Recommended Dietary Allowances (RDA) and Adequate Intakes

AGE (YR)	Thiamin RDA (mg/day)	Riboflavin RDA (mg/day)	Niacin RDA (mg/day)[a]	Biotin AI (μg/day)	Pantothenic acid AI (mg/day)	Vitamin B$_6$ RDA (mg/day)	Folate RDA (μg/day)[b]	Vitamin B$_{12}$ RDA (μg/day)	Choline AI (mg/day)	Vitamin C RDA (mg/day)	Vitamin A RDA (μg/day)[c]	Vitamin D AI (μg/day)[d]	Vitamin E RDA (mg/day)[e]	Vitamin K AI (μg/day)
Infants														
0–0.5	0.2	0.3	2	5	1.7	0.1	65	0.4	125	40	400	5	4	2.0
0.5–1	0.3	0.4	4	6	1.8	0.3	80	0.5	150	50	500	5	5	2.5
Children														
1–3	0.5	0.5	6	8	2	0.5	150	0.9	200	15	300	5	6	30
4–8	0.6	0.6	8	12	3	0.6	200	1.2	250	25	400	5	7	55
Males														
9–13	0.9	0.9	12	20	4	1.0	300	1.8	375	45	600	5	11	60
14–18	1.2	1.3	16	25	5	1.3	400	2.4	550	75	900	5	15	75
19–30	1.2	1.3	16	30	5	1.3	400	2.4	550	90	900	5	15	120
31–50	1.2	1.3	16	30	5	1.3	400	2.4	550	90	900	5	15	120
51–70	1.2	1.3	16	30	5	1.7	400	2.4	550	90	900	10	15	120
>70	1.2	1.3	16	30	5	1.7	400	2.4	550	90	900	15	15	120
Females														
9–13	0.9	0.9	12	20	4	1.0	300	1.8	375	45	600	5	11	60
14–18	1.0	1.0	14	25	5	1.2	400	2.4	400	65	700	5	15	75
19–30	1.1	1.1	14	30	5	1.3	400	2.4	425	75	700	5	15	90
31–50	1.1	1.1	14	30	5	1.3	400	2.4	425	75	700	5	15	90
51–70	1.1	1.1	14	30	5	1.5	400	2.4	425	75	700	10	15	90
>70	1.1	1.1	14	30	5	1.5	400	2.4	425	75	700	15	15	90
Pregnancy														
<18	1.4	1.4	18	30	6	1.9	600	2.6	450	80	750	5	15	75
19–30	1.4	1.4	18	30	6	1.9	600	2.6	450	85	770	5	15	90
31–50	1.4	1.4	18	30	6	1.9	600	2.6	450	85	770	5	15	90
Lactation														
<18	1.4	1.6	17	35	7	2.0	500	2.8	550	115	1200	5	19	75
19–30	1.4	1.6	17	35	7	2.0	500	2.8	550	120	1300	5	19	90
31–50	1.4	1.6	17	35	7	2.0	500	2.8	550	120	1300	5	19	90

NOTE: For all nutrients, values for infants are AI.

[a]Niacin recommendations are expressed as niacin equivalents (NE), except for recommendations for infants younger than 6 months, which are expressed as preformed niacin.

[b]Folate recommendations are expressed as dietary folate equivalents (DFE).

[c]Vitamin A recommendations are expressed as retinol activity equivalents (RAE).

[d]Vitamin D recommendations are expressed as cholecalciferol and assume an absence of adequate exposure to sunlight.

[e]Vitamin E recommendations are expressed as α-tocopherol.

Recommended Dietary Allowances (RDA) and Adequate Intakes (AI) for Minerals

AGE (YR)	Sodium AI (mg/day)	Chloride AI (mg/day)	Potassium AI (mg/day)	Calcium AI (mg/day)	Phosphorus RDA (mg/day)	Magnesium RDA (mg/day)	Iron RDA (mg/day)	Zinc RDA (mg/day)	Iodine RDA (μg/day)	Selenium RDA (μg/day)	Copper RDA (μg/day)	Manganese AI (mg/day)	Fluoride AI (mg/day)	Chromium AI (μg/day)	Molybdenum RDA (μg/day)
Infants															
0–0.5	120	180	400	210	100	30	0.27	2	110	15	200	0.003	0.01	0.2	2
0.5–1	370	570	700	270	275	75	11	3	130	20	220	0.6	0.5	5.5	3
Children															
1–3	1000	1500	3000	500	460	80	7	3	90	20	340	1.2	0.7	11	17
4–8	1200	1900	3800	800	500	130	10	5	90	30	440	1.5	1.0	15	22
Males															
9–13	1500	2300	4500	1300	1250	240	8	8	120	40	700	1.9	2	25	34
14–18	1500	2300	4700	1300	1250	410	11	11	150	55	890	2.2	3	35	43
19–30	1500	2300	4700	1000	700	400	8	11	150	55	900	2.3	4	35	45
31–50	1500	2300	4700	1000	700	420	8	11	150	55	900	2.3	4	35	45
51–70	1300	2000	4700	1200	700	420	8	11	150	55	900	2.3	4	30	45
>70	1200	1800	4700	1200	700	420	8	11	150	55	900	2.3	4	30	45
Females															
9–13	1500	2300	4500	1300	1250	240	8	8	120	40	700	1.6	2	21	34
14–18	1500	2300	4700	1300	1250	360	15	9	150	55	890	1.6	3	24	43
19–30	1500	2300	4700	1000	700	310	18	8	150	55	900	1.8	3	25	45
31–50	1500	2300	4700	1000	700	320	8	8	150	55	900	1.8	3	25	45
51–70	1300	2000	4700	1200	700	320	8	8	150	55	900	1.8	3	20	45
>70	1200	1800	4700	1200	700	320	8	8	150	55	900	1.8	3	20	45
Pregnancy															
<18	1500	2300	4700	1300	1250	400	27	12	220	60	1000	2.0	3	29	50
19–30	1500	2300	4700	1000	700	350	27	11	220	60	1000	2.0	3	30	50
31–50	1500	2300	4700	1000	700	360	27	11	220	60	1000	2.0	3	30	50
Lactation															
<18	1500	2300	5100	1300	1250	360	10	14	290	70	1300	2.6	3	44	50
19–30	1500	2300	5100	1000	700	310	9	12	290	70	1300	2.6	3	45	50
31–50	1500	2300	5100	1000	700	320	9	12	290	70	1300	2.6	3	45	50

About the Author

Judith E. Brown is Professor Emerita of Nutrition at the School of Public Health, and of the Department of Obstetrics and Gynecology at the University of Minnesota. She received her Ph.D. in human nutrition from Florida State University and her M.P.H. in public health nutrition from the University of Michigan. Dr. Brown has received competitively funded research grants from the National Institutes of Health, the Centers for Disease Control and Prevention, and the Maternal and Child Health Bureau and has over 100 publications in the scientific literature including the *New England Journal of Medicine,* the *Journal of the American Medical Association,* and the *Journal of the American Dietetic Association.* A recipient of the Agnes Higgins Award in Maternal Nutrition from the March of Dimes, Dr. Brown is a registered dietitian and the successful author of *Every-woman's Guide to Nutrition, Nutrition for Your Pregnancy,* and *What to Eat Before, During, and After Pregnancy.*

Judith E. Brown

University of Minnesota

Nutrition Now

FIFTH EDITION

THOMSON

WADSWORTH

Australia • Canada • Mexico • Singapore • Spain •
United Kingdom • United States

THOMSON

WADSWORTH

Acquisitions Editor: Peter Adams
Development Editor: Nedah Rose
Assistant Editor: Elesha Feldman
Editorial Assistant: Jean Blomo
Technology Project Manager: Ericka Yeoman-Saler
Marketing Manager: Jennifer Somerville
Marketing Assistant: Catie Ronquillo
Marketing Communications Manager: Jessica Perry
Project Manager, Editorial Production: Teresa L. Trego
Creative Director: Rob Hugel
Art Director: John Walker

Print Buyer: Rebecca Cross
Permissions Editor: Roberta Broyer
Production Service/Compositor: Newgen
Text Designer: Norman Baugher
Photo Researcher: Robin Sterling
Copy Editor: Margaret Pinette
Cover Designer: Larry Diodona
Cover Image: Main image: ImageShop/Corbis;
    Pears: Poisson d'Avril/photocuisine/Corbis
Cover Printer: Quebecor World
Printer: Quebecor World

Printed in the United States of America
2 3 4 5 6 7 11 10 09 08 07

For more information about our products, contact us at:
Thomson Learning Academic Resource Center
1-800-423-0563
For permission to use material from this text or product,
submit a request online at
http://www.thomsonrights.com.
Any additional questions about permissions can be submitted
by email to **thomsonrights@thomson.com.**

All Unit opening photos: Photo Disc

Library of Congress Control Number: 2007922480

ISBN-13: 978-0-495-11769-8

ISBN-10: 0-495-11769-2

Thomson Wadsworth
10 Davis Drive
Belmont, CA 94002-3098
USA

Asia
Thomson Learning
5 Shenton Way #01-01
UIC Building
Singapore 068808

Australia/New Zealand
Thomson Learning
102 Dodds Street
Southbank, Victoria 3006
Australia

Canada
Nelson
1120 Birchmount Road
Toronto, Ontario M1K 5G4
Canada

Europe/Middle East/Africa
Thomson Learning
High Holborn House
50/51 Bedford Row
London WC1R 4LR
United Kingdom

Latin America
Thomson Learning
Seneca, 53
Colonia Polanco
11560 Mexico D.F.
Mexico

Spain/Portugal
Paraninfo
Calle Magallanes, 25
28015 Madrid, Spain

CONTENTS IN BRIEF

CONTENTS

Nutrition Timeline →

1621
First Thanksgiving feast at Plymouth colony

Photo Disc

1702
First coffeehouse in America opens in Philadelphia

Photo Disc

1734
Scurvy recognized

1744
First record of ice cream in America at Maryland colony

Photo Disc

1747
Lind publishes "Treatise on Scurvy," citrus identified as cure

Photo Disc

1750
Ojibway and Sioux war over control of wild rice stands

1762
Sandwich invented by the Earl of Sandwich

Photo Disc

1771
Potato heralded as famine food

1774
Americans drink more coffee in protest over Britain's tea tax

Nutrition Timeline →

1775
Lavoisier ("the father of the science of nutrition") discovers the energy-producing property of food

Stefano Bianchetti/CORBIS

1816
Protein and amino acids identified, followed by carbohydrates and fats in the mid-1800s

1833
Beaumont's experiments on a wounded man's stomach greatly expand knowledge about digestion

1862
U.S. Department of Agriculture founded by authorization of President Lincoln

1871
Proteins, carbohydrates, and fats determined to be insufficient to support life; there are other "essential" components

1895

First milk station providing children with un-contami-nated milk opens in New York City

1896

Atwater publishes Proximate Composition of Food Materials

1906

Pure Food and Drug Act passed by President Theodore Roosevelt to protect consumers against contaminated foods

1910

Pasteurized milk introduced

1912

Funk suggests scurvy, beriberi, and pellagra caused by deficiency of "vitamines" in the diet

UNIT 8 CALORIES! FOOD, ENERGY, AND ENERGY BALANCE   8-1

UNIT 9 OBESITY TO UNDERWEIGHT: THE HIGHS AND LOWS OF WEIGHT STATUS   9-1

Nutrition Timeline →

1913
First vitamin discovered (vitamin A)

Photo Disc

1914
Goldberger identifies the cause of pellagra (niacin deficiency) in poor children to be a missing component of the diet rather than a germ as others believed

1916
First dietary guidance material produced for the public released; title is "Food for Young Children"

1917
First food groups published—the Five Food Groups: Milk and Meat; Vegetables and Fruits; Cereals; Fats and Fat Foods; Sugars and Sugary Foods

1921
First fortified food produced: iodized salt, needed to prevent widespread iodine-deficiency goiter in many parts of the United States

Morton Salt Co.

1928
American Society for Nutritional Sciences and the Journal of Nutrition founded

1929
Essential fatty acids identified

Photo Disc

1930s
Vitamin C identified in 1932, followed by pantothenic acid and riboflavin in 1933 and vitamin K in 1934

Photo Disc

1937
Pellagra found to be due to a deficiency of niacin

1941
First refined grain enrichment standards developed

Nutrition Timeline →

1941
First Recommended Dietary Allowances (RDAs) announced by President Franklin Roosevelt on radio

FDR Library

1946
National School Lunch Act passed

Photo Disc

1947
Vitamin B$_{12}$ identified

1953
Double helix structure of DNA discovered

Photo Disc

1956
Basic Four Food Groups released by the U.S. Department of Agriculture

1965
Food Stamp Act passed, Food Stamp program established

1966
Child Nutrition Act adds school breakfast to the National School Lunch Program

1968
First national nutrition survey in U.S. launched (The Ten State Nutrition Survey)

1970
First Canadian national nutrition survey launched (Nutrition Canada National Survey)

1972
Special Supplemental Food and Nutrition Program for Women, Infants, and Children (WIC) established

Nutrition Timeline →

1977

Dietary Goals for the U.S. issued

1978

First Health Objectives for the Nation released

1989

First national scientific consensus report on diet and chronic disease published

1992

The Food Guide pyramid is released by the USDA

1997

RDAs expanded to Dietary Reference Intakes (DRIs)

Photo Disc

1998

2003

2006

Folic acid fortification of refined grain products begins

Sequencing of DNA in the human genome completed; marks beginning of new era of research in nutrient–gene interactions

Obesity and diabetes recognized as gobal epidemics

PREFACE

"Everything should be made as simple as possible. But not simpler."
—Albert Einstein

I T IS WITH GREAT PLEASURE that we present to you the fifth edition of *Nutrition Now*. Athough the principles of the science of nutrition as developed for and presented in this text have not changed, much else has. Knowledge gains in nutrition are advancing at an incredibly rapid pace, and the implications of the advances to human health and well-being are impressive. Growth in our understanding of the role of nutrition in inflammation and the wide variety of disorders related to it is changing dietary guidance and previous conclusions about nutrition and disease relationships. Advances in knowledge about the effects of dietary components on gene function, and the long-term effects of nutrition during pregnancy and infancy on lifelong gene function and disease risk, are changing fundamental concepts about the origins and prevention of disease. Knowledge about the interrelated activities of antioxidants and other nutrients in foods (versus supplements) is refining recommendations for nutrient intake and supplement use. The expanding epidemics of obesity and type 2 diabetes are reaching into younger age groups, and much more public attention is being paid to the roles of diet and physical activity in their prevention and management.

Advances in the science of nutrition are increasingly being brought to us by a deeper level of understanding of the roles of nutrients and other components of food at the cellular level. To understand nutrition, as well as nutrition and health relationships, introductory courses in nutrition must now include topics such as visceral fat, endothelial function, inflammation, and nutrigenomics. This fifth edition covers these and other emerging topics directly related to the state of the science of nutrition now. It attempts to introduce these topics to nonmajors in a straightforward and clear way that keeps coverage as simple as possible—but no simpler than that.

What's New in the Fifth Edition?

This edition of *Nutrition Now* contains hundreds of small, factual updates and many extensive revisions to previous content. Pedagogical features have been updated and expanded as well. Particular attention has been paid to the development of additional interactive learning opportunities for students. You will find that the fifth edition includes:

- Review Questions and Answers to Review Questions for each unit

- Additional Health Action and Reality Check features

- A Media Menu of links to instructional videos, Powerpoint presentations, and other interactive learning activities

- In-depth coverage of MyPyramid and the Dietary Guidelines in the retitled Unit 6, Healthy Diets: The Dietary Guidelines, MyPyramid, and More.

- Expanded coverage of obesity and diabetes in youth and adults in the retitled Unit 9, Obesity to Underweight: The Highs and Lows of Weight Status, and in Unit 13, Diabetes Now

- New or revised dietary recommendations for heart disease, hypertension, diabetes, and cancer

- Integrated coverage of nutrition and chronic inflammation, endothelial function, the immune system, visceral and subcutaneous fat, and bariatrics

- Major changes in nutrition and physical performance content (from hydration to ergogenic aids)

- An updated and expanded Instructor's Activity Manual

Activities for students included in the Instructor's Activity Manual are intended to get students involved in the topics being covered in class through hands-on and interactive experiences. The activities can be undertaken in classes ranging in size from tens to hundreds. Activities include taste testing to identify genetically determined sensitivity to bitterness, developing a dietary behavioral change plan, anthropometry lab, designing fraudulent nutrition products, a physical activity assessment, and an assessment of three days of dietary intake. The Instructor's Activity Manual, as well as the forms used to conduct and submit activities, may be accessed by Instructors using *Nutrition Now* at the Web address http://nutrition.wadsworth.com.

Although a number of elements have changed in this new edition, many of the basic tenets of the text's approach have stayed the same. The text remains focused on meeting the needs of instructors offering introductory nutrition courses to nonmajors. The 33 units in the text stand alone and can be covered in the order of the instructor's choosing. Instructors may choose to customize their selection of units to be included in the text. The text remains heavily illustrated, and updated and revised figures, tables, and photographs have been added to the new edition.

Pedagogical Features

Nutrition Now continues to be oriented toward helping students build a foundation of knowledge about nutrition that will serve them well throughout life. Units are concise, focused on key facts and concepts, and provide ample real-life examples intended to enhance students' understanding of the material presented. Students are asked to apply their newly gained knowledge about nutrition in decision-making activities and exercises incorporated throughout the units.

Features included in this edition of *Nutrition Now* aimed at enhancing students' understanding of nutrition include:

- **Reality Check.** Reality Check presents brief, real-life scenarios and asks students to given a thumbs up or down to the optional solutions posed.

- **Nutrition Scoreboard.** Each unit begins with a three- to five-question pretest. Answers to the questions are given on the second page of the units.

- **Key Concepts and Facts.** Each unit begins with a listing of key concepts and facts related to the central topics covered.

- **On the Side.** Boxed inserts containing interesting facts related to nutrition topics covered in the units are sprinkled throughout the text.

- **Health Action.** Every unit has at least one Health Action box that relates to the personal application of information covered.

- **Margin definitions.** Unfamiliar terms are highlighted in bold in the text, and defined in nearby margins. Pronunciation guides are provided as needed.

- **Review questions and answers.** Students are invited to review their knowledge of key points made in each unit by answering eight to twelve review questions presented at the end of each unit.

- **Glossary.** Terms defined in the margins are listed in the Glossary near the end of the text. Approximately 20 new terms have been added to the glossary of the fifth edition.

- **Nutrition Up Close.** Each unit closes with an activity that gives students an opportunity to relate nutrition knowledge gained to their daily lives and experiences.

- **Media Menu.** Internet sources of reliable nutrition information are listed at the end of each unit. These have been thoroughly updated in the fifth edition and include interactive links.

- **Appended materials.** Resources included in the appendixes have been updated for the fifth edition of *Nutrition Now:* The Food Composition Table, the Reliable Sources of Nutrition Information list, and the Food Exchange System. The 2007 Healthy Eating Guidelines for Canada will be made available to instructors on their release.

- **Nutrition Timeline.** A Nutrition Timeline bannered across the bottom of the table of contents highlights major developments in the science of nutrition.

- **Daily Values for Food Labels and Glossary of Nutrient Measures.** These appear on the last book page.

Resources for the Instructor

- **Instructor's Manual and Test Bank.** This features lecture outlines, suggested classroom activities, Web resources, discussion questions, transparency masters, and chapter-by-chapter test questions linked to the ExamView Computerized Testing program.

- **ExamView® Computerized Testing.** An easy-to-use assessment and tutorial system facilitates creation, delivery, and customizing of tests and study guides, both print and online.

- **Multimedia Manager CD-ROM.** Book-specific Microsoft®, PowerPoint®, lecture slides with teaching points, graphics from the book, electronic versions of the Instructor's Manual and Test Bank, ABC video clips, and links to nutrition resources on the Web are included on this disk.

- **CNN Today: Nutrition Videos.** Three volumes of engaging video clips are available for launching lectures.

- **Instructor's Activity Book.** Classroom activities can be used by varying sizes of student groups.

- **Transparency Acetates.** This includes 80 full-color transparencies of key illustrations in the text.

- **Diet Analysis Plus 8.0.** This software enables students to track and assess their food choices and create personal nutrition and activity profiles. It is available on CD-ROM or online and may be packaged with the text.

Acknowledgments

This edition introduces Peter Adams as the new Executive Editor for *Nutrition Now* and Nedah Rose as the Senior Developmental Editor. Both have invested themselves fully in making *Nutrition Now* and its pedagogical aids effective teaching and learning tools. They follow in the footsteps of two remarkably talented professionals: Peter Marshall and Beth Howe. Pete has been this textbook's champion and guided its growth from the start, and Beth has kept the development of the last few editions on track and the process seamless. May you both read this preface and know I am thinking of you and wishing you the best in your new careers. Margaret Pinette copyedited this edition of *Nutrition Now* and brought a new eye for detail to the text.

My applause goes out to the sales representatives who are fans of *Nutrition Now* and who work hard to introduce faculty to its contents and features. You do a terrific job of communicating with instructors, as well as with me when you send instructors' thoughts my way.

It is said that instructors adopt a specific textbook but that students play a major role in instructors' decision to keep it. I am honored that you chose to adopt *Nutrition Now* and deeply pleased with the thought that students are helping you decide to keep it.

Reviewers' feedback is the lifeline of text writing, and the reviewers for the new edition conveyed very useful advice. The advice led me to some very interesting places on specific topics that changed my thinking and writing. May you remain students of *Nutrition Now* the textbook and keep your comments coming.

JUDITH E. BROWN

Reviewers of the Fifth Edition

Cassandra E. August
Baldwin-Wallace College

Laura Calderon
California State University,
Los Angeles

Jasia (Jayne) Chitharanjan
University of Wisconsin, Stevens Point

Dorothy Coltrin
De Anza College

Gary Fosmire
The Pennsylvania State University

Susan E. Helm
Pepperdine University

Linda S. Kolb
Wabash Valley College

Marcia Magnus
Florida International University

Lisa Sasson
New York University

Tasleem A. Zafar
Purdue University

UNIT 1

Key Nutrition Concepts and Terms

NUTRITION SCOREBOARD

	TRUE	FALSE
1 Calories are a component of food.		
2 Nutrients are substances in food that are used by the body for growth and health.		
3 Both high and low intakes of nutrients threaten health.		
4 Foods can basically be divided into two groups: those that are "good" for you and those that are "bad" for you.		

Key Concepts and Facts

- At the core of the science of nutrition are concepts that represent basic "truths" and serve as the foundation of our understanding about normal nutrition. (They are listed in Table 1.4 on p. 1-13.)

- Most nutrition concepts relate to nutrients.

Answers to
NUTRITION SCOREBOARD

		TRUE	FALSE
1	Calories are a measure of the amount of energy supplied by food. They're a property of food, not a substance present in food.		✔
2	That's the definition of nutrients.	✔	
3	Excessive as well as inadequate intake levels of vitamins, minerals, and other nutrients can be harmful to health.		✔
4	There are *no* good or bad foods, but there are healthy and unhealthy diets.	✔	

The Meaning of Nutrition

To be surprised, to wonder, is to begin to understand.
—José Ortega y Gasset

What is nutrition? It can be explained by situations captured in photographs as well as by words. This introduction presents a photographic tour of real-life situations that depict aspects of the study of nutrition.

Before the tour begins, take a moment to make yourself comfortable and clear your mind of clutter. Take a careful look at the photographs, pausing to mentally describe in two or three sentences what each shows.

Cary Conner/PhotoEdit

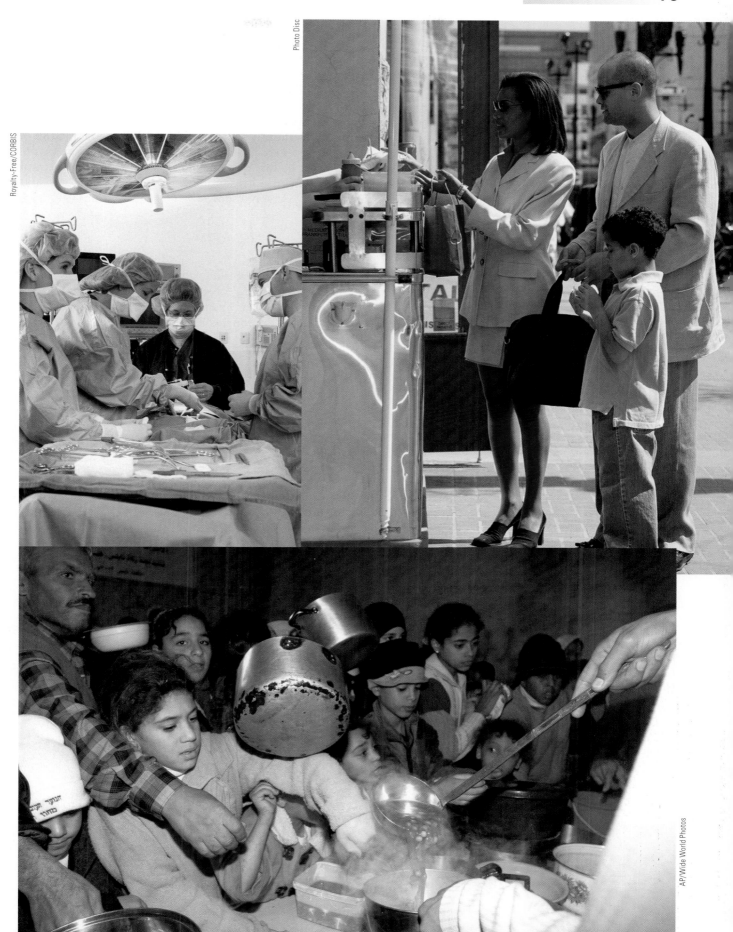

Not everyone who looks at the photographs will describe them in the same way. Reactions will vary somewhat due to personal experiences, interests, attitudes, and beliefs. An individual trying to gain weight will probably react differently to the photograph of the person on the scale than someone who is trying to lose weight. If you grew up in a family that farmed for a living, the picture of pesticides being sprayed on a crop may mean increased food production to you. But another person's reaction may be that pesticide residues on foods are harmful to health. Although knowledge about nutrition is generated by impersonal and objective methods, it can be a very personal subject.

Nutrition Defined

nutrition
The study of foods, their nutrients and other chemical constituents, and the effects of food constituents on health

In a nutshell, *nutrition* is the study of foods and health. It is a science that centers on foods, their nutrient and other chemical constituents, and the effects of food constituents on body processes and health. The scope of nutrition extends from food choices to the effects of specific components of foods on health.

Nutrition Is a "Melting Pot" Science

The broad scope of nutrition makes it an interdisciplinary science. Knowledge provided by the behavioral and social sciences, for example, is needed in studies that examine how food preferences develop and how they may be changed. Information generated by the biological, chemical, physical, and food sciences is required to propose and explain diet and disease relationships. Methods developed by quantitative scientists such as mathematicians and statisticians are needed to guide decision making about the significance of results produced by nutrition research. The study of nutrition will bring you into contact with information from a variety of disciplines (Illustration 1.1).

Nutrition Knowledge Is Applicable

As you study this science, you will discover answers to a number of questions about your own diet, health, and eating behavior. Is obesity primarily due to eating behaviors, physical inactivity, or your genes? How do you know whether new

Illustration 1.1
Nutrition is an interdisciplinary science.

Richard Anderson

information you hear about nutrition is true? Can sugar harm more than your teeth? Can the right diet or supplement give you a competitive edge? What is a healthful diet, how do you know if you have one, and, if improvements seem warranted, what's the best way to go about changing your diet for the better? These are just a few of the questions that nutrition and your course of study into it will address. If all goes well, you will take from this learning experience not only knowledge about nutrition and health, but skills that will keep the information and insights working to your advantage for a long time to come.

Foundation Knowledge for Thinking about Nutrition

Common sense requires a common knowledge.

You don't have to be a bona fide nutritionist to think like one. What you need is a grasp of the language and the basic concepts of the science. It's the purpose of this unit to give you this background. The essential topics covered here are explored in greater depth in units to come, and they build on this foundation knowledge. With a working knowledge of nutrition terms and the concepts that provide the foundation of our understanding about nutrition, you will have an uncommonly good sense of nutrition.

NUTRITION CONCEPT #1

Food is a basic need of humans.

> *If we could give individual the right amount of nourishment and exercise, we would have found the safest way to health.*
> —HIPPOCRATES

Humans need enough food to live and the right assortment of foods for optimal health. In the best of all worlds, the need for food is combined with the condition of **food security.** People who experience food security have access at all times to a sufficient supply of safe, nutritious foods needed for an active and healthy life. They are able to acquire acceptable foods in socially acceptable ways—without having to scavenge or steal food, for example, to eat or to feed their families. **Food insecurity** exists whenever the availability of safe, nutritious foods, or the ability to acquire them in socially acceptable ways, is limited or uncertain (Illustration 1.2).[1]

food security
Access at all times to a sufficient supply of safe, nutritious foods.

food insecurity
Limited or uncertain availability of safe, nutritious foods or the ability to acquire them in socially acceptable ways.

Illustration 1.2
"It is possible to go an entire lifetime without knowing about people's experiences with hunger."

—Meghan LeCates, Capitol Area Food Bank

Photo Disc

Soup kitchen

Ricin   Ricin, a deadly nerve toxin, comes from the seeds of the beautiful castor bean plant. The oil from the seeds (sometimes called beans) is used commercially in castor oil, motor oil, nylon, and paint. Castor plants are becoming increasingly abundant in southwestern parts of the United States, although their cultivation is discouraged. Ricin poisoning occurs when seeds with broken coatings or seed contents are ingested. 70 micrograms—the amount of a grain of salt—is enough to kill an adult. Exposure of the skin to the seeds causes irritation at the contact point. The U.S. Senate offices were closed for several days in 2004 when powdered ricin was discovered in a letter sent to Senator Bill Frist. No one was injured.

©W.P. Armstrong 2000

calorie
A unit of measure of the amount of energy supplied by food. (Also known as a kilocalorie, or the "large Calorie" with a capital C.)

nutrients
Chemical substances found in food that are used by the body for growth and health. The six categories of nutrients are carbohydrates, proteins, fats, vitamins, minerals, and water.

Illustration 1.3
Foods provide nutrients. "Please pass the complex carbohydrates, thiamin, and niacin . . . I mean, the bread!"

Brand X Pictures

Adults who live in food-insecure households are more likely to be overweight and to have heart disease or diabetes than adults who are poor but food secure.[2] Poor nutritional quality of food consumed, lack of exercise, and episodic food shortages may be partly responsible for the higher rates of overweight and chronic disease among food-insecure adults. High-calorie, filling, and inexpensive fast and convenience foods are often foods of choice when money is low. Although some children living in food-insecure households are nourished adequately and successful in school and develop high levels of social skills, as a group they are at higher risk of poor school performance and social and behavioral problems.[3,4]

Food insecurity exists in about 12% of U.S. and 4% of Canadian households.[4,5] It is most likely to occur in poor, female-headed households with young children living in inner-city areas. The rate in the United States is twice as high as the target of 6% established as a national health goal to be reached by the year 2010.[4]

Food Terrorism

The term *food security* now means more than it used to. Food is a potential weapon of bioterrorism. Although public health concerns related to bioterrorism center on disease threats such as smallpox and anthrax, food and water could also be used to intentionally spread illness. Simply recalling the latest announcement about a foodborne illness outbreak makes it easy to imagine the panic and health consequences that could result from intentional contamination of food or water supplies.

Toxic substances that could be introduced into food supplies include botulism toxin, ricin, radioactive particles, and microorganisms such as *Salmonella, E. coli 0517:H7,* and *Shigella.* Botulism toxin is the single most poisonous substance known. It takes at least 15 minutes of boiling to destroy this toxin. Most bacteria are killed when heated above 160°F or when municipal water supplies are treated with chlorine.[6,7] Ricin (pronounced rye-sin) is a widely available toxin found in the seeds of the castor plant, which is also widely available in tropical and warm areas of the world.[8] Castor plant seeds are sufficiently interesting to warrant being featured in the "On the Side" box nearby.

NUTRITION CONCEPT #2

Foods provide energy (calories), nutrients, and other substances needed for growth and health.

People eat foods for many different reasons. The most compelling reason is that we need the **calories, nutrients,** and other substances supplied by foods for growth and health.

Calories

A calorie is a unit of measure of the amount of energy in a food—and of how much energy will be transferred to the person who eats it. Although we often refer to the number of calories in this food or that one, calories are not a substance present in food. And, because calories are a unit of measure, they do not qualify as a nutrient.

Nutrients

Nutrients are chemical substances present in food that are used by the body (Illustration 1.3). Essentially everything that's in our body was once a component of the food we consumed.

There are six categories of nutrients (Illustration 1.4), and each category (except water) consists of a number of different substances used by the body for growth and health. The carbohydrate category includes simple sugars and complex carbohydrates (starches and dietary fiber). The protein category includes 20 amino acids, the chemical units that serve as the "building blocks" for protein. Several different types of fat are included in the fat category. Of primary concern are the saturated fats, unsaturated fats, essential fatty acids, *trans* fats, and cholesterol (read more about them in Illustration 1.4). The vitamin category consists of 13 vitamins, and the mineral category includes 15 minerals. Water makes up a nutrient category by itself.

Illustration 1.4
Nutrients are grouped into six categories. Here the major types of nutrients in each category are listed, along with a description of each.

SIX CATEGORIES OF NUTRIENTS

1. CARBOHYDRATES are substances in food that consist of a single sugar molecule or of multiples of them in various forms. They provide the body with energy.

Simple sugars are the most basic type of carbohydrates. Examples include glucose (blood sugar), sucrose (table sugar), and lactose (milk sugar). **Starches** are complex carbohydrates consisting primarily of long, interlocking chains of glucose units. **Dietary fiber** consists of complex carbohydrates found principally in plant cell walls. Because dietary fiber cannot be broken down by human digestive enzymes, it cannot be used for energy.

2. PROTEINS are substances in food made up of amino acids. Amino acids are specific chemical substances from which proteins are made. Of the 20 amino acids, 9 are "essential," or a required part of our diet.

3. FATS are substances in food that are soluble in fat and not water.

Saturated fats are found primarily in animal products, such as meat, butter, and cheese, and in palm and coconut oils. Diets high in saturated fat may elevate blood cholesterol. **Unsaturated fats** are found primarily in plant products, such as vegetable oil, nuts, and seeds, and in fish. Unsaturated fats tend to lower blood cholesterol levels. **Essential fatty acids** are two specific types of unsaturated fats that are required in the diet. *Trans* **fats** are a type of unsaturated fat present in hydrogenated oil, margarine, shortening, pastries, and some cooking oils that increase the risk of heart disease. **Cholesterol** is a fat-soluble, colorless liquid found in animals but not in plants. It can be manufactured by the liver.

4. VITAMINS are chemical substances found in food that perform specific functions in the body. Humans require 13 different vitamins in their diet.

5. MINERALS are chemical substances that make up the "ash" that remains when food is completely burned. Humans require 15 different minerals in their diet.

6. WATER is essential for life. Most adults need about 11–15 cups of water each day from food and fluids.

The distinctive odor garlic produces when chopped or crushed
is actually this plant's defense
against insect predators. Garlic
cloves contain an odorless,
sulfur-containing phytochemical
called "alliin." When the clove is
disrupted, alliin is released and
reacts with an enzyme located in
neighboring cells that converts the
alliin to the odoriferous "allicin."
Allicin is garlic's bug repellant—and
the "people repellant" that makes
many individuals shy
about eating it.[9]

Photo Disc

**phytochemicals (phyto =
plant)**
Chemical substances in plants,
some of which perform important
functions in the human body.
Phytochemicals give plants
color and flavor, participate in
processes that enable plants to
grow, and protect plants against
insects and diseases.

antioxidants
Chemical substances that prevent
or repair damage to cells caused
by exposure to oxidizing agents
such as environmental pollutants,
smoke, ozone, and oxygen.
Oxidation reactions are also a
normal part of cellular processes.

essential nutrients
Substances required for normal
growth and health that the body
cannot generally produce, or
produce in sufficient amounts;
they must be obtained in the diet.

Carbohydrates, proteins, and fats supply calories and are called the "energy nutrients." Although each of these three types of nutrients performs a variety of functions, they share the property of being the body's only sources of fuel. Vitamins, minerals, and water are chemicals the body needs for converting carbohydrates, proteins,
and fats into energy and for building and maintaining muscles, blood components,
bones, and other parts of the body.

Other Substances in Food

Food also contains many other substances, some of which are biologically active in
the body. The **phytochemicals** are a major type of these substances, and there are
over two thousand of them in plants. Illustration 1.5 presents examples of plant foods
that are particularly rich sources of phytochemicals. Phytochemicals provide plants
with color, give them flavor, foster their growth, and protect them from insects and
diseases. A specific example of how one phytochemical works is described in the
"On the Side" box on this page. In humans, consumption of phytochemicals in diets
high in vegetables and fruits is strongly related to a reduced risk of developing certain
types of cancer, heart disease, infections, and other disorders.[10]

Specific phytochemicals have names that are often hard to pronounce and difficult to remember. Nevertheless, here are a few examples. Plant pigments, such
as lycopene (like-o-peen), which helps make tomatoes red, anthocyanins (an-tho-
sigh-an-ins), which give blueberries their characteristic blue color, and beta-carotene
(bay-tah-kar-o-teen), which imparts a dark yellow color to carrots, are phytochemicals that act as **antioxidants.** They protect plant cells—and in some cases, human
cells, too—from damage that can make them susceptible to disease. Various types
of sulfur-containing phytochemicals are present in cabbage, broccoli, cauliflower,
brussels sprouts, and other vegetables of the same family. These substances may help
prevent a number of different types of cancer.[11]

Some Nutrients Must Be Provided by the Diet

Many nutrients are required for growth and health. The body can manufacture
some of these from raw materials supplied by food, but others must come as-
sembled. Nutrients that the body cannot make, or generally produce in sufficient
quantity, are referred to as *essential nutrients.* Here "essential" means "required
in the diet." Vitamin A, iron, and calcium are examples of essential nutrients. Table 1.1 lists all the known essential nutrients.

Nutrients used for growth and health that can be manufactured by the body
from components of food in our diet are considered nonessential. Cholesterol,

Illustration 1.5
Examples of good food sources of phytochemicals.

Photo Disc

TABLE 1.1

Essential nutrients for humans: A reference table

Energy Nutrients	Vitamins	Minerals	Water
Carbohydrates	Biotin	Calcium	Water
Fats[a]	Folate	Chloride	
Proteins[b]	Niacin (B_3)	Chromium	
	Pantothenic acid	Copper	
	Riboflavin (B_2)	Fluoride	
	Thiamin (B_1)	Iodine	
	Vitamin A	Iron	
	Vitamin B_6 (pyroxidine)	Magnesium	
	Vitamin B_{12}	Manganese	
	Vitamin C (ascorbic acid)	Molybdenum	
	Vitamin D	Phosphorus	
	Vitamin E	Potassium	
	Vitamin K	Selenium	
		Sodium	
		Zinc	

[a]Fats supply the essential nutrients linoleic and alpha-linolenic acid.
[b]Proteins are the source of 9 "essential amino acids": histidine, isoleucine, leucine, lysine, methionine, phenylalanine, threonine, tryptophan, and valine. The other 11 amino acids are not a required part of our diet; they are considered "nonessential."

creatine, and glucose are examples of nonessential nutrients. ***Nonessential nutrients*** are present in food and used by the body, but they are not required parts of our diet because we can produce them.

Both essential and nonessential nutrients are required for growth and health. The difference between them is whether we need a dietary source. A dietary deficiency of an essential nutrient will cause a specific deficiency disease, but a dietary lack of a nonessential nutrient will not. People develop scurvy (the vitamin C–deficiency disease), for example, if they don't consume vitamin C. But you could have zero cholesterol in your diet and not become "cholesterol deficient," because your liver produces cholesterol.

nonessential nutrients
Nutrients required for normal growth and health that the body can manufacture in sufficient quantities from other components of the diet. We do not require a dietary source of nonessential nutrients.

Our Requirements for Essential Nutrients

The amount of essential nutrients humans need each day varies a great deal, from amounts measured in cups to micrograms. (See Table 1.2 to get a notion of the amount represented by a gram, milligram, and other measures.) Generally speaking, adults need 11 to 15 cups of water from fluids and foods, 9 tablespoons of protein, one-fourth teaspoon of calcium, and only one-thousandth teaspoon (a 30-microgram speck) of vitamin B_{12} each day.

We all need the same nutrients, but not always in the same amounts. The amounts needed vary among people based on:

- Age

- Sex

- Growth status

- Body size

- Genetic traits

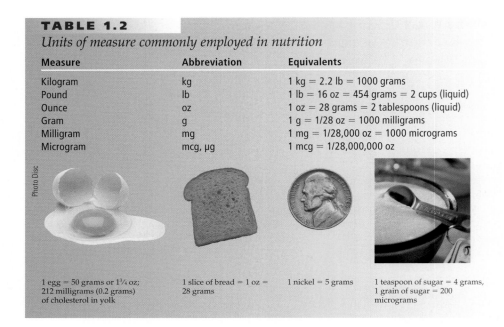

TABLE 1.2

Units of measure commonly employed in nutrition

Measure	Abbreviation	Equivalents
Kilogram	kg	1 kg = 2.2 lb = 1000 grams
Pound	lb	1 lb = 16 oz = 454 grams = 2 cups (liquid)
Ounce	oz	1 oz = 28 grams = 2 tablespoons (liquid)
Gram	g	1 g = 1/28 oz = 1000 milligrams
Milligram	mg	1 mg = 1/28,000 oz = 1000 micrograms
Microgram	mcg, μg	1 mcg = 1/28,000,000 oz

Photo Disc

1 egg = 50 grams or 1¾ oz; 212 milligrams (0.2 grams) of cholesterol in yolk

1 slice of bread = 1 oz = 28 grams

1 nickel = 5 grams

1 teaspoon of sugar = 4 grams, 1 grain of sugar = 200 micrograms

and the presence of conditions such as:

• Pregnancy

• Breastfeeding

• Illnesses

• Drug use

• Exposure to environmental contaminants

Each of these factors, and others, can influence nutrient requirements. General recommendations for diets that provide all the essential nutrients usually make allowances for major factors influencing the level of nutrient need, but they cannot allow for all of the factors.

Nutrient Intake Standards

Recommendations for levels of essential nutrient intake were first developed in the United States in 1943 and have been updated periodically since then. Called the Recommended Dietary Allowances (RDAs), these standards were established in response to the high rejection rate of World War II recruits due to underweight and nutrient deficiencies. Recommended levels of nutrient intake provided are based on age, gender, and condition (pregnant or breastfeeding). The most recent update, completed in 2004, was a major one. It gave the RDAs a whole new look and a new name—Dietary Reference Intakes (Illustration 1.6)

Dietary Reference Intakes (DRIs)   Previous editions of the RDAs were based on levels of essential nutrients that protected people from deficiency diseases such as scurvy and rickets. As the science of nutrition advances, it is becoming abundantly clear that other components of food besides essential nutrients affect

Illustration 1.6
One of the six reports on the DRIs.

Food & Nutrition Board

DRI

DIETARY REFERENCE INTAKES

For
Energy,
Carbohydrate,
Fiber,
Fat,
Fatty Acids,
Cholesterol,
Protein, and
Amino Acids
(Macronutrients)

health. It is also becoming apparent that levels of nutrient intake associated with deficiency disease prevention may be too low to help prevent cancer, heart disease, and osteoporosis. Excessively high intakes of nutrients from fortified foods and supplements are another problem that was not considered when previous editions of the RDAs were prepared.

The DRIs provide recommended intake levels of essential nutrients and safe upper levels of intake. The recommended daily levels of intake not only meet the nutrient needs of almost all healthy people (97 to 98%) but also promote health and help reduce the risk of chronic disease.[12] Table 1.3 provides examples of endpoints aimed at health promotion and disease prevention used by the DRI committees to estimate recommended levels of nutrient intake. Tables showing the DRIs begin on the inside front cover of this book

Recommended intakes of some nutrients, such as calcium, vitamin D, and fluoride, for which too little conclusive evidence on disease prevention exists, are represented by an "Adequate Intake," or AI, level (Table 1.4). Regardless of which label is attached, the RDAs and AIs represent the best estimates of nutrient intake

TABLE 1.3
Examples of primary endpoints used to estimate DRIs[12]

Carbohydrate: Amount needed to supply optimal levels of energy to the brain.

Total Fiber: Amount shown to provide the greatest protection against heart disease.

Folate: Amount that reduces the risk of heart disease and newborn abnormalities.

Iodine: Amount that corresponds to optimal functioning of the thyroid gland.

Selenium: Amount that maximizes its function in protecting cells from damage.

TABLE 1.4
Terms and abbreviations used in the 1997–2003 DRIs and a graphic representation of their meaning

Source: Adapted from Yates et al. J Am Diet Assoc 1998; 98:702

- **Dietary Reference Intakes (DRIs).** This is the general term used for the new nutrient intake standards for healthy people.
- **Recommended Dietary Allowances (RDAs).** These are levels of essential nutrient intake judged to be adequate to meet the known nutrient needs of practically all healthy persons while decreasing the risk of certain chronic diseases.
- **Adequate Intakes (AIs).** These are "tentative" RDAs. AIs are based on less conclusive scientific information than are the RDAs.
- **Estimated Average Requirements (EARs).** These are nutrient intake values that are estimated to meet the requirements of half the healthy individuals in a group. The EARs are used to assess adequacy of intakes of population groups.
- **Tolerable Upper Levels of Intake (ULs).** These are upper limits of nutrient intake compatible with health. The ULs do not reflect desired levels of intake. Rather, they represent total, daily levels of nutrient intake from food, fortified foods, and supplements that should not be exceeded.

Illustration 1.7
Schematic representation of the structure of a human cell.

levels that promote optimal health. These values are represented in the main DRI tables presented on the inside front cover of this book. The term Estimated Average Requirement (EAR) has been introduced in the new dietary intake standards. It represents the intake level of a nutrient that is estimated to meet the requirement of 50% of the individuals within a group.

Estimates of safe upper limits of nutrient intake are called Tolerable Upper Intake Levels, abbreviated ULs. The ULs do not represent recommended or desired levels of intake. Instead, they represent total daily levels of nutrient intake from food, fortified food products, and supplements that should not be exceeded. To reflect the fact that recommended nutrient intakes now include the RDAs, AIs, EARs, and ULs, the name of the recommendations has been broadened to Dietary Reference Intakes.

The DRIs are revolutionary in that they were developed by experts from the United States and Canada and apply to the people in both countries.

NUTRITION CONCEPT #3

Health problems related to nutrition originate within cells.

The main employers of nutrients are cells (Illustration 1.7). This is because all body processes required for growth and health take place within cells and the fluid that surrounds them. The human body contains more than one hundred trillion (100,000,000,000,000) cells. (Which type is the most common? See the "On the Side" box on the next page.) Functions of each cell are maintained by the nutrients it receives. Problems arise when a cell's need for nutrients differs from the available supply.

Nutrient Functions at the Cellular Level

Maximum health and lifespan require metabolic harmony.
—B. Ames[13]

Cells are the building blocks of tissues (such as muscles and bones), organs (such as the kidneys, heart, and liver), and systems (such as the respiratory, reproductive, circulatory, and nervous systems). Normal cell health and functioning are maintained when a nutritional and environmental utopia exists within and around the cells. Such circumstances allow *metabolism*—the chemical changes that take place within and outside of cells—to proceed flawlessly. Disruptions in the availability of nutrients, or the presence of harmful substances in the cell's environment, initiate diseases and disorders that eventually affect tissues, organs, and systems. Here are two examples of how cell functions can be disrupted by the presence of low or high levels of nutrients:

metabolism
The chemical changes that take place in the body. The formation of energy from carbohydrates is an example of a metabolic process.

- Folate, a B vitamin, participates in cellular reactions that keep the level of homocysteine (an amino-acid-like substance) normal. When too little folate is available, homocysteine accumulates in cells and spills into the blood. Once in the blood, high levels of homocysteine enhance plaque formation in arteries and increase the risk of heart disease.

- When too much iron is present in cells, the excess reacts with and damages cell components. If cellular levels of iron remain high, the damage spreads, impairing the functions of organs such as the liver, pancreas, and heart. Health problems in general begin with disruptions in the normal activity of cells. Humans are only as healthy as their cells.

NUTRITION CONCEPT #4

Poor nutrition can result from both inadequate and excessive levels of nutrient intake.

For each nutrient, every individual has a range of optimal intake that produces the best level for cell and body functions. On either side of the optimal range are levels of intake associated with impaired functions.[15] This concept is presented in Illustration 1.8. Inadequate essential nutrient intake, if prolonged, results in obvious deficiency diseases. Marginally deficient nutrient intakes generally produce subtle changes in behavior or physical condition. If the optimal intake range is exceeded, mild to severe changes in mental and physical functions occur, depending on the amount of the excess and the nutrient. Overt vitamin C deficiency, for example, produces bleeding gums, pain on being touched, and a failure of bone to grow. A marginal deficiency may cause delayed wound healing. On the excessive side, high intakes of vitamin C cause diarrhea.[12] Nearly all cases of vitamin and mineral overdose result from the excessive use of supplements or errors made in the level of nutrient fortification of food products. They are almost never caused by foods. For nutrients, "enough is as good as a feast."

Illustration 1.8
For every nutrient, there is a range of optimal intake that corresponds to the optimal functioning of that nutrient in the body.

Steps in the Development of Nutrient Deficiencies and Toxicities

Poor nutrition due to inadequate diets generally develops in the stages outlined in Illustration 1.9. To help explain the stages, this illustration includes an example of how vitamin A deficiency develops. After a period of deficient intake of an essential nutrient, the body's tissue reserves of the nutrient become depleted. Blood levels of the nutrient then decrease because there are no reserves left to replenish the blood supply. Without an adequate supply of the nutrient in the blood, cells get shortchanged. They no longer have the supply of nutrients needed to maintain normal function. If the dietary deficiency is prolonged, the malfunctioning cells cause sufficient impairment to produce physically obvious signs of a deficiency disease. Eventually, some of the problems produced by the deficiency may no longer be repairable, and permanent changes in health and function may occur. In most cases, the problems resulting from the deficiency can be reversed if the nutrient is supplied before this final stage occurs.

Excessively high intakes of many nutrients such as vitamin A and selenium produce toxicity diseases. The vitamin A toxicity disease is called "hypervitaminosis A," and the disease for selenium toxicity is called "selenosis." Signs of the toxicity disease stem from increased levels of the nutrient in the blood and the subsequent oversupply of the nutrient to the cells. The high nutrient load upsets

Illustration 1.9
The usual sequence of events in the development of a nutrient deficiency and an example of how vitamin A deficiency develops.

Source: Adapted with permission: Goodman AS, Vitamin A and retinoids in health and disease. (N Engl J Med 1984;310:1023–31). Copyright 1984. Massachusetts Medical Society. All rights reserved.

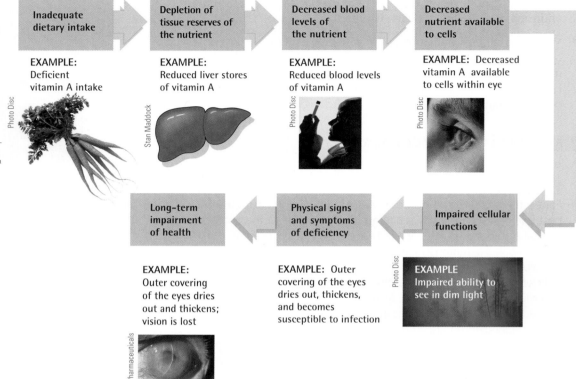

Inadequate dietary intake

EXAMPLE: Deficient vitamin A intake

Depletion of tissue reserves of the nutrient

EXAMPLE: Reduced liver stores of vitamin A

Decreased blood levels of the nutrient

EXAMPLE: Reduced blood levels of vitamin A

Decreased nutrient available to cells

EXAMPLE: Decreased vitamin A available to cells within eye

Long-term impairment of health

EXAMPLE: Outer covering of the eyes dries out and thickens; vision is lost

Physical signs and symptoms of deficiency

EXAMPLE: Outer covering of the eyes dries out, thickens, and becomes susceptible to infection

Impaired cellular functions

EXAMPLE Impaired ability to see in dim light

Illustration 1.10
This woman has iron deficiency anemia. She also has suboptimal blood levels of vitamin A, zinc, and iron. These nutrients are primarily found in animal products.

the balance needed for normal cell function. The changes in cell functions lead to the signs and symptoms of the toxicity disease.

For both deficiency and toxicity diseases, the best time to correct the problem is usually at the level of dietary intake, before tissue stores are adversely affected. In that case, no harmful effects on health and cell function occur—they are prevented.

Nutrient Deficiencies Are Usually Multiple Most foods contain many nutrients, so poor diets will affect the intake level of more than one nutrient (Illustration 1.10). Inadequate diets generally produce a spectrum of signs and symptoms related to multiple nutrient deficiencies. For example, protein, vitamin B_{12}, iron, and zinc are packaged together in many high-protein foods. The protein-deficient children you may see in news reports on television are rarely deficient just in protein. They likely have iron, zinc, and B vitamin deficiencies in addition to protein.

The "Ripple Effect"   Dietary changes affect the level of intake of many nutrients. Switching from a high-fat to a low-fat diet, for instance, generally results in a lower intake of calories, cholesterol, and vitamin E as well. So, dietary changes introduced for the purpose of improving the intake level of a particular nutrient produce a ripple effect on the intake of other nutrients.

NUTRITION CONCEPT #5

Humans have adaptive mechanisms for managing fluctuations in nutrient intake.

Healthy humans are equipped with a number of adaptive mechanisms that partially protect the body from poor health due to fluctuations in dietary intake. In the context of nutrition, adaptive mechanisms act to conserve nutrients when dietary supply is low and to eliminate them when they are present in excessively high amounts. Dietary surpluses of energy and some nutrients—such as iron, calcium, vitamin A, and vitamin B_{12}—are stored within tissues for later use. In the case of iron and calcium, the body regulates the amounts absorbed in response to its need for them. The body has a low storage capacity for other nutrients, for instance vitamin C and water, and eliminates any excesses through urine or stools.

Here are some examples of how the body adapts to changes in dietary intake:

- When caloric intake is reduced by fasting, starvation, or dieting, the body adapts to the decreased supply by lowering energy expenditure. Declines in body temperature and the capacity to do physical work also act to decrease the body's need for calories. When caloric intake exceeds the body's need for energy, the excess is stored as fat for energy needs in the future.

- The ability of the gastrointestinal tract to absorb dietary iron increases when the body's stores of iron are low. To protect the body from iron overdose, the mechanisms that facilitate iron absorption in times of need shut down when enough iron has been stored.

- The body protects itself from excessively high levels of vitamin C from supplements by excreting the excess in the urine.

Although these built-in mechanisms do not protect humans from all the consequences of poor diets, they do provide an important buffer against the development of nutrient-related health problems.

NUTRITION CONCEPT #6

Malnutrition can result from poor diets and from disease states, genetic factors, or combinations of these causes.

Malnutrition means "poor" nutrition and results from both inadequate and excessive availability of calories and nutrients in the body. Vitamin A toxicity, obesity, vitamin C deficiency (scurvy), and underweight are examples of malnutrition.

Malnutrition can result from poor diets and also from diseases that interfere with the body's ability to use the nutrients consumed. Diarrhea, alcoholism, cancer, bleeding ulcers, and HIV/AIDS, for example, may be primarily responsible for the development of malnutrition in people with these disorders.

malnutrition
Poor nutrition resulting from an excess or lack of calories or nutrients.

In addition, a percentage of the population is susceptible to malnutrition due to genetic factors. For example, people may be born with a genetic tendency to produce excessive amounts of cholesterol, absorb high levels of iron, or use folate poorly. Some cases of obesity, diabetes, and cancer appear to be related to a combination of genetic predisposition and dietary factors.[15]

NUTRITION CONCEPT #7

Some groups of people are at higher risk of becoming inadequately nourished than others.

Women who are pregnant or breastfeeding, infants, growing children, the frail elderly, the ill, and those recovering from illness have a greater need for nutrients than people who fit into "none of the above." As a result, they are at higher risk of becoming inadequately nourished than other people (Illustration 1.11). In cases of widespread food shortages, such as those induced by natural disasters or war, the health of these nutritionally vulnerable groups is compromised the soonest and the most.

Within the nutritionally vulnerable groups, certain people and families are at particularly high risk of malnutrition. These are people and families who are poor and least able to secure food, shelter, and high-quality medical services. The risk of malnutrition is not shared equally among all persons within a population.

NUTRITION CONCEPT #8

Poor nutrition can influence the development of certain chronic diseases.

Poor nutrition does not result only in nutrient deficiency or toxicity diseases. Faulty diets play important roles in the development of heart disease, hypertension, cancer, osteoporosis, and other **chronic diseases.** Diets high in animal fat, for example, are related to the development of heart disease; those low in vegetables and fruit to cancer; low-calcium diets to osteoporosis; and high-sugar diets to tooth decay. The harmful effects of negative dietary practices on the development of certain diseases and disorders may take years to become apparent.

NUTRITION CONCEPT #9

Adequacy, variety, and balance are key characteristics of a healthful diet.

Healthful diets contain many different foods that together provide calories and nutrients in amounts that promote the optimal functioning of the body. A variety of food is required to obtain all the different essential nutrients we need in our diet. No one food (except breast milk for young infants) contains all the nutrients we need, and most single foods don't come close.

Many different combinations of foods can make up a healthy diet. It could include milk, broccoli, chicken, and apples; or ants, snails, sea cucumbers, and burdock root. What's important is that the contributions of a variety of individual foods add up to an adequate and balanced diet.

Illustration 1.11
Women who are pregnant or breastfeeding and infants are among people who are at a higher risk of becoming inadequately nourished.

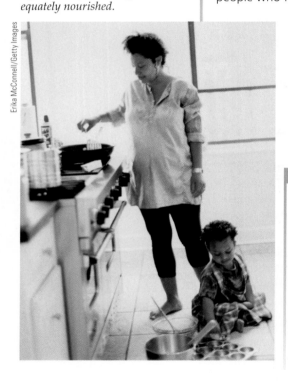

Erika McConnell/Getty Images

chronic diseases
Slow-developing, long-lasting diseases that are not contagious (for example, heart disease, diabetes, and cancer). They can be treated but not always cured.

Nutrient Density

Adequate diets are most easily achieved by including foods that are good sources of a number of nutrients but not packed with calories. Foods that provide multiple nutrients in appreciable amounts relative to calories are considered *nutrient-dense.* Those that provide calories and low amounts of nutrients are called *empty-calorie foods.* Illustration 1.12 shows an example of a nutrient-dense food and an empty-calorie food. It is obviously easier to build an adequate diet around nutrient-dense foods such as fruits, vegetables, breads, cereal, lean meats, and milk (Illustration 1.13) than around foods such as soft drinks, candy, pastries, chips, and alcoholic beverages.

nutrient-dense foods
Foods that contain relatively high amounts of nutrients compared to their calorie value. Broccoli, collards, bread, cantaloupe, and lean meats are examples of nutrient-dense foods.

empty-calorie foods
Foods that provide an excess of calories in relation to nutrients. Soft drinks, candy, sugar, alcohol, and fats are considered empty-calorie foods.

NUTRITION CONCEPT #10

There are no "good" or "bad" foods.

All things in nutriment are good or bad relatively.
—Hippocrates

Nearly eight of ten U.S. adults believe there are "good foods" and "bad foods."[16] Unless we're talking about spoiled stew, poison mushrooms, or something similar, however, no foods can be labeled as either good or bad. There are, however, combinations of foods that add up to a healthful or unhealthful diet. Consider the case of an adult who eats only foods thought of as "good"—for example, raw broccoli, apples, orange juice, boiled tofu, and carrots. Although all these foods are nutrient-dense, they do not add up to a healthy diet because they don't supply a wide enough variety of the nutrients we need. Or take the case of the teenager who occasionally eats fried chicken, but otherwise stays away from fried foods. The occasional fried chicken isn't going to knock his or her diet off track. But the person who eats fried foods every day, with few vegetables or fruits, and loads up on supersized soft drinks, candy, and chips for snacks has a bad diet. It's not any single food or group of foods that define a bad diet; it's the combination of foods generally consumed.

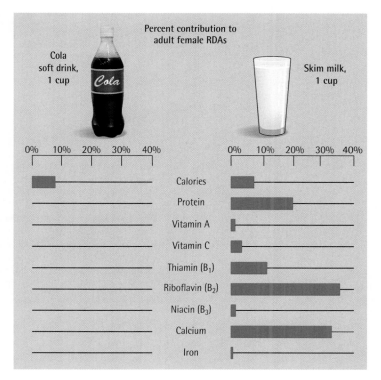

Illustration 1.12
An example of a food with a high nutrient density and a food with a low nutrient density. Percentages given represent percent contributions to adult female RDAs.

The basic nutrition concepts presented here are listed in Table 1.4. It may help you to remember the concepts and to start "thinking like a bona fide nutritionist," if you go back over each concept and give several examples related to it. If you understand these concepts, you will have gained a good deal of insight into nutrition.

Illustration 1.13
"Basic foods" are nutrient dense.

TABLE 1.4

Nutrition concepts

1. Food is a basic need of humans.
2. Foods provide energy (calories), nutrients, and other substances needed for growth and health.
3. Health problems related to nutrition originate within cells.
4. Poor nutrition can result from both inadequate and excessive levels of nutrient intake.
5. Humans have adaptive mechanisms for managing fluctuations in nutrient intake.
6. Malnutrition can result from poor diets and from disease states, genetic factors, or combinations of these causes.
7. Some groups of people are at higher risk of becoming inadequately nourished than others.
8. Poor nutrition can influence the development of certain chronic diseases.
9. Adequacy, variety, and balance are key characteristics of a healthy diet.
10. There are no "good" or "bad" foods.

Nutrition Up Close

Nutrition Concepts Review

Focal Point: Nutrition concepts apply to diet and health relationships. Write the number of the nutrition concept from Table 1.4 that applies to the situation described. Use each concept and do not repeat concept numbers in your responses.

Nutrition Concept Number	Situation
1. _____	The Irish potato famine caused thousands of deaths.
2. _____ and _____	Otis mistakenly thought that as long as he consumed enough calories, protein, vitamins, and minerals, he would stay healthy no matter what he ate.
3. _____	I feel guilty every time I eat potato chips. I wish they weren't bad for me.
4. _____	Phyllis was relieved to learn that her chronic diarrhea was due to the high level of vitamin C supplements she had been taking.
5. _____	A low amount of iron in Tawana's red blood cells was the reason for her loss of appetite and low energy level.
6. _____	Far more young children than soldiers died as a result of the 10-year civil war in Sudan.
7. _____	For the past 20 years, Don's idea of dinner was a big steak and potatoes. His recent heart attack changed his view of what's for dinner.
8. _____	During the two weeks they were backpacking in the Netherlands, Tomás and Ozzie ate very few vegetables and fruits. Their health remained robust, however.
9. _____	Zhang wasn't aware that he had the inherited condition hemochromatosis until he began taking iron supplements and developed iron overload symptoms.

FEEDBACK (answers to these questions) can be found at the end of Unit 1.

Key Terms

antioxidants, page 1-10
calorie, page 1-8
chronic diseases, page 1-18
empty-calorie foods, page 1-19
essential nutrients, page 1-10

food insecurity, page 1-7
food security, page 1-7
malnutrition, page 1-17
metabolism, page 1-14
nonessential nutrients, page 1-11

nutrients, page 1-8
nutrient-dense foods, page 1-19
nutrition, page 1-6
phytochemicals, page 1-10

Review Questions

Answers can be found on the last page of this unit.

	True	False
1. The word *nonessential* as in *nonessential nutrient* means that the nutrient is *not* required for growth and health.	☐	☐
2. Nutrition is defined as "the study of foods, their nutrients, and other chemical constituents and the effects of food contituents on health."	☐	☐
3. Food insecurity is a problem in developing countries, but it is *not* a problem in the United States or Canada.	☐	☐
4. Water is an essential nutrient.	☐	☐
5. The development of standards for nutrient intake levels was prompted		

	True	False
in part by the high rejection rate of World War II recruits due to underweight and nutrient deficiencies.	☐	☐
6. Tissue stores of nutrients decline after blood levels of the nutrients decline.	☐	☐
7. To maintain health, all essential nutrients must be consumed at the recommended level daily.	☐	☐
8. An individual's genetic traits play a role in how nutrient intake effects disease risk.	☐	☐
9. Nutritionally speaking, potato chips can be accurately described as a "bad" food and broccoli as a "good" one.	☐	☐

Media Menu

www.fns.usda.gov
The Food and Nutrition Service's main site provides information in English and Spanish on federal food and nutrition programs such as WIC and Food Stamps (including income eligibility and how to apply), responses to FAQs, links to the Dietary Guidelines, and nutrition education materials.

www.nutrition.gov
Go here for your guide to nutrition and health information. It includes information on food facts, food safety, life-cycle nutrition, food assistance, and nutrient databases.

www.nal.ars.usda.gov/ba/bhnrc/ndl
This terrific site from the National Agricultural Library is stuffed with information on nutrition labeling, composition of foods, lists

of food sources of nutrients, the DRI tables, dietary supplements, and answers to FAQs.

www.nal.usda.gov/fnic/etext/000/05.html
A list of DRI publications you can read online can be found on this site.

www.secondharvest.org
This site is a great source of information on food insecurity.

Notes

1. Anderson SA. Core indicators of nutritional state for difficult-to-sample populations. J Nutr 1990;120:1598.

2. Scheier LM. What is the hunger-obesity paradox? J Am Diet Assoc 2005;105:883–6.

3. Cook JT et al. Child food insecurity increases risks posed by household food insecurity to young children's health. J Nutr 2006;136:1073–6.

4. Wilde PE, Peterman JN. Individual weight change is associated with household food security status. J Nutr 2006;1395–1400.

5. Picciano MF et al., The National Nutrition Summit: history and continued commitment to the nutritional health of the U.S. population, J Nutr 2003;133:1949–52.

6. Bruemmer B. Food biosecurity. J Am Diet Assoc 2003:688–92.

7. Meinhardt PL. Water and bioterrorism: preparing for the potential threat to U.S. water supplies and public health. Ann Rev Public Health 2005;26:213–37.

8. Ricin toxin from the castor bean plant. www.ansci.cornell.edu/plants/toxicagents, and waynesword.palomar.edu, accessed 6/03.

9. Herber D. The stinking rose: oregano-sulfur compounds and cancer. Am J Clin Nutr 1997;66:425–6.

10. Herber D. Phytochemicals beyond anti-oxidation. J Nutr 2004;134:3175S–6S.

11. Standing Committee on the Scientific Evaluation of Dietary Reference Intakes (Food and Nutrition Board, National Academy of Sciences). Washington, DC: National Academy Press; 1997.

12. Recommended Dietary Allowances, www.iom.edu/fnb, accessed 5/04.

13. Ames BN. The metabolic tune-up: metabolic harmony and disease prevention. J Nutr 2003;133:1544S–48S.

14. Based on Mertz W. The essential trace elements. Science 1981;213:1332–8.

15. Trujillo E, Milner J. Nutrigenomics, proteomics, metabolomics, and the practice of dietetics. J Am Diet Assoc 2006;106:403–13.

16. Position of the American Dietetic Association: total diet approach to communicating food and nutrition information. J Am Diet Assoc 2002;102:100.

Nutrition Up Close

Nutrition Concepts Review

Feedback for Unit 1

The nutrition concepts apply to the situations as follows:

1. False (see page 1-11)
2. True (see page 1-6)
3. False (see page 1-8)
4. True (see page 1-11, Table 1.1)
5. True (see page 1-12)
6. False (see page 1-15 and Illustration 1.9)
7. False (see page 1-17)
8. True (see page 1-18)
9. False (see page 1-19)

1. 1
2. 2, 9
3. 10
4. 4
5. 3
6. 7
7. 8
8. 5
9. 6

The Inside Story about Nutrition and Health

NUTRITION SCOREBOARD

	TRUE	FALSE
1 How long people live and how healthy they are depend on four factors: lifestyle behaviors, the environment to which people are exposed, genetic makeup, and access to quality health care.		
2 Diet is related to the top two causes of death in the United States.		
3 The body of modern humans was designed over 40,000 years ago.		

Key Concepts and Facts

- Health and longevity are affected by diet. Other lifestyle behaviors, genetic traits, the environment to which we are exposed, and access to quality health care also affect health and longevity.

- Dramatic changes in the types of foods consumed by modern humans compared with early humans are related in some ways to the development of today's leading health problems.

- The diets and health of Americans are periodically evaluted by national studies.

- The health status of a population changes for the better or worse as diets change for the better or worse.

Answers to
NUTRITION SCOREBOARD

		TRUE	FALSE
1	There *are* no secrets to a long, healthy life.	✔	
2	Diet is associated with the development of heart disease and cancer (which, by the way, cause about half of all deaths in the United States)	✔	
3	Hairstyles may be different, but our bodies are the same as they were 40,000 years ago.	✔	

Nutrition in the Context of Overall Health

Think of your body as a machine. How well this machine performs depends on a number of related factors: the quality of its design and construction, the appropriateness of the materials used to produce it, and how well it is maintained.

A machine designed to produce 10,000 copies a day will break down sooner if it is used to make 20,000 copies a day. The repair call will, in all probability, come earlier if the machine is overused *and* poorly maintained or if it has a part that doesn't work well. On the other hand, chances are good the copy machine will function at full capacity if it is free from design flaws, skillfully constructed from appropriate materials, properly used, and kept in good shape through regular maintenance.

Although much more complex and sophisticated, the human body is like a machine in some important ways. How well the body works and how long it lasts depend on a variety of interrelated factors. The health and fitness of the human machine depend on genetic traits (the design part of the machine), the quality of the materials used in its construction (your diet), and regular maintenance (your diet, other lifestyle factors, and health care).

Lifestyles exert the strongest overall influence on health and longevity (Illustration 2.1).[1] Behaviors that constitute our lifestyle—such as diet, smoking habits, illicit drug use or excessive drinking, level of physical activity or psychological stress, and the amount of sleep we get—largely determine whether we are promoting health or disease.[2] Of the lifestyle factors that affect health, our diet is one of the most important.[3] In a sense, it is fortunate that diet is related to disease development and prevention. Unlike age, gender, and genetic makeup, our diets are within our control.

People have an intimate relationship with food—each year we put over a thousand pounds of it into our bodies! Food supplies the raw materials the body

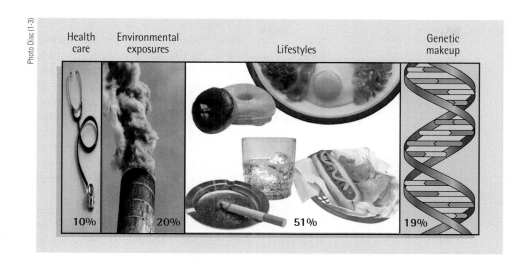

Photo Disc (1-3)

Health care | Environmental exposures | Lifestyles | Genetic makeup

10% | 20% | 51% | 19%

Illustration 2.1
Conditions that contribute to death among adults under the age of 75 years in the United States.
Health care refers to access to quality care; environmental exposures include the safety of one's surroundings and the presence of toxins and disease-causing organisms in the environment; lifestyle factors include diet, exercise, obesity, smoking, and alcohol and drug use; and genetic traits.

Source: Centers for Disease Control, Disease Prevention and Health, 1989 and 2003, www.cdc.gov/nchs, (accessed 6/06).

needs for growth and health, and these, in turn, are affected by the types of food we usually eat. The diet we feed the human machine can hasten, delay, or prevent the onset of an impressive group of today's most common health problems.

The Nutritional State of the Nation

> *Whosoever was the father of a disease, an ill diet was the mother.*
> —GEORGE HERBERT, 1660

Since early in the twentieth century, researchers have known that what we eat is related to the development of vitamin and mineral deficiency diseases, to compromised growth and impaired mental development in children, and to the body's ability to fight off infectious diseases. Seventy years ago in the United States, widespread vitamin deficiency diseases filled children's hospital wards and contributed to serious illness and death in adults (Illustration 2.2). Now, however, dietary excesses are filling hospital beds and reducing the quality of life for millions of Americans.

Photo Researchers, Inc.

Biophoto/Photoresearchers

Illustration 2.2
Vitamin D deficiency (rickets) and niacin deficiency (pellagra) were leading causes of hospitalization of children in the United States in the 1930s.

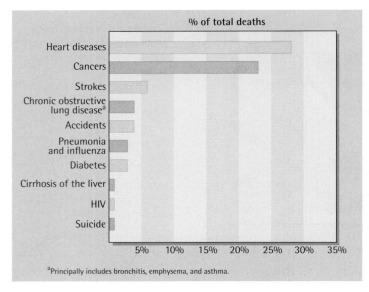

Illustration 2.3
Percentage of total deaths for the top ten leading causes of death in the United States, 2003

Source: Data from www.cdc.gov/nchs (accessed 6/06).

chronic diseases
Slow-developing, long-lasting diseases that are not contagious (for example, heart disease, cancer, diabetes). They can be treated but not always cured.

Today, the major causes of death in Americans are slow-developing, lifestyle-related *chronic diseases.* As Illustration 2.3 indicates, most Americans die from heart disease or cancer. Together these diseases account for 51% of all deaths. Diets high in saturated fat (the type found mainly in animal fat) and low in certain vitamins, vegetables, and fruits are linked to the development of heart disease. Certain types of cancer, such as breast and colon cancer, are related to low fruit and vegetable consumption and excessive levels of body fat.[4]

Diet is also related to four other leading causes of death: *diabetes, stroke, cirrhosis of the liver,* and accidents. Excessive body fat is closely associated with the development of diabetes in middle-aged adults, and excessive levels of alcohol intake contribute to stroke, cirrhosis of the liver, and accidents.[5]

Table 2.1 provides examples of the connections between diet and disease. Many of these relationships, which are highlighted in the following bulleted list, are presented in greater detail in other parts of this book.

- Obesity is the most common nutritional disorder in the United States. Approximately two-thirds of adults are overweight or obese, and one in five children is overweight, too.[6] Being overweight—or, more correctly, being overly fat—increases the likelihood that diabetes, heart disease, hypertension, and certain types of cancer will occur.[7]

- High sodium and low vegetable and fruit intakes and high intakes of sugary foods and refined grain products are associated with hypertension in many adults. Hypertension is one of the most common disorders in adults in the United States, and it is closely linked to heart disease and stroke.

TABLE 2.1

Examples of diseases and disorders linked to diet

Disease or Disorder	Dietary Connections
Heart disease	High saturated and trans fat and cholesterol intakes; low vegetable and fruit intakes; excessive body fat
Cancer	Low vegetable and fruit intakes; excessive body fat and alcohol intake
Stroke	Low vegetable and fruit intake; excessive alcohol intake, high-fat diets
Diabetes (primarily in adults)	Excessive body fat; low vegetable and fruit intake; high saturated fat, sugary foods, and refined grain products intake
Cirrhosis of the liver	Excessive alcohol consumption; poor overall diet
Hypertension (high blood pressure)	Excessive sodium (salt), sugary foods, refined grain products, and alcohol intake; low vegetable and fruit intake; excessive levels of body fat
Iron-deficiency anemia	Low iron intake
Tooth decay and gum disease	Excessive and frequent sugar consumption; inadequate fluoride intake
Osteoporosis	Inadequate calcium and vitamin D, low intakes of vegetables and fruits
Obesity	Excessive calorie intake
Overweight	Omitting breakfast
Chronic inflammation	Excessive calorie intake, high fat diets, low intake of vegetables and fruit, poor vitamin D status

Sources: Farshchi HR et al. Deleterious effects of omitting breakfast on insulin sensitivity and fasting lipid profiles in healthy lean women. Am J Clin Nutr 2005;81:388–96; Schulze MB et al. Dietary pattern, inflammation, and incidence of type 2 diabetes in women. Am J Clin Nutr 2005;82:675–84; Prynne CJ et al. Fruit and vegetable intakes and bone mineral status: a cross-sectional study in 5 age and sex cohorts. Am J Clin Nutr 2006;83:1420–8; Pavia M et al. Association between fruit and vegetable consumption and oral cancer: a meta-analysis of observational studies. Am J Clin Nutr 2006;83:1126–34; Kornman KS. Interleukin 1 genetics, inflammatory mechanisms, and nutrigenetic opportunities to modulate diseases of aging. Am J Clin Nutr 2006;83(suppl):475S–83S.

- Low iron intake is a common problem among women and young children in the United States. Consequently, iron deficiency remains a public health problem in the United States today.

- The frequent consumption of sugars, especially in the form of "sticky sweets," is related to the development of tooth decay and gum disease. Adequate fluoride intakes help prevent tooth decay.

- Low intake of calcium and poor vitamin D status, especially during the teen and young adult years, is associated with the development of *osteoporosis* and bone fractures. Nearly half of all women in the United States fail to get enough calcium in their diets.[8] Osteoporosis is also related to low intakes of vegetables and fruits.

- *Chronic inflammation* is increasingly cited as being a strong risk factor for heart disease, osteoporosis, diabetes, cancer, Alzheimer's disease, and other chronic health problems. Diets rich in colorful fruits and vegetables, moderate intake of fat, adequate vitamin D status, and normal body weight help prevent chronic inflammation and the disorders characterized by it.[9]

Nutrient–Gene Interactions and Health   Some diseases are promoted by interactions between nutrients and genes. Cancer-related genes, for example, can be activated or deactivated by certain components of food. Sulforaphane (pronounced sul-four-ah-phane) found in cabbage, broccoli, and brussel sprouts helps prevent colon cancer. It inactivates a gene that produces a substance that encourages cancer development.[10] About half of the U.S. population is genetically susceptible to cholesterol in the diet. For them, blood cholesterol levels increase with high cholesterol intake. People who lack the trait are considered "cholesterol nonresponders" because their blood cholestrol levels do not increase in response to a high cholesterol intake.[11]

The Importance of Food Choices

People are not born with a compass that directs them to select a healthy diet—and it shows.[12] If given access to a food supply like that available in the United States, people show a marked tendency to choose a diet that is high in animal fat and low in complex carbohydrates (Illustration 2.4). Such a diet also tends to

diabetes
A disease characterized by abnormally high levels of blood glucose.

stroke
The event that occurs when a blood vessel in the brain suddenly ruptures or becomes blocked, cutting off blood supply to a portion of the brain. Stroke is often associated with "hardening of the arteries" in the brain. (Also called a *cerebral vascular accident*.)

cirrhosis of the liver
Degeneration of the liver, usually caused by excessive alcohol intake over a number of years.

osteoporosis
A condition in which bones become fragile and susceptible to fracture due to a loss of calcium and other minerals.

chronic inflammation
Inflammation that lasts weeks, months, or years. Inflammation is the first response of the body's immune system to infection or irritation. Inflammation triggers the release of biologically active substances that promote oxidation and other potentially harmful reactions in the body.

Photo Disc

Illustration 2.4
Lopsided, all-American food choices.

include processed foods that are often high in salt or sugar and low in fiber, vegetables, and fruits. It is this type of diet that poses the greatest risks to the health of Americans.

Diet and Diseases of Western Civilization

Why is the U.S. diet—a "Western" style of eating—hazardous to our health in so many ways? What is it about a diet that is high in animal fat, salt, and sugar and low in vegetables, fruits, and fiber that promotes certain chronic diseases? A good deal of evidence indicates that the chronic diseases now prevalent in the United States and other Westernized countries have their roots in dietary and other changes that have taken place over centuries.

Our Bodies Haven't Changed

Biological processes that control what the body does with food today were developed more than 40,000 years ago. These hardwired, evolution-driven processes exist today because they are firmly linked to the genetic makeup of humans and genes change very little over great spans of time.[13]

Then . . .   For the first 200 centuries of existence, humans survived by hunting and gathering (Illustration 2.5). They were constantly on the move, pursuing wild game or following the seasonal maturation of fruits and vegetables. Meat, berries, and many other plant products obtained from successful hunting and gathering journeys spoiled quickly, so they had to be consumed in a short time. Feasts would be followed by famines that lasted until the next successful hunt or harvest.[14]

Illustration 2.5
Hunter-gatherers still exist in the world, but their numbers are diminishing. It is estimated that hunter-gatherers consume approximately 3000 calories daily due to their physically demanding way of life.

Illustration 2.6
The disconnect between high animal fat, high salt, high sugar, and processed foods in Western-type diets (right) and wild plants and animal foods consumed by our early ancestors (left).
Foods consumed by hunter-gatherers shown in the photograph include bird's eggs, wild cucumbers, roots, nut, and berries. Not shown are grubs and other insects, which might be consumed as quickly as they are discovered.

. . . and Now The bodies of modern humans, adapted to exist on a diet of wild game, fish, fruits, nuts, seeds, roots, vegetables, and grubs; to survive periods of famine; and to sustain a physically demanding lifestyle, are now exposed to a different set of circumstances. The foods we eat bear little resemblance to the foods available to our early ancestors (Illustration 2.6). Sugar, salt, alcohol, food additives, oils, margarine, dairy products, cereal grains, and refined and processed foods were not a part of their diets. These ingredients and foods came with Western civilization.[13] Furthermore, we do not have to engage in strenuous physical activity to obtain food, and our feasts are no longer followed by famines.

The human body developed other survival mechanisms that are not the assets they used to be. Mechanisms that stimulate hunger in the presence of excess body fat stores, conserve the body's supply of sodium, and confer an innate preference for sweet-tasting foods, as well as a digestive system that functions best on a high-fiber diet, were advantages for early humans. They are not advantageous for modern humans, however, because diets and lifestyles are now vastly different.

Although the human body has a remarkable ability to adapt to changes in diet, health problems of modern civilization such as heart disease, cancer, hypertension, and diabetes are thought to result partly from diets that are greatly different from those of our early ancestors. It appears that the human body was built to function best on a diet that is low in sugar and sodium, contains lean sources of protein, and is high in fiber, complex carbohydrates, and vegetables and fruits (Illustration 2.6).[12] Strong evidence for this conclusion is provided by studies that track how disease rates change as people adopt a Western style of eating.

TABLE 2.2

Life expectancy at birth for countries with high life expectancies

Country	Life expectancy (years)
Japan	81.5
Hong Kong	81.5
Switzerland	80.2
Sweden	79.9
Italy	79.8
Australia	79.7
Canada	79.7
Israel	79.4
Spain	79.3
France	79.2
Norway	78.9
Singapore	78.8
Austria	78.6
Germany	78.5
New Zealand	78.5
Netherlands	78.3
United Kingdom	78.3
Belgium	78.0
Finland	78.0
Greece	78.0
United States	77.1
Denmark	77.0
Portugal	76.9

Source: Health, United States, 2005, www.cdc.gov/nchs (accessed 6/06).

Changing Diets, Changing Disease Rates

Many countries are adopting the Western diet and the pattern of disease that accompanies it. People in Japan, for example, live longer than anyone else in the world—until they move to the United States (Table 2.2). In Japan, diets consist mainly of rice, vegetables, fish, shellfish, and meat (Illustration 2.7). When Japanese people move to the United States, their diets change to include, on average, more meat and fat and fewer complex carbohydrates. Japanese living in the United States are much more likely to develop diabetes (Illustration 2.8), heart disease, breast cancer, and colon cancer than people who remain in Japan.[15]

Dietary habits in Japan are rapidly becoming similar to those in the United States. Hamburgers, fries, steak, ice cream, and other high-fat foods are gaining in popularity. Rates of diabetes, heart disease, and cancer of the breast and colon are on the rise in Japan.[15] Similarly, the "diseases of Western civilization" are occurring at increasing rates in Russia, Greece, Israel, and other countries adopting the Western diet.[16, 17]

Eating more like our early ancestors did given today's food supply is a bit challenging. What types of foods would you choose if you wanted to shape a

Illustration 2.7

Typical Japanese foods. Compare these to the "all-American" food choices shown in Illustration 2.4.

Photo Disc

diet that is closer to that of hunter-gathers? Join Beth and Shandra in making decisions about which food choices would get them closer to the basic diets of early humans in the "Reality Check."

The Power of Prevention

Heart disease, cancer, and other chronic diseases are not the inevitable consequence of Westernization. High-animal-fat diets and lifestyle behaviors (such as smoking) that promote chronic disease can be avoided or changed. Although heart disease is still the leading cause of death in the United States, its rate has declined by 50% in the last 35 years. Reductions in animal fat and cholesterol intake, along with reductions in the proportion of people who smoke, are largely credited with this striking improvement in deaths from heart disease.[6]

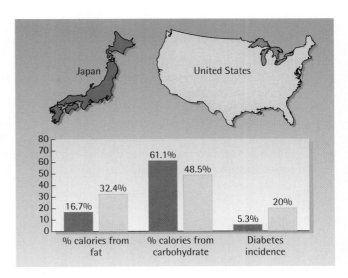

Illustration 2.8
An increased rate of diabetes in Japanese men immigrating to Seattle corresponds to dietary changes.

Source: CH Tsunehara, DL Leonetti, and WY Fujimoto, Diet of second-generation Japanese-American men with and without non-insulin-dependent diabetes (Am J Clin Nutr 1990;52:731–8).

Improving the American Diet

Many efforts are under way to improve the diet and health status of Americans. Like some other countries with high rates of "Western" diseases, the United States has set national health goals and implemented programs aimed at improving health. Goals for changes in health status in the United States are presented in the report, *Healthy People 2010: Objectives for the Nation*. Its objectives for improvements in nutrition are outlined in Table 2.3.

TABLE 2.3
A summary of nutrition objectives for the nation

Improvements in Health Status	Improvements in Health Practices	Improvements in Food Services and Programs
Reduce rates of: • Overweight • Heart disease and stroke • Hypertension • Cancer • Osteoporosis • Diabetes • Congenital (inborn) abnormalities • Growth retardation • Iron deficiency • Baby bottle tooth decay • Food allergy deaths	Increase rates of: • Breast-feeding • Safe and effective weight-loss practices (diet and exercise) • Healthy weight gain in pregnancy Reduce dietary intake of: • Fat • Saturated fat • Sodium Increase dietary intake of: • Vegetables • Fruits • Grain and grain products (especially whole grains) • Calcium	• Improve food safety practices • Increase food security • Increase nutrition education in schools • Improve nutritional quality of school meals and snacks • Increase work site programs in nutrition and weight management • Increase nutrition counseling for patients with heart disease, diabetes, and other nutrition- and diet-related conditions

Source: Data from Healthy People 2010—Objectives for the Nation, www.nutrition.gov (accessed 2/00).

Reality Check

Getting back to the basics—but how?

Beth and Shandra have been roommates for a year and usually shop for groceries together. For the next trip to the grocery store, they decide that each of them will make up a shopping list that includes foods that resemble what their early ancestors might have eaten. Here are the results:

Which list do you think comes closest to matching the basic foods consumed by our early ancestors?
Answers on page 2–11

Beth:
rice, yogurt, pork, honey, olive oil

Shandra:
carrots, nuts, asparagus, fish, blueberries

What Should We Eat?

Attempts to improve the U.S. diet have led to the development of a set of guidelines for healthy eating. Known as the "Dietary Guidelines for Americans" (Illustration 2.9), this document identifies nine key areas in need of improvement and provides practical suggestions for implementing dietary changes. The Dietary Guidelines, which are updated every five years, are a very useful tool for planning and evaluating dietary intake.

In addition to the Dietary Guidelines, recommendations for the selection of foods that contribute to a healthy diet are available. These recommendations are based on the food groups presented in "MyPyramid." MyPyramid is an updated version of the Basic Four Food Groups. The front page of this guide is shown in Illustration 2.10. Foods that don't appear in the MyPyramid groups are processed meats, salted and high-fat snack foods, soft drinks, and sugar-rich desserts. Overall, the Guide urges consumers to select a diet of basic foods at the appropriate caloric level and encourages people to engage in physical activity daily.

Illustration 2.9
Dietary Guidelines for Americans, 2005
www.nal.usda.gov/fnic/ etext/000105.html

USDA

Dietary Guidelines
for Americans
2005

U.S. Department of Health and Human Services
U.S. Department of Agriculture
www.healthierus.gov/dietaryguidelines

Nutrition Surveys: Tracking the American Diet

The food choices people make and the quality of the American diet are regularly evaluated by national surveys. Table 2.4 summarizes the major surveys conducted by the federal government. The first survey was started in 1936. It was conducted in conjunction with the original national program aimed at reducing hunger, poor growth in children, and vitamin and mineral deficiency diseases in the United States. Results of nutrition surveys are used to identify problem areas within the food supply, characteristics of diets consumed by people in the United States, and the nutritional health of the population. They provide information ranging from the amount of lead and pesticides in certain foods to the adequacy of diets of low-income families. Together with the results of studies conducted by university researchers and others, they provide the information needed to give direction to food and nutrition programs and efforts aimed at improving the availability and quality of the food supply.

TABLE 2.4

Periodic, national surveys of diet and health in the United States

Survey	Purpose
1. **National Health and Nutrition Examination Survey (NHANES)**[a]	• Assesses dietary intake, health, and nutritional status in a sample of adults and children in the United States on a continual basis
2. **Nationwide Food Consumption Survey (NFCS)**	• Performs regular surveys of food and nutrient intake and understanding of diet and health relationships among a national sample of individuals in the United States
3. **Total Diet Study** (sometimes called Market Basket Study)	• Ongoing studies that determine the levels of various pesticide residues, contaminants, and nutrients in foods and diets

[a]The Continuing Survey of Food Intakes by Individuals (CSF II) was integrated with NHANES in 2002.

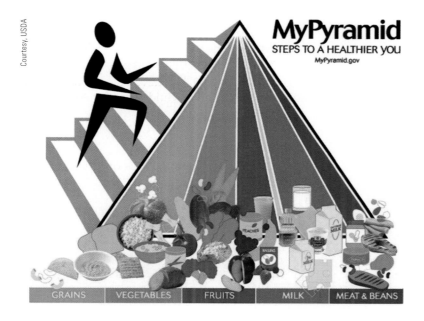

Courtesy, USDA

Illustration 2.10
Making daily food and exercise choices with the USDA MyPyramid.
www.nal.usda.gov/fnic/ etext/00010.5 html

ANSWERS TO
Reality Check

Getting back to the basics—but how?

Shandra's list contains unprocessed plant foods and fish, which would have been available to our early ancestors.

Beth:

Shandra:

Nutrition Up Close

Rate Your Nutrition Knowledge

Focal Point: How well informed are you? Compare your knowledge of nutrition against that of the typical American consumer.

Researchers at the Centers for Disease Control (CDC) contacted a sample of U.S. adults and asked them a set of questions about diet and health relationships. If you answer all of these questions correctly, you know more than most.

1. Which of the following choices represent the *two* best ways to lose weight and keep it off?
   a. Do not eat at bedtime.
   b. Eat fewer calories.
   c. Take diet pills.
   d. Increase physical activity.
   e. Do not eat any fat.
   f. Eat a grapefruit at each meal.

2. Eating a diet high in animal fat _____ the risk of heart disease.
   a. increases
   b. decreases
   c. does not affect

3. A high blood cholesterol level _____ the risk of heart disease.
   a. increases
   b. decreases
   c. does not affect

4. Which of the following substances in food is most often associated with high blood pressure?
   a. cholesterol
   b. sugar
   c. sodium (or salt)

5. Which of the following aspects of diet is related to the development of osteoporosis?
   a. not enough dietary fiber
   b. too much fat
   c. not enough calcium

FEEDBACK (answers to these questions) can be found at the end of Unit 2.

Source: Adapted from material from the U.S. Government—Centers for Disease Control and Prevention.

Key Terms

chronic diseases, page 2-4
chronic inflammation, page 2-5

cirrhosis of the liver, page 2-4
diabetes, page 2-4

osteoporosis, page 2-5
stroke, page 2-4

Review Questions

	True	False
1. Vitamin deficiency diseases remain a major health problem of poor children in the United States.	☐	☐
2. Our diets are directly related to most of the top ten causes of death.	☐	☐

	True	False
3. Low intake of vegetables and fruits is related to the development of heart disease, cancer, hypertension, and osteoporosis.	☐	☐

True False

4. Our genetic makeup is different from that of our earliest ancestors due to changes in the types of food we eat and our lower levels of physical activity. ☐ ☐

5. Although individuals cannot change their genetic makeup, they can change their risk for chronic disease development by making healthful eating and lifestyle changes. ☐ ☐

6. The incidence of chronic diseases such as diabetes, cancer, and heart

True False

disease increases as countries adopt a Western style of eating. ☐ ☐

7. National surveys in the United States assess the nutritional health of the population and the safety of the food supply. ☐ ☐

8. In the United States, the Basic Four Food groups is the most recently published guide to selection of a healthful diet. ☐ ☐

Media Menu

www.healthfinder.gov
A gateway for nutrition and health information can be found at this site.

www.nutrition.gov
This site provides one-stop shopping for information on government-sponsored food and nutrition programs, health statistics, and diet and health relationships.

www.nutrition.gov
You can go directly from this site to information on MyPyramid Food Guide, the Dietary Guidelines for Americans, nutrition and health issues, and Federal Nutrition Assistance programs.

www.nal.usda.gov/fnic/etext.000105.html
Get updates on the Food Guide Pyramid

(plus related educational materials) and the Dietary Guidelines for Americans at this Web site.

www.cdc.gov/nchs
Your gateway to information on Nutrition Monitoring in the United States can be found here.

www.healthypeople.gov
This site presents the Healthy People 2010 Objectives for the Nation.

www.hc-sc.gc.ca
Search the word *nutrition* and gain access to diet and health information, food and nutrition programs, and resources in Canada.

www.mayoclinic.com
Head to the Food and Nutrition center to obtain all sorts of information on nutrition, wellness, and disease.

www.nal.usda.gov/fnic/consumersite/justforyou.htm
This site presents interactive tools for dietary assessment and provides tools for making informed decisions about food, nutrition, and supplement use.

www.nlm.nih.gov/medlineplus
This site provides excellent information on diet–disease relationships.

Notes

1. Gold, M. HALY'S: Measuring lifestyle-related factors that contribute to premature death and disability. In: Estimating the Contributions of Lifestyle-Related Factors to Preventable Death: A Workshop Summary (2005). Washington, DC: National Academies Press, 2005:1–54.

2. Promoting Health: Intervention Strategies for Social and Behavioral Research, Institute of Medicine, National Academy of Sciences, July 2000.

3. German JB et al. Genomics and metabolomics as markers for the interaction

of diet and health: lessons from lipids. J Nutr 2003;133:2078S–83S.

4. Centers for Disease Control, Disease Prevention and Health Promotion Unit. Atlanta; 1989 and 2003, accessed 6/06.

5. Weisburger JH, Approaches for chronic disease prevention based on current understanding of underlying mechanisms, Am J Clin Nutr 2000;71:1710S–4S.

6. www.cdc.gov/nchs, accessed 6/06.

7. Mokdad AH et al. Prevalence of obesity, diabetes, and obesity-related

health risk factors, 2001. JAMA 2003;289:76–9.

8. Appendix D. Dietary Reference Intakes for Calcium through Fluoride. Institute of Medicine, Washington, DC: National Academy Press, 1998.

9. Kornman KS. Interleukin 1 genetics, inflammatory mechanisms, and nutrigenetic opportunities to modulate diseases of aging. Am J Clin Nutr 2006;83(suppl):475S–83S.

10. Go VLW et al. Nutrient-gene interaction: metabolic genotype-phenotype

relationships. J Nutr 2005;135:3016S–20S.

11. Greene CM et al. Maintenance of the LDL cholesterol: HDL cholesterol ratio in an elderly population given a dietary cholesterol challenge. J Nutr. 2005 Dec;135 (12):2793–8

12. Story M, Brown JE. Do young children instinctively know what to eat? N Engl J Med 1987;316:103–6.

13. Cordain L et al. Origins and evolution of the Western diet: health implications for the 21st century. Am J Clin Nutr 2005:81:341–54.

14. Milton K. Hunter-gatherer diets—a different perspective [editorial]. Am J Clin Nutr 2000;71:665–7.

15. Kato H, Haybuchi H. Study of the epidemiology of health and dietary habits of Japanese and Japanese Americans. Japan Journal of Nutrition 1989;47: 121–30.

16. Milton, Hunter-gatherer diets; Cooper R, Schatzkin A, Recent trends in coronary risk factors in the U.S.S.R., Am J Public Health 1982;72:431–40.

17. Sebastian et al., Net acid load of *Homo sapiens*.

18. Trichopoulou AD, Efstathiadis PP, Changes of nutrition patterns and health indicators at the population level in Greece, Am J Clin Nutr 1989;49(suppl):1042–7.

Answers to Review Questions

1. False (see page 2-3)
2. True (see page 2-4)
3. True (see page 2-4, 2-5)
4. False (see page 2-6)
5. True (see page 2-9)
6. True (see page 2-8)
7. True (see page 2-10)
8. False (see page 2-10)

Nutrition Up Close

Rate Your Nutrition Knowledge

Feedback for Unit 2

The number in parentheses indicates the percentage of U.S. adults in the sample who responded correctly.

b and d (73%)

a (80%)

a (86%)

c (58%)

c (43%)

UNIT 3

Ways of Knowing about Nutrition

NUTRITION SCOREBOARD	TRUE	FALSE
1 It is illegal to convey false or misleading information about nutrition in magazine and newspaper articles and on television.		
2 Knowledge about nutrition is gained by scientific studies.		
3 "Double-blind" studies are used to diminish the "placebo effect."		

Key Concepts and Facts

- For the most part, nutrition information offered to the public does not have to be true or even likely true.

- Nutrition information offered to the public ranges in quality from sound and beneficial to outrageous and harmful.

- Science is knowledge gained by systematic study. Reliable information about nutrition and health is generated by scientific studies.

- Misleading and fraudulent nutrition information exists primarily because of financial interests and personal beliefs and convictions.

Answers to NUTRITION SCOREBOARD

		TRUE	FALSE
1	Freedom of speech applies to information about nutrition in articles, speeches, pamphlets, and broadcasts. However, it is illegal to make false or misleading claims about nutrition in advertisements or on product labels and packaging.		✔
2	The "way of knowing" for the field of nutrition is science.	✔	
3	In a double-blind study, neither the research staff nor the research subjects know which subjects are getting the real treatment and which are receiving the fake treatment. This reduces the placebo ("sugar pill") effect, or changes in health that are due to the expectation that a specific treatment will have a particular impact on health.		✔

How Do I Know if What I Read or Hear about Nutrition Is True?

Maria, a college student: "You really ought to try this new Herbal Melt Down Diet I found on the Internet. I used the herbs for a week and lost five pounds!"

Jessie, a premed student: "If I were you, I'd treat that urinary tract infection with cranberry juice."

Newspaper headline: "Eating Cauliflower Daily Prevents Breast Cancer!"

Infomercial: "Lose fat, gain muscle and energy! Eat Complete Cereal Pro bars every day!"

Photo Disc

How do you know if what you read or hear about nutrition is true? With only the information given in these examples, you couldn't know. In reality, however, this is how much of the information we receive about nutrition comes to us—in bits and pieces. The nutrition information offered to the public is a mix of truths, half-truths, and gossip. The information does not have to meet any standard of truth before it can be represented as true in books, magazines, newspapers, TV and radio reports and interviews, pamphlets, the Internet, and speeches.

Opinions expressed about nutrition are protected by the freedom of speech provisions of the U.S. Constitution. Although a tabloid article or an Internet site announcing that 19 foods have negative calories is misleading and fraudulent, it is not illegal. The promotion of nutritional remedies that are not known to work, such as amino acid supplements for hair growth and beef extract for the treatment of cancer, is likewise protected by freedom of speech. It is illegal, however, to put false or misleading information about a product or service on a product label, in a product insert, or in an advertisement. In addition, the U.S. and Canadian mail systems cannot be used to send or to receive payments for products that are fraudulent.

With so much misinformation available, it can be difficult to know what to do when we hear or read something about nutrition that may benefit us personally or perhaps help a friend or relative. Why does such a mix of nutrition sense and nonsense exist? How can you separate sound information from the highly questionable? Where does nutrition information you can trust come from? These questions are addressed in this unit.

Why Is There So Much Nutrition Misinformation?

In addition to protections afforded to marketers by freedom of speech, there are other reasons why consumers are bombarded with nutrition misinformation and ineffective or untested products and services. These reasons can be grouped into two categories: profit, and personal beliefs and convictions. The first and the more important reason, not surprisingly, is the profit motive.

Motivation for Nutrition Misinformation #1: Profit As long as consumers seek quick and easy ways to lose weight, build muscle, slow aging, and reduce stress—goals that cannot be achieved quickly or easily—there will continue to be a huge market for nutritional products and services that offer assistance. The financial incentives are in place:

> . . . [join] a special company with 22nd century breakthrough
> nutritional products and unequaled compensation plan!
> —USA TODAY BUSINESS ADVERTISEMENT

> Teaming up nutritional science with network marketing makes an
> awesome twosome!
> —PROMOTIONAL ARTICLE FOR "NUTRITIONAL SYSTEM" IN AN ADVERTISING MAGAZINE

Not everyone who is in the business of nutrition has the goal of maintaining or improving people's health. Many are in the business to make money from people who are willing to believe their advertisements and buy products or services just in case they work.

People believe nutrition nonsense and buy fraudulent nutrition products for many reasons:

- Many people believe what they want to hear.

- People tend to believe what they see in print.

- Promotional materials are made to sound scientific and true.

- The products offer solutions to important problems that have few or no solutions in orthodox health care.

- Promotional materials often appeal to people who are disenchanted with traditional medical care and the side effects of many medications or to those who want a "natural" remedy.

Max, a 50-year old engineer: "If I hadn't been following a low-fat diet for the last 10 years, I would have had a heart attack by now—just like my dad did when he was 48."

Photo Disc

- Laws that govern truth in advertising are often not enforced.

- New bogus products and services find a ready market because existing bogus products and services don't work.

Instead of evidence and facts, profit-oriented companies use paid testimonials ("It worked for me, it will work for you!"), "medical experts," and Hollywood stars and sports heroes to promote their products. Their advertisements promote ideas like "a wonder to science" or "miraculous" that appeal to some people's inclinations toward the mysterious or the divine. Real remedies aren't advertised in such terms. Nor are they portrayed as being effective because they come from Europe, the Ecuadoran highlands, the ancient Orient, or organic algae ponds.

Illustration 3.1 offers a formula for developing and marketing a fraudulent nutrition product. Try following the steps and make up your own miraculous nutritional cure. Once you've devised your own product, you'll find it easy to detect when other people are doing it for real.

Controlling Profit-Motivated Nutrition Frauds   The Federal Trade Commission (FTC) has authority to remove from the airwaves and Internet advertisements that make false claims. In the past, the FTC has exerted its authority by removing blatantly false and misleading advertisements, including those for the "European Weight Loss Patch," shark cartilage, juices with herbal extracts, coral calcium, and bee pollen. Nevertheless, misleading and inaccurate advertisements still appear because enforcement efforts are weak and are concentrated on very dangerous

Illustration 3.1

Create Your Own Fraudulent Nutrition Product

A Step-by-Step Guide

1. Identify a common problem people really want fixed that cannot easily or quickly be fixed another way.
   EXAMPLES: Obesity, low energy, weak muscles.

2. Make up a nutritional remedy and connect it to a biological process in the body. Try to think of a remedy that probably won't harm anyone. While you are at it, create a catchy name for your product.
   EXAMPLES: An herb that burns fat by speeding up metabolism, vitamins that boost the body's supply of energy, protein supplements that go directly to your muscles.

3. Develop a scientific-sounding explanation for the effect the product has on the body. Refer to the results of scientific studies.
   EXAMPLES: "Research has shown that the combination of herbs in this product stimulates the production of chemicals that trigger the energy-production cycle in the body." "As nutritional scientists have known for a century, the body requires B vitamins to form energy. The more you consume of this special formulation of B vitamins, the more energy you can produce and the more energy you will have!" "The unique combination of amino acids in this product is the same as that found in the jaw muscles of African lions. It is well known that an African lion can lift a 600-pound animal in its teeth."

4. Dream up testimonials from bogus previous users of the product, before-and-after photos, or "expert" opinions to quote. Use terms like "magical," "miraculous," "suppressed by traditional medicine," "secret," or "natural" as much as possible.
   EXAMPLES: Jane Fondu, Indianapolis: "I didn't think this product was going to work. The other products I tried didn't. My doctor couldn't help me lose weight. Thank God for [insert name of product]! I miraculously lost 20 pounds a week while eating everything I wanted." Dr. J. R. Whatsit, DM, NtD, Director of Nutritional Research: "We discovered this secret herb in our laboratories after years of looking for the substance in plants that keeps them from getting fat. Voilà! We found it. This discovery may win us a Nobel Prize." Or, show photos of a skinny person and a beefed-up person. (The photos don't have to be of the same person; the people only need to look similar.)

5. Offer customers a money-back guarantee.

products. The FTC and other federal agencies do respond when several consumers register complaints about a nutritional product or advertisement.

Science for Sale   Here's a true story:

> Friday, 2:00 P.M. A call came into the Nutrition Department from a man who wanted to talk to a nutrition expert. He had heard on last night's news that zinc lozenges were good for treating cold symptoms. Since he had a cold and didn't think the zinc would hurt, he purchased and consumed a whole roll of lozenges. Now he had a horrible, metallic taste in his mouth that he couldn't get rid of, no matter how often he brushed his teeth or gargled with mouthwash. He was worried that the taste was going to stay in his mouth forever. The nutrition expert assured him that it wouldn't. He had overdosed on zinc, and the taste would go away slowly.

This man was not the only one who heard the newscast about zinc and consumed too many lozenges the next day. Sales of zinc lozenges skyrocketed after the newscast, and as it happened, the author of the research study profited handsomely from the increased sales. After completing the study, the author bought shares of stock in the company that made the lozenges.[1] Such a financial tie should have been reported in the article. It is possible that a financial incentive could have influenced how the author presented the study's results.

Many researchers have a vested interest in their research results (Illustration 3.2). A study of 1000 Massachusetts scientists who had published research articles found that one-third held a patent for the product tested, were paid industry consultants, or had another form of financial stake in the research.[2] Another study found that authors of research supportive of a new artificial fat were four times more likely to have financial affiliations with the company producing the product than were authors with no such company ties.[3]

Who Is Conducting the Research?   *Nutrition research is often conducted by people or companies that have a stake in the results.* Although this doesn't mean the research

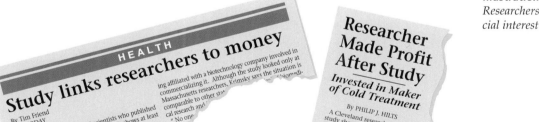

Illustration 3.2
Researchers may have a financial interest in their research.

A checklist for identifying nutrition misinformation

1. Is something being sold? ____ Yes ____ No

2. Does the product or service offer a new remedy for problems that are not easily or simply solved (for example, obesity, cellulite, arthritis, poor immunity, weak muscles, low energy, hair loss, wrinkles, aging, stress)? ____ Yes ____ No

3. Are such terms as "miraculous," "magical," "secret," "detoxify," "energy restoring," "suppressed by organized medicine," "immune boosting," or "studies prove" used? ____ Yes ____ No

4. Are testimonials, before-and-after photos, or expert endorsements used? ____ Yes ____ No

5. Does the information sound too good to be true? ____ Yes ____ No

6. Is a money-back guarantee offered? ____ Yes ____ No

Illustration 3.3
If a red flag comes up as you read about a product or service, beware.

results are invalid, it does mean that the study design should be carefully scrutinized before it is published in a scientific journal and broadcast on the news. People and companies with vested interests in the results of studies tend not to report findings that reflect negatively on a product, or their promotional materials may neglect to mention studies that produced different results. For these reasons, it is important to consider whether the study results or other presentations of nutrition facts may be tainted by a financial motive.

A Checklist for Identifying Nutrition Misinformation Illustration 3.3 lists some common features of fraudulent information for nutrition products and services. If you find any of these characteristics in articles, advertisements, Web sites, or pamphlets or hear them on infomercials or TV and radio interviews, beware of the product or service. People offering legitimate information are cautious and don't exaggerate nutritional benefits to health. They usually aren't selling anything but rather are trying to inform people about new findings so they can make better decisions about optimal nutrition.

The Business of Nutrition News Newspapers, TV stations, magazines, books, and other sources of information make money when the number of viewers or readers is high. To increase viewers or readers, the media attempt to present information that will pique people's interest. Nutrition breakthroughs tend to do that. The media have access to a large number of nutrition studies; more than eight thousand nutrition-related research articles are published each year (Illustration 3.4).

Illustration 3.4
On average, over 800 nutrition-related research articles are published monthly in scientific journals such as those shown here.

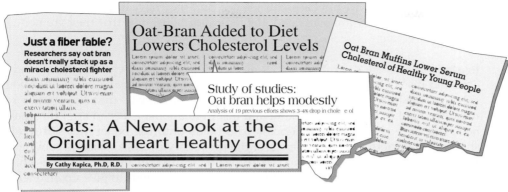

Just a fiber fable?
Researchers say oat bran doesn't really stack up as a miracle cholesterol fighter

Oat-Bran Added to Diet Lowers Cholesterol Levels

Oat Bran Muffins Lower Serum Cholesterol of Healthy Young People

Study of studies:
Oat bran helps modestly
Analysis of 19 previous efforts shows 3-4% drop in cholesterol

Oats: A New Look at the Original Heart Healthy Food
By Cathy Kapica, Ph.D, R.D.

Illustration 3.5
How to confuse the public: a lesson delivered by the headlines about oats and blood cholesterol.
Despite some conflicting reports, a consensus of scientific opinion about the effects of oat bran on blood cholesterol levels has been reached: Regular consumption of oat bran or oatmeal lowers LDL cholesterol (the so-called bad cholesterol) by 5% or more.[4]

Studies reporting new results related to obesity, cancer, diabetes, heart disease, vitamins, and food safety are particularly hot topics. To keep people interested, the media may sensationalize and oversimplify nutrition-related stories. They may report on one study one day and describe another study with opposite results the next. A classic example of headlines about nutrition that totally confused the public is shown in Illustration 3.5. Such study-by-study coverage of nutrition news leaves consumers not knowing what to believe or what to do. A Nutrition Trends Survey by the American Dietetic Association uncovered a strong consumer preference (81% of those sampled) for learning about the latest nutrition breakthroughs only *after* they had been generally accepted by nutrition and health professionals.

Nutrition studies are complex, and the results of one study are almost never enough to prove a point. Decisions about personal nutrition should be based on accumulated evidence that is broadly supported by nutritionists and other scientists.

Motivation for Nutrition Misinformation #2: Personal Beliefs and Convictions

Numerous alternative health practitioners (such as nutripaths, irridologists, electrotherapists, scientologists, and faith healers) use nutrition remedies. Although their approaches to and philosophies about health may vary, these practitioners often have strong beliefs in the benefits of what they do. These firm beliefs may be enough to "talk" some people out of their problems—problems that might have gone away anyway.

There's no way to say whether most of the remedies used by alternative nutrition practitioners work. (Some of the unproven remedies employed are shown in Illustration 3.6.) Strong belief in these remedies, rather than proof that they work or a desire for profits, may rule these practitioners' distribution of nutrition advice and products. The nutri-

Illustration 3.6
Supplements that may be used to treat a variety of health problems.
The safety and effectiveness of many of these products are yet to be proved.

tional cures offered are not based on science, and they have not been scientifically evaluated. Until they are evaluated, the logical conclusion is that they are neither safe nor effective.

Professionals with Embedded Beliefs Professionals who work in health care and research are not immune to the pull of deeply rooted convictions about diet and health relationships. Although they are scientists, they sometimes remain wedded to theories about diet and health relationships even after they have been disproved. The nurse or doctor who holds onto the belief that salt intake should be restricted in pregnancy and the university professor who is convinced that pesticides on foods pose no risk to health are two examples of professional sources of misinformation. The failure of such professionals to give up erroneous convictions about diet and health adds to the flow of nutrition misinformation as well as to consumer confusion and increased health risk.

How to Identify Nutrition Truths

The true method of knowledge is experiment.
—WILLIAM BLAKE, 1788

Where does nutrition information you can count on come from? There is only one way to identify sound nutrition information—you have to put it to the test and see if it survives the dispassionate, systematic examination dictated by science.

Science is a unique and powerful explanatory system. It produces information based on facts and evidence. Our understanding of nutrition is based on scientifically determined facts and evidence obtained from laboratory, animal, and human studies. These types of studies provide information that qualifies for use when developing public policies about nutrition and health (Illustration 3.7). Science delivers information eligible for inclusion in textbooks about nutrition. (Many faculty members review the information in nutrition textbooks for scientific accuracy.)

Sources of Reliable Nutrition Information

Reliable sources of information about nutrition use the standards of proof required by science. They report decisions about nutrition and health relationships that are based on multiple studies and arrived at by "scientific consensus." These decisions represent the majority opinion of scientists knowledgeable about a particular nutrition topic.

Nutrition recommendations made to the public, such as the Dietary Guidelines for Americans and the Dietary Reference Intakes, are based on the consensus of scientific opinion. The information is made available to the public not to sell a product or to pass on an ideology, but to inform consumers honestly about nutrition and to help people use the information to maintain or improve their health. Organizations and individuals offering reliable nutrition information are listed in Table 3.1.

Nutrition Information on the Web Web-based search engines are transforming the way we get information about foods, diets, supplements, and health. Eight in 10 Internet users, including scientists and consumers, look for health and nutrition information online.[5] As is the case for many other media sources, the accuracy of nutrition information made available on the Web varies considerably. In

Reality Check

Microwaves and Plastic

You've got mail: DO NOT microwave foods in plastic! Hot plastic releases highly toxic dioxins that can cause cancer . . . No source for the facts stated is given in the e-mail. Cyndi's and Scott's responses:

Who gets the "thumbs up"?
Answers on next page

Cyndi:
I'll check it out at the FDA Web site.

Scott:
I was afraid of that. Now I know microwaving food in plastic containers is dangerous.

The Warning Signs of poor nutritional health are often overlooked. Use this checklist to find out if you or someone you know is at nutritional risk.

Read the statements below. Circle the number in the yes column for those that apply to you or someone you know. For each yes answer, score the number in the box. Total your nutritional score.

DETERMINE YOUR NUTRITIONAL HEALTH

	YES
I have an illness or condition that made me change the kind and/or amount of food I eat.	2
I eat fewer than two meals per day.	3
I eat few fruits or vegetables, or milk products.	2
I have three or more drinks of beer, liquor, or wine almost every day.	2
I have tooth or mouth problems that make it hard for me to eat.	2
I don't always have enough money to buy the food I need.	4
I eat alone most of the time.	1
I take three or more different prescribed or over-the-counter drugs a day.	1
Without wanting to, I have lost or gained 10 pounds in the last 6 months.	2
I am not always physically able to shop, cook and/or feed myself.	2
TOTAL	

Total Your Nutritional Score. If it's . . .

0–2 **Good!** Recheck your nutritional score in 6 months.

3–5 **You are at moderate nutritional risk.** See what can be done to improve your eating habits and lifestyle. Your office on aging, senior nutrition program, senior citizens center, or health department can help. Recheck your nutritional score in 3 months.

6 or more **You are at high nutritional risk.** Bring this checklist the next time you see your doctor, dietitian, or other qualified health or social service professional. Talk with them about any problems you may have. Ask for help to improve your nutritional health.

These materials developed and distributed by the Nutrition Screening Initiative, a project of:

AMERICAN ACADEMY OF FAMILY PHYSICIANS   THE AMERICAN DIETETIC ASSOCIATION   NATIONAL COUNCIL OF THE AGING

Remember that warning signs suggest risk, but do not represent diagnosis of any condition.

Illustration 3.7
This evidence-based tool for assessing poor nutritional health in seniors was developed as part of the Nutrition Screening Initiative by the American Academy of Family Physicians, the American Dietetics Association, and the National Council on the Aging.

ANSWERS TO *Reality Check*

Microwaves and Plastic

Good thinking, Cyndi. Health scares in your inbox happen. The FDA Web site at www.cfsan.fda.gov reports that plastic does leach into food cooked in microwave ovens, but not enough of it enters food to cause harm. It also notes that plastic wraps and containers do not contain dioxins.

Cyndi:

Scott:

TABLE 3.1
Reliable sources of nutrition information[a]

Source of Nutritional Information	Examples
Nonprofit, professional health organizations	American Heart Association American Cancer Society National Kidney Foundation American Red Cross
Scientific organizations	National Academy of Sciences American Society for Clinical Nutrition American Society of Nutrition Scientists
Government publications: nutrition, diet, and health reports	National Institutes of Health Surgeon General Food and Drug Administration Centers for Disease Control U.S. Department of Agriculture
Registered dietitians	Hospitals Public health departments Extension service Universities
Nutrition textbooks	College and university nutrition courses Nutrition faculty of accredited universities

[a]See Appendix B for more details about reliable sources of nutrition information.

general, the most reliable sites are those developed by government health agencies (Web addresses that end in or include .gov), educational institutions (.edu), and professional health and science organizations (.org).[6] Specific Web sites that offer nutrition information you can count on are given on page 3-20 under "Media Menu."

Who Are Qualified Nutrition Professionals?

Many people refer to themselves as "nutritionists," but only some of them are qualified based on education and experience. These individuals are registered, licensed, or certified dietitians or nutritionists who meet qualifications established by national and state regulations. They have demonstrated a mastery of knowledge about the science of nutrition and appropriate clinical practices. Table 3.2 (p. 3–11) describes the qualifications of those who legitimately use the title *dietitian* or *nutritionist*. The specific titles vary somewhat depending on state regulations and laws governing the practice of nutrition and dietetics.

TABLE 3.2

Who's who in nutrition and dietetics:
Laws and regulations governing practice[7]

Title	Qualifications
Registered dietitian	Individual who has acquired knowledge and skills necessary to pass a national registration examination and participates in continuing professional education. Qualification is conferred by the Commission of Dietetic Registration.
Licensed dietitian/nutritionist	Individual (usually a registered dietitian) who is qualified to practice nutrition counseling based on education and experience. Nonlicensed individuals who practice nutrition counseling can be prosecuted for practicing without a license. Licenses are conferred by state law.
Certified dietitian/nutritionist	Individual who meets certain educational and experience qualifications. Unqualified individuals may practice nutrition counseling but have to call themselves something else (such as "nutritional counselor"). Certification is established by state regulation.
Dietitian/nutritionist	Individuals with any type of education or experience may register with the state of California as a dietitian or nutritionist

It is difficult for anyone to have all the information needed to make sound decisions about every nutrition question or issue that arises. When trying to make sense out of the information you read or hear, don't hesitate to get help. Visit the Web sites at the end of this unit, check out the index of this book for the relevant topic, or call your local health department or area Food and Drug Administration office. The more solid your information, the better your decisions will be.

The Methods of Science

Does the term *scientific method* conjure up an image of beady-eyed, white-coated scientists working diligently in windowless basement laboratories? Although we tend to assume that scientists are busy advancing our knowledge of nutrition and many other fields, for most people the methods of science are a mystery. They should not be (Illustration 3.8). Consumers are big users of nutrition information. To make sound judgments about nutrition and health, people need to know how to distinguish results produced by scientific studies from those generated by personal opinion, product promotions, and bogus studies.

Illustration 3.8
The scientific method is not mysterious at all, but a carefully planned process for answering a specific question.

There are many established methods of science. The specific methods employed vary from study to study depending on the type of research conducted. Nevertheless, all types of scientific studies have one feature in common: They are painstakingly planned. Planning is the *most* important, and often the most time-consuming, part of the entire research process.

Developing the Plan

The first part of the planning process entails clearly stating the question to be addressed—and, one hopes, answered—by the research. For the purposes of illustration, let's say our research will address the effects of vitamin *X* supplements on hair loss (Table 3.3 and Illustration 3.9). The idea for the research came from a

Photo Disc

Illustration 3.9
Do high doses of vitamin X increase hair loss?

TABLE 3.3

Overview of a human nutrition research study: Vitamin X supplements and hair loss[a]

A. **Pose a clear question:** "Do vitamin *X* supplements increase hair loss?"

B. **State the hypothesis to be tested:** "Vitamin *X* supplements of 25,000 IU per day taken for three months increase hair loss in healthy adults."

C. **Design the research:**

    1. *What type of research design should be used?* "In this study, a clinical trial will be used. The supplement and placebo will be allocated by a double-blind procedure."

    2. *Who should the research subjects be?* "The study will exclude subjects who may be losing or gaining hair due to balding, hair treatments, medications, or the current use of vitamin *X* supplements."

    3. *How many subjects are needed in the study?* "The required sample size is calculated to be 20 experimental and 20 control subjects."

    4. *What information needs to be collected?* "Information on hair loss, conditions occurring that might affect hair loss, and the use of supplements and placebos will be collected."

    5. *What are accurate ways to collect the needed information?* "The study will use the 'measure the hairs in a square inch of scalp' technique."

    6. *What statistical tests should be used to analyze the results?* "Appropriate tests identified."

D. **Obtain approval for the study from the committee on the use of humans in research:** "Approval obtained."

E. **Implement the study design:** "Implemented."

F. **Evaluate the findings:** "Subjects receiving the 25,000 IU of vitamin *X* for three months lost significantly more hair than subjects receiving the placebo. Hypothesized relationship found to be true."

G. **Submit paper on the research for publication in a scientific journal or other document.**

[a]This is a fictitious study used for illustrative purposes only.

---

study that hinted that users of vitamin *X* supplements had increased hair loss. In that study, adults were given 25,000 international units (IU) of vitamin *X* for three months to test its safety. Although the study found that vitamin *X* had no adverse effects on health and offered some benefits, several subjects who had received the supplements complained of hair loss. Accordingly, our study poses the question, "Do vitamin *X* supplements increase hair loss?"

The Hypothesis: Making the Question Testable

The question is then transformed into an explicit hypothesis that can be proved or disproved by the research. (*Hypothesis* and other research terms used in this unit are defined in Table 3.4.) The results of the research must provide a true or false response to the hypothesis and not an explanatory sentence or paragraph. So, in this example, "vitamin *X* increases hair loss" wouldn't do as a hypothesis because it leaves too many questions about the relationship unanswered. The effect of vitamin *X* on hair loss may depend on the amount of vitamin *X* given; how long it is taken; or on the research subjects' health, age, and other characteristics. The hypothesis "vitamin *X* supplements of 25,000 IU per day taken for 3 months by healthy adults increase hair loss" is concrete enough to be addressed by research.

TABLE 3.4
A short glossary of research terms

Association	The finding that one condition is correlated with, or related to another condition, such as a disease or disorder. For example, diets low in vegetables are associated with breast cancer. Associations do not prove that one condition (such as a diet low in vegetables) causes an event (such as breast cancer). They indicate that a statistically significant relationship between a condition and an event exists.
Cause and effect	A finding that demonstrates that a condition causes a particular event. For example, vitamin C deficiency causes the deficiency disease scurvy.
Clinical trial	A study design in which one group of randomly assigned subjects (or subjects selected by the "luck of the draw") receives an active treatment and another group receives an inactive treatment, or "sugar pill," called the placebo.
Control group	Subjects in a study who do not receive the active treatment or who do not have the condition under investigation. Control periods, or times when subjects are not receiving the treatment, are sometimes used instead of a control group.
Double blind	A study in which neither the subjects participating in the research nor the scientists performing the research know which subjects are receiving the treatment and which are getting the placebo. Both subjects and investigators are "blind" to the treatment administered.
Epidemiological studies	Research that seeks to identify conditions related to particular events within a population. This type of research does not identify cause-and-effect relationships. For example, much of the information known about diet and cancer is based on epidemiological studies that have found that diets low in vegetables and fruits are associated with the development of heart disease.
Experimental group	Subjects in a study who receive the treatment being tested or have the condition that is being investigated.
Hypothesis	A statement made prior to initiating a study of the relationship sought to be proved (found to be true) by the research.
Meta-analysis	An analysis of data from multiple studies. Results are based on larger samples than the individual studies and are therefore more reliable. Differences in methods and subjects among the studies may bias the results of meta-analyses.
Peer review	Evaluation of the scientific merit of research or scientific reports by experts in the area under review. Studies published in scientific journals have gone through peer review prior to being accepted for publication.
Placebo	A "sugar pill," an imitation treatment given to subjects in research.
Placebo effect	Changes in health or perceived health that result from expectations that a "treatment" will produce an effect on health.
Statistically significant	Research findings that likely represent a true or actual result and not one due to chance.

The Research Design: Gathering the Right Information

Poor research design, or the lack of a solid plan on how the research will be conducted, is often a weak link that renders studies useless. It's the "oops, we forgot to get this critical piece of information!" or the "how did all the measurements come out wrong?" at the end of a study that can ruin months or even years of work. Each step in the research process must be thoroughly planned. In research, there are no miracles. If something can go wrong because of incomplete planning, it probably will.

Research designs are often based on the answers to the following questions:

- What type of research design should be used?

- Who should the research subjects be?

- How many subjects are needed in the study?

- What information needs to be collected?

- What are accurate ways to collect the needed information?

- What statistical tests should be used to analyze the findings?

What Type of Research Design Should Be Used? Several different research designs can be used to test hypotheses. We could use an *epidemiological study* design to determine if hair loss is more common among people who take vitamin X supplements than among people who do not. Researchers commonly use this type of study to identify conditions that are related to specific health events in humans. To provide preliminary evidence, we could use animal studies that determine hair loss in supplemented and unsupplemented animals. Or we could use another design, such as a *clinical trial.* Since this design would work well for the proposed hypothesis about vitamin X supplements, we'll follow the rules for conducting a clinical trial in this example. (You can see a map of the research design used for this hypothetical study in Illustration 3.10.)

The purpose of clinical trials is to test the effects of a treatment or intervention on a specific biological event. In our example, we will test the effect of vitamin X supplements on hair loss.

■ **The Experimental and Control Groups** Clinical trials, as well as other research studies that address questions about nutrition, require an *experimental group* (the group of subjects who receive vitamin X in this example) and a *control group* (the comparison group that receives a *placebo* and not vitamin X). Never trust the results of a study that didn't employ both, and here's why. How do we know the effect of a certain treatment isn't due to something *other* than the treatment? What if we gave a group of adults vitamin X supplements and they lost hair? Does this mean the vitamin X caused the hair loss? Could it be that the subjects lost no more hair during the treatment period than they would have lost without the vitamin X? We can't know whether vitamin X supplements produce hair loss if we don't know how much hair is lost without vitamin X.

After measuring their usual hair loss, we will randomly assign (by the "luck of the draw") people in the study to serve in either the experimental or the control group. Individuals assigned to the control group will be given pills that look, taste, and feel like the vitamin X supplements but have no effect on hair loss or gain. Neither the research staff who will be assessing the hair loss nor the people in the study will know who is getting vitamin X and who is getting the placebo (Illustration 3.11). Scientists use this *double-blind procedure* because knowing which group is which may affect people's expectations and change the results. This important topic deserves additional attention.

■ **The "Placebo Effect"** The *placebo effect* can cause a good deal of confusion in research. That's because people tend to have expectations about what a treatment will do, and those expectations can influence what happens.

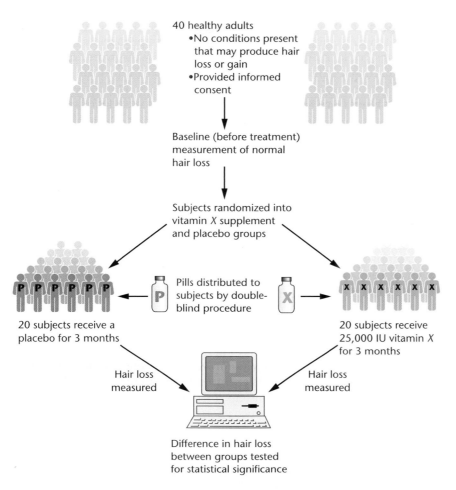

40 healthy adults
•No conditions present that may produce hair loss or gain
•Provided informed consent

Baseline (before treatment) measurement of normal hair loss

Subjects randomized into vitamin *X* supplement and placebo groups

Pills distributed to subjects by double-blind procedure

20 subjects receive a placebo for 3 months

20 subjects receive 25,000 IU vitamin *X* for 3 months

Hair loss measured

Hair loss measured

Difference in hair loss between groups tested for statistical significance

Illustration 3.10
Map of the design of the vitamin X supplement study.

A good example of the placebo effect occurred in a study that tested the effectiveness of a medication in reducing binge eating among people with bulimia.[8] After the usual number of binge-eating episodes was determined, 22 women with bulimia were given either the medication or a placebo in a double-blind fashion. The number of binge-eating episodes was then reassessed. At the end of the study period, a whopping 78% reduction in binge-eating episodes was found among women taking the medication. But binge-eating episodes dropped by 70% in the control group. Was the medication effective? No. Was the expectation that a medication would help effective? Yes, at least for the time covered by the study. Due to the placebo effect, a reduction of greater than 90% in episodes of binge eating would have been needed to conclude that the medication had a real effect on binge eating.

Who Should the Research Subjects Be?   Although expedient, it is not enough to say, "Well, let's use ten faculty members in the study." The type and number of subjects employed by the research are important considerations.

The type of subjects involved in research is important, because we need to exclude people who have conditions that might produce the problem the research is examining. For example, hair loss may result from periodic balding, a bad perm, the use of mousse or hair spray, or illnesses or medications. If we include people with these conditions and circumstances in the study, it will be difficult to determine whether hair loss is due to vitamin *X* or some-

Illustration 3.11
Which is the vitamin X supplement? Neither the subject nor the investigator should know which is vitamin X and which is the placebo.

Stan Maddock

thing else. In addition, it is important to exclude from the study people who already take vitamin X supplements or are bald. An inappropriate or biased selection of subjects could make the results useless or, in the worst case, make the study's results come out in a predetermined way.

How Many Subjects Are Needed in the Study?   The number of subjects needed for a study is a mathematical question, and we won't go into the formulas here. The number of subjects is based upon the number needed to separate true differences between the experimental and control groups from differences that are due to chance (Illustration 3.12). Suppose that hair loss normally varies from 40 to 80 strands per week. We'll need to include enough subjects in the study to make sure differences in hair loss between the groups are not due to normal fluctuations in hair loss. This point is important because studies employing too few subjects lead to inconclusive results and a great deal of controversy about nutrition. One or a few subjects are never enough to prove a point about nutrition.

Let's assume that the mathematical formulas applied to the vitamin X study show that 20 people are needed in both the experimental group and the control group.

What Information Needs to Be Collected?   Information recorded by researchers must represent an evenhanded approach to discovering the facts. We must identify not only the findings that may support the hypothesis, but *those that may refute it*. In order to know if vitamin X supplements increase hair loss, we need to know how much hair is lost, whether the subjects faithfully took the supplement, and whether something came up during the study (such as a bad perm) that might alter hair loss. The presence of such conditions should be determined by methods known to provide accurate results.

Photo Disc

Herbert Spichtinger/Zefa/Corbis

Illustration 3.12
How many subjects are needed in the study? It's important to get the number right.

Nutrition Up Close

Checking Out a Fat-Loss Product

Focal Point: How to make an informed decision about the truthfulness of an advertisement for a fat-loss product.

Read the accompanying advertisement for a fat-loss product. Then, check out the information using the checklist for identifying nutrition misinformation.

Obesity Research Ce...
P.O. Box 281752,
Miami, FL

**Lose Over 2 Inches
In 3 Weeks**

**A Scientifically Proven Fat
Reduction Cream That
Actually Works!**

A scientific, double-blind, placebo-controlled research study of both sexes demonstrated that LIPID MELT fat reduction cream reduces inches from the thigh area. Subjects lost an average of 2 inches from each thigh in only 3 weeks!

Subjects from a preliminary pilot study reported that LIPID MELT cream reduced inches from their abdomen. Both studies reported no skin rashes, discomfort or sensitivity. Formulated with patented liposome technology, LIPID MELT contains no drugs, only natural active ingredients. Initially sold only in salons and spas.

Six-week supply: $25 + $4.95 s/h. We guarantee that you will lose inches or your money will be refunded.

Obesity Research

1. Is something being sold?

   Yes   No

2. Is a new remedy for problems that are not easily or simply solved being offered?

   Yes   No

3. Are terms such as *miraculous, magical, secret, detoxify, energy restoring, suppressed by organized medicine, immune boosting,* or *studies prove* used?

   Yes   No

4. Are testimonials, before-and-after photos, or expert endorsements used?

   Yes   No

5. Does the information sound too good to be true?

   Yes   No

6. Is a money-back guarantee offered?

   Yes   No

Optional: Repeat the activity using a nutrition-related advertisement from the Internet.

FEEDBACK (answers to these questions) can be found at the end of Unit 3.

What Are Accurate Ways to Collect the Needed Information? "Garbage in—garbage out." If the information obtained on and from subjects is inaccurate, then the study is worthless. To avoid this problem, good research employs methods of collecting information known to produce accurate results.

Let's consider the problem of measuring hair loss. How can we do that accurately? First, we look for a method that has already been demonstrated by research to be accurate. Otherwise, we'll first have to prove that our method for measuring hair loss is accurate. For this example, assume an accurate method is used.

What Statistical Tests Should Be Used to Analyze the Findings? Before collecting the first piece of information, we'll select appropriate tests for identifying *statistically significant results* of the research. Many different statistical tests can be used to identify significant differences between the findings from the experimental and control groups. The trick here is to make sure to select the appropriate tests for proving or disproving the hypothesis. Without such tests, it is often very hard to decide what the findings mean. What if the study shows the group that took vitamin X lost an average of 5% more hair than the control group? Is that 5% difference due to the vitamin X or to something else, such as chance fluctuations in normal hair loss? Well-chosen statistical tests will tell us whether the differences between groups are in all probability real or due to chance occurrences or coincidence.

Obtaining Approval to Study Human Subjects

An important step in the planning process is applying for approval to conduct the proposed research on human subjects. Universities and other institutions that conduct research have formal committees, called institutional review boards, that scrutinize plans to make sure proposed studies follow the rules governing research on human subjects. For this study, we will have to show to the committee's satisfaction that the level of vitamin X employed is safe. As a part of this process, the committee usually requires that human subjects consent, in writing, to participate in the study.

Implementing the Study

With the design in place and the appropriate approvals obtained, it is time to implement the study. Assume that, due to the study's solid design and importance, we have been awarded a grant to fund the research. We can now recruit subjects for the study and enroll them if they are eligible and consent to participate. Measurements of hair loss are performed according to schedule, use of the pills by subjects is monitored, results are checked for errors and entered into a computerized data file, and statistical tests are applied. This process usually ends with computer printouts that exhibit the findings.

Making Sense of the Results

Let's say subjects who received the supplement lost 30% more hair than the control subjects did. Having applied the appropriate statistical test, we find that 30% is a highly significant difference. Does that mean that vitamin X supplements *cause* hair loss? *Cause* is a strong word in research terms. It implies that a **cause-and-effect relationship** (that the vitamin X supplements caused the hair loss) exists. But many factors can contribute to hair loss, and since many causes may be unknown or not measured by the study, it is difficult to conclude with absolute certainty that vitamin X by itself caused the hair loss. Assume 4 of the 20 subjects in the experimental group who were known to have taken the vitamin X supplement as indicated lost less hair than usual. The vitamin X supplements *didn't* cause them to lose hair.

We could conclude from this research (if, as a reminder, there was such a thing as vitamin X) that vitamin X supplements, given to healthy adults at a dose of 25,000 IU per day for three months, are strongly *associated* with hair loss. The term *associated* as used here means that the vitamin X supplements were strongly *related* to hair loss but may not have *caused* the hair loss. The hypothesized relationship between supplemental vitamin X and hair loss would be found to be true. Although the research strongly indicates a cause-and-effect relationship, additional studies would be needed to prove the relationship exists in other groups of people and at different doses of vitamin X.

After determining the results, we can write a paper describing the research and its conclusions and submit it for publication in a scientific journal. If judged to be acceptable after review by other scientists, the paper is published.

Complete coverage of how nutrition questions are addressed through research would take a three-course sequence. Nevertheless, it is hoped that the information presented here makes it easier to judge the likely accuracy of reports about nutrition studies in the popular media. There is a lot more to a "Vitamin X Supplements Linked to Hair Loss" story than meets the eye.

Science and Personal Decisions about Nutrition   Science is based on facts and evidence. The grounding ethic of scientists is that facts and evidence are more sacred than any other consideration. These characteristics of science and scientists are strong assets for the job of identifying truths.

Although imperfect because they are undertaken by humans in an environment of multiple constraints, only scientific studies produce information about nutrition you can count on. Evidence is the single best ingredient for decision making about nutrition and your health.

Key Terms

association, page 3–13
cause-and-effect relationship, page 3–18
clinical trial, page 3–14
control group, page 3–14

double-blind procedure, page 3–14
epidemiological study, page 3–14
experimental group, page 3–14
hypothesis, page 3–12
meta-analysis, page 3–13

peer review, page 3–13
placebo, page 3–14
placebo effect, page 3–15
statistically significant results, page 3–18

Review Questions

	True	False
1. By law, nutrition information presented in articles and on television has to be scientifically accurate.	☐	☐
2. The leading motivation underlying the presentation of nutrition misinformation is the profit motive.	☐	☐
3. The Food and Drug Administration (FDA) has the authority to remove from the airwaves and the Internet advertisements making false claims about nutrition.	☐	☐
4. Individuals can help combat the proliferation of bogus nutrition advertisements and products by notifying the Federal Trade Commission (FTC) of their existence.	☐	☐

	True	False
5. Nutrition products and services that offer a "money back guarantee" are most likely to work as advertised.	☐	☐
6. Nutrition information offered by a scientist or a medical doctor can be counted on as being accurate.	☐	☐
7. The term *registered dietitian* means that individuals with the title have acquired knowledge and skills necessary to pass a national examination on nutrition and participate in continuing education.	☐	☐
8. The association between vegetable and fruit intake and breast cancer development is an example of a cause-and-effect relationship.	☐	☐

	True	False

9. Research results may be unreliable if too few subjects are used in studies. ☐ ☐

10. Research articles published in scientific journals undergo peer review, a process by which experts in the areas under investigation evaluate articles prior to their acceptance for publication. ☐ ☐

Media Menu

pubmedcentral.nih.gov
PubMed Central is the National Institutes of Health's source of free, full-text journal articles for the biomedical and life sciences.

http://scholar.google.com
Google Scholar provides a simple way to search for peer-reviewed papers and articles from academic publishers, professional societies, and other scholarly organizations.

www.senseaboutscience.org.uk
Developed to respond to the misrepresentation of science and scientific evidence on issues that matter to society, this site covers and updates topics ranging from scares about plastic bottles to fluoride. The site helps people query the status of science reported on the internet.

site:gov, site:nih.gov, site:edu, and site:org
Using these terms in Google searches limits results to those from the government (gov), the National Institute of Health (nih), academic institutions (edu), or nonprofit organizations (org).

www.ftc.gov/ftc/consumer.htm
At this Web site you'll find publications with advice on avoiding scams and rip-offs and the button for filing a consumer fraud complaint.

www.altmedicine.com
This frequently updated site covers alternative health, diet and health, and other nutrition topics based on recent scientific reports.

www.eatright.org
The American Dietetic Association's Web site is an excellent source of information about diet, health, and qualified dietetics personnel in your area.

www.nlm.nih.gov/medlineplus
Maintained by the National Library of Medicine, this site is a sure bet for authoritative information on food and nutrition topics, ongoing clinical trials, and other information. It also offers a medical dictionary and glossary.

www.quackwatch.com
Uncovers health frauds, including nutrition scams, dubious allergy treatments, and suspect alternative medicines.

Notes

1. Who's minding the lab? Tufts University Health and Nutrition Letter 1997, May:4.

2. Who's minding the lab?

3. Levine J, Gussow JD, et al. Authors' financial relationships with the food and beverage industry and their published positions on the fat substitute Olestra. Am J Pub Health 2003;93:664–9.

4. Schneeman BO. Building scientific consensus: the importance of dietary fiber. Am J Clin Nutr 1999;69:1.

5. Steinbrook R. Searching for the right search—reaching the medical literature. N Engl J Med 2006;354:4–7

6. Al-Ubaydli M. Using search engines to find online medical information. PLoS Med 2(9):e228, 10/3/05.

7. Pathways to excellence. Chicago: The American Dietetic Association; 1997.

8. Alger SA, Schwalberg MD, Bigaouette JM, Michalek AV, Howard LJ. Effect of a tricyclic antidepressant and opiate antagonist on binge-eating behavior in normal weight, bulimic, and obese, binge-eating subjects. Am J Clin Nutr 1991;53:865–71.

Answers to Review Questions

1. False (see page 3-2)
2. True (see page 3-3)
3. False (see page 3-4)
4. True (see page 3-5)
5. False (see page 3-6)
6. False (see page 3-8)
7. True (see page 3-11)
8. False (see page 3-13)
9. True (see page 3-16)
10. True (see pages 3-13, 3-19)

Nutrition Up Close

Checking Out a Fat-Loss Product

Feedback for Unit 3

The product advertised earns five "yes" responses, making it highly unlikely that the product works. One or two "yes" responses provide a strong clue that advertised products may not work.

UNIT 4

Understanding Food and Nutrition Labels

NUTRITION SCOREBOARD

	TRUE	FALSE
1 Nutrition labels are required on all foods and dietary supplements sold in the United States.		
2 Nutrition labeling rules allow health claims to be made on the packages of certain food products.		
3 Nutrition labels contain all of the information people need to make healthy decisions about what to eat.		

Key Concepts and Facts

- People have a right to know what is in the food they buy.

- The purpose of nutrition labeling is to give people information about the composition of food products so they can make informed food-purchasing decisions.

- Nutrition labeling regulations cover the type of foods that must be labeled and set the standards for the content and format of labels.

Answers to
NUTRITION SCOREBOARD

	TRUE	FALSE
1 Labeling is now required for almost all processed foods and for dietary supplements, but remains largely voluntary for fresh fruits and vegetables and raw meats.[1]		✔
2 Health claims for food products are allowed on many food packages. The claims must be truthful and adhere to FDA standards.	✔	
3 Nutrition labels are necessarily short and can't tell the whole story about food and health. They help people make several key decisions about a food's composition.		✔

Nutrition Labeling

Misleading messages, hazy health claims, and the slippery serving sizes that characterized food labels of the past have led to a revolution in nutrition labeling. Consumers—especially those responsible for buying food for the family; people with weight concerns, food allergies, or diabetes; and health-conscious consumers—made it clear they wanted to end the mystery about what's in many foods. Passage of the 1990 Nutrition Labeling and Education Act by Congress indicated that their concerns had been heard. In 1993 the Food and Drug Administration published rules for nutrition labeling, and implementation and revisions of the new standards have been ongoing since then.

What Foods Must Be Labeled?

According to nutrition labeling regulations, almost all multiple-ingredient foods must be labeled with nutrition information.[1] Labeling of fresh vegetables and fruits and raw meats is voluntary because these foods are often sold individually and without packaging. It is recommended, however, that supermarkets make nutrition labeling information available to consumers through appropriately placed charts or pamphlets (Illustration 4.1).[2] Nutrition labeling is not required on foods sold by local bakeries and foods with packaging that is too small to fit a label (bite-sized candy bars, for example).

Labeling is not required on restaurant menus unless a nutrition claim is made for an item. When an item is labeled, for example "low fat" or "low calorie," then restaurants are required to display nutrition information for the nutrient for which the claim is made.[1]

Illustration 4.1

Many supermarkets make nutrition information available in posters, cards, or pamphlets in areas where produce and meats are sold.

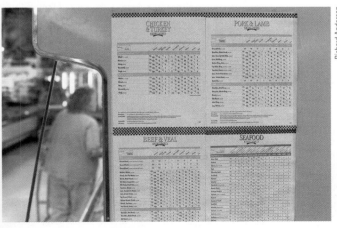

Richard Anderson

Due to the obesity and diabetes epidemics, and because people in the United States spend on average half of their food dollars in restaurants, the pressure on fast-food and other restaurants to include nutrition information on menus is increasing. Fast-food restaurants may provide nutrient composition information on packaging and Web sites. In general, however, this information is still not readily available. About two-thirds of major chain, table-service restaurants provide point-of-sale nutrition information for the healthiest items on their menus.[3]

What's on the Nutrition Label?

The nutrition information label is known as the "Nutrition Facts" panel. The panel, shown in Illustration 4.2, highlights a product's content of fat, saturated fat, *trans* fat, cholesterol, sodium, dietary fiber, two vitamins (A and C), and two

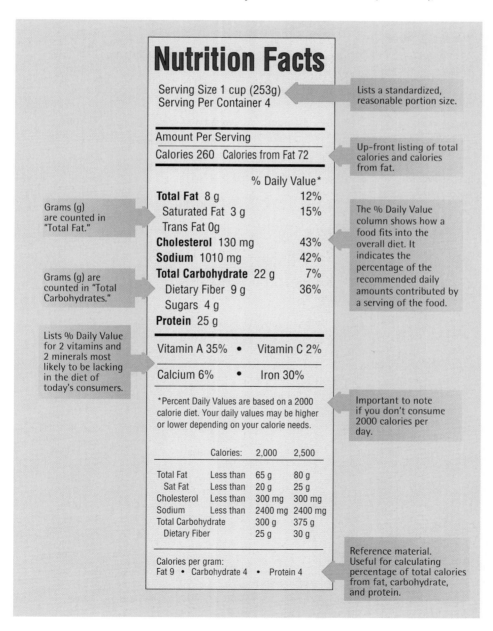

Illustration 4.2
What are you having for dinner? The Nutrition Facts panel lets you know.

Nutrition Facts

Serving Size 1 cup (253g)
Serving Per Container 4

Amount Per Serving

Calories 260   Calories from Fat 72

Lists a standardized, reasonable portion size.

Up-front listing of total calories and calories from fat.

	% Daily Value*
Total Fat 8 g	12%
Saturated Fat 3 g	15%
Trans Fat 0g	
Cholesterol 130 mg	43%
Sodium 1010 mg	42%
Total Carbohydrate 22 g	7%
Dietary Fiber 9 g	36%
Sugars 4 g	
Protein 25 g	

Grams (g) are counted in "Total Fat."

Grams (g) are counted in "Total Carbohydrates."

The % Daily Value column shows how a food fits into the overall diet. It indicates the percentage of the recommended daily amounts contributed by a serving of the food.

Vitamin A 35%   •   Vitamin C 2%

Calcium 6%   •   Iron 30%

Lists % Daily Value for 2 vitamins and 2 minerals most likely to be lacking in the diet of today's consumers.

*Percent Daily Values are based on a 2000 calorie diet. Your daily values may be higher or lower depending on your calorie needs.

Important to note if you don't consume 2000 calories per day.

	Calories:	2,000	2,500
Total Fat	Less than	65 g	80 g
Sat Fat	Less than	20 g	25 g
Cholesterol	Less than	300 mg	300 mg
Sodium	Less than	2400 mg	2400 mg
Total Carbohydrate		300 g	375 g
Dietary Fiber		25 g	30 g

Calories per gram:
Fat 9 • Carbohydrate 4 • Protein 4

Reference material. Useful for calculating percentage of total calories from fat, carbohydrate, and protein.

minerals (calcium and iron). The food's content of these nutrients must be based on a standard serving size as defined by the Food and Drug Administration (FDA). Standard serving sizes, which represent portion sizes generally consumed, are established for 131 types of food. For example, a serving of fruit juice is 6 ounces (¾ cup), and a serving of cooked vegetables is ½ cup.

All labeled foods must provide the information shown on the Nutrition Facts panel in Illustration 4.2. Additional information on specific nutrients can be added to the panel on a voluntary basis (Table 4.1). However, if the package makes a claim about the food's content of a particular nutrient not on the "mandatory" list, then information about that nutrient *must* be added to the Nutrition Facts panel. Nutrition labels contain a column headed **% *Daily Value (%DV)*.** Figures given in this column are intended to help consumers answer such questions as "Does a serving of this macaroni and cheese contain more fat than the other brand?" and "How much fiber is in this cereal?"

% Daily Value (%DV)
Scientifically agreed-upon standards of daily intake of nutrients from the diet developed for use on nutrition labels. The "% Daily Values" listed in nutrition labels represent the percentages of the standards obtained from one serving of the food product.

TABLE 4.1

Mandatory and voluntary components of the Nutrition Facts panel and assigned Daily Values (DVs)

Components are listed in the order in which they must appear on the nutrition panel; unapproved components may not be listed.[a]

Mandatory	DV	Voluntary	DV
Total calories	—	Calories from saturated fat	
Calories from fat	30%	Polyunsaturated fat	
Total fat	65 g	Monounsaturated fat	
Saturated fat	20 g	Stearic acid	
Trans fat	—[b]		
Cholesterol	300 mg	Insoluble fiber	
Sodium	2400 mg		
Total carbohydrate	300 g	Other carbohydrates	
Dietary fiber	25 g	Soluble, Insoluble fiber	
Sugars	—[b]	Sugar alcohols (xylitol,	
Protein	—[b]	mannitol, sorbitol)	
Vitamin A	5000 IU	Vitamin D	400 IU
Vitamin C	60 mg	Vitamin E	30 IU
Calcium	1000 mg	Vitamin K	80 mcg
Iron	18 mg	Thiamin	1.5 mg
		Riboflavin	1.7 mg
		Niacin	20 mg
		Vitamin B6	2.0 mg
		Folate	400 mcg
		Vitamin B12	6 mcg
		Biotin	300 mcg
		Pantothenic acid	10 mg
		Phosphorus	1000 mg
		Iodine	150 mcg
		Magnesium	400 mg
		Zinc	15 mg
		Selenium	70 mcg
		Copper	2 mg
		Manganese	2 mg
		Chromium	120 mcg
		Molybdenum	75 mcg
		Chloride	3400 mg
		Potassium	3500 mg

[a]If the food package makes a claim about any of the voluntary components, or if the food is enriched or fortified with any of them, they become a mandatory part of the nutrition panel.
[b]DVs are not shown on Nutrition Facts panels for *trans* fats, sugars, or protein. *Trans* fats and sugars have not been assigned DVs, and, although there is a DV for protein, it is not listed because most people have adequate protein intakes.[4]

Daily Values (DVs) are standard levels of dietary intake of nutrients developed specifically for use on nutrition labels (see Table 4.1). They are based on earlier editions of the Recommended Dietary Allowances and scientific consensus recommendations.[2] The %DV figures listed on labels represent the percentages of the standard nutrient amounts obtained from one serving of the food product. Standard values for total fat, saturated fat, and carbohydrate are based on a daily intake of 2000 calories. The %DV for total fat intake is based on 30% of total calories from fat (65 grams), the saturated fat standard on 10% of total calories (20 grams), and the standard for carbohydrate intake on 60% of total calories (300 grams). If, for example, a serving of a food product contained 10 grams of saturated fat, the %DV listed on the nutrition label would be 50% because the standard for saturated fat intake is 20 grams. In general, % Dietary Values of 5 or less are considered "low," and those listed as 20 or more "high."[4]

Food labels that make a claim about a particular nutrient or ingredient naturally contained in the food or added to it must include that nutrient or ingredient on the label. Other rules for including nutrition and health claims on food product labels exist.

Nutrient Content and Health Claims Must Be Truthful

Approximately 40% of food products sold in grocery stores make one or more nutrition-related claims on the packaging.[1] Can these claims be trusted? If they are among the nutrition claims approved by the FDA, the answer to this question is almost always yes.

Nutrient Content Claims   Statements used to label foods—such as "High fiber" and "Healthy"—must conform to standard definitions (Illustration 4.3). For example, foods labeled "low fat"—the most popular nutrient content claim made on food packages[1]—must contain 3 grams of fat or less per serving. Low-fat foods can be labeled with a "percent fat free" label, such as "98% fat free" turkey. This label means that the product contains approximately 2% fat on a weight basis. Meat products labeled "lean" must contain less than 10 grams of fat, 4.5 grams of saturated fat and *trans* fat combined, and 95 milligrams of cholesterol per serving. Some labels promote meat as lean based on the percentage of the meat's weight that consists of fat. So, a meatloaf mix that is 16% fat on a weight basis might be labeled as "16% lean." It would not be lean according to nutrition labeling standards. Illustration 4.4 provides an example of an appropriately and an inappropriately labeled meat to expand on this point.

A key feature of many nutrient content claims is that they cannot be used to make certain foods appear to be healthier for you than they really are. Foods often have to meet several standards for nutritional value before they qualify for a claim. You may see some food products labeled with claims like "pure," "10% fat," or "0 carbohydrates" that are not listed in the "Health Action" Chart. Because these claims have not been approved, they may not represent benefits and have no standard definition. These are the types of claims that are hard to trust.

Daily Values (DVs)
Scientifically agreed-upon daily nutrient intake standards. DVs are intended for use on nutrition labels only and are listed in the Nutrition Facts panel. "% Daily Value" on Nutrition Facts panels is calculated as the percentage of each DV supplied by a serving of the labeled food.

The "Jelly Bean Rule"
Foods high in fat, saturated fat, cholesterol, or sodium and low in certain vitamins, minerals, or fiber cannot be labeled with a nutrition claim. That means food manufacturers can't add, say, calcium to jelly beans, soft drinks, or similar foods to earn a claim that implies the product is "healthy."[5]

Illustration 4.3
Eye-catching nutrition claims on food labels must conform to standard definitions.

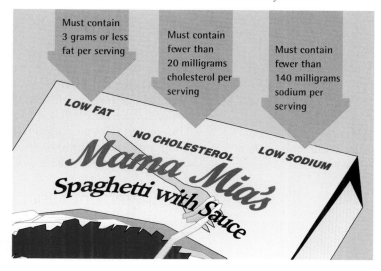

Must contain 3 grams or less fat per serving

Must contain fewer than 20 milligrams cholesterol per serving

Must contain fewer than 140 milligrams sodium per serving

LOW FAT

NO CHOLESTEROL

LOW SODIUM

Mama Mia's Spaghetti with Sauce

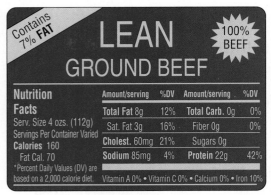

Health Claims   On approval by the FDA, foods or food components with scientifically agreed-upon benefits to disease prevention can be labeled with a health claim (Table 4.2). The health claim used on qualifying product labels, however, must be based on the FDA's "model claim" statements. For instance, scientific consensus holds that diets high in fruits and vegetables may lower the risk of cancer, and a health claim to this effect is allowed. The FDA's model claim for labeling fruits and vegetables is "Low fat diets rich in fruits and vegetables may reduce the risk of some types of cancer, a disease associated with many factors." The FDA approves health claims only for food products that are not high in fat, saturated fat, cholesterol, or sodium. See Table 4.3 for examples of model health claims approved by the FDA.

Labeling Foods as Enriched or Fortified   The vitamin and mineral content of foods can be increased by *enrichment* and *fortification.* Definitions for these terms were established more than 50 years ago. Enrichment pertains only to refined grain products, which lose vitamins and minerals when the germ and bran are removed during processing. Enrichment replaces the thiamin, riboflavin, niacin, and iron lost in the germ and bran. By law, producers of bread, cornmeal, pasta, crackers, white rice, and other products made from refined grains must use enriched flours. Beginning in 1998, federal regulations mandated that folate (a B vitamin) in the form of folic acid be added to refined grain products. This new regulation was put into effect to help reduce the risk of a particular type of inborn, structural problem in children (neural tube defects) related to low blood levels of folate early in pregnancy.

Any food product can be fortified with vitamins and minerals—and many are. One of the few regulations governing the fortification of foods is that the amount of vitamins and minerals added to the food must be listed in the Nutrition Facts panel. Illustration 4.5 shows some examples of enriched and fortified foods.

Food enrichment and fortification began in the 1930s to help prevent deficiency diseases such as rickets (vitamin D deficiency), goiter (from iodine deficiency), pellagra (niacin deficiency), and iron-deficiency anemia.[6] Today, foods are increasingly being fortified for the purpose of reducing the risk of chronic diseases such as osteoporosis, cancer, and heart disease. Other foods, such as energy bars, sports drinks and hydration fluids, and margarines are being fortified with an array of vitamins and minerals principally for sales appeal. Regular consumption of fortified foods increases the risk that people will exceed Tolerable Upper Intake Levels (ULs) of nutrients designated in the Recommended Dietary Allowances (RDAs). Regular use of multiple vitamin and mineral supplements along with liberal intake of fortified foods enhances the likelihood that excessive amounts of some nutrients will be consumed.[7]

Illustration 4.4
Two examples of labels claiming "lean" ground beef. Only the 7% fat beef is actually lean. "20% fat" ground beef is far from lean, actually providing 14 grams and 55% of total calories from fat per serving.

enrichment
The replacement of thiamin, riboflavin, niacin, and iron lost when grains are refined.

fortification
The addition of one or more vitamins and/or minerals to a food product.

The Ingredient Label

Still more useful information about the composition of food products is listed in ingredient labels. The label of any food that contains more than one ingredient must list the ingredients in order of their contribution to the weight of the food

HEALTH *Action* Some Examples of What "Front of the Package" Nutrient-Content Claims Must Mean

Term	Examples	Means That a Serving of the Product Contains:
Fresh	Fresh spinach	Foods are raw, not frozen or heated, and contain no preservatives.
Extra lean	Extra-lean pork, extra-lean hamburger	Fewer than 5 grams of fat, fewer than 2 grams of saturated fat and trans fat combined, *and* fewer than 95 milligrams of cholesterol (applies to meats only).
Lean	Lean beef, lean turkey	Fewer than 10 grams of fat, fewer than 4.5 grams of saturated fat and *trans* fat combined, *and* fewer than 95 milligrams of cholesterol (applies to meats only).
Extra, More	Bread with added fiber, fortified foods	At least 10% more of the Daily Value of a nutrient per serving than in a similar food.
Fat Free	Skim milk, no-fat salad dressing	Less than 0.5 grams of fat.
Free	Sugar-free, sodium-free	No—or negligible amounts—of sugars, sodium, or fat.
Good source	Good source of fiber, good source of calcium	From 10 to 19% of the Daily Value for a particular nutrient.
Healthy	Healthy burritos, canned vegetables	No more than 60 milligrams of cholesterol, 3 grams of fat, and 1 gram of saturated fat; and more than 10% of the Daily Value of vitamin A, vitamin C, iron, calcium, protein, or fiber. "Healthy" foods must also contain 360 milligrams or less of sodium.
High	High in iron, high in vitamin C	20% or more of the Daily Value for a particular nutrient.
Less	Less saturated fat, less cholesterol	25% less of a nutrient than a comparable food.
Light or lite	Light in sodium, lite in fat, light brown sugar, light and fluffy	33% fewer calories or half the fat as in the regular product, or 50% or less sodium than usual in a low-calorie, low-fat food. "Light" can also be used on labels to describe the texture or color of a food.
Low calorie	Low-calorie cookies, low-calorie fruit drink	40 calories or fewer.
Low cholesterol	Low-cholesterol egg product	20 milligrams or less cholesterol (applies to animal products only)
Low fat	Low-fat cheese, low-fat ice cream	3 grams or less fat.
Low saturated fat	Low-saturated-fat pancake mix, low-saturated-fat eggnog	1 gram or less saturated fat and 0.5 grams or less *trans* fat.
Low sodium	Low-sodium soup, low-sodium hot dogs	140 milligrams or less sodium.
Percent fat free	95% fat free, 98% fat free	The specified percentage of fat on a weight basis (only low-fat foods can use this label).
Reduced	Reduced calorie/cholesterol/ saturated fat bacon	25% less calories, cholesterol, or saturated fat and *trans* fat than the regular product.
***Trans* fat free**	*Trans* fat–free margarine	Less than 0.5 grams *trans* fat and less than 0.5 grams saturated fat.

TABLE 4.2
Approved health claims for food and disease prevention

Topics Covered by Approved Claims

1. Calcium and osteoporosis
2. Fats and cancer
3. Saturated fat and cholesterol and heart disease
4. Fruits and vegetables and heart disease, cancer
5. Sodium and hypertension
6. Folate or folic acid and neural tube defects (newborn malformations such as spina bifida)
7. Whole grains and heart disease
8. Soluble fiber in oats and psyllium seed husks and heart disease
9. Sugar alcohols (such as xylitol and sorbitol) and tooth decay
10. Soy protein and heart disease
11. Whole-grain foods and heart disease and certain cancers
12. Plant stanols and sterols (used in spreads such as Take Control and Benecol) and heart disease
13. Potassium and high blood pressure and stroke
14. Fruits, vegetables, high-fiber grain products and heart disease.

TABLE 4.3
Examples of model health claims approved by the FDA for labels of foods that qualify based on nutrient content; model claims are often abbreviated on food labels

Food and Related Health Issues	Model Health Claim
Whole-grain foods and heart disease, certain cancers	Diets rich in whole-grain foods and other plant foods and low in fat, saturated fat, and cholesterol may reduce the risk of heart disease and certain cancers.
Soy protein and heart disease	Diets low in saturated fat and cholesterol that include 25 grams of soy protein a day may reduce the risk of heart disease.
Sugar alcohols and tooth decay	Frequent between-meal consumption of food high in sugars and starches promotes tooth decay. The sugar alcohols in this food do not promote tooth decay.
Saturated fat, cholesterol, and heart disease	Development of heart disease depends on many factors. Eating a diet low in saturated fat and cholesterol and high in fruits, vegetables, and grain products that contain fiber may lower blood cholesterol levels and reduce your risk of heart disease.
Calcium and osteoporosis	Regular exercise and a healthy diet with enough calcium help maintain good bone health and may reduce the risk of osteoporosis later in life.
Fruits and vegetables and cancer	Low-fat diets rich in fruits and vegetables may reduce the risk of some types of cancer, a disease associated with many factors.
Folate and neural tube defects	Women who consume adequate amounts of folate daily throughout their childbearing years may reduce their risk of having a child with a brain or spinal cord defect.

(Table 4.4). The ingredient that makes up the greatest portion of the product's weight must be listed first.

The FDA now requires that ingredient labels include the presence of common food allergens in products. Potential food allergens consist of milk, eggs, fish, shellfish, tree nuts, wheat, peanuts, and soybeans. These foods, sometimes called the "Big Eight," account for 90% of food allergies. Precautionary statements, such as "may contain," or "processed in a facility with" are allowed.[8]

Food Additives on the Label Specific information about *food additives* must be listed on the ingredient label. Nearly 3000 chemical substances may be added to food to enhance its flavor, color, texture, cooking properties, shelf life, or nutrient content. Food additives on the FDA's GRAS (Generally Recognized As Safe) list can be used in food without preapproval. New additives must be approved by the FDA prior to use. Table 4.5 provides examples of functions of some additives used in food, and the nearby "On the Side" shows you an example of a highly preserved snack food. The most common food additives are sugar and salt, but trace amounts of polysorbate, potassium benzoate, and many other additives that are not so familiar are also included in foods. Although the new labeling

Illustration 4.5
Examples of foods that are enriched (left) and fortified (right).

food additives
Any substances added to food that become part of the food or affect the characteristics of the food. The term applies to substances added both intentionally and unintentionally to food.

In 1984 the Kellogg Company launched an ad campaign for All-Bran cereal that announced, "eating the right foods may reduce your risk of some kinds of cancer." Sales of the high-fiber cereal increased 37% in one year, but then Kellogg's had to withdraw the ads. The FDA ruled the All-Bran statement was equivalent to a claim for a drug. The campaign, however, started the nutrition and health claims revolution.[5]

ON THE SIDE

TABLE 4.4
All foods with more than one ingredient must have an ingredient label. Here are the rules.

1. Ingredients must be listed in order of their contribution to the weight of the food, from highest to lowest.

2. Beverages that contain juice must list the percentage of juice on the ingredient label.

3. The terms "colors" and "color added" cannot be used. The name of the specific color ingredients must be given (for example, caramel color, turmeric).

4. Milk, eggs, fish, and five other foods to which some people are allergic must be listed on the label.

TABLE 4.5
Solving the mystery of ingredient label terms

Cake Mix Ingredient Label	Additive	Function
Ingredients: Sugar, enriched flour bleached (wheat flour, niacin), iron thiamin Mononitrait, (vitamin B₁), riboflavin (vitamin B₂), vegetable shortening (contains partially hydrogenated soybean cottonseed oil), sodium aluminum phosphate, dextrose, leavening, (baking soda, monocalcium phosphate, diacalcium phosphate, aluminum sulphate), wheat starch, propylene glycol monoesters, modified corn starch, salt, egg white, vanilla, dried corn syrup, polysorbate 60, nonfat milk mono and diglycerides, sodium caseetrate, xanthan gum, soy lecithin.	Sodium aluminum phosphate	Gives baked products a light texture
	Propylene glycol monoesters	Helps blend ingredients uniformly, enhances moisture content and texture
	Mono- and diglycerides	Maintains product softness after baking
	Xanthan gum	Thickening agent, helps hold product together after baking

regulations are more comprehensive than the old ones, they don't help consumers understand what some additives do. Appendix D lists many of the most common additives and indicates their function in foods.

Trace amounts of substances such as pesticides; hormones and antibiotics given to livestock; fragments of packaging materials such as plastic, wax, aluminum, or tin; very small fragments of bone; and insects may end up in foods. These are considered "unintentional additives" and do not have to be included on the label.

Nutrition Labeling of Food in Other Countries   Nutrition labeling regulations exist in countries throughout Europe and elsewhere. In the United Kingdom, a nutritional traffic light labeling system is used on labels to help consumers identify the nutritional qualitites of food.[9] Canada's nutrition labeling requirements call for the inclusion of a Nutrition Facts table on food packages (Illustration 4.6). Canadian standards allow for nutrient content and health claims on food packages for qualifying foods. When grocery shopping in Canada, you'll see qualifying foods labeled as a "good source of calcium" or an educational message such as "ample intake of vegetables and fruits may reduce your risk of cancer" on packaging.[10]

Dietary Supplement Labeling

dietary supplement
Any product intended to supplement the diet, including vitamins, minerals, proteins, enzymes, herbs, hormones, and organ tissues. Such products must be labeled "Dietary Supplement."

structure/function claim
Statement appearing on a supplement label that describes the effects a dietary supplement may have on the structure or function of the body. Such statements *cannot* claim to diagnose, cure, mitigate, treat, or prevent disease.

A **dietary supplement** is a product taken by mouth that contains a "dietary ingredient" intended to supplement the diet. The "dietary ingredients" in these products include vitamins, minerals, proteins, enzymes, herbs, hormones, and organ tissues. Nutrition labeling regulations place dietary supplements in a special category under the general umbrella of "foods," and not drugs. Dietary supplements differ from drugs in that they do not have to undergo vigorous testing and obtain FDA approval before they are sold. In return, dietary supplement labels cannot claim that the products treat, cure, or prevent disease.[11]

According to FDA regulations, dietary supplements must be labeled as such and include a "Supplement Facts" panel that lists serving size, ingredients, and percent Daily Value (%DV) of essential nutrient ingredients (Illustration 4.7). Like foods, qualifying dietary supplements can be labeled wih nutrient claims ("high in calcium," for example), and health claims (for instance, "diets low in saturated fat and cholesterol that include sufficient soluble fiber may reduce the risk of heart disease"). Dietary supplement labels can also include **structure/function claims**. These claims are unique to dietary supplement labels.

Structure/Function Claims   Dietary supplements can be labeled with statements that describe effects the supplement may have on body structures or functions. Under this regulation, for example, vitamin E supplements can be labeled with the statement "Promotes healthy heart and circulatory function," Saint-John's-wort with "Helps maintain mental health," and zinc supplements with "Supports the immune system." If a structure/function claim is made on the label or package inserts, the label or insert must acknowledge that the FDA does not support the claim: *"This statement has not been evaluated by the FDA. This product is not intended to diagnose, treat, cure, or prevent any disease."* That's why you can see this acknowledgment printed on the labels of dietary supplements that make such claims. Structure and function claims don't have to be approved before they can appear on product labels. Dietary supplements labeled with misleading or untruthful information, and those that are not safe, can be taken off the market and the manufacturers fined.

Qualified Health Claims   In 2003, the FDA began allowing companies to use "qualified health claims" on dietary supplement and food packaging. These claims are not as strongly supported by scientific findings, as are health claims, and they carry qualifying language that indicates the lack of strong, scientific support. The qualified health claim related to walnuts and heart disease, for example, must read "There is supportive but not conclusive evidence that walnuts may reduce your risk of heart disease." Other examples of qualified health claims you may see on labels include those related to the potentially protective effects of tomatoes, selenium, nuts, or green tea on certain types of cancer and the possible relationship between intake of olive oil and heart disease.

Qualified health claims can be used only for a 9-month period while the FDA investigates the level of scientific agreement related to the claim. If at the end of the 9 months the claim doesn't pass scientific muster, the claim must be dropped from packaging.[12]

Irradiated Foods

Food irradiation is an odd example of a food additive. It is actually a process that doesn't add anything to foods. Irradiation uses X-rays, gamma rays, or electron beams to kill insects, bacteria, molds, and other microorganisms in food. Food irradiation enhances the shelf life of food products and decreases the risk of foodborne illness.[13] It must be performed according to specific federal rules.

Irradiated foods retain no radioactive particles. The process is like having your luggage X-rayed at the airport. Your luggage doesn't become radioactive, nor has anything in it changed. The process leaves no evidence of having occurred. Actually, this lack of change creates a challenge because inspection agencies can't determine whether a food has been irradiated or not. All irradiated foods except spices added to processed foods are required to display the international "radura" symbol and to indicate in writing that the food has been irradiated (Illustration 4.8). Irradiation is approved for use on chicken, turkey, pork, beef, eggs, grains, fresh fruits and vegetables, and other foods in the United States.[13]

Irradiated foods have a long history but have not yet won the trust of some American consumers.[14] The desire to limit radioactive processes of all sorts, as

Nutrition Facts
Per 125mL (87g)

Amount		% Daily Value
Calories 80		
Fat 0.5 g		1%
Saturated 0 g + Trans 0 g		0%
Cholesterol 0 mg		
Sodium 0 mg		0%
Carbohydrate 18 g		6%
Fibre 2 g		8%
Sugars 2 g		
Protein 3 g		
Vitamin A 2%	Vitamin C	10%
Calcium 0%	Iron	2%

Illustration 4.6
A Nutrition Facts table from a food package in Canada. Can you identify three ways in which this label differs from the label shown in Illustration 4.7?

†This statement has not been evaluated by the Food and Drug Administra... intended to diagnose, treat, cure...

Nutrition Facts
Serving Size 1 softgel

Amount Per Softgel	% Daily Value
Vitamin A 10,000 I.U.	200%
100% as Beta Carotene	
Vitamin C 250 mg	417%
Vitamin E 200 I.U.	667%
Zinc 7.5 mg	50%
Selenium 15 mcg	21%
Copper 1 mg	50%
Manganese 1.5 mg	75%

SUGGESTED USE: As a dieta... softgel daily, with a meal. Keep... in a cool, dry place, out of reach...
Do not use if imprinted seal under...
Our pure gelatin shell has no arti... you may notice a natural color va...
INGREDIENTS: Ascorbic Aci... Acetate, Gelatin, Cottonseed Oil... Calcium Phosphate, Partially H... and Soybean Oils, Lecithin Oil, Z... Beeswax, Manganese Sulfate,... Selenate.

☎ 1-800-276-2878 M-F 7am-4pm P...
Oil- and Water-soluble Vitamins with M...
Mfd. By **Nature Made Nutritional** ...
Mission Hills, CA 91346-9606 **Mad**...

Illustration 4.7
Dietary supplement label.

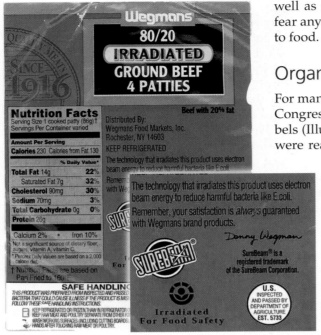

Illustration 4.8
The "radura," as the symbol is called, must be displayed on irradiated foods. The words "treated by irradiation, do not irradiate again" or "treated with radiation, do not irradiate again" must accompany the symbol.

Illustration 4.9
Foods certified as organic by the USDA can display this seal on food packages.

well as radioactive waste products, is widespread; many people fear anything that involves radioactivity, especially when it comes to food.

Organic Foods

For many years, consumers and producers of organic foods urged Congress to set criteria for the use of the term *organic* on food labels (Illustration 4.9). Consumers wanted to be assured that foods were really organic, and producers wanted to keep the business honest. Standards were needed because consumers cannot distinguish organically produced foods from others by looking at them, by tasting them, or by reading their nutrient values. The U.S. Department of Agriculture (USDA) has developed and is implementing standards for organic foods.[15]

Labeling Organic Foods   Rules that qualify foods as organic are shown in Illustration 4.10. If organic growers and processors qualify according to USDA approved certifying organizations, they can place the green-and-white USDA Organic seal on product labels. The USDA can impose financial penalties on companies that use the seal inappropriately. Organically grown and produced foods can be labeled in four other ways:

- "100% Organic" if they contain entirely organically produced ingredients

- "Organic" if they contain at least 95% organic ingredients

- "Made with organic ingredients" if they contain at least 70% organic ingredients

- "Some Organic Ingredients" if the product contains less than 70% organic ingredients

Many people choose organic foods not so much for what is in them, but for what is not in them: hormones and antibiotics, pesticide and herbicide residues (Illustration 4.11). Most organic foods, however, are not totally free of some of these ingredients. Due to "pesticide drift" from sprayed crops, past use of pesticides on farmland, and the leaching of chemicals used on crops into groundwater, organically grown plants may contain traces of pesticides. However, conventionally grown crops are six times more likely to have traces of several pesticides than are organic crops.[16]

Beyond Nutrition Labels: We Still Need to Think

Even with the new food labels, consumers can't be stupid.
—MAX BROWN

Understanding and applying the information relayed in nutrition labels calls for more, not less, nutrition knowledge on the public's part. As highlighted in the nearby Reality Check, people need to know a good bit about nutrition before they

1. Plants
   • Must be grown in soils not treated with synthetic fertilizers, pesticides, and herbicides for at least three years
   • Cannot be fertilized with sewer sludge
   • Cannot be treated by irradiation
   • Cannot be grown from genetically modified seeds, or contain genetically modified ingredients
2. Animals
   • Cannot be raised in "factorylike" confinement conditions
   • Cannot be given antibiotics or hormones to prevent disease or promote growth
   • Must be given feed products that are 100% organic

Illustration 4.10
USDA rules for qualifying foods as organic.

can understand nutrition labels and incorporate labeled foods appropriately into an overall diet. Good diets include more than just foods with nutrition labels. The ice cream cone from the stand in the mall; the orange, potato, or fish we buy in the store; and the pizza delivered to the dorm are unlabeled parts of many diets. We need to know enough about the composition of unlabeled foods to fit them into a healthy diet. Nutrition labels often list only two vitamins and two minerals, but many more are required for health, so it's particularly important for people to know if their diet is varied enough to supply needed vitamins and minerals. Use of the Food Guide Pyramid should go hand in hand with label reading and food choices.

Healthy diets consist of a wide variety of foods that may include high-fat or high-sodium foods on occasion. Not every food we eat has to have the "right" label profile. Serving mostly low-fat or low-calorie foods to children, for example, might have unintended, unhealthy effects. Young children need calories and fat for growth and development. If diets are severely restricted, growth and development will be impaired. Nutrition labels are an important tool for helping people make informed food-purchasing decisions. However, labels do not now—nor will they ever—provide all the information needed to make wise decisions about food. Only people who are well informed about nutrition can do that.

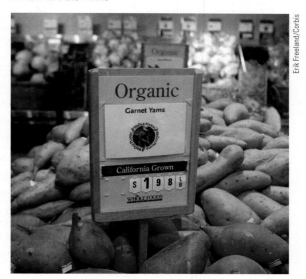

Illustration 4.11
Many people choose organic foods not so much for what is in them, but for what is not in them: hormones, antibiotics, and pesticide and herbicide residues.

Reality Check

Nutrition labeling

Foods labeled as "fat free" have few or no calories.

Who gets thumbs up?
Answers on page 4-15

Jared:
Right. You take the fat out of food, and calories go away.

Ronald:
Maybe, maybe not. "Fat-free" foods could still contain sugars, protein, and other ingredients that have calories.

Nutrition Up Close

Comparison Shopping

Focal Point: Comparing label information can help you choose the most nutritious product for your money.

Whole-Grain Cereal

Nutrition Facts

Serving Size **1 cup (50g)**
Serving Per Container 10

Amount Per Serving

Calories 170   Calories from Fat 5

% Daily Value*

Total Fat 0.5 g	1%
Saturated Fat 0 g	0%
Trans Fat 0 g	
Cholesterol 0 mg	0%
Sodium 0 mg	0%
Total Carbohydrate 41 g	14%
Dietary Fiber 5 g	21%
Sugars 0 g	
Protein 5 g	

Vitamin A 0%		Vitamin C 0%

Calcium 2%		Iron 8%

---

*Percent Daily Values are based on a 2000 calorie diet. Your daily values may be higher or lower depending on your calorie needs:
**Intake should be as low as possible.

		Calories:	2,000	2,500
Total Fat	Less than		65 g	80 g
Sat Fat	Less than		20 g	25 g
Cholesterol	Less than		300 mg	300 mg
Sodium	Less than		2400 mg	2400 mg
Total Carbohydrate			300 g	375 g
Dietary Fiber			25 g	30 g

Health Granola Cereal

Nutrition Facts

Serving Size **1 cup (50g)**
Serving Per Container 10

Amount Per Serving

Calories 210   Calories from Fat 30

% Daily Value*

Total Fat 3 g	5%
Saturated Fat 0 g	0%
Trans Fat 0 g	
Cholesterol 0 mg	0%
Sodium 120 mg	5%
Total Carbohydrate 43 g	14%
Dietary Fiber 3 g	12%
Sugars 16 g	
Protein 5 g	

Vitamin A 2%		Vitamin C 0%

Calcium 2%		Iron 10%

---

*Percent Daily Values are based on a 2000 calorie diet. Your daily values may be higher or lower depending on your calorie needs:
**Intake should be as low as possible.

		Calories:	2,000	2,500
Total Fat	Less than		65 g	80 g
Sat Fat	Less than		20 g	25 g
Cholesterol	Less than		300 mg	300 mg
Sodium	Less than		2400 mg	2400 mg
Total Carbohydrate			300 g	375 g
Dietary Fiber			25 g	30 g

Rate the nutrition value of these two cereals by completing the second paragraph below using information from the Nutrition Facts panels. The first paragraph has been done for you.

FEEDBACK (answers to these questions) can be found at the end of Unit 4.

The **Whole-Grain Cereal**, providing 170 **calories** per serving, contributes 5 **calories from fat** (or 1% of the % Daily Value). Each serving also provides 0 mg of **sodium** (or 0% of the % Daily Value), 5 g of **dietary fiber** (or 21% of the % Daily Value), and 0 g of **sugars**.

The **Health Granola Cereal**, providing _____ **calories** per serving, contributes _____ **calories from fat** (or _____ % of the % Daily Value). Each serving also provides _____ mg of **sodium** (or 5% of the % Daily Value), _____ g of **dietary fiber** (or _____ % of the % Daily Value), and _____ g of **sugars**.

Which cereal provides the better nutritional value?

ANSWERS TO
Reality Check

Nutrition labeling

Ronald got it. "Fat free" does not equal "calorie free." Fat-free Caesar salad dressing, for example, provides 40 calories in a 2-tablespoon serving.

Jared: 👎

Ronald: 👍

Key Terms

Daily Values (DVs), page 4-5
dietary supplement, page 4-10
enrichment, page 4-6

food additives, page 4-9
fortification, page 4-6
% Daily Value (%DV), page 4-4

structure/function claim, page 4-10

Review Questions

True False

1. Almost all multiple-ingredients foods must be labeled with nutrition information. ☐ ☐

2. Food manufactures can list any serving size they want on Nutrition Facts panels. ☐ ☐

3. In general, %DV listed for nutrients in Nutrition Facts panels of 10 or more are considered "low," and those listed as 50% or more are considered "high." ☐ ☐

4. An overriding principle of nutrition labeling regulations is that nutrient content and health claims made about a food on the packaging must be truthful. ☐ ☐

5. The term *enriched* on a food label means that extra vitamins and minerals have been added to the food to bolster its nutritional value. ☐ ☐

6. The ingredient that makes up the greatest portion of a food product's weight must be listed first on ingredients labels. ☐ ☐

True False

7. Food labeling regulations do *not* require food manufacturers to list the presence of major food allergens on ingredient labels. ☐ ☐

8. Dietary supplements, such as those for herbs and vitamins, are considered drugs and are not regulated by FDA's Nutrition Labeling rules. ☐ ☐

9. Foods can be labeled with nutrient content and health claims, whereas dietary supplements can be labeled with these two types of claims plus structure/function claims. ☐ ☐

10. Animals providing meats labeled "organic" cannot be given antibiotics or hormones. ☐ ☐

11. Foods bearing the USDA Organic seal are certified as organic by the U.S. Department of Agriculture. ☐ ☐

12. Nutrition labels provide all the nutrition information we need to make healthful food choices. ☐ ☐

Media Menu

www.cfsan.fda.gov/~dms/flquiz1.html
Test your food label knowledge and get answers to questions about identifying healthy food choices based on nutrition label information.

www.cfsan.fda.gov/~dms/foodlab.html
The FDA's main page is devoted to understanding and using food labels.

www.nutrition.gov
For quick access to high-quality information, search the terms "nutrition labels," "dietary

supplement labels," "organic foods," and "food irradiation" from this Web site.

www.hc-sc.gc.ca/fn-an/label-etiquet/nutrition/interactive/index_e.html
This address leads to a Health Canada site, information, and an interactive tool on nutrition labeling in Canada.

www.seekwellness.com/nutrition/new_food_label.htm
This is a very good, comprehensive site about the benefits and uses of nutrition labels.

http://www.ams.usda.gov/nop
This address leads to the USDA's organic standards home page. Search "National Organic Standards" on Google if needed; Web address will change. You can call USDA's National Organic Program to get detailed information on organic food regulations: 202-720-3252.

Notes

1. Brecher SJ et al. Status of nutrition labeling, health claims, and nutrient content claims for processed foods. J Am Diet Assoc 2000;100:1057–62.

2. Pennington JA, Hubbard VS. Derivation of daily values used for nutrition labeling. J Am Diet Assoc 1999;97:1407–12.

3. Harnack LJ. Availability of nutrition information on menus at major chain table-service restaurants. J Am Diet Assoc 2006;106:1012–15.

4. How to Understand Use the Nutrition Facts Label. www.cfsan.fda.gov/~dms/foodlab.html#nopercent, accessed 7/06.

5. Marquart L et al. Solid science and effective marketing for health claims. Nutr Today 2001;36:107–14.

6. Park YE et al. History of cereal-grain product fortification in the United States. Nutr Today 2001;36:124–36.

7. Position of the American Dietetic Association: food fortification and nutritional supplements. J Am Diet Assoc 2005;105:1300–11.

8. Food Allergen Labeling and Consumer Protection Act of 2004. www.cfsan.fda.gov/~dms/alrgact.html, accessed 7/06.

9. Armitage T. Confused Europeans snub nutrition label information. Nutra Ingredients.com/Europe, 4/3/05.

10. It's Your Health: Nutrition Labeling. www.hc-sc.gc.ca/iyh-vsv/life-vie/labelling-etiquetage_e.html, accessed 7/06.

11. Ross S. Functional foods: the Food and Drug Administration perspective. Am J Clin Nutr 2000;71(suppl):1735S–8S.

12. Qualified health claims subject to enforcement discretion, http://www.cfsan.fda.gov/~dms/qhc-sum.html, accessed 7/06.

13. Shea KM. Technical report: irradiation of food. Pediatrics 2000;106:1505–9.

14. McCally M. et al. Irradiation of food. N Engl J Med 2004,351:402.

15. The National Organic Program, http://www.ams.usda.gov/nop/indexNet.htm, accessed 7/06.

16. Consumer Union and the Organic Materials Review Institute study on pesticide residues in organic foods. Consumer Reports, August 2002.

Answers to Review Questions

1. True, see page 4-2.
2. False, see page 4-4.
3. False, see page 4-5.
4. True, see page 4-5.
5. False, see page 4-6.
6. True, see pages 4-6, 4-7.
7. False, see pages 4-8, 4-9.
8. False, see page 4-10.
9. True, see page 4-10.
10. True, see page 4-12.
11. True, see page 4-12.
12. False, see pages 4-12, 4-13.

Nutrition Up Close

Comparison Shopping

Feedback for Unit 4

The **Health Granola Cereal,** providing 210 **calories** per serving, contributes 30 **calories from fat** (or 5% of the % **Daily Value**). Each serving also provides 120 mg of **sodium** (or 5% of the % **Daily Value**), 3 g of **dietary fiber** (or 12% of the % **Daily Value**), and 16 g of **sugars.** The **Whole-Grain Cereal** provides the better nutritional value because it is lower in calories, total fat, sodium, and sugar and higher in fiber than the **Health Granola Cereal.** Don't be fooled by an attractive-sounding product. The proof of nutritional value is in the label, not the name.

Nutrition, Attitudes, and Behavior

NUTRITION SCOREBOARD

	TRUE	FALSE
1 Food preferences are genetically determined.		
2 Food habits never change.		
3 Skipping breakfast does not affect school performance in well-nourished children.		

Key Concepts and Facts

- Most food preferences are learned.

- The value a person assigns to eating right has more effect on dietary behaviors than does knowledge about how to eat right.

- Food habits can and do change.

- The smaller and more acceptable the dietary change, the longer it lasts.

- Behavior and mental performance can be affected by diet.

Answers to **NUTRITION SCOREBOARD**	TRUE	FALSE
1 Although genetics plays a role in food preferences, the predominant influences are environmentally determined.[1]		✔
2 The idea that food habits don't change is a myth.		✔
3 Well-nourished children who don't eat breakfast tend to score lower on problem-solving tests than children who eat breakfast.[2] (If you got this one wrong, answer one more question: Did you skip breakfast this morning?)		✔

Origins of Food Choices

Horse meat is a favorite food in a large area of north-central Asia. Pork, which is widely consumed in North and South America, Europe, and other areas, is rigidly avoided by many people in Islamic countries. Bone-marrow soup and sautéed snails are delicacies in France, while kidney pie is traditional in England. Dog is a popular food in Borneo, New Guinea, the Philippines and other countries, whereas snake is a delicacy in China. In some countries, people enjoy insects (Illustration 5.1) but consider corn and soybeans fit only for animal feed. And then there are steamed clams and raw oysters—food passions for some, but absolutely disgusting to others.[3]

When did you first think "yecck!"? The food choices just described would elicit that response among people from a variety of cultures. But they would not be responding to the same foods.

Why do people eat what they do? People learn from their culture and the society they live in which animals and plants are considered food and which are

Illustration 5.1
Maggots (right) and grasshoppers (left) are Mexican delicacies, served at Girasoles Restaurant in Mexico City.

Jorge Uzon/AFP/Getty Images

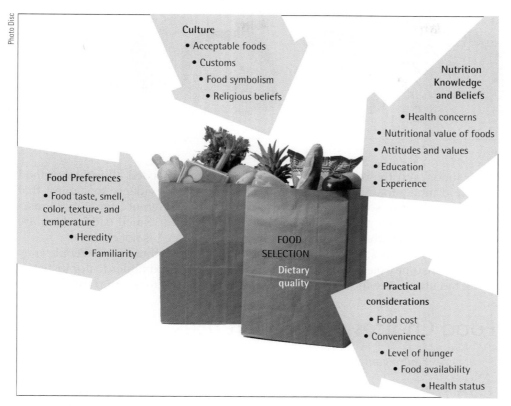

Culture
- Acceptable foods
- Customs
- Food symbolism
- Religious beliefs

Nutrition Knowledge and Beliefs
- Health concerns
- Nutritional value of foods
- Attitudes and values
- Education
- Experience

Food Preferences
- Food taste, smell, color, texture, and temperature
- Heredity
- Familiarity

FOOD SELECTION
Dietary quality

Practical considerations
- Food cost
- Convenience
- Level of hunger
- Food availability
- Health status

Illustration 5.2
Factors influencing food selection and dietary quality.
Each of these sets of factors interacts with the others.[6]

not.[4] Once items are identified as food, they develop a legacy of strong symbolic, emotional, and cultural meanings. Comfort foods, health foods, junk foods, fun foods, soul foods, fattening foods, mood foods, and pig-out foods, for example, have been identified in the United States. And all cultures have their "super food." In Russia and Ireland, it's potatoes. In Central America, corn and yucca (a starchy root, also called manioc) are super foods, and in Somalia it's rice. The designation refers to the cultural significance of the food and not to its nutritional value.[3]

In countries like the United States, where a wide variety of foods are available and people have the luxury of selecting which foods they will eat, food choices are influenced by a wide range of factors (Illustration 5.2). Of these factors, food preference has the largest impact on food choices for most people.[5] Food preferences vary a good deal among individuals, and, as addressed in the next "On the Side" feature, lead to a wide array of specific food choices. Rather than being inborn, food preferences are primarily learned.[1]

People Are Not Born Knowing What to Eat

The food choices people make are not driven by a need for nutrients or guided by food selection genes. People deficient in iron, for example, do not seek out iron-rich foods. If we're overweight, no inner voice tells us to reject high-calorie foods. Women who are pregnant don't instinctively know what to eat to nourish their growing fetuses. No evidence indicates that young children, if offered a wide variety of foods, would select and ingest a well-balanced diet.[7]

Humans are born with mechanisms that help them decide when and how much to eat, however.[8] An inborn attraction to sweet-tasting foods, a dislike for

Why do we prefer the foods we do?

J. E. Steiner

Illustration 5.3

Newborn infants respond to different tastes: (a) Baby at rest, (b) tasting distilled water, (c) tasting sugar, (d) tasting something sour, and (e) tasting something bitter.

Source: Taste-induced facial expressions of neonate infants from the classic studies of J. E. Steiner, in *Taste and Development: The Genesis of Sweet Preference,* ed. J. M. Weiffenbach, HHS Publication no. NIH 77-1068 (Bethesda, Md.: U.S. Department of Health and Human Services, 1977), pp. 173–189, with permission of the author.

bitter foods, and the response of thirst when water is needed all influence food and fluid intake to an extent (Illustration 5.3).[9] There is evidence to suggest that people deficient in sodium experience an increased preference for salty foods.[10] Most people have a strong aversion to foods they think "smell bad," which are likely spoiled. Whether this response is inborn or learned is not clear.

Inborn traits that affect food choices are not very fine-tuned. People may overdo (consume excessive sugar and salt, for example) and underdo (for instance, drink too little water in response to thirst) their response to internal cues.

Food Choices and Preferences

"You're going to eat that! Do you have any idea what's in that hot dog?"

"Yeah, I do. There's barbecue in the backyard, ball games with my mom and dad, and parties at my friend's house. The memories taste great!"

The strong symbolic, emotional, and cultural meanings of food come to life in the form of food preferences. We choose foods that, based on our cultural background and other learning experiences, give us pleasure.[10] Foods give us pleasure when they relieve our hunger pains, delight our taste buds, or provide comfort and a sense of security. We find foods pleasurable when they outwardly demonstrate our superior intelligence, our commitment to total fitness, or our pride in our ethnic heritage. We reject foods that bring us discomfort, guilt, and unpleasant memories and those that run contrary to our values and beliefs.

The Symbolic Meaning of Food

Food symbolism, cultural influences, and emotional reasons for food choices are broad concepts that may become clearer with concrete examples. Here are a few examples to consider.

Status Foods   Vance Packard, in his book *The Status Seekers,* provided a memorable example of the symbolic value of food:

As a lad, this man had grown up in a poor family of Italian origin. He was raised on blood sausages, pizza, spaghetti, and red wine. After completing high school, he went to Minnesota and began working in logging camps, where— anxious to be accepted—he soon learned to prefer beef, beer, and beans, and he shunned "Italian" food. Later, he went to a Detroit industrial plant, and eventually became a promising young executive. . . . In his executive role he found himself cultivating the favorite foods and beverages of other executives: steak, whiskey, and seafood. Ultimately, he gained acceptance in the city's upper class.

ON THE SIDE

Due to individual differences in preferences for coffee, Starbucks offers more than 19,000 ways a cup of it can be served.[12]

Michael Newman/Photo Edit

Now he began winning admiration from people in his elite social set by going back to his knowledge of Italian cooking, and serving them, with the aid of his manservant, authentic Italian treats such as blood sausage, spaghetti, and red wine![11]

Comfort Foods

The first thing I remember liking that liked me back was food.
—RITA RUDNER, COMEDIAN

Ice cream, apple pie, chicken noodle soup, boxed chocolates, meat loaf and mashed potatoes: these are the most popular comfort foods in the United States.[13] The feelings of security and love that came along with the tea and honey or chicken soup that your mother or father gave you when you had a cold, or with the ice cream and popsicles lovingly given to soothe a sore throat, are renewed with comfort foods. Once the symbolic value of a food is established, its nutritional value will remain secondary.[14] Food status is a strong determinant of food choices; and after all, as a noted nutritionist once said, "life needs a little bit of cheesecake."[15]

"Discomfort Foods"

"When I was a kid, my parents made me eat the brussels sprouts they put on my plate. I finally ate them, but then I threw up. Nobody has ever made me eat brussels sprouts again."

Memories of bad experiences with food, and expectations that certain foods will harm us in some way, each contribute to our learning about food and affect our food preferences. Eating a piece of blueberry pie right before an attack of the flu hits or overdosing on sweet pickles or olives, for example, may take these foods off your preferred list for a long time.

Cultural Values Surrounding Food

A team of scientists observed that the diet of certain groups in the Chin States of Upper Burma was seriously deficient in animal protein. After considerable study a way was found to improve the situation by cross-breeding the small, local black pigs raised by the farmers with an improved strain to obtain progeny giving a greater yield of meat. The entire operation, however, completely failed to benefit the nutrition of the population because of one fact which had been viewed as irrelevant. The cross-bred pigs were spotted. And it was firmly believed—as firmly as we believe that to eat, say, mice, would be disgusting—that spotted pigs were unfit to eat.[3]

Dietary change introduced into a culture for the purpose of improving health can be successful only if it is accepted by the culture. Cultural norms are not easily modified.

Other Factors Influencing Food Choices and Preferences

Food preferences and selections are also affected by the desire to consume foods considered healthy. Reducing fat intake, eating more fruits and vegetables, and cutting down on sweets bring rewards and pleasures such as weight loss and

According to the Rudd Center for Food Policy at Yale, Americans are introduced to approximately 20,000 new food items in grocery stores each year.

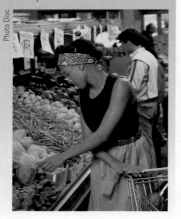

maintenance of the lost weight, an end to constipation, lower blood cholesterol level, and a newly discovered preference for basic foods.

Food Cost and Availability   Food choices are also affected by the cost and availability of food. Researchers found that college students eating in dining halls have better diets when they prepay for their meals for the entire term rather than paying at each meal.[16] Grocery shoppers tend to select more low-fat and high-fiber foods when presented with a wide selection of those foods rather than just a few.[17]

Genetic Influences   The primary role genetics plays in food preference development relates to taste. Some people are born with a strong sensitivity to bitterness and may reject foods such as brussels sprouts and broccoli (especially if they are overcooked!), bitter-tasting teas and wines, and tonic water. Some people with this genetic trait will continue to eat strong-flavored vegetables—particularly after they cover up the bitterness with seasonings or sauces.

Food Choices Do Change

Who says old dogs can't learn new tricks? Most Americans aren't eating the way they used to. Per person consumption of whole-grain products has increased 41% since 1970, while fresh egg intake has fallen from an average of 4½ eggs per week to 4. Low-fat milk (2%) sales have risen 112%, beef consumption dropped 18%, and broccoli consumption is skyrocketing. Per person consumption of broccoli has risen 1,007% since 1970—a bigger gain than for any other food. Americans are eating 168% more cheese and 20 to 25% more vegetables and fruits than in 1970.[18] Other examples of changes in food choices since 1990 are shown in Illustration 5.4.

Food choices are largely learned and do change as we learn more about foods. Perhaps your food choices have changed over time. How do the food choices you make now compare with the choices you made five years ago?

How Do Food Choices Change?

What are the ingredients for change in food choices? Why do some people succeed in improving their food choices while other people find that very hard to do? Nutrition knowledge, attitudes, and values have a lot to do with changing food choices for the better.

Nutrition Knowledge and Food Choices   Sound knowledge about good nutrition necessarily precedes the selection of a healthful diet. But is knowledge enough to ensure that healthy changes in diet will be made? The answer is "yes" for some

Illustration 5.4
Changes in Americans' food choices since 1990.[18]

Sugars +17%

Skim milk +25%

Chicken +40%

Margarine −42%

Oranges −13%

people and "no" for others. Students at James Madison University in Virginia improved their diets after taking a course in nutrition. At the end of the course, students reported eating, on average, less fat and sodium than at the beginning of the course (Illustration 5.5).[19] Changes toward healthier diets among students taking a nutrition course at the University of Texas at Austin have also been reported.[20] For some Americans, news that animal fats raise blood cholesterol levels was enough to convince them to cut down on beef and whole milk. Knowledge that diets containing an adequate amount of fiber reduce the risk of certain types of cancer led many Americans to switch to high-fiber breakfast cereals. Information about good nutrition leads some people to modify their eating behavior some of the time and is more likely to change the food choices of women than men.[21] Knowledge alone is often not enough, however.

Daily Tribune

Nutrition Course Improves Diets of College Students

The diets of college students improved after taking a basic nutrition course. Results indicated that students decreased the amount of total fat from 82 to 68 grams and made other changes putting their diets more in line with the Dietary Guidelines.

Illustration 5.5
A true story.[19]

When Knowledge Isn't Enough   Many people know far more about the components of a good diet than they put into practice. The problem is that between knowledge and practice lie multiple beliefs and experiences that act as barriers to change (Illustration 5.6). Change of any type is most likely to succeed when the benefits of making the change outweigh the disadvantages. This makes changes in food choices a very individual decision. Individuals decide whether a change is in their best interests. But what sorts of circumstances, in addition to increased knowledge, make changes in food choices worthwhile for individuals—and even highly desired?

Nutrition Attitudes, Beliefs, and Values   The value individuals place on diet and health is reflected in the food choices they make. A survey of restaurant patrons found that food choices varied according to the consumer's perceptions of the importance of diet to health:[22]

- "Unconcerned" consumers—people who are unconcerned about the connection between diet and health and who tend to describe themselves as "meat and potato eaters"—select foods for reasons other than health.

- "Committed" consumers believe that a good diet plays a role in the prevention of illness. They tend to consume a diet consistent with their commitment to good nutrition.

- "Vacillating" consumers—people who describe themselves as concerned about diet and health but who do not consistently base food choices on this

Illustration 5.6
Why knowledge about a good diet may not be enough to improve food choices.

"I feel guilty about eating the foods I like." | "Eating right is too expensive." | "I tried eating better, but I didn't stick with it." | "I don't have the time to eat right." | "The vegetables I like aren't available." | "I'm healthy now... Why should I worry about my diet?"

TABLE 5.1

Factors that enhance food changes

- Attitude that nutrition is important
- Belief that diet affects health
- Perceived susceptibility to diet-related health problems
- Perception that benefits of change outweigh barriers to change

concern—tend to vary their food choices depending on the occasion. These consumers are likely to abandon diet and health concerns when eating out or on special occasions, but they generally adhere to a healthy diet.

Avoiding illness and curing or diminishing current health problems are likewise strong incentives for changing food choices (Table 5.1).[23] One study found that nurses and dietitians based dietary changes primarily on the benefits of good diets to future health.[24] In almost all instances, the key to lasting improvements in diet is to make changes you can live with and know you can make. Changes that bring you more pleasure than inconvenience or discomfort have staying power, while those that require a lot of willpower to maintain rarely last.[25]

Successful Changes in Food Choices

Perhaps the key in making dietary changes is to determine which changes are the easiest.

The primary reason efforts to improve food choices fail is that the changes attempted are too drastic. Improvements that last tend to be the smallest acceptable changes needed to do the job.[26]

The Process of Changing Food Choices Assume you need to lose weight and want to keep it off by modifying your food choices. A promising plan to accomplish this goal would begin by identifying food choices you would like to change and lower-calorie food options you would be willing and able to eat (Table 5.2). Then you could make the plan more specific by identifying the changes that would be easiest to implement. For example, assume the low-calorie foods you like and would eat include frozen nonfat yogurt and oranges. You might decide to eat yogurt or an orange in place of your usual bedtime snack of ice cream. A specific dietary change such as this is much easier to implement than a broad

TABLE 5.2

Changing food choices for the better[25,26]

The Process	An Example
1. Identify a healthful change in your diet you'd like to make.	1. I'd like to lower my fat intake.
2. Identify two food choices you make that should change because they contribute to the need for the healthful change you identified.	2. I eat at fast-food restaurants five times a week. I always have a large order of fries, and several times a week I eat fried chicken.
3. Identify two or more specific, acceptable options for more healthful food choices than the ones identified under number 2.	3. Options identified: • Order tossed salad instead of fries. • Eat grilled chicken sandwiches once a week instead of fried chicken. • Eat Mexican fast-food more often.
4. Decide which option is easiest to accomplish and requires the smallest change to get the job done.	4. I love Mexican food. It would be easy to eat it more often.
5. Plan how to incorporate the change into your diet.	5. Mondays and Fridays, I'll eat Mexican.
6. Implement the change. Be prepared for midcourse corrections.	6. Midcourse correction: On Fridays, when I'm with my friends, it's easier to eat at the restaurant they like. I'll order the grilled chicken sandwich and a salad.

notion, such as "eat less." Although weight loss will take a while, such a small acceptable change has a much better chance of working than a drastic change in diet.

Planning for Relapses   When making a change in your diet, be prepared for relapses. Relapses happen for a number of reasons, and they don't mean the attempt has failed. People often return to old habits because the change they attempted was too drastic or because they tried to make too many changes at once. If the change undertaken doesn't work out, rethink your options and make a midcourse correction.

Does Diet Affect Behavior?

Food affects behavior in some rather striking ways. Irritable, crying infants rapidly change into cooing, sleepy angels after they are fed. Low-on-sleep employees perk up after their morning coffee. A high-calorie lunch makes many people feel calm and sleepy.[27] Not only do our behaviors affect our diet, but our diets can affect our behaviors. Behaviors shown to be affected by various components of the diet, as well as behaviors for which the effects of diet are unsubstantiated but nevertheless widely believed to exist, are listed in Table 5.3. One common belief about food and behavior is the subject of this unit's "Reality Check."

Malnutrition and Mental Performance

The World Health Organization estimates that 40 to 60% of the world's children suffer mild to moderate malnutrition.[30]

Like growth and health, mental development and intellectual capacity can be affected by diet. The effects range from mild and short term to serious and lasting,

TABLE 5.3
Examples of relationships between diet and behavior.[28]

***Established* relationships between diet and behavior:**
- Severe protein-calorie malnutrition early in life and impaired mental development
- Breakfast skipping and reduced problem-solving skills in children
- Fetal exposure to high levels of alcohol and mental retardation and behavioral problems
- Iron-deficiency anemia and reduced motor skills, attention span, and problem-solving skills
- Overexposure to lead and poor academic performance
- Folate deficiency and depression
- High-carbohydrate and high-calorie meals and calmness, sleepiness

***Unsubstantiated* claims about diet and behavior relationships:**
- Sugar intake, hyperactivity, and criminal behavior
- Vitamin supplements and increased intelligence
- Meat and aggressive behavior
- High-carbohydrate intake and increased brain serotonin level, decreased appetite, and weight loss
- Autism and food allergies

Reality Check

Can food be a love potion?

Toward the end of a friend's birthday party, Glenda and Cassell got into a heated debate about the existence of food aphrodisiacs. Part of the conversation went like this:

Is it all in Glenda's head?
Answers on next page

Glenda:
Chocolate and vanilla are natural love potions, there's no doubt about it. If I eat a chocolate truffle and spritz on vanilla flavoring like perfume before a date, the guy always goes nuts for me!

Cassell:
Are you kidding? No way! I've heard about oysters and this tree bark that are supposed to work miracles, too. It's all in your head.

Can food be a love potion?

The idea that food can act as an aphrodisiac has been around since ancient times. Although many have looked and others have tried, no one has found a food that acts like a love potion.[29]

Glenda: 👎

Cassell: 👍

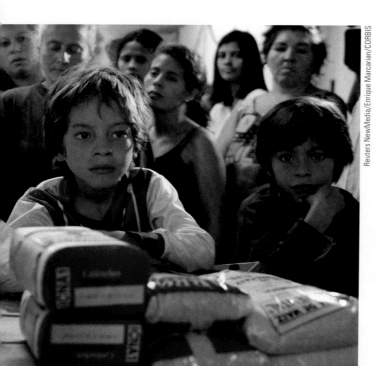

Illustration 5.7
Malnutrition in early childhood has long-lasting effects. Some children never fully recover.

depending on when the malnutrition occurs, how long it lasts, and how severe it is. The effects are most severe when malnutrition occurs while the brain is growing and developing.

Severe deficiency of protein, calories, or both early in life leads to growth retardation, low intelligence, poor memory, short attention span, and social passivity. When the nutritional insult is early and severe, some or all of these effects may be lasting (Illustration 5.7). In Barbados, for example, children who experienced protein-calorie malnutrition in the first year of life did not fully recover even with nutritional rehabilitation. Growth improved but academic performance did not. Compared to well-nourished children, those experiencing protein and calorie deficits during infancy were more likely to drop out of school and were four times more likely to have symptoms of ADHD (attention deficit hyperactivity disorder).[31]

Protein-calorie malnutrition that occurs later in childhood, after the brain has developed, produces behavioral effects that can be corrected with nutritional rehabilitation. Correction of other deficits that often accompany malnutrition, such as the lack of educational and emotional stimulation and harsh living conditions, hastens and enhances recovery.[31]

Protein-calorie malnutrition severe enough to cause permanent delays in mental development rarely occurs in the United States. When it does, malnutrition is usually due to neglect or inadequate caregiving. More common dietary events that impair learning in U.S. children are breakfast skipping, fetal exposure to alcohol, iron deficiency, and lead toxicity.

Does Breakfast Help You Think Better? Short-term fasting, such as skipping breakfast, reduces the late-morning problem-solving performance of children. About one in seven young children in the United States doesn't eat breakfast regularly.[32] (No one has yet studied the effects of breakfast skipping on college students' midterm exam scores. . . .) In general, children and adolescents who are breakfast eaters are less likely to be overweight than their peers who aren't.[2]

Early Exposure to Alcohol Affects Mental Performance Mental development can be permanently delayed by exposure to alcohol during fetal growth.

Although growth is also retarded, the most serious effects of fetal exposure to alcohol are permanent delays in mental development and behavioral problems associated with them. Women are advised not to drink if they are pregnant or may become pregnant.[33]

Iron Deficiency Impairs Learning

On a global basis, an estimated 20% of men, 35% of women, and 40% of all children are anemic, primarily due to iron deficiency.[34]

Most cases of iron-deficiency anemia in children result from inadequate intake of dietary iron. Iron-deficiency anemia in children is a widespread problem in developed and developing countries and likely is the most common single nutrient deficiency.[34] The potential impact of iron-deficiency anemia on the functional capacity of humans represents staggering possibilities.

Until recently, it was thought that the effects of iron-deficiency anemia on intellectual performance were short term and could be corrected by treating the anemia. It now appears that some of the effects may be lasting. Five-year-old children in Costa Rica treated for iron-deficiency anemia during infancy scored lower on hand–eye coordination and other motor skill tests than similar children without a history of anemia. Studies in the United States have detected shortened attention span and reduced problem-solving ability in iron-deficient children.[35,36]

Overexposure to Lead Our concern about exposure to lead has recently increased, due in large measure to information indicating that exposure to low levels of lead has long-term behavioral effects.[37] There are many opportunities for overexposure to lead. Approximately 84% of U.S. houses built before 1980 contain some lead-based paint.[38] Children living in or near these houses may eat the paint flakes (they taste sweet), or the old paint may contaminate the soil near the houses (Illustration 5.8). Lead also ends up in soil from industrial and agricultural chemicals, in water from lead-based pipes and solder, and in the air from the days when leaded gas was used.

Although the use of lead in cans, pipes, and gasoline has decreased dramatically, lead remains in the environment for long periods. It also stays in the body, stored principally in the bones, for a long time—20 years or more. The effects of excessive exposure to lead include increased absenteeism from school, longer reaction times, impaired reading skills, and higher dropout rates. Blood lead levels in children have dropped substantially in recent decades. Despite this drop, however, half a million young children in the United States still have elevated blood lead levels.[39]

Carbohydrates and Behavior

Carbohydrates are credited with improving mood, decreasing food intake, and inducing sleepiness. Simple sugars are blamed for causing hyperactivity in children and criminal behavior in adults. Where do the truths related to carbohydrate intake and behavior lie?

Carbohydrate Intake, Mood, and Appetite *Serotonin* is a chemical messenger produced from *tryptophan*, an essential amino acid. High blood levels of tryptophan increase the production of serotonin in the brain. Serotonin produced in the brain plays important roles in mood and the regulation of appetite. Due to

Illustration 5.8
There are many opportunities for overexposure to lead. Young children are especially vulnerable.

PhotoEdit, Felicia Martinez

serotonin (pronounced sare-uh-tone-in)
A neurotransmitter, or chemical messenger, for nerve cell activities that excite or inhibit various behaviors and body functions. It plays a role in mood, appetite regulation, food intake, respiration, pain transmission, blood vessel constriction, and other body processes.

tryptophan (pronounced trip-tuh-fan)
An essential amino acid that is used to form the chemical messenger serotonin (among other functions). Tryptophan is generally present in lower amounts in food protein than most other essential amino acids. It can be produced in the body from niacin, a B vitamin.

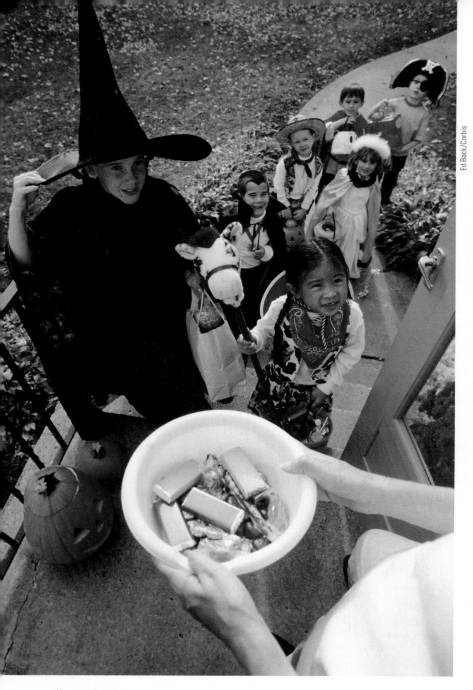

Ed Bock/Corbis

Illustration 5.9
Does a high sugar intake make children hyperactive? It's unlikely.[41]

its influence on appetite regulation, serotonin is a target of action for a number of weight-loss drugs and popular weight-loss diets. Consumed by itself, carbohydrate increases blood levels of tryptophan and serotonin production in the brain.[40]

For years it was thought that high-carbohydrate meals may decrease appetite and food intake by increasing levels of serotonin in the brain. Recent research indicates, however, that increases in serotonin related to carbohydrate intake are small, and that any increase is quickly blocked by the simultaneous intake of small amounts of protein. Although eating food can erase a bad mood in hungry people, it does not appear that high-carbohydrate foods decrease appetite by increasing brain levels of serotonin. High-carbohydrate diets are also purported to decrease fat intake, but this notion is controversial. High-carbohydrate meals may induce drowsiness or calmness.[40]

Sugar, Hyperactivity, and Criminal Behavior Sugar has been wrongly blamed for all sorts of behavioral problems. A number of studies have investigated the relationship of sugar intake to hyperactivity (see Illustration 5.9) and criminal behavior,[40,41] and none has shown an effect. In fact, studies generally indicate that sugars and other carbohydrates *decrease* activity level in children and adults.[41]

The myth that sugar causes hyperactivity in children lingers on in the minds of many parents, however. The excitement that often accompanies high-sugar eating occasions such as Halloween and birthday parties, or the expectation that children will be more active after they eat sugary foods, may sustain the myth. It has also been claimed that food additives, especially artificial food colors, cause hyperactivity in children. Well-controlled studies do not support this claim, either.[42,43]

The Future of Diet and Behavior Research

Identifying the effects of nutrition on behavior is a tricky business. Many factors in addition to diet influence behavior, making it difficult to separate dietary from social, economic, educational, and genetic influences. We still have much to learn, and many assumptions about diet and behavior must await confirmation through research.

Nutrition Up Close

Improving Food Choices

Focal Point: Developing a plan for healthier eating.

Identify a change in your diet that you would like to make. Then develop a plan for making the change by thinking through and responding to each element of the dietary change process listed. (Refer to Table 5.2 for examples of responses.)

Dietary Change Process   **Your Response**

1. Identify a healthful change in your diet you'd like to make.   1. _____

2. Identify two food choices you make that should change because they
   contribute to the need for the healthful change you identified.   2. _____

3. Identify two specific, acceptable options for food choices more healthful
   than the ones identified under number 2.   3. _____

4. Decide which option is easiest to accomplish and requires the smallest
   change to get the job done.   4. _____

5. Plan how to incorporate the change into your diet.   5. _____

FEEDBACK (answers to these questions) can be found at the end of Unit 5.

Key Terms

serotonin, page 5-11   *tryptophan, page 5-11*

Review Questions

	True	False		True	False
1. Food preferences are universal—everyone likes the same foods.	☐	☐	5. Nutrition knowledge is an important prerequisite for making healthful food choices.	☐	☐
2. Foods will be rejected by a population, no matter how nutritious they may be, if the foods don't fit into a culture's definition of what foods are appropriate to eat.	☐	☐	6. The primary reason efforts to improve food choices fail is that many people lack the willpower to stick to the change.	☐	☐
3. Food choices are driven largely by a person's need for nutrients.	☐	☐	7. Broad dietary changes, such as a decision to simply "eat less," are more likely to produce lasting behavioral changes than are small changes in diets, such as snacking on favorite fruits rather than candy.	☐	☐
4. Potatoes, rice, corn, and yucca are examples of "super foods" in specific countries due to their cultural significance.	☐	☐			

True False

True False

8. Changes in dietary intake that are acceptable to an individual and easy to implement are the types of changes that are most likely to last. ☐ ☐

9. Changing food choices for the better takes planning and includes individual decisions on specifically how the change in food choices will be implemented. ☐ ☐

10. Individuals need to plan for modifying their approach to improving food choices because even the best

planned changes in food choices sometimes fail. ☐ ☐

11. Examples of ways in which dietary intake affects behavior include the relationships between sugar intake and hyperactivity in children and between food allergies and autism. ☐ ☐

12. Results of recent research studies make it clear that the consumption of breakfast makes little difference to children's academic performance or weight status. ☐ ☐

Media Menu

www.nimh.nih.gov
The National Institute of Mental Health provides information on attention deficit hyperactivity and other disorders.

healthfinder.gov
Use this site to find reliable information about iron deficiency, fetal alcohol syn-

drome, and other disorders presented in this unit.

www.ncemch.org
This URL connects to the National Center for Education in Maternal and Child Health and provides information on lead poisoning and searchable databases.

www.ipm.iastate.edu/misc/insectsasfood.html
Go to this site for insect recipes from Iowa State University.

Notes

1. van den Bree MBM et al. Genetic and environmental influences on eating patterns of twins aged ≥50 y. Am J Clin Nutr 1999;70:456–65.

2. Rampersaud GC et al. Breakfast habits, nutritional status, body weight, and academic performance in children and adolescents. J Am Diet Assoc 2005:105:743–60.

3. Pyke M. Man and food. New York: McGraw-Hill; 1972.

4. Beauchamp GK, Mennella JA. Sensitive periods in the development of human flavor perception and preference. Ann Nestle 1998;56:19–31.

5. Rolls B. Obesity and Weight Control Symposium, Experimental Biology Annual Meeting, San Diego, CA, April 14, 2003.

6. Birch LL, et al. Effects of a nonenergy fat substitute on children's energy and macronutrient intake (Am J Clin Nutr 1993;58:326–33).

7. Story M, Brown JE. Do young children instinctively know what to eat? The studies of Clara Davis revisited. N Engl J Med 1987;316:103–6.

8. Birch LL et al. Effects of a nonenergy fat substitute on children's energy and macronutrient intake. Am J Clin Nutr 1993;58:326–33.

9. Steiner JE. The gustofacial response: observation on normal and anencephalic newborn infants. In: Bosma JF, ed. Fourth symposium on oral sensation and perception: development in the fetus and infant. Bethesda (MD): National Institute of Dental Research, DHEW Publication No. 73-546, US Dept HEW, NIH; 1973:254.

10. Factors influencing food choices in humans. Nutr Rev 1990;48:442.

11. V Packard, The status seekers (New York: Random House, 1959), 146.

12. Horovitz B. You want it your way. USA Today, Mar. 5, 2004, p. 1A.

13. Yankelovich Survey results reported in USA Today, 2000 Feb. 7:D1.

14. Parraga IM. Determinants of food consumption. J Am Diet Assoc 1990;90:661–3.

15. Zallen DT, Brown ML. Food fights: deciding about diet and disease. Am J Clin Nutr 1990;52:944–5.

16. Beerman KA. Variation in nutrient intake of college students: a comparison by students' residence. J Am Diet Assoc 1991;91:343–4.

17. Cheadle A et al. Community-level comparisons between the grocery store environment and individual dietary practices. Prev Med 1991;20:250–6.

18. Economic Research Service, Food Guide Pyramid Servings 1990 and 2004, www.ers.usda.gov/data/foodconsumption, accessed 7/06.

19. Brevard PB. Nutrition education improves the diet of college students. J Am Diet Assoc 1991, abs. A-47.

20. Davis JN et al. Is nutrition knowledge reflected in healthier dietary practices of college students? Experimental Biology 2003: Meeting Abstract 709.8, San Diego, April 2003.

21. Ankeny K et al. Dietary fat: sources of information used by men. J Am Diet Assoc 1991;91:1116–7.

22. Shepherd R, Stockley L. Nutrition knowledge, attitudes, and fat consumption. J Am Diet Assoc 1987;87:615–9.

23. Becker MH et al. The health belief model and prediction of dietary compliance: a field experiment. J Health Soc Behav 1977;18:348–66.

24. Holdt CS, et al. Knowledge and behaviors of allied health professionals regarding meat (abs.). J Am Diet Assoc 1991;91S, abs. A-13.

25. Handley B et al. Using action plans to help primary care patients adopt healthy behaviors: a descriptive study. J Am Board Fam Med 2006;19:224–31.

26. Gould L. Consumers taking small steps toward big lifestyle changes. Results of Datmonitor's survey of diet trends in Europe and the US. Reported in nutraingredients.com/Europe, March 18, 2005.

27. Young SN. Nutrition 3. The fuzzy boundary between nutrition and psychopharmacology. Can Med Assoc J, Jan. 22, 2002, vol. 155.

28. Young SN, Nutrition 3: The fuzzy boundary between nutrition and psychopharmacology (Can Med Assoc J, vol. 155, Jan. 22, 2002); Wolraich ML et al., The effect of sugar on behavior or cognition in children: a meta-analysis (JAMA 1995;274:1617); I deAndraca et al., Psychomotor development and behavior in iron-deficient anemic infants (Nutr Rev 1997;55:125); Fernstrom JD, Can nutrient supplements modify brain function? (Am J Clin Nutr 2000; 71(suppl.):1669S–73S); Jackson K. Nutrition and autism, are they linked? Today's Dietitian 2005;Jan:31–4.

29. Parker G, et al. Mood state effects of chocolate. J Affect Disord 2006; 92:149–59.

30. Bellamy C. The state of the world's children. Oxford (UK): Oxford University Press; 1996.

31. Allen LH. Functional indicators of nutritional status of the whole individual or the community. Clin Nutr 1984;3:169–75.

32. Nicklas TA et al. Breakfast consumption affects adequacy of total daily intake in children. J Am Diet Assoc 1993;93:886–91.

33. Tittman HG. What's the harm in just a drink? Alcohol and Alcoholism 1990;25:287–91; and Dietary Guidelines for Americans, US Dept of HHS and USDA, 2000.

34. Lazoff B. Nutrition and behavior. Am Psychologist 1989;44:231–6.

35. Lazoff B, et al. Long-term developmental outcome of infants with iron deficiency, N Engl J Med 1991;325: 687–94.

36. Hurade EK et al., Early childhood anemia and mild or moderate mental retardation, Am J Clin Nutr 1999; 69:115–9.

37. Canfield L et al, Intellectual impairment in children with blood lead concentrations below 10 μg per deciliter. N Engl J Med 2003;34:1517–26.

38. Child Health USA, 1997. US Maternal and Child Health Bureau, PHS, Washington, DC, p. 29.

39. Rogan WJ, Ware JH. Exposure to lead—how low is low enough? N Engl J Med 2003;348:1515–16.

40. Gans DA, et al. Sucrose and delinquency: oral sucrose tolerance test and nutritional assessment. Pediatrics 1990;86:254–62.

41. Young, Fernstrom JD. Can nutrient supplements modify brain function? Am J Clin Nutr 2000;71(suppl):1669S–73S.

42. Ferguson HB et al. Double-blind challenge studies of behavioral and cognitive effects of sucrose-aspartame ingestion in normal children. Nutr Rev 1986;44:144–50.

43. Regalado M. Busting the sugar-hyperactivity myth. WebMD, www.webMD.com, 1999, accessed 1/01.

Answers to Review Questions

1. False, see page 5-2.
2. True, see pages 5-2, 5-4, 5-5.
3. False, see pages 5-3, 5-4.
4. True, see pages 5-3.
5. True, see pages 5-6, 5-7.
6. False, see page 5-8.
7. False, see page 5-8.
8. True, see page 5-8.
9. True, see page 5-8.
10. True, see page 5-9.
11. False, see page 5-9.
12. False, see page 5-10.

Nutrition Up Close

Improving Food Choices

Feedback for Unit 5

There are no right or wrong answers to this unit's "Nutrition Up Close." Plans for dietary change that are specific, easy to accomplish, and highly acceptable are most likely to work in the long run.

UNIT 6

Healthy Diets, Dietary Guidelines, MyPyramid, and More

This Is Just to Say
I have eaten
the plums
that were in
the icebox
and which
you were probably
saving
for breakfast
Forgive me
they were delicious
so sweet
and so cold
—WILLIAM CARLOS WILLIAMS

NUTRITION SCOREBOARD

	TRUE	FALSE
1 Nearly one in two U.S. adults fails to consume three servings of vegetables daily.		
2 The basic food groups include a "healthy snack" food group.		
3 "Supersized" fast-food meals often provide two to three times more calories than regular-sized fast food meals.		

Key Concepts and Facts

- Healthful diets are characterized by adequacy and balance.

- There are many types of healthful diets.

- There are no good or bad foods, only diets that are healthful or unhealthful.

- The Dietary Guidelines for Americans and MyPyramid Food Guide provide foundation information for healthful diets and physical activity levels.

Answers to **NUTRITION SCOREBOARD**	TRUE	FALSE
1 One in two U.S. adults fails to consume at least three servings of vegetables daily.[1]	✔	
2 The basic food groups do not include a "healthy snack" food group.		✔
3 Supersizing fast-food meals piles on calories.	✔	

Healthy Eating: Achieving the Balance between Good Taste and Good for You

May I have your attention please? For a moment, think about the foods in Illustration 6.1. If your mouth is watering and you're ready to go out and buy some ripe peaches, you have found the balance between good taste and good for you. Who said foods that taste good aren't good for you?

Illustration 6.1
Can you smell it? Can you taste it?
A plump, golden peach. It's so ripe that juice spurts from it and drips down your chin when you take a bite. . . . A golden brown turkey just taken out of the oven. The wonderful smell fills the kitchen. A steaming loaf of homemade bread just set out to cool. A perfect ripe tomato just picked from the garden. It melts in your mouth.

Characteristics of Healthful Diets

Healthful diets come in a variety of forms. They may be based on bread, olives, nuts, fruits, beans, vegetables, lamb, and chicken (as in Greece); rice, vegetables, and small amounts of fish and other meats (as in China); or black beans, rice, meat, and tropical fruits (as in Cuba and Costa Rica). Illustration 6.2 gives examples of the diverse foods that can be part of a healthy diet.

Although the types of foods that go into them can vary substantially, healthful diets all share two characteristics: They are adequate and balanced. *Adequate diets* include a wide variety of foods that together provide sufficient levels of calories and *essential nutrients.* What's sufficient? For calories, it's the number

Illustration 6.2
Foods that contribute to healthy diets in different countries.
a. Dinner in Italy might be linguine primavera (pasta with vegetables). b. Pad Thai, rice noodles and vegetables, is a favorite dish in Thailand. c. Tamales are a celebration food in many Latin cultures. d. Dal, curry dishes, vegetables, and chicken are popular parts of the cuisine of India.

a.

b.

c.

d.

adequate diet
A diet consisting of foods that together supply sufficient protein, vitamins, and minerals and enough calories to meet a person's need for energy.

essential nutrients
Substances the body requires for normal growth and health but cannot manufacture in sufficient amounts; they must be obtained in the diet.

balanced diet
A diet that provides neither too much nor too little of nutrients and other components of food such as fat and fiber.

macronutrients
The group name for the energy-yielding nutrients of carbohydrate, protein, and fat. They are called macronutrients because we need relatively large amounts of them in our daily diet.

unsaturated fats
The type of fat that tends to lower blood cholesterol level and the risk of heart disease. They are liquid at room temperature and found in foods such as nuts, seeds, fish, shellfish, and vegetable oils.

Illustration 6.3
Acceptable Macronutrient Distribution Ranges (AMDRs) for percent of total calories in the diet that should consist of carbohydrate, protein, and fat.[3]

that maintains a healthy body weight. For essential nutrients, sufficiency corresponds to intakes that are in line with recommended intake levels represented by the *Recommended Dietary Allowances (RDAs)* and *Adequate Intakes (AIs)*. (These are shown in the Dietary Reference Intake tables inside the front cover of this book.) Recommended amounts of essential nutrients should be obtained from foods to reap the benefits offered by the variety of naturally occurring substances in foods that promote health. Following the pattern of food intake recommended in the MyPyramid Food Guide helps ensure that the need for essential nutrients as well as the need for other beneficial components of food are met.[2]

A *balanced diet* provides calories, nutrients, and other components of food in the right proportion—neither too much nor too little. Diets that contain too much sodium or too little fiber, for example, are out of balance. Diets that provide more calories than needed to maintain a healthy body weight are also out of balance.

Current dietary intake standards include guidelines to help individuals balance their intake of carbohydrate, protein, and fat—the *macronutrients.* Called the "Acceptable Macronutrient Distribution Ranges," or AMDRs, the guidelines indicate percentages of total caloric intake that should consist of carbohydrate, protein, and fat. These recommendations apply to individuals over the age of 4 years. Ranges are recommended because the prevention of heart disease, diabetes, and obesity as well as the assurance of dietary adequacy are related to various levels of intakes rather than to one specific intake level of these macronutrients. In the previous edition of the Dietary Reference Intakes (DRIs), recommendations were given for the percentage of total calories that should consist of *unsaturated* and *saturated fats.* These recommendations have been removed from the new standards. It is now recommended that diets be kept low in saturated and *trans* fat and provide sufficient amounts of linoleic acid and alpha-linolenic acid. Both of these fatty acids are unsaturated and considered essential nutrients.[3]

How Balanced Is the American Diet?

Results of assessments of what Americans actually consume compared to recommendations for healthy diets given in the DRIs is shown in Illustration 6.3. Compared to earlier studies, results reported in this illustration show that the American diet is improving in some respects. Over the past few years, intake of vegetables and fruits has increased, soft drink consumption has declined, and we are eating more dairy and grain products.[4,6]

On average, the U.S. diet remains overloaded with added sugars and low in essential fatty acids, dairy products, and vegetables and fruits. Continuing high

Macronutrient	CARBOHYDRATE	Added sugars	PROTEIN	FAT	Linoleic acid	Alpha-linolenic acid
AMDR	45–65%	no more than 25%	10–35%	20–35%	5–10%	0.6–1.2%
Average Intake	52%	>28%	15%	33%	5.7%	0.6%

Photo Disc

intakes of saturated and *trans* fat place many Americans at risk for heart disease, and low intake of dairy products increases the risk of osteoporosis.[3]

Under-consumption of Vegetables and Fruits   Nearly one in two U.S. adults fails to consume at least three servings of vegetables daily; and three out of four adults consume less than two servings of fruits.[1] Risk of heart disease, stroke, and cancer decrease as vegetable and fruit intakes increase to 5–9 servings daily.[5] Approximately 6% of adults regularly consume the types of vegetables most closely associated with reduced risk of cancer: broccoli, cauliflower, brussels sprouts, and dark, leafy, green vegetables. The most commonly consumed vegetable in the United States is french fries.[6]

Guides to Healthy Diets

Due to the impact of food choices on the health of individuals and population groups, many countries have established recommendations for dietary intake. Such recommendations are periodically updated as new discoveries about diet and health emerge. Many of the guidelines include recommendations for physical activity as well.

National recommendations for diet and physical activity usually apply to children over the age of 2 years and aren't appropriate for every individual in a population. General guidelines for diet and physical activity may not completely match the needs of individuals who, for example, are strict vegetarians, disabled, or have disorders such as hypertension or diabetes. A basic premise of national dietary guidelines is that nutrient needs should be met primarily through food consumption.

Benefits of population-based dietary and physical activity recommendations are multiple. The information is science based, free, and made widely available to the public on the Web. Adherence to the information can help people stay healthy and lowers their risk of developing disorders such as cancer, osteoporosis, heart disease, and obesity.

Some of the national guidelines also give credit to the cultural and social importance of food. "Enjoy your food!" is listed as Ireland's first dietary guideline, and the guidelines for Japanese people include "Happy eating makes for happy family life; sit down and eat together and talk; treasure family taste and home cooking."[7] Table 6.1 provides a summary of the dietary guidelines established for Canada and Japan. Web addresses that will link you to more examples of national dietary guidelines are given at the end of this unit. Additional information on dietary guidance for Canadians is given in Appendix F.

National diet and physical activity guidance in some countries is accompanied by information on how to select a healthful diet and achieve recommended levels of physical activity. In the United States, the national guidelines for diet and physical activity are called the "Dietary Guidelines for Americans," and the major how-to guide for consumers is "MyPyramid."[8]

Dietary Guidelines for Americans

The Dietary Guidelines for Americans provide science-based recommendations to promote health and to reduce the risk for major chronic diseases through diet and physical activity. Due to this credibility and focus on health promotion and

saturated fats
The type of fat that tends to raise blood cholesterol levels and the risk for heart disease. They are solid at room temperature and are found primarily in animal products such as meat, butter, and cheese.

Mexican is the most popular ethnic cuisine in the United States, followed by Italian and Chinese.

Photo Disc

ON THE SIDE

TABLE 6.1
Good advice from the dietary guidelines of Canada and Japan

Canadian Dietary Guidelines

A. **Summary of Nutrition Recommendations**
The Canadian diet should:
- provide energy consistent with the maintenance of body weight within the recommended range.
- include essential nutrients in amounts specified in the Recommended Nutrient Intakes.
- include no more than 30% of energy as fat (33 g/1000 kcal or 39 g/5000 kJ) and no more than 10% as saturated fat (11 g/1000 kcal or 13 g/5000 kJ).
- provide 55% of energy as carbohydrate (138 g/1000 kcal or 165 g/5000 kJ) from a variety of sources.
- be reduced in sodium content.
- include no more than 5% of total energy as alcohol, or 2 drinks daily, whichever is less.
- contain no more caffeine than the equivalent of 4 cups of regular coffee per day.
- Community water supplies containing less than 1 mg/litre should be fluoridated to that level.

B. **Guidelines for Healthy Eating**
- Enjoy a variety of foods.
- Emphasize cereals, breads, other grain products, vegetables, and fruits.
- Choose low-fat dairy products, lean meats, and foods prepared with little or no fat.
- Achieve and maintain a healthy body weight by enjoying regular physical activity and healthy eating.
- Limit salt, alcohol, and caffeine.

Japan's Dietary Guidelines for Health Promotion

1. Enjoy your meals
   - Have delicious and healthy meals that are good for your mind and body.
   - Enjoy communication at the table with your family or other people and participate in the preparation of meals.
2. Establish a healthy rhythm by keeping regular hours for meals.
3. Eat well-balanced meals with staple food, as well as main and side dishes.
4. Eat enough grains, such as rice and other cereals.
5. Combine vegetables, fruits, milk products, beans, and fish in your diet.
6. Avoid too much salt and fat.
7. Learn your healthy body weight and balance the calories you eat with physical activity.
8. Chew your food well and do not eat too quickly.
9. Enjoy nature's bounty and the changing seasons by using local food products and ingredients in seasons and by enjoying holiday and special-occasion dishes.
10. Reduce leftovers and waste through proper cooking and storage methods.
11. Assess your daily eating habits.
    - Learn and practice healthy eating habits at school and at home.
    - Promote appreciation of good eating habits from an early stage of life.

trans fats
A type of unsaturated fat present in hydrogenated oils, margarine, shortening, pastries, and some cooking oils that increase the risk of heart disease.

cholesterol
A fat-soluble, colorless liquid found in many animal products but not in plants. High cholesterol intake is related to the development of heart disease.

disease prevention for the public, the Dietary Guidelines form the basis of federal food and nutrition education programs and policies.

The 2005 edition of the Dietary Guidelines stresses the importance of selecting nutrient-dense foods, balancing caloric intake with output, and increasing physical activity. This document includes the promise that the health of most individuals will be enhanced if the recommendations are followed. Each recommendation is part of an integrated whole—all the recommendations should be implemented for best results.

Focus Areas and Key Recommendations

The Dietary Guidelines for 2005 include nine "Focus Areas" and 23 "Key Recommendations." These are highlighted in the Health Action presented on page 6-7. "Key Recommendations for Specific Population Groups" are also provided by the

Dietary Guidelines. Special population groups addressed are infants and young children, pregnant women, older adults, and people with weakened immune systems. These and other recommendations included in the Dietary Guidelines are available online as part of the full report (refer to the Media Menu section at the end of this unit).

Under legislative mandate, the Dietary Guideline for Americans must be updated every five years. This requirement was established over 20 years ago to ensure that up-to-date, scientifically based information on diet, physical activity,

HEALTH *Action* The 2005 Dietary Guidelines for Americans: Focus Areas and Examples of Key Recommendations

Find your way to a healthier you.

Adequate Nutrients within Calorie Needs

- Consume a variety of nutrient-dense foods and beverages within and among the basic food groups while choosing foods that limit the intake of saturated and *trans* **fats, cholesterol,** added sugars, salt, and alcohol.
- Meet recommended intakes within energy needs by adopting a balanced eating pattern, such as the USDA MyPyramid Food Guide or the DASH Eating Plan.

Weight Management

- To maintain body weight in a healthy range, balance calories from foods and beverages with calories expended.
- To prevent gradual weight gain over time, make small decreases in food and beverage calories and increase physical activity.

Physical Activity

- Engage in regular physical activiy and reduce sedentary activites to promote health, psychological well-being, and a healthy body weight.
- Engage in at least 30 minutes of moderate-intensity physical activity, above usual activity, at work or home on most days of the week.
- To help manage body weight engage in approximately 60 minutes of moderate- to vigorous-intensity activity on most days of the week.

Food Groups to Encourage

- Consume a sufficient amount of fruits and vegetables while staying within energy needs.
- Choose a variety of fruits and vegetables each day. In particular, select from all five vegetable subgroups (dark green, orange, legumes, starchy vegetables, and other vegetables) several times a week.
- Consume three or more ounce-equivalents of whole-grain products per day.
- Consume three cups per day of fat-free or low-fat milk or equivalent milk products.

Fats

- Consume less than 10% of calories from saturated fatty acids and less than 300 mg/day of cholesterol and keep *trans* fatty acid consumption as low as possible.
- Keep total fat intake between 20 to 35% of calories.

Carbohydrates

- Choose fiber-rich fruits, vegetables, and whole grains often.
- Choose and prepare foods and beverages with little added sugars or caloric sweeteners.

Sodium and Potassium

- Consume less than 2,300 mg (approximately 1 tsp of salt) of sodium per day.

Dietary Guidelines
for Americans
2005

- Choose and prepare foods with little salt. At the same time, consume potassium-rich foods, such as fruits and vegetables.

Alcoholic Beverages

- Those who choose to drink alcoholic beverages should do so sensibly and in moderation—defined as the consumption of up to one drink per day for women and up to two drinks per day for men.

Food Safety

- To avoid microbial foodborne illness:
  - Clean hands, food contact surfaces, and fruits and vegetables.
  - Cook foods to a safe temperature to kill microorganisms.

and health will be available to the U.S. public, health professionals, and policy makers.

Implementation of the Dietary Guidelines

The MyPyramid Food Guide is the major how-to tool intended to help the public implement the 2005 Dietary Guidelines. In addition, the Dietary Guidelines report identified the DASH (*Dietary Approaches to Stop Hypertension*) Eating Plan as being consistent with the dietary recommendations. MyPyramid covers both food selection and physical exercise, while the DASH Eating Plan addresses only dietary intake. Both are valuable tools that provide the framework for planning nutrient-dense, calorically appropriate diets that diminish the risk of chronic disease.

MyPyramid Food Guide

Food group guides to the intake of healthful diets have been available in the United States since 1916. Known by names such as "Basic Four Food Groups" and "Food Guide Pyramid," the guides have been periodically updated since then. New releases of food group guides reflect existing scientific knowledge about nutrition and health and are modified to address emerging health problems.

The U.S. Department of Agriculture (USDA) released its newest version of the food guide in 2005. Called "MyPyramid," it is very different from previous guides (Illustration 6.4). It is Internet based and resource filled, and some of the educational resources offered are interactive. It is a popular Web site. So many people (48 million) logged on in the first day it was available that it temporarily crashed the site.[11]

What's New in MyPyramid?

The last food guide was published by USDA in 1992, leaving the new guide with some catching up to do. National recommendations for nutrient intake levels (the Dietary Reference Intakes), rising rates of obesity and diabetes, and new knowledge about diet and health relationships reported in the 2005 Dietary Guidelines all had to be addressed in the new food guide. It was also time to jump into the information age with both feet in order to make the information widely available. Consequently, MyPyramid differs in many ways from previous editions of food guides.

To help stem the tide of increasing rates of obesity, MyPyramid guidance stresses the importance of lower calorie intakes and increased physical activity levels. Because low intakes increase the risk of certain chronic diseases, the new food guide emphasizes intake of whole-grain products, colorful vegetables, low-fat dairy products, and lean meats.

No More Standard Serving Sizes This version of the food guide does not include standard serving sizes for food portions. They were a source of major confusion for the public previously because serving sizes were perceived as being bigger than they actually were.[12] MyPyramid uses cups and ounces as the primary measures of how much food to consume daily and gives the recommended number of cups or ounces for food within each group.

Illustration 6.4
MyPyramid's major graphic for the food guide is much different from those used in the past.

MyPyramid.gov
STEPS TO A HEALTHIER YOU

MyPyramid.gov: The Web Site

Log on to mypyramid.gov and be prepared to be amazed at what the site offers.

- Click on "MyPyramid Plan" and enter your age, sex, and activity level. You will quickly be rewarded with an estimate of what and how much food you need to eat. Web pages that correspond to the assessments displayed are shown Illustration 6.5.

- Use the advice located under "Inside MyPyramid" to make smart choices from every food group, to find your balance between food and physical activity, and to identify nutrient-dense foods.

- Follow the lead to "Tour MyPyramid" and take an animated trip through MyPyramid.

- Go to "MyPyramid for kids" to play the MyPyramid Blast-Off game and find other materials designed for elementary school–aged childern.

- The "For Professionals" link will connect you to detailed information about using MyPyramid educational materials and provides seven days of sample menus that correspond to the MyPyramid recommendation for a 2000-calorie food pattern. Four of the seven days included in the menus, and the nutrient analysis for the menus, are shown in Illustration 6.6.

- The "Tips & Resources" link heads you to a wealth of ideas that can help you get started toward a healthy diet. Included are tips for each food group, physical activity, and eating out.

- "Inside the Pyramid" explains each food group, discretionary calories, and physical activity recommendations. It provides information on which foods are within the various groups and food measure equivalents so you can convert food amounts into cups and ounces, and oils into teaspoons. Table 6.2 lists food measure equivalents for common foods by food group shown in this section of Inside the Pyramid.

- "MyPyramid Tracker" is a dietary and physical activity assessment tool that provides information on your diet quality, physical activity status, related nutrition messages, and links to nutrient and physical activity information. Although the MyPyramid Food Guide is not designed for weight-loss diets, the MyPyramid Tracker tool can be used to assess and monitor calorie intake and physical activity levels.

Food group recommendations generated by MyPyramid are based on an individual's calorie need and provide adequate amounts of essential nutrients. People who exceed the recommended intake levels of fats and sweets may consume too many calories, however, and gain weight as a result.[2]

Limitations of MyPyramid   Materials available for MyPyramid are almost entirely made available on the Web, making the information inaccessible to people who do not use the Internet. This situation leaves primarily low-income families without access to MyPyramid.[11] MyPyramid does not provide specific recommendations for infants, pregnant or breastfeeding women, or for strict vegetarians.

Menus offered as examples on MyPyramid may not correpond to individual food preferences and contain relatively few ethnic foods. As with past food guides,

Illustration 6.5
MyPyramid Plan

Sample Menus for a 2000 Calorie Food Pattern

Averaged over a week, this seven day menu provides all of the recommended amounts of nutrients and food from each food group. (Italicized foods are part of the dish or food that preceeds it.)

Day 1	Day 2	Day 3	Day 4
BREAKFAST	**BREAKFAST**	**BREAKFAST**	**BREAKFAST**
Breakfast burrito	Hot cereal	Cold cereal	1 whole wheat English muffin
1 flour tortilla (7" diameter)	*1/2 cup cooked oatmeal*	*1 cup bran flakes*	*2 tsp soft margarine*
1 scrambled egg (in 1 tsp soft margarine)	*2 tbsp raisins*	*1 cup fat-free milk*	*1 tbsp jam or preserves*
*1/3 cup black beans**	*1 tsp soft margarine*	*1 small banana*	1 medium grapefruit
2 tbsp salsa	1/2 cup fat-free milk	1 slice whole wheat toast	1 hard-cooked egg
1 cup orange juice	1 cup orange juice	*1 tsp soft margarine*	1 unsweetened beverage
1 cup fat-free milk		1 cup prune juice	
			LUNCH
LUNCH	**LUNCH**	**LUNCH**	White bean-vegetable soup
Roast beef sandwich	Taco salad	Tuna fish sandwich	*1 1/4 cup chunky vegetable soup*
1 whole grain sandwich bun	*2 ounces tortilla chips*	*2 slices rye bread*	*1/2 cup white beans**
3 ounces lean roast beef	*2 ounces ground turkey, sauteed in*	*3 ounces tuna (packed in water, drained)*	2 ounce breadstick
2 slices tomato	*2 tsp sunflower oil*	*2 tsp mayonnaise*	8 baby carrots
1/4 cup shredded romaine lettuce	*1/2 cup black beans**	*1 tbsp diced celery*	1 cup fat-free milk
1/8 cup sauteed mushrooms (in 1 tsp oil)	*1/2 cup iceberg lettuce*	*1/4 cup shredded romaine lettuce*	
1 1/2 ounce part-skim mozzarella cheese	*2 slices tomato*	*2 slices tomato*	**DINNER**
1 tsp yellow mustard	*1 ounce low-fat cheddar cheese*	1 medium pear	Rigatoni with meat sauce
3/4 cup baked potato wedges*	*2 tbsp salsa*	1 cup fat-free milk	*1 cup rigatoni pasta (2 ounces dry)*
1 tbsp ketchup	*1/2 cup avocado*		*1/2 cup tomato sauce tomato bits**
1 unsweetened beverage	*1 tsp lime juice*	**DINNER**	*2 ounces extra lean cooked ground beef (sauteed in 2 tsp vegetable oil)*
	1 unsweetened beverage	Roasted chicken breast	*3 tbsp grated Parmesan cheese*
DINNER		*3 ounces boneless skinless chicken breast**	Spinach salad
Stuffed broiled salmon	**DINNER**	1 large baked sweetpotato	*1 cup baby spinach leaves*
5 ounce salmon filet	Spinach lasagna	1/2 cup peas and onions	*1/2 cup tangerine slices*
1 ounce bread stuffing mix	*1 cup lasagna noodles, cooked (2 oz dry)*	*1 tsp soft margarine*	*1/2 ounce chopped walnuts*
1 tbsp chopped onions	*2/3 cup cooked spinach*	1 ounce whole wheat dinner roll	*3 tsp sunflower oil and vinegar dressing*
1 tbsp diced celery	*1/2 cup ricotta cheese*	*1 tsp soft margarine*	1 cup fat-free milk
2 tsp canola oil	*1/2 cup tomato sauce tomato bits**	1 cup leafy greens salad	
1/2 cup saffron (white) rice	*1 ounce part-skim mozzarella cheese*	*3 tsp sunflower oil and vinegar dressing*	**SNACKS**
1 ounce slivered almonds	1 ounce whole wheat dinner roll		
1/2 cup steamed broccoli	1 cup fat-free milk		1 cup low-fat fruited yogurt
1 tsp soft margarine		**SNACKS**	
1 cup fat-free milk	**SNACKS**	1/4 cup dried apricots	
		1 cup low-fat fruited yogurt	
SNACKS	1/2 ounce dry-roasted almonds*		
	1/4 cup pineapple		
1 cup cantaloupe	2 tbsp raisins		

Illustration 6.6
Sample menus and nutrient analysis from the "For Professionals" link

planning and evaluating how mixed dishes (such as stews, soups, salads, and various types of pizza) fit into the foods groups still can be perplexing.

In MyPyramid, total amounts of food recommended from each food group are given in cups or ounces, servings of oils are given in teaspoons, and a calorie number is given for the "Discretionary Calories" allotment. People are not used to thinking of serving sizes in ounces of bread or rice or as teaspoons of oils. They may not know the calorie value of sweets and fats covered by the Discretionary Calorie allotment.

The DASH Diet

Originally published as a diet that helps control mild and moderate **hypertension** in experimental studies, the DASH Eating Plan also reduces the risk of cancer, osteoporosis, and heart disease. Improvements in blood pressure are generally seen within two weeks of starting this dietary pattern.[13]

The DASH dietary pattern emphasizes fruits, vegetables, low-fat dairy foods, whole-grain products, poultry, fish, and nuts. Only low amounts of fats, red meats, sweets, and sugar-containing beverages are included. This dietary pattern provides ample amounts of potassium, magnesium, calcium, fiber, and protein and limited amounts of saturated and *trans* fats.

The DASH Eating Plan is available as part of the 2005 Dietary Guidelines. Recommendations for types and amounts of foods included in this eating plan by calorie need are shown in Table 6.3 Although two calorie need levels are shown in the Illustration, a DASH eating plan is available for 12 levels (1600 to 3200 calories) at the Web site listed in Table 6.3.

hypertension
High blood pressure. It is defined as blood pressure exerted inside blood vessel walls that typically exceeds 140/90 millimeters of mercury.

TABLE 6.2

MyPyramid food measure equivalents

Grains		
	bagel	1 mini bagel = 1 oz
		1 large bagel = 4 oz
	biscuit	1–2" diameter = 1 oz
		1–3" diameter = 2 oz
	bread	1 slice = 1 oz
	cooked cereal	½ cup = 1 oz
	crackers	5 whole wheat = 1 oz
		7 square/round = 1 oz
	English muffin	½ muffin = 1 oz
	muffin	1–2½" diameter = 1 oz
		1–3½" diameter = 3 oz
	pancake	1–4½" diameter = 1 oz
		2–3" diameter = 1 oz
	popcorn	3 cups = 1 oz
	breakfast cereal	1 cup flakes = 1 oz
		1¼ cups = 1 oz
	rice	½ cup = 1 oz
	pasta	½ cup = 1 oz
	tortilla	1–6" diameter = 1 oz
		1–12" diameter = 4 oz

Vegetables		
	cooked	1 cup = 1 cup
	carrots	2 medium = 1 cup
		½ baby = 1 cup
	celery	1 large stalk = 1 cup
	corn on the cob	1–6" long = ½ cup
		1–9" long = 1 cup
	green/red peppers	1 large = 1 cup
	potatoes	1 medium (3" diameter) = 1 cup
	raw, leafy	2 cups = 1 cup
	tomato	1 large = 1 cup

Milk		
	milk	1 cup = 1 cup
	yogurt	1 cup = 1 cup
	cheese	1½ oz hard = 1 cup
		⅓ cup shredded = 1 cup
		2 oz processed = 1 cup
		½ cup ricotta = 1 cup
		2 cups cottage cheese = 1 cup
	pudding	1 cup = 1 cup
	frozen yogurt	1 cup = 1 cup
	ice cream	1½ cup = 1 cup

fruits		
	apple	1 small = 1 cup
		½ large = 1 cup
	banana	1 large = 1 cup
	cantaloupe	⅛ = 1 cup
	grapes	32 = 1 cup
	grapefruit	1 = 1 cup
	orange	1 = 1 cup
	peach	1 = 1 cup
	pear	1 = 1 cup
	plums	3 = 1 cup
	strawberries	8 large = 1 cup
	watermelon	1" wedge = 1 cup
	dried fruit	½ cup = 1 cup
	fruit juice	1 cup = 1 cup

Meats and Beans		
	steak	1–3½" × 2½" × ½" = 3 oz
	hamburger	1 small = 2 oz
		1 medium = 4 oz
		1 large = 6 oz
	chicken	½ breast = 3 oz
		1 thigh = 2 oz
		1 leg = 3½ oz
	pork chops	1 medium = 3 oz
	fish	1 small can tuna = 3½ oz
		1 small fish = 3 oz
		1 salmon steak = 5 oz
	seafood	5 large shrimp = 1 oz
		10 medium clams = 3 oz
		½ cup lobster = 2½ oz
	eggs	1 small = 1 oz
		1 large = 2 oz
	nuts	12 almonds = 1 oz
		1 Tbsp peanut butter = 1 oz
	beans	¼ cup cooked, dry beans = 1 oz
		¼ cup tofu = 1 oz
		2 Tbsp hummus = 1 oz

Food Guides from Other Countries   As is the case for dietary guidelines, many countries and regions of the world have developed food intake guides. Of these, the most well known is the "Mediterranean Diet Pyramid" shown in Illustration 6.7. The World Health Organization (WHO) developed this guide in 1994 to help popularize a type of diet associated with reduced risk of heart disease and cancer.[25] It emphasizes olive oil, breads, whole grain cereals, nuts, fish, dried beans, vegetables, and fruits—and wine in moderation. Intake of red meats is limited to monthly intake, and sweets and poultry to weekly intake. The Asian Food Guide Pyramid (Illustration 6.8) shares a number of elements in common with the Mediterranean Diet Pyramid. This guide separates food into monthly, weekly, optional daily, and daily intakes. It focuses on plant foods while de-emphasizing intake of red meat, eggs, poultry, and other animal products. The Asian Food Guide

TABLE 6.3
The DASH Eating Plan

Food Group	2000 Calories	2600 Calories	Serving Size
	NUMBER OF SERVINGS PER DAY		
Grains[a]	6–8	10–11	1 slice bread 1 oz dry cereal 1/2 cup cooked rice, pasta, cereal
Vegetables	4–5	5–6	1 cup raw leafy vegetable 1/2 cup cut-up or raw or cooked vegetable 1/2 cup vegetable juice
Fruits	4–5	5–6	1 medium fruit 1/4 cup dried fruit 1/2 cup fresh, frozen, or canned fruit 1/2 cup fruit juice
Fat-free or low-fat milk and milk products	2–3	3	1 cup milk 1 cup low-fat or fat-free yogurt 1 1/2 oz low-fat or fat-free cheese
Lean meats, poultry, fish	2 or fewer	2	3 oz cooked meat, poultry, or fish
Nuts, seeds, legumes	4–5/week	1/day	1/3 cup or 1 1/2 oz nuts 2 Tbsp peanut butter 2 Tbsp or 1/2 oz seeds 1/2 cup cooked dry beans or peas
Fats and oils	2–3	3	1 tsp soft margarine 1 Tbsp low-fat mayonnaise 2 Tbsp light salad dressing 1 tsp vegetable oil
Sweets	5/week	2 or fewer/day	1 Tbsp sugar, jelly, jam 1/2 cup sorbet and ices 1 cup lemonade

[a]whole grain products primarily

Source: Dietary Guidelines for Americans, Appendix A-1: The DASH Eating Plan at 1,600-, 2,000-, 2,600-, and 3,100- Caloric Levels. Available at www.health.gov/dietaryguidelines/dga2005/document/html/appendixA.htm.

Illustration 6.7
The Mediterranean Diet Pyramid

Illustration 6.8
Asian Food Guide Pyramid

Pyramid was developed by the Cornell-China-Oxford Project from results of studies on more than 10,000 families in China and Taiwan.[26]

Canada has chosen to stress grain products, fruits, and vegetables in its food guide. The Canadian Food Guide to Healthy Eating is depicted as a rainbow with grains, fruits, and vegetables located on the outer bands to emphasize these foods. The Canadian format gives consumers a visual display of a wide variety of food choices within each group.

Portion Distortion

> *As portion sizes have grown, people don't really know what a cup of spaghetti looks like on a plate. We have no conception of what portions are anymore.*
>
> —ELLEN SCHUSTER, PH.D.[16]

Although food measure equivalents needed to estimate how much food should be consumed daily from each food group are described in MyPyramid (see Table 6.2), the problem of estimating portion sizes remains. Individual ideas of what a normal serving amount is vary based on past experiences at family meals, the size of portions provided by restaurants, and packaged food and beverage sizes. Supersizing meals at fast-food restaurants, large portions served by other restaurants, expanding sizes of bakery products, and larger cups of soft drinks are contributing to the problem of portion distortion. Table 6.4 provides examples of how food portion sizes are expanding.

Is Supersizing Leading to Supersized Americans?   Supersizing fast foods can double or triple the caloric content of the foods compared to their regular-sized counterparts. Larger portions don't cost restaurants much more than smaller portions, they increase sales volume, and they also encourage people to eat more. Many Americans are already eating plenty, and it is suspected that rising rates of obesity may be partly related to increased portion sizes.[17] Some health activists and legislators think it's time for fast-food and other restaurants to let consumers know the caloric cost, as well as the price, of menu items (Illustration 6.9).[18]

A report recently commissioned by the Food and Drug Administration (FDA) concluded that restaurants are in a prime position to help improve the nation's

TABLE 6.4

Typical portion sizes in the marketplace

Marketplace Portion	Size
Bagel	5.8 oz
Muffin	6.5 oz
Hamburger bun	2.2 oz
Hamburger	3.9 oz
Steak	8.1 oz
French fries	5.3 oz
Pasta	2.9 C

The standard bakery size for a muffin used to be 1.5 ounces. It has grown to become the 6.5-ounce muffin.

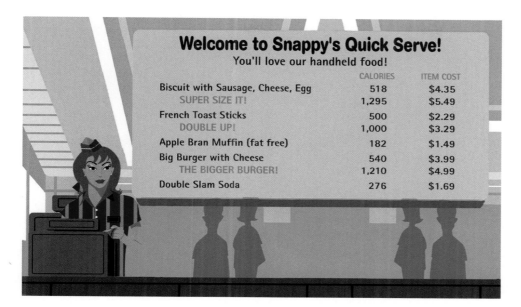

Illustration 6.9
Should fast-food and chain restaurants be required to disclose the caloric content of menu items?

diet and combat obesity. There are nearly a million restaurants and other eating establishments in the United States and a minority of them provide readily available nutrition information about the products they serve. How many calories are in the large portions of high-fat food served up by many resturants remains a mystery to most diners, and chances are good they are being served more calories than they realize. Simply letting consumers know how many calories are in the menu items offered may help them make better-informed decisions about what to eat.[18]

Can You Still Eat Right When Eating Out?

The question about what to eat often boils down to choosing a restaurant. According to recent USDA data, 50% of Americans eat out every day. In general, foods eaten away from home have lower nutrient content and are higher in fat than foods eaten at home (see Illustration 6.10). In addition, children and teenagers who eat dinner with their families most days tend to have more healthful diets—including more vegetables, fruits, vitamins, minerals, and fiber, and less fat—than others who never or occasionally eat dinner with the family.[19]

Reality Check

Portion Distortion

Mohammad and Kevin decided to eat at an Italian restaurant after soccer practice. Just for the fun of it, they agreed to guess how many ounces of spaghetti they would consume if they ate the portion served to them.

Answers on page 6-17

Mohammad: Let me see . . . I'd guess it would be 6 ounces.

Kevin: I bet it's 16 ounces. It looks like it weighs a pound.

Illustration 6.10
Hamburgers, french fries, and pizza are the top-selling food items in U.S. restaurants.[18]

Staying on Track while Eating Out   You'll find it easier to stick to a healthy diet if you decide what to eat before you enter a restaurant and look over the menu (Illustration 6.11). You could make the decision to order soup and a salad, broiled meat, a half-portion of the entrée, or no dessert *before* entering the restaurant. "Impulse ordering" is a hazard that can throw diets out of balance. If you're going to a party or business event where food will be served, decide before you go what types of food you will eat and what you will drink. If only high-calorie foods are offered, plan on taking a small portion and stopping there.

Illustration 6.11
"No, no thank you. Extra cheese isn't part of what I planned to eat."

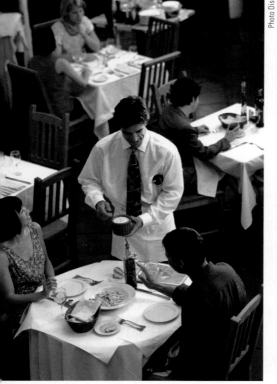

Can Fast Foods Be Part of a Healthy Diet?   Many fast foods deserve their reputation for being high in calories, fat, saturated fat, and salt. (Table 6.5 lists the calorie and fat contributions of a variety of foods sold in fast-food restaurants). Tostadas, bean burritos, grilled chicken sandwiches, and baked potatoes are among the lowest-fat options available at fast-food restaurants. In response to negative publicity about the caloric and nutritional value of foods traditionally offered, many fast-food chains have added specialty salads to their menus. The salads offer nutrient-dense vegetables; but additions like crispy chicken, bacon, and salad dressing can increase their caloric value to around 500 and increase fat calories to 40 to 50% of total calories.

You can eat an occasional cheeseburger, fries, or specialty sandwich at your favorite fast-food restaurant and still maintain a healthy diet. Usual dietary intake is more important to long-term health than are occasional dinners at a fast-food restaurant.

The Slow Food Movement

Forsake fast food and convenience cuisine. Savor the enjoyment of ripe, locally grown produce and freshly prepared foods that have made a short, simple field-to-plate journey.
—Sharon Palmer, RD, 2005.[22]

An interesting trend in food preparation and consumption is making its way across the globe. The trend is away from fast food and junk food, and

ON THE SIDE

Do you recognize these vegetables?

a. Parsnips, b. Turnips, c. kale, d. Leeks, e. Swiss chard

TABLE 6.5

The fat content of some fast foods[a]

Food	Calories	% of Calories from Fat	Food	Calories	% of Calories from Fat
Sausage Breakfast Croissanwich	538	69	Burrito Supreme	457	43
Fried chicken thigh	278	62	Fillet-o-fish	373	43
Fried chicken wing	181	60	Egg McMuffin	340	42
Bacon cheeseburger	610	58	Hamburger	350	41
Chicken McNuggets® (6)	323	56	Beef tostada	239	41
Fried chicken breast	276	55	Beef n' Cheddar	490	39
Fried chicken drumstick	147	55	Taco	186	39
Quarter Pounder with cheese	525	55	Thin-n-Crispy Cheese Pizza (2 slices)	398	38
Big Mac	570	55	Beef burrito	357	38
Sausage McMuffin	427	55	Roast beef sandwich	353	38
Whopper with cheese	711	54	Pan Pizza with pepperoni (2 slices)	540	37
Fried onion rings	274	53	Pan Pizza with cheese (2 slices)	492	33
Whopper	628	52	Chili, regular	240	30
Coleslaw	103	52	Tostada	179	30
French fries, regular	227	52	Grilled chicken breast on bun	320	28
Hot dog	280	51	Pizza with pepperoni (2 slices)	380	28
Whaler Fish Sandwich®	488	50	Bean burrito	357	25
Quarter Pounder®	427	48	Pizza with cheese (2 slices)	376	24
Pan Supreme Pizza® (2 slices)	589	46	Vanilla shake	280	19
Cheeseburger	318	45	Chocolate shake	447	16
Ham and Cheese Specialty Sandwich	471	44	Mashed potatoes, without gravy	62	15
Thin-n-Crispy Pizza with pepperoni (2 slices)	413	44	Baked potato, plain	250	7
Thin-n-Crispy Pizza Supreme (2 slices)	459	43			

[a]Calories and fat content may vary somewhat depending on the fast-food restaurant or chain.

ANSWERS TO
Reality Check

Portion Distortion

The average portion size of pasta served by restaurants is nearly 3 cups—or 6 ounces.[14] For people who need 2000 calories, that's a day's allotment for the foods from the grain group.

Mohammad:

Kevin:

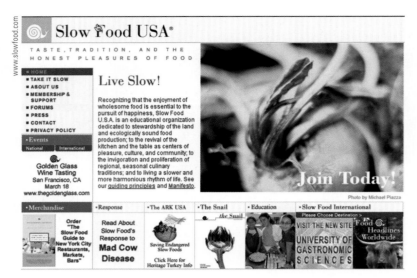

Illustration 6.12
Backlash to fast foods and fast lives?
New food trends are under way, and here's an example.

source: www.slowfood.com

toward locally grown "homespun" foods eaten in a casual atmosphere.[20] The "Slow Food USA" (Illustration 6.12) movement represents some aspects of this trend. Part of an international group, Slow Food USA is an educational organization that supports ecologically sound food production; the revival of the kitchen and the table as centers of pleasure, culture, and community; and living a slower and more harmonious rhythm of life.[24] The trend is placing the topic of healthy eating in a new light for many individuals, and it may help bring people closer to family, friends, and the environment.

What If You Don't Know How to Cook?

I can be warmed by the memory of cooking something for somebody.
—MAYA ANGELOU, 2004

Illustration 6.13
Some good starter cookbooks and recipe sources.
Most bookstores offer a wide variety of basic cookbooks.

With so many convenience foods available and time at a premium, there is growing concern that we're becoming a nation of cooking illiterates. Cooking at home gives you control over what you eat and how it's prepared. Some people immensely enjoy cooking and get a real thrill out of making their specialties for friends and family. It's becoming a popular leisure-time activity: 43% of U.S. adults have taken it up for enjoyment.[23]

If you don't know how to cook, there are several ways to learn. You could start on your own by using the recipes on food packages like pasta, tomato sauce, or dried beans. Or you could buy a basic cookbook and make simple dishes like salads, tacos, shish kebab, and lentil soup. You could even take a community education course. Illustration 6.13 shows some examples of good starter cookbooks. Read the sections on basic cooking skills and learn what types of equipment and utensils you need to prepare basic dishes. Get the foods and other supplies you need. Select the recipes that look doable and be sure they pass your taste and nutritional standards tests. Voilà! You're cooking.

Bon Appétit!

Dietary guidelines from some other countries contain one other rule of healthy eating that would serve Americans well: "Enjoy your meals." Eating a healthy diet should be enjoyable. If it's too much of a struggle, the healthy diet won't last. The best diets are those that keep us healthy and enhance our sense of enjoyment and well-being. The trick is to remember the broad array of nutritious foods we like that give us good taste and enjoyment when we eat them. And remember not to feel guilty when you occasionally eat hamburgers or ice cream. Enjoy them to the utmost—as much as ripe oranges, papaya, homemade soups, roast turkey, hummus, and countless other nutritious delicacies.

Nutrition Up Close

MyPyramid: A Balancing Act

Focal Point: Translating dietary intake into MyPyramid Food Group amounts

Planning a diet that supplies recommended amounts of foods from each food group requires that you be able to determine portion sizes and know how to use food measure equivalents. Here's an activity that will give you practice doing that.

Breakfast: 1 cup corn flakes
              ½ cup 2% milk

Lunch: ¼ pound cheeseburger on bun
              1 cup potato chips
              16 ounces iced tea (no sugar)

Dinner: 10″ diameter white pizza with a thick crust
              16 ounces cola

Snack: 2 cups ice cream

Using the information on Food Measure Equivalents in Table 6.2, and your best judgment, estimate how many cups or ounces of food are provided by the following one-day's diet for each MyPyramid food group.

Food Group

Grains	_____ ounces
Fruits	_____ cups
Vegetables	_____ cups
Meat and Beans	_____ ounces
Milk	_____ cups

FEEDBACK (answers to these questions) can be found at the end of Unit 6.

Key Terms

adequate diet, page 6-4
balanced diet, page 6-4
cholesterol, page 6-6

essential nutrients, page 6-4
macronutrients, page 6-4
hypertension, page 6-11

saturated fats, page 6-5
trans fats, page 6-6
unsaturated fats, page 6-4

Review Questions

True   False

1. Adequate diets are defined as those that provide sufficient calories to relieve hunger and maintain a person's body weight. ☐ ☐

2. The diets of Americans tend to be out of balance in a number of ways. ☐ ☐

3. Consumption of five to nine servings of vegetables and fruit daily confers more health benefits than consuming five. ☐ ☐

4. The 2005 Dietary Guidelines for Americans emphasize the importance of selecting nutrient-dense foods, balancing calorie intake with output, and increasing physical activity. ☐ ☐

True   False

5. "MyPyramid " is the major tool used in the United States to help people implement the Dietary Guidelines. ☐ ☐

6. The DASH Eating Plan represents a dietary pattern that is consistent with the 2005 Dietary Guidelines recommendations for dietary intake. ☐ ☐

7. Reduction in salt (sodium) consumption is no longer recommended by the Dietary Guidelines. ☐ ☐

8. "Key Recommendations" are included for food safety in the Dietary Guidelines. ☐ ☐

9. To help stem the tide of increasing rates of obestiy in the United States,

	True	False			True	False
MyPyramid stresses the importance of lower caloric intake and increased physical activity.	☐	☐	11. "Food Measure Equivalents" are used to identify standard serving sizes of foods with food groups.		☐	☐
10. An advantage, as well as a limitation, of MyPyramid Food Guide is that it is principally available on the Internet.	☐	☐	12. Fast-food resturants are now required to label food wrappers with the calorie value of the food in the wrapper.		☐	☐

Media Menu

www.kraftfoods.com/kf/HealthyLiving/ articles/dietaryguidelines.htm
Developed by Kraft Foods, this site summarizes "Dietary Guidelines from around the World" including those for Ireland, India, the Philippines, and Korea.

www.nutrition.gov
This site provides links to MyPyramid and the Dietary Guidelines.

www.dietitians.ca
Go from this Dietitians of Canada home page to find consumer information on advice for healthy eating and food preparation, visit a virtual grocery store, assess your nutrition profile, and get menu-planning help.

www.mypyramid.gov
This is the U.S. Department of Agriculture's MyPyramid Food Guide main page, with links to dietary and physical activity assessments, menu planning information, interactive learning tools, and much more. Don't miss "My Pyramid Tracker"—it's fun.

www.dietitian.or.jp/english/jp_health_ nutrition/dietary_guidelines-s.html
The source of the Japanese dietary guidelines is at this site.

www.slowfoodusa.org, wwwslowfoods .com
Slow Food is an international association that promotes the pleasures of the table and defends food and agricultural biodiversity worldwide.

www.ars.usda.gov/foodsearch
This site presents USDA's What's In The Foods You Eat Search Tool. You can use it to view nutrient profiles of 13,000 foods.

Notes

1. Food consumption and spending. Food Review, vol. 23, no. 3, www.ers.usda.gov/data/foodcomposition, accessed 8/03.

2. Gao X et al. The 2005 Food Guide Pyramid is associated with more adequate nutrient intakes within energy constraints than the 1992 Pyramid. J Nutr 2006;136:1341–6.

3. Dietary Reference Intakes. Energy, carbohydrate, fiber, fat, fatty acids, cholesterol, protein, and amino acids. Institute of Medicine, National Academy of Sciences. Washington, DC: National Academies Press, 2002.

4. ERS/USDA Food Consumption Data, www.ers.usda.gov/Data/foodconsumption, accessed 7/06.

5. Eat 5 to 9 day for better health, www.5aday.gov, accessed 7/06.

6. Johnston CS et al. More Americans are eating "5-A-Day" but intake of dark green cruciferous vegetables remains low. J Nutr 2000;130:3063–67.

7. Asian Food Information Centre, Dietary Guidelines, www.afic.org, accessed 7/06.

8. Dietary Guidelines Executive Summary, www.health.gov/dietaryguidelines/dga2005/document/html/executivesummary.htm, accessed 7/06.

9. New Canadian Food Guide references published at Health Canada's Web site, hc-sc.gc.ca/hpfb-dgpsa/onpp-bppn/revision_food_guide_e.html

10. Source of Japanese DGs: http://www.dietitian.or.jp/english/jp_health_nutrition/dietary_guidelines-s.html, accessed 7/06.

11. Johnston CS. Uncle Sam's diet sensation: My Pyramid—an overview and commentary. Medscape General Medicine 2005;7:10/21/05, accessed at www.medscape.com.

12. Meerschaert CM. One size does not fit all. The new Pyramid food guidance system. Today's Dietitian 2005;Aug:42–6.

13. Nowson CA. et al. Blood pressure response to dietary modifications in free-living individuals J. Nutr. 134:2322–2329.

14. Young LR, Nestle M. Expanding portion sizes in the US marketplace: implications for nutrition counseling. J Am Diet Assoc 2003;103:231–4.

15. Tillotson JE. Fast-casual dining: our next eating passion? Nutr Today 2003; 38:91–4.

16. Ellen Schuster is a professor of nutrition at Oregon State University. The quote was taken from a *Wall Street Journal* article published 5/20/03, p. D1.

17. Young and Nestle, Expanding portion sizes; and Fisher JO et al. Children's bite size and intake of an entrée are greater with large portions than with age-appropriate or self-selected portions. Am J Clin Nutr 2003;77:1164–70.

18. Bridges A. Restaurants should shrink portions. cnn.netscape.com/news/story, accessed 6/3/06.

19. Guthrie, Role of food prepared away from home; and Gillman MW et al. Family dinner and diet quality among older children and adolescents. Arch Fam Med 2000;9:235–40.

20. Sloan AE. Top 10 trends to watch and work on: 3rd biannual report. Food Tech 2001;55:38–58.

21. Slow Food International Organizations, www.slowfood.com., accessed 7/06.

22. Palmer S. The slow food movement picks up speed. Today's Dietitian 2005;Nov.:36–9.

23. Riley NS. The joys of cooking class. Wall Street Journal, March 9, 2006, page D9.

24. Cooking. CNI Nutrition Newsletter 1999;April 12:8.

25. de Lorgeril M, et al. Mediterranean diet, traditional risk factors, and the rate of cardiovascular disease complications after myocardial infarction. Circulation 1999;99:779–85.

26. Asian Food Guide Pyramid, www.nal.usda.gov/fric/etext/000023.html, accessed 8/03.

Nutrition Up Close

MyPyramid: A Balancing Act

Feedback for Unit 6

If you came up with numbers close to the following you did a good job of breaking down a one-day diet into MyPyramid food group amounts:

Grains	20–21 ounces
Fruits	0 cups
Vegetables	½ cup
Meat and beans	3 ounces
Milk	4½ cups

Note: MyPyramid Tracker calculates that this day's diet provides 3125 calories.

Answers
to Review Questions

1. False, see page 6-3.
2. True, see pages 6-4 and 6-5.
3. True, see page 6-5.
4. True, see page 6-6.
5. True, see page 6-8.
6. True, see page 6-8.
7. False, see page 6-7.
8. True, see page 6-7.
9. True, see page 6-8.
10. True, see page 6-9.
11. False, see page 6-8.
12. False, see page 6-14.

How the Body Uses Food: Digestion and Absorption

NUTRITION SCOREBOARD

	TRUE	FALSE
1 Almost all of the carbohydrate and fat you consume in foods is absorbed by the body, but only about half of the protein is.		
2 Disorders of the digestive system are a leading cause of hospitalizations and medical visits in the United States and Canada.		
3 Lactose maldigestion is a common digestive disorder.		

Key Concepts and Facts

- Our bodies are in a continuous state of renewal. Materials used to renew body tissues come from the food we eat in the form of nutrients.

- Digestion and absorption are processes that make nutrients in foods available for use by the body.

- Digestive disorders are common and often related to dietary intake.

Answers to
NUTRITION SCOREBOARD

		TRUE	FALSE
1	Over 90% of all the carbohydrates, fats, *and* proteins consumed in food are absorbed and become part of the body.		✔
2	Digestive disorders are the leading cause of hospitalizations among 20- to 44-year-olds in the United States and Canada. They account for over 70 million medical visits in the United States each year.[1,2]	✔	
3	Over half of the world's population digests lactose from milk and milk products incompletely or not at all.[3]	✔	

Photo Disc

My Body, My Food

La vie est une fonction chimique. ("Life is a chemical process.")
—ANTOINE LAVOISIER, LATE 18TH CENTURY

You are not the same person you were a month ago. Although your body looks the same and you don't notice the change, the substances that make up the organs and tissues of your body are constantly changing. Tissues we generally think of as solid and permanent, such as bones, the heart and blood vessels, and nerves, are continually renewing themselves. The raw materials used in the body's renewal processes are the nutrients you consume in foods.

Each day, about 5% of our body weight is replaced by new tissue. Existing components of cells are renewed, the substances in our blood are replaced, and body fluids are recycled. Taste cells, for example, are replaced about every 7 days, and cells lining the intestinal tract every 1 to 3 days. All of the cells of the skin are replaced every month. Red blood cells turn over every 120 days. If you thought it was hard to maintain a car, an apartment, or a house, just imagine what the maintenance is on a body! Maintenance is just one of the body's ongoing functions that require nutrients as raw material.

How Do Nutrients in Food Become Available for the Body's Use?

The human body is an amazing machine. Mine is, anyway. For example, I regularly feed my body truly absurd foods, such as cheez doodles, and somehow it turns them into useful body parts, such as glands. At least I assume it turns them into useful body parts.
—DAVE BARRY, 1987

The components of food that make "useful body parts" are nutrients. Through the processes of *digestion* and *absorption,* they are made available for use by every cell in the body.

The Internal Travels of Food: An Overview   The "food processor" of the body is the digestive system shown in Illustration 7.1. It consists of a 25- to 30-foot-long

digestion
The mechanical and chemical processes whereby ingested food is converted into substances that can be absorbed by the intestinal tract and utilized by the body.

absorption
The process by which nutrients and other substances are transferred from the digestive system into body fluids for transport throughout the body.

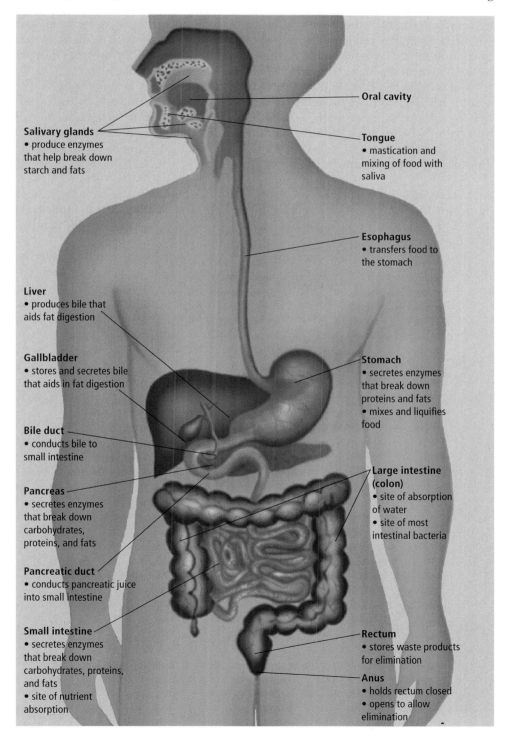

Oral cavity

Salivary glands
• produce enzymes that help break down starch and fats

Tongue
• mastication and mixing of food with saliva

Esophagus
• transfers food to the stomach

Liver
• produces bile that aids fat digestion

Gallbladder
• stores and secretes bile that aids in fat digestion

Stomach
• secretes enzymes that break down proteins and fats
• mixes and liquifies food

Bile duct
• conducts bile to small intestine

Pancreas
• secretes enzymes that break down carbohydrates, proteins, and fats

Large intestine (colon)
• site of absorption of water
• site of most intestinal bacteria

Pancreatic duct
• conducts pancreatic juice into small intestine

Small intestine
• secretes enzymes that break down carbohydrates, proteins, and fats
• site of nutrient absorption

Rectum
• stores waste products for elimination

Anus
• holds rectum closed
• opens to allow elimination

Illustration 7.1
The digestive system.

monosaccharides
(*mono* = one, *saccharide* = sugar) Simple sugars consisting of one sugar molecule. Glucose, fructose, and galactose are monosaccharides.

enzymes
Protein substances that speed up chemical reactions. Enzymes are found throughout the body but are present in particularly large amounts in the digestive system.

starch
Complex carbohydrates made up of complex chains of glucose molecules. Starch is the primary storage form of carbohydrate in plants. The vast majority of carbohydrate in our diet consists of starch, monosaccharides, and disaccharides.

disaccharide
Simple sugars consisting of two sugar molecules. Sucrose (table sugar) consists of a glucose and a fructose molecule, lactose (milk sugar) consists of glucose and galactose, and maltose (malt sugar) consists of two glucose molecules.

bile
A yellowish-brown or green fluid produced by the liver, stored in the gallbladder, and secreted into the small intestine. It acts like a detergent, breaking down globs of fat entering the small intestine to droplets, making the fats more accessible to the action of lipase.

muscular tube and organs such as the liver and pancreas that secrete digestive juices. The digestive juices break foods down into very small particles that can be absorbed and used by the body. The absorbable forms of carbohydrates are ***monosaccharides,*** such as glucose and fructose. Proteins are absorbed as amino acids and fats as fatty acids and glycerol. Vitamins and minerals are not broken down before they are absorbed; they are simply released from foods during digestion.

Much of the work of digestion is accomplished by ***enzymes*** manufactured by components of the digestive system such as the salivary glands, stomach, and pancreas. Enzymes are complex protein substances that speed up reactions that break down food. A remarkable feature of enzymes is that they are not changed by the chemical reactions they affect. This makes them reusable.

Carbohydrates, proteins, and fat each have their own set of digestive enzymes. All together, over a hundred different enzymes participate in the digestion of carbohydrates, proteins, and fat. Table 7.1 presents information on some of the enzymes involved in digestion and highlights their specific roles. In Table 7.2 you will see these enzymes cited in the summary of the processes involved in the digestion of carbohydrate, fat, and protein.

A Closer Look   As you chew food, glands under the tongue release saliva that lubricates food so that it can be swallowed and pass easily along the intestinal tract. Saliva also gets food digestion started. It contains salivary amylase and lipase that begin to break down carbohydrates and fats.

After food is chewed, it is swallowed and passes down the esophagus to the stomach. Muscles that act as valves at the entrance and exit of the stomach ensure that the food stays there until it's liquefied, mixed with digestive juices, and ready for the digestive processes of the small intestine. Solid foods tend to stay in the stomach for over an hour, whereas most liquids pass through it in about

TABLE 7.1

Primary function of some digestive enzymes

Enzyme	Enzyme Function	Enzyme Source
A. Carbohydrate Digestion		
Amylase	Breaks down **starch** into smaller chains of glucose molecules	Produced in the salivary glands (salivary amylase) and the pancreas (pancreatic amylase)
Sucrase	Separates the **disaccharide** sucrose into glucose and fructose	Produced in the small intestine
Lactase	Splits the disaccharide lactose into glucose and galactose	Produced in the small intestine
Maltase	Separates maltose into two molecules of glucose	Produced in the small intestine
B. Fat Digestion		
Lipase	Breaks down fats into fragments of fatty acids and glycerol	Produced in salivary glands (lingual lipase) and the pancreas (pancreatic lipase). The action of lipase is enhanced by **bile.**
C. Protein Digestion		
Pepsin	Separates protein into shorter chains of amino acids	Produced by the stomach
Trypsin	Splits short chains of amino acids into molecules containing, one, two, or three amino acids	Produced by the pancreas

TABLE 7.2

Summary of the digestion of carbohydrates, fats, and proteins

	Mouth	Stomach	Small Intestine, Pancreas, Liver, and Gall Bladder	Large Intestine (Colon)
Carbohydrates (excluding fiber)	The salivary glands secrete saliva to moisten and lubricate food; chewing crushes and mixes it with salivary amylase that initiates starch digestion.	Digestion of starch continues while food remains in the stomach. Some alcohol (a carbohydrate-like substance) is absorbed. Acid produced in the stomach aids digestion and destroys bacteria in food.	Pancreatic amylase continues starch digestion. Sucrase, lactase, and maltase break down disaccharides into monosaccharides. Some alcohol is absorbed here.	Undigested carbohydrates reach the colon and are partly broken down by intestinal bacteria.
Fiber	The teeth crush fiber and mix it with saliva to moisten it for swallowing.	No action.	Fiber binds cholesterol and some minerals.	Most fiber is excreted with feces; some fiber is digested by bacteria in the colon.
Fat	Fat-rich foods are mixed with saliva. Small amounts of lingual lipase accomplish some fat breakdown.	Fat tends to separate from the watery stomach fluid and foods and float on top of the mixture. Only a small amount of fat is digested. Fat is last to leave the stomach.	Bile readies fat for the action of lipase from the pancreas. Lipase splits fats into fatty acids and glycerol fragments.	A small amount of fatty materials escapes absorption and is carried out of the body with other wastes.
Protein	In the mouth, chewing crushes and softens protein-rich foods and mixes them with saliva.	Stomach acid works to uncoil protein strands and to activate the stomach's protein-digesting enzyme. Pepsin breaks the protein strands into smaller fragments.	Trypsin splits protein into molecules containing one, two, or three amino acids.	The large intestine carries undigested protein residue out of the body. Normally, almost all food protein is digested and absorbed.

20 minutes. When the stomach has finished its work, it ejects 1 to 2 teaspoons of its liquefied contents into the small intestine through the muscular valve at its end. Stomach contents continue to be ejected in this fashion until they are totally released into the small intestine. These small pulses of liquefied food stimulate muscles in the intestinal walls to contract and relax; these movements churn and mix the food as it is digested by enzymes. When the diet contains a lot of fiber and sufficient fluids, the bulge of digesting food in the intestine tends to be larger. Larger food bulges stimulate a higher level of intestinal muscle activity than do smaller food bulges. Thus, high-fiber meals pass through the digestive system somewhat faster than low-fiber meals.

Digestion, as well as the absorption of nutrients, is greatly enhanced by the structure of the intestines (Illustration 7.3). Fingerlike projections called "villi" line the inside of the intestinal wall and increase its surface area tremendously. If laid flat, the surface area of the small intestine would be about the size of a baseball infield, or approximately 675 square feet. This large mass of tissue requires a high level of nutrients for maintenance. Much of this need (50% in the small intestine and 80% in the large intestine) is met by foods that are being digested.[5]

Digestion is completed when carbohydrates, proteins, and fats are reduced to substances that can be absorbed and when vitamins and minerals are released from food. Most nutrients are absorbed in the small intestine. Water, sodium, and some of the end products of bacterial digestion, however, are absorbed from the large intestine. The large intestine is home to many strains of bacteria that consume undigested fiber and other types of complex carbohydrates that are not broken down by human digestive enzymes. These bacteria excrete gas as well as

Get Your Juices Flowing

You don't have to actually eat food to start your digestive juices flowing. You just have to think about food or see it.[4] Put this information to the test. Clear your mind, turn the page, and take a close look at Illustration 7.2.

ON THE SIDE

Illustration 7.2
Testing, testing.
This is a test of your salivary secretions. Did the lemon speak directly to your salivary glands?
If you want to turn the digestive processes off, quit thinking about food. Finish your reading assignment!

fatty acids that are partly absorbed in the large intestine. Substances in food that cannot be absorbed collect in the large intestine and are excreted in the stools.

Absorption   Absorption is the process by which the end products of digestion are taken up by the *lymphatic system* (Illustration 7.4) and the *circulatory system* (Illustration 7.5) for eventual distribution to cells of the body. Lymph vessels and blood vessels infiltrate the villi that line the inside of the intestines (Illustration 7.6) and transport absorbed nutrients toward the major branches of the lymphatic and circulatory systems. The breakdown products of fat digestion are largely absorbed into lymph vessels, whereas carbohydrate and protein breakdown products enter the blood vessels.

The nutrient-rich contents of the lymphatic system are transferred to the bloodstream at a site near the heart where vessels from both systems merge into one vessel. From there the lymph and blood mixture is sent to the heart and subsequently throughout the body by way of the circulatory system. The circulatory system reaches every organ and tissue in the body, thereby supplying cells with nutrients obtained from food.

lymphatic system
A network of vessels that absorb some of the products of digestion and transport them to the heart, where they are mixed with the substances contained in blood.

circulatory system
The heart, arteries, capillaries, and veins responsible for circulating blood throughout the body.

Illustration 7.3
Scanning electron micrographs of cross sections of the small intestine (left) and the large intestine (right).
Note the high density of villi in the small intestine and the relative flatness of the lining of the large intestine.

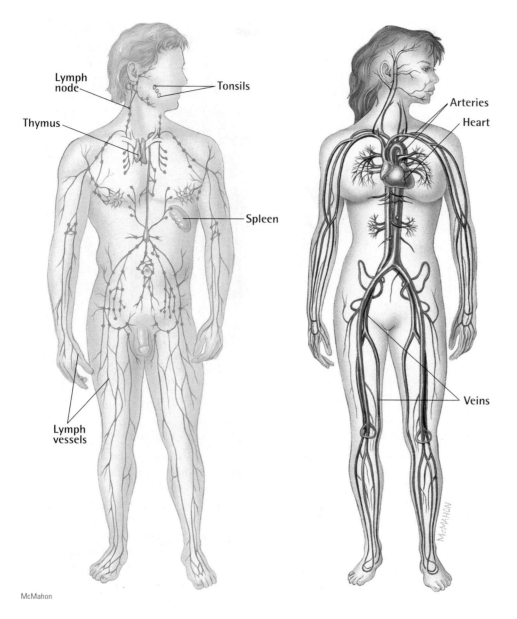

Illustration 7.4
(left) The lymphatic system.

Illustration 7.5
(right) The circulatory system includes the heart and blood vessels.
This system serves as the nutrient transportation system of the body.

McMahon

■ **Beyond Absorption** Cells can use nutrients directly for energy, body structures, or the regulation of body processes or convert them into other usable substances. For example, glucose delivered to cells can be used "as is" for energy formation or converted to glycogen and stored for later use. Fatty acids, an end product of fat digestion, can be incorporated into cell membranes or used in the synthesis of certain hormones. Vitamins and minerals freed from food by digestion can be used by cells to control enzyme activity or can be stored for later use. The body has a limited storage capacity for some vitamins and minerals. Consequently, excessive amounts of certain vitamins and minerals such as vitamin C, thiamin, and sodium are largely excreted in urine.

■ **Digestion and Absorption Are Efficient** Of our intake of energy nutrients, approximately 99% of the carbohydrate, 92% of the protein, and 95% of the fat we

Illustration 7.6
Structure of villi, showing blood
and lymph vessels.

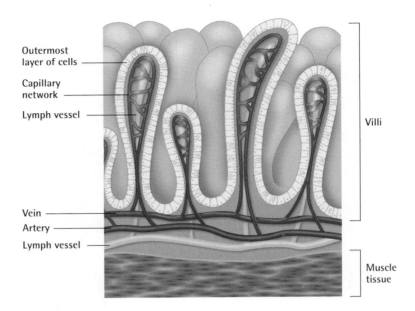

Outermost
layer of cells

Capillary
network

Lymph vessel

Villi

Vein

Artery

Lymph vessel

Muscle
tissue

consume in food are digested and absorbed. Dietary fiber, however, leaves the digestive system in much the same form as it entered. Humans don't have enzymes that break down fiber. It should be noted, however, that some fiber is digested in the large intestine by bacteria.

Digestive Disorders Are Common

heartburn
A condition that results when acidic stomach contents are released into the esophagus, usually causing a burning sensation.

*hemorrhoids
(hem-or-oids)*
Swelling of veins in the anus or rectum.

*irritable bowel syndrome
(IBS)*
A disorder of bowel function characterized by chronic or episodic gas; abdominal pain; diarrhea, constipation, or both.

*duodenal (do-odd-en-all)
and stomach ulcers*
Open sores in the lining of the duodenum (the uppermost part of the small intestine) or the stomach.

Excluding childbirth, digestive disorders such as ***heartburn, hemorrhoids, irritable bowel syndrome,*** and ***duodenal and stomach ulcers*** are the leading cause of hospitalization among U.S. and Canadian adults aged 20 to 44 years. They account for over 70 million medical visits yearly in the United States alone.[1] Table 7.3 shows the percentages of U.S. adults who have common digestive disorders. Digestive disorders are common in children as well as adults. At least one-third of U.S. adults experience heartburn, and up to 28% of school children experience constipation.[6]

Constipation and Hemorrhoids   Both constipation and hemorrhoids are often due to diets that provide too little fiber. Fiber intakes by adults of 25 to 30 grams per day along with plenty of fluids (8 to 12 cups a day) can help prevent constipation and hemorrhoids in healthy people. Some good sources of dietary fiber are shown in Illustration 7.7.

Ulcers and Heartburn   Ulcers develop when the protective barrier formed by cells lining the stomach and duodenum (the uppermost part of the small intestine) is damaged. This allows stomach acid and digestive enzymes to erode the lining of the stomach and duodenum and cause an "ulcer." Duodenal ulcers are ten times more common than stomach ulcers and are closely associated with the presence of *Helicobacter pylori (H. pylori)* bacteria.[7] *H. pylori* infects and irritates the lining of the stomach. The infection is acquired by the ingestion of foods and other substances contaminated with saliva, vomit, or feces from people harboring the bacteria in their stomach. Rates of *H. pylori* infection are highest in countries with poor sanitary conditions.[8] Excessive production of stomach acid can lead to ulcers, too, and may also cause heartburn. Heartburn is unrelated to heart conditions. It is called that because acid that escapes from the stomach causes a burning sensation in an area of the esophagus located near the heart.

Stress, anxiety, and frequent use of aspirin, ibuprofen, naproxen, and other medications appear to be related to the development of ulcers and heartburn in some people.[9] Fatty foods, coffee, alcohol, citrus fruits, soft drinks, and a variety of other foods have been implicated in the development of ulcers and heartburn. Rather than cause these disorders, certain foods appear to aggravate the symptoms of ulcers and heartburn. High-fiber diets, on the other hand, appear to be protective against the development of ulcers and heartburn.[10] Recently, weight gain in women[11] and eating within three hours of going to bed[12] have been found to increase heartburn symptoms.

A number of medications are available for the treatment of *H. pylori* and the excessive production of stomach acid. Reduction of stress, judicious elimination of offending foods and beverages from the diet, and small, frequent meals also help relieve ulcers and heartburn in some people.[9]

Irritable Bowel Syndrome   Irritable bowel syndrome, abbreviated IBS, is not considered a disease but rather a persistent disorder in the way the colon functions. For some reason, the colon (or bowel) in people with IBS spasms, and that leads to painful cramps. Gas production, a feeling of being bloated, diarrhea, constipation, or both, are common features of IBS. Most commonly, mild IBS is treated with a low-fat diet (25% of calories), counseling, and over-the-counter fiber powders, antidiarrhea pills, and pain medications. More serious cases may respond to stress reduction, eating in relaxed surroundings, antidepressants, and drugs that reduce colon muscle spasms. *Probiotics,* or "friendly" bacteria that help colonize the colon with microorganisms that help protect and heal the colon lining, are being increasingly used for this disorder.[13]

TABLE 7.3	
Common digestive disorders[1]	
U.S. Adults Affected[a]	
Heartburn	33.0%
Hemorrhoids	12.8
Irritable bowel syndrome	6–12%
Ulcers	3.5
Chronic constipation	3.0
Chronic diarrhea	1.2

[a]Noninstitutionalized adults experiencing the disorder in the past 12 months.

probiotics
Nonharmful bacteria and some yeasts that help colonize the intestinal tract with beneficial microorganisms and that sometimes replace colonies of harmful microorganisms. Most common probiotic strains are *Lactobacilli* and *Bifidobacteria.*

Richard Anderson

Illustration 7.7
Food sources of dietary fiber. *Together, the foods shown provide 29 grams of dietary fiber, an amount that helps prevent constipation.*

1 mango (4 g)

All Bran Cereal, 1/3 c. (10 g)

1/2 c. lima beans (5 g)

1 pear with skin (4 g)

1/2 c. corn (3 g)

1/2 c. carrots (2.8 g)

diarrhea
The presence of three or more liquid stools in a 24-hour period.

Diarrhea   *Diarrhea* is a common problem in the United States and a leading public health problem in developing countries. Most cases of diarrhea are due to bacterial- or viral-contaminated food or water, lack of immunizations against infectious diseases, and vitamin A, zinc, and other nutrient deficiencies that make children particularly susceptible to diarrhea. Diarrhea can deplete the body of fluid and nutrients and produce malnutrition as well. If it lasts more than two weeks or is severe, diarrhea can lead to dehydration, heart and kidney malfunction, and death. An estimated 3.5 million deaths from diarrhea diseases occur each year to the world's population of children 5 years of age or under.[14]

The vast majority of cases of diarrhea can be prevented through food and water sanitation programs, immunizations, and adequate diets. The early use of oral rehydration fluids (for example, the formula provided by the World Health Organization and commercial formulas such as Pedialyte and Rehydralyte) shortens the duration of diarrhea. Rehydration generally takes 4 to 6 hours after the fluids are begun.[15]

Rather than "resting the gut" during diarrhea as used to be recommended, children and adults, once rehydrated, should eat solid foods. Foods such as yogurt, lactose-free or regular milk, chicken, potatoes and other vegetables, dried beans, and rice and other cereals are generally well tolerated and provide nutrients needed for the repair of the intestinal tract. It is best to avoid sugary fluids such as soft drinks. High-sugar beverages tend to draw fluid into the intestinal tract rather than increase the absorption of fluid.[16]

flatulence (flat-u-lens)
Presence of excess gas in the stomach and intestines.

Flatulence   Everyone experiences *flatulence*—it's normal. Gas can occur in the esophagus, stomach, small intestine, and large intestine due to swallowed air or bacterial breakdown of food in the large intestine. Air may be swallowed along with food and beverages or while chewing gum. Eating and drinking while in a rush generally increases air ingestion. Bacterial production of gas in the large intestine may be related to the ingestion of dried beans, broccoli, cauliflower, brussels sprouts, onions, corn, and other vegetables containing a type of complex carbohydrate that bacteria, but not humans, can break down. Fructose, which is used to sweeten a variety of food products and beverages, and sorbitol (used in some types of candy and gum) may lead to gas formation by bacteria that produce gas as a waste product of carbohydrate digestion. Heartburn and other gastrointestinal tract disorders and medications such as antibiotics are also associated with gas production.[1]

People often think they produce too much gas, even when they don't. The amount of gas swallowed and produced by gut bacteria varies a good deal among individuals and within the same individual. Gas production changes depending on what foods are eaten, the types of bacteria populating the large intestine, the medications used, and the presence of gastrointestinal tract disorders. Severe and painful symptoms related to gas production may signal the presence of a digestive disorder.[1]

Stomach Growling   Gas in the stomach can make your stomach growl. When your stomach growls, you know that gas and food or fluids are mixing in your stomach. The growling tends to be louder when your stomach is empty, when there's no food to muffle the noise. It can occur anytime, whether your stomach is empty or full, but is more likely to happen when your stomach is empty.[17]

Ancient Perspectives on Flatulence

Photo Disc

Passing gas is necessary to well-being.
—HIPPOCRATES

All Roman citizens shall be allowed to pass gas whenever necessary.
—CLAUDIUS

Lactose Maldigestion and Intolerance

I can't drink milk. It tears me up inside.

A very common digestive disorder is *lactose maldigestion.* The lactose found in milk and milk products presents a problem for most of the world's adults, who cannot digest it, either partially or completely (Table 7.4).[3] The condition occurs more commonly in population groups that have no historical links to dairy farming and milk drinking.[18] Early humans in central and northwestern Europe and the regions of Africa and China highlighted in Illustration 7.8 tended to raise dairy animals and drink milk.

Lactose maldigestion is caused by a genetically determined low production of lactase, the enzyme that digests lactose. People who lack this enzyme end up with free lactose in their large intestine after they consume milk or milk products that contain lactose. The presence of lactose in the large intestine produces the symptoms of *lactose intolerance.* These symptoms include a bloated feeling and diarrhea due to fluid accumulation and gas and abdominal cramping caused by the excretion of gas by bacteria that digest lactose.

Lactose maldigestion is rare in young children and affects adults to various degrees. Some adults produce little or no lactase and develop symptoms of lactose intolerance when they consume only small amounts of milk or milk products. Others produce some lactase and can tolerate limited amounts of lactose-containing milk and milk products, such as a cup of milk at a time or two cups of milk consumed with meals during the day. Regular consumption of milk may

TABLE 7.4

Estimated incidence of lactose maldigestion among older children and adults in different population groups[19]

Incidence of Lactose Maldigestion

Asian Americans	90%
Africans	70
African Americans	70
Asians	65 or more
American Indians	62 or more
Mexican Americans	53 or more
U.S. adults (overall)	25
Northern Europeans	20
American Caucasians	15

lactose maldigestion
A disorder characterized by reduced digestion of lactose due to the low availability of the enzyme lactase.

lactose intolerance
The term for gastrointestinal symptoms (flatulence, bloating, abdominal pain, diarrhea, and "rumbling in the bowel") resulting from the consumption of more lactose than can be digested with available lactase.

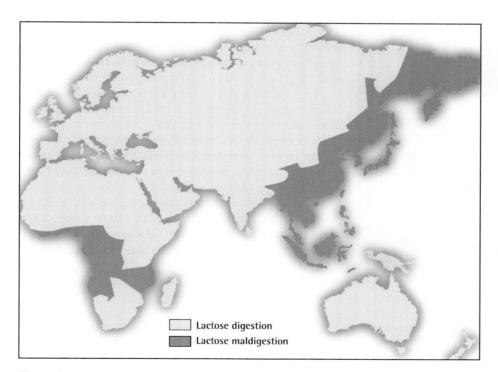

Lactose digestion
Lactose maldigestion

Illustration 7.8
Lactose maldigestion is less common among descendants of people who consumed milk from domesticated animals during prehistoric times (light areas) than among people whose early ancestors did not drink milk (dark areas).[19]

Photo Disc

O N   T H E   S I D E

"Brain freeze," or that splitting headache you can get from eating ice cream too fast, is caused by the quick drop in temperature in the back of your mouth. That causes vessels to constrict, and the result is an "ice cream headache."[21]

improve lactose digestion due to enhanced bacterial breakdown of lactose in the gut.[20]

Many people who are lactose maldigesters have no trouble eating yogurt and other fermented milk products such as cultured buttermilk, kefir, and aged cheese. The bacteria used to culture yogurt can digest half or more of the lactose. This reduction in lactose content is sufficient to prevent adverse effects in many people with lactose maldigestion.[22]

Milk solids, milk, and other lactose-containing components of milk may be added to foods you wouldn't expect. Consequently, it's best to examine food ingredient labels when in doubt. Milk, for instance, is a primary ingredient in some types of sherbet, and milk solids are added to many types of candy.

Do You Have Lactose Maldigestion?   The single most reliable indicator of lactose maldigestion is the occurrence of lactose intolerance within hours after consuming lactose.[18] If you consistently experience these symptoms (described earlier), visit your health care provider for a diagnosis. The symptoms could be due to lactose, other substances in milk, or another problem.

How Is Lactose Maldigestion Managed?   Lactose maldigestion should *not* be managed by omitting milk and milk products from the diet! Doing so would exclude a food group that contributes a variety of nutrients that cannot easily be replaced by other foods. The omission of milk and milk products from the diet of people with lactose intolerance promotes the development of osteoporosis.[20] Rather, fortified soy milk, low-lactose cow's milk, milk pretreated with lactase drops, and yogurt and other fermented milk products (if tolerated) should be consumed. Illustration 7.9 shows a variety of dairy products that are generally well tolerated by people with lactose maldigestion.

Lactase tablets are also available and should be taken within 30 minutes of consuming lactose.[20] Lactase is an enzyme, and enzymes are made of protein. Protein is partially digested in the stomach, so some of the lactase ingested in the tablets may not reach the small intestine where it is needed if the tablets are taken too far in advance of eating.

Digestion is a remarkably complex and efficient process that could be covered in much more detail than has been presented here. Readers are encouraged to consult the Web sites listed at the end of the unit for additional information.

Illustration 7.9
Dairy products generally well tolerated by people with lactose maldigestion.

Richard Anderson

Key Terms

absorption, page 7–3
bile, page 7–4
circulatory system, page 7–6
diarrhea, page 7–10
digestion, page 7–3
disaccharide, page 7–4
duodenal and stomach ulcers, page 7–8

enzymes, page 7–4
flatulence, page 7–10
heartburn, page 7–8
hemorrhoids, page 7–8
irritable bowel syndrome (IBS),
    page 7–8
lactose intolerance, page 7–11

lactose maldigestion, page 7–11
lymphatic system, page 7–6
monosaccharides, page 7–4
probiotics, page 7–9
starch, page 7–4

Nutrition Up Close

Personal History of Digestive Upsets

Focal Point: Digestive disorders are common.

To bring you closer to your digestive system, review the following list and check the upsets you have experienced in the past month:

Digestive Upset	Experienced in the Past Month?
Heartburn or indigestion	_____
Diarrhea	_____
Constipation	_____
Stomach cramps	_____
Vomiting	_____
Abdominal bloating	_____
Hemorrhoids	_____

FEEDBACK: Most bouts of digestive upsets are brief and only bothersome. Some will be related to diet (inadequate or excessive dietary fiber or bacterially contaminated food, for example) whereas others will be initiated by illness, stress, or another condition. Painful or prolonged episodes of digestive disorders should, of course, be brought to the attention of your health care provider.

Media Menu

www.healthfinder.gov
Search digestive diseases by name, or search "digestion."

digestive.niddk.nih.gov
NIH's National Digestive Disease Informa-tion Clearinghouse homepage provides links to information about specific digestive diseases, clinical trials, statistics, and answers to questions such as "Why do I have gas?"

www.nlm.nih.gov/medlineplus
Find out more about digestion, absorption, organs of the gastrointestinal tract, and digestive diseases at this site.

Review Questions

	True	False
1. Cells lining the intestinal tract are replaced every 30 days.	☐	☐
2. Enzymes are protein substances that speed up chemical reactions and are reusable.	☐	☐
3. Digestion of proteins begins with the action of salivary amylase.	☐	☐
4. The end products of digestion are taken up by the lymphatic and circulatory systems.	☐	☐
5. The nutrient-rich contents of the lymphatic system do not mix with blood. These nutrients are distributed		

	True	False
throughout the body by a separate system of lymph vessels.	☐	☐
6. The body has a limited capacity to store some vitamins and minerals.	☐	☐
7. Only about 75% of the protein, carbohydrate, and fat we consume in food is digested and absorbed.	☐	☐
8. Weight gain in women and eating within three hours of going to bed increase heartburn symptoms.	☐	☐
9. Most of the world's adult population has lactose maldigestion.	☐	☐

Notes

1. Digestive disease statistics, National Digestive Disease Clearinghouse, http://digestive.niddk.nih.gov/statistics/statistics.htm, accessed 8/03.

2. Leading causes of hospitalization in Canada. Health Canada, Population and Public Health Branch, www.hc-sc.gc.ca, accessed 8/03.

3. Hertzler SR, Clancy SM. Kefir improves lactose digestion and tolerance in adults with lactose maldigestion. J Am Diet Assoc 2003;103:582–74.

4. Schneeman B. Nutrition and gastrointestinal function. Nutr Today 1993; Jan/Feb:20–24.

5. Bengmark S. Econutrition and health maintenance: a new concept to prevent GI inflammation, ulceration, and sepsis. Clin Nutr 1996;15:1–10.

6. Borowitz SM et. al. Precipitants of constipation during early childhood. J Am Fam Pract 2003;16:213–8.

7. Current medical diagnosis and treatment, 35th ed. Stamford (CT): Appleton & Lange; 1997.

8. Suerbaum S, Michetti P. Helicobacter pylori infection. N Engl J Med 2002; 347:1175–86.

9. Kurata AN, et al. Dyspepsia in primary care: perceived causes, reasons for improvement, and satisfaction with care. J Fam Practice 1997;44:281–8.

10. The Merck manual of medical information. Whitehouse (NJ): Merck Research Laboratory; 2000; and Suerbaum and Michetti, Helicobacter pylori infection.

11. Jacobson BC et al. Weight gain in women and GERD. N Engl J Med 2006; 354:2340–8.

12. Fujiwara Y et al. Dinner-to-bed time and GERD. Am J Gastroenterol 2005; 100:2633–6.

13. Chang HY et al. Current gut-directed therapies for irritable bowel syndrome. Curr Treat Options Gastroenterol 2006; 9:314–23.

14. Kilgore PE et al. Trends in diarrheal disease associated mortality in US children, 1968 through 1991. JAMA 1995; 274:1143–8.

15. Kilgore et al. Trends in diarrheal disease; Goepp JG, Katz SA, Oral rehydration therapy, Am Fam Phys 1993;47: 843–8; and Meyers A. Oral rehydration therapy: what are we waiting for? Am Fam Phys 1993;47:740–2.

16. Lima AAM, Guerrant RL. Persistent diarrhea in children: epidemiology, risk factors, pathophysiology, nutritional impact, and management. Epidemiol Rev 1992;34:222–42; and Goepp and Katz, Oral rehydration therapy.

17. Stomach growling. Scientific American. Com ask the expert. www.scientificamerican.com, accessed 8/03.

18. Simmons FJ. Primary adult lactose intolerance and the milking habit: a problem in biological and cultural interrelationships. I. Review of the medical research. Am J Digestion 1981;14:819.

19. Scrimshaw NS, Murray, EB. Prevalence of lactose maldigestion. Am J Clin Nutr 1988;48(suppl):1086–98; and Inman-Felton AE. Overview of lactose maldigestion (lactose nonpersistence). J Am Diet Assoc 1999;99:481–9.

20. Lactose intolerance: a self-fulfilling prophecy leading to osteoporosis. Nutr Rev 2003;61:221–3.

21. Ice cream headache. www.mayoclinic.com/health/ice-creamheadaches/DS00640, accessed 3/06.

22. Hertzler and Clancy, Kefir improves lactose digestion; and Martini MC, Smith DE, Savaiano DA. Lactose digestion from flavored and frozen yogurts, ice milk, and ice cream by lactase-deficient persons. Am J Clin Nutr 1987;46:636–40.

Answers
to Review Questions

1. False, see page 7-2.
2. True, see page 7-4.
3. False, see pages 7-4, 7-5.
4. True, see page 7-6.
5. False, see page 7-6.
6. True, see page 7-7.
7. False, see pages 7-7, 7-8.
8. True, see page 7-9.
9. True, see page 7-11.

Calories! Food, Energy, and Energy Balance

NUTRITION SCOREBOARD

	TRUE	FALSE
1 Gram for gram, carbohydrates provide the body with more energy than protein does.		
2 A teaspoon of butter has a higher caloric value than a teaspoon of margarine.		
3 Energy can be neither created nor destroyed. It can, however, change from one form to another.		

Key Concepts and Facts

- A calorie is a unit of measure of energy.

- The body's sources of energy are carbohydrates, proteins, and fats (the "energy nutrients").

- Fats provide over twice as many calories per unit weight as carbohydrates and proteins do.

- Most foods contain a mixture of the energy nutrients as well as other substances.

- Weight is gained when caloric intake exceeds the body's need for energy. Weight is lost when caloric intake is less than the body's need for energy.

Answers to
NUTRITION SCOREBOARD

		TRUE	FALSE
1	Carbohydrates and protein both provide 4 calories per gram. Fat, on the other hand, provides 9 calories per gram, and alcohol 7 calories per gram.		✔
2	Butter and margarine contain the same amount of fat, so they provide the same number of calories (35 per teaspoon).		✔
3	Stated is the conservation of energy principle, the first law of thermodynamics.	✔	

Energy!

*It's high in energy, you say? Sure, I'll eat one, thanks.
I thought those snack bars were loaded with calories!*

When you think of calories, do you think of energy? That is the "scientifically correct" way to think about them. Energy is what calories are all about.

Illustration 8.1

A calorie is the amount of energy needed to raise the temperature of 1 kilogram of water (about 4 cups) from 15°C to 16°C (59°F to 61°F).

calorie (*calor* = **heat**)
A unit of measure used to express the amount of energy produced by foods in the form of heat. The calorie used in nutrition is the large "Calorie," or the "kilocalorie" (kcal). It equals the amount of energy needed to raise the temperature of 1 kilogram of water (about 4 cups) from 15 to 16°C (59 to 61°F). The term **kilocalorie,** or "calorie" as used in this text, is gradually being replaced by the "kilojoule" (kJ) in the United States; 1 kcal = 4.2 kJ.

Calories Are a Unit of Measure

The *calorie* is like a centimeter or pound in that it is a unit of measure. Rather than serving as a measure of length or weight, the calorie is used as a measure of energy. Specifically, a calorie is the amount of energy needed to raise the temperature of 1 kilogram of water (about 4 cups) from 15°C to 16°C (59°F to 61°F) (Illustration 8.1). This amount of energy is used as a standard for assigning caloric values to foods. Because calories are a unit of measure, they are not a component of food like vitamins or minerals. When we talk about

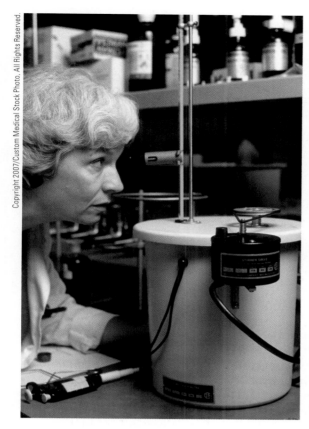

Illustration 8.2
A bomb calorimeter used to measure the calorie value of foods.
A food's calorie value is determined by the amount of heat released and transfered to water when the food is completely burned.

the caloric content of a food, we're really talking about the caloric value of the food's energy content.

The caloric value of food is determined in a "bomb calorimeter" by burning it completely in a container surrounded by a specific amount of water (Illustration 8.2). The energy released by the food in the form of heat raises the temperature of the surrounding water. The rise in temperature indicates how many calories were released from the portion of food. Although the body doesn't literally burn food, the amount of heat released by food while burning is approximately the same as the amount of energy it supplies the body.

The Body's Need for Energy

The body uses energy from foods to fuel muscular activity, growth, and tissue repair and maintenance; to chemically process nutrients; and to maintain body temperature (to name a few examples). These needs for energy are subdivided into three categories: basal metabolism, physical activity, and dietary thermogenesis (Table 8.1 and Illustration 8.3).

TABLE 8.1
The three energy-requiring processes of the body.

Basal metabolism
- Energy required to maintain normal body functions while at rest

Physical activity
- Energy needed for muscular work

Dietary thermogenesis
- Energy use related to food ingestion. (The process gives off heat.)

basal metabolism

physical activity

dietary thermogenesis

Illustration 8.3
Examples of the three types of energy-requiring processes in the body.

Reality Check

Mel:
I've got a slow metabolism. It doesn't matter how little I eat, I still can't lose weight.

Juanita:
Some of my friends who have trouble losing weight say that, too, but it can't be right. "Slow metabolism" is a myth.

basal metabolism
Energy used to support body processes such as growth, health, tissue repair and maintenance, and other functions. Assessed while at rest, basal metabolism includes energy the body expends for breathing, the pumping of the heart, the maintenance of body temperature, and other life-sustaining, ongoing functions.

basal metabolic rate (BMR)
The rate at which energy is used by the body when it is at complete rest. BMR is expressed as calories used per unit of time, such as an hour, per unit of body weight in kilograms. Also commonly called *resting metabolic rate*, or *RMR*.

The largest single contributor to energy need is basal metabolism or "resting metabolism," as it is also called. It accounts for 60 to 75% of the total need for calories in the vast majority of people.[1] Energy-requiring processes of *basal metabolism* include breathing, the beating of the heart, maintenance of body temperature, renewal of muscle and bone tissue, and other ongoing activities that sustain life and health. Growth is considered a component of basal metabolism. The proportion of total calories needed for basal metabolism is particularly high during the growing years.

Some organs and tissues in the body are more metabolically active, or require more energy to sustain their functions, than others. Body fat is less metabolically active than other tissues and accounts for under 20% of basal metabolic calorie expenditure in most people. The brain, liver, kidneys, and muscle are metabolically active. Together, these four body components account for 80% or more of the energy used for basal metabolism.[2]

Only a small percentage of people have unusually low or high *basal metabolic rates (BMR).* In healthy individuals, slight differences in BMR rarely account for the ease with which weight is gained, maintained, or lost. Uncommon metabolic disorders, such as underactive thyroid and Cushing's syndrome, can modify basal metabolic processes and lead to changes in weight status.[3]

The Reality Check for this unit addresses the topic of basal metabolism and weight loss. Check it out.

Energy-using activities of basal metabolic processes require no conscious effort on our part; they are continuous activities that the body must perform to sustain life. The energy needed to carry out basal metabolic functions is assessed when the body is in a state of complete physical and emotional rest.

How Much Energy Do I Expend for Basal Metabolism?   You can quickly estimate the calories needed for basal metabolic processes:

• For men: Multiply body weight in pounds by 11.

• For women: Multiply body weight in pounds by 10.[4]

Thus, a man who weighs 170 pounds needs approximately 170 × 11, or 1870 calories per day for basal metabolic processes. A 135-pound woman needs 135 × 10, or 1350 calories.

TABLE 8.2

Energy expenditure by usual level of activity[6]

Activity Level	Percentage of Basal Metabolism Calories
Inactive. Sitting most of the day; less than 2 hours of moving about slowly or standing	30%
Average. Sitting most of the day; walking or standing 2 to 4 hours, but no strenuous activity	50
Active. Physically active 4 or more hours each day; little sitting or standing; some physically strenuous activities	75

  This formula gives an estimate of calories used for basal metabolism in adults based on sex and weight. Physical activity level, muscle mass, height, health status, and genetic traits, also influence calorie expenditure for basal metabolism to some extent. Consequently, results obtained using this quick formula may be 10 to 20% lower or higher than the true number of calories required for basal metabolism.[5]

How Much Energy Do I Expend in Physical Activity? The caloric level needed for physical activity can vary a lot, depending on how active a person is. It usually accounts for the second highest amount of calories we expend. The energy cost of supporting a physically inactive lifestyle (Table 8.2) is about 30% of the number of calories needed for basal metabolism. An "average" activity level requires roughly 50%, and an "active" level requires approximately 75% of the calories needed for basal metabolism.[6] A physically inactive person needing 1500 calories a day for basal metabolism, for example, would require about 450 calories (1500 calories × 0.30—for 30%) for physical activity.

■ **A Common Source of Error** People have a tendency to overestimate time spent in physical activity. This in turn tends to overestimate calories needed for physical activity, and depending on the amount of overestimation, can lead to highly inaccurate estimates of total calorie need. Physical activity level should be based on time spent actually engaged in physical activity and not include time spent getting ready for the activity, on breaks, or between activities.

How Many Calories Does Dietary Thermogenesis Take? A portion of the body's energy expenditure is used for digesting foods, absorbing, utilizing, and storing nutrients, and transporting nutrients into cells. Some of the energy involved in such activities escapes as heat. These processes are referred to as *dietary thermogenesis.* Calories expended for dietary thermogenesis are estimated

dietary thermogenesis
Thermogenesis means "the production of heat." Dietary thermogenesis is the energy expended during the digestion of food and the absorption, utilization, storage, and transport of nutrients. Some of the energy escapes as heat. It accounts for approximately 10% of the body's total energy need. Also called *diet-induced thermogenesis* and *thermic effect of foods or feeding.*

ANSWERS TO
Reality Check

Is there any such thing as a slow metabolism?
Mel's basal metabolism may be lower than average BMR for his weight, but his metabolism wouldn't be "slow." Calories needed for basal metabolism can be somewhat lower in people with relatively high amounts of body fat relative to muscle mass.[3]

Felicia Martinez/Photo Edit

Michael Newman/Photo Edit

Mel:

Juanita:

TABLE 8.3

Summary of calculations for estimating total calorie need of a 130-pound, inactive woman

	Calories
1. **Basal metabolism** Multiply body weight in pounds by 10. (For men the figure is 11.)	$130 \times 10 = 1300$
2. **Physical activity** Multiply basal metabolism calories by 0.30 (for 30%) based on the usual energy expenditure level of "inactive" (see Table 8.2).	$1300 \times 0.30 = 390$
3. **Dietary thermogenesis** Add calories needed for basal metabolism and physical activity together: $1300 + 390 = 1690$ Multiply the result by 0.10 (for 10%).	$1690 = 0.10 = 169$
4. **Total calorie need** Add calories needed for basal metabolism, physical activity, and dietary thermogenesis together. $1300 + 390 + 169 =$	**Total calorie need = 1859**

TABLE 8.4

Foods that do have calories

1. A candy bar eaten with a diet soda
2. Celery and grapefruit
3. Hot chocolate, cheesecake, or soft drinks consumed to make you feel better
4. Cookie pieces
5. Foods "taste-tested" during cooking
6. Foods you eat while on the run
7. Foods you eat straight from their original containers (like ice cream from the carton, milk from a jug, or chips from a large bag)

TABLE 8.5

Caloric values of the energy nutrients and alcohol

	cals/gm
Carbohydrate	4
Protein	4
Fat	9
Alcohol	7

as 10% of the sum of basal metabolic and usual physical activity calories. For instance, say a person's basal metabolic need is 1500 calories and 450 calories are required for usual activity: 1500 calories + 450 calories = 1950 calories. Calories expended for dietary thermogenesis would equal approximately 10% of the 1950 calories, or 195 calories.

Adding It All Up　Your estimated total daily need for calories is the sum of calories used for basal metabolism, physical activity, and dietary thermogenesis. In the preceding example, total calorie need would be 2145 calories (1500 + 450 + 195). A complete example of calculations involved in estimating a person's total calorie need is provided in Table 8.3. Although the caloric level calculated won't be exactly right, it should provide a good estimate of your total caloric need.

Where's the Energy in Foods?

Any food that contains carbohydrates, proteins, or fats (the "energy nutrients") supplies the body with energy. (That *includes* the foods listed in Table 8.4!) Carbohydrates and proteins supply the body with 4 calories per gram, and fat provides 9 calories per gram. Alcohol also serves as a source of energy. There are 7 calories in each gram of alcohol in the diet (Table 8.5).

If you enjoy grilling foods on an outdoor barbecue, you have probably observed firsthand the high level of stored energy in fats (Illustration 8.4). Unlike the drippings of low-fat foods such as shrimp or vegetables, drips from high-fat foods cause bursts of flames to shoot up from the grill. The high-energy content of alcohol can be seen in the alcohol-fueled flames that adorn cherries jubilee and bananas Foster.

If you know the carbohydrate, protein, fat, and (if present) alcohol content of a food or beverage, you can calculate how many calories it contains. For example, say a cup of soup contains 15 grams of carbohydrate, 10 grams of protein, and 5 grams of fat. To calculate the caloric value of the soup, multiply the number of

Illustration 8.4
If you have observed the flames produced by fat dripping from a steak or hamburger on a grill, you have seen the powerhouse of energy stored in fat.
The carbohydrate and protein contents of grilled foods don't burn with nearly the same intensity. They have less energy to give.

grams of carbohydrate and protein by 4 and the number of grams of fat by 9. Then add the results together:

$$
\begin{aligned}
15 \text{ grams carbohydrate} \times 4 \text{ calories/gram} &= 60 \text{ calories} \\
10 \text{ grams protein} \times 4 \text{ calories/gram} &= 40 \text{ calories} \\
5 \text{ grams fat} \times 9 \text{ calories/gram} &= \underline{45 \text{ calories}} \\
&\ 145 \text{ calories}
\end{aligned}
$$

You can calculate the percentage of total calories from carbohydrate, protein, and fat in the soup by dividing the number of calories supplied by each nutrient by the total number of calories and then multiplying by 100:

- Carbohydrate: $\dfrac{60 \text{ calories}}{145 \text{ calories}} = 0.41 \times 100 = 41\%$

- Protein: $\dfrac{40 \text{ calories}}{145 \text{ calories}} = 0.28 \times 100 = 28\%$

- Fat: $\dfrac{45 \text{ calories}}{145 \text{ calories}} = 0.31 \times 100 = \underline{31\%}$

$$100\%$$

Given this information about the caloric value of carbohydrates, proteins, and fats, which of the items in Illustration 8.5 would you expect to be highest in calories?

Illustration 8.5
Which contains the most calories—a tablespoon of margarine, sugar, or pork? (Answers are on page 8-8.)

Richard Anderson

Answer to Illustration 8.5:
Margarine.
*There are about 101 calories in
one tablespoon of margarine,
46 in a tablespoon of sugar, and
40 in a tablespoon of relatively
lean pork.*

College students have been found to miss this question 47% of the time.[7] It's the margarine! Margarine contains the most fat; it is made primarily from oil. *High-fat foods provide more calories ounce for ounce than foods that contain primarily carbohydrate or protein.* That means bread and potatoes (rich in carbohydrates), catsup (rich in water and a low-calorie vegetable), sugar, and other low-fat or no-fat foods provide fewer calories than equal amounts of foods that contain primarily fat.

Most Foods Are a Mixture

Some foods (such as oil and table sugar) consist almost exclusively of one energy nutrient, but most foods contain many substances in varying amounts.

Bread is high in complex carbohydrates ("starch"), but it also contains many other substances such as protein, water, vitamins, minerals, and fiber. Likewise, steak is not all protein. Although protein constitutes about 32% of the total weight of lean sirloin steak, fats make up about 8%, and the largest single ingredient is water—60% of the weight. Yet we don't think of steak as a "high-liquid food"; instead, it's often thought of as pure protein.

So, even though some foods provide relatively more carbohydrate, protein, or fat than other foods, most foods contain a mixture of energy nutrients.

The Caloric Value of Foods: How Do You Know?

Some people say you can estimate the caloric value of food by how it tastes or by how appetizing it looks. Actually, it's not that simple. How many calories do you think are contained in a half-cup of peanuts; a boiled, medium-sized potato; and a cup of rice (Illustration 8.6)?

Taste, appearance, and reputation do not make good criteria for determining the caloric value of foods. Here's an example:

> I used to order the fish sandwich at fast-food restaurants because I was trying to avoid all the calories in hamburgers. Then I found out the fish sandwich had about the same number of calories as the quarter-pound hamburger! Where did I get the idea that fried fish loaded with tartar sauce has fewer calories than a hamburger?

This doesn't have to happen to you! By referring to Appendix A, you could determine that a typical fast-food fish sandwich and a quarter-pound hamburger

Illustration 8.6
What's the caloric value of these foods?
One contains 420 calories, another 205, and the third 118 calories. Try matching the caloric values with the foods, and then check your answers on p. 8-10.

weigh in at about 400 calories each. If you're interested in the caloric value of foods you frequently eat, look them up in this appendix.

How Is Caloric Intake Regulated by the Body?

Because energy is critical to survival, the body has a number of mechanisms that encourage regular caloric intake. It has less effective means of discouraging an excessive intake of calories.[8]

Mechanisms that encourage food intake don't depend on weight status.[9] Rather, they are keyed to encouraging eating on a regular basis so that food, if available, will be consumed and carry the body through times when food isn't around. Humans developed on a schedule of "feast and famine." Those who could store enough fat to see them through the times when food was scarce had an advantage. So, no matter how thin or fat a person is, she or he experiences *hunger* if food is not consumed several times throughout the day. The "hungry" signal is thought to be sent by a series of complex mechanisms when cells run low on energy nutrients supplied by the last meal or snack.[9]

When we eat, we reach a point when we feel full and are no longer interested in eating. The signal is due to hormones and internal sensors in the brain, stomach, liver, and fat cells that indicate *satiety*—the feeling that we've had enough to eat.[11]

For most people, hunger and satiety mechanisms adjust energy intake to match the body's need for energy. However, internal signals that urge us to eat or to stop eating can be overridden. People can resist eating, no matter how strong the hunger pains. Even after the "I'm full" siren has sounded, people can go on eating.[12] And sometimes people eat because they have an *appetite* for specific foods and desire the pleasure foods can bring.

Appetite may or may not be related to being hungry. It can be triggered when we smell or see a tasty food right after a meal or when we're really hungry. Ever see those TV commercials for a juicy burger and fries that air about 11:00 P.M.? How many people who jump into their cars and head to the carry-out window are actually hungry? Or have you noticed how appealing the idea of eating is and how good food tastes when you're very hungry? That's appetite at work.

The Question of Energy Balance   Unless you are currently losing or gaining weight, the number of calories you need is the number you usually consume in your diet. Adults who maintain their weight are in a state of energy balance (Illustration 8.7). Because they are not losing weight (using fat and other energy stores) or gaining it (storing energy), their body's expenditure of energy and its intake of energy are balanced.

When energy intake is less than the amount of energy expended, people are in negative energy balance. In this case, energy stores are used and people lose weight. When a positive energy balance exists, weight and fat stores are gained because more energy is available from foods than is needed by the body. Slight changes in caloric need may not amount to fluctuations in body weight on a day-to-day basis. Over the course of a year, however, a *positive* or *negative* daily energy balance of 50 calories would result in a weight gain or loss of approximately 5 pounds.

Sometimes, a positive energy balance is a healthy and normal circumstance. For example, a positive energy balance is normal when growth is occurring, as in childhood or pregnancy, or when a person is regaining weight lost during an illness.

hunger
Unpleasant physical and psychological sensations (weakness, stomach pains, irritability) that lead people to acquire and ingest food.

satiety
A feeling of fullness or of having had enough to eat.

appetite
The desire to eat; a pleasant sensation that is aroused by thoughts of the taste and enjoyment of food.

People expend an average of 11 calories an hour chewing gum. If you chewed gum for an hour every day and didn't change any other component of energy balance, you'd lose about a pound a year.[10]

Photo Disc

ON THE SIDE

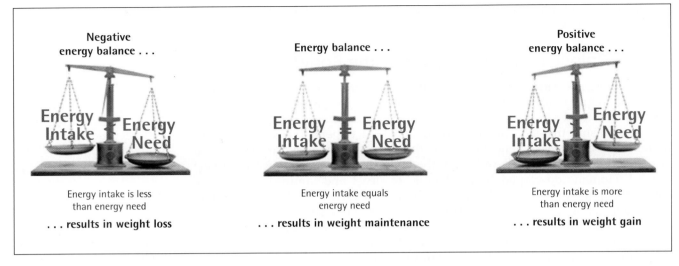

Illustration 8.7
The body's energy status.

Keep Calories in Perspective

Answer to Illustration 8.6:

	Calories
½ cup of peanuts	420
1 medium boiled potato	118
1 cup of white rice	205

Calories is not a word that means "fattening" or "bad for you." Calories are a life and health-sustaining property of food. Diets compatible with health contain a mixture of foods providing various amounts of calories. It's not the caloric content of individual foods that makes them good or not. It's the sum of calories and nutrients in foods that make up our total diets.

Key Terms

appetite, page 8-9
basal metabolic rate (BMR), page 8-4
basal metabolism, page 8-4

calorie, page 8-2
dietary thermogenesis, page 8-5
hunger, page 8-9

resting metabolic rate (RMR), page 8-4
satiety, page 8-9

Media Menu

www.ars.usda.gov/ba/bhnrc/ndl
At the homepage for USDA's food composition information, you can obtain nutrient data files and get answers to commonly asked questions about the calorie value of foods from this huge site.

http://exercise.about.com/od/ fitnesstoolscalculators/
This site offered by "About Health and Fitness," provides BMI and calorie expenditure

for physical calculators based on data you input. The site has lots of ads, but the calculators are easy to use.

http://www-users.med.cornell.edu/~spon/ picu/calc/beecalc.htm
This address brings you to Cornell University's Medical Calculations site. It includes a tool for estimating basal energy expenditure.

www.cyberdiet.com
Go directly to "Fast Food Quest" to look up and compare caloric and nutrient values of foods from fast-food restaurants.

Nutrition Up Close

Food as a Source of Calories

Focal Point: Examine the distribution of calories in the foods you eat.

What are the sources of calories in food? Determine the caloric contribution of the fat, carbohydrate, and protein content of the following snack foods using the composition data and calorie conversion factors listed. Then answer the two questions that follow.

Which snack is lower in total calories?
_____

Which snack is lower in fat calories?
_____

Calorie conversion factors

1g fat	= 9 calories
1g carbohydrate	= 4 calories
1g protein	= 4 calories

Potato Chips

Serving size: 1 oz (about 20 chips)

Fat: 10g × ____ = ____ calories from fat
Carbohydrate: 15g × ____ = ____ calories from carbohydrate
Protein: 2g × ____ = ____ calories from protein
____ Total calories

% of calories from fat: ____ = _____ = ____

Mini Pretzels

Serving size: 1 oz (about 17 pieces)

Fat: 0g × ____ = ____ calories from fat
Carbohydrate: 24g × ____ = ____ calories from carbohydrate
Protein: 3g × ____ = ____ calories from protein
____ Total calories

% of calories from carbohydrate: ____ = _____ = ____

FEEDBACK (answers to these questions) can be found at the end of Unit 8.

Review Questions

True False

1. A calorie is also referred to as a kilo-calorie (kcal). ☐ ☐

2. Calories are a component of food that provide the body with energy. ☐ ☐

3. A person's calorie need is based on his or her energy expenditure from basal metabolism, physical activity, and dietary thermogenesis. ☐ ☐

4. Basal metabolic rate varies substantially among people due to differences in genetic traits and disease history. ☐ ☐

5. George, a city bus driver, weighs 220 pounds. The number of calories he expends for basal metabolism daily would equal about 2420 calories. ☐ ☐

6. A common source of error in estimations of physical activity level is underestimation of the amount of time spent in physical activities. ☐ ☐

True False

7. Dietary thermogenesis accounts for approxiamtely 30% of a person's total calorie need. ☐ ☐

8. Say you ate a medium serving of french fries that contained 3 grams of protein, 38 grams of carbohydrate, and 14 grams of fat. The calorie value of the french fries would be 290. ☐ ☐

9. The serving of french fries referred to in question 8 would provide 60% of calories from fat. ☐ ☐

10. The composition of lean steak is nearly 100% protein. ☐ ☐

11. *Appetite* is related to the desire for food, whereas *hunger* refers to a physiological need for food. ☐ ☐

12. A person who is gaining weight is in positive energy balance. ☐ ☐

Notes

1. Poehlman ET, Horton ES. Energy needs: assessment and requirement in humans. In: Shils ME et al. Modern Nutrition in Health and Disease, 9e, Philadelphia: Lippincott, Williams, and Wilkins, 1998, p.96.

2. Hsu A et al. Larger mass of high-metabolic rate organs does not explain higher resting energy expenditure in children. Am J Clin Nutr 2003;77: 1506–11.

3. Slow metabolism: Is there any such thing? www.mayoclinic.com/health/slow-metabolism/AN00618, accessed 7/06.

4. Harris JA, Benedict FG. A biometric study of basal metabolism in men. Publication no. 279 of the Carnegie Institute of Washington, 1919.

5. Sjodin AM, et al., The influence of physical activity on BMR, Med Sci Sports Exercise 1996;28:85–91; Klausen B, et al., Age and sex effects on energy expenditure, Am J Clin Nutr 1997;65:895–907; and Frankenfield DC, et al., The Harris-Benedict studies of human basal metabolism: history and limitations, J Am Diet Assoc 1998;98:439–45.

6. Boothby WM, Berksen J, Dunn HL. Studies of the energy of metabolism of normal individuals: a standard for basal metabolism, with a nomogram for clinical application. Am J Physiol 1936;116:468–84.

7. Melby CL, Femea PL, Sciacca JP. Reported dietary and exercise behaviors, beliefs and knowledge among university undergraduates. Nutr Res 1986;6:799–808.

8. Jacobs DR, Jr. Fast food and sedentary lifestyle: a combination that leads to obesity. Am J Clin Nutr 2006;83: 189–90.

9. Raben A et al. Meals with similar energy densities but rich in protein, fat, carbohydrate, or alcohol have different effects on energy expenditure and substance metabolism but not on appetite and energy intake. Am J Clin Nutr 2003;77:91–100.

10. Levine J et al. The energy expended in chewing gum. N Eng J Med 1999;341: 2100.

11. McCrory MA et al. Dietary determinants of energy intake and weight regulation in healthy adults. J Nutr 2000; 130:276S–9S.

12. Marmonier C et al. Snacks consumed in a nonhungry state have poor satiating efficiency. Am J Clin Nutr 2002;76: 518–28.

Answers to Review Questions

1. True, see page 8-2.
2. False, see page 8-2.
3. True, see page 8-3.
4. False, see page 8-4.
5. True, see page 8-4.
6. False, see page 8-5.
7. False, see pages 8-5, 8-6.
8. True, see pages 8-6, 8-7.
9. False, see page 8-7.
10. False, see page 8-8.
11. True, see page 8-9.
12. True. see pages 8-9, 8-10.

Nutrition Up Close

Food as a Source of Calories

Feedback for Unit 8

Potato Chips

Serving size: 1 oz (about 20 chips)

Fat: 10 g 3 __9__ = __90__ calories from fat

Carbohydrate: 15 g 3 __4__ = __60__ calories from carbohydrate

Protein: 2 g 3 __4__ = __8__ calories from protein

__158__ Total calories

% of calories from fat: $\dfrac{90}{158}$ = 0.57×100 = __57%__

Mini Pretzels

Serving size: 1 oz (about 17 pieces)

Fat: 0 g 3 __9__ = __0__ calories from fat

Carbohydrate: 24 g 3 __4__ = __96__ calories from carbohydrate

Protein: 3 g 3 __4__ = __12__ calories from protein

__108__ Total calories

% of calories from carbohydrate: $\dfrac{96}{108}$ = 0.89×100 = __89%__

The pretzels are a better snack choice because they are lower in total calories and fat calories than the potato chips.

Obesity to Underweight: The Highs and Lows of Weight Status

NUTRITION SCOREBOARD

	TRUE	FALSE
1 There is usually little difference between the ideal body weights defined by cultural norms and those defined by science.		
2 Fat stores concentrated around the stomach present a greater hazard to health than do fat stores located around the hips and thighs.		
3 High-calorie diets are the only cause of obesity.		
4 Healthy underweight people often find it nearly impossible to gain weight.		

Key Concepts and Facts

- Ideal body weight and shape are defined by culture and by health measures. The operating definition should be based on health.

- Rates of overweight and obesity are increasing worldwide. Diseases and disorders related to excess body fat are also increasing.

- The location of body fat stores, as well as the amount of body fat, are important to health.

- The causes of obesity are complex and not completely understood. Diet, physical activity, environmental exposures, and genetic factors influence the development of obesity.

- Underweight in the United States usually results from a genetic tendency to be thin or from poverty, illness, or the voluntary restriction of food intake.

Answers to
NUTRITION SCOREBOARD

		TRUE	FALSE
1	Cultural norms of ideal body weights are often at odds with the "healthy" weights defined by science.		✔
2	Health is affected by the location of body fat stores as well as by obesity.[1] Adults who store body fat in the stomach, or central area of the body, are at higher risk for a number of health problems than are adults who store fat primarily in their hips and thighs.	✔	
3	Obesity has many causes, and high-calorie diets are just one of them.		✔
4	Gaining weight is as difficult for many underweight people as losing weight is for many overweight people.	✔	

Variations in Body Weight

Wouldn't it be terrific if your body adjusted your food intake based on a healthy level of body fat stores? Then you wouldn't have to worry about being too thin or getting too fat. Why doesn't that happen?

Photo Disc

For humans, one size does not fit all. Human bodies come in a range of sizes that vary from heavy and tall to thin and short. Over recent decades, however, the distribution of body sizes has become lopsided. The proportion of people in the United States and other countries classified as overweight or obese is increasing, and rates of underweight are dropping. What lies underneath these trends? Why are humans so susceptible to gaining weight?

Obesity appears to represent a weak link in the biological evolution of humans. Body processes that regulate food intake developed over 40,000 years ago when "feast and famine" cycles were common. Being underweight was a distinct disadvantage. The more body fat people could store after a feast, the better their chances of surviving the subsequent famines. Consequently, multiple mechanisms that favored food intake and body fat storage developed.[2] These body mechanisms continue to encourage food intake even when food is constantly available and obesity poses a greater threat to survival than does famine. In the words of Theodore Van Itallie, a noted obesity researcher, in environments with an abundant supply of food and no requirement for vigorous physical activity, "perhaps thin people are the ones who are abnormal."[3]

Circumstances today are very different from those existing when the human genetic endowment and its food intake regulatory mechanisms were established.

These changes may go a long way to explaining why obesity is a major problem in many countries.

How Is Weight Status Defined?

Culture and science define the appropriateness of body size. In the 15th century, moon-faced and pear-shaped women were considered beautiful.[4] In the 1930s, full-figured women such as Jean Harlow were box-office stars, and Marilyn Monroe was an icon of the 1950s. Then came Twiggy, Calista Flockhart, and Keira Knightley. Thinness beyond health boundaries became the standard of beauty. With "too thin" as the cultural ideal, many people regard weights within the normal range as being too high. Men feel the pressure, too. For them, visible body fat and weak-looking bodies are culturally taboo.[5]

Science defines standards for body weight for adults primarily based on the risk of death from all causes. Death rates are highest among adults who have very high body weights for height, and lowest among adults who are normal weight for height (Illustration 9.1).[6] You can get an idea of what a normal weight for height is by following the steps detailed in the "Health Action."

In the past, standards used to identify whether a person was underweight, normal weight, overweight, or obese were based on weight for height tables developed by the insurance industry to estimate life expectancy and the risk of death of adults. These tables have been replaced by standards that employ **body mass index**.[8]

Body Mass Index

Most commonly referred to as BMI, body mass index is a measure of weight for height that provides a fairly good estimate of body fat content.[9] Ranges of BMI are used to define weights for height that correspond to underweight, normal weight, and obesity in adults (Table 9.1). BMI has the advantage of being calculated the same way for adult males and females. Calculating BMI involves dividing weight

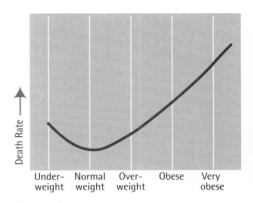

Illustration 9.1
The relationship between body weight status and deaths from all causes for adults.[6]

TABLE 9.1
Classifying weight status by body mass index.[10]

	Body Mass Index
Underweight	under 18.5 kg/m²
Normal weight	18.5–24.9 kg/m²
Overweight	25–29.9 kg/m²
Obese	30 kg/m² or higher[a]

[a]Obesity is subdivided for some purposes into the BMI groups of moderate obesity (30–34.9 kg/m²), severe obesity (35–39.9 kg/m²), and very severe obesity (40+ kg/m²).

body mass index (BMI)
An indicator of the status of a person's weight for their height. It is calculated by dividing weight in kilograms by height in meters squared. It can also be calculated using inches and pounds as shown in Table 9.2.

HEALTH *Action* Estimating Normal Weight for Height

There is a quick way to estimate within ±10% what a normal, or healthy, weight for height in adults is. It's the Hamwi[8] method, and it is done as follows.

Women

Begin with 5 feet equals 100 pounds, and then add 5 pounds for each additional inch of height. Here's an example of how you would estimate a healthy weight for a woman who is 5 feet, 7 inches tall:

$$5 \text{ feet} = 100 \text{ pounds}$$
$$7 \text{ inches} \times 5 \text{ pounds} = 35 \text{ pounds}$$
$$100 \text{ pounds} + 35 \text{ pounds} = 135 \text{ pounds}$$

Men

For men, five feet equals 106 pounds, and each additional inch of height adds on 6 pounds. So, for example, we would estimate a healthy weight for a man who is 5 feet 10 inches tall the following way:

$$5 \text{ feet} = 106 \text{ pounds}$$
$$10 \text{ inches} \times 6 \text{ pounds} = 60 \text{ pounds}$$
$$100 \text{ pounds} + 66 \text{ pounds} = 166 \text{ pounds}$$

TABLE 9.2

Calculating BMI: An example

Say Chris weighs 140 pounds and is 5 feet, 3 inches tall. To calculate BMI:

Figure out how many inches tall Chris is:

    5 feet × 12 inches per foot = 60 inches

    60 inches + 3 inches = 63 inches

Divide Chris's weight by his height in inches:

$$\frac{140\ \text{pounds}}{63\ \text{inches}} = 2.22$$

Divide the result (2.22) by Chris's height in inches again:

$$\frac{2.22}{63\ \text{inches}} = 0.035$$

Multiply this result by 703:

    0.035 × 703 = 24.8   This is Chris's BMI.

in pounds by height in inches, then dividing this result by height in inches again, and then multiplying that result by 703. An example of BMI calculation is given in Table 9.2. You can use the chart shown in Illustration 9.2 to identify weight status based on BMI from information on weight and height.

Assessing Weight Status in Children and Adolescents Standards used to assess weight status in children and adolescents employ BMI percentile ranges for girls and boys (Illustration 9.3). Percentiles of BMI are based on the proportion of children and adolescents who have different levels of BMI at given ages. For example, if a child's BMI is at the 50th percentile for his or her age, then half of children will have BMIs that are below and half will have values above this child's BMI level. Table 9.3 shows BMI percentile ranges that correspond to underweight, at risk of underweight, normal weight, at risk of overweight, and overweight in children and adolescents. Obesity is not included in this table. Body fat content, rather than BMI percentile ranges, should be used to diagnose obesity in children and adolescents.[11]

BMI percentile ranges for children and adolescents do not provide information on growth progress in terms of height. Consequently, growth in height cannot be assessed using BMI. Other standards for assessment of height in children and adolescents are available.

Illustration 9.2
BMI Chart

Body Mass Index (BMI)

Height	18	19	20	21	22	23	24	25	26	27	28	29	30	31	32	33	34	35	36	37	38	39	40
4'10"	86	91	96	100	105	110	115	119	124	129	134	138	143	148	153	158	162	167	172	177	181	186	191
4'11"	89	94	99	104	109	114	119	124	128	133	138	143	148	153	158	163	168	173	178	183	188	193	198
5'0"	92	97	102	107	112	118	123	128	133	138	143	148	153	158	163	168	174	179	184	189	194	199	204
5'1"	95	100	106	111	116	122	127	132	137	143	148	153	158	164	169	174	180	185	190	195	201	206	211
5'2"	98	104	109	115	120	126	131	136	142	147	153	158	164	169	175	180	186	191	196	202	207	213	218
5'3"	102	107	113	118	124	130	135	141	146	152	158	163	169	175	180	186	191	197	203	208	214	220	225
5'4"	105	110	116	122	128	134	140	145	151	157	163	169	174	180	186	192	197	204	209	215	221	227	232
5'5"	108	114	120	126	132	138	144	150	156	162	168	174	180	186	192	198	204	210	216	222	228	234	240
5'6"	112	118	124	130	136	142	148	155	161	167	173	179	186	192	198	204	210	216	223	229	235	241	247
5'7"	115	121	127	134	140	146	153	159	166	172	178	185	191	198	204	211	217	223	230	236	242	249	255
5'8"	118	125	131	138	144	151	158	164	171	177	184	190	197	203	210	216	223	230	236	243	249	256	262
5'9"	122	128	135	142	149	155	162	169	176	182	189	196	203	209	216	223	230	236	243	250	257	263	270
5'10"	126	132	139	146	153	160	167	174	181	188	195	202	209	216	222	229	236	243	250	257	264	271	278
5'11"	129	136	143	150	157	165	172	179	186	193	200	208	215	222	229	236	243	250	257	265	272	279	286
6'0"	132	140	147	154	162	169	177	184	191	199	206	213	221	228	235	242	250	258	265	272	279	287	294
6'1"	136	144	151	159	166	174	182	189	197	204	212	219	227	235	242	250	257	265	272	280	288	295	302
6'2"	141	148	155	163	171	179	186	194	202	210	218	225	233	241	249	256	264	272	280	287	295	303	311
6'3"	144	152	160	168	176	184	192	200	208	216	224	232	240	248	256	264	272	279	287	295	303	311	319
6'4"	148	156	164	172	180	189	197	205	213	221	230	238	246	254	263	271	279	287	295	304	312	320	328
6'5"	151	160	168	176	185	193	202	210	218	227	235	244	252	261	269	277	286	294	303	311	319	328	336
6'6"	155	164	172	181	190	198	207	216	224	233	241	250	259	267	276	284	293	302	310	319	328	336	345

Height — Body Weight (pounds)

Underweight (<18.5)   Healthy Weight (18.5-24.9)   Overweight (25-29.9)   Obese (30)

Find your height along the left-hand column and look across the row until you find the number that is closest to your weight. The number at the top of that column identifies your BMI. The area shaded in green represents healthy weight ranges.

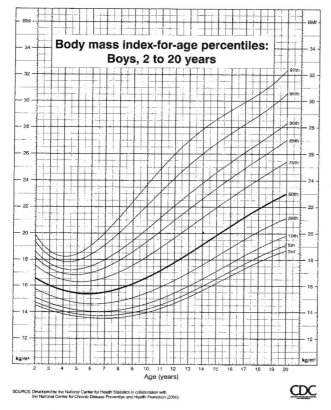

Illustration 9.3
An example of the CDC's 2000 Growth Charts: BMI-for-age.
Copies of the growth charts can be obtained from www.cdc.gov/growthcharts

TABLE 9.3
Weight status standards for 2- to 20-year-olds based on BMI-for-age.[11]

	Percentile Range(s)
Underweight	5th or less
At risk of underweight	5th-15th
Normal weight	15th-85th
At risk of overweight	85th-95th
Overweight	95th or higher

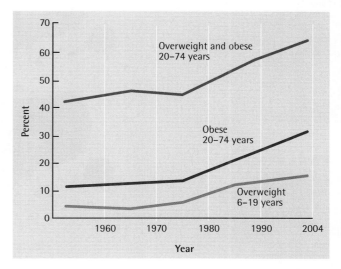

Illustration 9.4
Overweight and obesity by age in the United States, 1960-2004.

Source: Centers for Disease Control and Prevention, National Center for Health Statistics, CDC Fast Stats, 2006.

overweight
A high weight-for-height.

obese
Characterized by excess body fat.

Most Adults in the United States Weigh Too Much

Being *overweight* or *obese* is the norm in the United States. The combined incidence of overweight and obesity among adults is nearly 66% and is on its way up (Illustration 9.4).

Over 1 in 7 children and adolescents in the United States are overweight, and this percentage may be rising as you read this.[15] Overweight and obesity are becoming America's number one health problem. Obesity, which represents the largest risks to health, varies considerably among U.S. states (Illustration 9.5). In 2005, for example, rates of obesity were highest in Mississippi, West Virginia, and Alabama and lowest in Colorado and Vermont.[12]

The trend toward higher rates of overweight and **obesity** in the United States is now shared by almost all other countries of the world. In Europe, 10 to 25% of adults are obese, and over a billion adults worldwide are overweight or obese.[13] A sampling of overweight plus obesity rates by country is shown in Table 9.4. Overweight and obesity are major, worldwide public health problems. The high and growing incidence of overweight and obesity in the United States and around the world represents a time bomb for a future explosion in obesity-related disease rates.

TABLE 9.4

Prevalence of overweight plus obesity in a sampling of countries[14,15]

Country	Percent Overweight and Obese
Albania	79
Czech Republic	65
United States	65
Iran	64
England	61
Costa Rica	59
Germany	55
Canada	49
Netherlands	46
Australia	42
France	42
Israel	40
China	34
Japan	24

TABLE 9.5

The mismatch between obesity and health. All these problems are associated with obesity[16,17]

Type 2 diabetes
Chronic inflammation
Metabolic syndrome
Hypertension
Stroke
Elevated cholesterol level
Low HDL-cholesterol level
Heart disease
Certain types of cancer
Gallbladder disease
Shortened life expectancy
Discrimination
Depression
Infertility
Accidents
Skin disorders
Sleep disorders

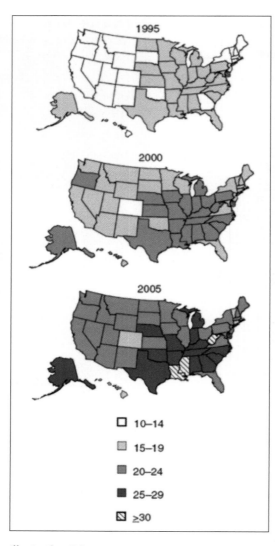

Illustration 9.5
Percent of obese (BMI ≥ 30kg m²) adults by state in 1995, 2000, and 2005.

Source: Centers for Disease Control and Prevention, MMWR 2006;55:985–988.

The Influence of Obesity on Health True obesity is not a healthy state. It carries an increased risk of diabetes, hypertension, stroke, heart disease, elevated total cholesterol levels, low HDL-cholesterol levels (the "good cholesterol"), certain types of cancer, and other health problems (Table 9.5).[16] Life expectancy is also influenced by body weight. Contrary to the popular view that underweight people live the longest, adults with BMIs between 22 and 25 kg/m² live longer on average than do underweight adults (BMI < 18.5 kg/m²).[18] Life expectancy in overweight adults is approximately 3 years, and in obese persons 6 years, shorter than average.[19] The more obese a person is, the more common health problems related to obesity become.[16] In the category of good news is the fact that loss of 5 to 10% of body weight in obese adults decreases risks to health substantially. Weight loss lowers blood pressure and LDL cholesterol, increases HDL cholesterol (the one you want to be high), and decreases the risk of type 2 diabetes or high glucose levels in people with diabetes.[20,21]

Obesity and Psychological Well-Being Another consequence of obesity is the ingrained cultural prejudice to which obese people are subjected. Children who are obese are more likely to suffer unfair or indifferent treatment from teachers than are nonobese children. They experience more isolation, rejection, and feelings of inferiority than other children. Obese adults are likely to be discriminated against in hiring and promotion decisions and to be thought of as lazy or lacking in self-control, even by health professionals who care for them.[22] Society's prejudice against people who don't conform to the cultural ideal of body size may be the most injurious consequence of obesity.

Body Fat and Health: Location, Location, Location

It is becoming increasingly clear that some of the health problems associated with obesity are directly related to where excess fat is stored. Humans store fat in two major locations: under the skin over the hips, upper arms, and thighs, and in the

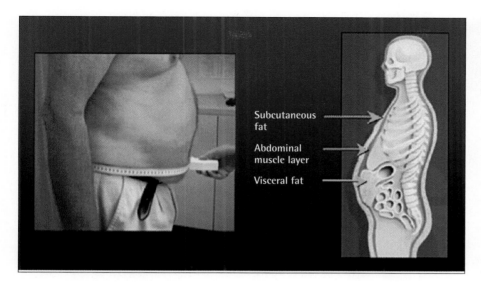

Illustration 9.6
A diagram of the location of subcutaneous and visceral fat.

Source: Caballero E. Dyslipidemia and vascular function: The rationale for early and aggressive intervention. www.medscape.com, 4/18/06

subcutaneous fat
(Pronounced sub-q-tain-e-ous) Fat located under the skin.

visceral fat
(Pronounced vis-sir-el) Fat located under the skin and muscle of the abdomen.

metabolism
The chemical changes that take place in the body. The conversion of glucose to energy or to body fat is an example of a metabolic process.

chronic inflammation
Inflammation that lasts weeks, months, or years. Inflammation is the first response of the body's immune system to infection or irritation. Inflammation triggers the release of biologically active substances that promote oxidation and other potentially harmful reactions in the body.

abdomen. Fat stored under the skin is called ***subcutaneous fat*** and that stored in the abdomen ***visceral fat*** (see Illustration 9.6). People who store fat primarily in their hips, upper arms, and thighs are said to have a "pear shape" and those who store fat principally in the abdomen an "apple shape." These body shapes are shown in Illustrations 9.7 and 9.8. You may be better off healthwise if you're a "pear" rather than an "apple."

Visceral fat is much more ***metabolically*** active than subcutaneous fat, and some of its metabolic effects are hazardous to health. Metabolic reactions taking place within visceral fat cells initiate inflammatory reactions that lead to the production of unstable substances that cause oxidation. Over time, these oxidation reactions lead to a state of ***chronic inflammtion***. Chronic inflammation causes damage to molecules within other cells in the body, including those that line blood

PEAR
Fat stores around hips predominate

APPLE
Fat stores around waist predominate

llustration 9.7
Basic body shapes.
The pear normally has narrow shoulders, a small chest, and an average-size waist. Fat is concentrated in the hips, upper arms, and thighs. Weight loss in these areas is usually difficult.

The apple looks round in the middle. Fat is concentrated in the waist and can be lost with diet and exercise. The apple shape is riskier than the pear shape due to the presence of visceral fat.

Illlustration 9.8
The man on the left is 6 feet tall and weighs 240 pounds; the man on the right is 6 feet 1 inch tall and weighs 230 pounds. Although similar in weight and height, the man on the left is at higher risk for diabetes, hypertension, and heart disease due to his "apple" shape and body fat content.

insulin resistance
A condition in which cells "resist" the action of insulin in facilitating the passage of glucose into cells.

metabolic syndrome
A constellation of metabolic abnormalities generally characterized by insulin resistance, abdominal obesity, high blood pressure and triglyceride levels, low levels of HDL cholesterol, and impaired glucose tolerance. Metabolic syndrome predisposes people to the development of heart disease, hypertension, and other disorders. It is common—one in five U.S. adults has metabolic syndrome.[17]

type 2 diabetes
A disease characterized by high blood glucose levels due to the body's inability to use insulin normally or to produce enough insulin (previously called adult-onset diabetes).

vessels. Metabolic consequences of chronic inflammation promote the development of plaque in arteries, osteoporosis, *insulin resistance, metabolic syndrome, type 2 diabetes,* elevated blood pressure, cancer, Alzheimer's disease, and other disorders.[23] Loss of abdominal fat reduces the risk of developing disorders related to chronic inflammation.[20]

Visceral Fat and Waist Circumference   In the recent past, the ratio of waist-to-hip circumference was used to assess body fat distribution, or whether a person is an "apple" or a "pear." This method is being replaced by the use of waist circumference only (Illustration 9.9). Waist circumference is closely related to the

Illustration 9.9
Determining your waist-circumference.

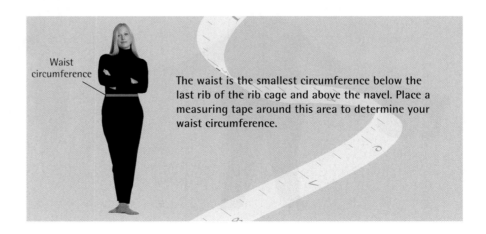

Waist circumference

The waist is the smallest circumference below the last rib of the rib cage and above the navel. Place a measuring tape around this area to determine your waist circumference.

size of visceral fat stores.[24] In men, waist circumferences over 40" (102 cm), and in women, over 35" (88 cm), indicate an apple shape and excessive storage of visceral fat.[25]

Assessment of Body Fat Content

Body mass index is often used to approximate body fat content because the two measures correspond closely in groups of people. This is not always the case for individuals. Take a 165-pound, 30-year-old woman who is 5 feet 6 inches tall and weight trains heavily. Her body weight for height would indicate obesity, but she may have a low body fat content and be healthy. If BMI were used to assess body fat levels in professional football players, most all of them would be wrongly categorized as overly fat.[26] Sometimes people who are classified as normal weight or underweight by BMI standards have too much body fat because they are physically inactive. Certain medications make people retain fluid. Their weight-for-height may qualify them as overweight, but their body fat content may actually be very low. Obviously, measures of body fat content are better estimators of health status than are measures of weight-for-height.

Methods for Assessing Body Fat Content   Tools for assessing body fat content that are easy to use, accurate, and inexpensive are now available, and their use is spreading. This development is starting a trend for the determination of weight status based on body fat content.[25] Standards for classifying percent body fat, or the percent of weight that consists of fat, have been developed (Table 9.6) and will be refined as additional studies on the relationships between body fat and health risk are done.

Here are the most common methods for determining body fat content:

- skinfold thickness measures
- bioelectrical impedance analysis (BIA)
- underwater weighing
- magnetic resonance imaging (MRI)
- computerized axial tomography (CT or CAT scans)
- dual-energy X-ray absorptiometry (DEXA)
- whole body air displacement (BOD POD)

The theories underlying each of these methods and its advantages and limitations are presented in Table 9.7. Each body fat test has its advantages and drawbacks. The tests are most likely to provide accurate results when they are performed by skilled, experienced technicians using proper, well-maintained equipment and when the measurements are converted into percent body fat by the appropriate formulas.

Everybody Needs Some Body Fat   A certain amount of body fat—3 to 5% for men and 10 to 12% for women—is needed for survival. Body fat serves essential roles in the manufacture of hormones; it's a required component of every cell in the body; and it provides a cushion for internal organs. Fat that serves these purposes is not available for energy formation no matter how low energy reserves become. Low body fat levels are associated with delayed physical maturation during adolescence, infertility, accelerated bone loss, and problems that accompany starvation.[28]

ON THE SIDE

Between 25 and 40% of domestic cats seen in veterinary practices are overweight or obese.[54]

Photo Disc

TABLE 9.6
Percentages of body weight as fat considered low, average, and high[26,27]

	% Body Fat		
	Low	Average	High
Women	less than 12	32	35 or more
Men	less than 5	22	25 or more

ON THE SIDE

We're getting bigger. According to the CDC, men now weigh 191 pounds and are 5'9½" tall on average. In 1960, the average weight of men was 166 pounds and height was 5'8". Women are now 5'4" tall and weigh 164 pounds on average. In 1960, those figures were 5'3" and 140 pounds.

TABLE 9.7

Commonly used methods for assessing body fat content

Skinfold Measurement

Body fat content can be estimated by measuring the thickness of fat folds that lie underneath the skin. Calipers are used to measure the thickness of fat folds, preferably over several sites on the body. Body fat content is estimated by "plugging" the thicknesses into the appropriate formula.

Advantages: Calipers are relatively inexpensive (they cost about $250–$350). The procedure is painless if done correctly and can yield a good estimate of percent body fat.

Limitations: This technique is often performed by untrained people. Skinfolds may be difficult to isolate in some individuals.

Bioelectrical Impedance Analysis (BIA)

Because fat is a poor conductor of electricity and water and muscles are good conductors, body fat content can be estimated by determining how quickly electrical current passes from the ankle to the wrist.

Advantages: The equipment required is portable, and the test is easy to do and painless. The results are fairly accurate for people who are not at the extremes of weight-for-height.

Limitations: Equipment may be expensive; inferior equipment produces poor results. Hydration status and meal ingestion may affect electrical conductivity and produce inaccurate results, as may inaccurate formulas used to calculate body fat content from test results.

Underwater Weighing

The subject is first weighed on dry land; next he or she is submerged in water and exhales completely; then his or her weight is measured. The less the person weighs under water compared to the weight on dry land, the higher the percent body fat. (Fat, but not muscle or bone, floats in water.)

Advantages: If undertaken correctly and if appropriate formulas are used in calculations, this technique gives an accurate value of percent body fat.

Limitations: The equipment required is expensive and not easily moved. The test doesn't work well for people who don't swim or who are ill or disabled in some way.

Magnetic Resonance Imaging (MRI)

MRI and CT or CAT scans provide similar results. Using this technology, a person's body fat and muscle mass can be photographed from cross-sectional images obtained when the body, or parts of it, is exposed to a magnetic field (or, in the case of CAT scans, to radiation). Based on the volume of fat and muscle observed, total fat and muscle content can be determined.

Advantages: Provides highly accurate assessment of fat and muscle mass.

Limitations: The test is expensive ($500–$1000 per assessment) and largely used for research purposes.

TABLE 9.7

Commonly used methods for assessing body fat content (continued)

Dual-energy X-ray Absorpiotmetry (DEXA)

DEXA (or DXA) is based on the principle that various body tissues can be differentiated by the level of X-ray absorption. The measure is made by scanning the body with a small dose of X-rays (similar to the level of exposure from a transcontinental flight) and then calculating body fat content based on the level of X-ray absorption.

Advantages: Provides highly accurate results when measurements are undertaken correctly. DEXA is safe and "user friendly" for people being measured and can also be used to assess bone mineral content and lean tissue mass.

Limitations: DEXA machine is expensive, and so is the cost of individual assessments ($150 per person). Machine must be operated by a trained and certified radiation technologist in many states.

Courtesy Life Measurement, Inc. www.bodpod.com

Whole Body Air Displacement

This is an established method that has become practical for broader use due to development of the BOD POD. The method is similar to that of underwater weighing but uses air displacement for determining percent body fat. Individuals sit in an enclosed "cabin" for about 5 minutes while wearing a tight-fitting swim suit and cap. Computerized sensors determine body weight and the amount of air that is displaced by the body.

Advantages: This method provides a quick, comfortable, automated, and reasonably accurate way to assess body fat content. It is suitable for disabled individuals, the elderly, and children.

Limitations: Results can be modified by drinking, eating, or exercising before testing; a full bladder or failure to adhere to test procedures may lead to error in results. The test is costly but less expensive than DEXA. A BOD POD machine costs around $36,000.

What Causes Obesity?

Simply stated, obesity results when the intake of calories exceeds caloric expenditure. But the cause of obesity is not that simple. Whether people accumulate excess body fat or not is due to complex and interacting factors that include

- diet
- physical activity
- environmental exposures
- genetic background[29]

Some medications such as antipsychotics, antidepressants, insulin, and beta-blockers used for hypertension are also associated with the development of obesity.[30] Their contribution to the overall incidence of obesity is small (although meaningful to the affected people).

Are Some People Born to Be Obese?

Accumulating evidence indicates that genetics plays important roles in the development and maintenance of obesity. Heredity may account for approximately 50% of the risk for developing obesity. Genetic influences on obesity take several paths. Some people are born with errors in metabolism that produce obesity (this appears to be rare). Others are born with multiple genetic traits that predispose them to becoming obese (probably common). A predisposing genetic trait is expressed when the right *environmental trigger* exists. For example, a person with a genetic predisposition to becoming obese may maintain normal weight as long as she or he is physically active or consumes a low-fat diet. If activity level becomes low or the diet changes to one high in fat (two types of environmental triggers), the genetically susceptible person then gains weight. Genetically based differences in biological processes affecting food intake and energy utilization appear to influence a person's susceptibility to obesity. Individuals lacking leptin, a hormone involved in appetite regulation, or who secrete high levels of insulin may be at high risk for obesity.[31]

environmental trigger
An environmental factor, such as inactivity, a high-fat diet, or a high sodium intake, that causes a genetic tendency toward a disorder to be expressed.

The propensity to gain weight when excessive levels of calories are consumed may be affected by genetic traits. Claude Bouchard and his colleagues at Lavel University in Canada fed adult identical twins 840 more calories a day than each twin needed to maintain his or her weight.[32] At the end of the 100-day feeding period, weight gain among the twins ranged from 9 to 29 pounds! (The average weight gain was 24 pounds.) Related twins tended to gain similar amounts of weight, but the weight gains varied widely even though all subjects consumed the same number of excess calories. Another study showed that identical twins reared apart are less likely to have similar body weights than identical twins who are reared together.[33]

Other Potential Genetic Influences   Other hereditary factors have also been theorized to be related to obesity. The "set-point theory," for example, suggests that individuals are programmed to weigh a certain amount and that body weight will return to that level after weight is lost or gained. It has also been proposed that the number and size of fat cells predispose some people to obesity. Neither of these two popular theories has been verified by research.[29]

Recent increases in the incidence of obesity in the United States and other developed countries indicate that genetic influences in combination with environmental factors, such as inactivity, high-calorie diets, and a continuous supply of large servings of highly palatable foods, play key roles in the development of obesity.[34] We are just beginning to understand the nature of the interactions between environmental factors and genes. The day will come, however, when people with inborn tendencies toward obesity can be identified by an examination of their genes.

Do Obese Children Become Obese Adults?

The link between early and later obesity is weak for young children under the age of three years.[16] It becomes stronger among older children, especially if one or both parents are obese. Only 8% of obese children who are heavy at one to two years of age and who do not have an obese parent are obese as adults. However, nearly 80% of children who are obese between the ages of 10 and 14 and have at least one obese parent are obese as adults.[35]

The Role of Diet in the Development of Obesity

Regardless of the cause of obesity, weight gain results when more energy is consumed than expended. Americans are consuming more energy than in previous years: Caloric intake is up by an average of 340 per day while physical activity level has remained about the same. Although Americans are consuming a lower proportion of total calories from fat, total fat intake is up due to increased caloric intake. Fruits, vegetables, and fiber are often missing from diets that provide too many high-calorie foods.[34]

The generous availability of inexpensive, calorie-dense foods; eating out regularly at fast-food restaurants, and large portions of favorite foods tend to increase caloric intake. Adults served large portions of food consume an average of 30 to 50% more food than when presented with small food portions.[36] When offered a lot of tasty food, people tend to eat beyond the feeling of satiety and past the point where food continues to taste good.[37]

Low Levels of Physical Activity Promote Obesity Low levels of physical activity are related to the high and increasing incidence of obesity among Americans.[38] To a large extent, physical activity has become voluntary. Many farmers now plow, sow, and reap in air-conditioned tractors with power steering. Lawn mowers propel themselves and sometimes their operators, too. Instead of walking or biking two and a half miles to the store and back, we drive. An average-size adult driving 2.5 miles burns about 17 calories. Biking that distance would use seven times more calories (122). But walking 2.5 miles would burn around 210 calories, more than 12 times the calories it takes to drive! We have traded physical activity for convenience, time, and, perhaps, personal fat stores. Too much television watching, in particular, has been blamed for obesity in children and adults; but it is not clear which comes first—obesity, sedentary lifestyles, or too much TV.

Obesity: The Future Lies in Its Prevention

Whether obesity is related to genetic predisposition, environmental factors, or a combination of both, certain steps can be taken to help prevent it.

Preventing Obesity in Children

For children, the prevention of obesity includes the early development of healthy eating and activity habits. Parents should offer a nutritious selection of food, but children themselves should be allowed to decide how much they eat.[55] Physical activities that are fun for every child—not just those who show athletic promise—should be routine in schools and summer programs.

Interactions between parents and children around eating and body weight can set the stage for the prevention or the promotion of overweight in children. Parents who overreact to a child's weight by focusing on it, restricting food access, and making negative comments to the child may increase the likelihood that eating and weight problems will develop or endure. Lifestyle changes for the whole family—such as incorporating fun physical activities into daily schedules; making a wide assortment of nutritious foods available in the home; and decreasing a focus on eating, foods, and weight—are some of the positive changes families can make to promote healthy eating and exercise habits, and normal weight in children.[39]

Preventing Obesity in Adults

Action needs to be taken to prevent weight gain during the adult years. Many adults gain weight at a slow pace (about a pound per year) as they age, whereas others gain substantial amounts of weight over short periods of time.[40] Data from a national nutrition and health survey indicate that major gains in weight are most likely to occur in adults between the ages of 25 and 34 years.[41] Regular, vigorous exercise may prevent or lessen the amount of weight gain that occurs with age, as may decreased portion sizes at home and in restaurants.[37] Paying attention to the "I'm hungry" and "I'm full" feelings can help moderate food intake.[42] For more on this topic, visit the "Reality Check" on page 9-15.

Regular intake of 2 to 4 servings per day of milk and milk products and getting enough sleep may also help prevent weight gain. A number of studies have shown that people who routinely consume milk, yogurt, and other dairy products weigh less and have less body fat than people who don't.[43] People who sleep 6 hours or less a night are more likely to be overweight than those who get 8 hours or more.[44]

Some People Are Underweight

underweight
Usually defined as a low weight-for-height. May also represent a deficit of body fat.

Illustration 9.10
Some underweight people are genetically thin and are healthy.

Photo Researchers, Richard Hucthings

Worldwide, underweight due to inadequate diets is much more common than obesity and a good deal more life-threatening.[45] There are many preventable reasons why people fail to get enough to eat. The lows of body weight discussed here relate to underweight as it occurs in many economically developed nations.

In contrast to the desperate situations faced by many people in economically emerging countries, underweight in developed nations largely results from illnesses such as HIV/AIDS, pneumonia, and cancer; an eating disorder (anorexia nervosa); or the voluntary restriction of food. An important and preventable cause of underweight in the United States and some economically developed nations is poverty.

Underweight Defined

People who are ***underweight*** have too little body fat, or less than 12% body fat in adult females and 5% body fat in males.[27] They have BMIs below 18.5 kg/m^2 as indicated in Table 9.1. A portion of the 2% of U.S. adults classified as underweight by BMI will not actually be underweight, just as a subset of people categorized as obese by BMI are not really obese.[12]

Some people assessed as underweight for height are healthy and have a healthy body composition. Like the person in Illustration 9.10, they are probably genetically thin. People who are naturally thin often have as much difficulty gaining weight as obese people have losing it.[46] People who are thin and unhealthy are more likely than others to experience apathy, fatigue, and illnesses frequently, and they take longer to recover from illness. They tend to have reduced bone mineral density and more bone fractures, be intolerant of cold temperatures, and have impaired concentration.[41,46]

The "I'm full" feeling

How do you know when you've had enough to eat?
Answers on next page

Carlos
I'm on automatic pilot for eating until my plate is clean.

Sandra
I stop eating when, like boom!—the "I'm full" signal hits.

Underweight and Longevity in Adults

Longevity can be extended in adult mice and monkeys by feeding them a nutritious, calorie-restricted diet that produces underweight.[47] Although it is not known whether caloric restriction and underweight would serve as a fountain of youth for humans, a growing number of adults believe it will. Devotees of calorie restriction for longer life tightly control their food intake and activity level to maintain a lean body. Calorie restrictors tend to consume 1,600 to 1,700 calories per day, emphasize nutrient-rich foods, and exercise regularly.[48] They in general appear thin but healthy (see Illustration 9.11).

Adults following calorie-restricted diets tend not to have chronic inflammation nor the health problems associated with it. They are at risk of iron deficiency, osteoporosis, infertility, infections, and of becoming irritable. The theory that life expectancy would be increased by calorie restriction is challenged by data on humans that show increased disease and death rates among underweight compared to normal weight people.[47] Several long-term studies of health outcomes related to calorie restriction in men are now under way. Relatively few women follow this diet and lifestyle; little is known about its effects on their health.

Toward a Realistic View of Body Weight

A widespread belief among Americans is that individuals can achieve any body weight or shape they desire if they just diet and exercise enough. It's a myth. People naturally come in different weights and shapes, and these can only be modified so much (Illustration 9.12).[49] Half of all women in the United States wear sizes 14 to 26, yet many clothing models are very underweight. Many men, no matter how hard they work out, will never have a washboard stomach or fit into "slim jeans."

Size Acceptance

The U.S. obsession with body weight and shape is spreading to other industrialized countries as part of popular culture. Ironically, strong societal bias against certain body sizes may contribute broadly to weight and health problems. Intolerance of overweight and obese children and

Illustration 9.11
The appearance of one man who follows a calorie-restricted diet.

Illustration 9.12
People come in many different sizes and shapes.

adults tends to increase discrimination against them. This type of societal bias lowers the individual's feeling of self-worth and may promote eating disorders, including the consumption of too much food. Females are hardest hit by negative attitudes about body size. Although the incidence of overweight and obesity tends to be higher in males, obesity in females carries with it many more negative stereotypes.[50] Overreactive parents make things worse.

Of all the things societies can do to prevent obesity, acceptance of people of different sizes and a more realistic view of obtainable body weights and shape may be two of the most important.

The Health at Every Size Program   A kinder and effective approach to health improvement in obese people has been developed. Called "Health at Every Size," the program is gaining acceptance among consumers and health professionals in the U.S. and Canada.[51] The program emphasizes eating when hungry and stopping when full; enjoyable, life-enhancing physical activities; body size acceptance; healthy eating behaviors; a peaceful relationship with food; self-esteem; and social support networks. The program reduces a number of health problems related to obesity even though it does not lead to weight loss in the short term. Health at Every Size program participation is associated with reduced blood pressure and LDL-cholesterol, increased HDL-cholesterol levels, improved self-esteem, decreased disordered eating behaviors, and improved body image.[52] The Health at Every Size philosophy and program components could help future generations of children, men, and women achieve healthful eating patterns and high levels of well-being and quality of life, as well as improved long-term health.[53]

ANSWERS TO
Reality Check

The "I'm full" feeling

Even favorite foods lose some or all of their appeal after we're full. The "I'm full" feeling will let you know when that happens—if you pay attention.[60]

Carlos

Sandra:

Nutrition Up Close

Are You an Apple?

Focal Point: Determining your waist circumference.

Follow the directions given in Illustration 9.9 for determining your waist circumference. If you don't have a measuring tape, use a string. Mark the string where the two ends intersect when placed around your waist. Then use a ruler to measure the distance between the end and the mark on the string.

FEEDBACK can be found at the end of Unit 9.

Key Terms

body mass index (BMI), page 9-3
chronic inflammation, page 9-7
environmental trigger, page 9-12
insulin resistance, page 9-8
metabolic syndrome, page 9-8

metabolism, page 9-7
obese, page 9-5
obesity, page 9-5
overweight, page 9-5
subcutaneous fat, page 9-7

type 2 diabetes, page 9-8
underweight, page 9-14
visceral fat, page 9-7

Review Questions

	True	False
1. Death rates are lowest among adults who remain underweight as they age.	☐	☐
2. According to the Hamwi method for estimating normal weight in adults, a 6'1" male should weigh approximately 204 pounds.	☐	☐
3. A child with a body mass index higher than the 95th percentile on the CDC growth charts is considered obese.	☐	☐
4. The rate of overweight and obesity in the United States is higher than that reported for any other country.	☐	☐
5. Excess visceral fat poses higher risks to health than excess subcutaneous fat.	☐	☐
6. Key features of metabolic syndrome include excess abdominal fat and insulin resistance.	☐	☐

	True	False
7. Women but not men require a minimal amount of body fat stores for health.	☐	☐
8. If undertaken correctly, and if appropriate formulas are used, underwater weighing produces accurate values of a person's percent body fat.	☐	☐
9. The use of medications that cause weight gain is a leading reason for the obesity epidemic in the United States.	☐	☐
10. Vegetables, fruits, and fiber are often consumed in low amounts by people who regularly eat high-calorie foods.	☐	☐
11. Adults tends to eat 30 to 50% more food when given large portions of food than when given smaller portions.	☐	☐
12. Calorie-restricted diets have been shown to prolong life in mice, monkeys, and humans.	☐	☐

Media Menu

www.cdc.gov/nchs/fastats/overwt.htm
Rates of overweight and obesity are increasing fast enough to make it difficult to keep textbooks updated between revisions. For the latest statistics, go to this site.

www.nhlbi.nih.gov/subsites/index.htm
This site provides information on how to "Aim for Healthy Weight." It includes interactive BMI calculator, menu planner, information on portion sizes, and other applications.

www.naafa.org
The National Association to Advance Fat Acceptance is dedicated to improving the

quality of life for fat people. The organization takes on issues of size discrimination and policies against the interest of fat people, and it promotes fat acceptance through media and legislative routes.

www.sizewise.com
This site provides information, inspirational stories, updates, and resources related to size acceptance.

www.healthfinder.gov
Search "obesity" to find clinical guidelines for diagnosis and treatment, the role of genes in the development of obesity,

national efforts aimed to reduce obesity, and more.

www.shapeup.org
This site offers information, advice, and products for weight management. Provides a BMI calculator for kids and information on "Diabesity."

www.cdc.gov/growthcharts
The 2000 growth charts for children and adolescents 2 to 20 years old, including BMI-for-age graphs, are available at this site.

Notes

1. Lean MJ, Han TS. Waist worries. Am J Clin Nutr 2002;76:699–70.

2. Brown PJ, Konner M. An anthropological perspective on obesity. Annals of NY Acad Sci 1987;499:29–46.

3. Van Itallie TB. Obesity: the American disease. Food Technol 1979;(Dec):43–47.

4. Rossner S. Ideal body weight—for whom? (editorial). Acta Medica Scand 1984;216:241–2.

5. Fallon A, Rozin P. Sex differences in perception of desirable body shape. J Abnormal Psychology 1985;94:102–5.

6. Stevens J et al. Evaluation of WHO and NHANES II standards for overweight using mortality rates. J Am Diet Assoc 2000;100:825–7.

7. Roth JH. What is optimum body weight? (letter) J Am Diet Assoc 1995;95:856–7.

8. Defining overweight and obesity, www.cdc.gov/nccdphp/dnpa/obesity/defining.htm, accessed 7/03.

9. Fernández JR et al. Is percentage body fat differentially related to body mass index? Am J Clin Nutr 2003;77:71–5.

10. Defining overweight and obesity (www.cdc.gov/nccdphp/dnpa/obesity/defining.htm).

11. CDC growth charts: United States. Advance Data, No. 314, 12/4/00 (Revised), www.cdc.gov/nchs.

12. State-specific prevalence of obesity among adults—United States, 2005. Centers for Disease Control and Prevention, MMWR. 2006;55:985–988.

13. European nations failing obese patients—survey. Survey results from health professionals at the 12th European Congress in Obesity in Helsinki, www.medscape.com, accessed 6/13/03.

14. National Center for Health Statistics. Faststats A to Z. http://www.cdc.gov/nchs/fastats/overwt.htm, accessed 8/06.

15. Global rates of obesity. International Obesity Task Force, http://www.iotf.org/media/globalprev.htm, prevalence rates updated 3/05, site accessed 8/06.

16. Bray GA. Physiology and consequences of obesity. Medical Education Collaborative, Diabetes and endocrinology clinical management, www.medscape.com/Medscape/endocrinology/Clinical Mgmt/Cm.v03/public/index.CM.v03.html, accessed 1/8/01.

17. Broom I. Thinking about abdominal obesity and cardiovascular risk. Br J Vasc Dis 2006;6:58–61.

18. Calle EE et al., Body mass index and mortality in a prospective cohort of U.S. adults. N Engl J Med 1999;341:1097–1105.

19. Peeters A et al., Obesity in adulthood and its consequences for life expectancy. Ann Int Med 2003;138:24–32.

20. Weinbrenner T et al. Circulating oxidized LDL is associated with increased waist circumference in women. Am J Clin Nutr 2006;83:30–5.

21. Harrard WR. Weight control and exercise. JAMA 1995;274:1964–5.

22. Price J et al. Family practice physicians' beliefs, attitudes and practices regarding obesity, Am J Prev Med 1987;3:339–45.

23. Kornman KS. Interleukin 1 genetics, inflammatory mechanisms, and nutrigenetic opportunities to modulate diseases of aging. Am J Clin Nutr 2006;83(suppl):475S–83S.

24. Wang Y et al. Comparison of abdominal adiposity and overall obesity in predicting risk of type 2 diabetes among men. Am J Clin Nutr 2005;81:555–63.

25. Farin HMK et al. Body mass index and waist circumference both contribute to differences in insulin-mediated glucose disposal in nondiabetic adults. Am J Clin Nutr 2006;83:47–51.

26. Adams R. Gut check: the NFL watches its weight. New York Times, Oct. 8–9, 2005; p. P3.

27. Appendix H, Table H-1. Body measurement summary statistics. In: Dietary Reference Intakes for Energy through Amino Acids, Washington, DC: The National Academies Press, 1999.

28. Gibson RS, Principles of nutritional assessment, New York: Oxford University Press; 1990.

29. Purness JQ, Obesity, www.medscape.com/viewarticle/45726, accessed 6/03.

30. Malone M et al. Medication associated with weight gain may influence outcome in weight management program. Ann pharmacotherapy 2005;39:1204–13.

31. Lyon HN et al. Grnetics of common forms of obesity: a brief overview. Am J Clin Nutr 2005;82(suppl):215S–7S. Eisenberg D et al. Update on obesity surgery, World J Gastroenterol 2006;12:3196–3203.

32. Bouchard C et al. The response to long-term overfeeding in identical twins. N Engl J Med 1990;322:1477–82.

33. Macdonald A, Stunkard AJ. Body-mass indexes of British separated twins (letter). N Engl J Med 1990;322:1530.

34. Ledikwe JH et al. Dietary energy density is associated with energy intake and weight status in US adults. Am J Clin Nutr 2006;83:1362–8.

35. Whitaker JA et al. Predicting obesity in young adulthood from childhood parental obesity. N Engl J Med 1997; 337:869–73.

36. Rolls BJ et al. Larger portion sizes lead to sustained increases in energy intake over 2 days. J Am Diet Assoc 2006;106:543–9.

37. Rolls BJ. The super-sizing of America. Nutr Today 2003;38:42–53.

38. McCrory MA et al., Biobehavioral influences on energy intake and adult weight gain, J Nutr 2002;132:383OS–34S.

39. Position of the American Dietetic Association: Individual-, family-, school-, and community-based interventions for pediatric overweight. J Am Diet Assoc 2006;106:925–45.

40. Jeffery RW et al, Prevalence and correlates of large weight gains and losses in adults, Int J Obes 2002;26:969–72.

41. Costanzo PR, Schiffman SS. Thinness—not obesity—has a genetic component. Neuroscience and Biobehavioral Reviews 1989;13:55–58.

42. Mello MM et al. Obesity—the new frontier of Public Health law. N Engl J Med 2006;354:2601–10.

43. Teegarden D. The influence of dairy product intake on body composition. J Nut 2005;135:2749–52.

44. Lie D. Getting more sleep may reduce weight gain. Medscape Medical News, www.medscape.com/viewarticle/495410, assessed 1/05.

45. Onis M et al. Prevalence and trends of overweight among preschool children in developing countries. Am J Clin Nutr 2000;72:1032–9.

46. Bouchard C. Heredity and the path to overweight and obesity. Med Sci Sports 1991;23:285–91.

47. Shapses SA, Riedt CS. Bone, body weight, and weight reduction: What are the concerns? J Nutr 2006;136:1453–6.

48. Meyer TE t al. Long-term calorie restriction ameliorates the decline in diastolic functions in humans. J Am Coll Cardio 2006:47:398–402.

49. Satter EM. Internal regulation and the evaluation of normal growth as the basis for prevention of obesity in children. J Am Diet Assoc 1996;96:860–4.

50. UK panel calls for media to end pressure on girls to be thin, Report of the UK's Cabinet's Body Image Summit, Reuter Health 2000; June 22.

51. Chapman GE et al. Canadian dietitians' approaches to counseling adult clients seeking weight-management advice. J Am Diet Assoc, 2005;105:1275–9.

52. Bacon L et al. Size acceptance and intuitive eating improve health for obese, female chronic dieters. J Am Diet Assoc 2005;105:929–36.

53. Robinson J. Health at every size: toward a new paradigm of weight and health. www.medscape.com-viewarticle/506299, accessed 9/2/05.

54. Scarlett JM et al. Overweight cats: prevalence and risk factors. Int J Obes Relat Metab Disord. 1994;18 Suppl 1: S22–8.

Answers
to Review Questions

1. False, see page 9-3.
2. False (it's 184 pounds), see page 9-3.
3. False, see page 9-4.
4. False, see Table 9.4, page 9-6.
5. True, see page 9-7.
6. True, see page 9-8.
7. False, see page 9-9.
8. True, see page 9-10.
9. False, see page 9-12.
10. True, see page 9-13.
11. True, see page 9-13.
12. False, see page 9-15.

Nutrition Up Close

Are You an Apple?

Feedback for Unit 9

If your waist circumference is over 35" (88 cm) and you're a female, and over 40" (120 cm) if male, you're an apple. (You are all a peach for doing this exercise.)

Weight Control:
The Myths and
Realities

NUTRITION SCOREBOARD

	TRUE	FALSE
1 Anybody who really wants to can lose weight and keep it off.		
2 Weight loss is the cure for obesity.		
3 Weight loss can be accomplished using many types of popular diets. Such diets rarely help people maintain the weight loss.		
4 Weight-loss products and services must be shown to be safe and effective before they can be marketed.		
5 Small and acceptable improvements in eating and exercise behaviors are more likely to produce weight loss and weight loss maintenance than are large and unpleasant changes in behaviors.		

Key Concepts and Facts

- The effectiveness of weight-control methods should be gauged by their ability to prevent weight regain.

- That a weight-loss product or service is widely publicized and utilized doesn't mean it works.

- Successful weight control is characterized by gradual weight loss from small, acceptable, and individualized changes in eating and activity.

Answers to NUTRITION SCOREBOARD

		TRUE	FALSE
1	If this claim were true, hardly anybody would be obese.		✔
2	Maintenance of weight loss is the cure for obesity.		✔
3	Many popular diets can lead to weight loss, but none successfully help people prevent weight regain.[1]	✔	
4	Unfortunately, many weight-loss products and services on the market have not been shown to be safe or effective. Laws and regulations do not fully protect the consumer from the introduction of bogus products and services.		✔
5	Small and acceptable behavioral changes are easier to live with over time than are drastic and disliked changes.[1]	✔	

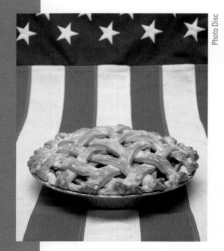

Photo Disc

Baseball, Hot Dogs, Apple Pie, and Weight Control

Americans are preoccupied with their weight. On any given day, 40% of adults, and a majority of individuals who are overweight or obese, are trying to lose weight. To help get the weight off, Americans spend over *$33 billion* annually (an average of approximately $120 per person each year) on weight-loss products and services. Yet Americans are gaining weight faster than they are losing it, and the incidence of obesity in the United States is on the rise.[2] Roughly 5 to 10% of people who lose weight keep it off.[3] Consumers are paying handsomely for weight loss without experiencing the desired cosmetic changes or health benefits that come when weight loss is maintained.

Why do so many Americans fail at weight control? For some people, achieving permanent weight reduction on their own may be truly impossible.[1] For others, the problem is the ineffective methods employed, not the people who use them.

Weight Loss versus Weight Control

"Jess—did I hear you say you wanted to lose some weight? It happens I know about a terrific diet! A couple of months ago I went on Dr. Quick's Amino Acid Diet and lost 15 pounds! You eat nothing but fish, papaya, and broccoli, and the pounds just melt away!"

Jess was shocked. Selma looked heavier than she did last year when they were in class together.

"But Selma," Jess inquired, "what happened to the weight you lost?"

"Oh, I gained it back. I ran out of willpower and started eating everything in sight. I'm going to get back on track, though. Tomorrow I start the Slim Chance Diet."

Dietary torture and weight loss are not the cure for overweight. If losing weight was all it took to achieve the cultural ideal of thinness, nobody would be overweight. Any popularized approach to weight loss (and some get pretty spectacular) that calls for a reduction in caloric intake can produce weight loss in the short run. These methods fail in the long run, however, because they become too unpleasant. Feelings of hunger, deprivation, and depression that often occur while on a weight-loss diet eventually lead to a breakdown of control, eating binges, a return to previous habits, and weight regain.[4] Humans are creatures of pleasure and not pain. Any painful approach to weight control is bound to fail. Improved and enjoyable eating and exercise habits are needed to keep excess weight off, and quick weight-loss approaches don't change habits.

Unfortunately, many dieters think weight-control methods are successful if they lead to a rapid loss of weight.[5] When the lost weight is regained, dieters tend to blame themselves and not the faulty method. Blaming themselves for the failure, many people are ready to try other quick weight-loss methods. But they usually fail, too.[6]

The Business of Weight Loss   More than 29,000 weight-loss products and services are available. Some of these are shown in Illustration 10.1. Most of them either don't work at all or don't prevent weight regain. "Quick-fix" weight-

Illustration 10.1
There is no lack of weight-loss books and products.
There is a lack of popular approaches that help people keep weight off.

Do some foods have negative calories?

You're looking for a way to shed 10 pounds you gained during freshman year and are seriously thinking about adding grapefruit and vinegar to your diet. You have heard these foods have "negative calories" because they make the body burn fat.

Answers on next page

Juanita:
I know vinegar cleans the grease off windows. Maybe it will melt away my fat, too.

Chuck:
Vinegar and grapefruit are so sour—I bet they rev up the metabolism so you burn fat.

loss approaches that don't lead to long-term changes in behavior are the primary reason approximately 90 to 95% of people who lose weight gain it back. The demand for such products and services is so great, however, that many are successful—financially.

One reason so many weight-loss products and services are available is that almost none of them work. If any widely advertised approach helped people lose weight and keep it off, manufacturers of bogus methods would go out of business. The weight-loss industry also thrives because of the social pressure to be thin. Many people try new weight-loss methods even though they sound strange or too good to be true. Often people believe that a product or service must be effective, or it wouldn't be allowed on the market. Although reasonable, this belief is incorrect.

The Lack of Consumer Protection The truth is that general societal standards for consumer protection do not apply to the weight-loss industry.[6] No laws require a product to be effective: In most cases, companies do not have to show that weight-loss products or services actually work before they can be sold. That's why products like herbal remedies, forks with stop and go lights, weight-loss skin patches, colored "weight-loss" glasses, electric cellulite dissolvers, mud and plastic wraps, and inflatable pressure pants are available for sale. These, like many other weight-loss products, do not work, and there is no reason why they should. Promotions for the products just have to make it *seem* as though they might work wonders on appetite or body fat.

Furthermore, weight-loss products and services are usually not tested for safety before they reach the market. The history of the weight-loss industry is littered with abject failures: fiber pills that can cause obstructions in the digestive tract, very-low-calorie diets and intestinal bypass surgeries that lead to nutrient-deficiency diseases and serious health problems, amphetamines that can produce physical addiction, diet pills that cause heart valve problems, and liquid protein diets that have led to heart problems and death from heart failure.[10] A brief history of weight-loss method failures is chronicled in Table 10.1.

Pulling the Rug Out Fraudulent weight-loss products and services may be investigated and taken off the market. Currently, the Federal Trade Commission (FTC) monitors deceptive practices and weight-loss claims on a case-by-case basis. It has filed suits against companies making exaggerated claims, and has identified the most common dubious claims made by the industry (Table 10.2). The

TABLE 10.1

A brief history of discontinued weight-loss methods

Year	Method	Reason for Discontinuation
1940s-1960s	Amphetamines	Highly addictive, heart and blood pressure problems
	Vibrating machines	Did not work
1970s	Jejunoileal; bypass surgery	Often caused chronic diarrhea, vitamin and mineral deficiencies, kidney stones, liver failure, arthritis
	Liquid protein diet	Poor-quality protein caused heart failure, deaths
1980s	Intestinal bypass surgery	Excessive risk of serious health problems
1990	Oprah Winfrey liquid diet success (she lost 67 pounds)	
1991	Oprah Winfrey gains 67 pounds and declares "No more diets!"	
1997	Phen-Fen (Redux)	Heart valve defects, hypertension in lung vessels
2004	Ephedra	Excessive risk of stroke, heart attack, and psychiatric illnesses

FTC performs most investigations in response to consumer complaints. Illustration 10.2 shows three examples of products that were taken off the market and explains why.

Requirements for truth in labeling keep many bogus products from including false or misleading information on their labels. These laws, however, do not keep outrageous claims from being made on television or printed in pamphlets, books, magazines, and advertisements. Few companies get caught if they include false or misleading information in weight-loss product ads.

What if the Truth Had to Be Told?   Suppose the weight-loss industry had to inform consumers about the results of scientific tests of the effectiveness and safety of weight-loss products and services. What if this information had to be routinely included on weight-loss product labels (Illustration 10.3)? What do you think would happen to consumer choices and the weight-loss industry? Increased federal enforcement of truth-in-advertising laws may help put an element of honesty into promotions for many weight-loss products. In addition, the FTC has proposed that claims about long-term weight loss "must be based on the experience of patients followed for at least two years after they complete the [weight-loss] program."[12] If this proposal ever becomes law, it will change the weight-loss industry in the United States.

Diet Pills   Two types of prescription diet pills are currently approved for long-term use in obese people in the United States. One is Meridia (sibutramine), and

TABLE 10.2

Loads of false and misleading weight-loss advertisements appear in the media. Here is a list of the top six features of weight-loss ads that make false or misleading claims.[11]

1. Use testimonials, before-and-after photos.

2. Promise rapid weight loss.

3. Require no special diet or exercise.

4. Guarantee long-term weight loss.

5. Include a "clinically proven" or "doctor approved" statement.

6. Make a "safe," "natural," or "easy" claim.

ANSWERS TO
Reality Check

Do some foods have negative calories?

The vinegar, grapefruit, negative calorie myth lives on because it strikes many people as reasonable. Neither vinegar nor grapefruit, nor any other food has negative calories or causes the body to gear up metabolism and burn fat.[9]

Juanita

Chuck

Stan Maddock

Fat magnet pills were purported to break into thousands of magnetic particles once swallowed. When loaded with fat, the particles simply flushed themselves out of the body. The FTC found the product too hard to swallow. The company took the product off the market and made $750,000 available for customer refunds.

"Blast" away 49 pounds in less than a month with Slim Again, Absorbit-All, and Absorbit-AllPlus pills, claimed ads in magazines, newspapers, and on the Internet. The company got blasted by the FTC to the tune of $8 million for making fraudulent claims.

Take a few drops of herbal liquid, put them on a bandage patch, and voilà—a new weight-loss product! The herbal liquid was supposed to reach the appetite center of the brain and turn the appetite off. Federal marshals weren't impressed. They seized $22 million worth of patch kits and banned the sale of others.

Illustration 10.2
A few examples of bogus weight-loss products that were removed from the market.

Illustration 10.3
Just imagine what would happen if weight-loss approaches were required to divulge their effectiveness and risks.

BEE POLLEN-GRAPEFRUIT DIET

The Amazing
BEE POLLEN-GRAPEFRUIT DIET
Dr. Quickfix

MELTS AWAY POUNDS OF CELLULITE IN JUST 10 DAYS!

Consumer Protection Label
Effectiveness: Ineffective for 95% of users
Risks: Bee pollen may cause adverse reactions in allergic people. Weight regain may exceed weight loss. If the diet fails, people may blame themselves, thereby precipitating anxiety and self-doubt.

the other is Xenical (orlistat).[13] Meridia works by enhancing satiety (which reduces food intake), and Xenical works by partially blocking fat absorption in the intestines. Both have side effects. Meridia may increase blood pressure and heart rate, cause headaches, and lead to dry mouth. Xenical's primary side effect is oily stools, which are particularly bothersome if high-fat meals are consumed. Malabsorption of fat caused by Xenical reduces absorption of a number of fat-soluble nutrients such as vitamin D, E, and beta-carotene.[14]

It is recommended that Meridia and Xenical be used in conjunction with reduced-calorie diets and exercise. Both effectively lower weight by an average of about 10% over a period of six months to a year. Weight regain after pill use stops is common.[14]

Many other diet drugs are under development or testing. Much activity currently centers around vaccines for obesity prevention, and hormone treatments that modify the sensations of hunger or satiety. Even though people using prescribed diet drugs are carefully screened and monitored by medical professionals, serious side effects can develop. No diet drug is absolutely safe, and none are known to cure obesity forever.[14]

Popular Diets

There is a never-ending supply of diet books on the market. More than 1000 of them are available at any one time. Although the safety and effectiveness of almost all of the diets described in popular weight loss books have not been evaluated, a few have been selected for examination due to their popularity. Five popular diets that have undergone scrutiny by researchers are Dr. Atkins's New Diet Revolution, Weight Watches Pure Points Program, the Slim-Fast Plan, the

Zone Diet, and Rosemary Conley's Eat-Yourself-Slim Diet (Illustration 10.4). Studies have compared weight loss, adherence to the diet, and health effects of these popular diets.[15-17]

Approaches to weight loss described in these books represent a mix of balanced, low-carbohydrate, high-protein diets (see Illustration 10.5); and low-fat diets. All of the diets were found to produce a reduction in calorie intake if followed and weight loss. Weight loss varied by how long a person stayed on the diet and averaged about 7 pounds. Most of the people in the studies were unable to stick to the diet for a year, primarily because it was too hard to follow.[15,16]

Health Effects of Popular Diets

A number of improvements in health status have been reported for individuals who lost weight by following the advice given in popular diet books. Body fat is reduced as part of the weight lost. Approximately 80% of the weight lost consists of fat, and the rest is lean body tissue such as muscle. Blood levels of insulin, C-reactive protein (a marker of chronic inflammation), and total cholesterol go down, and levels of beneficial HDL cholesterol head up.[15,16] These are positive changes that benefit health if weight is not regained. But, as we know, weight is almost always regained after the diet ends.

Weight Regain: A Shared Characteristic of Popular Diets The behavioral changes required to sick to dietary and exercise prescriptions dictated by popular diets turn out to be too difficult and unpleasant for most people to follow in the long run. Differences in the amounts of weight lost for this or that diet and the health benefits achieved disappear over time as weight is regained. The situation has lead James Hill, one of the directors of the National Weight Control Registry, to conclude, "Our real challenge is not helping people lose weight but is helping them keep it off."[1]

There are indications that consumers are growing tired of popular, money-making approaches to weight loss. A survey undertaken by the food industry

The high-protein, low-carbohydrate diet for weight loss has been around for over a century. The first "Dr. Atkins" was William Banting, a London undertaker. He published a book on his terrific new diet in 1864. It was so popular that diet-ing to this day is referred to as "Banting" in England.[23]

Photo Disc

Susan Van Etten/PhotoEdit

Illustration 10.4
Examples of popular weight loss diets evaluated by research studies.

Illustration 10.5
An example meal from a low-carbohydrate, high-protein diet.

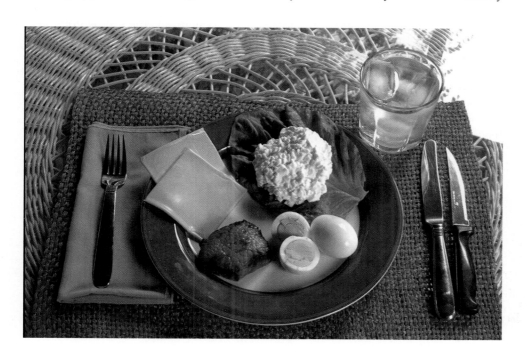

in the United Kingdom and the United States found that 29% and 44% of the adults studied were aware that extreme diets *cannot* produce sustained weight loss. Instead of popular weight-loss diets, these consumers are deciding that the path to weight control is paved by small, easy changes to their diets and physical activities.[18]

Organized Weight-Loss Programs

The mark of successful weight-loss programs is maintenance of weight loss in the long term. Whether organized weight-loss programs such as Weight Watchers, Jenny Craig, Overeaters Anonymous, or services provided by many medically supervised weight-loss clinics hit this mark is not known. Results of studies of the effectiveness of privately run weight-loss programs are rarely published. Programs requiring the largest changes in diet and lifestyle would be expected to have the lowest success rates in terms of weight-loss maintenance. No weight-loss product, diet, program, or clinic that substantially alters usual eating and activity patterns has been shown to result in a long-term loss of excess body fat in more than 5 to 10% of people who use it.[19]

Drastic Measures

Fasting and surgery represent extreme approaches to body fat reduction. These approaches are riskier than weight-loss methods that produce a gradual loss in weight.

Fasting Sometimes people fast to lose weight, but this method is not recommended unless the fast is closely supervised by medical professionals. Abruptly stopping food intake not only does not foster long-term weight-loss maintenance, but it produces a high level of wear and tear on the gastrointestinal tract. Fasting causes "intestinal starvation" because it removes a primary source of nutrients from the intestines—foods that are being digested. Without sufficient nutrients to maintain the gastrointestinal tract, mucus and cells lining the tract are lost, and infection is more likely. Far from "cleansing the system," as the saying goes, fasting can clog up the lower intestines. Fasting can also lead to a preoccupation with food and eating and a rapid rebound in weight once the fast ends.[20]

Weight-Loss Surgery Weight-loss surgery, referred to as *bariatric* surgery in the medical field, is a weight control method of last resort. Weight-control surgery is reserved for people with BMIs over 40 or for people with BMIs of 35 to 40 who have serious health problems related to their weight. Although several methods of weight-loss surgery exists, gastric bypass surgery is the most frequently performed worldwide.[21]

Gastric bypass surgery is the most effective method for weight loss and weight maintenance available.[21] On average, individuals undergoing this surgery weigh 30% less than they did before the surgery, and they often maintain the loss over the long term.[22] Health status generally improves dramatically as a result of the weight loss. Resolution of type 2 diabetes, hypertension, sleep disorders, and elevated LDL cholesterol blood levels often follow gastric bypass surgery.[21]

This type of bariatric surgery is becoming increasing popular. In 2005, 170,000 gastric bypass operations were performed in the United States. Some people who don't qualify are gaining weight to get past the BMI thresholds.[23]

In gastric bypass surgery, most of the stomach is stapled shut, leaving a pouch at the top that can hold about 2 tablespoons of food. A section of the small

---

ON THE SIDE

Newspaper headline in 2035: "Diet and exercise are the keys to weight loss."

—GRABBED OFF THE INTERNET

"The second day of a diet is always easier than the first. By the second day, you're off it."

—JACKIE GLEASON

"My doctor told me to stop having intimate dinners for four, unless there are three other people."

—ORSON WELLES

bariatrics
The field of medicine concerned with weight loss.

intestine that connects to the bottom of the stomach is cut off and attached as a drain for the stomach pouch. The loose end of the small intestine is then reattached to the section of the small intestine that leads from the pouch. This way, digestive juices from the stomach and the upper part of the small intestine are made available for digesting food. People who have their stomachs stapled can eat only a small amount of food. If they eat too much, they feel nauseous and dizzy and vomit.

■ **Concerns Related to Gastric Bypass Surgery** Gastric bypass surgery is not a panacea for weight control. It comes at a high cost ($25,000 to $30,000) and is accompanied by high rates of postsurgery complications. There is a 2% chance of dying from this operation, but that risk is considered worth it because obese people who receive the surgery live significantly longer than do those who stay obese. Rates of complications from the surgery are 22% during the postsurgery hospital stay and 40% within the 6 months that follow surgery. Readmissions for complications can increase costs associated with gastric bypass surgery to over $60,000.[24]

Some of the complications arising from the surgery are related to nutrient deficiencies. The small stomach and changes in absorption that results from the surgey leads to malabsorption of a number of vitamins and minerals, most notably vitamin B_{12}, folate, calcium, and iron. Multivitamin and mineral supplements are a routine component of care after bypass surgery.[22] Other complications related to adverse consequences of anesthesia, infection, nauses, vomiting, dehydration, and gallstones may also develop.[25]

Some amount of lost weight is usually regained in the years following gastric bypass surgery. The reason for the gain is increased volume of the stomach pouch. Increasing food intake can stretch people's stomachs. The 2-tablespoon pouch left after the surgery can be enlarged to hold 1/2 to 2/3 cup of food.[21] (A regular-sized stomach has a capacity of about four cups.)

About half of the individuals who undergo bypass surgery struggle with emotional issues around food when old habits begin to come back. People who undergo this surgery have to be committed to long-term lifestyle changes and follow-up care.[22] Because weight loss from bypass surgery averages 30% of initial body weight, most people will still be overweight.

■ **Body Contouring Surgery** Gastric bypass surgery leaves some people with folds of excess skin where fat stores were lost (see Illustration 10.7). Skin tissue previously stretched by high levels of fat stores does not retract on its own after fat stores are lowered. The only known way of removing it is "body contouring" surgery. The surgery may have to be repeated a number of times and can come at a cost of tens of thousands of dollars.

Liposuction At a cost of over $3000 per surgery (prices vary by fat deposit site, inflation, and surgeon), fat deposits in the thighs, hips, arms, back, or chin can be partially removed by liposuction. The procedure is the most common type of cosmetic surgery performed in the United States. Considered cosmetic, it is not intended for weight loss (Illustration 10.8). Surgical standards require that no more than 8 pounds of fat be removed by liposuction.[26] If a person gains a good deal of weight after liposuction, fat will be deposited to some extent in the breasts and

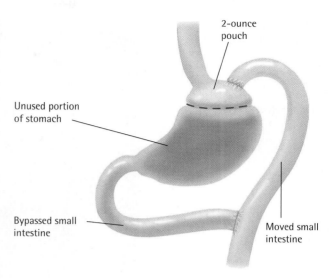

Illustration 10.6
Gastric bypass surgery.

Illustration 10.7
The photo on the left shows excess skin folds that developed after weight loss from gastric bypass surgery. The one on the right shows the results of body contouring surgery.

other areas not operated on. These deposits can lead to a return of an undesired body shape.[27] In addition, surgery always carries a risk of infection and other complications, so it cannot be taken lightly.

Weight Loss: Making It Last

An approach to weight loss can be considered successful only if it is safe, healthful, and prevents weight regain. Successful approaches focus on healthy eating and exercise for a lifetime, rather than on "dieting." People who lose weight and maintain their weight afterward have found physical activities they enjoy and ex-

Illustration 10.8
The effects of liposuction.
The photo on the left shows the thighs of a woman before liposuction, and the photo on the right shows the same woman's thighs after liposuction.

Illustration 10.9
Small, acceptable changes are the key to a successful weight-loss/weight-maintenance program. Try to find activities that you enjoy; get out there and play!

ercise regularly (Illustration 10.9). That is the major commonality among people who maintain their weight after weight loss.[1] Other characteristics of people who lose weight and keep it off and characteristics of people who regain lost weight are listed in the nearby Health Action.

Safe, healthful, and effective methods of weight loss and maintenance described here are not the subject of most popular diet books and weight loss programs. The approach that works in the long run depends on individual decisions about changes in diet and physical activiy that can be made and kept for the rest of one's life. Changes in diet and physical activity most likely to be maintained are small, easy to implement, and acceptable, even preferable, to existing behaviors.[1,28]

HEALTH *Action* Weight-Loss Maintainers versus Weight Regainers[1,29]

Weight-Loss Maintainers

- Exercise regularly.
- Make small and comfortable changes in diet and physical activity.
- Eat breakfast.
- Choose low-fat foods.
- Keep track of their weight, dietary intake, and physical activity level.

Weight Regainers

- Exercise little.
- Use popular diets.
- Make drastic and unpleasant changes in their diets and physical activity levels.
- Take diet pills.
- Cope with problems and stress by eating.

Small, Acceptable Changes

For most people, excess body fat accumulates slowly over time. Consuming just an extra 50 calories a day, for example, will lead to a gain of approximately 5 pounds in a year. (Approximately 3500 excess calories will produce a weight gain of 1 pound, and a 3500-calorie deficit will produce a loss of 1 pound in body weight in many people.) Excess weight is rarely all gained over the course of a few weeks or months. Fat is put on slowly, and that's the best way to take it off.

Gradual losses in body fat do not require dramatic changes in diet or activity level. Only small changes in diet and activity are needed. By cutting back food intake by 100 calories per day, a person could lose 10 pounds in a year. Table 10.3 provides examples of small changes in diet and physical actvity level that, if all

TABLE 10.3
Small changes in diet and physical activity worth about 100 calories[a]

Diet	Physical Activity
1. Consume 1 cup of fat-free yogurt with fruit instead of a cup of low-fat yogurt with fruit	1. Walk an additional 20 minutes.
2. Drink $1\frac{1}{2}$ cups of skim milk rather than $1\frac{1}{2}$ cups of whole milk.	2. Lift arm weights for 15 minutes.
3. Eat a roasted or grilled chicken sandwich rather than a fried chicken sandwich.	3. Garden for 15 minutes.
4. Eat $\frac{1}{2}$ cup rather than a cup of rice.	4. Play Frisbee for 30 minutes.
5. Order a regular rather than a large fish sandwich.	5. Clean house for 30 minutes.

[a]Physical activity calories are based on a 150-pound person.

HEALTH *Action* Small Changes Can Make a Big Difference[31, 32]

Eight rules for getting portion sizes right:

1. Listen carefully to your internal hunger and satiety cues and follow them.
2. Prepare less food for meals, or store extra food before eating.
3. Start with a small serving, then get more if you're still hungry.
4. Use small dishes, spoons, and glasses.
5. Slow down and enjoy every bit of food you eat.
6. Don't eat out of serving containers or packages. You won't know how much you're eating.
7. Divide portions given in resturants on your plate into food you will eat now and food you will take home, share, or not eat.
8. At fast-food restaurants, order small or regular-sized foods and beverages.

Some ideas on how to choose and prepare lower-energy dense foods:[33,34]

1. Use nonfat yogurt for making fruit smoothies.
2. Add more vegetables to soup and salads.
3. Add a generous helping of fruit to your cereal.
4. Add extra chopped celery, pickles, and sweet onion to your tuna and egg salad.
5. Enjoy thick slices of cucumbers on your sandwiches.
6. Serve yourself larger portions of vegetables and fruits at meals and keep portions of meat, fried foods, and high-calorie foods small.
7. Choose high-fiber foods like legumes and cooked, whole grains.
8. Load up your pizza with mushrooms, sliced tomatoes, spinach, onions, and broccoli.

else about a person's diet and physical activity remain the same, would produce about a 100-calorie deficit a day.

Identifying Small, Acceptable Changes   To identify changes in eating and activity that have staying power, first list the weak points in your diet and activity. Dietary weak points might include the consumption of high-fat foods due to eating out often, or relying on high-fat convenience foods that can be heated up in seconds. Another weak point might be skipping breakfast and overeating later in the day because of extreme hunger. Weak points in physical activity might include driving instead of walking, not engaging in sports, or spending too little time playing outside.

For each weak point, identify options that seem acceptable and enjoyable. A person who enjoys broiled chicken with barbecue sauce might not mind eating that at restaurants instead of fried chicken. That's a change people can make if they plan ahead. A person who gets too full from a large serving of fries might be happier ordering a small serving and not eating so many. A breakfast skipper might find grabbing a piece of fruit and a slice of cheese for breakfast acceptable and doable. People who enjoy walking may not mind leaving the car or bus behind and letting their feet carry them to class, the grocery store, or a friend's house.

Many acceptable options for making small improvements in diet and activity may be available (see the "Health Action—Small Changes Can Make a Big Difference"). The easiest changes to accomplish are the ones that should be incorporated into the overall lifestyle improvement plan. Some people, for example, lose weight and keep it off simply by consciously cutting down on portion sizes. Others avoid eating too much at any meal and walk more. Simply adding breakfast helps some people lose weight and maintain the loss.[30] Increasingly, people are losing weight and keeping it off by eating more energy-dense foods. This change doesn't require that people eat less food but rather select and prepare foods that are nutrient-dense rather than calorie-dense. The easier the changes are to follow, the more likely they are to succeed.

Individualized plans that don't work out often include unacceptable or unenjoyable changes. The changes may be too large or too different from the usual. In that case, go back to the drawing board and modify the plan to include small changes that are acceptable in the long run. Perhaps the original plan included jogging, but it turns out that jogging isn't any fun. In that case, take jogging out of the plan! Replace it by any other physical activity that would be enjoyed. Midcourse corrections should be expected. Some experimentation may be required to identify the small changes that will last.

What to Expect for Weight Loss   If you're happy with the changes and have found the right levels of calorie intake and physical activity, weight loss will be gradual but lasting. The pattern of loss should be somewhat like that graphed in Illustration 10.10, where the person lost 18 pounds over 30 weeks. The pattern will include peaks, valleys, and plateaus—not a straight downward curve. Sometimes a bit of weight will be gained, and other times more weight than expected will be lost. It's more important to enjoy and continue improved eating and activity patterns than to concentrate on the number of pounds lost.

If diet and exercise behaviors are improved in acceptable ways, there may be little need to become preoccupied with the number of calories consumed, the number of calories burned off in a bout of exercise, or the number of pounds lost last week. The goal is reached when the small changes become an enjoyable part of life on a day-to-day basis. Improved eating and activity patterns offer many benefits. Weight loss is only one of them.

Illustration 10.10
People who lose weight gradually are more likely to keep it off than those who lose weight rapidly. The weight loss graphed here averages half a pound per week.

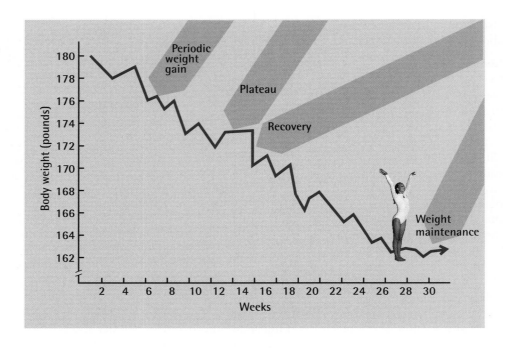

Key Term

bariatrics, page 10-8.

Review Questions

	True	False
1. Claims made for the effectiveness of weight loss products in advertisements must, by federal regulation, be true.	☐	☐
2. Commerical weight-loss services and products must be tested for safety before they are made available to the public.	☐	☐
3. Some foods, including celery and grapefruit, have negative calories and help overweight people burn fat.	☐	☐
4. It is recommended that people using prescription diet pills for weight loss should also follow a reduced-calorie diet and exercise.	☐	☐
5. Many different types of popular diets lead to weight loss if followed.	☐	☐

	True	False
6. Health benefits that result from following advice given in a popular diet book are usually due to the specific dietary recommendations made rather than the weight loss achieved.	☐	☐
7. The major problem shared by all popular weight-loss approaches is that they do not help people maintain weight in the long run.	☐	☐
8. A person who weighs 300 pounds before gastric bypass sugery and loses the average amount of initial body weght years later would be expected to weigh 170 pounds.	☐	☐
9. The best way to lose weight and keep it off is by making small and acceptable changes in diet and physical activity.	☐	☐

Nutrition Up Close

Will You Maintain or Regain?

Focal Point: People who lose weight and keep the pounds off share similar successful behaviors.

Have you recently lost weight? Answer the following questions to find out if your current eating and activity behaviors promote making the weight loss permanent.

	YES	NO

1. Do you usually eat breakfast?
2. Do you avoid skipping meals?
3. Do you plan ahead for snack and mealtime foods?
4. Do you prepare most of your meals at home?
5. Do you limit consumption of high-fat foods like butter, margarine, mayonnaise, and fried foods?
6. Do you consume calorie-dense foods like desserts infrequently and in small portions?
7. Do you walk instead of drive whenever possible?
8. Do you engage in regular physical activities that you enjoy?
9. Do you seek social support to help maintain your weight loss?
10. Do you eat when you're hungry and stop eating when you are full?

Feedback (answers to these questions) can be found at the end of Unit 10.

Media Menu

www.thedietchannel.com/faddiets.htm
Read up on how popular diets compare to the Dietary Guidelines, specific strengths and weaknesses of selected fad diets, and guidance on how to lose weight and keep it off here.

http://nutrition.about.com/library/bl_fad_diets.htm
Take the nutrition quiz on fad diets at this site developed by About Fitness & Health.

www.nhlbi.nih.gov/health/public/heart/obesity/lose_wt/behavior.htm
A Guide to Behavior Change for weight loss from the National Heart, Blood, and Lung Institute is available at this site. The site leads you to a BMI calculator and menu planner.

www.ars.usda.gov/ba/bhnrc/ndl
At the homepage for USDA's food composition information, you can obtain nutrient data files and get answers to commonly asked questions about the calorie value of foods.

http://exercise.about.com/od/fitnesstoolscalculators/
This site, offered by "About Health and Fitness," generates BMI and calorie expenditure data for physical activity based on data you input. The site has lots of ads, but the calculators are easy to use.

www.dietbites.com/article0159.html
This site, brought to you by dietbites.com, provides a calorie index you can use to quickly look up notes on the nutrient and calorie values basic foods. Watch out for the ads.

www.nwcr.ws
The National Weight Control Registry was developed to identify and investigate the characteristics of individuals who succeed at long-term weight loss. It is tracking over 5000 individuals who have lost significant amounts of weight and kept it off for long periods of time. Learn more about characteristics of long-term weight losers or share your story.

www.ediets.com
This commercial site offers personalized diet plans and other resources developed largely by dietitians.

www.shapedown.com
This commercial site provides background information and methods for the tested Shapedown Program for weight management in children and adolescents.

www.quackwatch.com
Check out "quacky" weight-loss products and services.

Notes

1. Hill JO et al. Weight maintenance: what's missing? J Am Diet Assoc 2005; 105:S63–S66.

2. Obesity Trends, www.cdc.gov/nccdphp/–dnpa/obesity/trend/prev_bmi.htm, accessed 9/03.

3. Sarlio-Lahteenkorva SS, et al. A descriptive study of weight loss maintenance: 6 and 15 year follow-up of initially overweight adults. Int J Obesity 2000;24:116–25.

4. Goodrick GK, Foreyth JP. Why treatments for obesity don't last. J Am Diet Assoc 1991;91:234–47.

5. Womble L et al. Unrealistic weight loss goals and feelings of failure. Presentation at the North American Association for the Study of Obesity, Long Beach, CA, 2000 Nov.

6. Wooley SC, Garner DM. Obesity treatment: the high cost of false hope. J Am Diet Assoc 1991;91:1248–51.

7. National Center for Health Statistics. Faststats A to Z. http://www.cdc.gov/nchs/fastats/overwt.htm, accessed 8/06.

8. Jacobs DR, Jr. Fast food and sedentary lifestyle: a combination that leads to obesity. Am J Clin Nutr 2006;83:189–90.

9. Cunningham A, Marcason W. Is it possible to burn calories by eating grapefruit or vinegar?

10. Wadden TA, Stunkard AJ, Brownell KD. Very low calorie diets: their efficacy, safety, and future. Ann Intern Med 1983;99:675–84.

11. Current trends in weight-loss advertising. J Am Diet Assoc 2003;103:150.

12. Taubes G. Dietary approaches to weight control. What works? Experimental Biology Annual Meeting, San Diego, April 14, 2003.

13. Broom I. Thinking about abdominal obesity and cardiovascular risk. Br J Vasc Dis 2006;6:58–61.

14. Yanovski SZ, Yanovski JA. Obesity. N Engl J Med 2002;346:591–602.

15. Dansinger MG et al. Comparisons of the Atkins, Ornish, Weight Watchers, and Zone diets for weight loss and heart disease risk reduction: a randomized trial. JAMA 2005;293:43–50.

16. Arterburn D et al. Comparison of commercial weight loss diets. JAMA 2006;292:101–3,

17. Buchholz AC, Schoeller DA. Is a calorie a calorie? Am J Clin Nutr 2004;79(suppl):899S–906S.

18. Gould L. Consumers taking small steps toward big lifestyle changes. Datmonitor survey of diet trends and advertising in the United Kingdom and the United States. NUTRAingredients.com/europe, 3/18/05.

19. Goodrick and Foreyth, Why treatments for obesity don't last. J Am Diet Assoc 1991;91:234–47.

20. Polivy J. Psychological consequences of food restriction. J Am Diet Assoc 1996;96:589–92.

21. Eisenberg D et al. Update on obesity surgery, World J Gastroenterol 2006;12:3196–3203.

22. Scheier L, Bariatric surgery: Life-threatening risk of life-saving procedure? J Am Diet Assoc 2004; 104:1338–40.

23. Parker-Pope T. A desperate diet: putting on pounds to qualify for weight loss surgery. Wall Street Journal, 7/8/03, p. D1.

24. Encinosa WWE et al. Late complication rates of bariatric surgey. Med Care 2006; July, www.medscape.com/viewarticle/541389, accessed 8/06.

25. Marcason W. What are the dietary gudelines following bariatric surgery? J Am Diet Assoc 2004;104:487–8.

26. Liposuction comes of age, and New liposuction technique faster, less invasive. www.healthfinder.gov, accessed 9/03.

27. Liposuction comes of age (www.healthfinder.gov); and Cohen S. Lipolysis: pitfalls and problems in a series of 1246 procedures. Aesthetic Plast Surg 1985;9:207.

28. Wing RR, Phelan S. Long-term weight maintenance. Am J Clin Nutr 2005;82(suppl):222S–5S.

29. Weinsier RL et al., Free-living energy expenditure in women successful and unsuccessful at maintaining a normal body weight, Am J Clin Nutr 2002.

30. Schlundt DG, Hill JO, Sbrocco T, Pope-Cordle J, Sharp T. The role of breakfast in the treatment of obesity: a randomized trial. Am J Clin Nutr 1992;55:645–51.

31. 12 smart ways to right-size your portions. Today's Dietitian 2005;May:60.

32. Mello MM et al. Obesity—the new frontier of Public Health law. N Engl J Med 2006;354:2601–10.

33. Orr J, Davy B. Dietary influences on peripheral hormones and regulating energy intake: potential applications for weight management. J Am Diet Assoc 2005;105:1115–22.

34. Ledikwe JH et al. Dietary energy density is associated with energy intake and weight status in US adults. Am J Clin Nutr 2006;83:1362–8.

Answers to Review Questions

1. False, see page 10-4.
2. False, see page 10-4.
3. False, see page 10-5.
4. True, see page 10-6.
5. True, see page 10-7.
6. False, see page 10-7.
7. True, see page 10-7.
8. False, it's 210 pounds. See page 10-8.
9. True, see page 10-12.

Nutrition Up Close

Will You Maintain or Regain?

Feedback for Unit 10

The more checks you made in the "Yes" column, the greater the chances that you will maintain your weight loss. Reflect for a moment upon any checks you made in the "No" column. Enacting small, acceptable changes in your eating and activity plan now may make the difference between maintaining your weight loss or regaining the pounds.

UNIT 11

Disordered Eating: Anorexia Nervosa, Bulimia, and Pica

NUTRITION SCOREBOARD

	TRUE	FALSE
1 The United States has one of the world's highest rates of anorexia nervosa.		
2 Eating disorders result from psychological, and not biological, causes.		
3 People in many different cultures may consume clay, dirt, and other nonfood substances.		

Key Concepts and Facts

- Anorexia nervosa, bulimia nervosa (bulimia), binge-eating disorder, and pica are four specific eating disorders. They may seriously threaten health.

- Eating disorders are much more common in females than males.

- The incidence of eating disorders in a society is related to the value placed on thinness by that society.

- An important route to the prevention of anorexia nervosa and bulimia is to change a society's cultural ideal of thinness and to eliminate biases against people (especially women) who are not thin.

Answers to
NUTRITION SCOREBOARD

	TRUE	FALSE
1 Anorexia nervosa is most common in the United States and other Westernized countries.[1]	✔	
2 The cause (or causes) of eating disorders is not yet known with certainty. Both psychological and biological factors may play a role.		✔
3 Although not recommended for health reasons, people in many different cultures practice pica—the regular ingestion of nonfood items such as clay and dirt.	✔	

The Eating Disorders

Three square meals a day, an occasional snack or missed meal, and caloric intakes that average out to match the body's need for calories—this set of practices is considered "orderly" eating. Self-imposed semistarvation, feast and famine cycles, binge eating, *purging*, and the regular consumption of nonfood substances such as paint chips and clay—these behaviors are symptoms of disordered eating.

purging
The use of self-induced vomiting, laxatives, or diuretics (water pills) to prevent weight gain.

Four specific types of disordered eating patterns are officially recognized as eating disorders and have been assigned diagnostic criteria. They are (1) anorexia nervosa, (2) bulimia nervosa, (3) binge-eating disorder, and (4) pica.[2] Other forms of disordered eating such as compulsive overeating and purging disorder have been observed, but too little research exists to establish criteria for diagnosis.

Anorexia Nervosa

It's about 9:30 on a Tuesday night. You're at the grocery store picking up sandwich fixings and some milk. Although your grocery list contains only four items, you arrive at the checkout line with a half-filled cart. The woman in front of you has only five items: a bag with about 10 green beans, an apple, a bagel, a green pepper, and a 4-ounce carton of nonfat yogurt. As she carefully places each item into her shopping bag, you notice that she is dreadfully thin.

The woman is Alison. She has just spent half an hour selecting the food she will eat tomorrow. Alison knows a lot about the caloric value of foods and makes only low-calorie choices. Otherwise, she will never get rid of her excess fat. To Alison, weight is everything—she cannot see the skeleton-like appearance others see when they look at her.

Alison has an intense fear of gaining weight and of being considered fat by others. She is annoyed when her parents and friends express their concern about her weight.

Richard Anderson

Illustration 11.1
A day's diet?
For a person with anorexia nervosa, it was. The foods shown provide approximately 562 calories.

anorexia nervosa
An eating disorder characterized by extreme weight loss, poor body image, and irrational fears of weight gain and obesity.

You didn't know this about Alison when you saw her. There is much more to anorexia nervosa than meets the eye.

Individuals with ***anorexia nervosa*** starve themselves (Illustration 11.1). They can never be too thin—no matter how emaciated they may be. As shown in Illustration 11.2, people with anorexia nervosa look extraordinarily thin from the neck down. The face and the rest of the head usually look normal because the head is the last part of the body to be affected by starvation.

People with anorexia nervosa have little fat (7 to 18% of body weight). They keep their body fat content low by consuming between 300 and 700 calories a day. They become cold easily and have unusually low heart rates and sometimes an irregular heartbeat, dry skin, and low blood pressure. Women with anorexia nervosa experience absent or irregular menstrual cycles, infertility, and poor pregnancy outcomes (Table 11.1). Low testosterone levels in males with this eating disorder produce diminished sexual drive and impaired fertility. Approximately 9 in 10 women with anorexia nervosa have significant bone loss, and 38% have osteoporosis. The extent of bone loss correlates strongly with undernutrition: the lower the body weight, the lower the bone density. Males with anorexia nervosa lose bone mass, too. Improving calcium and vitamin D intake is recommended along with nutritional rehabilitation experience.[3,4,5]

The Female Athlete Triad Pediatricians, nutritionists, and coaches are beginning to be on the lookout for eating disorders, menstrual cycle dysfunction, and decreased bone mineral density in young female athletes. Low caloric in-

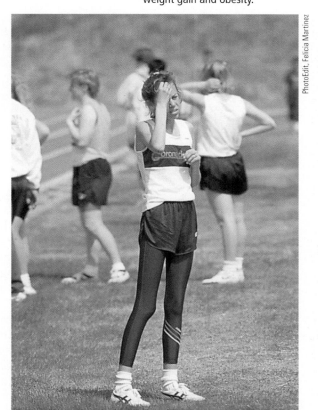

PhotoEdit, Felicia Martinez

Illustration 11.2
Eating disorders occur in males as well as females, but females make up approximately 90% of all cases.

TABLE 11.1

Features of anorexia nervosa

A. Essential Features

1. Refusal to maintain body weight at or above 85% of normal weight for age and height
2. Intense fear of gaining weight or becoming fat, despite being underweight
3. Disturbance in the way in which body weight or shape is experienced, undue influence of body weight or shape on self-evaluation, or denial of the seriousness of current low body weight
4. Lack of menstrual periods in teenage females and women (missing at least three consecutive periods)

Restricting type: Person does not regularly engage in binge-eating or purging behavior.
Binge-eating type: Person regularly engages in binge-eating or purging behavior (self-induced vomiting; laxative, diuretic, or enema use).

B. Common Features in Females

1. Low-calorie diet, extensive exercise, low body fat
2. Soft, thick facial hair, thinning scalp hair
3. Loss of heart muscle, irregular, slow heartbeat
4. Low blood pressure
5. Increased susceptibility to infection
6. Anemia
7. Constipation
8. Low body temperature (hypothermia)
9. Dry skin
10. Depression
11. History of physical or sexual abuse
12. Low estrogen levels
13. Low bone density
14. Infertility, poor pregnancy outcome

C. Common Features in Males

1. Most of the common features in females
2. Substance abuse
3. Mood and other mental disorders, self-loathing
4. Decreased testosterone level, sex drive, and fertility

Source: Reprinted with permission from the Diagnostic and Statistical Manual of Mental Disorders, Fourth Edition, Text Revision. Copyright 2000 American Psychiatric Association.

ON THE SIDE

The average size of female gymnasts on the U.S. Olympic team shrank from 5 feet 3 inches tall and 105 pounds in 1976 to 4 feet 9 inches tall and 88 pounds in 1992.

Photo Disc

takes and underweight related to eating disorders can lower estrogen levels and disrupt menstrual cycles. The lack of estrogen decreases calcium deposition in bones and reduces bone density at a time when peak bone mass is accumulating.

Irregular or absent menstrual cycles used to be thought of as "no big deal." That attitude has changed, however, due to research results indicating that abnormal cycles in young females are related to delayed healing of bone and connective tissue injuries, and to bone fractures and osteoporosis later in life.[6]

Motivations Underlying Anorexia Nervosa   The overwhelming desire to become and remain thin drives people with anorexia nervosa to refuse to eat, even when ravenously hungry, and to exercise intensely. Half of the people with anorexia turn to *binge eating* and purging—features of bulimia nervosa—in their efforts to lose weight.[2] Preoccupied with food, people with anorexia may prepare wonderful meals for others but eat very little of the food themselves. Family members and friends, distressed by their failure to persuade the person with anorexia to eat, report high levels of anxiety. Although adults often describe people with anorexia as "model students" or "ideal children," their personal lives are usually marred by low self-esteem, social isolation, and unhappiness.[7]

binge eating
The consumption of a large amount of food in a small amount of time.

Richard Anderson

Illustration 11.3
There is a need for a more real-istic body shape on television and in fashion magazines.

—Vivienne Nathanson, British physician, 2000

What Causes Anorexia Nervosa?   The cause of anorexia nervosa isn't yet clear. It is likely that many different conditions, both psychological and biological, pre-dispose an individual to become totally dedicated to extreme thinness.[8] The value that Western societies place on female thinness, the need to conform to society's expectations of acceptable body weight and shape, low self-esteem, and a need to control some aspect of one's life completely are commonly offered as potential causes for this disorder (Illustration 11.3).[9]

How Common Is Anorexia Nervosa?   It is estimated that 1% of adolescent and young women in the Western world and less than 0.1% of young males have anorexia nervosa. The disorder has been reported in girls as young as 5 and in women through their forties;[10] however, it usually begins during adolescence. It is estimated that one in ten females between the ages of 16 and 25 has "subclinical" anorexia nervosa, or exhibits some of the symptoms of the disorder.[4]

Certain groups of people are at higher risk of developing anorexia nervosa than others (Table 11.2). People at risk come from all segments of society but tend to be overly concerned about their weight and food and have attempted weight loss from an early age.[4]

Treatment   There is no "magic bullet" treatment that cures anorexia nervosa quickly and completely. In all but the least severe cases, the disorder generally takes 5 to 7 years and professional help to correct. Treating the disorder is of-ten difficult because few people with anorexia believe their weight needs to be increased.[2]

Treatment programs for anorexia nervosa generally focus on the prompt res-toration of nutritional health and body weight, psychological counseling to im-prove self-esteem and attitudes about body weight and shape, antidepressant or other medications, family therapy, and normalizing eating and exercise behav-iors. These programs are successful in 50% of people and partially successful in

TABLE 11.2
Risk groups for anorexia nervosa[4]

- Dieters
- Ballet dancers
- Competitive athletes (gymnasts, figure skaters)
- Fitness instructors
- Dietetics majors
- People with type 1 (insulin depen-dent) diabetes

most other cases.[11] One-third of people with full recovery from anorexia nervosa will relapse within 7 years or less. By 8 years after diagnosis, 3% of people with anorexia nervosa will have died from the disorder, and it claims the lives of 18% 33 years later. Results of treatment are often excellent when the disorder is treated early.[5] Unfortunately, many people with the condition deny that problems exist and postpone treatment for years. Initiation of treatment is often prompted by a relative, coach, or friend.[12]

Bulimia Nervosa

Finally home alone, Lisa heads to the pantry and then to the freezer. She has carefully controlled her eating for the last day and a half and is ready to eat everything in sight.

It's a bittersweet time for her. Lisa knows the eating binge she is preparing will be pleasurable, but that she'll hate herself afterward. Her stomach will ache from the volume of food she'll consume, she'll feel enormous guilt from losing control, and she'll be horrified that she may gain weight and will have to starve herself all over again. Lisa is so preoccupied with her weight and body shape that she doesn't see the connection between her severe dieting and her bouts of uncontrolled eating. To get rid of all the food she is about to eat, she will do what she has done several times a week for the last year. Lisa avoids the horrible feelings that come after a binge by "tossing" everything she ate as soon as she can.

In just 10 minutes, Lisa devours 10 peanut butter cups (the regular size), a 12-ounce bag of chocolate chip cookies, and a quart of mint chocolate chip ice cream (Illustration 11.4). Before 5 more minutes have passed Lisa will have emptied her stomach, taken a few deep breaths, thrown on her shorts, and started the 5-mile route she jogs most days. As she jogs, she obsesses about getting her 138-pound, 5-foot 5-inch frame down to 115 pounds. She will fast tomorrow and see what news the bathroom scale brings.

bulimia nervosa
An eating disorder characterized by recurrent episodes of rapid, uncontrolled eating of large amounts of food in a short period of time. Episodes of binge eating are often followed by purging.

Lisa is not alone. **Bulimia nervosa** occurs in 1 to 3% of young women and in about 0.5% of young males in the United States.[3] The disorder is characterized by regular episodes of dieting, binge eating, and attempts to prevent weight gain by purging; use of laxatives, diuretics, or enemas; dieting; and sometimes exercise. In most cases, bulimia nervosa starts with voluntary dieting to lose weight. At

Illustration 11.4
Bulimia nervosa is characterized by the consumption of a large amount of food (such as shown here) followed by purging, and dieting.

TABLE 11.3
Features of bulimia nervosa[2]

A. Essential Features

1. Recurrent episodes of binge eating. An episode of binge eating is characterized by both of the following:
   a. Eating an amount of food within a two-hour period of time that is definitely larger than most people would eat in a similar amount of time and under similar circumstances.
   b. A sense of lack of control over eating during the episode; a feeling that one cannot stop eating or control what or how much one is eating.
2. Recurrent inappropriate compensatory behavior in order to prevent weight gain, such as self-induced vomiting; misuse of laxatives, diuretics, enemas, or other medications; fasting; or excessive exercise.
3. The binge eating and inappropriate compensatory behaviors both occur, on average, at least twice a week for three months.
4. Self-evaluation is unduly influenced by body weight and shape.
5. The disturbance does not occur exclusively during episodes of anorexia nervosa.

Purging type: The person regularly engages in self-induced vomiting or the misuse of laxatives, diuretics, or enemas.
Nonpurging type: The person regularly engages in fasting or excessive exercise but does not regularly engage in self-induced vomiting or the misuse of laxatives, diuretics, or enemas.

B. Common Features

1. Weakness, irritability
2. Abdominal pain, constipation, bloating
3. Dental decay, tooth erosion
4. Swollen cheeks and neck
5. Binging on high-calorie foods
6. Eating in secret
7. Normal weight or overweight
8. Guilt and depression
9. Substance abuse
10. Dehydration
11. Impaired fertility
12. History of sexual abuse

Source: Reprinted with permission from the Diagnostic and Statistical Manual of Mental Disorders, Fourth Edition, Text Revision. Copyright 2000 American Psychiatric Association.

some point, voluntary control over dieting is lost, and people feel compelled to engage in binge eating and vomiting.[13] The behaviors become cyclic: Food binges are followed by guilt, purging, and dieting. Dieting leads to a feeling of deprivation and intense hunger, which leads to binge eating, and so on. Once a food binge starts, it is hard to stop.

Table 11.3 lists the features of bulimia nervosa. Approximately 86% of people with this condition vomit to prevent weight gain and avoid postbinge anguish. A smaller proportion of people use laxatives, diuretics (water pills), or enemas alone or in combination with vomiting.[2] Laxatives, enemas, and diuretics do not prevent weight gain, however, and their regular use can be harmful. The habitual use of laxatives and enemas causes "laxative dependency"—these products become necessary for bowel movements. Diuretics can cause illnesses by depleting the body of water and certain minerals and disturbing its fluid balance.[11]

The lives of people with bulimia nervosa are usually dominated by conflicts about eating and weight. Some affected individuals are so preoccupied with food that they spend days securing food, bingeing, and purging. Others experience only occasional episodes of binge eating, purging, and fasting.[2]

Unlike those with anorexia nervosa, people with bulimia usually are not underweight or emaciated. They tend to be normal weight or overweight.[4] Like anorexia nervosa, bulimia nervosa is more common among athletes (including gymnasts, weight lifters, wrestlers, jockeys, figure skaters, physical trainers, and distance runners) and ballet dancers than in other groups.[14]

Bulimia nervosa leads to major changes in metabolism. The body must constantly adjust to feast and famine cycles and mineral and fluid losses. Salivary glands become enlarged, and teeth may erode due to frequent vomiting of highly acidic foods from the stomach.[2]

Is the Cause of Bulimia Nervosa Known?

"Do you follow a special diet?" asks the dietitian at the eating disorder clinic.

"Yes," answers the client with bulimia. "Feast or famine."

The cause of bulimia nervosa is not known with certainty, but the scientific finger is pointing at depression, abnormal mechanisms for regulating food intake, and feast-and-famine cycles as possible causes. Fasts and *restrained eating* may prompt feelings of deprivation and hunger that may trigger binge eating.[15] The ideal thinness may become more and more difficult to achieve as the feast-and-famine cycles continue.

Treatment   The cornerstone of bulimia treatment is nutrition and psychological counseling to break the feast-and-famine cycles. Replacing the disordered pattern of eating with regular meals and snacks often reduces the urge to binge and the need to purge. Psychological counseling aimed at improving self-esteem and attitudes toward body weight and shape goes hand in hand with nutrition counseling. In many cases, antidepressants are a useful component of treatment.[16] The full recovery of women with bulimia nervosa is higher than that for anorexia nervosa. Nearly all women with bulimia achieve partial recovery, but one-third will relapse into bingeing and purging within 7 years.[15] Bulimia nervosa usually improves substantially during pregnancy; about 70% of women with the condition will improve their eating habits for the sake of their unborn baby.[17]

Binge-Eating Disorder

Psychiatrists now recognize an eating disorder called *binge-eating disorder* (Table 11.4). People with this condition tend to be overweight or obese, and one-third are male.[9] Like individuals with bulimia nervosa, people with binge-eating disorder eat several thousand calories' worth of food within a short period of time during a solitary binge, feel a lack of control over the binges, and experience distress or depression after the binges occur. People must experience eating binges twice a week on average over a period of 6 months to qualify for the diagnosis. Unlike individuals with bulimia nervosa, however, people with binge-eating disorder don't vomit, use laxatives, fast, or exercise excessively in an attempt to control weight gain.[18]

It is estimated that 9 to 30% of people in weight-control programs and 30 to 90% of obese people have binge-eating disorder.[19] The condition is far less common (2 to 5%) in the general population.[17] Stress, depression, anger, anxiety, and other negative emotions appear to prompt binge-eating episodes. Preliminary

---

restrained eating
The purposeful restriction of food intake below desired amounts in order to control body weight.

binge-eating disorder
An eating disorder characterized by periodic binge eating, which normally is not followed by vomiting or the use of laxatives. People must experience eating binges twice a week on average over a period of 6 months to qualify for the diagnosis.

TABLE 11.4

Features of binge-eating disorder[2]

1. Rapid consumption of extremely large amounts of food (several thousand calories) in a short period of time
2. Two or more such episodes of binge eating per week over a period of 6 months
3. Binge eating by oneself
4. Lack of control over eating or an inability to stop eating during a binge
5. Post binge-eating feelings of self-hatred, guilt, and depression or disgust
6. Purging, fasting, excessive exercise, or other compensation for high-calorie intakes not present

Close to home

Although she hides it, you are sure your sister has bulimia nervosa and that she is not getting help. You are deeply concerned for her health and well-being but don't know what to do about it.

Here's what Heather and Crystal say they would do:

Who do you think has the better idea?
Answers on next page

Heather
I'd talk with her about getting help.

Crystal
I'd spend more time with her to let her know I love her.

evidence indicates that binge-eating disorder aggregates in families and has both genetic and environmental orgins.[20]

The Treatment Approach to Binge-Eating Disorder The treatment of binge-eating disorder focuses on both the disordered eating and the underlying psychological issues.[21] Persons with this condition will often be asked to record their food intake, indicate bingeing episodes, and note feelings, circumstances, and thoughts related to each eating event (Illustration 11.5). This information is used to identify circumstances that prompt binge eating and practical alternative behaviors that may prevent it. Individuals being treated for binge-eating disorder are usually given information about it, attend individual and group therapy sessions, and receive nutrition counseling on normal eating, hunger cues, and meal planning. Antidepressants may be part of the treatment. Treatment is successful in 85% of women treated for binge-eating disorder.[18]

Illustration 11.5
Example of a food dairy of a person with binge-eating disorder.

Resources for Eating Disorders

Information and services related to eating disorders are available from a variety of sources. Services are best delivered by health care teams specializing and experienced in the treatment of eating disorders. Contact with a primary care physician, dietitian, or nurse practitioner is often a good start to the process of identifying qualified health care teams. Reliable sources of information about eating disorders, support groups, nearby treatment centers, and hot lines can be found on the Internet. (See "Media Menu" at the end of the unit.)

One of the most important resources for people with an eating disorder may be a trusted friend or relative. This unit's "Reality Check" explores this resource in a very personal way—by putting you in the shoes of a person whose sister has bulimia.

Daily Food Record

Date_____

Time	Type and amount of food and beverage	Meal, Snack, Binge?	Eating triggers (feelings, situation)
7:30 am	coffee, 2 cups sugar 2 tsp cornflakes, 2 cups skim milk, 1 cup	M	Hunger!
11:30 am	tuna sandwich ice tea, 2 cups	M	Bored, hungry
7:30 pm	3 hamburgers 2 large fries 24 oreos 1/2 gallon ice cream	B	Stressed out, angry at my coach

Close to home

Both ideas are admirable and deserve a thumbs-up. Heather's idea is aimed directly at helping her sister consider treatment and may sometimes be the appropriate action to take.

There is a way to talk to a relative or friend about your concerns for them that may help both of you. Learn more about it from the information presented in Table 11.5.

Heather 👍   Crystal 👍

Undieting: The Clash between Culture and Biology

There is a saying that women underreport their weight and men overreport their height. Clearly there are cultural norms at work here.
—L. COHEN, 2001[22]

The pressure to conform to society's standard of beauty and acceptability is thought to be a primary force underlying the development of eating disorders.[1] Children acquire prevailing cultural values of beauty before adolescence. As early as age 5, American children learn to associate negative characteristics with people who are overweight and positive characteristics with those who are thin.[10] Standards of beauty defined by models and movie and television stars often include thinness, but the body shape portrayed as best is often unhealthfully thin and unattainable by many. The disparity between this ideal and what people normally weigh has led to widespread discontent. Approximately 50% of normal-weight adult women are dissatisfied with their weight; and many diet, binge, and purge occasionally or fast in an attempt to reach the standard of beauty set for them.[23]

TABLE 11.5
Helping a family member or friend with an eating disorder

Whether at work, home, or play, many of us experience anxiety and a sense of helplessness when someone we love is living with an eating disorder. We may feel compelled to take action to help but aren't sure what to do or how to do it. Here are some tips on how to express your concerns to a friend or relative with an eating disorder:

1. Gather information about services for people with eating disorders to share with your friend or relative.
2. Talk with your friend or relative privately when there is enough time to fully discuss the issue. Tell them you are worried and that they may need to seek help.
3. Encourage your friend or relative to express his or her feelings and then listen intently. Be accepting about the feelings that are expressed. Be ready to talk to the friend or relative more about it in the future.
4. Do not argue with your friend or relative about whether she or he has an eating disorder. Let your friend or relative know you heard what was said but that you are concerned that he or she may not get better without treatment.
5. Seek emergency medical help in life-threatening situations.

Only individuals with an eating disorder can make the decision to get help. Knowledge that people who love them will be around to support them and their decision to seek treatment may help encourage people with an eating disorder to take action.

Illustration 11.6
The trend toward size acceptance.
Acceptance of a realistic standard of body weight and shape—one that corresponds to health and physical fitness—and respect for people of all body sizes may be the most effective measures that can be taken to prevent anorexia nervosa, bulimia nervosa, and binge-eating disorder.

A movement toward acceptance of body size, fashionable attire for larger people, full-size models, and a more realistic view of individual differences in body shapes is emerging slowly in America (Illustration 11.6). Acceptance of a realistic standard of body weight and shape—one that corresponds to health and physical fitness—and respect for people of all body sizes may be the most effective measures that can be taken to prevent anorexia nervosa, bulimia nervosa, and binge-eating disorder.

Pica

> When did I start eating clay? I know it might sound strange to you, but I started craving clay in the summer of '58. It was a beautiful spring morning—it had just rained. I smelled something really sweet in the breeze coming in my bedroom window. I went outside and knew instantly where the sweet smell was coming from. It was the wet clay that lies all around my house. I scooped some up and tasted it. That's when and how I started my craving for that sweet-smelling clay. I keep some in the fridge now because it tastes even better cold.

A most intriguing type of eating disorder, *pica* has been observed in chimpanzees and in humans in many different cultures since ancient times.[24] The history and persistence of pica might suggest that the practice has its rewards. Nevertheless, important health risks are associated with eating many types of nonfood substances.

The characteristics of pica are summarized in Table 11.6. Young children and pregnant women are most likely to engage in the practice; for unknown reasons, it rarely occurs in men.[25] It most commonly takes the form of *geophagia* (clay or dirt eating), *pagophagia* (ice eating), *amylophagia* (laundry starch and cornstarch eating), or *plumbism* (lead eating). A potpourri of nonfood substances, listed in Table 11.7, may be consumed.

TABLE 11.6
Characteristics of pica

A. **Essential features:** Regular ingestion of nonfood substances such as clay, paint chips, laundry starch, paste, plaster, dirt, or hair.
B. **Other common features:** Occurs primarily in young children and pregnant women in the southern United States.

pica (pike-eh)
The regular consumption of nonfood substances such as clay or laundry starch.

geophagia (ge-oh-phag-ah)
Clay or dirt eating.

pagophagia (pa-go-phag-ah)
Ice eating.

amylophagia (am-e-low-phag-ah)
Laundry starch or cornstarch eating.

plumbism
Lead (primarily from old paint flakes) eating.

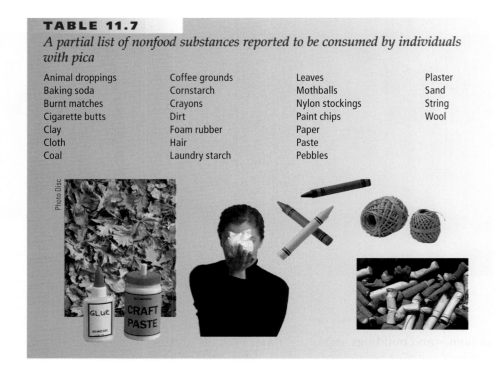

TABLE 11.7

A partial list of nonfood substances reported to be consumed by individuals with pica

Animal droppings	Coffee grounds	Leaves	Plaster
Baking soda	Cornstarch	Mothballs	Sand
Burnt matches	Crayons	Nylon stockings	String
Cigarette butts	Dirt	Paint chips	Wool
Clay	Foam rubber	Paper	
Cloth	Hair	Paste	
Coal	Laundry starch	Pebbles	

Pica has a forceful calling card:

> *Pica permits the mind no rest until it is satisfied.*[26]

It is not clear why pica exists, although several theories have been proposed.

Geophagia

Some people very much like to eat certain types of clay or dirt. Those who do often report that the clay or dirt tastes or smells good, quells a craving, or helps relieve nausea or an upset stomach. The belief that certain types of clay provide relief from stomach upsets may have some validity: A component of some clays is used in nausea and diarrhea medicines. There is no evidence that geophagia is motivated by a need for minerals found in clay or dirt, however.[27]

Although the reasons given for clay and dirt ingestion make the practice understandable and helps explain its acceptance in some cultures, the consequences to health outweigh the benefits. Clay and dirt consumption can block the intestinal tract and cause parasitic and bacterial infections.[2] The practice is also associated with iron-deficiency and sickle-cell anemia in some individuals.[28,29]

Pagophagia

Have you ever known somebody who constantly crunches on ice? That person may have a 9-in-10 chance of being iron deficient. Regular ice eating, to the extent of one or more trays of ice cubes a day, is closely associated with an iron-deficient state. Ice eating usually stops completely when the iron deficiency is treated.[29]

Ice eating may be common during pregnancy. In one study of women from low-income households in Texas, 54% of pregnant women reported eating large amounts of ice regularly. Ice eaters had poorer iron status than other pregnant women who did not eat ice.[30]

Amylophagia

The sweet taste and crunchy texture of flaked laundry starch are attractive to a small number of women, especially during pregnancy. If the laundry starch preferred is not available, cornstarch may be used in its place. Laundry starch is made from unrefined cornstarch. The taste for starch almost always disappears after pregnancy.[25]

Laundry starch and cornstarch have the same number of calories per gram as do other carbohydrates (4 calories per gram). Consequently, starch eating provides calories and may reduce the intake of nutrient-dense foods. In addition, starch may contain contaminants because it is not intended for consumption. Starch eaters' diets are generally inferior to the diets of pregnant women who don't consume starch, and their infants are more likely to be born in poor health.[25]

Plumbism

The consumption of lead-containing paint chips poses a major threat to the health of children in the United States and many other countries (Illustration 11.7). Many older homes and buildings, especially those found in substandard housing areas, are covered with lead-based paint and its dried-up flakes. Children may develop lead poisoning if they eat the sweet-tasting paint flakes or inhale lead from contaminated dust and soil near the buildings. An estimated 1 million young children in the United States have elevated blood lead levels.[31]

High levels of exposure to lead can cause profound mental retardation and death in young children. Low levels of exposure can lead to hearing problems, growth retardation, reduced intelligence, and poor classroom performance. Children with lead poisoning are more likely to fail or drop out of school than children not exposed to lead in their environment.[31]

Eating disorders affect the health and well-being of over a million people in the United States. Although there are treatment strategies, such as counseling and the removal of lead-based paints from old houses and apartments, the solution to eating disorders lies in their prevention. With the exception of certain types of pica discussed here, the most effective way to prevent eating disorders may be to adjust our expectations and cultural norms to reflect reality.

Proposed Eating Disorders   Several other patterns of abnormal eating behaviors have been described and may become official diagnoses in the future.[32,33] These tentative eating disorders include:

- Compulsive overeating—the uncontrolled ingestion of large amounts of food, as found in binge-eating disorder

- Purging disorder—frequent purging without binge eating

- Restrained eating—the consistent limitation of food intake to avoid weight gain

- Orthorexia nervosa (pronounced ortho-rex-e-ah)—an unhealthy fixation with the health value and purity of food

These disorders exist, and experimental prevention and treatment services are available. Web sites listed under "Media Menu" will lead you to more information on these lesser-known, proposed eating disorders.

PhotoEdit

Illustration 11.7
The regular consumption of lead-based paint chips from old houses is a major cause of lead poisoning in young children.

Nutrition Up Close

Eating Attitudes Test

Focal Point: Discover if your eating attitudes and behaviors are within a normal range.

Date _____ Age _____ Gender _____
Height _____ Present weight _____ How long at present weight? _____
Highest past weight _____ How long ago? _____
Lowest past weight _____ How long ago? _____

Answer the following questions using these responses:

A = always   S = sometimes   U = usually   R = rarely   O = often   N = never

_____ 1. I am terrified of being overweight.
_____ 2. I avoid eating when I am hungry.
_____ 3. I find myself preoccupied with food.
_____ 4. I have gone on eating binges where I feel that I may not be able to stop.
_____ 5. I cut my food into very small pieces.
_____ 6. I am aware of the calorie content of the foods I eat.
_____ 7. I particularly avoid foods with a high-carbohydrate content.
_____ 8. I feel that others would prefer that I ate more.
_____ 9. I vomit after I have eaten.
_____ 10. I feel extremely guilty after eating.
_____ 11. I am preoccupied with a desire to be thinner.
_____ 12. I think about burning up calories when I exercise.
_____ 13. Other people think I am too thin.
_____ 14. I am preoccupied with the thought of having fat on my body.
_____ 15. I take longer than other people to eat my meals.
_____ 16. I avoid foods with sugar in them.
_____ 17. I eat diet foods.
_____ 18. I feel that food controls my life.
_____ 19. I display self-control around food.
_____ 20. I feel that others pressure me to eat.
_____ 21. I give too much time and thought to food.
_____ 22. I feel uncomfortable after eating sweets.
_____ 23. I engage in dieting behavior.
_____ 24. I like my stomach to be empty.
_____ 25. I enjoy trying new rich foods.
_____ 26. I have the impulse to vomit after meals.

Feedback (including scoring) can be found at the end of Unit 11.

Source: The EAT-26 has been reproduced with permission. Garner, D. M., Olmsted, M. P., Bohr, Y., and Garfinkel, P. E. (1982). The Eating Attitudes Test: Psychometric features and clinical correlates. Psychological Medicine, 12, 871–878.

Key Terms

amylophagia, page 11-11
anorexia nervosa, page 11-3
binge eating, page 11-4
binge-eating disorder, page 11-8

bulimia nervosa, page 11-6
geophagia, page 11-11
pagophagia, page 11-11
pica, page 11-11

plumbism, page 11-11
purging, page 11-2
restrained eating, page 11-8

Review Questions

True False

1. A critical aspect of anorexia nervosa is an intense fear of gaining weight or becoming fat. ☐ ☐

2. One of the most common health consequences of anorexia nervosa is blurred vision. ☐ ☐

3. The "female athlete triad" refers to female athletes who have an eating disorder, abnormal menstrual function, and decreased bone mineral density. ☐ ☐

4. More individuals have anorexia nervosa than any other recognized eating disorder. ☐ ☐

5. An essential feature of bulimia nervosa is binge eating followed by periods of intense physical activity. ☐ ☐

True False

6. Obese people are more likely to experience binge-eating disorder than are people who are not obese. ☐ ☐

7. Treatment is effective for 85% of women treated for binge-eating disorder. ☐ ☐

8. *Pica* refers to the regular consumption of raw whole grains, grasses, and other foods generally used to feed livestock. ☐ ☐

9. People who consume lead, such as from lead-based paints, have a condition called "geophagia." ☐ ☐

Media Menu

www.edap.org
The National Eating Disorders Association is the largest not-for-profit organization in the United States. The organization works to prevent eating disorders and provide treatment referrals to those suffering from them.

www.anad.org
The National Association of Anorexia Nervosa and Associated Disorders (ANAD) provides help for those struggling with eating disorders. Resources include a hotline (847-831-3438, Monday–Friday, 9A.M.–5P.M, Central Time), support groups, referrals to health care professionals, and education and prevention programs to promote self-acceptance and healthy lifestyles. All services are free.

www.anred.com
On the Anorexia Nervosa and Related Eating Disorders home page, this nonprofit organization provides information on the major and less well-known eating disorders. Site includes self-help tips and information about recovery and prevention.

www.bulimiaguide.org
A resource guide for families and friends of individuals with an eating disorder is presented in the Web site.

www.nedic.ca
The National Eating Disorder Info Center in Toronto provides information and resources on eating disorders and weight preoccupation and a telephone support line, information on support groups, and listings of Canada-wide treatment resources.

www.edreferral.com
This site includes basic information on eating disorders along with specific information on treatment and recovery for men, pregnant women, and others with eating disorders.

www.naafa.org
From the National Association to Advance Fat Acceptance, Inc., this site is dedicated to improving the quality of life for fat people. It takes on policies and practices related to size discrimination and promotes size acceptance by individuals and society.

Notes

1. Lake AJ et al. Effect of Western culture on women's attitudes to eating and perceptions of body shape. Int J Eat Disord 2000;27:83–9.

2. American Psychiatric Association. Diagnostic and statistical manual of mental disorders: DMS-IV, 4th ed., Text Revision. Washington, DC: 2000.

3. Eating disorders. Facts about eating disorders and the search for solutions. National Institute of Mental Health, Bethesda, MD, 2001. www.nimh.nih.gov.

4. Grinspoon S et al. Prevalence and predictive factors for regional osteopenia in women with anorexia nervosa. Ann Intern Med 2000;133:790–4.

5. Yager J et al. Anorexia nervosa. N Engl J Med 2005;353:1481–8.

6. Beals KA, Manore MM. Disorders of the female athlete triad among collegiate athletes. Int J Sports Nutr Exer Metab 2002;12:281–93.

7. Omizo SA, Oda EA. Anorexia nervosa: psychological considerations for nutrition counseling. J Am Diet Assoc 1988; 88:49–51.

8. Tamburrino MB, McGinnis RA. Anorexia nervosa: a review. Panminerva Med 2002;44:301–11.

9. Eating disorders. www. nimh.nih.gov, accessed 8/03.

10. Feldman W, et al. Culture versus biology: children's attitudes toward thinness and fatness, Pediatrics 1988;81: 190–4.

11. Robinson PH, Recognition and treatment of eating disorders in primary and secondary care, Alimentary Pharmacol & Thera 2000;14:367–77; and Patel et al., eating disorders.

12. Treating eating disorders, Harvard Women's Health Watch 1996 May:4–5; and When eating goes awry: an update on eating disorders, Food Insight 1997 Jan/Feb:35.

13. Faris PL et al. Effect of decreasing afferent vagal activity with ondansetron on symptoms of bulimia nervosa: a randomized, double-blind trial. Lancet 2000;355:792–70.

14. Edell D, Beware of personal trainers with their looks, IDEA Health and Fitness Source. www.healthcentral.com/ drdean/ DeanFullTextTopics.cfm?ID= 41044&src=n2, accessed 9/2000.

15. Herzog DB, Copeland PM. Bulimia nervosa—psyche and satiety (editorial). N Engl J Med 1988;319:716–8.

16. Eating disorders III, Disease definition, epidemiology, and natural history, www.mentalhealth.com, accessed 12/2000.

17. Hohlstein LA. Eating disorders program. American Dietetic Association annual meeting, Boston, 1997 Oct 27.

18. Fairburn CG et al. Distinctions between binge eating disorder and bulimia nervosa. Arch Gen Psychiatry 2000;57:659–65.

19. Branson R et al. Binge eating as a major phenotype of melanocortin 4 receptor gene mutations. N Engl J Med 2003;348:1096–103.

20. Hudson JI et al. Binge eating disorder may be a familial disorder. Arch Gen Psychiatry 2006;63:313–19.

21. Hay PJ, Bacaltchuk J. Psychotherapy for bulimia nervosa and bingeing (Cochrane Review). In: The Cochrane Library, Issue 2, 2003.

22. Cohen, LA, Nutrition and cancer prevention. Nutr Today 2001;36:78–9.

23. Zuckerman DM, et al. The prevalence of bulimia among college students. Am J Public Health 1986;76:1135–7.

24. Cooper M. Pica: a survey of the historical literature as well as reports from the fields of veterinary medicine and anthropology. Springfield (IL): Charles C. Thomas; 1957.

25. Edwards CH et al. Clay- and cornstarch-eating women. J Am Diet Assoc 1959;35:810–15.

26. Craign FW. Observations on cachexia Africana or dirt-eating. Am J Med Sci 1935;17:365.

27. Johns T, Duquette M. Detoxification and mineral supplementation as functions of geophagy. Am J Clin Nutr 1991;53: 448–56.

28. Hackworth SR, Williams LL. Pica for foam rubber in patients with sickle-cell disease. South Med J 2003; 96:81–3.

29. Coltman CA Jr., Pagophagia and iron lack, JAMA 1969;207:513–16.

30. Rainville AJ. Pica practices of pregnant women are associated with lower maternal hemoglobin level at delivery. J Am Diet Assoc 1998;98:293–6.

31. Child health USA. Washington, DC: Maternal and Child Health Bureau, U.S. PHS;2000.

32. Mathieu J. What is orthorexia? J Am Diet Assoc 2005;105:1510–2.

33. Keel PK et al. Purging disorder as a distinct eating disorder. Int J Eat Disord 2005;38:191–9.

Nutrition Up Close

Eating Attitudes Test

Feedback for Unit 11

Answers to Review Questions

1. True, see page 11-3.
2. False, see page 11-3.
3. True, see page 11-3.
4. False, see pages 11-5, 11-6, 11-8.
5. False, see pages 11-6, 11-7.
6. True, see page 11-8.
7. True, see page 11-9.
8. False, see page 11-11.
9. False, see pages 11-11, 11-12.

Never   = 3
Rarely   = 2
Sometimes   = 1
Always, usually, and often = 0

A total score under 20 points may indicate abnormal eating behavior. If you think you have an eating disorder, it is best to find out for sure. Careful evaluation by a qualified health professional is necessary to exclude any possible underlying medical reasons for your symptoms. Contacting a physician, nurse practitioner, dietitian, or the student health center is an important first step. You may wish to show your Eating Attitudes Test to the health professional.

Useful Facts about Sugars, Starches, and Fiber

NUTRITION SCOREBOARD | TRUE | FALSE

1 Pasta, bread, and potatoes are good sources of complex carbohydrates.

2 Ounce for ounce, presweetened breakfast cereals and unsweetened cereals provide about the same number of calories.

3 A 12-ounce can of soft drink contains about 3 tablespoons (9 teaspoons) of sugar.

4 Lettuce, onions, and celery are high in dietary fiber.

5 Cooking vegetables destroys their fiber content.

6 Sugar consumption is the leading cause of tooth decay.

Key Concepts and Facts

- Simple sugars, the "starchy" complex carbohydrates, and dietary fiber are members of the carbohydrate family.

- Ounce for ounce, sugars and the starchy complex carbohydrates supply fewer than half the calories of fat.

- Tooth decay and poor-quality diets are related to high sugar intake.

- Fiber benefits health in a number of ways.

Answers to
NUTRITION SCOREBOARD

		TRUE	FALSE
1	If you get this right, you may be in the minority. In one study of college students, only 38% could identify good sources of complex carbohydrates. Rice, crackers, grits, dried beans, corn, peas, tortillas, biscuits, and oatmeal are also good sources of complex carbohydrates.	✔	
2	That's true! Both sweetened and unsweetened cereals consist primarily of carbohydrate. A gram of carbohydrate provides 4 calories whether the source is sugar or flakes of corn.	✔	
3	That's a lot of sugar!	✔	
4	Although these are healthy food choices, they do not contain very much dietary fiber. (Not all that goes "crunch" is a good source of fiber.)		✔
5	Cooking doesn't destroy dietary fiber.		✔
6	Rates of tooth decay increase in populations as sugar intake increases.[1]	✔	

The Carbohydrates

carbohydrates
Chemical substances in foods that consist of a simple sugar molecule or multiples of them in various forms.

Carbohydrates are the major source of energy for people throughout the world. They are the primary ingredient of staple foods such as pasta, rice, cassava, beans, and bread. On average, Americans consume less carbohydrate than people in much of the world: approximately 51% of total calories. This level of intake is on the low end of the recommended range of carbohydrate intake of 45 to 65% of total calories.[2]

The carbohydrate family consists of three types of chemical substances:

1. Simple sugars

2. Complex carbohydrates ("starch")

3. Total fiber

Some food sources of these different types of carbohydrate are shown in Illustration 12.1. With the exception of fiber, members of the carbohydrate family perform one role in common in the body: They provide energy. Simple sugars and starchy complex carbohydrates supply 4 calories per gram. Dietary fiber cannot be digested by enzymes produced by humans and is not considered a source of

Felicia Martinez/PhotoEdit

Illustration 12.1
The carbohydrate family.
Some food sources of simple sugars (left), and some food sources of starch and dietary fiber (right) are shown.

energy. Nevertheless, dietary fiber plays a number of important roles in maintaining health.

Carbohydrates have two sets of "chemical relatives" of significance to human nutrition. Both are derived from carbohydrates and have an alcohol group as part of their chemical structure. The first of these, the *alcohol sugars,* are presented in this unit. *Alcohol* (or, more precisely, ethanol), the other chemical relative of carbohydrates, supplies 7 calories per gram. It is covered in a separate unit. Increasingly, carbohydrates and carbohydrate-containing foods are being classified by their "glycemic index," or the extent to which they increase blood glucose levels. This topic is introduced in this unit and explored further in Unit 13 on diabetes.

Simple Sugar Facts

Simple sugars are considered "simple" because they are small molecules that require little or no digestion before they can be used by the body. They come in two types: *monosaccharides* and *disaccharides.* The monosaccharides consist of one molecule and include glucose ("blood sugar" or "dextrose"), fructose ("fruit sugar"), and galactose. Disaccharides consist of two molecules. The combination of a glucose molecule and a fructose molecule makes sucrose (or "table sugar"); maltose ("malt sugar") is made from two glucose molecules; and lactose ("milk sugar") consists of a glucose molecule plus a galactose molecule. Honey, by the way, is a disaccharide. It is composed of glucose and fructose just as sucrose is, but it's a liquid rather than a solid because of the way the two molecules of sugar are chemically linked together. Disaccharides are broken down into their monosaccharide components during digestion; only glucose, fructose, and galactose are absorbed into the bloodstream.

High-fructose corn syrup—a liquid sweetener used in many soft drinks, fruit drinks, and other products—and alcohol sugars are also considered simple sugars.[3] Most of the simple sugars have a distinctively sweet taste.

simple sugars
Carbohydrates that consist of a glucose, fructose, or galactose molecule, or a combination of glucose and either fructose or galactose. High-fructose corn syrup and alcohol sugars are also considered simple sugars. Simple sugars are often referred to as "sugars."

monosaccharides
(*mono* = one, *saccharide* = sugar): Simple sugars consisting of one sugar molecule. Glucose, fructose, and galactose are monosaccharides.

disaccharides
(*di* = two, *saccharide* = sugar): Simple sugars consisting of two molecules of monosaccharides linked together. Sucrose, maltose, and lactose are disaccharides.

Illustration 12.2
Glycogen in a liver cell. *The black "rosettes" are aggregates of glycogen molecules. This cell was photographed under an electron microscope at a magnification of 65,000X.*

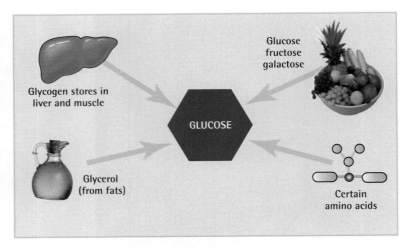

Illustration 12.3
The body's sources of glucose.

glycogen
The body's storage form of glucose. Glycogen is stored in the liver and muscles.

The only simple sugar the body can use to form energy is glucose. That's not a problem because both fructose and galactose are readily converted by the body to glucose. When the body has more glucose than it needs for energy formation, it converts the excess to fat and to *glycogen,* the body's storage form of glucose. Glycogen is a type of complex carbohydrate. It consists of chains of glucose units linked together in long strands. Glycogen is produced only by animals and is stored in the liver (Illustration 12.2) and muscles. When the body needs additional glucose, glycogen is broken down, making glucose available for energy formation. Glucose can also be derived from certain amino acids and the glycerol component of fats. Illustration 12.3 shows the various ways glucose becomes available to the body. A constant supply of glucose is needed because the brain, red blood cells, white blood cells, and specific cells in the kidneys require glucose as an energy source.[2]

Where's the Sugar? Sugars enter our diets primarily from foods that have added sugar.[4] Although used primarily because of their sweet taste, simple sugars improve the appearance, consistency, and cooking properties of many food products. Consequently, they are the most commonly used food additive. On average, Americans consume 124 grams (31 teaspoons) of added sugar per day, and that equals about 16% of the average caloric intake. Sugar consumption has increased 23% in the past 30 years. The biggest source of sugar in many diets is soft drinks,[4] but sucrose is added to a lot of other foods, too. Sugar is a very common ingredient of breakfast cereals, for example (Table 12.1), and high-fructose corn syrup sweetens many types of breakfast cereal and beverages. Simple sugars are also present in fruits, and small amounts occur in some vegetables (Table 12.2). Unlike many foods with added sugar, both fruits and vegetables provide an array of nutrients. With the exception of milk, which contains the simple sugar lactose, animal products are nearly devoid of simple sugars.

TABLE 12.1

The simple sugar content of a 1-ounce serving of some breakfast cereals

Cereal	Simple Sugars (grams)	% Total Calories from Simple Sugars
Apple Jacks	13	52%
Corn Pops	14	47
Frosted Cheerios	13	43
Raisin Bran	19	40
Frosted Flakes	11	40
All Bran	5	29
Frosted MiniWheats	6	24
Bran Flakes	5	22
Life	6	20
Rice Krispies	3	13
Grape-Nuts	3	11
Product 19	3	11
Special K	3	11
Wheaties	3	11
Wheat Chex	2	8
Cornflakes	2	7
Cheerios	1	4
Shredded Wheat	0	0

TABLE 12.2

The simple sugar content of some common foods

	Amount	Simple Sugars (grams)[a]	% Total Calories from Simple Sugars
Sweeteners:			
Corn syrup	1 tsp	5	100%
Honey	1 tsp	6	100
Maple syrup	1 tsp	4	100
Table sugar	1 tsp	4	100
Fruits:			
Apple	1 medium	16	91
Peach	1 medium	8	91
Watermelon	1 wedge (4" × 8")	25	87
Orange	1 medium	14	86
Banana	1 medium	21	85
Vegetables:			
Broccoli	1/2 cup	2	40
Corn	1/2 cup	3	30
Potato	1 medium	1	4
Beverages:			
Fruit drinks	1 cup	29	100
Soft drinks	12 oz	38	100
Skim milk	1 cup	12	53
Whole milk	1 cup	11	28
Candy:			
Gumdrops	1 oz	25	100
Hard candy	1 oz	28	100
Caramels	1 oz	21	73
Fudge	1 oz	21	73
Milk chocolate	1 oz	16	44

[a]4 grams sucrose = 1 teaspoon.

TABLE 12.3

Added sugars

Brown sugar
Confectioner's sugar
Corn syrup
Corn syrup solids
Crystal corn syrup
Dextrose
Fructose
High-fructose corn syrup
Honey
Liquid fructose
Malt syrup
Maple syrup
Molasses
Pancake syrup
Powdered sugar
Raw sugar
White sugar

Added Sugars The major source of simple sugars in most diets is sugar added to food during food processing or preparation (Table 12.3). Nutrition labels must list the total amount of sugar per serving under the heading "sugars" (Illustration 12.4). In addition, in the ingredient list, all simple sugars contained in the product must be lumped together under "sweeteners" followed by a parenthetical list

Illustration 12.4
Labeling the sugar content of breakfast cereals: two examples.

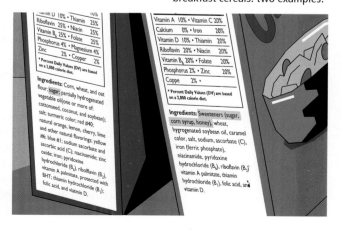

itemizing each type in order of weight.[5] Labels contain information on total sugars per serving and do not distinguish between sugars naturally present in foods and added sugars.

What's So Bad about Sugar?   Foods to which simple sugars have been added are often not among the top sources of nutrients. Simple sugars are among the few foods that provide only calories. Many foods high in simple sugars, such as cake, sweet rolls, cookies, pie, candy bars, and ice cream, are also high in fat. The likelihood that diets will provide insufficient amounts of vitamins and minerals increases along with sugar intake. Furthermore, it is perfectly clear that the frequent consumption of sticky sweets causes tooth decay.[6]

Simple sugars have been blamed for far more problems than tooth decay. Sugar has been accused of depleting the body of nutrients (it doesn't), causing violent and criminal behavior (this has been disproved by a number of studies), and producing diabetes (also untrue).[7,8] There is growing evidence, however, that high intakes of sugars may increase weight and the risk of obesity.[9,10] Nor is there any evidence that sugar intake causes hyperactivity in children.[11,12] Actually, high sugar intakes may not only decrease children's activity level but also tend to make adults sleepy.[13,14] Sugary foods have a dampening effect on appetite in the short term.[13] In other words, your parents were right when they told you not to eat that cookie before dinner.

Advice on Sugar Intake   What's the bottom line on eating sugary foods? Enjoy them in limited amounts. The Dietary Reference Intakes call for limiting intake of added sugars to less than 25% of total caloric intake. Other health authorities, including the World Health Organization and those of 26 countries, recommend limiting added sugars to 10% of calories or less.[15] Brush your teeth after eating sticky sweets and think about replacing regular soft drinks with water. And don't let sweet, high-fat foods like candy bars, pastries, and ice cream replace nutrient-dense foods such as vegetables, fruits, and whole grain products in your diet!

The Alcohol Sugars—What Are They?

alcohol sugars
Simple sugars containing an alcohol group in their molecular structure. The most common are xylitol, mannitol, and sorbitol.

Nonalcoholic in the beverage sense, the *alcohol sugars* are like simple sugars except that they include a chemical component of alcohol. Like simple sugars, the alcohol sugars have a sweet taste. Xylitol is by far the sweetest alcohol sugar—it's much sweeter than the other two common alcohol sugars, mannitol and sorbitol.

Alcohol sugars are found naturally in very small amounts in some fruits. Their presence in foods is most often due to their use as sweetening agents in gums and candy (Illustration 12.5). Unlike the simple sugars, xylitol, mannitol, and sorbitol do not promote tooth decay. The extensive use of alcohol sugars in food is limited by their tendency to cause diarrhea if consumed in large amounts.[16]

Artificial Sweetener Facts

Unwanted calories in simple sugars, the connection of sucrose with tooth decay, the need for a sugar substitute for people with diabetes, and sugar shortages such as occurred during the two world wars have all provided incentives for developing sugar substitutes. Four artificial sweeteners are currently on the market in the United States, and a fifth has been approved by the FDA. Many more artificial sweeteners are being developed.

Artificial sweeteners, also known as "intense sweeteners," are contained in a wide assortment of prepared foods (Illustration 12.6). None of these sweeteners

Illustration 12.5
Examples of products sweetened with xylitol, mannitol, and sorbitol.
Alcohol sugars provide 4 calories per gram.

Illustration 12.6
Some of the thousands of foods that contain artificial sweeteners.

are significant sources of calories, and they do not promote tooth decay or raise blood glucose levels. But none developed so far provide the optimal taste qualities of sucrose.[17] Whether the use of artificial sweeteners leads to weight loss or the prevention of weight gain is much debated. Research results can be cited that support or refute a role of artificial sugars in weight loss.[18]

Saccharin Saccharin was the first artificial sweetener developed. Did you know that it was discovered in a laboratory in the late 1800s? That's right—saccharin is over 100 years old. The availability of this artificial sweetener, which is 300 times as sweet as sucrose, helped relieve the sugar shortages that occurred during World Wars I and II.

In 1977 saccharin was taken off the market after very high doses were found to cause cancer in laboratory animals. At that time, however, saccharin was the only no-calorie, artificial sweetener available, and its removal sparked a public outcry. After many people complained to Congress, saccharin was returned to the market by congressional mandate. Saccharin was deemed safe in 2000 after scientists concluded there was no clear evidence that it causes cancer in humans.[19]

A few drops of sucrose solution on an infant's tongue have a pain-relieving effect that persists for several minutes. Sugar solution is being increasingly used in hospitals to control pain in infants undergoing heel pricks for blood collection and other minor procedures.[12]

ON THE SIDE

phenylketonuria (feen-ol-key-tone-u-re-ah), PKU
A rare genetic disorder related to the lack of the enzyme phenylalanine hydroxylase. Lack of this enzyme causes the essential amino acid phenylalanine to build up in blood.

NEOTAME

SUCRALOSE

SACCHARIN

ACESULFAME K

ASPARTAME

FRUCTOSE

SUCROSE

XYLITOL

GLUCOSE

SORBITOL

MANNITOL

GALACTOSE

MALTOSE

LACTOSE

Illustration 12.7
A ranking of various types of artificial sweeteners and naturally occurring sugars in order of sweetess.

Aspartame　Early in the 1980s, the artificial sweetener aspartame was approved for use in the United States and more than 90 other countries. Known as Nutrasweet, this artificial sweetener is about 200 times sweeter than sucrose. Aspartame is used in more than 4000 products worldwide, including soft drinks, whipped toppings, jellies, cereals, puddings, and some medicines.[20] Products containing aspartame must carry a label warning people with *phenylketonuria* (an inherited disease) and others with certain liver conditions about the presence of phenylalanine. People with these disorders are unable to utilize the amino acid phenylalanine, causing it to build up in the blood. Because high temperatures tend to break down aspartame, it is not used in baked or heated products.

■ **Aspartame Was Developed by Chemists, Not Nature**　Aspartame is made from two amino acids. Both are found in nature, but it took chemists to arrange their chemical partnership. Because aspartame is made from amino acids (the building blocks of protein), it supplies 4 calories per gram. Aspartame is so sweet, however, that very little is needed to sweeten products. Illustration 12.7 shows the relative sweetening power of various artificial sweeteners and naturally occurring sugars.

■ **Is Aspartame Safe?**　A safe level of aspartame intake is defined as 50 milligrams per kilogram of body weight per day in the United States and as 40 milligrams per kilogram of body weight in Canada.[21] In food terms, the limit in the United States is equivalent to approximately 20 aspartame-sweetened soft drinks or 55 desserts per day (Illustration 12.8). The average intake of aspartame in the United States, Canada, Germany, and Finland, for example, ranges from 2 to 10 milligrams per day, well below the level of intake considered safe.[21] A small proportion of individuals, however, report that they are sensitive to aspartame and develop headaches, dizziness, or anxiety when they consume small amounts. Studies have failed to confirm these effects.[22]

Sucralose　This noncaloric, intense sweetener is made from sucrose, is safe, is very sweet (600 times sweeter than sucrose), and does not leave a bitter aftertaste. Known as "Splenda" on product labels, it is used in both hot and cold food products, including soft drinks, baked goods, frosting, pudding, and chewing gum.

Acesulfame Potassium　The fourth artificial sweetener currently available in the United States is acesulfame potassium, also known as acesulfame K, "Sunette" and "Sweet One." Approved by the FDA in 1988, acesulfame K is added to at least 4000 foods and is used in food production in about 90 countries. It is 200 times as sweet as sucrose, provides zero calories, and does not break down when heated. The FDA has approved "neotame," an artificial sweetener deemed safe. It delivers over 7000 times the sweetness of sucrose, so very small amounts are needed in foods.[23]

Cyclamate: Still Banned in the USA　Cyclamate was widely used in food products in the United States from the 1940s through most of the 1960s. It was banned in 1969 because large doses of a cyclamate-saccharin mixture caused cancer in laboratory mice. Cyclamates are available in over 50 countries, some of which (including Canada) have banned the use of saccharin! Several national groups of experts have recently concluded that cyclamates are safe.[24]

Illustration 12.8
You would have to consume more than 20 cans of soft drinks sweetened with aspartame (Nutrasweet) a day to exceed the safe limit set for this artificial sweetener.

Complex Carbohydrate Facts

Starches, glycogen, and dietary fiber constitute the ***complex carbohydrates*** known as ***polysaccharides.*** Only plant foods such as grains, potatoes, dried beans, and corn that contain starch and dietary fiber are considered dietary sources of complex carbohydrates (Table 12.4). Very little glycogen is available from animal products.

Which Foods Have Carbohydrates?   Food sources of complex carbohydrates include whole-grain breads, cereals, pastas, and crackers, as well as these same foods produced from refined grains. Whole grain products provide more fiber and

complex carbohydrates
The form of carbohydrate found in starchy vegetables, grains, and dried beans and in many types of dietary fiber. The most common form of starch is made of long chains of interconnected glucose units.

polysaccharides
(*poly* = many, *saccharide* = sugar): Carbohydrates containing many molecules of sugar linked together. Starch, glycogen, and dietary fiber are the three major types of polysaccharides.

TABLE 12.4
The complex carbohydrate content of some common foods

	Amount	Complex Carbohydrate (grams)	% Total Calories from Complex Carbohydrates
Grain and grain products:			
Rice (white), cooked	$^1/_2$ cup	21	83%
Pasta, cooked	$^1/_2$ cup	15	81
Cornflakes	1 cup	11	76
Oatmeal, cooked	$^1/_2$ cup	12	74
Cheerios	1 cup	11	68
Whole wheat bread	1 slice	7	60
Dried beans (cooked):			
Lima beans	$^1/_2$ cup	11	64
White beans	$^1/_2$ cup	13	63
Kidney beans	$^1/_2$ cup	12	59
Vegetables:			
Potato	1 medium	30	85
Corn	$^1/_2$ cup	10	67
Broccoli	$^1/_2$ cup	2	40

Reality Check

Carbohydrate craving

Have you ever pined or whined for something sweet? Is a meal not complete unless there's a "starch"? Do you think you can get addicted to carbohydrates?

Who gets thumbs up?
Answers on page 12-13

Terry:
Maybe you could really love to eat them, but addicted? I don't think so.

Wolfgang:
The more I don't eat candy, the more I want to eat some. I swear, I'm addicted.

beneficial substances naturally present in grains than do refined grain products. Whole grain foods reduce the risk of heart disease and some types of cancer.[25]

Are They Fattening?   Starchy foods are caloric bargains (Illustration 12.9, next page). A medium baked potato weighs in at only 122 calories, a half-cup of corn at 85 calories, and a slice of bread at 70 calories. You can expand the caloric value of complex carbohydrates quite easily by adding fat, sauces, and cheese. One cup of macaroni (about 200 calories) gains around 180 calories when it comes as macaroni and cheese. Adding a quarter-cup of gravy to potatoes elevates calories by 150.

Dietary Fiber Facts

What has no calories; prevents constipation; may lower the risk of heart disease, obesity, and diabetes; and is generally underconsumed by people selecting a Western-type diet? The answer is dietary fiber.

Total fiber intake by U.S. children and adults is well below the amount recommended (Table 12.5).[2] People who consume the recommended amount of fiber tend to select whole-grain breads, high-fiber cereal, and dried beans most days and eat at least five servings of vegetables and fruits daily.[26] Food sources of dietary fiber are listed in Table 12.6. It doesn't matter whether the fiber foods are mashed, chopped, cooked, or raw. They retain their fiber value through it all. Note, though, that fast foods tend to be poor sources of fiber.

Types of Fiber   A new classification system for defining edible fibers based on source of fiber and effects on body processes has been developed.[2] Fibers are classified as *functional fiber, dietary fiber,* and *total fiber.* All fiber shares the property of not being digested by human digestive enzymes.

Functional fibers perform specific, beneficial functions in the body, including decreasing food intake by providing a feeling of fullness, reducing postmeal rises in blood glucose levels, preventing constipation, and decreasing fat and cholesterol absorption.[2] This type of fiber can be extracted from foods or produced commercially for use in fortifying foods with fiber, as well as in fiber supplements. Connective tissues from animals included in some meat products, psyllium, pectin, gels, and seed and plant gums are classified as functional fibers.

Dietary fiber consists of nondigestible carbohydrates found in plant foods. Because functional fibers are components of plant foods, this type of fiber comes with all of the other beneficial nutrients and other substances found in plants.

functional fiber
Specific types of nondigestible carbohydrates and connective tissues that have beneficial effects on health. Two examples of functional fibers are psyllium and pectin.

dietary fiber
Naturally occurring, intact forms of nondigestible carbohydrates in plants and "woody" plant cell walls. Oat and wheat bran, and raffinose in dried beans, are examples of this type of fiber.

total fiber
The sum of functional and dietary fiber.

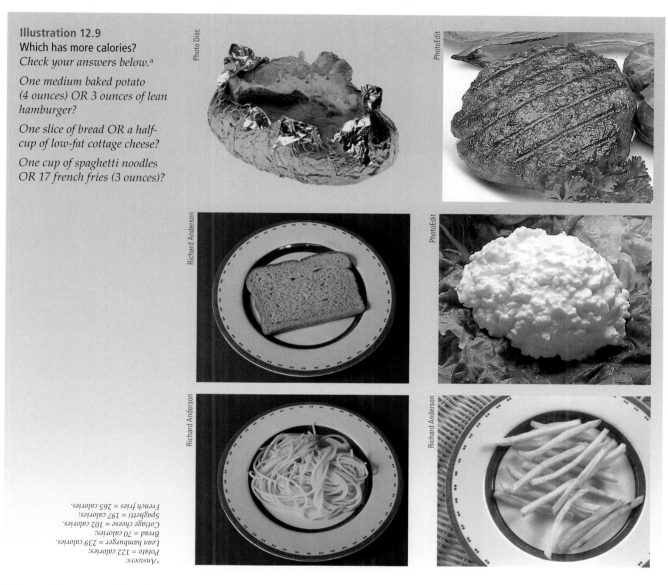

Illustration 12.9
Which has more calories?
Check your answers below.[a]

One medium baked potato (4 ounces) OR 3 ounces of lean hamburger?

One slice of bread OR a half-cup of low-fat cottage cheese?

One cup of spaghetti noodles OR 17 french fries (3 ounces)?

[a]*Answers:
Potato = 122 calories;
Lean hamburger = 239 calories.
Bread = 70 calories;
Cottage cheese = 102 calories.
Spaghetti = 197 calories;
French fries = 265 calories.*

Dietary fibers are found in the bran component of oats and wheat (Illustration 12.10), in cellulose (a rigid component of plant cell walls), in vegetables and fruits, and in the nondigestible starch components of dried beans. The recommended daily intake of fiber is based on total fiber, which is the sum of functional plus dietary fiber intake.[2]

■ **Soluble Fiber**   The term *soluble fiber* is still used by the Food and Drug Administration for nutrition labeling. Soluble fiber is a functional fiber because it benefits health in several ways. Soluble fiber slows glucose absorption, thereby lowering peak blood levels of glucose, and reduces fat and cholesterol absorption. Soluble fibers are found in oats, barley, fruit pulp, and psyllium. They are called "soluble" because of their ability to combine chemically with water.[25]

TABLE 12.5
Average and recommended daily intakes of total fiber[2]

Age-Group	Average Total Fiber Intake (GM)	Recommended Intake of Total Fiber (gm)
4–8 years	12.0	19
men, 19–50 years	13.7	38
women, 19–50 years	13.2	25

Have you ever had buckwheat pan-cakes (3 grams of fiber per serving)? Did you know that buckwheat is a member of the rhubarb family—not a grain?

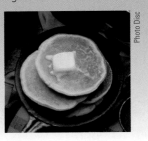

Photo Disc

TABLE 12.6
Examples of good sources of dietary fiber

The recommended intake of total fiber for men is 38 grams per day and 25 grams per day for women.

	Amount	Dietary Fiber (grams)
Grain and grain products:		
Bran Buds	1/2 cup	12.0
All Bran	1/3 cup	10.0
Raisin Bran	1 cup	7.0
Granola (homemade)	1/2 cup	6.0
Bran Flakes	3/4 cup	5.0
Oatmeal	1 cup	4.0
Spaghetti noodles	1 cup	4.0
Shredded Wheat	1 biscuit	2.7
Whole wheat bread	1 slice	2.0
Bran (dry; wheat, oat)	2 tbs	2.0
Fruits:		
Avocado	1/2 medium	7.0
Raspberries	1 cup	5.0
Mango	1 medium	4.0
Pear (with skin)	1 medium	4.0
Orange (no peel)	1 medium	3.0
Apple (with skin)	1 medium	2.4
Peach (with skin)	1 medium	2.3
Strawberries	10 medium	2.1
Banana	6" long	1.8
Vegetables:		
Lima beans	1/2 cup	5.0
Green peas	1/2 cup	4.0
Potato (with skin)	1 medium	3.5
Broccoli	1/2 cup	3.4
Brussels sprouts	1/2 cup	3.0
Collard greens	1/2 cup	3.0
Corn	1/2 cup	3.0
Carrots	1/2 cup	2.8
Green beans	1/2 cup	2.7
Cauliflower	1/2 cup	2.0
Nuts:		
Almonds	1/4 cup	4.5
Peanuts	1/4 cup	3.3
Peanut butter	2 tbs	2.3
Dried beans (cooked):		
Pinto beans	1/2 cup	10.0
Black beans (turtle beans)	1/2 cup	8.0
Black-eyed peas	1/2 cup	8.0
Kidney or navy beans	1/2 cup	6.2
Peas	1/2 cup	5.7
Lentils	1/2 cup	5.0
Fast foods:		
Big Mac	1	3
French fries	1 regular serving	3
Whopper	1	3
Cheeseburger	1	2
Taco	1	2
Chicken sandwich	2	1
Egg McMuffin	1	1
Fried chicken, drumstick	1	1

Reality Check

Carbohydrate craving

The popular theory about carbohydrate craving and addiction goes like this: Carbohydrates increase blood levels of glucose, then insulin level surges to drop blood glucose levels. Resulting low blood glucose levels send a message to your brain telling you that carbohydrates are needed to bring your blood glucose levels back up. The truth is there are no biological processes that make healthy people crave carbohydrate. (There may be psychological cravings, however.) Any source of calories will relieve hunger.[25]

Terry   Wolfgang

Be Cautious When Adding More Fiber to Your Diet Newcomers to adequate fiber diets often experience diarrhea, bloating, and gas for the first week or so of increased fiber intake. These side effects can be avoided. They occur when too much fiber is added to the diet too quickly. Although humans do not produce enzymes that can break down fiber, bacteria that live in the lower intestines do. The bacteria consume dietary fiber as food and excrete gas and other products as wastes. The gas produced by the bacteria causes the bad feelings that can accompany unusually high fiber intakes. Adding sources of dietary fiber to the diet gradually can prevent these side effects. In addition, dietary fiber can be constipating if consumed with too little fluid. Your fluid intake should increase along with your intake of dietary fiber. You know you've got the right amount of fiber in your diet when stools float and are soft and well formed.

The "Health Action" on the next page suggests some food choices that will put fiber into your meals and snacks. Fiber should be added carefully to children's diets, because the large volume of some high-fiber diets can fill the children up before they consume enough calories.[27]

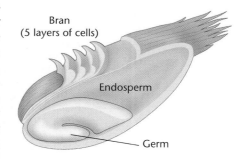

Bran
(5 layers of cells)

Endosperm

Germ

Illustration 12.10
Diagram of a grain of wheat showing the bran that is a rich source of dietary fiber.
The germ contains protein, unsaturated fats, thiamin, niacin, riboflavin, iron, and other nutrients. (The bran and germ are removed in the refining process.) The endosperm primarily contains starch, the storage form of glucose in plants.

Glycemic Index of Carbohydrates and Carbohydrates in Foods

In the not-too-distant past it was assumed that "a carbohydrate is a carbohydrate is a carbohydrate." It was thought that all types of carbohydrates had the same effect on blood glucose levels and health, so it didn't matter what type was consumed. As is the case with many untested assumptions, this one fell by the wayside. It is now known that some types of simple and complex carbohydrates in foods elevate blood glucose levels more than do others. Such differences are particularly important to people with disorders such as *insulin resistance* and *type 2 diabetes.*[28]

Carbohydrates and cabohydrate-containing foods are now being classified by the extent to which they increase blood glucose levels. This classification system is called the *glycemic index.* Carbohydrates that are digested and absorbed quickly have a high glycemic index and raise blood glucose levels to a higher extent than do those with lower glycemic index values. Carbohydrates and carbohydrate-containing foods with high glycemic index values include glucose, white bread, baked potatoes, and jelly beans. Fructose, xylitol, hummus, apples, and all-bran

insulin resistance
A condition is which cell membranes have a reduced sensitivity to insulin so that more insulin than normal is required to transport a given amount of glucose into cells.

type 2 diabetes
A disease characterized by high blood glucose levels due to the body's inability to use insulin normally or to produce

glycemic index
A measure of the extent to which blood glucose is raised by a 50-gram portion of a carbohydrate-containing food compared to 50 grams of glucose or white bread.

cereal are examples of carbohydrates and carbohydrate-containing foods with low glycemic indexes.

Diets providing low glycemic index carbohydrates have been found to improve blood glucose control in people with diabetes, to reduce elevated levels of blood cholesterol and triglycerides, increase levels of beneficial HDL cholesterol, and decrease the risk of developing type 2 diabetes, some types of cancer, and heart disease.[29]

HEALTH *Action* Putting the Fiber into Meals and Snacks

HIGH-FIBER OPTIONS FOR BREAKFAST

Whole grain toast ————————— 2 g per slice
Bran cereal:
  Bran flakes ——————— 1 cup ——— 7 g
  All Bran ——————— 1/3 cup ——— 10 g
  Raisin bran ——————— 3/4 cup ——— 5 g
  Oat bran ——————— 1/3 cup ——— 5 g
Bran muffin, with fruit: ——— 1 small ——— 3 g
Strawberries ——————— 10 ——— 2 g
Raspberries ——————— 1/2 cup ——— 3 g
Bananas ——————— 1 medium — 2 g

LUNCHES THAT INCLUDE FIBER

Whole grain bread ————————— 2 g per slice
Baked beans ——————— 1/2 cup ——————— 10 g
Carrot ——————— 1 medium ——— 2 g
Raisins ——————— 1/4 cup ——————— 2 g
Peas ——————— 1/2 cup ——————— 4 g
Peanut butter ——————— 2 tablespoons — 2 g

FIBER ON THE MENU FOR SUPPER

Brown rice ——————— 1/2 cup ——— 2 g
Potato ——————— 1 medium — 3 g
Dried cooked beans ——— 1/2 cup ——— 8 g
Broccoli ——————— 1/2 cup ——— 3 g
Corn ——————— 1/2 cup ——— 3 g
Tomato ——————— 1 medium — 2 g
Green beans ——————— 1/2 cup ——— 3 g

FIBER-FILLED SNACKS

Peanuts ——————— 1/4 cup ——————— 3 g
Apple ——————— 1 medium ——— 2 g
Pear ——————— 1 medium ——— 4 g
Orange ——————— 1 medium ——— 3 g
Prunes ——————— 3 ——————— 2 g[a]
Sunflower seeds ——— 1/4 cup ——————— 2 g
Popcorn ——————— 2 cups ——————— 2 g

[a]Prunes contain fiber, but their laxative effect is primarily due to a naturally occurring chemical substance that causes an uptake of fluid into the intestines and the contraction of muscles that line the intestines.

Carbohydrate-Related Disorders

Carbohydrates are related to a number of health problems. Carbohydrate-related disorders include diabetes, insulin resistance, and hypoglycemia (Unit 13), lactose intolerance (Unit 7), and tooth decay and other oral health problems.

Carbohydrates and Your Teeth

We tried the white man's food, and we liked it. Now we have toothaches.
—Chief Senebura, Xavante people in Pimentel Barbosa, Brazil[30]

The relationship between sugar and **tooth decay** is very close, and the history of tooth decay closely parallels the availability of sugar. The incidence of tooth decay is estimated to have been very low (less than 5%) among hunter-gatherers who had minimal access to sugars.[31] Tooth decay did not become a widespread problem until the late 17th century, when great quantities of sucrose were exported from the New World to Europe and other parts of the world. When sugar shortages occurred in the United States and Europe during World War I and World War II, rates of tooth decay declined; they rebounded when sugar became available again.[31] Rates of tooth decay in children vary substantially among countries, but the highest rates are in countries where sugar is widely available in processed foods and beverages. Tooth decay is spreading rapidly in developing countries where sugar, candy, soft drinks, and fruit drinks are becoming widely available.[32]

Sweets are not the only culprit. Simple sugars that promote tooth decay can also come from starchy foods, especially pretzels, crackers, and breads that stick to your gums and teeth. Some of the starch is broken down to simple sugars by enzymes in the mouth.

To reduce the incidence of tooth decay, a number of countries have developed campaigns to help inform consumers about cavity-promoting foods. Switzerland, for example, labels foods safe for the teeth with a "happy tooth" symbol (Illustration 12.11) and encourages the use of alcohol sugars (which don't promote tooth decay) in sweets. Other countries recommend that sweets be consumed with meals or that teeth be brushed after sweets are eaten.

There's More to Tooth Decay Than Sugar Per Se   How frequently sugary and starchy foods are consumed and how long they stick to gums and teeth make a difference in their tooth-decay-promoting effects. Marshmallows, caramels, and taffy, for example, are much more likely to promote tooth decay than are apples and milk chocolate. Nevertheless, all of the foods listed in Table 12.7 can promote tooth decay if allowed to remain in contact with the gums and teeth. Drinking coffee or tea with sugar throughout the day or consuming three or more regular soft drinks between meals hastens tooth decay (more than if these beverages are consumed with meals) Candy, cookies, and crackers eaten between meals are much more likely to promote tooth decay than are the same foods consumed as part of a meal. Chewing as few as two sticks of sugar-containing gum a day also significantly increases tooth decay.[34,35]

Why Does Sugar Promote Tooth Decay?   Sugar promotes tooth decay because it is the sole food for certain bacteria that live in the mouth and excrete acid that dissolves teeth. In the presence of sugar, bacteria in the mouth multiply rapidly and form a sticky, white material called ***plaque***. Tooth areas covered by plaque are

tooth decay
The disintegration of teeth due to acids produced by bacteria in the mouth that feed on sugar. Also called dental caries or cavities.

Illustration 12.11
Switzerland's "happy tooth" symbol.

Bacteria on the tongue have recently been identified as the primary cause of bad breath. Mouthwash gets rid of bad breath for about an hour; brushing teeth with toothpaste takes care of it in 25% of people, but brushing the tongue eliminates it in 70 to 80%.[33]

ON THE SIDE

plaque
A soft, sticky, white material on teeth; formed by bacteria.

TABLE 12.7

The "stickiness" value of some foods

The stickier the food, the worse it is for your teeth.

Very Sticky	Sticky	Somewhat Sticky	Barely Sticky
Caramels	Doughnuts	Bagels	Apples
Chewy cookies	Figs	Cake	Bananas
Crackers	Frosting	Cereal	Fruit drinks
Cream-filled cookies	Fudge	Dry cookies	Fruit juices
Granola bars	Hard candy	Milk chocolate	Ice cream
Marshmallows	Honey	Rolls	Oranges
Pretzels	Jelly beans	White bread	Peaches
Taffy	Pastries		Pears
	Raisins		
	Syrup		

prime locations for tooth decay because they are dense in acid-producing bacteria. Acid production by bacteria increases within 5 minutes of exposure to sugar. It continues for 20 to 30 minutes after the bacteria ingest the sugar.[36] If teeth are frequently exposed to sugar, the acid produced by bacteria may erode the enamel, producing a cavity. If the erosion continues, the cavity can extend into the tooth and allow bacteria to enter the inside of the tooth. That can cause an infection and the loss of the tooth. It can be prevented if the plaque is removed before the acid erodes much of the enamel. Teeth are capable of replacing small amounts of minerals lost from enamel.[37]

The Remarkable Success of Water Fluoridation   In the early 1930s, lower rates of tooth decay were observed among children living in areas where water naturally contained fluoride. This provided the initial evidence that led to the fluoridation of many community water supplies. Fluoridated water reduces the incidences of tooth decay by 50% or more and is primarily responsible for declining rates of tooth decay and loss.[38]

Credit for declines in tooth decay and tooth loss in the United States is also shared by fluoride supplements, toothpastes, rinses and gels, protective sealants, and improved dental hygiene and care.[1] Further improvements in rates of dental caries will occur with reduced intake of sugars and sticky carbohydrates. Fluoridation is a safe, effective, and cheap method of controlling dental disease; providing fluoridated community water supplies costs about 50 cents per person per year.[39] Despite the advantages of fluoride, more than a million Americans consume water from a less than optimally fluoridated water supply.[40]

Illustration 12.12

"Baby bottle caries" (also called "nursing bottle syndrome") occurs in infants who habitually receive sweet fluids or milk in bottles when they go to sleep. Cavities occur first in the upper front teeth because that's where fluid pools when babies sleep.

Custom Medical Stock

Baby Bottle Caries   A startling example of the effect that frequent and prolonged exposure to sugary foods can have is "baby bottle caries" (Illustration 12.12). Infants and young children who routinely fall asleep while sucking a bottle of sugar water, fruit drink, milk, or formula—or while breast-feeding—may develop severe decay. After the child falls asleep, the fluid may continue to drip into the mouth. A pool of the fluid collects between the tongue and the front teeth, bathing the teeth in the sweet fluid for as long as the child sleeps. The upper front teeth become decayed first because the tongue protects the lower teeth. Baby bottle caries occur in 5 to 10% of infants and young children and can lead to the destruction of all baby teeth.[39]

Foods with little carbohydrate such as cheese and peanut butter, protein (eggs, meats, yogurt), and dietary fiber (vegetables and whole grain products, for example) inhibit tooth decay (Table 12.8).[38]

TABLE 12.8
Foods that don't promote tooth decay.[2]

Artificial sweeteners	Gum and candy sweetened	Peanut butter
Cheese	with alcohol sugars	Tea
Coffee (no sugar)	Meats	Water
Eggs	Milk	Yogurt (plain)
Fats and oils	Nuts	

Key Terms

alcohol sugars, page 12-6
carbohydrates, page 12-2
complex carbohydrates, page 12-9
dietary fiber, page 12-10
disaccharides, page 12-3
functional fiber, page 12-10

glycemic index, page 12-13
glycogen, page 12-4
insulin resistance, page 12-13
monosaccharides, page 12-3
phenylketonuria (PKU), page 12-8
plaque, page 12-15

polysaccharides, page 12-9
simple sugars, page 12-3
tooth decay, page 12-15
total fiber, page 12-10
type 2 diabetes, page 12-13

Review Questions

True False

1. Excluding fiber, one gram of carbohydrate provides 4 calories. ☐ ☐

2. Fructose is a monosaccharide, maltose is a diasaccharide, and glycogen is a trisaccharide. ☐ ☐

3. When the body has more glucose than it needs for energy formation, the excess glucose is converted to glycogen and fat. ☐ ☐

4. Sugars are the most commonly used food additive. ☐ ☐

5. Excess sugar intake causes hyperactivity activity in children and predisposes youth to acts of violence. ☐ ☐

6. Xylitol and sorbitol are examples of alcohol sugars. ☐ ☐

True False

7. Excess intake of aspartame (also called Nutrasweet) causes headaches in a majority of children and adults. ☐ ☐

8. Complex carbohydrates are also known as polysaccharides. ☐ ☐

9. All types of fiber share the characteristic of providing 4 calories per gram. ☐ ☐

10. Carbohydrates and carbohydrate-containing foods are classified by their glycemic index, or the extent to which they increase blood glucose levels. ☐ ☐

11. Declining rates of dental caries in the United States are primarily related to increased access to fluoridated water supplies. ☐ ☐

Nutrition Up Close

Does Your Fiber Intake Measure Up?

Focal Point: Approximate the amount of fiber your diet contains.

Are you meeting your fiber quota, or do you consume the typical low-fiber American diet? To determine if your fiber intake is adequate, award yourself the allotted number of points for each serving of the following foods that you eat in a typical day. For example, if you normally eat 1 slice of whole grain bread each day, give yourself 2 points. If you eat 2 slices daily, give yourself 4 points. After tallying your score, refer to the Feedback section for the results.

High-fiber food choices	**2 points for each serving**
Fruits:	
1 whole fruit (e.g., apple, banana)	_____
½ cup cooked fruit	_____
¼ cup dried fruit	_____
Grains:	
½ cup cooked brown rice	_____
1 whole grain slice of bread, roll, muffin, or tortilla	_____
½ cup hot whole-grain cereal (e.g., oatmeal)	_____
¾ cup cold whole-grain cereal (e.g., Cheerios)	_____
2 cups popcorn	_____
Nuts and seeds:	
¼ cup seeds (e.g., sunflower)	_____
2 tablespoons peanut butter	_____
Vegetables: **3 points for each serving**	
1 whole vegetable (e.g., potato)	_____
½ cup cooked vegetable (e.g., green beans)	_____
Bran cereals: **7 points for each serving**	
½ cup cooked oat bran cereal	_____
1 cup cold bran cereal	_____
Legumes: **8 points for each serving**	
½ cup cooked beans (e.g., baked beans, pinto beans)	_____
Total score	_____

Special note: You can also calculate your fiber intake using the Diet Analysis Plus software. Input your food intake for one day. Then go to the Analyses/Reports section to view the total number of grams of fiber in your diet on that day.

FEEDBACK (including scoring) can be found at the end of Unit 12.

Media Menu

www.cancer.gov/cancertopics/factsheet/
Risk/artificial-sweeteners
The National Cancer Institute provides information on artificial sweeteners and cancer from this site.

www.nlm.nih.gov/medlineplus/
dietaryfiber.html
This site offers specific suggestions on how to fit more fiber into your diet, fiber and health information, fiber content of foods, and the latest research on fiber.

www.mendosa.com/gilists.htm
This site provides a table of the glycemic index of foods.

www.ada.org
It's not the American Dietetic Association, it's the American Dental Association; they have a site called "The Public" that provides information on oral health, finding a dentist, and tips for teachers.

www.nlm.nih.gov/medlineplus/
dentalhealth.html
The long menu of topics on this site includes answers to FAQs about caring for your teeth and gums; information on diet and dental health, gum chewing and caries prevention, causes of periodontal disease, facts about fluoride, and an atlas of teeth.

www.adha.org
The American Dental Hygienists Association Web site provides dental health education tools and information on dental health and care here.

www.ific.org
This is an industry-sponsored site offering information on sweeteners, artificial sweeteners, and oral health.

www.mchoralhealth.org
The Bureau of Maternal and Child Health offers information on practices and programs aimed at improving oral health in children.

Notes

1. Touger-Decker R et al. Sugars and dental caries. Am J Clin Nutr 2003;78(suppl):881S–92S.

2. Dietary reference intakes: energy, carbohydrate, fiber, fat, fatty acids, cholesterol, protein, and amino acids. Institute of Medicine, National Academies of Sciences. Washington, DC: National Academies Press, chapter 11, 2002.

3. Oberrieder HK, Fryer EB. College students' knowledge and consumption of sorbitol. J Am Diet Assoc 1991;91: 715–7.

4. Position of the American Dietetic Association: Use of nutritive and non-nutritive sweeteners. J Am Diet Assoc 2004;104:255–58.

5. Oberrieder and Fryer, College students' knowledge; and Food and Drug Administration (Department of Health and Human Services), Nutrition labeling, Federal Register 1991 Nov 27.

6. Dietary reference intakes, NAS, chap. 11; and National Institutes of Health Consensus Development Conference Statement: Diagnosis and management of dental caries throughout life. March 28, 2001; 18:1–24.

7. Bachorowski J-A et al. Sucrose and delinquency: behavioral assessment, Pediatrics 1990;86:244–53.

8. Janket S-J et al., A prospective study of sugar intake and the risk of type 2 diabetes in women, Diab Care 2003;26:1008–15.

9. Malik VS et al. Intake of sugar-sweetened beverages and weight gain: a systematic review. Am J Clin Nutr 2006;84:274–88.

10. Bray GA et al. Consumption of high-fructose corn syrup in beverages may play a role in the epidemic of obesity. Am J Clin Nutr 2004;79:537–43.

11. Wolraich M et al., The effect of sugar on behavior or cognition in children: a meta-analysis, JAMA 1995;274:1617–21.

12. Young SN. Nutrition 3. The fuzzy boundary between nutrition and psychopharmacology. Can Med Assoc J, Jan. 22, 2002; 1:66.

13. Jones JM, Elam K. Sugars and health: Is there an issue? J Am Diet Assoc 2003; 103:1058–60.

14. Wolraich et al., The effect of sugar; and Liberman HR, Sugars and behavior, Clin Nutr 1985;4:195–9; and Pivonka EEA, Grunewald KK, Aspartame- or sugar-sweetened beverages: effects on mood in young women, J Am Diet Assoc 1990;90:250–4.

15. Dietary reference intakes, NAS, chap. 11; and Joint WHO/FAO Expert Consultation on Diet, Nutrition, and the Prevention of Chronic Diseases, Geneva, Switzerland; 2002.

16. Badiga MS et al. Diarrhea in diabetics: the role of sorbitol, J Am Coll Nutr 1990;9:578–82.

17. Coulston AM et al. Sugar and sugars: myths and realities. J Am Diet Assoc 2002;102:351–53.

18. Usefulness of artificial sweeteners for body weight control. Nutr Rev 2003;61:219–220.

19. Saccharin deemed safe. Community Nutrition Institute, 2000; June 2:8.

20. Verni C. Artificial sweeteners: review and update for practitioner. SCAN's Pulse, Summer 1999.

21. Butchko HH, Kotsonis FN. Acceptable daily intake vs. actual intake: the aspartame example. J Am Coll Nutr 1991;10:258–66.

22. Spiers PA et al. Aspartame: neuro-psychologic and neurophysiologic evaluation and chronic effects. Am J Clin Nutr 1998;68:531–7.

23. Neotame: a new sweetener. Nutr Today 2002;37:84.

24. Cyclamate. http://en.wikipedia.org/wiki/Cyclamate, accessed 8/06.

25. American Association of Cereal Chemists Report: All fibers are essen-

tially functional, Cereal Foods World 2003;48:128.

26. Marlett JA, Slavin JL. Position of the American Dietetic Association: health implications of dietary fiber. J Am Diet Assoc 1997;97:1157–9.

27. Williams CL. A summary of conference recommendations on dietary fiber in childhood. Pediatrics 1995; 96:1023–8.

28. Rizkella SW et al. Low glycemic index diet and glycemic control in type 2 diabetes. Diabetes Care 2004;27:1866–72.

29. Kendall CW et al. The glycemic index: methodology and use. Nestle Nutr Workshop Ser Clin Perform Programme. 2006;11:43–56.

30. Jordan M. Colgate brings dental care to Brazilian Indian tribes; ravages of tobacco and rice. Wall Street Journal, 7/23/02, page B1.

31. Cornero S, Puche RC. Diet and nutrition of prehistoric populations at the alluvial banks of the Parana River. Medicina 2000;60:109–14.

32. Parajas IL. Sugar content of commonly eaten snack foods of school children in relation to their dental health status. J Phillipine Dent Assoc 1999;51:4–21.

33. Bad breath. Nutr Today 2000;35:6.

34. Ismail A, Food cariogenicity in Americans aged from 9 to 29 years accessed in a national cross-sectional study, 1971–1974, J Dental Res 1986; 65:1435–40.

35. Edgar WM, Sugar substitutes, chewing gum, and dental caries—a review, British Dental J 1998;184:29–32.

36. Schachtele CF, Harlander SK. Will the diets of the future be less cariogenic? J Canadian Dental Assoc 1984;3:213–9.

37. Palmer CA, Watson LM. Position of the American Dietetic Association: the impact of fluoride on dental health. J Am Diet Assoc 2005;105: 1620–28.

38. Position of the American Dietetic Association: oral health and nutrition. J Am Diet Assoc 2003;103:615–25.

39. Trends in children's oral health. National Maternal and Child Oral Health Resource Center, www .ncemch.org, 1999.

40. Populations receiving optimally fluoridated public drinking water—United States, 2000. MMWR 2002;51:144–7.

Nutrition Up Close

Does Your Fiber Intake Measure Up?

Feedback for Unit 12

The total number of points you scored approximates the **grams** of total fiber you typically consume daily.[a] Use this scale to find out if your fiber intake meets the recommended goal:

- **0-10 grams:** You consume less than the average American. Increase your fiber intake by including more fruits, vegetables, whole grains, and legumes in your diet overall.
- **11-15 grams:** Like other Americans, you consume too little fiber. Increase the number of servings of high-fiber foods you already enjoy, while substituting more high-fiber foods for refined food products. A quick way to add fiber to your diet is to consume more of the two fiber powerhouses: legumes and bran cereal.
- **15-20 grams:** You currently consume more fiber than the average American. Make sure you're including 5 or more servings of fruits and vegetables. Eat 6 to 11 servings of bread, cereal, rice, and pasta daily; choose whole-grain versions of these foods often.
- **20-40 grams:** Congratulations! Your dietary fiber intake is in the vicinity of that recommended. Keep up the good work.

[a]Because different foods within a food group contribute varying amounts of dietary fiber, the point values have been average. Make sure to check the Nutrition Facts panel on bran cereals because these cereals vary in the amount of dietary fiber they contain.

Answers to Review Questions

1. True, see page 12-2.
2. False, see pages 12-3, 12-9.
3. True, see page 12-4.
4. True, see page 12-4.
5. False, see page 12-6.
6. True, see page 12-6.
7. False, see page 12-8.
8. True, see page 12-9.
9. False, see page 12-10.
10. True, see page 12-13.
11. True, see page 12-16.

Diabetes Now

NUTRITION SCOREBOARD

	TRUE	FALSE
1 Excess sugar consumption causes diabetes.		
2 Diabetes generally develops over the course of many years.		
3 Glycemic index is a measure of the simple sugar content of foods.		

Key Concepts and Facts

- Diabetes is related to abnormal utilization of glucose by the body.

- The three main forms of diabetes are type 1, type 2, and gestational diabetes.

- Rates of type 2 diabetes increase as obesity does.

- Weight loss and physical activity can prevent or delay the onset of type 2 diabetes in many people.

Answers to NUTRITION SCOREBOARD	TRUE	FALSE
1 Simple sugar intake does not cause diabetes.[1]		✔
2 Disease processes underlying the development of diabetes exist for years before the onset of diabetes in most cases.[2]	✔	
3 Glycemic index is a measure of the potential of carbohydrate-containing foods to raise blood glucose level. Only a few simple sugars have a high glycemic index.		✔

The Diabetes Epidemic

What AIDS was in the last 20 years of the 20th century, diabetes is to be in the first 20 years of this century.
—Paul Zimmet, International Diabetes Institute[4]

diabetes
A disease characterized by abnormal utilization of carbohydrates by the body and elevated blood glucose levels. There are three main types of diabetes: type 1, type 2, and gestational diabetes. The word *diabetes* in this unit refers to type 2, which is by far the most common form of diabetes.

It's not the plague, yellow fever, or heart disease. The latest worldwide disease epidemic is **diabetes**, and the rising rates are directly related to the global increase in obesity (Illustration 13.1). Diabetes affects 6% of adults worldwide,[5] and 9.6% of U.S. adults.[6] Less than 1% of U.S. adults were diagnosed with type 2 diabetes in 1960. Over a third of the 9.6% of U.S. adults with diabetes have yet to be diagnosed and are not receiving health care services for the disorder.[7]

There are three major forms of diabetes: *type 1, type 2,* and **gestational diabetes.** Type 2 diabetes is the most common by far and is fueling the diabetes epidemic. Table 13.1 summarizes key features of type 1 and type 2 diabetes. Both types are diagnosed when fasting levels of blood glucose are 126 milliliters/deciliter (mg/dL) and higher; these types generally take years to develop.[2] In all cases of diabetes, the central defect is elevated blood glucose level caused by an inadequate supply of insulin, an ineffective utilization of insulin, or both.[8]

Insulin is a hormone produced by the pancreas that performs many functions, one of which is to reduce blood glucose levels after meals. By facilitating the passage of glucose into cells, insulin keeps a steady supply of glucose going into cells. Glucose is needed by cells as a source of energy for thousands of chemical reactions that participate in the maintenance of ongoing body functions and health. If insulin is produced in

Illustration 13.1
Diabetes headlines say it all.

TABLE 13.1

Key characteristics of type 1 and type 2 diabetes[6]

Characteristic	Type 1 Diabetes	Type 2 Diabetes
Insulin deficiency?	Yes	Possible in advanced stages of the disease
Proportion of cases	5–10%	90–95%
Risk factors	Viral infection early in life (or other triggers in genetically sensitive individuals) that destroys part of the pancreas. Young birth, age, certain medications.	Obesity (especially abdominal fatness), sedentary lifestyle, insulin resistance, low weight at birth, certain ethnicities, family history, older age
Treatment	Insulin, individualized diet and exercise	Weight loss (in most cases), increased physical activity, individualized diet, sometimes oral medications, and/or insulin

insufficient amounts, or if cell membranes are not sensitive to the action of insulin, cells become starved for glucose. Functional levels of multiple tissues and organs in the body degrade as a result. High levels of blood glucose are related to adverse side effects on the body, too, such as elevated blood levels of triglycerides, *chronic inflammation,* increased blood pressure, and hardening of the arteries.[9]

Health Consequences of Diabetes

Health effects of diabetes vary depending on how well blood glucose levels are controlled and on the presence of other health problems such as hypertension or heart disease. In the short run, poorly controlled and untreated diabetes produces blurred vision, frequent urination, weight loss, increased susceptibility to infection, delayed wound healing, and extreme hunger and thirst. In the long run, diabetes may contribute to heart disease, hypertension, nerve damage, blindness, kidney failure, stroke, and the loss of limbs due to poor circulation. The number one cause of death among people with diabetes is heart disease. Many of the side effects of diabetes can be prevented or delayed if blood glucose levels are maintained within the normal range.[7]

Type 2 Diabetes

The development of this common type of diabetes is most likely to occur in overweight and obese, inactive people (Illustration 13.2). The term *diabesity* is being used to describe the close relationship between obesity and type 2 diabetes, and the current surge in worldwide rates of both is being called the "diabesity epidemic."[10] Approximately 60% of people who develop type 2 diabetes are obese, and 30% are overweight.[11] Illustration 13.3 shows the close relationship between rising rates of obesity and type 2 diabetes in the United States.

Although most often diagnosed in people over the age of 40, type 2 diabetes is becoming increasingly common in children and adolescents.[13] There is a genetic component to this disease, as evidenced by the fact that it tracks in families and is more likely to occur in certain groups (Hispanic American, African American, Asian and Pacific Islanders, and Native Americans) than others.[7] Rather than inheriting genes that cause diabetes, certain individuals inherit or acquire very

type 1 diabetes
A disease characterized by high blood glucose levels resulting from destruction of the insulin-producing cells of the pancreas. This type of diabetes was called juvenile-onset diabetes and insulin-dependent diabetes in the past, and its official medical name is type 1 diabetes mellitus.

type 2 diabetes
A disease characterized by high blood glucose levels due to the body's inability to use insulin normally, or to produce enough insulin. This type of diabetes was called adult-onset diabetes and non-insulin-dependent diabetes in the past, and its official medical name is type 2 diabetes mellitus.

gestational diabetes
Diabetes first discovered during pregnancy.

chronic inflammation
Inflammation that lasts weeks, months, or years. Inflammation is the first response of the body's *immune system* to infection or irritation. Inflammation triggers the release of biologically active substances that promote oxidation and other potentially harmful reactions in the body.

immune system
Body tissues that provide protection against bacteria, viruses, foreign proteins, and other substances identified by cells as harmful.

insulin resistance
A condition in which cell membranes have reduced sensitivity to insulin so that more insulin than normal is required to transport a given amount of glucose into cells. It is characterized by elevated levels of serum insulin, glucose, and triglycerides, and increased blood pressure.

early in life multiple genetic traits that increase the likelihood that diabetes will develop. These genetic traits lead to metabolic changes that promote the development of diabetes when a person is exposed to certain environmental triggers, such as a diet high in saturated fats or excess calories.[14]

Type 2 diabetes takes years to develop. Its development is characterized by a number of metabolic changes that can be identified before blood levels of glucose become high enough to qualify for the diagnosis. Individuals on their way to developing type 2 diabetes tend to have somewhat higher that normal blood glucose levels and a disorder called *insulin resistance*.

Prediabetes and Insulin Resistance

Elevated fasting blood glucose levels that are somewhat below the cut point used to diagnose type 2 diabetes characterize *prediabetes*. Approximately 26% of U.S. adults, and 314 million people worldwide, are at risk of type 2 diabetes due to this condition. [6] The presence of prediabetes increases a person's odds of developing type 2 diabetes by 10% per year.[15] Most people diagnosed with prediabetes have insulin resistance. Obesity, low levels of physical activity, and genetic predispositions are common risk factors for insulin resistance. Abdominal obesity, or high levels of central-body fat, is a particularly potent risk factor for insulin resistance.[16]

Insulin resistance is due to abnormalities in the way the body uses insulin. Normally, insulin is able to lower blood glucose levels after meals by binding to receptors on cell membranes. These receptors are activated by insulin and allow glucose to pass into cells. With insulin resistance, cell membranes "resist" the effects of insulin, and that lowers the amount of glucose transported into cells. Excessive amounts of abdominal fat contribute to insulin resistance by releasing "free fatty acids" into the blood. These fatty acids accumulate in muscles, the liver, and other organs and further increase insulin resistance.[16] To prevent blood glucose levels from becoming too high, the pancreas produces and secretes additional insulin. Higher-than-normal levels of insulin are generally sufficient to keep blood glucose levels under control for a number of years. Cells in the pancreas may become exhausted, however, from the years of overwork. In such cases, production of insulin slows down or stops, and glucose accumulates in the blood as a result. When fasting blood glucose levels consistently reach 126 mg/dL or higher, type 2 diabetes is under way.[6]

Insulin resistance is also related to the development of a spectrum of metabolic abnormalities that have far-reaching effects. Collectively, the adverse effects of insulin resistance are included in a disorder called *metabolic syndrome*.

Metabolic Syndrome Physicians have known for decades that obese people with hypertension and type 2 diabetes are at high risk of heart disease. What they didn't know is why. Over time, research studies discovered that part of the answer to the "why" question was insulin resistance.[17] There is currently no

Illustration 13.2
Obesity characterized by central-body fat stores and physical inactivity is a strong risk factor for type 2 diabetes.

Illustration 13.3
The incidence of type 2 diabetes increases as rates of obesity increase.[12]

Diagnosed diabetes, percent of adults

Obesity, percent of adults

Percent

Percent

Year

simple and inexpensive test for insulin resistance. Clinically, it is assumed to be present in people with high waist circumferences.[19] Insulin resistance is related to a cluster of metabolic abnormalities included in the metabolic syndrome that increase the risk of type 2 diabetes as well as heart disease. Symptoms related to the metabolic syndrome include:

- High levels of central-body fat

- Waist circumference 40″ or over in males, 35″ or over in females

- High blood pressure (130/85 mm/Hg or higher)

- Elevated blood triglycerides levels (150 mg/dL or higher)

- Low levels of protective HDL cholesterol (less than 50 mg/dL in women and 40 mg/dL in men)

- Elevated fasting blood glucose levels (110 mg/dL or higher)

The diagnosis of metabolic syndrome is made when three or more abnormalities are identified. Individuals with four or five metabolic abnormalities have a 3.7 times greater risk of heart disease, and a 25 times higher risk for diabetes than do people with no abnormalities.[9] It is estimated that 20% of men and women in the United States have metabolic syndrome, with the bulk of cases being made up of overweight and obese inactive adults.[18] Weight loss, exercise, adequate vitamin D status, adequate magnesium intake, and the other factors listed in Illustration 13.4 are the key components of preventing and managing metabolic syndrome.[7,20]

Managing Type 2 Diabetes

Diet and exercise are the cornerstones of the treatment of type 2 diabetes.[22] Modest weight loss alone (5 to 10% of body weight) has been repeatedly shown to significantly improve blood glucose control in overweight and obese people with type 2 diabetes.[23]

In general, diets developed for diabetes emphasize

- Complex carbohydrates, including whole-grain breads and cereals, and other high-fiber foods, vegetables, fruits, low-fat milk and meats, and fish

- Unsaturated fats

- Regular meals and snacks[24]

prediabetes
A condition in which blood glucose levels are higher than normal but not high enough for the diagnosis of diabetes. It is characterized by impaired glucose tolerance, or fasting blood glucose levels between 110 and 126 mg/dL.

metabolic syndrome
A constellation of metabolic abnormalities that increase the risk of heart disease and type 2 diabetes. Metabolic syndrome is characterized by insulin resistance, abdominal obesity, high blood pressure and triglycerides levels, low levels of HDL cholesterol, and impaired glucose tolerance. It is also called *Syndrome X* and *insulin resistance syndrome.*

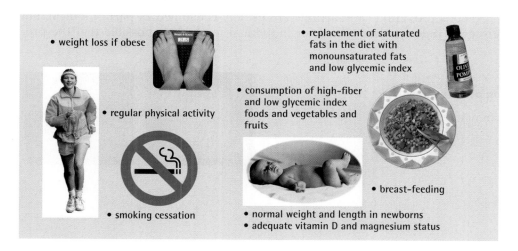

- weight loss if obese
- regular physical activity
- smoking cessation
- replacement of saturated fats in the diet with monounsaturated fats and low glycemic index
- consumption of high-fiber and low glycemic index foods and vegetables and fruits
- breast-feeding
- normal weight and length in newborns
- adequate vitamin D and magnesium status

Illustration 13.4
Key components of the prevention and management of metabolic syndrome.

Chromium supplements (500 to 1000 µg per day) are sometimes recommended. It appears the essential mineral chromium improves blood glucose and lipid levels in many people with type 2 diabetes.[25]

Dietary management of diabetes should focus on heart disease risk reduction as well as blood glucose control. Food sources of monounsaturated fats, such as vegetable oils, nuts, seeds, and lean meats and seafoods are recommended over foods high in saturated or *trans* fats. Monounsaturated fats raise blood levels of HDL cholesterol while reducing LDL cholesterol level, improve insulin resistance, and lower blood glucose levels somewhat.[26] Diet and weight-loss interventions may be supplemented by oral medications that decrease insulin resistance and blood lipids and by insulin if needed.[27]

Knowledge of the role of diet, exercise, insulin, and other factors in the management of diabetes is expanding so rapidly that the American Diabetes Association updates management recommendations yearly. Dietary recommendations are currently not consistent across developed countries, indicating that scientific consensus is yet to be reached on a number of important issues related to diet and diabetes.

Glycemic Index and Glycemic Load   Carbohydrate-containing foods have a range of effects on blood glucose levels—some cause a rapid rise, and others do not. Foods that increase blood glucose to relatively high levels require more insulin to move glucose into cells than do foods that produce lower levels of glucose. Over the past 25 years, many carbohydrate-containing foods have been tested for their effect on blood glucose level and assigned a *glycemic index* value. Compared to high glycemic index (GI) carbohydrate sources, foods with low GI values increase blood glucose levels to a lower extent, and decrease insulin need.[28]

The glycemic index of carbohydrate-containing foods is determined by assessing the elevation in blood glucose level caused by ingestion of 50 grams of a carbohydrate-containing food compared to the rise in blood glucose level that results from consuming 50 grams of glucose. (Sometimes the standard for comparison is white bread). Table 13.2 shows the glycemic index of a number of foods, using 50 grams of glucose as the standard for comparison. As you read over the list of foods and their glycemic index values, you may find some surprises. Sucrose, honey, fructose, and other simple sugars are not high glycemic index foods, and many fruits we think of as sweet do not cause a relatively high rise in blood glucose level. Coarse-ground whole-wheat breads and dried beans have medium-to-low glycemic index values. It probably comes as no surprise that glucose has a glycemic index of 100, but it is infrequently found by itself in foods.

Foods providing carbohydrate are usually consumed as part of a meal, but glycemic index is determined for individual foods. It is possible that the pro-

glycemic index (GI)
A measure of the extent to which blood glucose level is raised by a 50-gram portion of a carbohydrate-containing food compared to 50 grams of glucose or white bread.

Reality Check

Will the real whole grain please stand up?

Which bread is made from whole grains?

Who gets thumbs up?
Answers on page 13-8

Wheat Bread

Whole-Wheat Bread

TABLE 13.2
Glycemic index (GI) of selected foods[3]

High GI	(70–100)	Medium GI	(56–69)	Low GI	(55 or lower)
glucose	100	orange soda	68	honey	55
French bread	95	sucrose	68	oatmeal	54
scone	92	croissant	67	corn	53
potato, baked	85	Cream of Wheat	66	cracked wheat bread	53
potato, instant mashed	85	couscous	65	orange juice	52
Corn Chex	83	chapati	62	banana	52
pretzel	83	sweet potato	61	mango	51
Rice Krispies	82	muffin, blueberry	59	potato, boiled	50
cornflakes	81	Coca-Cola	58	green peas	48
Corn Pops	80	rice, white or brown	60	pasta	48
Gatorade	78	breadfruit	69	carrots, raw	47
jelly beans	78	taco shells	68	lactose	46
doughnut, cake	76	angel food cake	67	milk chocolate	43
waffle, frozen	76	Quaker Quick Oats	65	orange	42
french fries	75	French bread with butter and jam	62	peach	42
Shredded Wheat	75	couscous	61	apple juice	40
Cheerios	74	Raisin bran	61	apple	38
popcorn	72	bran muffin	60	pear	38
watermelon	72	Just Right cereal	60	tomato juice	38
Grape Nuts	71	power bar	56	yam	37
wheat bread	70			dried beans	25
white bread	70			grapefruit	25
				fructose	19
				xylitol	8
				hummus	6

tein and fat content of other foods in a meal affect glycemic index values of the carbohydrate-containing foods. To determine if this is the case, scientists assessed the glycemic index of carbohydrate foods consumed in mixed meals. The results of the study showed that the protein and fat content of a meal has little effect on the glycemic index of carbohydrate-containing foods consumed. Glycemic index values for the carbohydrate-containing foods in the meals were within 90% of the values reported for the individual foods.[29]

Some high-GI foods such as baked potatoes, French bread, and cornflakes are good sources of a number of nutrients. Just because a food has a high glycemic index doesn't mean it should not be consumed as part of a balanced diet. Adjusting food choices toward selection of mainly low-GI foods is most helpful for people attempting to prevent or control type 2 diabetes or to diminish the effects of insulin resistance.[30]

Because of the way glycemic index is calculated, it is difficult to know the extent to which blood glucose will be raised by consumption of a particular amount of a food. Another index of the blood-glucose-raising potential of carbohydrate-containing foods has been developed to help straighten out this confusion. It's called **glycemic load**, and it represents the blood-glucose-raising potential of the specific amount of food consumed. Glycemic load is calculated by multiplying the grams of carbohydrate in a specific amount of food times the food's glycemic index. This result is then divided by 100 to calculate glycemic load. A carrot provides approximately 7 grams of carbohydrate and has a glycemic index of 47. Its glycemic load would be calculated as:

$$7 \times 47 = 329$$
$$329/100 = 3.29$$
$$\text{Glycemic load} = 3.29$$

glycemic load (GL)
A measure of the extent to which blood glucose level is raised by a given amount of a carbohydrate-containing food. GL is calculated by multiplying a food's GI by its carbohydrate content.

Reality Check

Will the real whole grain please stand up?

You can't tell by looking, but the slice of bread on the right contains more fiber (4 vs. 2 grams). Want to find whole-grain breads? Look for the term "whole grain" or "whole wheat" on the label. Bread products carrying that label contain 51% or more whole grains. The ingredient label on whole-grain products will list one or more whole grains before listing enriched flour.

Wheat Bread

Whole Wheat Bread

The blood-glucose-raising effect of one carrot doesn't amount to much. If you consumed 4 slices (4 ounces) of French bread, the result would tell a different story. The glycemic load supplied by this amount of French bread is 49.4, a level that raises blood glucose and insulin levels far more than a carrot or 1 slice of French bread.

Consumption of low-GI foods is a recommended component of the dietary management of type 2 and gestational diabetes in a number of countries, and is becoming increasingly recommended in the United States. Consumption of low-GI foods is viewed as a useful part of the management of insulin for resistance and metabolic syndrome, and as a secondary aid to blood glucose control for people with all forms of diabetes.[31]

Sugar Intake and Diabetes

Does sugar intake cause diabetes? The answer to this reasonable and often asked question is no. Results of multiple studies indicate that high intakes of simple sugars do not cause diabetes.[1] Regular consumption of high glycemic index foods, however, may be linked to type 2 diabetes in genetically susceptible people.[30] Should people with diabetes exclude sugars from their diet? Sugar does not have to be eliminated from the diet of people with diabetes. Intake of total carbohydrates and the glycemic index of the carbohydrate-containing foods, rather than sugar intake specifically, are most strongly related to blood glucose levels.[31] High-sugar diets increase blood triglyceride levels in people with metabolic syndrome, and that may increase the risk of heart disease.[24]

Prevention of Type 2 Diabetes

The effects of weight loss and exercise in preventing type 2 diabetes can be quite dramatic. In one large study that took place over a 3-year period, people with pre-diabetes reduced their risk of developing type 2 diabetes by over 50% with losses in body weight of around 7% and 150 minutes a week of exercise.[32]

Diets rich in whole-grain and high-fiber foods are protective against the development of type 2 diabetes and appear to aid weight loss. Components of high-fiber, whole grain foods raise blood glucose levels marginally and appear to provide nutrients and other biologically active substances that lessen the risk of this disease.[33] Consumption of regular or decaffeinated coffee (1 to 4 cups daily) and moderate alcohol intake (1 to 2 drinks per day if appropriate) also appear to decrease the risk of developing type 2 diabetes.[34,35] Finally, type 2 diabetes may be prevented or postponed by not gaining weight during the young adult years. Weight gains of 10 to 15 pounds between the ages of 25 and 40 years have been

PhotoDisc

Illustration 13.7
Whole-grain products, other high-fiber foods, and ample servings of vegetables and fruits can help prevent type 2 diabetes.

HEALTH *Action* Preventing Type 2 Diabetes

Are You at Risk?

Check each category that applies to you:

___ I have a brother, sister, mother, or father with type 2 diabetes.

___ My BMI is 25 or higher.

___ My waist cicrumfrence is over 35 inches (if female) or 40 inches (if male).

___ I have been diagnosed with pre-diabetes or insulin resistance.

___ I am habitually inactive.

___ I have been diagnosed with high blood pressure.

___ I have been diagnosed as having high bood triglyceride levels.

If you checked one or more of the categories above, you are at increased risk of developing type 2 diabetes.

What to Do Now?

1. Recognize that there are no quick fixes for the prevention of type 2 diabetes.
2. If you are overweight or obese, gradually cut back on your calorie intake by making small and acceptable changes in your diet. Aim for a reduction in caloric intake of 250 calories a day.
3. Gradually increase your physical activity level until you are moderately to vigorously active for at least 150 minutes per week. Select activities you enjoy and will keep doing in the long run.
4. Choose low glycemic index foods for your carbohydrate-containing food choices as often as possible. (You can use Table 13.2 to check out your options for low glycemic index foods).
5. Consume enough fiber (25 to 35 grams daily).
6. Eat your favorite vegetables and fruits often.
7. Choose low-fat dairy products.
8. Select and consume whole-grain products rather than refined grain products.[28,34]
9. See your health care provider if you think you have a problem with your blood glucose levels.

Royalty-Free/Corbis

found to increase the risk of type 2 diabetes sevenfold compared to not gaining weight or gaining it after the age of 40.[36]

It is anticipated that 800,000 new cases of type 2 diabetes may develop each year in the United States due to rising rates of obesity.[7] The additional burdens such an increase would place on individuals and health care costs clearly convey the message that prevention is urgently needed. Public health campaigns are now under way to encourage people to lose weight if overweight, to exercise regularly, and to select whole-grain products and other high-fiber foods along with ample intake of vegetables and fruits.

Are you at risk of developing type 2 diabetes? Find out if you are, and what people at risk can do to help prevent it in the Health Action.

Type 1 Diabetes

Type 1 diabetes results from a deficiency of insulin and accounts for 5 to 10% of all cases. The diagnosis of type 1 peaks around the ages of 11 to 12 years, and it usually occurs before the age of 40.[6]

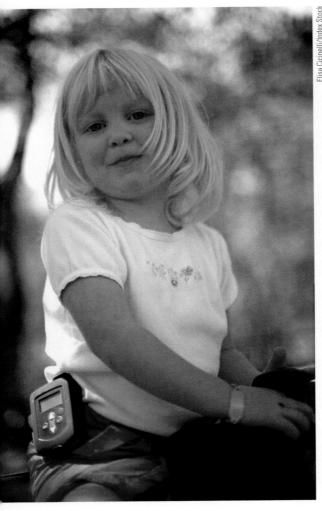

Illustration 13.5
The first insulin pumps in the 1960s were heavy and cumbersome—so large they had to be worn as a backpack. Today's pumps allow much greater freedom of movement.

1500 B.C Early healers noticed that ants and files were attracted to the urine of people with a mysterious, emaciating disease. The attraction was later found to be glucose and the emaciating disease, diabetes.[46]

Managing Type 1 Diabetes

The main goal guiding the management of type 1 diabetes is blood glucose control. Individuals who keep their blood glucose levels within the normal range experience the fewest complications over time. Blood glucose levels are primarily managed by regular, healthful meals consumed at planned times and in specified amounts to match insulin needs. Physical activity is recommended, and levels of activity have to be accounted for when calculating food and insulin need for blood glucose control.[42]

People with type 1 diabetes are instructed to measure blood glucose levels several times daily and to adjust insulin dose according to the results. Both the skin pricks required to test blood and delivery of insulin by injection may be painful, however, and are a feared part of the disease.[6] Technological advances such as the insulin pump (Illustration 13.5) are taking out some of the sting. The insulin pump technology uses the amount of carbohydrate in a meal or snack and the blood glucose level a user enters and calculates and injects the required amount of insulin. Although it is a bit complicated to use, the insulin pump can cut down on the number of insulin injections people have to take, and it gives people more flexibility about when and what they eat, and the timing of their physical activity. Insulin pump technology is advancing rapidly and soon will include insertable glucose sensors that will accurately measure blood glucose levels.[43]

Other advances in methods for managing blood glucose levels have arrived and include inhaled insulin and pancreatic islet cell transplant. Inhaled insulin was approved in 2006 and can be used to replace some insulin injections, as the pump does.[44] Insulin-producing pancreatic cell transplants have been tested for decades and offer superior blood glucose control. Some people with type 1 diabetes receiving the transplant no longer require insulin injections. The current problem with the transplanted pancreatic cells is that they tend to lose their ability to produce insulin after several years.[45]

Insulin deficiency appears to develop when a person's own immune system destroys beta cells in the pancreas that produce insulin.[47] The destruction is likely triggered by a viral infection (such as mumps, rubella, or the flu) or allergic reactions in genetically susceptible people. Medications used to treat high blood pressure, arthritis, and other disorders may also contribute to the development of type 1 diabetes. Breastfeeding for the first 4 months of life or so and adequate vitamin D status confer elements of protection against type 1 diabetes.[49,50] Weight loss, even as little as 10 to 15 pounds, benefits blood glucose control in overweight and obese individuals with type 1 diabetes.[51]

Environmental factors appear to be more important than genetic background in the development of this type of diabetes.[50] The incidence of type 1 diabetes varies 36-fold among countries (Illustration 13.6), suggesting that environmental factors play a key role in its development.[42] For yet unknown reasons, the worldwide incidence of type 1 diabetes is increasing by 3% a year.[47]

Gestational Diabetes

Approximately 5 to 6% of women develop gestational diabetes during pregnancy, but the incidence varies a good deal based on age, body weight, and ethnicity. Native Americans, African Americans, women over the age of 35 years, obese women, and those with habitually low levels of physical activity are at higher risk than other women.[53] Gestational diabetes is likely caused by the same basic environmental and genetic predisposition factors as is type 2 diabetes.[54] Infants born to women with poorly controlled diabetes may be excessively fat at birth, require a cesarean delivery, and have blood glucose control problems after delivery. They are at greater risk for developing diabetes later in life.[53]

As is the case for type 2 diabetes, women with gestational diabetes are insulin resistant and can often control their blood glucose levels with an individualized diet and exercise plan. Some women require daily insulin injections for blood glucose control.[55]

Gestational diabetes disappears after delivery, but type 2 diabetes may appear later in life. Exercise, maintenance of normal weight, and consumption of a healthy diet reduce the risk that diabetes will return.[54]

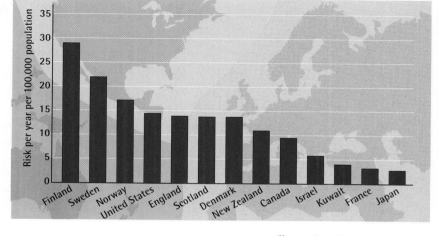

Illustration 13.6
Incidence of type 1 diabetes in various countries.[46,52]

Hypoglycemia: The Low Blood Sugar Blues?

Hypoglycemia is due to abnormally low blood glucose levels. It is thought to be a rare disorder because it is not often diagnosed. The diagnosis is tricky—blood tests for glucose should be conducted when the symptoms of hypoglycemia are present rather than during a scheduled appointment.[37] The true incidence of hypoglycemia is not known.

Hypoglycemia is most often caused by an excessive availability of insulin in the blood. The oversupply of insulin may be caused by certain tumors that secrete insulin, by other health problems, by high alcohol intake on an empty stomach, or, in people with diabetes, by an insulin dose that is too high. Hypoglycemia also occurs during prolonged starvation or fasting.[38]

Symptoms of hypoglycemia include weakness, sweating, nervousness, confusion, and irritability. They may appear before blood glucose levels are low enough to be clinically considered abnormal. Apparently, people vary considerably in their response to low-normal glucose levels.[38] Symptoms of hypoglycemia disappear within 5 to 15 minutes after glucose, candy, orange juice, or similar food is consumed. Increased blood glucose levels need to be maintained by food after the drop in blood glucose level has been corrected.[24]

The notion that high sugar intakes by themselves cause hypoglycemia has not been verified by research.[40] People who suspect that they have this disorder should consult a physician and have their blood glucose level checked at a time when the symptoms normally occur.[41]

hypoglycemia
A disorder resulting from abnormally low blood glucose levels. Symptoms of hypoglycemia include irritability, nervousness, weakness, sweating, and hunger. These symptoms are relieved by consuming glucose or foods that provide carbohydrate.

Standard diet therapy for hypoglycemia calls for people to consume five to six small meals that include a variety of food sources of complex carbohydrates, protein, and fat and to avoid alcohol and snacks high in simple sugars.[38]

Diabetes in the Future

The anticipated surge in the worldwide incidence of type 2 diabetes is not inevitable. It could be lowered substantially by environmental and lifestyle changes that reduce the risk for, and incidence of, overweight and obesity. Increased awareness of the connection between diabetes and body weight may help. Only a small proportion of people are aware of the connection now.[56] The hoped-for future of diabetes would be the one that negates the dire forecasts of the experts.

Nutrition Up Close

Calculating Glycemic Load

Focal Point: To gain an appreciation of the effect of source and amount of carbohydrate consumed on blood glucose levels.

Glycemic index provides an estimate of the rise in blood glucose level expected from consuming 50 grams of a carbohydrate containing food. Glycemic load, on the other hand, is a measure of the expected rise in blood glucose related to the ingestion of other amounts of this food. Consequently, glycemic load estimates the blood-glucose-raising potential of the amount of carbohydrate containing food actually consumed.

1. Use Appendix A to look up the carbohydrate content of each food for the serving size listed.
2. Use Table 13.2 to identify the glycemic index for each food.
3. Calculate glycemic load based on this formula:

   Glycemic load = grams carbohydrate × glycemic index of the food, divided by 100

Here's an example for 2 tsp sucrose:

$$\text{grams carbohydrate in 2 tsp sucrose} = 8$$
$$\text{glycemic index of sucrose} = 68$$
$$8 \times 68 = 544$$
$$\text{glycemic load} = 544 \div 100 = 5.44$$

Food	Serving Size	grams carbohydrate	×	Glycemic Index	÷	Glycemic Load 100
cola beverage	36 oz (3 cans)	_____		_____		_____
potato, baked, no skin	1	_____		_____		_____
apple juice, bottled	1 cup	_____		_____		_____
milk chocolate, plain	1 oz	_____		_____		_____
hummus	½ cup	_____		_____		_____

FEEDBACK Glycemic load answers can be found at the end of Unit 13.

Key Terms

chronic inflammation, page 13-3
diabetes, page 13-2
gestational diabetes, page 13-2
glycemic index, page 13-6

glycemic load, page 13-8
hypoglycemia, page 13-11
immune system, page 13-3
insulin resistance, page 13-4

metabolic syndrome, page 13-4
prediabetes, page 13-4
type I diabetes, page 13-2
type 2 diabetes, page 13-2

Review Questions

True   False   True   False

1. The incidences of type 2 diabetes and of gestational diabetes increase as rates of overweight and obesity increase. ☐ ☐

2. Insulin facilitates the passage of vitamins into cells. ☐ ☐

3. Individuals with type 2 diabetes and gestational diabetes generally have the condition known as insulin resistance. ☐ ☐

4. The most effective strategies for the prevention and control of type 2 diabetes are weight loss and exercise. ☐ ☐

5. Prediabetes is defined as a condition in which fasting blood glucose levels are between 110 and 126 mg/dL. ☐ ☐

6. A female with a waist circumference of 29 inches, high blood pressure, and low levels of LDL cholesterol would be diagnosed as having "metabolic syndrome." ☐ ☐

7. Gestational diabetes appears to be caused by the same environmental and genetic predisposition factors that apply to type 2 diabetes. ☐ ☐

8. All foods high in complex carbohydrates have low glycemic indices. ☐ ☐

9. Excess sugar intake from soft drinks is a direct cause of type 2 diabetes. ☐ ☐

10. Hypoglycemia is most often caused by an excess availability of insulin. ☐ ☐

Media Menu

www.webmd.com
Go to "Diabetes" or to "Conditions A–Z," select diabetes, and hit "go." View a video, get the latest information on the treatment or a risk assessment for diabetes, and test your diabetes IQ.

www.diabetes.com
Part of WebMD, provides videos, audios, and print material on diabetes facts, tips for eating well, information on insulin resistance, and more.

www.niddk.nih.gov
Search diabetes and receive information for people newly diagnosed with diabetes, the low-down on diabetes treatment, diet planning, and a lot more.

http://diabetes.niddk.nih.gov/dm/pubs/ hypoglycemia/index.htm
From the National Institutes of Health, this site describes hypoglycemia and its causes and care.

www.healthfinder.gov
Search insulin resistance, diabetes, metabolic syndrome, and hypoglycemia to find reliable reports.

www.diabetes.org
The online home of the American Diabetes Association provides lots of resources and information on the various types of diabetes.

www.ndep.nih.gov
The home page of the National Diabetes Education Program leads you to information and resources on diabetes, prediabetes, diabetes in children and youth, prevention strategies, programs, and more.

http://diabetes.niddk.nih.gov/dm/pubs/ statistics/index.htm#7
Get the latest statistics on the incidence of diabetes and its health complications as well as.information on its prevention and treatment on this site.

Notes

1. Liu S et al. Sugar intake and diabetes risk. Diabetes Care 2003;26:1008–15.

2. Nathan DM. Initial management of glycemia in type 2 diabetes mellitus. N Engl J Med 2002;347:1242–50.

3. Foster-Powell K et al. International table of glycemic index and glycemic load values: 2002. Am J Clin Nutr 2002;76:5–56.

4. World seen facing diabetes catastrophe, impact may outpace AIDS. International Diabetes Federation Conference, Paris 2003. Reported in www.medscape.com, 9/6/03.

5. www.nlm.nih.gov/medlineplus/news/fullstory_34711.html, 7/06. Type 2 affects 6% of adults worldwide.

6. Diabetes. http://diabetes.niddk.nih.gov/dm/pubs/statistics/index.htm#7, accessed 8/06.

7. Health, United States, 2005. www.cdc.gov, accessed 8/06.

8. What is gestational diabetes? www.nichd.nih.gov/publications/pubs/gdm/sub1.htm, accessed 8/06.

9. Alexander CM et al. NCEP-defined metabolic syndrome, diabetes, and prevalence of coronary heart disease among NHANES III participants age 50 years and older. Diabetes 2003;52:1210–14.

10. Jones V. The "Diabesity" epidemic: Let's rehabilitate America. Medscape Gen Med 2006; 8:34.

11. Incidence of diagnosed diabetes rising in the US. www.cdc.gov, 10/05.

12. Graph produced based on information in: The CDC's Database Modeling Project: developing a new tool for chronic disease prevention and control, www.cdc.gov, accessed 8/06

13. Vivian EM. Type 2 diabetes in children and adolescents—the next epidemic? Curr Med Res Opin 2006;22:297–306.

14. Marin, C et al. The Ala54Thr polymorphism of the fatty-acid binding protein 2 gene is associated with a change in insulin sensitivity after a change in the type of dietary fat. Am J Clin Nutr 2005; 82:196–200.

15. Cowie CC, MMWR, CDC Surveillance System, 2003:52:833–7.

16. Ronti T et al. The endocrine function of adipose tissue. Clin Endocrinol 2006;64:355–65.

17. Reaven GM. The metabolic syndrome: is this diagnosis necessary? Am J Clin Nutr 2006;83:1237–47.

18. Park Y-W et al. The metabolic syndrome: prevalence and associated risk factors. Arch Int Med 2003;163:427–36.

19. Ehrmann DA. Polycystic ovary syndrome. N Engl J Med 2005;352:1223–36.

20. Pittas AG et al. High levels of vitamin D, calcium may lower type 2 diabetes risk. Diabetes Care 2006;29:650–56.

21. Huerta MG et al. Magnesium deficiency and insulin resistance in children. Diabetes Care 2005;28:1175–81.

22. Nathan DM et al, Guidelines for management of hypoglycemia of type 2 diabetes. Diabetes Care 2006;29:1963–72.

23. American Diabetes Association. The prevention or delay of type 2 diabetes. Diabetes Care 2002;25:vol.4.

24. Franz MJ et al. Evidence-based nutrition principles and recommendations for the treatment and prevention of diabetes and related complications. Diabetes Care 2002;25:148–66.

25. Martin J et al. Chromium picolinate supplementation attenuates body weight gain and increases insulin sensitivity in subjects with type 2 diabetes. Diabetes Care. 2006;29:1826–32.

26. Solomon CG, Reducing cardiovascular risk in type 2 diabetes, N Engl J Med 248:2003;457–60.

27. Hamdy O et al. New "lifestyle" guidelines for type 2 diabetes. Medscape Med News, www. medscape.com .viewarticle/502699, accessed 4/05.

28. Rizkella SW et al. Low glycemic index diet and glycemic control in type 2 diabetes. Diabetes Care 2004;27:1866–72.

29. Wolever TMS et al. Food glycemic index, as given in Glycemic Index tables, is a significant determinant of glycemic response elicited by a composite breakfast meal. Am J Clin Nutr 2006;83:1306–12.

30. Brand-Miller JC. Postprandial glycemic, glycemic index, and the prevention of type 2 diabetes. Am J Clin Nutr 2004;80:243–4.

31. Ludwig DS, Glycemic load comes of age. J Nutr 2003;133:2728–32.

32. Diabetes Prevention Program Research Group, Reduction in the incidence of type 2 diabetes with lifestyle intervention or metaformin, N Engl J Med 2002.

33. McKeown N et al., Whole-grain intake is favorably associated with metabolic risk factors for type 2 diabetes and cardiovascular disease, Am J Clin Nutr 2002;76:390–8.

34. Gruber A et al. Diabetes prevention: is there more to it than lifestyle change? Int J Clin Pract 2006;60:590–4.

35. Van Dam RM et al. Coffee and the risk of type 2 diabetes in women. Diabetes Care 2006;29:398:403.

36. Schienkiewitz A et al. Body mass index history and risk of type 2 diabetes: results from the European Prospective Investigation into Cancer and Nutrition (EPIC)-Potsdam Study. Am J Clin Nutr 2006;84:427–33.

37. Amiel S. Reversal of unawareness of hypoglycemia. N Engl J Med 1993; 329:876–7.

38. Blumberg S. Should hypoglycemia patients be prescribed a high-protein diet? J Am Diet Assoc 2005;105:196–7.

39. Palard J et al. Blood glucose measurements during symptomatic episodes in patients with suspected postprandial hypoglycemia, N Engl J Med 1989;321:1421.

40. Nelson RL. Hypoglycemia: fact or fiction? Mayo Clin Proc 1985;60:844–50.

41. Service FJ. Hypoglycemia and the postprandial syndrome. N Engl J Med 1989;321:1472–4.

42. Studies aim to preserve insulin production in type 1 diabetes. www.nih.gov/news/pr/jun2006/niddk-09.htm, accessed 8/06.

43. Baltz JC et al. Brave new world: insulin pump technology. Today's Dietitian 2004;Nov:48–50.

44. First inhaled insulin product approved. FDA Consumer Magazine 2006;Mar–Apr.

45. Shapiro AMJ. Transplanting islets for diabetes. The Scientist 2006; May:43–8.

46. MacCraken J et al. From ants to analogs. Postgrad Med 1997;101.

47. McNally RJQ et al. Space-time cluster analysis of type one diabetes among 0–29 year-olds in Yorkshire, UK. Diabetological 2006;49:900–4.

48. LaPorte RE, Taijima N, Akerblom HR, et al. Geographic differences in the risk of insulin-dependent diabetes mellitus. Updated in 2000. Diabetes Care 1985; 8(suppl): 101–7.

49. Virtanen SM, Rasanen L, Aro A et al. Infant feeding in Finnish children over the age of 4 years with newly diagnosed IDDM. Diabetes Care 1991;14:415–7.

50. Harris SS. Vitamin D in type 1 diabetes. J Nutr 2005;135:323–5.

51. Intensive diabetes self-management: 2000 update, American Association of Clinical Endocrinologists, Endocrine Practice 2000;6:42–67.

52. LaPorte et al., Geographic differences; and Muntoni S et al., Nutritional factors and worldwide incidence of childhood type 1 diabetes, Am J Clin Nutr 2000;71:1525–9.

53. What is gestational diabetes? www.nichd.nih.gov/publications/pubs/gdm/sub1.htm, accessed 8/06.

54. Buchanan TA et al. Gestational diabetes mellitus. J Clin Invest 2005;115: 485–91.

55. Reader D et al. Impact of gestational mellitus nutrition practice guidelines implemented by registered dietitians on pregnancy outcomes. J Am Diet Assoc. 2006;106:1426–33.

56. Few overweight and obese people aware of diabetes risk. www.medscape.com, accessed 9/03.

Answers to Review Questions

1. True, see pages 13-3, 13-11.
2. False, see page 13-2.
3. True, see pages 13-4, 13-11.
4. True, see pages 13-8, 13-11.
5. True, see page 13-5.
6. False, see page 13-5.
7. True, see page 13-11.
8. False, see page 13-6.
9. False, see page 13-8.
10. True, see page 13-11.

Nutrition Up Close

Calculating Glycemic Load

Feedback for Unit 13

Food	Glycemic Load
cola beverage	69.6
baked potato	28.9
apple juice	11.6
milk chocolate	7.3
hummus	1.5

Effects of carbohydrate-containing foods on blood glucose levels vary depending on the glycemic index of these foods and the amount of them we consume.

Alcohol: The Positives and Negatives

NUTRITION SCOREBOARD

	TRUE	FALSE
1 Alcohol is produced from carbohydrates.		
2 You can protect your body from the harmful effects of consuming excessive amounts of alcohol by eating a nutritious diet.		
3 Alcohol abuse plays a major role in injuries and deaths in the United States.		
4 Dark beer provides more calories than the same amount of regular beer.		

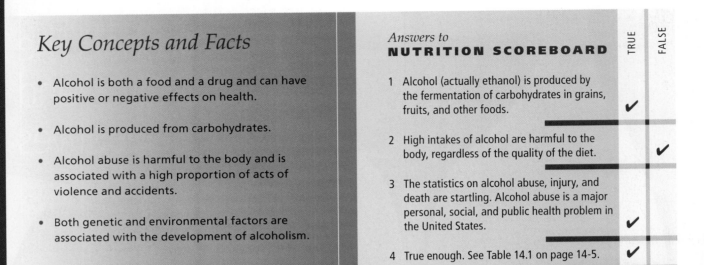

Key Concepts and Facts

- Alcohol is both a food and a drug and can have positive or negative effects on health.

- Alcohol is produced from carbohydrates.

- Alcohol abuse is harmful to the body and is associated with a high proportion of acts of violence and accidents.

- Both genetic and environmental factors are associated with the development of alcoholism.

Answers to
NUTRITION SCOREBOARD

		TRUE	FALSE
1	Alcohol (actually ethanol) is produced by the fermentation of carbohydrates in grains, fruits, and other foods.	✔	
2	High intakes of alcohol are harmful to the body, regardless of the quality of the diet.		✔
3	The statistics on alcohol abuse, injury, and death are startling. Alcohol abuse is a major personal, social, and public health problem in the United States.	✔	
4	True enough. See Table 14.1 on page 14-5.	✔	

Alcohol Facts

Alcohol is both a food and a drug. It's a food because alcohol is made from carbohydrates, and the body uses it as an energy source. Alcohol is a drug because it modifies various body functions.

The type of alcohol people consume in beverages is ethanol. (We refer to ethanol by the broader term *alcohol* in this unit.) Alcohol is produced from carbohydrates in grains, fruits, and other foods by the process of *fermentation.* Wines, brews made from grains, and other alcohol-containing beverages are a traditional part of the food supply of many cultural groups. In high doses, however, alcohol is harmful to the body and can cause a wide variety of nutritional, social, and physical health problems.

The Positive

Whether alcohol has harmful effects on health depends on how much is consumed. The consumption of moderate amounts of alcohol by healthy adults who are not pregnant appears to cause no harm. In fact, moderate alcohol consumption is associated with a significant level of protection against heart disease. A moderate level of alcohol consumption is considered to be one standard-sized drink per day for women and two drinks for men (Illustration 14.1).[1] Alcohol increases the body's production of high-density lipoprotein cholesterol (HDL).[3] HDL is known as the "good cholesterol" because it helps eliminate cholesterol from the body, and that reduces plaque buildup in arteries.[2]

All types of alcohol-containing beverages reduce the risk of heart disease. But is there something special about red wine that reduces the risk of heart disease more than other alcohol-containing beverages do? The answer appears to be yes. Red wine includes pigments that make grapes deep red, blue, or purple. A number of these pigments act as antioxidants and may decrease the tendency of blood to clot (which may happen during a heart attack) and may decrease the ability of

fermentation
The process by which carbohydrates are converted to ethanol by the action of the enzymes in yeast.

LDL cholesterol to stick to plaque in the arteries. Purple grape juice contains these antioxidants, too, and may also help prevent heart disease.[3]

People don't have to consume alcohol to reduce their risk of heart disease. Diets low in saturated and *trans* fat, liberal intakes of vegetables and fruits, ample physical activity, and not smoking also reduce the risk of heart disease.[4]

The Negative

Heavy drinking, often defined as the consumption of five or more drinks per day, poses a number of threats to the health of individual drinkers and often to other people as well. Although health can be damaged by the regular consumption of high amounts of alcohol, the ill effects of alcohol are most obvious in people with *alcoholism.*

Habitually high alcohol intakes and alcoholism increase the risk of developing high blood pressure, stroke, and cirrhosis of the liver; throat, stomach, and bladder cancer; central nervous system disorders; and vitamin and mineral deficiency diseases. Alcohol abuse is associated with a high proportion of deaths from homicide, drowning, fires, traffic accidents, and suicide (Illustration 14.2). It is also involved in a large proportion of rapes and assaults. *Alcohol poisoning* from the consumption of a large amount of alcohol in a short period of time can cause death—and does to a number of college students each year.[5] You can read more about the important topic of alcohol poisoning in the Health Action.

Illustration 14.1
Standard serving sizes of alcohol-containing beverages.
Standard servings shown each contain 0.6 oz (17g) of alcohol.

alcoholism
An illness characterized by a dependence on alcohol and by a level of alcohol intake that interferes with health, family and social relations, and job performance.

alcohol poisoning
A condition characterized by mental confusion, vomiting, seizures, slow or irregular breathing, and low body temperature due to the effects of excess alcohol consumption. It is life threatening and requires emergency medical help.

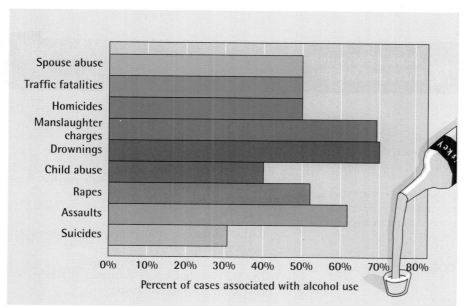

Illustration 14.2
Violence and injuries associated with alcohol.

Source: National Institute on Alcohol Abuse and Alcoholism, 2001, 2006.

Illustration 14.3
Children with fetal alcohol syndrome experience growth and mental retardation, in addition to specific facial characteristics. Alcohol-containing beverages must show a warning statement on labels.

Drinking during pregnancy may harm the fetus. Women who binge drink or drink regularly during pregnancy are at risk of delivering an infant with signs of fetal alcohol syndrome (Illustration 14.3). Children with fetal alcohol syndrome experience long-term growth and mental retardation. The severity of the condition depends on how much alcohol was consumed during pregnancy, whether the mother is genetically susceptible to adverse effects of alcohol, and if excessive intake occurred early or late in pregnancy. Although small amounts of alcohol do not appear to be hazardous, it is recommended that women who are, or may become, pregnant not drink.[6]

Alcohol Intake and Diet Quality

Alcohol provides 7 calories per gram, making alcohol-containing beverages rather high in caloric content (Table 14.1). Because many alcohol-containing beverages provide calories and few or no nutrients, they are considered "empty-calorie" foods. On average, alcohol accounts for 3 to 9% of the caloric intake of U.S. adults who drink. The average goes up to around 50% among heavy drinkers.[7] Although beer, wine, and mixed drinks are known to contain alcohol and to provide calories, there exists some confusion about whether calories from alcohol contribute to weight gain. This issue is addressed in the "Reality Check."

As caloric consumption from alcohol-containing beverages increases, the quality of the diet generally decreases. Diets of heavy drinkers frequently provide too little thiamin, vitamins A and C, calcium, and iron.[7] Deficiencies of nutrients, as well as direct, toxic effects of high levels of alcohol ingestion, produce

---

HEALTH *Action* Alcohol Poisoning: Facts and Action[18]

Facts

- Alcohol poisoning is a serious and sometimes deadly result of drinking excessive amounts of alcohol, usually in an episode of binge drinking.
- A high level of alcohol in the blood depresses breathing, promotes choking and vomiting, and slows heart rate.
- Blood alcohol levels continue to rise after a person passes out.

CRITICAL SIGNS of alcohol poisoning include:

1. Mental confusion and loss of consciousness; sometimes the person cannot be roused.
2. Vomiting
3. Seizures
4. Slow or irregular breathing (10 seconds or more between breaths)
5. Off-colored skin (paleness or bluish skin color)

ACTION

1. CALL 911 for help if there is any suspicion of an alcohol overdose.
2. DO NOT wait for all the symptoms to appear.
3. DO NOT assume a person can safely "sleep it off" or that a cold shower will help.

most of the physical health problems associated with alcoholism. The lack of thiamin, for example, impairs the brain's utilization of glucose. When people with alcoholism initially withdraw from alcohol, the thiamin deficiency may result in "delirium tremens," a condition called the "DTs" by people who staff detoxification centers. People with delirium tremens experience convulsions and hallucinations and are severely confused. Thiamin injections are a key component of treatment for delirium tremens.[8] Because alcohol in excess is directly toxic to body tissues, consuming an adequate diet protects heavy drinkers from only some of the harmful effects of alcoholism.[9]

TABLE 14.1
Caloric value of common alcohol-containing beverages

	Serving	Calories
Beer, regular	12 oz	150
Beer, light	12 oz	110
Beer, dark	12 oz	168
Malt beverage	12 oz	225
80-proof liquor	1.5 oz	100
Wine, red	5 oz	105
Wine, white	5 oz	100
Wine cooler	12 oz	215

How the Body Handles Alcohol

Alcohol is easily and rapidly absorbed in the stomach and small intestine. Within minutes after it is consumed, alcohol enters the circulatory system and is on its way to the liver, brain, and other tissues throughout the body. Alcohol remains in blood and body tissues until it is broken down and used for energy or is converted into fat and stored. The process of converting alcohol into a source of energy takes several hours or more to complete, depending on the amount of alcohol consumed. Because of the lag time between alcohol intake and utilization, blood levels of alcohol build up as drinking continues (Table 14.2).

TABLE 14.2
Alcohol doses and estimated percent blood alcohol levels

Number of Drinks[a]	Percent Blood Alcohol by Body Weight					
	100 LB	120 LB	140 LB	160 LB	180 LB	200 LB
1	0.04	0.03	0.03	0.02	0.02	0.02
2	0.04	0.03	0.03	0.03	0.02	0.02
3	0.07	0.06	0.05	0.05	0.04	0.04
4	0.11	0.09	0.08	0.07	0.06	0.06
5	0.14	0.12	0.10	0.09	0.08	0.07
6	0.18	0.15	0.13	0.11	0.10	0.09

[a]Taken within an hour or so; each drink equal to ½ ounce pure ethanol. Affects may vary based on food intake, sex, and other factors.
Source: University of Oklahoma Police Dept., Blood Alcohol Calculator, www.ou.edu/oupd/bac.htm, accessed 8/06.

Do alcohol calories count?

Maybe you've never seen an obese person with alcoholism, so you're tempted to think alcohol calories don't count. Chronic alcohol abuse is associated with weight loss and muscle wasting, even though the calorie intake of heavy drinkers is high. The effect appears to be due to an inhibition of fat tissue accumulation. The calories do count for light and moderate drinkers, however.[12]

Pedro:

Erik:

Illustration 14.4
The legal limit for intoxication is 0.08%—beyond the point where driving is impaired.

The term **proof** was derived hundreds of years ago by testing whether an alcohol containing liquid would ignite. If it did, that was "proof" of a high enough content of alcohol.[17]

The intoxicating effects of alcohol correspond to blood alcohol levels. A drink or two in an hour raises blood levels of alcohol to approximately 0.03% in most people who weigh about 140 pounds. Blood alcohol levels of 0.03% correspond to mild intoxication. At this level, people lose some control over muscle movement and have slowed reaction times and impaired thought processes. A person's ability to drive or operate equipment in a safe manner is decreased at this level of blood alcohol content (Illustration 14.4). Blood alcohol levels of around 0.06% are associated with an increased involvement in traffic accidents. The legal limit for intoxication according to all states' highway safety ordinances is 0.08%—beyond the point where driving is impaired. When blood alcohol content increases to 0.13%, speech becomes slurred, "double vision" occurs, reflexes are dulled, and body movements become unsteady. If blood alcohol level continues to increase, drowsiness occurs and people may lose consciousness. Levels of blood alcohol above 0.6% can cause death.[10]

A given amount of alcohol intake among women produces higher blood levels of alcohol than for men of the same body weight. Pound for pound, women's bodies contain less water than men's bodies, so blood alcohol levels in women increase faster than in men.[13]

Over 150 medications, including sleeping pills, antidepressants, and painkillers, interact harmfully with alcohol. Combining three or more drinks per day with aspirin or nonaspirin pain relievers (acetaminophen, ibuprofen) may cause stomach ulcers or liver damage.[6]

How to Drink Safely if You Drink

Many of the problems related to alcohol intake can be prevented by not drinking or by drinking responsibly. That means:

- Not drinking if you are or could become pregnant.
- Not drinking on an empty stomach (which can make you intoxicated surprisingly fast).
- Slowly sipping rather than gulping drinks.
- Limiting alcohol to an amount that doesn't make you lose control over your mind and body.
- Never driving a car or boat, hunting, or operating heavy equipment while under the influence of alcohol.

What Causes Alcoholism?

One in 13 adults in the United States abuse alcohol or has alcoholism.[13] Alcoholism tends to run in families, and there is a documented genetic component to the disease.[14] Its development is also influenced by environmental factors. In general, the younger individuals are when they begin to drink, the greater likelihood that they will develop a drinking problem at some point in life. Individuals who begin drinking before the age of 15, for example, are four times more likely to become alcohol dependent than are people who do not drink before age 21 (Illustration 14.5). Close association with friends or peers who drink, high levels of stress, and availability of alcohol may also increase the risk of alcoholism. Television ads depicting youth-oriented parties, fun, and beer may increase underage drinking.[15]

Alcohol Use Among Adolescents

Alcohol use among adolescents is increasing, and the age when teens begin drinking is going down. Underage drinking accounts for 20% of all the alcohol consumed in the United States. The average age when teens begin drinking is now 14 years. These trends are particularly disturbing because they may lead to higher rates of alcoholism and alcohol-related problems in the near future.[16] Reduction in alcohol intake by adolescents is a major public health initiative of the Year 2010 Health Objectives for the Nation.

Much remains to be learned about the prevention of alcohol abuse and alcoholism. Until then treatment programs, such as Alcoholics Anonymous and educational programs that stress safe drinking, will be relied upon to lessen the impact of alcohol abuse on personal and public health.

Illustration 14.5
The younger a person is when drinking begins, the higher the probability that a drinking problem will develop.

Nutrition Up Close

Effects of Alcohol Intake

Focal Point: Estimating blood alcohol levels and side effects.

Scenario: Ligia and Mark attend a wedding reception. Prior to the meal, they both drink a glass of champagne to toast the bride. Fifteen minutes later, they drink another glass to toast the groom. Ligia weighs 140 pounds and Mark 180.

Feedback (answers to these questions) can be found at the end of Unit 14.

Questions: Using the information in Table 14.2 and the information on "How the Body Handles Alcohol" (p. 14-5), answer the following questions:

A. After two glasses of champagne, what would be Ligia's estimated blood alcohol level? ____% blood alcohol. What would be Mark's? ____% blood alcohol

B. List three side effects of 0.03% blood alcohol content:
   1)   2)   3)

Key Terms

alcoholism, page 14-3   *alcohol poisoning, page 14-4*   *fermentation, page 14-2*

Review Questions

	True	False
1. Moderate consumption of alcohol-containing beverages decreases the risk of heart disease.	☐	☐
2. A "moderate" intake of alcohol-containing beverages is considered to be two standard servings daily for men and one for women.	☐	☐
3. Alcohol is considered a food because it is a good source of a variety of vitamins and minerals.	☐	☐

	True	False
4. Alcohol is used by the body for energy or is converted to glycogen.	☐	☐
5. Alcohol poisoning represents an emergency situation requiring medical care.	☐	☐
6. The idea that alcohol is absorbed more slowly if you drink after you eat or while eating is a myth.	☐	☐
7. Some people are genetically susceptible to developing alcoholism.	☐	☐

Media Menu

www.niaaa.nih.gov
The National Institute on Alcohol Abuse and Alcoholism offers this site for exploration of alcohol and health issues, quick facts, college drinking prevention programs, current research projects, and answers to common questions on alcohol abuse.

www.MayoClinic.com
Search "alcohol" and get the latest information on alcoholism, treatment programs, taking control, pros and cons of alcohol use, and an alcohol quiz.

www.alcoholics-anonymous.org
Information about services for alcoholism available from Alcoholics Anonymous World Service is available at this address.

www.al-anon.alateen.org
Al-Anon/Alateen alcohol treatment services can be found at this address.

www.ou.edu/oupd/bac.htm
The University of Oklahoma Police Department has developed an online tool for estimating blood alcohol level based on alcohol consumption and other factors.

www.e-chug.com
The e-CHUG is an evidence-based, online alcohol intervention and personalized feedback tool developed by counselors and psychologists at San Diego State University. It provides personalized information about an individual's drinking level and risk factors.

www.nlm.nih.gov/medlineplus
Search the word *alcohol* and be greeted by a large selection of topics related to alcohol: women and alcohol, fetal alcohol syndrome, Alcoholics Anonymous, and alcohol and youth.

Notes

1. Chapter 9 Alcoholic Beverages, Dietary Guidelines for Americans, www.health.gov/dietaryguidelines/dga2005/document, accessed 8/06.

2. Maraldi C et al. Relationships among alcohol, cardiac events, and death. Arch Intern Med 2006;166:1490–7.

3. Szmitk PE et al. Red wine and your heart. Circulation 2005;111:e10–e11.

4. Goldberg IJ. To drink or not to drink? N Engl J Med 2003;348:163–64.

5. Markel H. Dying for a drink: alcohol on campus. Medscape Public Health and Prevention 2006;4(1), 6/07/2006.

6. Alcohol and your health. Weighing the pros and cons. www.mayoclinic.com/health/alcohol/SC00024, accessed 8/06.

7. Kesse E et al., Do eating habits differ according to alcohol consumption? Am J Clin Nutr 2001;74:322–7.

8. Lieber CS. The influence of alcohol on nutritional status. Nutr Rev 1988;46:241–54.

9. Grant LP et al. Nutrition education is positively associated with substance abuse treatment program outcomes. J Am Diet Assoc 2004;104:604–10.

10. Charness ME, Simon RP, Greenberg DA. Ethanol and the nervous system. N Engl J Med 1989;321:442–53.

11. Levine JA, et al. Energy expenditure in chronic alcohol abuse. Euro J Clin Invest 2000;30:779–86.

12. Alcohol Alert. National Institute on Alcohol Abuse and Alcoholism. www.niaaa.nih.gov, accessed 6/03.

13. Alcoholism: getting the facts. National Institute on Alcohol Abuse and Alcoholism. www.niaa.nih.gov, accessed 6/03.

14. Mulligan MK et al. Toward understanding of genetics of alcohol drinking through transcriptome meta-analysis. Proc Natl Acad Sci 2006;103:6368–73.

15. Alcohol Alert. Underage drinking. www.niaaa.nih.gov, accessed 1/06.

16. Alcohol consumption and expenditures for underage drinking and adult excessive drinking. Alcoholism & Drug Abuse Weekly 2003;15(9):3–4.

17. Franzen, J. How do you determine the alcohol proof of a given liquid? MadSci Network: Chemistry, 2/98.

18. College drinking—changing the culture. www.collegedrinkingprevention.gov, National Institute of Alcohol Abuse and Alcoholism, accessed 8/06.

Answers to Review Questions

1. True, see page 14-2.
2. True, see page 14-2.
3. False, see page 14-4.
4. The second part of this statement is false; see page 14-5.
5. True, see pages 14-3, 14-4.
6. False, see page 14-6.
7. True, see page 14-7.

Nutrition Up Close

Effects of Alcohol Intake

Feedback for Unit 14

A. Ligia's estimated % blood alcohol = 0.03%.
   Mark's estimated % blood alcohol = 0.02%.
B. Three side effects of these blood alcohol levels:
   1. Loss of some control over muscle movements
   2. Slowed reaction time
   3. Impaired thought processes

Proteins and Amino Acids

NUTRITION SCOREBOARD

	TRUE	FALSE
1 The primary function of protein is to provide energy.		
2 "Nonessential amino acids" are not required for normal body processes. Only "essential amino acids" are.		
3 High-protein diets and amino acid supplements by themselves increase muscle mass and strength.		

Key Concepts and Facts

- Proteins are made of amino acids. Some amino acids are "essential" (required in the diet), and some are "nonessential" (not a required part of diets).

- Although protein can be used for energy, its major functions in the body involve the construction, maintenance, and repair of protein tissues.

- Protein tissue construction in the body proceeds only when all nine essential amino acids are available.

- Appropriate combinations of plant foods can supply sufficient quantities of all the essential amino acids.

Answers to
NUTRITION SCOREBOARD

		TRUE	FALSE
1	Energy is a function of protein, but it's not the primary one.		✔
2	"Nonessential amino acids" are required by the body, but they are not required components of our diet. (Yes, it is confusing.)		✔
3	Muscles contain protein, but you can't increase muscle mass by consuming a high-protein diet or amino acid supplements.		✔

---

protein
Chemical substance in foods made up of chains of amino acids.

hormone
A substance, usually a protein or steroid (a cholesterol-derived chemical), produced by one tissue and conveyed by the bloodstream to another. Hormones affect the body's metabolic processes such as glucose utilization and fat deposition.

Illustration 15.1
The protein perception.

Protein

Other nutrients

Protein's Image versus Reality

The term *protein* is derived from the Greek word *protos*, meaning "first." The derivation indicates the importance ascribed to this substance when it was first recognized. An essential structural component of all living matter, protein is involved in almost every biological process in the human body. Protein has a very positive image (Illustration 15.1). It's so positive that you don't have to talk about the importance of protein—people are already convinced of it.

Rich or poor, nearly all people in the United States get enough protein in their diets. Actually, most people consume more protein than they need. Average intakes of protein exceed the Recommended Dietary Allowance (RDA) level for all age and sex groups. Approximately 15% of total calories in the average U.S. adult diet are supplied by protein.[2]

High-protein intakes are generally accompanied by high-fat and low-fiber intakes. That's because foods high in protein such as hamburger, cheese, nuts, and eggs are high in fat and contain little or no fiber. Even lean meats provide a considerable proportion of their total calories as fat (Illustration 15.2).

Functions of Protein

Proteins perform four major functions in the body (Table 15.1). They are an integral structural component of skeletal muscle, bone, connective tissues (skin, collagen, and cartilage), organs (such as the heart, liver, and kidneys), red blood cells and hemoglobin, hair, and fingernails. Proteins are the basic substance that make up thousands of enzymes in the human body; they are a major component of *hormones* such as insulin and serve as other substances that perform important biological functions. Tissue maintenance and the repair of organs and tissues dam-

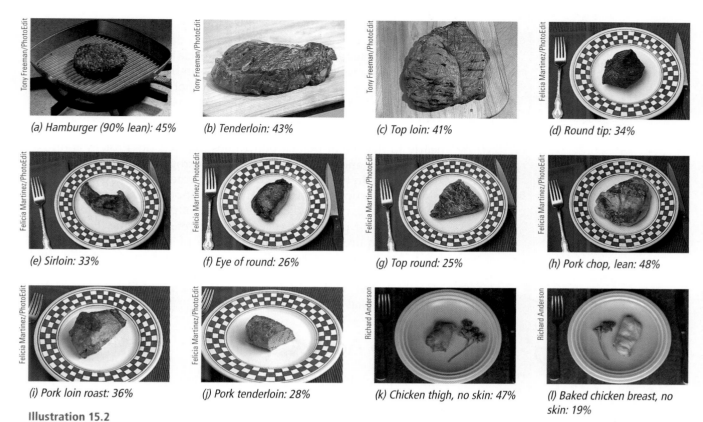

(a) Hamburger (90% lean): 45%
(b) Tenderloin: 43%
(c) Top loin: 41%
(d) Round tip: 34%
(e) Sirloin: 33%
(f) Eye of round: 26%
(g) Top round: 25%
(h) Pork chop, lean: 48%
(i) Pork loin roast: 36%
(j) Pork tenderloin: 28%
(k) Chicken thigh, no skin: 47%
(l) Baked chicken breast, no skin: 19%

Illustration 15.2
The fat content of 3-ounce portions of "lean" meats. *The percentage of calories from fat is indicated for each portion. (A 3-ounce portion of meat is about the size of a deck of cards.) Each portion of meat provides approximately 21 grams of protein*

aged due to illness or injury are functions of different types of protein. Finally, protein serves as an energy source.

The body of a 154-pound man contains approximately 24 pounds of protein. Nearly half of the protein is found in muscle, while the rest is present in the skin, collagen, blood, enzymes, and *immunoproteins*; organs such as the heart, liver, and intestines; and other body parts. All protein in the body is continually being turned over, or broken down and rebuilt. This process helps maintain protein tissues in optimal condition so they continue to function normally. The process

immunoproteins
Blood proteins such as phagocytes (pronounced *fag-o-sites*) and antibodies that play a role in the functioning of the immune system (the body's disease defense system). White blood cells are a type of phagocyte that engulfs and absorbs foreign bodies and harmful organisms in blood and tissues. Antibodies attack foreign proteins.

TABLE 15.1
Functions of protein

1. Serves as a structural material in muscles, connective tissue, organs, and hemoglobin
2. Serves as the basic component of enzymes, hormones, and other biologically important chemicals
3. Maintains and repairs protein-containing tissues
4. Serves as an energy source

Illustration 15.3
Red blood cells enmeshed in fibrin in a color-enhanced microphotograph.
Red blood cells and fibrin (which helps stop bleeding by causing blood to clot) are made primarily from protein.

Glycine
|
Isoleucine
|
Valine
|
Glutamic acid
|
Glutamine
|
Cysteine
|
Cysteine ────── Cysteine
|
Alanine
|
Serine
|
Valine
|
Cysteine
|
Serine
|
Leucine
|
Tyrosine
|
Glutamine
|
Leucine
|
Glutamic acid
|
Asparagine
|
Tyrosine
|
Cysteine ──────── Cysteine
|
Asparagine

Phenylalanine
|
Valine
|
Asparagine
|
Glutamine
|
Histidine
|
Leucine
|
Glycine
|
Serine
|
Histidine
|
Leucine
|
Valine
|
Glutamic acid
|
Alanine
|
Leucine
|
Tyrosine
|
Leucine
|
Valine
|
Glycine
|
Glutamic acid
|
Arginine
|
Glycine
|
Phenylalanine
|
Phenylalanine
|
Tyrosine
|
Threonine
|
Proline
|
Lysine
|
Alanine

Illustration 15.4
Amino acid chains in the protein insulin and diagram of the structure of insulin.

DNA (deoxyribonucleic acid)
Genetic material contained in cells that initiates and directs the production of proteins in the body.

essential amino acids
Amino acids that cannot be synthesized in adequate amounts by humans and therefore must be obtained from the diet. They are sometimes referred to as "indispensable amino acids."

nonessential amino acids
Amino acids that can be readily produced by humans from components of the diet. Also referred to as "dispensable amino acids."

of protein turnover utilizes roughly 9 ounces of protein each day. Yet, we consume only 2 to 3 ounces of protein daily. Most of the protein used for maintenance is recycled from muscle and other protein tissues being turned over. Proteins play key roles in the repair of body tissues by serving as substances such as fibrin that helps blood clot (Illustration 15.3) and by replacing tissue proteins damaged by illness or injury.[3]

Protein serves as a source of energy in healthy people, but not nearly to the extent that carbohydrates and fats do. Protein is unlike carbohydrate and fat in that it contains nitrogen and does not have a storage form in the body. In order to use protein for energy, amino acids that make up proteins must first be stripped of their nitrogen. The free nitrogen can be used as a component of protein formation within the body; or, if present in excess, it is excreted in urine. Excretion of nitrogen requires water, so high intake of protein increases water need. Amino acids missing their nitrogen component are converted to glucose or fat that then can be used to form energy. A small amount of protein (1%) can be obtained from the liver and blood and used to cover occasional deficits in protein intake.[4]

Amino Acids

The "building blocks" of protein are amino acids. Protein consumed in food is broken down by digestive enzymes and absorbed into the bloodstream as amino acids. There are 20 common amino acids (Table 15.2) that form proteins when linked together. Every protein in the body is composed of a unique combination of amino acids linked together in chains (Illustration 15.4). The organization of amino acids into the chains is orchestrated by *DNA*, the genetic material within each cell that directs protein synthesis. Once formed, the chains of amino acids may fold up into a complex shape. Some proteins are made of only a few amino acids, while other proteins contain hundreds. Whatever the number of amino acids, the specific amino acids involved and their arrangement determine whether the protein is an enzyme, a component of red blood cells, a muscle fiber, or another tissue made from protein.

Nine of the 20 common amino acids are considered *essential*, and 11 are *nonessential*. Despite the labels, all 20 amino acids are required to build and maintain protein tissues. The essential amino acids are called "essential" because the body cannot produce them, or produce enough of them, so they must be provided by the diet. Our bodies can produce nonessential amino acids, so we don't require

TABLE 15.2
Essential and nonessential amino acids

Essential		Nonessential	
Histidine	Threonine	Alanine	Glutamine
Isoleucine	Tryptophan	Arginine	Glycine
Leucine	Valine	Asparagine	Proline
Lysine		Aspartic acid	Serine
Methionine		Cysteine	Tyrosine
Phenylalanine		Glutamic acid	

a dietary source of them. Proteins in foods contain both essential and nonessential amino acids.

Proteins Differ in Quality

The ability of proteins to support protein tissue construction in the body varies depending on their content of essential amino acids. How well dietary proteins support protein tissue construction is captured by tests of the protein's "quality."

Proteins of high quality contain all the essential amino acids in the amounts needed to support protein tissue formation by the body. If any of the essential amino acids are missing in the diet, proteins are not formed—even those proteins that could be produced from available amino acids. Shutting off all protein formation for want of an amino acid or two may appear inefficient; but if the body did not cease all protein formation, cells would end up with an imbalanced assortment of proteins. This would seriously affect cell functions. When the required level of an essential amino acid is lacking, the remaining amino acids are primarily used for energy.

Amino acids cannot be stored very long in the body, so we need a fresh supply of essential amino acids daily. This means we need to consume foods that provide a sufficient amount of all essential amino acids every day.

Complete Proteins Food sources of high-quality protein (meaning they contain all the essential amino acids in the amount needed to support protein formation) are called *complete proteins.* Proteins in this category include those found in animal products such as meat, milk, and eggs (Illustration 15.5). *Incomplete proteins* are deficient in one or more essential amino acids. Proteins in plants are "incomplete," although soybeans are considered a complete source of protein for adults.[4] (Soybeans may not meet the essential amino acid requirements of young infants.) You can "complement" the essential amino acid composition of plant sources of protein by combining them to form a "complete" source of protein. Illustration 15.6 shows a few complementary plant food combinations that produce complete proteins.

Vegetarian Diets Diets consisting only of plant foods can provide an adequate amount of complete proteins. The key to success is eating a variety of complementary sources of protein each day. Vegetarian diets have been practiced for centuries by some religious and cultural groups, bearing testimony to their general adequacy and safety. (Unit 16 on vegetarian diets expands on this topic.)

Amino Acid Supplements

Because amino acids occur naturally in foods, people often assume they are harmless, no matter how much is taken (Illustration 15.7). Researchers have known for decades, however,

Illustration 15.5
Animal sources of protein supply "complete proteins."
Each food shown is a source of complete protein.

complete proteins
Proteins that contain all of the essential amino acids in amounts needed to support growth and tissue maintenance.

incomplete proteins
Proteins that are deficient in one or more essential amino acids.

Illustration 15.6
Each of these combinations of plant foods provides complementary protein sources.

that high intakes of individual amino acids can harm health. High amounts may disrupt normal protein production by overwhelming cells with a surplus of some amino acids and a relative deficit of others.[5] Excess consumption of the amino acid methionine, in particular, causes a host of problems. Intake levels two to five times above the amount normally consumed from foods worsen the symptoms of schizophrenia, promote hardening of the arteries, impair fetal and infant development, and lead to nausea, vomiting, bad breath, and constipation.[6]

Not all individual amino acid supplements are hazardous to health. Safe doses of certain amino acids are now being used to treat specific clinical conditions. The best example of clinically useful amino acids is tryptophan and its derivatives 5-HT and melatonin (Illustration 15.8).

Tryptophan, 5-HT, and Melatonin Supplements   In 1990, tryptophan supplements were banned from the market due to concern that they caused the devastating illness eosinophilia-myalgia syndrome (EMS). EMS was eventually determined to be caused by a contaminant in some tryptophan supplements and not by tryptophan itself.[7] Although the ban on tryptophan sales continues, supplements of this amino acid are available on the Web and by prescription.

The tryptophan ban spurred the growth of sales of its derivatives, 5-HT (5-hydroxytryptophan) and melatonin. 5-HT is converted in the body to serotonin, a chemical messenger that influences mood, sleep, appetite, anxiety, and pain sensation. Melatonin plays an important role in triggering sleep and wakefulness. Supplements of melatonin have been found to trigger sleep and to help reset the sleep clock in shift workers, pilots, jet-lagged travelers, and people with sleep disorders. 5-HT appears to help induce sleep in many people, as do nighttime snacks high in tryptophan and carbohydrate.[8,9] Some of the best food sources of tryptophan are listed in Table 15.3.

As is the case for other dietary supplements, the purity, dose, and safety of supplements available on the Web and in stores is not guaranteed. Long-term effectiveness and safety studies are needed to take all of the worry out of using these supplements.

Can Amino Acid Supplements and Protein Powders Build Muscle?   You can't just consume amino acids or protein powders and watch your muscles grow (no matter how convincing the ads that sell such products are). If that happened, everyone who wanted a rippled stomach and bulging triceps could have them. Neither essential amino acids nor protein supplements by themselves increase

Illustration 15.7
A wide array of amino acid supplements are available over the counter, but the safety of these supplements is unknown.

Richard Anderson

ON THE SIDE

On average, U.S. adults consume 2,232 grams (5 pounds) of protein each year from turkey.
Are these animals prone to holding their heads up in the rain and drowning? No, they're not. Normal turkeys don't do that.[12]

Taxi/Getty Images

Illustration 15.8
Tryptophan supplements, although banned in the United States and other countries, are available on the Web and by prescription. 5-HT and melatonin are derivatives of tryptophan and can be sold.

Richard Anderson

TABLE 15.3
Some food sources of tryptophan

		Tryptophan, mg
Veal	3 oz	370
Ham	3 oz	370
Lamb	3 oz	340
Beef	3 oz	319
Turkey	3 oz	290
Crab	3 oz	250
Salmon	3 oz	245
Flounder	3 oz	280
Tuna	3 oz	277
Cashews	20–26 nuts	215
Cottage cheese	½ cup	168
Skim milk	1 cup	124

muscle size and strength.[1] Muscle size and strength are built slowly from the raw ingredients of a healthy diet and resistance training (Illustration 15.9).[10] It has been shown that you can build muscle size and strength while consuming a low-protein diet. It's not the protein that build muscles; it takes training and the calories and nutrients provided by a healthy diet.[11]

Food as a Source of Protein

The average intake of protein in the United States is 75 grams per day, exceeding the RDA for men of 56 grams and that for women of 46 grams. Approximately 70% of the protein consumed by Americans comes from meats, milk, and other animal products.[3] Dried beans and grains are not as well known for their protein, but are nevertheless good sources (Table 15.4). Plant sources of protein are

TABLE 15.4
Food sources of protein

The adult RDA is 46 grams for women aged 19 to 24 years and 58 grams for men aged 19 to 24 years.

Food	Amount	Protein Content Grams	Percentage of Total Calories
Animal products			
Tuna (water packed)	3 oz	24	89%
Shrimp	3 oz	11	84
Cottage cheese (low-fat)	1/2 cup	14	69
Beef steak (lean)	3 oz	26	60
Chicken (no skin)	3 oz	24	60
Pork chop (lean)	3 oz	20	59
Beef roast (lean)	3 oz	23	45
Skim milk	1 cup	9	40
Fish (haddock)	3 oz	19	38
Leg of lamb	3 oz	22	37
Yogurt (low-fat)	1 cup	13	34
Hamburger (lean)	3 oz	24	34
Egg	1 medium	6	32
Swiss cheese	1 oz	8	30
Sausage (pork links)	3 oz	17	28
2% milk	1 cup	8	26
Cheddar cheese	1 oz	7	25
Whole milk	1 cup	8	23
Dried beans and nuts			
Tofu	1/2 cup	14	38
Soybeans (cooked)	1/2 cup	10	33
Split peas (cooked)	1/2 cup	5	31
Lima beans (cooked)	1/2 cup	6	27
Dried beans (cooked)	1/2 cup	8	26
Peanuts	1/2 cup	9	17
Peanut butter	1 tbs	4	17
Grains			
Corn	1/2 cup	3	29
Egg noodles (cooked)	1/2 cup	4	25
Oatmeal (cooked)	1/2 cup	3	15
Whole-wheat bread	1 slice	2	15
Macaroni (cooked)	1/2 cup	3	13
White bread	1 slice	2	13
White rice (cooked)	1/2 cup	2	11
Brown rice (cooked)	1/2 cup	2	10

Photo Disc

Illustration 15.9
Building muscles like these takes time, a good diet, and lots of resistance exercise.

Pure protein

You've got a nutrition exam coming up, so you and your classmate Carole have gotten together to study. You get into a discussion about food sources of protein that goes like this:

Who gets the "thumbs up"?
Answers on next page

Carole:
Lean meats are the best source. They're pure protein!

Lauren:
Pure protein? Even the driest, toughest meats contain more than protein.

TABLE 15.5

Iron content in a 3-ounce serving of various meats

The RDA for women aged 19 to 24 years is 15 milligrams. The RDA for men aged 19 to 24 years is 10 milligrams.

Meat	Iron Content (mg)
Pork chop (lean)	3.4
Round steak (lean)	3.1
Hamburger (lean)	3.0
Shrimp	2.6
Tuna	1.6
Baked chicken (no skin)	1.4
Lamb (lean)	1.3

generally low in fat, making them a wise choice for consumers who are trying to limit their intake of fat. Nearly all food sources of protein provide an assortment of vitamins and minerals as well. Beef and pork are particularly good sources of iron, a mineral often lacking in the diets of women (Table 15.5).

What Happens When a Diet Contains Too Little Protein?

Protein deficiency can occur by itself or in combination with a deficiency of calories and nutrients. Because food sources of protein generally contain essential nutrients such as iron, zinc, vitamin B_{12}, and niacin, diets that produce protein deficiency usually cause a variety of other deficiencies, too. Protein does not generally serve as an important source of energy, but body protein will be used as a major energy source during starvation. To meet the need for energy, the body will extract protein from the liver, intestines, heart, muscles, and other organs and tissues. Loss of more than about 30% of body protein results in reduced body strength for breathing, susceptibility to infection, abnormal organ functions, and death.[13]

Protein-Energy Malnutrition   Protein deficiency contributes to the development of the two major forms of protein-energy malnutrition (PEM): *kwashiorkor* and *marasmus* (the more common form). These two diseases are related to severe undernutrition and include deficiencies of protein, calories, and an assortment of vitamins and minerals. Children affected by PEM tend to be easily irritated, apathetic, small, and highly vulnerable to infection.

PEM generally develops in infants and young children when insufficient amounts of inferior foods are given. Children with a predisposition toward abnormal liver and metabolic functions develop kwashiorkor as the form of PEM. Those with normal metabolic adaptations to starvation tend to develop marasmus. Kwashiorkor is the more serious form of PEM and is accompanied by edema (Illustration 15.10) and many abnormal changes in the body's metabolic processes.

Marasmus is characterized by a skin-and-bone appearance (Illustration 15.11) and fewer metabolic abnormalities than is the case for kwashiorkor. Children with marasmus, for example, can use proteins from muscle and other tissues to help meet their need for protein. Children with kwashiorkor are less able to use body proteins. Marasmus is more easily treated than kwashiorkor and has higher survival rates.[13, 14]

Cases of kwashiorkor and marasmus rarely occur in developed countries. Instead, malnutrition in the western world is largely due to anorexia nervosa, disease, nutritional ignorance, social chaos, or food faddism.[15]

kwashiorkor (kwa-she-or-kor)
A deficiency disease primarily caused by a lack of protein and calories in the diet and abnormal metabolic adaptations to starvation.

marasmus
A condition of severe body wasting due to deficiencies of both protein and calories.

Illustration 15.10
Children suffering from a severe form of protein deficiency called kwashiorkor experience swelling in the arms, legs, and stomach area. At right, the same child after nutritional therapy.

Illustration 15.11
Individuals with marasmus look as starved as they are. At right is the same child after nutritional therapy.

How Much Protein Is Too Much?

Adults can consume a substantial amount of protein—approximately 35% of total calories—for months at a time without ill effects. This observation is based on studies of the diets of Eskimos, explorers, trappers, and hunters in northern America. The very high-protein diets would generally contain a good deal of fat in the form of whale blubber, lard, or fat added to dried meat. Consumption of 45% of total calories from protein is considered too high. Consumption of this level of protein is related to nausea, weakness, and diarrhea. Diets very high in protein result in death after several weeks. The disease resulting from excess protein intake was termed "rabbit fever" after it occurred in trappers attempting to exist on wild rabbit only.[3]

High-protein diets have been implicated in the development of weak bones, kidney stones, cancer, heart disease, and obesity. The National Academy of Sciences has concluded that the risk of such disorders does not appear to be increased among individuals consuming 10 to 35% of total calories from protein, and on average adults consume 15%.[2]

A Tolerable Upper Intake Level (UL) for protein has not been established. Because information on the effects of high-protein intakes is limited, people are cautioned not to consume high levels of protein from foods or supplements.

ANSWERS TO
Reality Check

Pure protein

Some people think muscle and lean meat consist only of protein. They don't. By weight, lean cooked sirloin steak is 29% protein, 8% fat, and 62% water. Lean pork is 29% protein, 9% fat, and 61% water, for example.

Carole 👎

Lauren 👍

Nutrition Up Close

My Protein Intake

Focal Point: Determine the amount of protein in your diet yesterday.

For *each serving* of a food item you ate yesterday, write the grams of protein the food contains in the corresponding blank. For example, a standard serving of meat is 3 ounces (about the size of the palm of your hand or a deck of cards). If you had *one* 3-ounce pork chop yesterday, write *20 grams* in the corresponding blank. If you had *two* 3-ounce pork chops, write *40 grams*. If a protein food you ate yesterday is not included, choose the item on the list closest to it. Then, total the grams of protein you ate yesterday from both plant and animal sources. Finally, compare your protein intake with the RDA of 46 grams for women or 56 grams for men.

Food	One Serving	Protein in One Serving (grams)	Protein You Ate (grams)
Animal products			
Milk (whole)	1 c (8 oz)	8	_____
Yogurt	1 c (8 oz)	13	_____
Cottage cheese	½ c (4 oz)	14	_____
Hard cheese	1 oz	7	_____
Hamburger (lean)	3 oz	24	_____
Beef steak (lean)	3 oz	26	_____
Chicken (no skin)	3 oz	24	_____
Pork chop (lean)	3 oz	20	_____
Fish	3 oz	19	_____
Hot dog	1	6	_____
Sausage	3 oz	17	_____
		Subtotal from animal foods:	_____
Plant products			
Bread	1 slice	2	_____
Rice	½ c (4 oz)	2	_____
Pasta	½ c (4 oz)	3	_____
Cereals	½ c (4 oz)	3	_____
Vegetables	½ c (4 oz)	2	_____
Peanut butter	1 tbs	4	_____
Nuts	¼ c (2 oz)	7	_____
Cooked beans (legumes)	½ c (4 oz)	8	_____
		Subtotal from plant foods:	_____
		Total grams of protein from plant and animal foods:	_____
		Amount above/below RDA:	_____

Special note: You can also calculate your protein intake using Thomson's Diet Analysis Plus software. Input your food intake for one day. Then go to the Analyses/Reports section to view the total number of grams of protein in your diet.

FEEDBACK can be found at the end of Unit 15.

Key Terms

complete proteins, page 15-5
DNA (deoxyribonucleic acid), page
    15-4
essential amino acids, page 15-4
hormone, page 15-2

immunoprotein, page 15-3
incomplete proteins, page 15-5
kwashiorkor (kwa-she-or-kor), page
    15-8
marasmus, page 15-8

nonessential amino acids, page 15-4
protein, page 15-2

Review Questions

True False

1. The only known function of protein is to serve as a structural component of muscle, bones, and other solid tissues. ☐ ☐

2. Low protein intake is a fairly common problem among adults and children in the United States. ☐ ☐

3. Immunoproteins are a part of the body immune system. ☐ ☐

4. Fibrin is a protein that helps blood clot. ☐ ☐

5. The nitrogen component of amino acids must be removed before they can be used for energy. ☐ ☐

6. There are 22 amino acids, and half of them are "essential." ☐ ☐

True False

7. DNA provides the blueprint for the production of specific proteins by the body. ☐ ☐

8. People need to eat animal sources of protein in order to consume enough high-quality "complete" proteins. ☐ ☐

9. Individual amino acids supplements are uniformly hazardous to health. ☐ ☐

10. Amino acid supplements plus resistance exercise are the keys to building muscle mass and strength. ☐ ☐

11. Kwashiorkor is the name of the protein deficiency disease. ☐ ☐

12. Protein intake should constitute 10 to 35% of total calorie intake. ☐ ☐

Media Menu

www.healthfinder.gov
Search topics related to protein on this site.

www.nemsn.org
The National EMS Network provides information on symptoms, current issues, and answers to questions about EMS, tryptophan, and melatonin.

www.iom.edu
Select "Food & Nutrition" to gain access to the 2002 Dietary Reference Intake report on macronutrients and the chapter on Protein.

Notes

1. Williams AG et al. Is glucose/amino acid supplementation after exercise an aid to strength training? Br J Sports Med 2001;35:109–13.

2. What We Eat in America. http://www.ars.usda.gov/Services/doc.htm?docid=7674, accessed 9/06.

3. Dietary Reference Intakes. Energy, carbohydrate, fiber, fat, fatty acids, cho-lesterol, protein, and amino acids. Institute of Medicine, National Academy of Sciences, Washington, DC: National Academies Press; 2002.

4. Matthews DE. Proteins and amino acids. In: Modern nutrition in health and disease, 9th edition, Shils ME et al., eds. Philadelphia: Lippincott, Williams & Wilkins, 1998, pp. 11–48.

5. High protein intake harms the kidneys. www.healthfinder.gov/high protein diet, accessed 9/03.

6. Garlick PJ. Toxicity of methionine in humans. J Nutr 2006;136:1722S–25S.

7. Information paper on 5-hydroxytryptophan and 5-hydroxy-L-tryptophan, www.nal.usda.gov/fnic/etext/macronut.html, accessed 7/03.

8. Kay LK. Melatonin. Today's Dietitian 2004: Nov., p. 54–6.

9. Markus CR et al. Evening intake of alpha-lactalbumin increases plasma tryptophan availability and improves alertness and brain measures of attention. Am J Clin Nutr 2005;81:1026–33.

10. Phillips SM et al. Dietary protein to support anabolism with resistance exercise in young men. AM Coll Nutr 2005;24:134S–139S.

11. Castaneda C et al. Resistance training to counteract the catabolism of a low-protein diet in patients with chronic renal insufficiency. Ann Int Med 2001;135:965–76.

12. Quote by Tom Savage, Animal Science Professor, Oregon State University in Corvallis, reported in Lewis, R. Gobbler, Genome tapped. The Scientist 2003;Nov. 3:37.

13. Pascal N et al. Serum concentrations of sex hormone binding globulin are elevated in kwashiorkor and anorexia nervosa but not in marasmus. Am J Clin Nutr 2002;76:239–44.

14. Badaloo AV et al. Lipid kinetic differences between children with kwashiorkor and those with marasmus. Am J Clin Nutr 2006;83:1283–8.

15. Liu T et al. Kwashiorkor in the US: fad diets, perceived and true milk allergy, and nutritional ignorance. Arch Dermatol 2001;137:630–7.

Answers to Review Questions

1. False, see pages 15-2, 15-3.
2. False, see page 15-2.
3. True, see page 15-3.
4. True, see page 15-3.
5. True, see page 15-4.
6. False, see page 15-4.
7. True, see page 15-4.
8. False, see page 15-5.
9. False, see page 15-6.
10. False, see pages 15-6, 15-7.
11. False, see page 15-8.
12. True, see page 15-9.

Nutrition Up Close

My Protein Intake

Feedback for Unit 15

Compare your subtotals to find out which protein source you prefer, plant or animal. Protein from animal products is often accompanied by fat. If you are concerned about calories and fat in your diet, choose plant protein sources more often. And, if you are similar to many Americans, your intake of protein will exceed the RDA by quite a bit.

Vegetarian
Diets

NUTRITION SCOREBOARD

	TRUE	FALSE
1 The human body developed to function best on a vegetarian diet.		
2 People who don't eat meat have more health problems than people who do.		
3 Macrobiotic diets cure some types of cancer.		
4 In order to consume enough high-quality protein, vegetarians need to consume combinations of plant foods that provide a complete source of protein at every meal.		

Key Concepts and Facts

- Vegetarianism is more than a diet. It is often part of a value system that influences a variety of attitudes and behaviors.

- Appropriately planned vegetarian diets are health promoting.

- Vegetarian diets that lead to caloric and nutrient deficiencies generally include too narrow a range of foods.

Answers to
NUTRITION SCOREBOARD

		TRUE	FALSE
1	Early humans developed on an omnivorous diet (meat and plant diet). However, health can be fostered on an omnivorous or a vegetarian diet.		✔
2	Persons who consume appropriately planned vegetarian diets do not have more health problems than other people. Furthermore, such diets may reduce the risk of developing several chronic diseases.		✔
3	No type of vegetarian or quasi-vegetarian diet, including macrobiotic and vegan diets, has been found to cure cancer.		✔
4	Vegetarians do need to plan their diets to ensure that they obtain enough high-quality protein, but they need to consume complete sources of protein daily, not at every meal. Appropriately planned vegetarian diets provide sufficient amounts of high-quality protein.		✔

Perspectives on Vegetarianism

. . . unless vegetarianism comes from the soul, it will just be a passing fad.
—JANIS BARKAS, 1978

It wouldn't be any fun to live without meat . . . . I do not believe in this meatless, dairyless, butterless society.
—JULIA CHILD, 1991

Photo Disc

Vegetarianism in the United States, Canada, and other economically developed countries is moving from the realm of counter-culture to the mainstream.[1] Yet even with the increased acceptance of vegetarian diets, people tend to be for vegetarianism or against it, often without knowing much about it. Few of those opposed to vegetarianism have tried to learn about the vegetarian way of life or understand that vegetarian diets can be healthful. A small percentage of health professionals are vegetarians, and those who are not are often skeptical about how healthy a vegetarian diet can be. The possibility that something so different from the customary diet can be nourishing to the body may be rejected out of hand.

An objective look at vegetarianism reveals that both appropriately planned vegetarian diets and the lifestyle often followed by vegetarians can be very good for health.[2] Vegetarians tend to be health conscious; often avoid using alcohol, tobacco, and illicit drugs; and engage in regular physical activity.[3] Diet is usually one of several characteristics shared by people practicing particular types of vegetarianism.

Reasons for Vegetarianism

Worldwide, vegetarians number in the hundreds of millions.[4] Most of the world's population subsists on vegetarian diets because meat and other animal products are scarce or too expensive (Table 16.1). In other societies, people have the luxury of choosing a healthy assortment of food from an abundant and affordable food supply that includes a wide variety of items acceptable to vegetarians (Illustration 16.1). When food availability is not an issue, people tend to adopt vegetarian diets because of a desire to cause no harm to animals, a personal commitment to preserve the environment and the world's food supply by "eating low on the food chain," or a belief that animal products are unhealthful or unsafe. They may avoid animal products as part of a value or religious belief system. Others follow vegetarian diets to keep their weight down or to lower the risk of developing specific diseases such as diabetes or heart disease.[5]

Vegetarian Diets Come in Many Types

> Yeah, I guess I'm kind of a vegetarian, too. I try not to eat meat—
> well, I love chicken but I don't eat red meat. I mean, I eat hamburgers.
> I can't remember the last time I had a steak.
> —A 17-YEAR-OLD "VEGETARIAN"

There is no one vegetarian diet. People who consider themselves to be vegetarians range from those who eat all foods except red meat (mainly beef) to those who exclude all foods from animal sources, including honey. According to the American Vegetarian Society, a person is a vegetarian only if she or he eats *no* meat.

Vegetarian Diet Options   The types of foods included in common vegetarian diets are summarized in Table 16.2. The least restrictive form of vegetarian diet has been unofficially labeled "far vegetarian" because it excludes only red meats. This diet is very much like that consumed by omnivores, or meat and plant eat-

TABLE 16.1
Reasons for vegetarianism[3, 5]

- Lack of availability or affordability of animal products
- Desire not to cause harm to animals
- Religious beliefs
- Desire to "eat low on the food chain"
- Desire to preserve the world's food supply
- Health
- Desire to omit hormones, antibiotics, and possible contaminants in meats

Illustration 16.1
The growing selection of vegetarian foods in supermarkets.

TABLE 16.2
Vegetarian diets come in many types

	Foods Included					
Type of Diet	Beef Lamb, Pork ("Red Meat")	Poultry	Fish	Eggs	Milk and Milk Products	Plant Foods
"Far" vegetarian		x	x	x	x	x
Quasi-vegetarian[a]			x	x	x	x
Lacto-ovo vegetarian[b]				x	x	x
Lacto-vegetarian					x	x
Macrobiotic						x
Vegan						x
Raw food						x

[a]Quasi-vegetarian diets (also called semivegetarian) may include poultry; they tend to vary in the type of animal products consumed.
[b]Macrobiotic diets may include fish and other animal products.

What's in the bowl? It depends on what type of vegetarian diet you are following.

ers. Quasi-vegetarian diets (also called semivegetarian and flexitarian diets) vary somewhat but generally exclude beef, pork, and poultry while including fish, eggs, dairy products, and plant foods. The lacto-ovo vegetarian diet includes only dairy foods, eggs, and plant foods. Individuals practicing this type of diet exclude all meats. The lacto-vegetarian diet, as the name implies, includes only milk and milk products and plant foods. Vegan (pronounced *vee-gun*) and raw food diets are a more restrictive type of vegetarian diet. Vegans and raw food dieters eat only plant foods; in addition, vegans may avoid honey and clothes made from wool, leather, or silk.

■ **Macrobiotic Diets** Macrobiotic diets fall somewhere between quasi-vegetarian and vegan diets. The formulation of macrobiotic diets has changed dramatically over the past few decades. In addition to brown rice, other grains, and vegetables, these diets now include fish, dried beans, spices, fruits, and many other types of foods. No specific foods are prohibited, and locally grown and whole foods are emphasized.

Persons adhering to the macrobiotic philosophy place value on consuming organic foods and balancing the intake of "yin" and "yang" foods. Foods are classed as yin or yang based on beliefs about the food's relationship to the emotions and the physical condition of the body. Yin foods such as corn, seeds, nuts, fruits, and leafy vegetables are considered negative, dark, cold, and feminine. Yang foods represent opposing positive forces of light, warmth, and masculinity. Poultry, fish, eggs, and cereal grains such as buckwheat are yang foods.[6]

■ **Raw Food Diets** Diets consisting primarily of raw foods have come in and out of fashion for hundreds of years. Raw food diets currently in rogue center on the health benefits of plant-based diets and the belief that raw foods provide enzymes helpful for digestion. Enzymes in raw foods do not promote normal digestion, however. Enzymes in food are inactivated by stomach acid and by digestive system enzymes that break down proteins.[7]

Although there is no formal definition, raw food diets are usually described as an uncooked vegan diet. The diet consists of vegetables, fruits, nuts, seeds, sprouted grains, and beans. Uncooked foods usually make up between 50 and 100% of the diet.[7]

Extremely restrictive raw food diets are associated with impaired growth in children. Raw food diets, in general, lower beneficial HDL cholesterol levels and

Reality Check

Reintroducing meat

Larry has been a vegetarian for the last 7 years and wants to eat a hamburger again to see if it tastes as good as he remembers. He's a bit nervous about doing it because he thinks meat might make him feel sick.

Who gets thumbs up?
Answers on next page

Susan: Larry should eat the hamburger if he wants to. It's a food, not an indigestion time bomb.

Doug: Larry's stomach isn't used to the heaviness of meat. It will make him nauseated.

may lead to vitamin B_{12} deficiency and loss of bone mineral density. On the positive side, they are related to low body weights for height and healthy blood levels of triglycerides and total cholesterol.[8,9]

■ **Other Vegetarian Diets** For a number of other vegetarian regimes, the spiritual or emotional importance assigned to certain foods supersedes consideration of their contribution to an adequate diet.[10] Vegetarians adhering to the "living foods diet" consume uncooked and fermented plant foods only. This diet is inadequate in a number of nutrients, including vitamin B_{12}.[11] *Fruitarians* consume only fruit and olive oil. People adopting this type of diet rarely stick with it for long—it does not sustain health.

fruitarian
A form of vegetarian diet in which fruits are the major ingredient. Such diets provide inadequate amounts of a variety of nutrients.

Vegetarian Diets and Health

Appropriately planned vegetarian diets have been shown to be healthful, nutritionally adequate, and beneficial in the prevention and treatment of certain diseases.
—POSITION OF THE AMERICAN DIETETIC ASSOCIATION AND DIETITIANS OF CANADA, 2003[12]

Long-standing vegetarian dietary regimes that promote health in India and China have been tested by time (if not by science) and found to be adequate in essential nutrients.[1] In fact, vegetarian diets are generally healthier than those of nonvegetarians.[13] On the other hand, highly restrictive vegetarian diets—such as fruitarian regimes, the raw food diet, and other popular vegetarian diets that stray from established regimes—can be harmful to health.[1] In general, the more restrictive the diet, the more likely it is to be inadequate and lead to health problems. This is especially true for pregnant women, children, and persons who are ill, all of whom have relatively high needs for nutrients. Reports of caloric and nutrient deficiencies in young children of vegan parents are more common than is the case with less restrictive vegetarian diets.[1]

Availability of a large assortment of vegan foods in the United States and other developed countries is making it less likely that vegetarian diets of children and adults will be inadequate.[12] Vegetarians in economically developed countries are not at greater risk for iron deficiency than nonvegetarians. They generally obtain sufficient iron from plants and have ample intake of vitamin C, which enhances absorption of iron from plants. In contrast, the diets of vegetarians in developing countries often include too few iron-rich plants, making iron deficiency

ANSWERS TO
Reality Check

Reintroducing meat

The thought that meat could make you sick may be enough to trigger indigestion. Many self-defined vegetarians consume meat on rare occasion without reported ill effects.[12]

Susan: 👍

Doug: 👎

common. Both the quality and the quantity of protein in vegetarian diets can also be a source of concern. However, vegetarians in developed countries generally have adequate protein intakes.[12] Vitamin B_{12} has consistently been shown to be the most likely nutrient to be lacking in the diets of individuals who do not consume animal products.[12] Increasingly, vegetarian diets are being recognized for their beneficial effects on health and disease prevention.[14]

What Are the Health Benefits of Vegetarian Diets?

I am a great eater of beef, and I believe that does great harm to my wit.
—WILLIAM SHAKESPEARE, 1589

Compared to the usual American diet and lifestyle, vegetarianism is associated with a lower risk of developing heart disease, stroke, hypertension, type 2 diabetes, chronic bronchitis, and gallstones and kidney stones. People who follow a vegetarian diet rarely become obese or develop high blood cholesterol levels.[14,15] For children and adults at risk of early heart disease, the low-fat and high-fiber vegetable and fruit content of vegetarian diets may help reduce blood cholesterol levels and the risk of heart disease.[16] Benefits of vegetarian diets, macrobiotic diets, and other plant-based diets in the prevention and treatment of cancer have not been demonstrated.[17]

Dietary Recommendations for Vegetarians

Because vegetarian diets exclude one or more types of foods, it's important that the foods included provide sufficient calories and the assortment and quantity of nutrients needed for health. No matter what the motivation underlying the assortment of foods included, vegetarian diets that fail to provide all the nutrients humans need in the required amounts will not sustain health.

Well-Planned Vegetarian Diets

There are a number of types of vegetarian diets, so there is no one set of dietary recommendations that is appropriate for all of them. Dietary guidelines for vegan diets generally recommend a variety of foods that includes grains, legumes, vegetables, fruits, fats and oils, and sweets (Illustration 16.2). Additional nutrition guidance, such as that given by the American and Canadian Dietetic Associations (Table 16.3) is offered to help vegetarians make decisions about specific types of foods to consume within each food group and good ways to meet key nutrient needs.

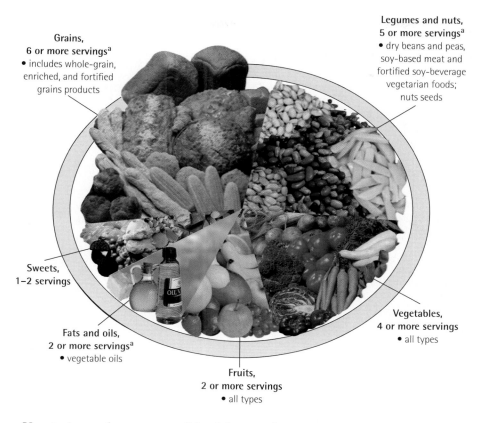

Grains,
6 or more servings[a]
• includes whole-grain, enriched, and fortified grains products

Legumes and nuts,
5 or more servings[a]
• dry beans and peas, soy-based meat and fortified soy-beverage vegetarian foods; nuts seeds

Vegetables,
4 or more servings
• all types

Fruits,
2 or more servings
• all types

Fats and oils,
2 or more servings[a]
• vegetable oils

Sweets,
1–2 servings

Illustration 16.2
Foundation foods for well-planned vegetarian diets.[18–21]
VEGAN DIETS
GRAINS, 6 or more servings[a]
• includes whole-grain, enriched, and fortified grain products
LEGUMES AND NUTS, 5 or more servings[a]
• dry beans and peas, soy-based meat and fortified soy-based beverage vegetarian foods; nuts, seeds
VEGETABLES, 4 or more servings
• all types
FRUITS, 2 or more servings
• all types
FATS AND OILS, 2 or more servings[a]
• vegetable oils
SWEETS, 1–2 servings

[a]Number of serving depends on calorie need.

Vegetarians who consume fish, dairy products, or eggs can reduce their reliance on legumes, nuts, and grains as protein and nutrient sources. Animal products are a major source of **complete proteins,** vitamin B_{12}, vitamin D, calcium, and EPA (eicosapentaenoic acid) and DHA (docosahexaenoic acid). EPA and DHA are two important fatty acids primarily found in fish and seafood. Vitamin B_{12}, vitamin D, calcium, and EPA and DHA are the nutrients most likely to be lacking in the diets of vegetarians.[17, 21]

A specific example of the food and nutrient composition of one day of a vegetarian's diet are shown in Table 16.4. The diet provides low amounts of cholesterol and saturated fat, is high in fiber, and provides adequate amounts of protein and most vitamins and minerals. These are fairly typical results for a vegetarian diet.[14] Of the four key nutrients most likely to be missing in the diet of vegetarians, this day's diet provides the recommended amount of vitamin D, but not of vitamin B_{12}, calcium, or EPA/DHA. Vitamin D intake reaches 6 mcg (240 IU) and B_{12} levels 1.9 mcg in this diet due to the inclusion of fortified soy milk.

complete proteins
Proteins that contain all of the nine essential amino acids in amounts sufficient to support protein tissue construction by the body.

TABLE 16.3

The American and Canadian dietetic associations' guidelines for vegetarian meal planning[12]

1. Choose a variety of foods, including whole grains, vegetables, fruits, legumes, nuts, seeds, and, if desired, dairy products and eggs.
2. Choose whole, unrefined foods often and minimize intake of highly sweetened, fatty, and highly refined foods.
3. Choose a variety of fruits and vegetables.
4. If animal foods such as dairy products and eggs are used, choose lower-fat versions of these foods. Cheeses and other high-fat dairy foods and eggs should be limited in the diet because of their saturated fat content and because their frequent use displaces plant foods in some vegetarian diets.
5. Vegans should include a regular source of vitamin B_{12} in their diets along with a source of vitamin D if sun exposure is limited.
6. Do not restrict dietary fat in children younger than two years. For older children, include some foods higher in unsaturated fats (for example, nuts, seeds, nut and seed butters, avocado, and vegetable oils) to help meet nutrient and energy needs.

TABLE 16.4

Example of food intake and nutrient content in one day for the vegetarian diet of a 31-year-old, 130-pound female

One day's diet:

Oatmeal, 1 cup	Salad dressing, 3 tablespoons
Banana, 1 medium	Soy milk, 1 cup
Soy milk, 1 cup	Veggie burger, 1 patty
Brown sugar, 1 Tablespoon	Bun, 1
Almonds, 1 ounce (22 almonds)	French fries, 1 cup
Black beans, 2 cups	Herbal tea, 1 cup
Brown rice, 1 cup	Hummus, 1/2 cup
Lettuce salad with mixed vegetables, 1 cup	Baby carrots, 10

Selected nutrient analysis results

NUTRIENT	AMOUNT CONSUMED	RECOMMENDED INTAKE
Protein, g	87	46
Total fiber, g	53	25
Total fat, g	100	53–92
Saturated fat, g	20	<26
Cholesterol, mg	27	<300
EPA + DHA, mg[a]	50	300–600
Vitamin A, mcg	723	700
Vitamin C, mg	54	75
Vitamin E, mg	14	15
Vitamin B_6, mg	3.2	1.3
Vitamin B_{12}, mcg	1.9	2.4
Thiamin, mg	3.1	1.1
Riboflavin, mg	1.6	1.1
Niacin, mg	19	14
Folate, mcg	561	400
Vitamin D, mcg[a]	6	5
Calcium, mg	491	1000
Magnesium, mg	614	320
Iron, mg	18.5	18
Zinc, mg	12.2	8
Selenium, mg	37	55

[a]EPA and DHA availability are estimated based on alpha-linolenic acid intake of 2.8 g; vitamin D intake is estimated based on fortified soymilk intake. Analysis is from mypyramidtracker.gov, 9/06.

Vegetarians can meet their needs for vitamin B_{12}, vitamin D, calcium, and EPA/DHA by consuming fortified foods (such as those shown in Table 16.5) or by use of B_{12}-fortified yeast, DHA algae supplements, or fortified plant foods (such as DHA-fortified peanut butter or margarine). EPA can be obtained from sources of DHA. EPA and DHA are closely related chemicals, and the body can convert one to the other to an extent.[22] Direct exposure of the skin to sunlight for about 15 minutes a day is one of the best ways to get vitamin D.

Vitamin and mineral supplements can be, and often are, used by vegetarians to meet specific nutrient needs. If needed, supplementary intake levels of vitamins B_{12} and D, calcium, and EPA/DHA should approximate those shown in Table 16.6. The best way to know if supplemental nutrients are needed is to carefully perform a dietary assessment that covers several typical days of food intake. (The Thomson's Diet Analysis Plus Program or USDA's dietary analysis program available at mypyramidtracker.gov can be used for the dietary assessment.)

Complementary Protein Foods Most well-planned vegetarian diets provide adequate amounts of protein, but the quality of the protein varies depending on which plant foods are consumed.

TABLE 16.5
Key nutrient fortification levels of some vegetarian foods[a]

	Key Nutrients		
	VITAMIN B$_{12}$	VITAMIN D	CALCIUM
Recommended intake for adults:	2.4 mcg	5 mcg	1000 mg
Food			
Rice Dream, 1 cup	1.5 mcg	2.5 mcg	300 mg
Soy milk, 1 cup	3 mcg	3 mcg	300 mg
Meat analogs[b]	0–1.4	0	0
Fully fortified breakfast cereals	6 mcg	1 mcg	1000 mg
Other breakfast cereals	1.5 mcg	1 mcg	0–100 mg

[a]Check product nutrition information lablels for product-specific information. None of the foods listed are yet fortified with EPA and DHA.
[b]Includes burgers, hot dogs, mixed dishes, and other foods made with plant foods.

TABLE 16.6
Amounts of key nutrients that may be needed by vegetarians daily[a]

Vitamin B$_{12}$	2–6 mcg
Vitamin D	200–400 IU (10 mcg)
Calcium	500 mg
EPA and DHA	300 mg

[a]Fortified foods may also supply these nutrients.

Animal products such as meat, eggs, and milk provide all of the nine ***essential amino acids*** in sufficient quantity to qualify as complete sources of protein. In addition, tests have shown soy proteins to be complete protein sources for children and adults.[23] The diet must include complete proteins because the body needs a sufficient supply of each essential amino acid if it is to build and replace protein substances such as red blood cells and enzymes. If any of the essential amino acids are missing in the diet, protein tissue construction stops, and the available amino acids will be used for energy instead. Essential amino acids consumed in foods are not stored for long, so the body needs a fresh supply each day or so.

Vegetarians who don't consume animal products can meet their need for essential amino acids by combining plant foods to yield complete proteins. This is done by consuming plant foods that *together* provide all the essential amino acids although each individual food is missing some of these essential nutrients. The goal is to "complement" plant sources of essential amino acids, or to consume ***complementary protein sources*** from plant foods regularly.[5]

Many different combinations of plant foods yield complete proteins. Basically, complete sources of protein can be obtained by combining grains such as rice, bulgur (whole wheat), millet, or barley with dried beans, tofu, or green peas, or corn with lima beans or dried beans, or seeds with dried beans. Some examples of complementary sources of plant proteins are shown in Illustration 16.3. Milk, meat, and eggs contain complete proteins and will complement the essential amino acids profile of any plant source of protein.

Where to Go for More Information on Vegetarian Diets

You can find information about vegetarian diets in a variety of sources. World Wide Web resources appear at the end of this unit, and several cookbooks that offer information about vegetarian nutrition and delicious recipes are pictured in Illustration 16.4. Unfortunately, some of the information available on vegetarianism is wrong or misconstrued. Some vegetarian organizations are more committed to selling particular beliefs, memberships, or books and magazines than to promoting healthy vegetarian diets. Beware of vegetarian diets that claim to cure cancer, AIDS, or other serious illnesses or promise you'll experience inner peace or spiritual renewal. Appropriately planned vegetarian diets are health promoting, but they are not magic bullets that will cure all the ills of body and soul.

essential amino acids
Amino acids that cannot be synthesized in adequate amounts by the human body and must therefore be obtained from the diet.

complementary protein sources
Plant sources of protein that together provide sufficient quantities of the nine essential amino acids.

There are more than 70 varieties of dried beans. They are an important part of the cuisine of almost all cultures. Bean cuisine includes chili, baked beans, beans and rice, hummus, and refried beans.

Photo Disc

ON THE SIDE

Rice and black beans.

Hummus and bread.

Corn and black-eyed peas.

Bulgur (whole wheat) and lentils.

Tofu and rice.

Corn and lima beans (succotash).

Tortilla with refried beans (e.g., a bean burrito).

Pea soup and bread.

Illustration 16.3
Some combinations of plant foods that provide complete protein:

- *Rice and dried beans*
- *Rice and green peas*
- *Bulgur (wheat) and dried beans*
- *Barley and dried beans*

- *Corn and dried beans*
- *Corn and green peas*
- *Corn and lima beans (succotash)*

- *Soybeans and seeds*
- *Peanuts, rice, and dried beans*
- *Seeds and green peas*

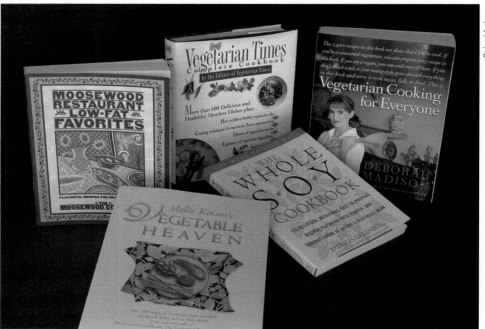

Illustration 16.4
Many excellent vegetarian cookbooks are available; most provide information on diet as well as recipes. The books pictured here are Moosewood Restaurant Low-Fat Favorites: Flavorful Recipes for Healthful Meals, *edited by Pam Krauss;* Vegetarian Times Complete Cookbook, *by the editors of* Vegetarian Times; Vegetarian Cooking for Everyone, *by Deborah Madison;* The Whole Soy Cookbook, *by Patricia Greenberg; and* Vegetable Heaven, *by Mollie Katzen.*

Nutrition Up Close

Vegetarian Main Dish Options

Focal Point: Serving up vegetarian alternatives to meat dishes.

Assume you are a member of the "vegetarian option" planning committee for your college. You are asked to identify three meatless main dishes that could be served in the dining halls for lunch and another three that could be served for dinner. The one stipulation is that they should be main dishes you would enjoy eating.

What meatless dishes would you identify as options for lunch and for dinner?

Lunch Dishes

1.

2.

3.

Feedback (answers to these questions) can be found at the end of Unit 16.

Dinner Dishes

1.

2.

3.

Key Terms

complementary protein sources, page 16-9

complete proteins, page 16-7
essential amino acids, page 16-9

fruitarian, page 16-5

Review Questions

True False

1. The emotional and spiritual importance of certain foods in some vegetarian regimes supersedes any contribution the foods may make to an adequate diet. ☐ ☐

2. Raw food diets supply ample amounts of all vitamins and required minerals and improve indigestion in people with digestive problems. ☐ ☐

3. Compared to omnivore diets, vegetarian diets are associated with lower rates of obesity, heart disease, hypertension, and type 2 diabetes. ☐ ☐

4. Vitamin B_{12} and EPA/DHA are largely found in animal foods. ☐ ☐

True False

5. The four nutrients most likely to be lacking in the diets of vegetarians are vitamin B_{12}, EPA/DHA, magnesium, and vitamin E. ☐ ☐

6. Well-planned vegetarian diets provide adequate amounts of iron and protein. ☐ ☐

7. When forming a new protein tissue, the body will respond to a deficiency of an essential amino acid by replacing it with another essential amino acid that is available. ☐ ☐

8. Corn combined with cooked, dried beans is an example of a complementary protein source. ☐ ☐

Media Menu

http://www.eatright.org/cps/rde/xchg/ada/hs.xsl/governance_5105_ENU_HTML.htm You'll find the "New Food Guide for North American Vegetarians" at this address. It was developed by dietitians in the United States and Canada.

www.nal.usda.gov/fnic/pubs/bibs/gen/vegetarian.htm
An extensive listing of, and links to, information on vegetarian diets and health and weight status, education materials, scientific reports, and other resources are available at this site.

www.mypyramid.gov/tips_resources/vegetarian_diets.html
USDA's site for vegetarian diet information

is based on MyPyramid food group recommendations.

www.vrg.org
Contact the Vegetarian Resource Group for dietary advice, healthy cooking and shopping tips, and other information for the vegetarian community.

www.kushiinstitute.org
Learn more about the macrobiotic philosophy, conferences, and events.

www.vegweb.com
This site provides a guide to vegetarianism, recipes and books, and much more (including ads).

Notes

1. Sabate J. The contribution of vegetarian diets to health and disease: a paradigm shift? Am J Clin Nutr 2003;78(suppl):501S–7S.

2. Cade JE et al. The UK Women's Cohort Study: comparison of vegetarians, fish-eaters and meat-eaters. Public Health Nutr. 2004;7:871–8.

3. Buffonge I. The vegetarian/vegan lifestyle. West Indian Med J 2000; 49:17–19.

4. Antony AC. Vegetarianism and vitamin-12 (cobalamin) deficiency. Am J Clin Nutr 2003;78:3–6.

5. Barr SL et al. Perceptions and practices of self-defined current vegetarian, former vegetarian, and nonvegetarian women. J Am Diet Assoc 2002;102:354–6.

6. Kushi LH et al. The macrobiotic diet and cancer. J Nutr 2001. 131:3056S–64S.

7. Cunningham E. What is a raw food diet and are there any risks or benefits associated with it? J Am Diet Assoc 2004;104:1623.

8. Koebnick C. Long-term consumption of a raw food diet is associated with favorable serum LDL cholesterol and triglyc-erides but also with elevated plasma homocysteine and low serum HDL cholesterol in humans. J Nutr. 2005;135:2372–8.

9. Fontana L et al. Low bone mass in subjects with long-term raw vegetarian diet. Arch Int Med 2005;165:684–9.

10. Mutch PB. Food guides for the vegetarian. Am J Clin Nutr 1988;48:913–19.

11. Rauma A-L et al. Vitamin B-12 status of long-term adherents of a strict, un-cooked vegan diet ("living foods diet") is compromised. J Nutr 1995;125:2511–15.

12. Position of the American Dietetic Association and Dietitians of Canada: vegetarian diets. J Am Diet Assoc 2003;103:748–65.

13. Haddad EH, Tanzman JS. What do vegetarians in the United States eat? Am J Clin Nutr 2003;78(suppl):626S–32S.

14. Aronson D. Vegetarian nutrition: What every dietitian should know. Today's Dietitian 2005;Mar:31–7.

15. Berkow SE et al. Vegetarian diets and weight status. Nutr Rev. 2006;64:175–88.

16. Dunham L et al. Vegetarian eating for children and adolescents. J Pediatr Health Care 2006;20:27–34.

17. Key TJ et al. Health effects of vegetarian and vegan diets. Proc Nutr Soc. 2006;65:35–41.

18. Messina V et al. A new food guide for North American vegetarians. J Am Diet Assoc 2003;103:771–5.

19. Haddad EH et al. Vegetarian food guide pyramid: a conceptual framework. Am J Clin Nutr 1999;70(suppl):615S–9S.

20. Vegetarian food guide pyramid, http://fnic.nal.usda.gov, accessed 9/06.

21. Leitzmann C. Vegetarian diets: what are the advantages? Forum Nutr. 2005;57:147–56.

22. Arterburn LM et al. Distribution, inter-conversion, and dose response of n-3 fatty acids in humans. Am J Clin Nutr 2006;83:(suppl):1467S–76S.

23. Dietary reference intakes for carbohydrates, fiber, fat, fatty acids, cholesterol, protein, and amino acids. Washington, DC: National Academies Press; 1999.

Answers to Review Questions

1. True, see page 16-5.
2. False, see page 16-4.
3. True, see page 16-6.
4. True, see page 16-7.
5. False, see page 16-7.
6. True, see pages 16-5.
7. False, see page 16-9.
8. True, see page 16-9.

Nutrition Up Close

Vegetarian Main Dish Options

Feedback for Unit 16

Vegetarians definitely had the advantage in completing this exercise! For the meat eaters, were the main dishes you identified meat-free? Here are six vegetarian options for lunch and dinner main dishes:

- Eggplant parmesan
- Falafel
- Red beans and rice
- Pasta with mixed vegetables
- Melted cheese, tomato, and sprout sandwiches
- Bean burritos
- Veggie burgers

Food Allergies and Intolerances

NUTRITION SCOREBOARD

	TRUE	FALSE
1 About one in every three Americans is allergic to at least one food.		
2 Food ingredient labels must indicate the presence of major food allergens in food products.		
3 Skin prick tests are an accurate way to diagnose food allergies.		
4 Food intolerances cause less severe reactions than food allergies do.		

Key Concepts and Facts

- True food allergies involve a response by the body's immune system to a particular substance in food.

- Food allergies may be caused by hundreds of different foods. However, 90% of food allergies are due to eight foods: nuts, eggs, wheat, milk, peanuts, soy, shellfish, and fish.[1]

- Food intolerances cover an array of adverse reactions that do *not* involve the body's immune system.

- The most accurate method of identifying food allergies and intolerances is the double-blind, placebo-controlled food challenge.

Answers to
NUTRITION SCOREBOARD

		TRUE	FALSE
1	About one in three Americans *believes* he or she is allergic to at least one food. Probably 2 in 100 adults and 5 in 100 infants and young children actually are.[1]		✔
2	Listing potential, relatively common food allergens on food ingredient labels is required for packaged foods.	✔	
3	Skin prick tests for food allergies are often inaccurate.[2]		✔
4	Food allergies and food intolerances have different causes, but both can produce life-threatening reactions.		✔

Food Allergy Mania

Talk about your controversial subjects. This one is very hot. People use the term *food allergy* to refer to virtually any type of problem they have with food. At one extreme are people who believe allergies to milk, wheat, and sugar are to blame for hyperactivity and a host of other behavioral problems in children. At the other extreme are some health professionals who think people who complain of food allergies need to have their heads examined rather than their bodies' reactions to foods. Despite the controversies, food allergies have become a major health concern in the United States and many other Westernized countries.[3]

Food allergies are real and can be very serious. At the minimum, true food allergies can cause a rash or an upset stomach. At the maximum, they can result in death. Unreal food allergies can cause problems, too. They can lead people to eliminate nutritious foods from their diet unnecessarily, resulting in inadequate diets and eventually in health problems. One of the most intriguing aspects of food allergies is the frequency and ease with which foods are falsely blamed for a variety of mental and physical health problems.

Adverse Reactions to Foods

People experience adverse reactions to food for three primary reasons. One is food poisoning. The other two involve the body's reaction to substances in food that are normally harmless. With *food allergies,* the body's *immune system* reacts to a substance in food (almost always a protein) that it identifies as harmful (Illustration 17.1). The immune system is not involved in *food intolerances,* which encompass other adverse reactions to normally harmless substances in food.

Why Do Food Allergies Involve the Immune System? It may seem odd that a system that helps the body conquer bacteria and viruses is involved in "protect-

---

food allergy
Adverse reaction to a normally harmless substance in food that involves the body's immune system. (Also called *food hypersensitivity.*)

immune system
Body tissues that provide protection against bacteria, viruses, and other substances identified by cells as harmful.

food intolerance
Adverse reaction to a normally harmless substance in food that does *not* involve the body's immune system.

PhotoEdit, Felicia Martinez

Illustration 17.1
What do these foods have in common?
Each food shown, plus many more not shown, may cause adverse reactions in some individuals.

Illustration 17.2
The development of allergic reactions to foods.

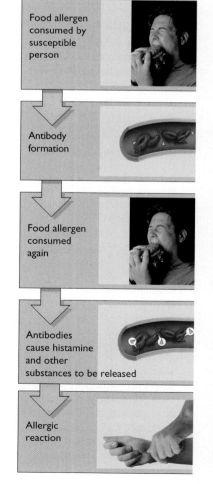

Food allergen consumed by susceptible person

Antibody formation

Food allergen consumed again

Antibodies cause histamine and other substances to be released

Allergic reaction

ing" us from normal constituents of food. In people with allergies, however, the cells recognize some food components as harmful—just as they recognize bacteria and viruses. Components of food that trigger the immune system are called *food allergens.*

For some susceptible people, exposure to trace amounts of an allergen triggers an allergic reaction. Highly sensitive people can develop an allergic reaction simply by being in the room when the food to which they are allergic is prepared or by touching the food. Allergic reactions to peanuts have been triggered by kissing someone who has recently consumed them and possibly by breathing peanut vapors on airplanes.[4]

In response to the allergen, the immune system forms *antibodies* (Illustration 17.2). The antibodies attach to cells located in the nose, throat, lungs, skin, eyes, and other areas of the body. When the allergen appears again, the body is ready. The previously formed antibodies recognize the allergen, attach to it, and signal the body to secrete *histamine* and other substances that cause the physical signs of allergic reactions. Most commonly, allergic reactions cause a rash, diarrhea, congestion, or wheezing (Table 17.1), but the symptoms may be much more serious.

TABLE 17.1

The three most common types of symptoms caused by food allergies

Some people experience more than one type of reaction.[5]

Symptom	Percentage of People with Food Allergies Who Develop the Symptom
• Skin eruptions: rash, hives	84%
• Gastrointestinal upsets: diarrhea, vomiting, cramps, nausea	52
• Respiratory problems: congestion, runny nose, cough, wheezing, asthma	32

Illustration 17.3
These foods can cause anaphylactic shock, a severe allergic reaction, in highly sensitive people.
(Most food allergies cause milder symptoms.)

food allergen
A substance in food (almost always a protein) that is identified as harmful by the body and elicits an allergic reaction from the immune system.

antibodies
Proteins the body makes to combat allergens.

histamine
(hiss-tah-mean) A substance released in allergic reactions. It causes dilation of blood vessels, itching, hives, and a drop in blood pressure and stimulates the release of stomach acids and other fluids. Antihistamines neutralize the effects of histamine and are used in the treatment of some cases of allergies.

anaphylactic shock
(an-ah-fa-lac-tic) Reduced oxygen supply to the heart and other tissues due to the body's reaction to an allergen in food. Symptoms of anaphylactic shock include paleness, a flush discoloration of the lips and fingertips, a "faceless" expression, weak and rapid pulse, and difficulty breathing.

Exposure to trace amounts of an allergen in peanuts, nuts (peanuts are actually a legume, not a nut), fish, and shellfish can cause *anaphylactic shock* (Illustration 17.3). This massive reaction of the immune system can result in death, caused by the cutoff of the blood supply to tissues throughout the body. People experiencing anaphylactic shock can be revived with an injection of epinephrine. The incidence of peanut allergy in the United States is increasing—about 1 in 150 people are affected.[2] Although it's not clear why peanut allergy is increasing, some scientists suspect the reasons are related to expanded use of soy infant formulas, skin oils and creams that contain peanut oil, or consumption of peanuts very early in life.[6]

Symptoms of food allergies may occur within seconds after the body is exposed to the food allergen, or they may take 2 hours or more to occur. The symptoms often disappear within 2 hours after they begin.[7]

What Foods Are Most Likely to Cause Allergic Reactions? Many foods may cause allergic reactions in susceptible individuals (Illustration 17.4). Nevertheless, approximately 90% of all food allergies are caused by eight foods: nuts, eggs, wheat, milk, peanuts, soy, shellfish, and fish. The nearby "On the Side" feature gives you an idea for how to remember these foods. Together, these are considered "the big eight" food allergies.[1]

Passage of the Food Allergen Labeling and Consumer Protection Act of 2004 is helping consumers with food allergies find offending foods and ingredients. Food ingredient labels are now required to state clearly whether the food contains one or more of the "big eight" allergenic foods.[1]

■ A Closer Look at Wheat Allergy

Minnie is 46 years old and has three sisters and two brothers, all of whom are taller than she is. (She considers herself the "runt of the litter.") Since childhood, Minnie has had problems with diarrhea and cramps. Over the past 2 years, these problems have become worse, and she has been chronically tired. Assuming that she was lactose intolerant, Minnie stopped eating all dairy products. Her problems persisted, however. When she almost lost her job as a receptionist due to her frequent bathroom breaks and her obvious fatigue, she decided to see a doctor again.

Diagnosing anemia, the doctor sent Minnie to a registered dietitian. Hearing Minnie's story, the dietitian suspected celiac disease caused by an allergy to a component of wheat and other grains. After two weeks on an allergen-free diet, Minnie started to regain her strength and spent much less time in the bathroom. Diagnostic tests subsequently confirmed that Minnie had celiac disease, a condition that likely had existed since childhood.

An allergy to wheat, or more specifically to a component of gluten found in wheat and other grains such as barley, rye, and triticale has been underdiagnosed in the past. Called *celiac disease* (and also celiac sprue and gluten-sensitive enteropathy), it affects one in 120 to 300 people in many countries, including the United States.[8] It differs somewhat from traditional immunoglobulin E-mediated allergies in that the immune system reaction to gluten is localized in the lining of the small intestine. Celiac disease can be difficult to diagnose because the symptoms (diarrhea, weight loss, cramps, anemia) are similar to those of other diseases. With

Photo Disc

Illustration 17.4
The "big eight" foods that can cause allergies.[1]

celiac disease
A disease characterized by inflammation of the small intestine lining resulting from a genetically based intolerance to gluten. The inflammation produces diarrhea, fatty stools, weight loss, and vitamin and mineral deficiencies. (Also called *celiac sprue and gluten-sensitive enteropathy*.)

new awareness about the disorder, doctors are ordering diagnostic tests more frequently, and the rate of confirmed cases of celiac disease is increasing.[9]

Celiac disease is treated with a lifelong gluten-free diet. Because gluten is included in many food additives as well as grain products, it generally takes the help of an experienced, registered dietitian to design and monitor gluten-free diets.[10]

The term *gluten-free* will be standardized and allowed on food labels for qualifying foods beginning in 2008.[1] This new regulation is voluntary but is expected to be widely employed.

Do Genetically Modified Foods Contain Allergens?   It's possible but unlikely. Genetically modified foods contain proteins from different types of plants or animals. Consequently, a protein that causes allergic reactions in some people could possibly end up in a genetically modified food. This possibility is recognized by manufacturers, and genetically modified foods that may contain allergens are tested for safety before they are sold. Pretests of soybeans with added Brazil-nut proteins, for example, found that people allergic to Brazil nuts were allergic to the genetically modified soybeans. Plans to make the soybeans commercially available were dropped, and the manufacturer halted further production.[11] Due to manufacturers' testing procedures and FDA regulations related to these foods, the possibility that genetically modified foods contain hidden allergens is limited to some extent. However, the true risks of allergic reactions from genetically modified food are unknown.[12]

How Common Are Food Allergies?

Infants and young children experience food allergies more often than adults do. Approximately 5% of infants and young children develop a food allergy, which most often is due to a protein in cow's milk.[13] To prevent allergic reactions in infants and young children at risk of allergies, it is recommended that cow's milk and dairy products be excluded from the infant's diet during the first year of life. Eggs should be introduced after two years, and peanuts, nuts, fish, and shellfish should not be given for the first 3 years.[14] Most infants outgrow their allergy to cow's milk, eggs, and wheat later in childhood. They often do not outgrow allergies to peanuts, nuts, fish, and shellfish.[2]

NEW Milk. Please Serve Safe Food.
This image and saying may help you remember the top foods related to allergies: You're standing in front of a refrigerator and have just taken a big gulp of milk straight from the carton. In that second, it hits you that the milk was sour. To avoid that ever happening again, you think of the saying, "NEW Milk. Please Serve Safe Food." Translation:

N = NUTS
E = EGGS
W = WHEAT

MILK
PLEASE = PEANUTS
SERVE = SOY
SAFE = SHELLFISH
FOOD = FISH

ON THE SIDE

The incidence of food allergies in adults is estimated to be 2% and appears to be increasing.[15] Nevertheless, around one-third of Americans believe they are allergic to one or more foods. About 60% of complaints of food allergy fail to be confirmed by testing.[16]

Why the Difference between Reality and Perception?

One reason food allergies seem more widespread than they actually are is that popular wisdom endows various foods with certain properties. The following are some of the more widely held beliefs:

- Allergies to sugar or food colors lead to hyperactivity in children.

- Yeast causes chronic fatigue syndrome.

- Dairy products, especially milk, cause phlegm.

None of these associations has been documented by scientific studies.

Bad timing and mental attitudes about certain foods can also lead to erroneous diagnoses of food allergies. Here are two examples:

1. Isaiah, a lover of blueberries and blueberry pie, hasn't touched a blueberry since 1995. That year, Isaiah had a piece of blueberry pie in a restaurant and later became violently sick to his stomach. Bingo! Isaiah thought he must be allergic to blueberries.

   Actually, Isaiah gave up blueberries for no good reason. He was coming down with the flu when he ate the pie. Nevertheless, to this day when he thinks of blueberries, he gets a bit queasy.

2. For 11 years, Emilia rigidly avoided even small amounts of cow's milk. She was convinced that just a few drops of cow's milk would cause pressure in her head, blurred vision, dizziness, cramps, and nausea.

   Finally, Emilia's presumed allergy to cow's milk was put to the test. She was given liquid through a dark tube inserted into her stomach from her mouth. When she was told the fluid was cow's milk—even though it was actually water—the familiar symptoms appeared within 10 minutes. When she was given milk but told it was water, no symptoms appeared.[17]

   Emilia was shocked. For 11 years she had scrutinized nearly everything she ate and had wasted hundreds of dollars on special food products and supplements. Finishing a frosted brownie and a glass of milk, Emilia contemplated the practical realities of the power of suggestion.

Diagnosis: Is It a Food Allergy?

It is essential to realize that the [allergic] reactions occur in human beings and are therefore enmeshed in the interplay between body, mind and soul. . . . Only through experience with blind food challenges can the amazing power of self-deception be appreciated.
—CHARLES MAY, 1980[17]

double-blind, placebo-controlled food challenge A test used to determine the presence of a food allergy or other adverse reaction to a food. In this test, neither the patient nor the care provider knows whether a suspected offending food or a placebo is being tested.

A variety of tests are used to diagnose food allergies, but there is only one "gold standard." That test is the **double-blind, placebo-controlled food challenge**.[18]

Suppose you suspect you are allergic to grapes because you got an annoying rash twice after eating them. Your doctor arranges an appointment for you at an

Reality Check

I think I'm allergic to . . .

You and your coworkers decide to go to a Chinese restaurant for lunch. Trevor wants to go but hesitates because, as he said, "I must be allergic to Chinese food. Every time I eat it I feel dizzy, sweat, and get this ringing in my ears."

Who's got the better idea?
Answers on next page

Haley:
Trevor, that's not an allergy. It's all in your head.

Darrel:
Have you ever asked them to leave out the MSG?

allergy clinic. At the clinic, you will be given grapes, either in a concentrated pill form or blended into a liquid, and placebo pills (or a liquid) without being told which is the grapes and which the placebo. After an appropriate amount of time, your reaction to the grapes and to the placebo will be noted. If the rash appears after the grapes but not after the placebo, you have an adverse reaction to grapes that may be an allergy.

Food challenges are best undertaken under medical supervision. True food allergies can cause serious reactions, and immediate help may be needed. Foods suspected of causing anaphylactic shock in the past should not be given in food challenge tests. They should be totally omitted from the diet.[18]

The Other Tests for Food Allergy

Other tests for food allergy are reliable to varying degrees. None of them is always accurate, and the results of bogus tests are not to be trusted at all.

Skin Prick Tests Skin prick tests (also called prick skin tests) for food allergy are commonly employed and are useful for identifying *the absence* of a food allergy. A positive test result isn't proof positive that an allergy exists, however, because positive test results are inaccurate about 50% of the time.[28]

For this test, a few drops of food extract are placed on the skin, and the skin is then pricked with a needle. At approximately the same time, another area of skin is pricked using only water. If the area around the food skin prick becomes redder and more swollen than the area pricked with water (Illustration 17.5), it is concluded that the person *might* be allergic to the food tested. Skin prick tests are useful for ruling out allergies and for reducing the number of foods that may have to be tested by double-blind, placebo-controlled food challenges.

The RAST Test The acronym RAST stands for "radioallergosorbent test" (pronounced *radio-all-er-go-sor-bent*). In this test, a sample of blood is examined for immune substances formed in response to an allergen. The test is not specific for individual food allergens, nor does it reliably indicate the presence of a food allergy.[28]

Bogus Tests A number of companies offer food allergy tests through the mail. In one study, researchers sent five of these

Illustration 17.5
A positive reaction to a skin prick test for a food allergy usually looks like this.
The result indicates that an allergy to the food tested may be present but doesn't guarantee it.

I think I'm allergic to . . .

Some people, like Trevor may be sensitive to MSG. It doesn't elicit an immune system reaction, so it's not a food allergy.

Haley

Darrel

companies the required samples, nine from adults who were allergic to fish and nine from adults who had no allergies. All five companies missed the fish allergy. When duplicate samples from the same adults were sent separately to each company, reports came back with different results for each of the pairs of samples. In addition, the companies reported their laboratories had identified various food allergies that none of the adults actually had.[17] Mail-order allergy tests are a waste of money; they do not use reliable assessment techniques. Other unreliable methods for identifying food allergies include hair analysis, cytotoxic blood tests, iridology, and sublingual (under the tongue) and food injection provocation tests.[20]

What's the Best Way to Treat Food Allergies?

After a food allergy is confirmed, the food is eliminated from the diet. This is the only treatment available for food allergies. "Allergy" shots for food allergens are not yet available but may be in the future.[2] If the eliminated food is an important source of nutrients or is found in many food products (such as milk or wheat), consultation with a registered dietitian is recommended.

Some food allergies don't last forever. Many infants and young children outgrow allergies to cow's milk, wheat, and eggs by late childhood. Children and adults with a severe allergy to peanuts, nuts, fish, or shellfish may have to eliminate the food from their diet for a lifetime.[2]

Food Intolerances

Food intolerances produce some of the same reactions as food allergies, but the reactions develop by different mechanisms. Food intolerance reactions do not involve the immune system; they are due to a missing enzyme or other cause.[15]

It is clear that some people are intolerant of lactose, sulfite, histamine (a component of red wine and aged cheese), and other foods (Table 17.2).[15] It is also clear that not all food intolerances are real. Just as with food allergies, the best way to separate the real from the unreal is the double-blind, placebo-controlled food challenge. True food intolerances produce predictable reactions. Problems such as headache, diarrhea, swelling, or stomach pain will occur every time a person consumes a sufficient amount of the suspected food.

Lactose Maldigestion and Intolerance

Lactose maldigestion, or the inability to break down lactose in dairy products due to the lack of the enzyme lactase, results in the condition known as *lactose intolerance.* Symptoms of lactose intolerance, such as flatulence, bloating, abdominal

TABLE 17.2

Foods and substances in them linked to food intolerance reactions

- Aged cheese
- Anchovies
- Beer
- Catsup
- Chocolate
- Dried beans
- Food coloring
- Lactose
- Mushrooms
- Pineapple
- Red wine
- Sausage (hard, cured)
- Soy sauce
- Spinach
- Tomatoes
- Yeast

lactose maldigestion
A disorder characterized by reduced digestion of lactose due to the low availability of the enzyme lactase.

lactose intolerance
The term for gastrointestinal symptoms (flatulence, bloating, abdominal pain, diarrhea, and "rumbling in the bowel") resulting from the consumption of more lactose than can be digested with available lactase.

pain, diarrhea, and "rumbling in the bowel," occur in lactose maldigesters within several hours of consuming more lactose than can be broken down by the available lactase. These symptoms are due to the breakdown of undigested lactose by bacteria in the lower intestines (which produce gas as a by-product of lactose ingestion) and by fluid accumulation.[21]

Reduced intake of lactose-containing dairy products (milk, ice cream, and cottage cheese, for example) reduces lactose intolerance. However, foods such as hard cheese, low- or no-lactose milk, buttermilk, and yogurt without added milk solids contain low amounts of lactose and can generally be consumed. Small amounts of milk or other lactose-containing dairy products are generally well tolerated by people with lactose maldigestion.[21] Care should be taken to ensure adequate intake of calcium and vitamin D if milk and dairy product intake is restricted.

Lactose maldigestion is a very common disorder. It occurs in about 25% of U.S. adults, but the incidence varies dramatically by population group. By one set of estimates, 62 to 100% of Native American adults, 90% of Asian Americans, 80% of African Americans, 53% of Mexican Americans, and 15% of Caucasian Americans are lactose maldigesters.[22]

Sulfite Sensitivity

Sulfite is a food additive used by food manufacturers in the past to keep vegetables and fruits looking fresh and to prevent mold growth. It is also added to some beers, wines, processed foods, and medications as a preservative (see Table 17.3). Very small amounts of sulfite can cause anaphylactic shock and bring on an asthma attack in sensitive people.[23] Because many foods have added sulfite, people sensitive to sulfite should read food ingredient labels carefully. The FDA requires that processed foods containing sulfites list sulfites on the ingredient label. It also prohibits the use of sulfite on fresh vegetables and fruits.

Red Wine, Aged Cheese, and Migraines

Some people develop migraine headaches when they consume red wine. The presence of histamine in wine has been blamed for the headaches. People with this intolerance are unable to break down histamine during digestion, so it accumulates in the blood and causes headaches.[24] Histamine is also found in beer, sardines, anchovies, hard cured sausage, pickled cabbage, spinach, and catsup. Tyramine, a compound closely related to histamine and found in aged cheese, soy sauce, and other fermented products, causes migraine headaches in sensitive people. Chocolate, too, may bring on migraine headaches in sensitive individuals, although this finding is controversial.[25]

MSG and the "Chinese Restaurant Syndrome"

Some people appear to be sensitive to MSG (monosodium glutamate), a flavor enhancer used on meats and in soups, stews, and many Chinese food items. Because symptoms are reported to occur shortly after sensitive individuals eat Chinese food, sensitivity to MSG has been dubbed the "Chinese restaurant syndrome." Symptoms associated with MSG sensitivity include dizziness, sweating, flushing, a rapid heartbeat, and a ringing sound in the ears.[19]

It is reported that some Chinese people do not believe in the existence of the Chinese restaurant syndrome.[26] Nonetheless, many Chinese restaurants are willing to prepare foods without MSG if requested.

TABLE 17.3
Some foods that may contain sulfite

- Wine
- Beer
- Hard cider
- Tea
- Fruit juices
- Vegetable juices
- Guacamole
- Dried fruit
- Potato products
- Canned vegetables
- Baked goods
- Spices
- Gravy
- Soup mixes
- Jam
- Trail mix
- Fish and seafood

Foods may be cross contaminated with potential allergens during production or preparation. Yogurt-covered peanuts have been accidentally mixed with yogurt-covered raisins, and milk chocolate has been found to contain peanut fragments.[27]

Photo Disc

Precautions

People with food allergies or intolerances have to be very careful about what they eat and should have a plan of action ready in case they develop a serious reaction. Becoming highly knowledgeable about which foods and food products contain ingredients that cause an adverse reaction is essential. When eating out, people with allergies and intolerances should ask a lot of questions to make sure they know what is being served. They have to become students of food ingredient labels.

Importantly, people who develop anaphylactic shock should always carry a preloaded syringe of epinephrine, or "epipen" and know how to use it. They should also know where to call for emergency help or be ready to dial 911.[2] People whose reactions to foods lessen with the use of asthma inhalers and antihistamines should keep them readily available.

One final thought about food allergies and intolerances—they are never caused by studying about them.

Key Terms

anaphylactic shock, *page 17-4*
antibodies, *page 17-3*
celiac disease, *page 17-4*
double-blind, placebo-controlled food
    challenge, *page 17-6*

food allergen, *page 17-3*
food allergy, *page 17-2*
food intolerance, *page 17-2*
histamine, *page 17-3*

immune system, *page 17-2*
lactose intolerance, *page 17-8*
lactose maldigestion, *page 17-8*

Review Questions

	True	False
1. Peanuts, shellfish, cow's milk, and eggs are among the "big eight" foods that can cause food allergies.	☐	☐
2. Adverse reactions to sulfite in foods are prompted by immune system-mediated processes.	☐	☐
3. The prevalence of food allergies appears to be increasing in the United States.	☐	☐
4. The Food Allergen Labeling and Consumer Protection Act requires that manufacturers remove all potentially allergic ingredients from their foods.	☐	☐
5. Skin prick tests definitively identify allergies to specific foods.	☐	☐

	True	False
6. Symptoms of food allergy can be triggered by exposure to a very small amount of a food allergen.	☐	☐
7. Food allergies can be treated with "allergy shots."	☐	☐
8. Use of the term *gluten-free* on labels of qualifying food will be allowed starting in 2008.	☐	☐
9. Allergies to peanuts, nuts, fish, and seafood are likely to last a lifetime.	☐	☐
10. Food allergies can be prevented to an extent in young children by delaying the introduction of eggs, cow's milk, dairy products, peanuts, nuts, fish, and shellfish until after the age of 1 to 3.	☐	☐

Prevention of Food Allergies

Focal Point: Foods that may cause allergies in the first year of life.

Scenario: Darla and Mitch have a 6-month-old son who has been breastfed exclusively. The time has come to add semisolid food to their son's diet. Mitch had food allergies as a child, so he and Darla are concerned that their son may develop allergies, too.

Question: If you were Darla or Mitch, what seven foods would you omit from your son's diet for the first one to three years of life?

1. _____

2. _____

3. _____

4. _____

5. _____

6. _____

7. _____

FEEDBACK (answers to these questions) can be found at the end of Unit 17.

Media Menu

www.aaaai.org
This is the home page for the American Academy of Allergy, Asthma, and Immunology. Go from there to tips to remember, information on diagnosis, treatment, and prevention.

www.foodallergy.org
The Food Allergy and Anaphylaxis Network

provides lots of information on food allergies and intolerances, how to avoid food allergens, advice on how to deal with school food service, support services, and recipes.

www.mayoclinic.com
This site is a great resource for scientific information and practical tips on food allergies.

www.healthfinder.gov
Search food allergy and food intolerance to find facts, specific information on peanut allergy, avoiding allergies, and treatment.

Notes

1. Bren L. Food labels identify allergens more clearly. FDA Consumer Magazine, 2006;Mar–Apr.

2. Chapman JA et al. New guidelines for food allergy. Ann Allergy Asthma Immunol 1006;96:S1–S68.

3. Allergic diseases—How big is the problem? An expert interview with Mark T O'Hollaren, MD. Medscape Allergy & Clinical Immunology 2006;6(1):4/24/06.

4. Maloney JM. Peanut allergen via kissing a real possibility. www.medscape.com/viewarticle/525036, accessed 9/06.

5. Sampson HA. State of the art theories on food allergies. Annual meeting of the American Dietetic Association, Boston, 1997 Oct. 28.

6. Leung et al., Effect of anti-IgE therapy; and Kuwayama SP. Peanut allergy. N Engl J Med 2002;347:1534.

7. Burks AW, Food hypersensitivity, Annual meeting of the American Dietetic Association, Boston, 1977 Oct. 28.

8. Farrell RJ, Kelley CP. Celiac sprue. N Engl J Med 2002;346:180–8.

9. Diagnosing gluten-sensitive enteropathy. WebMD Scientific American Online, www.WebMD.com, vol. 26, no. 9, Sept. 2003.

10. Murray JA. The widening spectrum of celiac disease. Am J Clin Nutr 1999;69:354–65.

11. Taylor SL et al. Food allergies and avoidance diets. Nutr Today 1999;34:15–22.

12. Falk MC et al. Food biotechnology: benefits and concerns. J Nutr 2002;132:1384–90.

13. Norris JM et al. Early cereal exposure and wheat allergy. Pediatrics 2006;117:2175–82.

14. Fiocchi A et al. Consensus document for introducing solid foods into an infant's diet to avoid food allergies. Ann Allergy Asthma Immunol 2006;97:10–21.

15. Kline DA. Food allergy symptoms and causes. Today's Dietitian 2005;Aug:10–14.

16. Allergy statistics (www.nal.usda.gov/fnic); and Food allergies at a glance, p. 47, accessed 8/03.

17. May CD. Food sensitivity: facts and fancies. Nutr Rev 1984;42:72–8.

18. Dutau G. Food allergies and diagnostic alternatives: labial food challenge, oral food challenge. Revue Francaise d'Allergologie et d'Immunologie Clinique. 2000;40:728–41.

19. Dietary Reference Intakes. Energy, carbohydrate, fiber, fat, fatty acids, cholesterol, protein, and amino acids. Institute of Medicine, National Academy of Sciences; Washington, DC: National Academies Press, 2002.

20. Ferguson A. Food sensitivity or self-deception? N Engl J Med 1990;323:476–8.

21. Tolstoi LG. Adult-type lactase deficiency. Nutr Today 2000;35:134–42.

22. Inman-Felton AE. Overview of lactose maldigestion (lactose non-persistence). J Am Diet Assoc 1999;99:481–89.

23. Yang WH, Purchase EC. Adverse reactions to sulfites. Can Med Assoc J 1985;133:865–7.

24. Jarisch R, Wantke F. Wine and headache. Int Arch Allergy Immunol 1996;110:7–12.

25. Hannington E, Preliminary report on tyramine headache, Br Med J 1967;2:550–1; and Chandra, Food hypersensitivity and allergic disease.

26. Definition of Chinese restaurant syndrome, MedicineNet.com, http://www.medterms.com/script/main/art.asp?articlekey=15584, accessed 9/06.

27. What people mistake as food allergies, www.intelihealth.com, accessed 8/03

Nutrition Up Close

Prevention of Food Allergies

Answers to Review Questions

1. True, see page 17-4.
2. False, see pages 17-8, 17-9.
3. True, see page 17-6.
4. False, see page 17-4.
5. False, see page 17-7.
6. True, see page 17-3.
7. False, see page 17-8.
8. True, see page 17-5.
9. True, see pages 17-5, 17-8.
10. True, see page 17-5.

Feedback for Unit 17

Although it is not known whether their son would develop food allergies, Darla and Mitch would be wise to delay the introduction of these seven foods until after the first one to three years of life.

1. nuts
2. eggs
3. milk
4. milk products
5. peanuts
6. shellfish
7. fish

Fats and Cholesterol in Health

NUTRITION SCOREBOARD

	TRUE	FALSE
1 It is currently recommended that adults consume diets providing less than 30% of calories from fat.		
2 The types of fat consumed are more important to health than is total fat intake.		
3 Cholesterol is present in every food.		
4 Saturated fat intake has a much stronger influence on blood cholesterol levels than does cholesterol intake.		

Key Concepts and Facts

- Fats are our most concentrated source of food energy. They supply 9 calories per gram.

- Dietary fats "carry" the essential fatty acids, fat-soluble vitamins, and healthful phytochemicals along with them in foods.

- Fats are not created equal. Some types of fat have positive effects, and some have negative effects on health.

- Saturated fats and trans fats raise blood cholesterol levels more than does dietary cholesterol or any other type of fat.

Answers to
NUTRITION SCOREBOARD

	TRUE	FALSE
1 The acceptable range of fat intake is now 20 to 35% of total calories.		✔
2 The types of fat consumed are more important to health than the total amount of fat.[1]	✔	
3 Cholesterol is present only in animal products.		✔
4 Saturated fat, which is found primarily in animal products, raises blood cholesterol levels a good deal more than does dietary cholesterol.	✔	

Changing Views about Fat Intake and Health

Scientific evidence and opinions related to the effects of fat on health have changed substantially in recent years—and so have recommendations about fat intake. In the past, it was recommended that Americans aim for diets providing less than 30% of total calories from fat. Evidence indicating that the type of fat consumed is more important to health than is total fat intake has changed this advice. The watchwords for thinking about fat have become "Not all fats are created equal: Some are better for you than others." American adults are being urged to select food sources of "healthy" or "good" fats while keeping fat intake within the range of 20 to 35% of total caloric intake. Concerns that high-fat diets encourage the development of obesity have been eased by studies demonstrating that excessive caloric intakes—and not just diets high in fat—are related to weight gain.[2]

New recommendations regarding fat intake do not encourage increased fat consumption. Rather, they emphasize that healthy diets include certain types of fat and that total caloric intake and physical activity are the most important components of weight management. Diets providing as low as 20% of calories from fat, and those providing 30 to 35%, can be healthy—depending on the types of fat consumed and the quality of the rest of the diet.[3] This unit provides facts about fats, explains the reasons behind recent changes in recommendations for fat intake, and addresses the practical meaning of it all.

Facts about Fats

Fats are a group of substances found in food. They have one major property in common: They are not soluble (or, in other words, will not dissolve) in water. If you have ever tried to mix vinegar and oil when making salad dressing, you have observed the principle of water and fat solubility firsthand.

Fats are actually a subcategory of the fat-soluble substances known as *lipids.* Lipids include fats, oils, and cholesterol. Dietary fats such as butter, margarine, and shortening are often distinguished from oils by their property of being solid at room temperature. This physical difference between fats and oils is due to their chemical structures.

Functions of Dietary Fats

Fats in Foods Supply Energy and Fat-Soluble Nutrients Dietary fats are a concentrated source of energy. Each gram of fat consumed supplies the body with 9 calories worth of energy. That's enough energy for a 160-pound person to walk casually for a minute and a half or to jog at a slow pace for about a minute. Fats in food supply the *essential fatty acids* (linoleic acid and alpha-linolenic acid) and provide the fat-soluble vitamins D, E, K, and A (the "deka" vitamins). So, part of the reason we need fats in our diet is to get a supply of the essential nutrients they contain (Table 18.1). Diets containing little fat (less than 20% of total calories) often fall short on delivering adequate amounts of essential fatty acids and fat-soluble vitamins.[4]

Fats Increase the Flavor and Palatability of Foods Although "pure" fats by themselves tend to be tasteless, they absorb and retain the flavor of substances that surround them. Thus, fats in meats and other foods pick up flavors from their environment and give those flavors to the food. This characteristic of fat is why butter, if placed next to the garlic in the refrigerator, tastes like garlic.

Fats Contribute to the Sensation of Feeling Full As they should, at 9 calories per gram! Fats tend to stay in the stomach longer than carbohydrates or proteins and are absorbed over a longer period of time. That's why foods with fat "stick to your ribs."[5]

Fats are a Component of Cell Membranes, Vitamin D, and Sex Hormones Some types of fats give cell membranes flexibility and help regulate the transfer of nutrients into and out of cells.[6] Others serve as precursors to vitamin D and sex hormones, such as estrogen and testosterone.

Excess Dietary Fat, Carbohydrates, and Protein Are Stored as Fat

Fats in the body include those consumed in the diet and those produced from carbohydrates and proteins. Humans eat only a few times a day, but we need energy throughout the day. To ensure a constant supply of energy, the body converts carbohydrates and proteins that have been supplied from foods and are not used to meet immediate needs to storage forms of energy. Some of the excess carbohydrate and protein is converted to glycogen, the storage form of glucose under normal circumstances, but most of the excess is changed to fat and stored in fat cells (Illustration 18.1).

Body fat is not just skin deep. Fat is also located around organs such as the kidneys and heart. It's there to cushion and protect the organs and keep them insulated. Cold-water swimmers can attest to the effectiveness of fat as an insulation material. They purposefully build up body fat stores because they need the extra layer of insulation (Illustration 18.2).

TABLE 18.1
Roles of dietary fat

- Provides a concentrated source of energy
- Carries the essential fatty acids, the fat-soluble vitamins, and certain phytochemicals
- Increases the flavor and palatability of foods
- Provides sustained relief from hunger
- Serves as a component of cell membranes, vitamin D, sex hormones

lipids
Compounds that are insoluble in water and soluble in fat. Triglycerides, saturated and unsaturated fats, and essential fatty acids are examples of lipids, or "fats."

essential fatty acids
Components of fats (linoleic acid, pronounced *lynn-oh-lay-ick,* and alpha-linolenic acid, pronounced *lynn-oh-len-ick*) required in the diet.

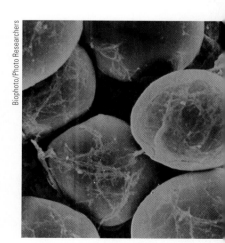

Illustration 18.1
A close look at fat cells (color-enhanced microphotograph).

AP Photo

Fats Come in Many Varieties

There are many types of fat in food and our bodies (Table 18.2). Of primary importance are *triglycerides* (or "triacyl-glycerols"), *saturated and unsaturated fats, cholesterol,* and *trans* fats (for definitions, see Table 18.3). The different types of fats have different effects on health.

Triglycerides, which consist of one *glycerol* unit (a glucoselike substance) and three fatty acids (Illustration 18.3), make up 98% of our dietary fat intake and the vast majority of our body's fat stores. Triglycerides are transported in blood attached to protein carriers and are used by cells for energy formation and tissue maintenance. A minority of fats take the form of *diglycerides* (glycerol plus two fatty acids) and *monoglycerides* (glycerol and one fatty acid). Diglycerides are present in some oils and small amounts are used in food products as emulsifiers—or to increase the blending of fat- and water-soluble substances. Monoglycerides are present in small amounts in some oils; we don't consume very much of them in foods.

Illustration 18.2
Although their body fat stores don't fit the image of the superathlete, cold-water swimmers need the fat to help stay warm.
Pictured here is the English swimmer Mike Read, who had swum the English Channel 20 times by age 39. The narrowest width of the English Channel is 22 miles, or 35 km.

As far as health is concerned, the glycerol component of fat is relatively unimportant. It's the fatty acids that influence what the body does with the fat we eat; and they are responsible, in part, for how fat affects health. Many different types of fatty acids are found in triglycerides. You've heard of the major ones: those that make fat "saturated" or "unsaturated."

Saturated and Unsaturated Fats   Fatty acids found in fats consist primarily of hydrogen atoms attached to carbon atoms (Illustration 18.4). When the carbons are attached to as many hydrogens as possible, the fatty acid is "saturated"—that is, saturated with hydrogen. Saturated fats tend to be solid at room temperature. Except for palm and coconut oil, only animal products are rich in saturated fats (Illustration 18.5).[7] Fatty acids that contain fewer hydrogens than the maximum are "unsaturated." They tend to be liquid at room temperature. By and large, plant foods are the best sources of unsaturated fats.

Unsaturated fats are classified by their degree of unsaturation. If only one carbon–carbon bond in the fatty acid is unsaturated, the fat is called "monounsaturated." If two or more carbon–carbon bonds are unsaturated with hydrogen, the fat qualifies as "polyunsaturated."

The Omega-6 and Omega-3 Fatty Acids

The essential fatty acids linoleic acid and alpha-linolenic acid are members of the fatty acid families of omega-6 (also called n-6 fatty acids) and omega-3 fatty

Glycerol + 3 fatty acids = Triglyceride

Illustration 18.3
A triglyceride.

TABLE 18.2

Basic facts about the types of fat

Fats can be:
- Monoglycerides
- Diglycerides
- Triglycerides

Fats can be:
- Saturated
- Monounsaturated
- Polyunsaturated

Unsaturated fats come in:
- "Cis" forms
- "Trans" forms

TABLE 18.3
A glossary of fats

Triglycerides: Fats in which the glycerol molecule has three fatty acids attached to it; also called *triacyl-glycerol*. Triglycerides are the most common type of fat in foods and in body fat stores.

Saturated fats: Molecules of fat in which adjacent carbons within fatty acids are linked only by single bonds. The carbons are "saturated" with hydrogens; that is, they are attached to the maximum possible number of hydrogens. Saturated fats tend to be solid at room temperature. Animal products and palm and coconut oil are sources of saturated fats.

Unsaturated fats: Molecules of fat in which adjacent carbons are linked by one or more double bonds. The carbons are not saturated with hydrogens; that is, they are attached to fewer than the maximum possible number of hydrogens. Unsaturated fats tend to be liquid at room temperature and are found in plants, vegetable oils, meats, and dairy products.

Glycerol: A syrupy, colorless liquid component of fats that is soluble in water. It is similar to glucose in chemical structure.

Cholesterol: A fat-soluble, colorless liquid found in animals but not in plants. Cholesterol is used by the body to form hormones such as testosterone and estrogen and is a component of animal cell membranes.

Diglyceride: A fat in which the glycerol molecule has two fatty acids attached to it; also called diacylglycerol.

Monoglyceride: A fat in which the glycerol molecule has one fatty acid attached to it; also called monoacylglycerol.

Monounsaturated fats: Fats that contain a fatty acid in which one carbon-carbon bond is not saturated with hydrogen.

Polyunsaturated fats: Fats that contain a fatty acid in which two or more carbon–carbon bonds are not saturated with hydrogen.

***Trans* fats:** Fats containing fatty acids in the *trans* form. Also called *trans* fatty acids.

Two hydrogens are missing from each of these carbon–carbon links, making the fatty acid polyunsaturated. With fewer hydrogens to attach to, these carbons are doubly bonded to each other. Monounsaturated fatty acids have only one carbon–carbon bond that is "unsaturated" with hydrogen atoms.

Illustration 18.4
A look at the difference between a saturated and an unsaturated fatty acid.

Saturated | Monounsaturated | Polyunsaturated

PERCENTAGE FATTY ACID COMPOSITION

DAIRY PRODUCTS

Food	Saturated	Monounsaturated	Polyunsaturated
Brick cheese	67	30	3
Cheddar cheese	67	30	3
Cottage cheese	67	30	3
American cheese	67	30	3
Whole milk	65	31	4
Butter	65	31	4

HUMAN MILK

	Saturated	Monounsaturated	Polyunsaturated
HUMAN MILK	48	40	12
EGG (medium)	37	48	15

OILS AND MARGARINES

Food	Saturated	Monounsaturated	Polyunsaturated
Canola oil	6	58	36
Coconut oil	92	6	2
Corn oil	13	25	62
Corn oil margarine	17	61	22
Flaxseed oil	9	18	73
Olive oil	14	77	9
Palm oil	52	38	10
Peanut oil	18	48	34
Soybean oil	15	25	60
Soybean oil margarine	22	50	28
Sunflower oil	11	21	68

MEATS

Food	Saturated	Monounsaturated	Polyunsaturated
Hamburger (21% fat)	53	43	4
Chicken, roasted (no skin)	26	46	28
Pork chop	39	49	12
Turkey, roasted	40	25	35

NUTS AND SEEDS

Food	Saturated	Monounsaturated	Polyunsaturated
Cashews	21	62	17
Macadamia nuts	16	46	2
Peanuts, dry-roasted	15	52	33
Peanut butter	18	51	32
Sunflower seeds (dried)	11	20	69

Illustration 18.5
Fat profiles of selected foods.

acids (also known as n-3 fatty acids), respectively. Both are polyunsaturated, can be used as a source of energy, and are stored in fat tissue. Because they are essential, both linoleic and alpha-linolenic acid are required in the diet.

Linoleic acid is required for growth, maintenance of healthy skin, and normal functioning of the reproductive system. It is a component of all cell membranes and is found in particularly high amounts in nerves and the brain. A number of biologically active compounds produced in the body that participate in regulation of blood pressure and blood clotting are derived from linoleic acid. The major food sources of linoleic acid are sunflower, safflower, corn, and soybean oils.

Alpha-linolenic acid is a structural component of all cell membranes and is found in high amounts in the brain and other nervous system tissues. It also forms biologically active compounds used in the regulation of blood pressure and blood clotting; but these compounds have the opposite effects on blood pressure and blood clotting as do derivatives of linoleic acid.[4] Omega-3 fatty acids are found in walnuts, dark, leafy green vegetables; and flaxseed, canola, and soybean oils in the form of alpha-linolenic acid.

Other, biologically important omega-3 fatty acids exist, and the two primary ones are EPA (eicosapentaenoic acid, pronounced *e-co-sah-pent-tah-no-ick*) and DHA (docosahexaenoic acid, pronounced *dough-cos-ah-hex-ah-no-ick*). These omega-3 fatty acids can be produced from alpha-linolenic acid, but the conversion process is slow and results in the availability of relatively small amounts of EPA and DHA. EPA and DHA can be converted to each other, however.[5]

EPA, DHA, and Fish Oils

Omega-3 fatty acids are not just good fats, they affect heart health in positive ways.
—Penny Kris-Etherton, Distinguished Professor of Nutrition

Fishy eggs? An odd source of fish oil has become part of North Americans' diets. Hens given feed with added cod liver oil or fish meal produce eggs with yolks containing 100 to 350 milligrams of DHA.[11]

Eric Risberg/AP Photo

ON THE SIDE

DHA and EPA are primarily ingredients of fish oils and perform a number of important functions in the body. DHA is a structural component of the brain and is found in high amounts in the retina of the eye. During the last three months of pregnancy and during infancy, DHA accumulates in these tissues and promotes optimal intellectual and visual development.[6]

EPA serves as a precursor of a number of biologically active compounds involved in blood pressure regulation, blood clotting, and anti-inflammatory reactions. Inflammation is a central component of many chronic diseases, including heart disease, type 2 diabetes, osteoporosis, cancer, Alzheimer's disease, and rheumatoid arthritis. It is part of the body's response to the presence of infectious agents or irritants. A by-product of the body's inflammatory processes, however, are oxidation reactions that can harm cells and tissues. Derivatives of EPA limit the harmful effects of inflammatory and oxidation reactions.[7]

Adequate intake of EPA and DHA for adults is considered to be 650 mg per day.[8] This level of intake can be achieved by consuming 12 ounces of fish and shellfish meals weekly. Fish, shellfish, and other foods that provide EPA and DHA are listed in Table 18.4. The table also highlights fish and shellfish that contain relatively low amounts of mercury. Mercury can cause nerve tissue damage and learning problems if consumed in excessive amounts. Pregnant and breast-feeding women are advised to limit their intake to 12 ounces per week of low-mercury fish and shellfish.[9]

Consumption of two fish meals a week reduces the risk of heart disease, heart attack, sudden death, and stoke and improves fetal and infant development.[6] Higher amounts of EPA and DHA (1 to 3 grams), generally provided by fish oils,

TABLE 18.4
Omega-3 fatty acid (EPA and DHA) content of selected seafood[12, 13]

Seafood (3½-Ounce Serving)	EPA and DHA (Grams)	Seafood (3½-Ounce Serving)	EPA and DHA (Grams)
Sardines in sardine oil[a]	3.3	Oysters[a]	0.6
Mackerel	2.6	Catfish[a]	0.5
Salmon, Atlantic, farmed[a]	2.2	Shrimp[a]	0.5
Lake trout[a]	2.0	Halibut	0.5
Herring[a]	1.8	Pollock[a]	0.5
Salmon, Atlantic, wild[a]	1.8	Scallops[a]	0.5
Tuna, white, canned	1.7	Whiting	0.5
Salmon, sockeye[a]	1.5	Carp	0.3
Whitefish, lake[a]	1.5	Crab[a]	0.3
Anchovies[a]	1.4	Pike, walleye	0.3
Salmon, chinook[a]	1.4	Tuna, fresh	0.3
Bluefish	1.2	Catfish, wild[a]	0.2
Halibut[a]	1.2	Clams[a]	0.2
Trout, rainbow, farmed[a]	1.2	Fish sticks	0.2
Salmon, pink[a]	1.0	Flounder[a]	0.2
Trout, rainbow, wild[a]	1.0	Haddock[a]	0.2
Bass, striped	0.8	Lobster	0.2
Swordfish	0.8	Salmon, red[a]	0.2
		Snapper, red	0.2
		OTHER	
		Human milk, 3½ oz	0.2
		DHA egg, 1	0.2

[a] Mercury content < 0.2 ppm.

are being used clinically to lower blood triglyceride levels, reduce the incidence of heart disease in people at risk, and reduce the need for anti-inflammatory drugs in people with rheumatoid arthritis.[7] The Food and Drug Administration recommends that consumers ingest no more than a total of 3 grams of EPA and DHA daily and limit supplementary intakes to 2 grams a day. Fish liver oils should be used with caution because they contain relatively high amounts of vitamins A and D. Fish oils, made from the body of the fish, do not.[6]

Balancing Intake of Omega-6 and Omega-3 Fatty Acids   The ratio of omega-6 to omega-3 fatty acid intake appears to be important because the functions of one are adversely modified by the presence of disproportionately high amounts of the other. Although an exact ratio has not been agreed upon, it is thought that people should consume omega-6 fatty acids and omega-3 fatty acids in a proportion of roughly 4 (or less) to 1. Many Americans regularly consume vegetable oils but eat fish infrequently. Consequently, the ratio between the intake of omega-6 and omega-3 fatty acids is over 9 to 1, indicating a need to increase intake of omega-3 fatty acids.[14]

Modifying Fats

Unsaturated fats aren't as stable as saturated fats. They are more likely to turn rancid with time and exposure to air and heat than are saturated fats. Additionally, solid fats are preferable to oils for some cooking applications. These problems with unsaturated fats have a solution. It's called hydrogenation.

hydrogenation
The addition of hydrogen to unsaturated fatty acids.

What's Hydrogenation?   *Hydrogenation* is a process that adds hydrogen to liquid unsaturated fats, thereby making them more saturated and solid. The shelf life, cooking properties, and taste of vegetable oils are improved in the process. Hydrogenation has two drawbacks, however. Hydrogenated vegetable oils contain more saturated fat than the original oil. Corn oil, for example, contains only 6% saturated fats; but corn oil margarine has 17%. The other negative is that hydrogenation causes a change in the structure of the unsaturated fatty acids. Specifically, hydrogenation converts some unsaturated fats into **trans fats**.

Trans Fatty Acids The bulk of *trans* fats in our diets comes from hydrogenated vegetable oils. Hydrogenation causes some of the unsaturated fatty acids to be converted from their naturally occurring *cis* to the *trans* form. Ruminant animals like cows, goats, and sheep form a small amount of *trans* fats in their stomachs. Consequently, milk and milk products from these animals will contain *trans* fats.[15]

Repositioned hydrogen molecules in *trans* fats change the way the body uses the fat. *Trans* fats raise blood cholesterol levels more than any other type of fat. They increase the risk of heart disease, stroke, sudden death from heart disease, and type 2 diabetes, and they promote inflammation. Intake of *trans* fats as low as 1% of total calories per day strongly increases the risk of heart disease.[16] An intake of 2.2 grams of *trans* fat daily would place a person consuming 2000 calories a day at increased risk of heart disease. As indicated by the *trans* fat content of food given in Table 18.5, it's not difficult to consume more than 2.2 grams of *trans* fats daily if fried and snack foods and bakery products are consumed often. On average, Americans consume 2.3 g of *trans* fat daily.[16]

It is recommended that Americans consume as little *trans* fats as possible,[1] and nutrition information labeling requirements that began in 2006 are making that easier to accomplish. Nutrition Facts panels must include the *trans* fat content of food products (Illustration 18.6). The %DV column (for percent of Daily Value) is not be used for *trans* fats because there is no recommended level of intake. Products labeled "*trans* fat-free" must contain less than 0.5 gram of both *trans* and saturated fats. The requirement to label the *trans* fat content of food products is increasing the number of foods labeled "*trans* fat-free" and "no *trans* fat" (Illustration 18.7) and decreasing its use in foods. Food companies are busy developing other ways to produce foods without them.[17]

Nutrition Facts		
Serving Size 1 Entree		
Serving Per Container 1		
Amount Per Serving		
Calories 380	Calories from Fat 170	
		%Daily Value
Total Fat 19g		**29%**
Saturated Fat 10g		**50%**
Trans Fat 2g		
Cholesterol 85mg		**28%**
Sodium 810mg		**34%**
Total Carbohydrate 33g		**11%**
Dietary Fiber 3g		**12%**
Sugars 5g		
Protein 20g		
Vitamin A 10%		Vitamin C 0%
Calcium 10%		Iron 15%

Percent Daily Values are based on a 2000 calorie diet. Your daily values may be higher or lower depending on your calorie needs:

		Calories	2000	2500
Total Fat	Less Than		65g	80g
Sat Fat	Less Than		20g	25g
Cholesterol	Less Than		300mg	300mg
Sodium	Less Than		2400mg	2400mg
Total Carbohydrate			300g	375g
Dietary Fiber			25g	30g

Illustration 18.6
Trans fat: the newest addition to nutrition facts panels.

TABLE 18.5

Trans *fatty acid content of food products*

Values may change as companies lower the trans fat content of foods.

Food	Trans Fatty Acids (Grams)
Kentucky Fried Chicken pot pie	8
French fries, large serving	4-7
Cake doughnut	6
Breaded fish sticks, 3	5
Margarine, 1 tbs	4
Dutch apple pie	3-4
Wheat crackers	3
Chocolate chip cookie	2
Snack crackers, one 2 oz	2
Shortening, 1 tbs	1
Tub margarine, 1 tbs	1
Butter, 1 tbs	0
"No *trans* fat" margarine	0
Olive oil	0

trans fats
Unsaturated fatty acids in fats that contain atoms of hydrogen attached to opposite sides of carbons joined by a double bond:

H
—C=C—
　　　H

　　　　　　H H
　　　　　—C=C—

Trans fatty   *Cis* fatty
   acid   acid

Fats containing fatty acids in the *trans* form are generally referred to as *trans* fats. *Cis* fatty acids are the most common, naturally occurring form of unsaturated fatty acids. They contain hydrogens located on the same side of doubly bonded carbons.

Illustration 18.7
Products that feature "no *trans* fats" and "*trans* fat-free" labels.

Checking Out Cholesterol

Cholesterol is a lipid found *only* in animal products. It is tasteless and odorless and contained in both the lean and fat parts of animal products. Table 18.6 lists some sources of cholesterol. Plants don't contain cholesterol because they can't produce it and don't need it to function and grow normally.

Sources of Cholesterol

The cholesterol used by the body comes from two sources. Most (about two-thirds) of the cholesterol available to the body is produced by the liver. The rest comes from the diet (Illustration 18.8). Because the liver produces cholesterol from other substances in our diet, it does not qualify as an essential nutrient.

The Contributions of Cholesterol

Would you be surprised to learn that cholesterol:

- is found in every cell in your body?
- serves as the building block for estrogen, testosterone, and the vitamin D that is produced in your skin on exposure to sunlight?
- is a major component of nerves and the brain?
- cannot be used for energy (so it provides no calories)?

The body has many uses for cholesterol (Table 18.7). It doesn't just accumulate in arteries!

Fat Substitutes

The fat content of processed foods can be partially or fully replaced by fat substitutes or, as they are also called, "fat replacers." Fat substitutes attempt to imitate the taste, texture, and cooking properties of our favorite fats but with fewer calories. Table 18.8 lists some of the more than 60 products that are already on the market; others are under development. The degree to which these products succeed in imitating the qualities of fat varies. Making the perfect fat substitute is not an easy task. Not all of the fat substitutes that have entered the marketplace have met the test of consumer acceptability.[18]

Are Fat Substitutes Safe?

Consumers sometimes wonder about the safety of fat substitutes. Because most fat substitutes are derived from ingredients of food such as carbohydrates, protein, and vegetable oils, they are often assumed to be safe. Fat replacers that are made from substances that do not occur naturally in foods must be tested for safety. Olean, a nonnaturally occurring fat replacer that cannot be digested, was tested extensively before being approved for use. It can cause diarrhea and oily stools if consumed in large amounts.[19]

TABLE 18.6
Food sources of cholesterol

Note that all the sources are animal products. Cholesterol in foods is a clear, oily liquid found in the fat and lean portions of many animal products.

Animal product	Amount	Cholesterol (Milligrams)
Brain	3 oz	1746
Liver	3 oz	470
Egg	1	212
Veal	3 oz	128
Shrimp	3 oz	107
Prime rib	3 oz	80
Chicken (no skin)	3 oz	75
Turkey (no skin)	3 oz	65
Hamburger, regular	3 oz	64
Pork chop, lean	3 oz	60
Fish, baked (haddock, flounder)	3 oz	58
Ice cream	1 cup	56
Sausage	3 oz	55
Hamburger, lean	3 oz	50
Milk, whole	1 cup	34
Crab, boiled	3 oz	33
Lobster	3 oz	29
Cheese (cheddar)	1 oz	26
Milk, 2%	1 cup	22
Yogurt, low-fat	1 cup	17
Milk, 1%	1 cup	14
Butter	1 tsp	10
Milk, skim	1 cup	7

TABLE 18.7
How the body uses cholesterol

- Cholesterol is a component of all cell membranes, the brain, and nerves.
- It is needed to produce estrogen, testosterone, and vitamin D.

TABLE 18.8
Some fat substitutes

Carbohydrate-Based (0–4 Cal/g)	Protein-Based (1–2 Cal/g)	Fat-Based (0–5 Cal/g)
Avicel	Simplesse	Benefat
Just fiber	Dairy-Lo	Caprenin
N-lite		Olean
Oat fiber		
Splendid		
Z Trim		

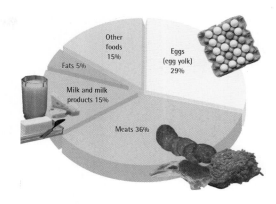

Illustration 18.8
Food sources of cholesterol in the U.S. diet.[21]
Percentages indicate the proportion of cholesterol each type of food contributes to the diet.

Can Fat Substitutes Benefit Health?

When substituted for foods containing saturated fats, fat substitutes may benefit health by lowering blood cholesterol levels. Studies suggest that the use of foods containing fat substitutes reduces total fat consumption, but whether they lead to weight loss is controversial. So, although they may not be the answer to weight control, fat substitutes may provide part of the solution to high fat intakes and the health problems related to them.[20]

Finding Out about the Fat Content of Food

Not all of the fat in food is visible. To avoid being fooled, it helps to use a reference on the fat composition of foods. Table 18.9 lists the fat content of common food sources of fat, including candy. Vegetables and fruits (except avocado and coconut) and grains are not listed because they contain relatively little fat. Other

TABLE 18.9
The fat content of some foods

Food	Amount	Grams	Percentage of Total Calories From Fat
Fats and oils			
Butter	1 tsp	4.0	100%
Margarine	1 tsp	4.0	100
Oil	1 tsp	4.7	100
Mayonnaise	1 tbs	11.0	99
Heavy cream	1 tbs	5.5	93
Salad dressing	1 tbs	6.0	83
Meats and fast foods			
Hot dog	1 (2 oz)	17.0	83
Bologna	1 oz	8.0	80
Sausage	4 links	18.0	77
Bacon	3 pieces	9.0	74
Salami	2 oz	11.0	68
Pork steak	3 oz	18.0	62
Hamburger, regular (20% fat)	3 oz	16.5	62
Burger King Double Whopper with Cheese	14 oz	57	56
Burger King Ultimate Breakfast Roll	7 oz	30	53
Chicken, fried with skin	3 oz	14.0	53
Big Mac	6.6 oz	31.4	52
Quarter Pounder with cheese	6.8 oz	28.6	50
Whopper	8.9 oz	32.0	48

(continued)

TABLE 18.9

The fat content of some foods (continued)

Food	Amount	Grams	Percentage of Total Calories From Fat
Meats and fast foods (*continued*)			
Steak (rib eye)	3 oz	9.9	47
Hamburger, lean (10% fat)	3 oz	9.5	45
Steak (T-bone), lean	3 oz	8.9	44
Rabbit	3 oz	7.0	38
Veggie pita	1	17.0	38
Ranch chicken pita	1	18.0	34
Steak (round), lean only	3 oz	5.2	29
Chicken, baked without skin	3 oz	4.0	25
Turkey wrap	1	9.0	24
Hamburger, extra lean (4% fat)	3 oz	2.3	23
Venison	3 oz	2.7	18
Subway, club	6"	5.0	14
Flounder, baked	3 oz	1.0	13
Subway, veggie	6"	3.0	11
Shrimp, boiled	3 oz	1.0	10
Milk and milk products			
Cheddar cheese	1 oz	9.5	74
American cheese	1 oz	6.0	66
Milk, whole	1 cup	8.5	49
Cottage cheese, regular	½ cup	5.1	39
Milk, 2%	1 cup	5.0	32
Milk, 1%	1 cup	2.7	24
Cottage cheese, 1% fat	½ cup	1.2	13
Milk, skim	1 cup	0.4	4
Yogurt, frozen	¾ cup	0.0-6.6	0-3
Other			
Olives	4 medium	1.5	90
Avocado	½	15.0	84
Almonds	1 oz	15.0	80
Sunflower seeds	¼ cup	17.0	77
Peanut butter	1 tbs	8.0	76
Peanuts	¼ cup	17.5	75
Cashews	1 oz	13.2	73
Egg	1	6.0	61
Potato chips	1 oz (13 chips)	11.0	61
Chocolate chip cookies	4	11.0	54
French fries	20 fries	20.0	49
Taco chips	1 oz (10 chips)	6.2	41
Candy			
Mr. Goodbar	1.7 oz	15.0	56
Peanut butter cups, 2 regular	1.6 oz	15.0	54
Milk chocolate	1.6 oz	14.0	53
Almond Joy	1.8 oz	14.0	50
Kit Kat	1.5 oz	12.0	47
M & M's, peanut	1.7 oz	13.0	47
Nestle's Crunch	1.6 oz	11.0	45
Twix	2.0 oz	14.0	45
Baby Ruth	2.1 oz	14.0	43
Pay Day	1.9 oz	12.0	43
Snickers	2.1 oz	13.0	42
Butterfinger	2.1 oz	12.0	39
M & M's, plain	1.7 oz	10.0	39
Milky Way	2.2 oz	11.0	35
3 Musketeers	2.1 oz	9.0	31
Tootsie Roll	2.3 oz	6.0	21

references can also be used, such as the food composition tables in Appendix A, the Diet Analysis Plus Program software, and the nutrition labels on food products.

Knowledge of the caloric and fat content of a food can be used to calculate the percentage of calories provided by fat. For example, suppose that a slice of cherry pie provides 350 calories and 15 grams of fat. To calculate the percentage of fat calories, multiply 15 grams by 9 (the number of calories in each gram of fat), divide the result by 350 calories, and then multiply by 100:

$$15 \text{ grams fat} \times 9 \text{ calories/gram} = 135 \text{ calories}$$
$$135 \text{ calories}/350 = 0.39$$
$$0.39 \times 100 = 39\% \text{ of total calories from fat}$$

Fat Labeling

Nutrition labeling regulations for fat require food manufacturers to adhere to standard definitions of "low fat," "fat-free," and related terms used on food labels. Similarly, claims made about the cholesterol content of food products must comply with standard definitions (Table 18.10). If a claim is made about the fat content of a food, the nutrition facts panel must specify the food's fat, saturated fat, *trans* fat, and cholesterol content. If a claim is made about cholesterol content (and claims can be made only for products that normally contain cholesterol), the nutrition panel must also reveal the product's fat and saturated fat content. To prevent the use of unrealistically small serving sizes as a way to appear to cut down on a product's fat content, standard serving sizes must be used on food labels.

Recent Changes in Recommendations for Fat and Cholesterol Intake

In the past, recommendations for fat intake were based on the percent of total calories consumed from fat. These recommendations have changed due to evolving evidence about relationships between dietary fat intake and health. Certain types of diets that are relatively high in total calories from fats, such as the Mediterranean diet highlighted in Illustration 18.9, do not raise the risk of heart disease, stroke, obesity, or a number of other diseases. They tend to promote health and prevent or delay the onset of certain chronic diseases.[22] Evidence established on the healthful effects of the Mediterranean diet and other examples of dietary fat intake and health in different populations prompted the development of new recommendations based primarily on the healthfulness of various types of fats.[1]

Good Fats, Bad Fats Fats come in many types in foods, and with few exceptions, they serve as a source of energy and provide a number of essential functions in the body. With regard to raising or lowering the risk of heart disease

TABLE 18.10

What claims about the cholesterol content of foods that normally contain cholesterol must mean

- No cholesterol or cholesterol-free: Contains less than 3 milligrams of cholesterol per serving
- Low cholesterol: Contains 20 milligrams or less of cholesterol per serving
- Reduced cholesterol: Contains at least 75% less cholesterol than normal
- Less cholesterol: Contains at least 25% less cholesterol than normal; the percentage less must be stated on the label

and stroke, however, fats differ. Those that elevate total cholesterol and LDL-cholesterol levels are regarded as "bad" or "unhealthy" fats. Those that lower total cholesterol and LDL-cholesterol and raise blood levels of HDL-cholesterol (the one that helps the body get rid of cholesterol in the blood) are considered "good" or healthy."[23]

The list of unhealthy fats includes *trans* fats, saturated fats, and cholesterol. Fats labeled "bad" are generally solid at room temperature and are included in foods such as high-fat meats and dairy products, hard margarines, shortening, and crispy snack foods.[23] Monounsaturated fats, polyunsaturated fats, alpha-linolenic acid, DHA, and EPA are considered healthy fats and are present in food in the form of oils (Table 18.11).

Illustration 18.9
A look at the cuisine of the Mediterranean diet.

Recommendations for Fat and Cholesterol Intake

Current recommendations for adults call for consumption of 20 to 35% of total calories from fat. The AIs (Adequate Intakes) for the essential fatty acid linoleic acid is set at 17 grams a day for men and 12 grams for women. AIs for the

TABLE 18.11
Healthy and unhealthy fats and examples of food sources

Healthy Fats	Unhealthy Fats
DHA, EPA (omega-3 fatty acids) fish and seafood	***Trans* fats** Snack and fried foods, bakery goods
Monounsaturated fats Olive and peanut oil, nuts, avocados	**Saturated fats** Animal fats
Polyunsaturated fats Vegetable oils	**Cholesterol** eggs, seafood, and meat
Alpha-linolenic acid Soybeans, walnuts, flaxseed	

Good fats, bad fats

What foods provide "healthy" fats?

Who gets thumbs up?
Answers below

Kristen:
How can I be wrong? Low-fat food products are best for healthy fat because they contain almost no fat!

Butch:
I'm thinking foods like fish, peanut butter, and *trans* fat–free margarine contain healthy fats.

other essential fatty acid, alpha-linolenic acid, are 1.6 grams per day for men and 1.1 grams for women. It is recommended that intake of *trans* fats and saturated fats be as low as possible while consuming a nutritionally adequate diet. Only a small proportion of Americans consume too little linoleic acid, but intakes of alpha-linolenic acid tend to be low. Americans are being encouraged to increase consumption of EPA and DHA by eating fish more often. In addition, saturated fat intake averages 11 to 12% of calories, an amount that increases the risk of heart disease.[1]

There is no recommended level of cholesterol intake, because there is no evidence to indicate that cholesterol is required in the diet. The body is able to produce enough cholesterol, and people do not develop a cholesterol deficiency disease if it is not consumed. Because blood cholesterol levels tend to increase somewhat as consumption of cholesterol increases, it is recommended that intake should be minimal. Although cholesterol intake averages around 250 mg per day in the United States, a more health-promoting level of intake would be less than 200 mg a day.[24]

Recent recommendations for fat intake represent an unusually large but necessary change in dietary intake guidance. Much remains to be understood about the effects of dietary fats on health, and how other components of diet, lifestyle, and genetic traits modify relationships between fat intake and health.

Good fats, bad fats

Low-fat foods contain less fat than the regular version of the foods. But that doesn't mean the products contain no fat, or only good fats. Food sources of fish oils, unsaturated fat, and *trans* fat–free products provide the healthy fats. As always, healthy diets aren't based on individual foods, they are based on overall diets. You can emphasize foods providing healthy fats without feeling bad about occasionally eating foods branded with the "bad fat" label.

Kristen:
👎

Butch:
👍

Nutrition Up Close

The Healthy Fats in Your Diet

FOCAL POINT: Identify your healthy fat food choices.

Are the fats in your diet the healthy type? Check it out by answering these questions:

How Often Do You Eat:	Seldom or Never	1-2 Times per Week	3-5 Times per Week	Almost Daily
1. Sausage, hot dogs, ribs, and luncheon meats?	☐	☐	☐	☐
2. Heavily marbled steaks or roasts and chicken with the skin?	☐	☐	☐	☐
3. Soybean products such as tofu or soynuts?	☐	☐	☐	☐
4. Nuts or seeds?	☐	☐	☐	☐
5. Whole milk, cheese, or ice cream?	☐	☐	☐	☐
6. Soft margarine or olive oil?	☐	☐	☐	☐
7. French fries, snack crackers, commercial bakery products?	☐	☐	☐	☐
8. Rich sauces and gravies?	☐	☐	☐	☐
9. Fish or seafood?	☐	☐	☐	☐
10. Peanut butter?	☐	☐	☐	☐

FEEDBACK (including scoring) can be found at the end of Unit 18.

Key Terms

essential fatty acids, page 18-3
hydrogenation, page 18-8

lipids, page 18-3

trans fats, page 18-9

Review Questions

	True	False
1. Animal products are by far the leading source of saturated fats in the American diet.	☐	☐
2. Flaxseed oil and dark, leafy green vegetables are excellent sources of the essential fatty acid alpha-linolenic acid.	☐	☐
3. The best food sources of EPA and DHA are fish and shellfish.	☐	☐
4. Cholesterol is not a required nutrient because it serves no essential, or life-sustaining, function in the body.	☐	☐
5. Due to new nutrition labeling requirements and the subsequent reduction of *trans* fats in food products, most Americans are not consuming too much *trans* fat.	☐	☐

	True	False
6. *Trans* fats raise blood cholesterol levels more than any other type of fat.	☐	☐
7. If a claim is made on a food package about the product's fat content, the nutrition facts panel must list the food's fat, saturated fat, *trans* fat, and cholesterol content.	☐	☐
8. The type of fat consumed is more important to health than is total fat consumption.	☐	☐
9. Unhealthful, or "bad," fats include *trans* fats and saturated fats.	☐	☐
10. It is recommended that adults consume 20 to 25% of total calories from fat.	☐	☐

Media Menu

www.nlm.nih.gov/medlineplus/dietaryfat.html
This site provides a menu that connects you to other sites such as good and bad fats, interpreting blood lipid profiles, benefits of omega-3 fatty acids, and fat substitutes.

www.nal.usda.gov/fnic
Find out more about fats and fat replacers from the extensive list of topics covered under the search term "fats."

www.healthfinder.gov
Here is a good source of information on fats, *trans* fat, and cholesterol through search terms such as dietary fat and healthy fats.

www.mayoclinic.org
Healthy fats, bad fats, know your fats, fats and heart disease, and other topics are intelligently covered in sites available through the Mayo Clinic's home page.

www.epa.gov/ost/fish
The Environmental Protection Agency's site is good for looking up national and local advisories on fish contamination and consumption.

Notes

1. Dietary Reference Intakes, Energy, carbohydrate, fiber, fat, fatty acids, cholesterol, protein, and amino acids. Institute of Medicine, National Academy of Sciences, Washington, DC: National Academies Press; 2002

2. Connor WE. Importance of n-3 fatty acids in health and disease. Am J Clin Nutr 2000;71(suppl):171S–5S.

3. Hepburn FN et al. Provisional tables on the content of omega-3-fatty acids and other fat components of selected foods. J Am Diet Assoc 1986;86:788–93.

4. Arterburn LM et al. Distribution, interconversion, and dose response of n-3 fatty acids in humans. Am J Clin Nutr 2006;83(suppl):1467S–76S.

5. Gebauer SK et al. n-3 fatty acid dietary recommendations and food sources to achieve essentiality and cardiovascular benefits. Am J Clin Nutr 2006;83:1526S–35S.

6. Oh R. Practical applications of fish oil (omega-3 fatty acids) in primary care. J Am Board Fam Pract 2005;18:28–36.

7. Kornman KS. Interleukin 1 genetics, inflammatory mechanisms, and nutrigenetic opportunities to modulate diseases of aging. Am J Clin Nutr 2006;83(suppl):475S–83S.

8. American Heart Association. Guidelines for the secondary prevention of cardiovascular disease. Circulation, posted online May 15, 2006, www.medscape.com/viewarticle/532327.

9. FDA affirms position on mercury in fish. www.fda.gov, issued 6/7/06.

10. Stephensen CP et al. The innate immune system: friend and foe. Am J Clin Nutr 2006;93:187–8.

11. Everything you always wanted to know about those newfangled eggs. Nutr Today 2003;38:75.

12. Mercury levels in commercial fish and shellfish. www.cfsan.fda.gov/seafood1.html, released June 6, 2006.

13. Addendum A: EPA and DHA Content of Fish Species, www.health.gov/dietaryguidelines/dga2005/report/HTML/table_g2_adda2.htm, accessed 9/06.

14. Kris-Etherton PM et al. Polyunsaturated fatty acids in the food chain in the United States. Am J Clin Nutr 2000;71(suppl):179S–88S.

15. Stuppy P. Transitioning away from trans fatty acids, Today's Dietitian 2003;Jan.:12–14; and Dietary Reference Intakes.

16. Mozaffarian D et al. Trans fatty acids and cardiovascular disease. N Engl J Med 2006;345:1602–12.

17. Tarrago-Trani MT et al. New and existing oils and fats used in products with reduced trans-fatty acids content. J Am Diet Assoc 2006;106:867–80.

18. Mattes RD, Position of the American Dietetic Association: fat replacers, J Am Diet Assoc 1998;98:463–8.

19. Bray GA et al. A 9-month randomized controlled trial comparing fat-substituted and fat-reduced diets in healthy obese men: the Ole Study. Am J Clin Nutr 2002;76:928–34.

20. Position of the American Dietetic Association: Fat replacers. J Am Diet Assoc 2005;105:266–75.

21. Sabar AF et al. Dietary sources of nutrients among U.S. adults. J Am Diet Assoc 1998;98:537–47.

22. Trichopoulou A et al. Adherence to a Mediterranean diet and survival in a Greek population. N Engl J Med 2003;348:2599–608.

23. Fats: the good and the bad (www.mayoclinic.com/invoke.cfm?id=NU00262); and Kris-Etherton et al., New guidelines.

24. Krauss RM et al., Revision 2000: a statement for healthcare providers from the Nutrition Committee of the American Heart Association, J Nutr 2001;131:132–46.

Nutrition Up Close

The Healthy Fats in Your Diet

Feedback for Unit 18

Give yourself a point for each time you checked the "3-5 Times per Week" or "Almost Daily" columns for numbers 3, 4, 6, 9, and 10. These foods are sources of healthy unsaturated fats or DHA and EPA. Take a point away for each time your answer ended up in the same columns for foods listed in numbers 1, 2, 5, 7, and 8. These foods provide saturated or *trans* fats. If you have any points left, your selection of food sources of fat regularly include healthy fats.

Answers to Review Questions

1. True, see page 18-4.
2. True, see page 18-7.
3. True, see page 18-7.
4. False, see page 18-10.
5. False, see page 18-9.
6. True, see page 18-9.
7. True, see page 18-13.
8. True, see page 18-13.
9. True, see page 18-14.
10. False, see page 18-14.

Nutrition and Heart Disease

NUTRITION SCOREBOARD	TRUE	FALSE
1 Heart disease is the leading cause of death among men and women in the United States.		
2 The most effective way to lower blood cholesterol is to exercise regularly.		
3. Plaque that causes harderning of the arteries develops on the inside surface of artery walls.		
4. Chronic inflammation is now considered a major contibutor to the development and progression of heart disease.[5]		
5 The risk factors for heart disease are the same for women as for men.		

Key Concepts and Facts

- Heart disease is the leading cause of death in men and women in the United States.[1]

- Dietary and lifestyle factors are among the most important contributors to heart disease.[2]

- Moderate-fat diets that provide primarily "healthy fats" decrease heart disease risk to a greater extent than do low-fat, high-carbohydrate diets.[3]

- Lowering high blood LDL-cholesterol and raising HDL-cholesterol levels reduces the risk of heart disease.[4]

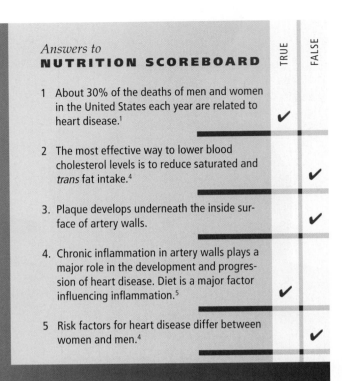

Answers to NUTRITION SCOREBOARD

		TRUE	FALSE
1	About 30% of the deaths of men and women in the United States each year are related to heart disease.[1]	✔	
2	The most effective way to lower blood cholesterol levels is to reduce saturated and *trans* fat intake.[4]		✔
3.	Plaque develops underneath the inside surface of artery walls.		✔
4.	Chronic inflammation in artery walls plays a major role in the development and progression of heart disease. Diet is a major factor influencing inflammation.[5]	✔	
5	Risk factors for heart disease differ between women and men.[4]		✔

Photo Disc

The Diet–Heart Disease Connection

Suspicions that dietary fat may be related to heart disease were first raised over 200 years ago. During the late 18th century, physicians noted that people who died of heart attacks had fatty streaks and deposits in the arteries that led to the heart. Unlike many early theories about the causes of diseases, this one has basically held up over time. It is clearly established that diets high in saturated fat and *trans* fat (rather than total fat as originally suspected) are a major risk factor for heart disease in the United States.[3] Other risk factors for heart disease have been identified and, except for genetic tendencies, sex, and age, can be modified. That makes heart disease a highly preventable cause of death and disability.

Declining Rates of Heart Disease

No disease that can be treated by diet should be treated with any other means.
—Maimonides, 1135–1204

As Americans have become more aware of the relationship between diet and heart disease, food choices have changed, and the average cholesterol level in adults has dropped from 240 milligrams in the early 1970s to 203 in recent years. As cholesterol levels have declined, so has the rate of heart disease in the United States (Illustration 19.1). The 60% drop in deaths from heart disease since 1950 appears to be primarily related to declines in blood cholesterol levels, reduced rates of smoking, improved blood pressure control, and advances in medications and medical care.[6]

Despite the general decline in cholesterol levels and rates of heart disease, the average cholesterol level of 203 milligrams among U.S. adults is still too high.

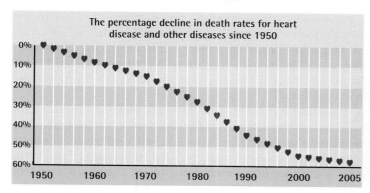

Illustration 19.1
The percentage decline in death rates for heart disease and other diseases since 1972.

Source: National Center for Health Statistics, 2006. Data for 2005 preliminary.

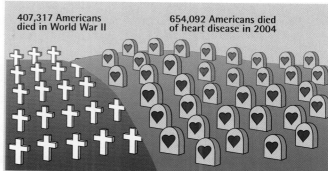

Illustration 19.2
The impact of heart disease.
If deaths from stroke and other diseases related to atherosclerosis are included, the annual death toll reaches 804,239.

Source: National Center for Health Statistics, 2006.

Every 1% drop in average cholesterol level in countries with high rates of heart disease produces a 2% decline in heart disease rate. Raising levels of beneficial HDL cholesterol by 1 milligram per deciliter reduces rates by 2 to 3%.[7]

A Primer on Heart Disease

There is no bigger health problem in the United States and other developed countries than heart disease (Illustration 19.2). Worldwide, heart disease accounts for one out of every four deaths.[8] It is an "equal opportunity" disease, striking as many women as men, although women on average die 10 years later from heart disease than do men. It is an *un*equal opportunity disease in that African Americans are more likely to die from it than are Caucasian Americans.[1]

What Is Heart Disease?

Heart disease, or more correctly "coronary" heart disease, is a term referring to several disorders that result from inadequate blood circulation to parts of the heart. Heart disease is almost always due to a narrowing of the arteries leading to the heart. Arteries become narrow due to a buildup of **plaque** (Illustration 19.3). People with narrowed arteries have **atherosclerosis,** or "hardening of the arteries" as it is often called. Heart disease develops silently over time (usually decades).

heart disease
One of a number of disorders that result when circulation of blood to parts of the heart is inadequate. Also called *coronary heart disease.* ("Coronary" refers to the blood vessels at the top of the heart. They look somewhat like a crown.)

plaque
Deposits of cholesterol, other fats, calcium, and cell materials in the lining of the inner wall of arteries.

atherosclerosis
"Hardening of the arteries" due to a build-up of plaque.

Illustration 19.3
The progression of atherosclerosis.
As plaque builds up, arteries narrow, reducing or stopping the supply of blood to the heart, brain, muscle, or other affected parts of the body.

ICI Pharmaceutical Division, Cheshire, England

cardiovascular disease
Disorders related to plaque buildup in arteries of the heart, brain, and other organs and tissues.

chronic inflammation
Inflammation that lasts weeks, months, or years. Inflammation is the first respose of the body's immune system to infection or irritation. It triggers the release of biologically active substances that promote oxidation and other potentially harmful reactions in the body.

Illustration 19.4
The heart after a heart attack.
The dark portion at the base of the heart is affected by the blockage in blood flow.

When arteries are narrowed by 50% or more, the shortage of blood to the heart can produce chest pain (called "angina"). A heart attack occurs when an artery leading to the heart becomes clogged by a piece of plaque released by a ruptured portion of the artery wall or by a blood clot.[2] (Illustration 19.4). Although heart disease primarily affects individuals over the age of 55, it's a progressive disease that may begin in childhood.[2]

Arteries leading to the heart aren't the only ones affected by atherosclerosis. Plaque can also build up in the arteries in the legs, neck, brain, and other body parts. If the blood supply to the legs is reduced, pain and muscle cramps may result after brief periods of exercise. Plaque buildup in arteries of the brain contributes to stroke—an event that occurs when the blood supply to a part of the brain is inadequate. Health problems due to atherosclerosis in arteries of the heart, brain, neck, and legs are collectively referred to as *cardiovascular disease.*

What Causes Atherosclerosis?

A number of conditions are known to increase plaque formation in arteries. The two most influential identified so far are elevated blood levels of cholesterol and *chronic inflammation* in the inner walls of arteries leading to the heart. High blood cholesterol levels and inflammation are interrelated and act together to increase atherosclerosis.[5]

Blood Cholesterol Levels and Heart Disease   In general, the higher the blood cholesterol level, the more likely it is that plaque will build up in the arteries and heart disease will occur (Illustration 19.5). (Some people with atherosclerosis don't have high blood levels of cholesterol, however.)[2] A person's blood cholesterol level is determined by a number of factors, including dietary intake, smoking, exercise, and genetic traits. Diets high in saturated fat elevate cholesterol levels in most people. Such diets are characterized by the regular consumption of high-fat milk and cheese, eggs, and beef. One type of unsaturated fat raises blood cholesterol levels more than saturated fat do, and that is *trans* fat.[3] This type of fat is produced when vegetable oils are hydrogenated—made solid by the addition of hydrogen. Blood cholesterol levels can also be raised by high cholesterol intakes. But blood cholesterol responds less to dietary cholesterol intake than to *trans* or saturated fat intake.[3]

All Blood Cholesterol Is Not Equal   Cholesterol is soluble in fat, but blood is mostly water. Therefore, cholesterol must be bound to compounds that mix with water, or it would float in blood. For this reason, cholesterol present in blood is bound to protein, which is soluble in water. The resulting combination is called a "lipoprotein," and there are a number of different types. Two lipoproteins have gained notoriety by virtue of their coverage in the popular press and their role in the development of heart disease. One is HDL cholesterol (for "high-density lipoprotein" cholesterol—it could be nicknamed "Heart-Disease-Lowering" cholesterol).

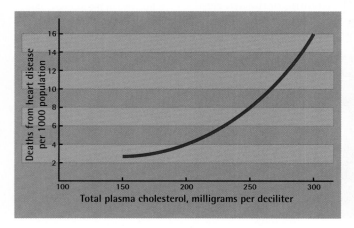

Illustration 19.5
The relationship between blood cholesterol level and death from heart disease.

Source: Grundy SM. Cholesterol and coronary heart disease: a new era. JAMA 1986;256:2849–58. Copyright 1986, American Medical Association.

Reality Check

Is shrimp off the menu for people with high blood cholesterol levels?

Who gets thumbs up?
Answers on next page

Lupe:
Shrimp has some cholesterol, but it also has good fats like DHA. You can still eat it on a cholesterol-lowering diet.

Sharon:
Save your money and bring your cholesterol down. Forget about eating shrimp.

It is the "good" type of cholesterol, and you want high levels of it in your blood. LDL cholesterol (low-density lipoprotein cholesterol) is the villain. The composition and roles of both types of lipoproteins are summarized in Illustration 19.6.

■ **Understanding HDL and LDL** HDL gets its reputation as the good cholesterol because it helps remove cholesterol from the blood. An avid cholesterol acceptor, HDL escorts cholesterol to the liver for its eventual excretion from the body. High HDL-cholesterol levels (over 40 milligrams per deciliter in men and over 50 in women) are protective against heart disease. LDL cholesterol carries more cholesterol than does HDL, and its cholesterol can be incorporated into plaque. The higher the LDL-cholesterol level, the more likely it is that atherosclerosis will develop and progress into heart disease.[2]

LDL- and HDL-cholesterol are components of total blood cholesterol measures. High levels of beneficial HDL-cholesterol may elevate total cholesterol values and make it appear that a person is at risk of heart disease. High levels

Seven out of 10 Americans over the age of 18 do not know the difference between "good" cholesterol (HDL) and "bad" cholesterol (LDL).

ON THE SIDE

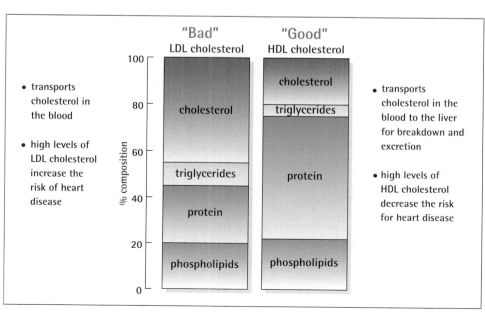

"Bad"
LDL cholesterol

"Good"
HDL cholesterol

- transports cholesterol in the blood

- high levels of LDL cholesterol increase the risk of heart disease

% composition

cholesterol

triglycerides

protein

phospholipids

cholesterol

triglycerides

protein

phospholipids

- transports cholesterol in the blood to the liver for breakdown and excretion

- high levels of HDL cholesterol decrease the risk for heart disease

Illustration 19.6
The "bad" cholesterol (LDL) and the "good" cholesterol (HDL) found in blood.

ANSWERS TO
Reality Check

Is shrimp off the menu for people with high blood cholesterol levels?

Shrimp contains cholesterol—about 100 mg in 3 ounces. But that doesn't mean you should not eat it if you're watching your cholesterol intake. Shrimp also contains DHA and EPA, which are heart healthy, and only a trace of saturated fat. Blood cholesterol level is much less responsive to dietary cholesterol than to saturated fat intake.[11]

Lupe:

Sharon:

of HDL-cholesterol (particularly those over 60 mg/dL) are strongly protective against heart disease.[10] Risk of heart disease is more accurately predicted based on individual measures of LDL- and HDL-cholesterol than by total blood cholesterol results.

■ **Triglycerides and Heart Disease Risk**   Triglycerides are transported in blood attached to VLDL (very-low-density lipoprotein) cholesterol. Until recently, the risk to heart disease posed by elevated blood levels of triglycerides had taken a back seat to blood cholesterol. Newer study results confirm that high blood levels of triglycerides increase heart disease risk and that efforts to prevent and treat heart disease should include a focus on blood triglyceride levels.[2] In addition to increasing the risk of heart disease, elevated triglyceride levels may signal the development or presence of metabolic syndrome or type 2 diabetes. These conditions are characterized by elevated blood levels of triglycerides and glucose, insulin resistance, abdominal obesity, and low levels of HDL cholesterol. People with metabolic syndrome are at particularly high risk for heart disease.[9]

■ **Genetic Effects on Blood Cholesterol Levels**   Genetic traits play important roles in how diet and exercise affect blood lipid levels. For example, some people are born with one or more genetic traits that reduces cholesterol absorption and therefore blood cholesterol levels. People who absorb cholesterol poorly do not experience the usual increase in blood cholesterol levels when they consume eggs or other high-cholesterol foods.[11] Other individuals have genetic traits that lower improvements in HDL-cholesterol levels that generally come with exercise or that keep blood triglyceride levels high even though ample amounts of EPA and DHA are consumed.[12]

Nutritional deprivation early in life can permanently modify the function of certain genes. Men born with low birth weight, for instance, tend to develop lower levels of HDL-cholesterol when they consume saturated fats then do men born with normal birth weights.[13] Most of the differences in genetic traits and functions identified so far are represented in a minority of the American population. Knowledge of individual genetic traits are, however, being increasingly used to guide decisions about dietary and physical activity approaches to the prevention and treatment of heart disease.

Chronic Inflammation and Heart Disease   The link between chronic inflammation and heart disease appears to be related to oxidation reactions that occur as a result of prolonged inflammatory processes.[5] Inflammatory reactions take place within the *endothelium*—the layer of cells that line the inside of artery walls (Illustration 19.7). Some of the LDL-cholesterol in people with high blood levels of this lipoprotein enters the endothelium and triggers inflammation.

endothelium
(Pronounced *n-dough-theil-e-um*)
The layer of cells lining the inside of blood vessels.

Inflammatory reactions lead to oxidation of LDL-cholesterol. Oxidized LDL-cholesterol is reactive; it damages cells within artery walls and is far more likely to become part of plaque than is unoxidized LDL-cholesterol. As shown in Illustration 19.8, plaque accumulates within artery walls and causes a thickening, hardening, and narrowing of the arteries.[14] Antioxidants consumed from plant foods and those produced in the body help lower inflammation by quenching oxidation reactions or by repairing their harmful effects.

Nutrients and other components of foods that decrease inflammation are summarized in Table 19.1. Antioxidants are principally found in plants foods such as fruits, vegetables, and whole grains. Some of the strongest natural antioxidants are the phytochemicals that add color to fruits and vegetables. Nutrients that function as antioxidants include vitamins C and E, selenium, and beta-carotene.[5]

Illustration 19.7
The endothelium consists of cells that form the inside surface of arteries.

Who's at Risk for Heart Disease?

Frankly, you may be. Take a look at the Health Action feature and see. Most Americans are at risk for heart disease because they have high blood cholesterol levels, hypertension, or a family history of heart attack before the age of 55; or because they smoke, are physically inactive, are obese, or have diabetes or high blood pressure (Illustration 19.9).

Some of the risk for heart disease may arise very early in life. Less-than-optimal availability of energy and nutrients during critical periods of fetal growth and development and early in childhood can modify gene functions in ways that increase the risk of heart disease later in life.[19] Infants who weigh less than 5.5 pounds at birth due to the mother's poor nutrition during pregnancy may be at increased risk for heart disease. Although researchers have long believed that heart disease begins in childhood, the risk for heart problems may begin before birth.[13]

Illustration 19.8
Plaque (shown in yellow) forms under the endothelium and thickens, hardens, and narrows the artery.

Are the Risks the Same for Women as for Men?

It is generally assumed that risk factors for heart disease identified in men apply to women as well. This assumption, along with the notion that heart disease represents a major health problem only for men, led to the exclusion of women from research studies. Both of these assumptions have been shown to be incorrect. Heart disease is a major health problem of women. Indeed, it is the leading cause of death among women in many countries.[21] Furthermore, risk factors for heart disease identified in studies of men do not always apply to women.

Results of recent studies involving women reveal that high blood cholesterol levels are a weak predictor of heart disease in older women and that low HDL-cholesterol levels are a stronger predictor than in men. Diabetes, obesity, high blood triglyceride levels, and increasing age are particularly strong risk factors for heart disease in women.[22]

The risk for heart disease in women increases substantially after menopause, which occurs around the age of 50 years. Menopause is characterized by declines in the levels of the hormone estrogen and HDL cholesterol and increases in the levels of LDL-cholesterol. It is not clear, however, whether these increases in LDL cholesterol or other factors increase heart-disease risk, or whether lowering LDL-cholesterol levels significantly reduces risk in older women. Compared to men, it

TABLE 19.1

Nutrients and other substances that reduce chronic inflammation[5, 15, 16]

- EPA and DHA
- Vitamin D
- Nuts, tea, most fruits and vegetables, wine, alcohol, whole grains
- Statins (the primary type of drug used to treat heart disease)
- Antioxidant nutrients

HEALTH *Action* Leading Risk Factors for Heart Disease[18]

- High blood cholesterol level (particularly LDL-cholesterol level)
- Low HDL-cholesterol level
- Diet high in saturated fat, cholesterol, and *trans* fats
- Family history of early heart attack
- Inadequate folate intake
- Diet low in fruits, vegetables, and whole grains

- High blood triglyceride level
- Hypertension
- Smoking
- Physical inactivity
- Obesity (especially central fat)
- Diabetes
- Age over 45 for men, over 55 for women

Walnuts are one of the oldest foods known to humans. Historical records related to walnut consumption date back to 7000 B.C.[17]

Courtesy of California Walnuts/ Walnut Marketing Board

appears that LDL-cholesterol levels are a much weaker predictor of heart disease in women.[23] Women normally have higher levels of HDL cholesterol throughout their lives than men, and their total blood cholesterol level is a bit higher, too.[24]

Diet and Lifestyle in the Management of Heart Disease

Many factors are involved in the development of heart disease, so approaches to treatment need to be broad. Treatment begins and continues with dietary and lifestyle modifications and includes reduction of high blood pressure and body weight (if needed), drugs in some cases, and smoking cessation. The goals of heart-disease treatment are improved overall health and blood lipid profiles.[10]

Diets and lifestyles that reduce the risk of heart disease benefit overall health in many ways. For most people, heart-healthy diets and lifestyles reduce the risk of diabetes, stroke, some types of cancer, osteoporosis, dementia, and obesity. When implemented, dietary recommendations for the treatment of heart disease

Illustration 19.9
How to have a heart attack.

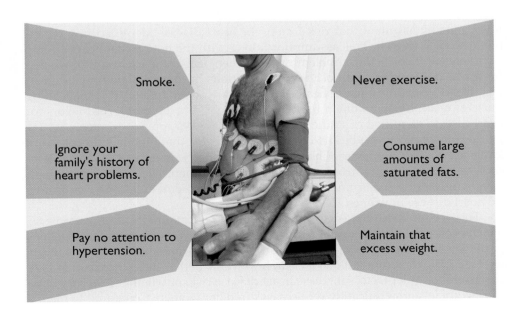

Smoke.

Ignore your family's history of heart problems.

Pay no attention to hypertension.

Never exercise.

Consume large amounts of saturated fats.

Maintain that excess weight.

TABLE 19.2

Risk categories for total, LDL-, and HDL-cholesterol levels and triglyceride levels in adults.[20]

	Total Cholesterol (mg/dL)	LDL Cholesterol (mg/dL)	HDL Cholesterol (mg/dL)	Triglyceride (mg/dL)
Desirable/optimal	<200	<100	>50 in women, >40 in men	<150
Near optimal	—	100–129	<40	—
Borderline high	200–239	130–159	(low) <40	150–200
High[a]	240+	160–189	—	200–499
Very high	—	190+	—	500+

[a]High cholesterol levels among women and people over the age of 70 may be a poor predictor of heart disease risk.

lower plaque formation and improve health and physical fitness levels in a large majority of people.[25]

Modification of Blood Lipid Levels

Blood levels of total, LDL, and HDL cholesterol, and triglycerides considered desirable and abnormal have been defined (Table 19.2) and are used to assess heart disease risk and treatment progress. Total cholesterol represents the sum of the levels of LDL, HDL, and VLDL cholesterol (the lipoprotein that transports triglycerides in blood).

Table 19.3 summarizes the effects of dietary components and lifestyle factors on blood levels of LDL cholesterol, HDL cholesterol, and triglycerides. Because some types of fat have both good and bad effects on blood lipid levels, dietary recommendations for heart disease focus on consumption of specific types of fat and foods that lower levels of LDL cholesterol and triglycerides and raise levels of HDL cholesterol (Table 19.4).

High levels of LDL cholesterol are reduced by limiting saturated fat intake to less than 7% of total calories and by excluding processed foods that contain trans fats. These fats are replaced in the diet by unsaturated fats and in particular by good food sources of monounsaturated fats like canola oil, olive oil, safflower oil, nuts, and avocados. Monounsaturated fats are preferred for lowering LDL cholesterol because they do not decrease HDL-cholesterol levels the way polyunsaturated fats do.

Whole-grain products, fiber, vegetables, fruits, and **plant stanols** and **sterols** also lower LDL cholesterol without decreasing HDL levels. Plant stanols and sterols lower LDL-cholesterol levels by reducing cholesterol absorption. Spreads containing stanols and sterols, used to replace margarine and butter, are widely available in supermarkets (Illustration 19.10).[27] Dietary recommendations for LDL-cholesterol reduction include limiting cholesterol intake to less than 200 mg daily. Improvements in weight status and increased physical activity result in lower levels of this risky lipoprotein.[10]

HDL-cholesterol levels can be increased by exercise, weight loss, and moderate alcohol consumption (1 to 2 drinks per day if appropriate) and by including nuts in the diet. It's better to increase HDL cholesterol through these means than by consuming foods with saturated fats because saturated fats also increase LDL cholesterol.[3]

Low-fat, high-carbohydrate diets were a mainstay of heart-disease prevention and treatment approaches in the past, principally because this type of diet lowers LDL-cholesterol levels. However, because low-fat, high-carbohydrate diets

plant stenols or sterols
Substances in corn, wheat, oats, rye, olives, wood, and some other plants that are similar in structure to cholesterol but that are not absorbed by the body. They decrease cholesterol absorption.

Illustration 19.10
Examples of spreads containing LDL-cholesterol-lowering plant stanols and sterols.

Keith Weller/ARS/USDA

TABLE 19.3

Effects of dietary components and lifestyle factors on blood lipid level[26]

Blood Lipid	Dietary Components	Lifestyle Factors
LDL cholesterol (low levels are heart healthy)	**Increases levels** • *trans* fats • saturated fats • dietary cholesterol	
	Decreases levels • mono- and polyunsaturated fats • whole-grain products • fiber • vegetables and fruits • plant stanols/sterols • nuts	**Decreases levels** • weight loss (if needed) • physical activity
HDL cholesterol (high levels are heart healthy)	**Increases levels** • moderate fat intake • saturated fats • alcohol (1–2 drinks per day if appropriate) • moderate carbohydrate intake	**Increases levels** • physical activity • weight loss (if needed)
	Decreases levels • polyunsaturated fats • high-carbohydrate diets • *trans* fats	
Triglycerides (low levels are heart healthy)	**Increases levels** • high-carbohydrate diets • high alcohol intake	**Decreases levels** • weight loss (if needed) • physical activity
	Decreases levels • moderate fat diets • DHA and EPA	

TABLE 19.4

Characteristics of diets that lower heart-disease risk[20]

- Provide 20–35% of total calories from fat, 10–25% from protein, and 45–65% from carbohydrates
- Emphasize healthy fats
  > plant sources of mono- and polyunsaturated fats
  > DHA and EPA from fish and shellfish (650 mg per day/12 ounces of fish/shellfish per week)
- Limit intake of unhealthy fats
  > *trans* fat in processed foods (< 1% of calories)
  > saturated fat from animal products, palm and coconut oil (< 7% of calories)
  > dietary cholesterol (< 200 mg per day)
- Are high in fiber; 35 grams per day for men and 25 grams per day for women
- Contain over 3 servings of vegetables and 2 servings of fruit daily
- Include whole-grain products and nuts
- Are moderate in carbohydrates and contain limited amounts of added sugars
- Include spreads with plant stanols or sterols (see Illustration 19.10)
- Contain alcohol in moderation (1–2 drinks per day if appropriate)

TABLE 19.5

Food choices that promote health and lower heart-disease risk

- Oils: canola, peanut, olive, safflower, flaxseed
- Fish
- Nuts
- Vegetables
- Fruits
- Whole-grain products
- Lean, unfried, and unprocessed meats
- Low-fat dairy products
- Dried beans
- Spreads containing plant stanols or sterols

also raise triglyceride levels and lower HDL-cholesterol levels, the recommendation has changed. Advice to consume low-fat, high-carbohydrate diets is being replaced by recommendations to consume healthy fats and foods that lower LDL cholesterol while maintaining triglyceride and HDL cholesterol at acceptable levels.[3] Food choices that are compatible with current recommendations for the dietary treatment of heart disease are shown in Table 19.5.

Cholesterol-lowering drugs are indicated for the treatment of heart disease if blood lipid changes achieved by diet and lifestyle improvements are insufficient.[3]

Modification of Chronic Inflammation and Oxidation

Table 19.1, presented earlier in this unit, summarizes nutrients, foods, and other substances that reduce chronic inflammation. The omega-3 fatty acids EPA and DHA, found in fish and shellfish, reduce inflammation because they lead to the formation of anti-inflammatory chemicals. Lower levels of inflammation that result from the presence of these substances reduce plaque formation and the progression of heart disease. EPA and DHA also lower the risk of heart disease by decreasing blood triglyceride levels.[30] Recent evidence indicates that vitamin D has anti-inflammatory properties[31] Chronic inflammation is also reduced by the loss of excess body fat, by regular exercise, and by foods such as fruits, vegetables, nuts, teas, and whole-grain products.[5, 15, 16] Antioxidants present in these plant foods appear to contribute to the prevention of artery wall damage and that reduces the accumulation of plaque.[5]

In the recent past it was thought that diets rich in folic acid; vitamins B_{12}, C, and E; and soy products reduced plaque buildup by reducing inflammation or oxidation. Their roles in prevention of heart disease are now in some doubt.[18, 32] It is not recommended that they be used to specifically help prevent or treat heart disease.[20]

The Statins

Statins have been hailed as wonder drugs for treating heart disease. These drugs, known by names such as Lipitor, Zocor, and Mevacor, markedly reduce cholesterol production in the liver; they also combat the ill effects of heart disease in other ways. Their use is related to a 30% drop in LDL-cholesterol levels and a 30 to 40% reduction in heart attack and stroke in both women and men.[12] Statins are not a substitute for diet and lifestyle changes. They improve blood lipid levels more when combined with dietary and lifestyle changes than when used alone.[2]

Statins are used widely but are expensive and have side effects such as muscle pain and weakness, liver disease, and kidney failure.[28] The cost and side effects of statins have prompted researchers to take a close look at alternatives—like extreme cholesterol-lowering diets. Table 19.6 shows an example of the "portfolio diet," used to decrease LDL-cholesterol levels. A menu for the type of diet more typically used to lower LDL cholesterol is provided in the table for comparison. The portfolio diet is vegetarian and based on soy milk and soy-based foods, fiber, oat bran, nuts, plant

Illustration 19.11
Heart-healthy food

TABLE 19.6

Type of food included in the portfolio diet for substantial reduction in LDL cholesterol [29]

Preventing Heart Disease	LDL Cholesterol Lowering
Breakfast	
breakfast cereal (fortified)	oat bran cereal with added psyllium fiber
skim milk	soy milk
fruit	oat bran toast
whole-grain toast	plant-sterol enriched margarine
soft margarine	jam, strawberries
coffee/tea	
Lunch	
turkey chili and onions	bean soup with rice
whole-grain roll	oat bran bread
soft margarine	plant-sterol enriched margarine
cole slaw	soy milk
skim milk	
Dinner	
salmon steak	spicy sautéed tofu
tartar sauce	ratatouille
green peas and corn	cooked barley
whole-grain pasta	broccoli, cauliflower
tossed salad with avocado and Italian dressing	plant-sterol enriched margarine
angel food cake with fruit	soy milk
wine/coffee/tea	almonds
Snacks	
nuts, popcorn,	nuts, soy bar,
skim milk, soy milk,	psyllium in juice,
fruit, raw vegetables,	fruit, raw vegetables
peanut butter on whole-grain bread	

sterols and stanols, and vegetables and fruits. Far from representing usual food preferences, the diet may be a challenge to follow in the long run. The diet does, however, reduce LDL cholesterol to levels that are achieved by statins. The results of one study showing dramatic effects of this diet on blood lipids were widely covered as a "wonder diet" in national newspapers. The end result appears to be that therapeutic diets are being increasingly viewed as one way to dramatically lower LDL-cholesterol levels while keeping the dose of statin required as low as possible. [29]

Looking toward the Future

Approaches to the prevention and treatment of heart disease have changed rather dramatically over recent years and will continue to evolve. Concerns about the cost of cholesterol-lowering drugs and their side effects, and the availability of low-cost preventive and treatment approaches, will factor into these changes. Approaches that use diet and lifestyle modification, changes in the quality of the food supply in stores and restaurants, and increased consumer involvement in risk reduction may lead the way to higher rates of decline in heart disease. An end to escalating rates of obesity and physical inactivity would serve our collective heart especially well.

Score Your Diet for Fat

Focal Point: Assess your chances for developing heart disease.

To estimate your risk for developing heart disease, circle the one number in each column that best describes you. Then, total the numbers from all categories to determine if your lifestyle encourages heart health or heart disease.

Heredity	Exercise	Age	Weight	Habits of Tobacco	Eating Fat
1 No known familial history of heart disease	**1** Intensive exercise at work and during recreation	**1** 15–23	**0** More than 5 lb below standard weight	**0** Nonuser	**1** No animal fat
2 One immediate family member who developed heart disease over age 55	**2** Moderate exercise at work and during recreation	**2** 24–34	**1** standard weight	**1** Cigar or pipe	**2** Very little animal fat (<10% of total calories)
3 Two immediate family members who developed heart disease over age 55	**3** Sedentary work with intensive exercise during recreation	**3** 35–44	**2** 6–20 lb overweight	**2** 10 cigarettes or fewer per day	**3** Some animal fat (11–20% of total calories)
4 One immediate family member who developed heart disease under age 55	**5** Sedentary work with moderate exercise during recreation	**4** 45–54	**4** 21–35 lb overweight	**4** 20 cigarettes or more per day	**4** Moderate animal fat (21–30% of total calories)
6 Two immediate family members who developed heart disease under age 55	**6** Sedentary work with light exercise during recreation	**6** 55+ years	**6** 36–50 lb overweight	**6** 30 cigarettes or more per day	**6** Excessive animal fat (>30% of total calories)

Total of all categories _____

FEEDBACK (including scoring) can be found at the end of Unit 19.

Source: Adapted from Heart Health Quiz (Loma Linda University).

Key Terms

atherosclerosis, page 19-3
cardiovascular disease, page 19-4
chronic inflammation, page 19-4

endothelium, page 19-6
heart disease, page 19-3
plant stanols or sterols, page 19-9

plaque, page 19-3

Review Questions

	True	False
1. Heart disease is the leading cause of death in women and men in developed countries.	☐	☐
2. The two most important conditions known to increase the risk of atherosclerosis are chronic inflammation in the walls of arteries and obesity.	☐	☐
3. The term *cardiovascular disease* refers to disorders related to plaque buildup in arteries of the heart, brain, and other organs and tissues.	☐	☐
4. High levels of LDL cholesterol increase the risk of heart disease just as low levels of HDL cholesterol do.	☐	☐
5. Vitamin E and C supplements do *not* appear to decrease the risk of heart disease.	☐	☐

	True	False
6. High levels of total cholesterol in the blood always represent a major risk factor for heart disease.	☐	☐
7. Diabetes, obesity, low HDL-cholesterol levels, and high blood triglyceride levels are stronger risk factors for heart disease in women than in men.	☐	☐
8. The major drawback to heart healthy diets is that they help prevent heart disease only.	☐	☐
9. Loss of excess body weight and physical activity decrease blood levels of HDL-cholesterol and triglycerides.	☐	☐
10. Heart healthy diets include ample fiber, whole-grain products, fish, and lean, unprocessed meats.	☐	☐

Media Menu

www.nlm.nih.gov/medlineplus/coronarydisease.html
Links to information on coronary heart disease, coronary artery disease, risk factor reduction, current news, and latest study results can be found on this site.

www.nlm.nih.gov/medlineplus/cholesterol.html
Get the latest news about good and bad cholesterol, *trans* fats, statins, dietary and blood cholesterol, and heart disease at this site—or visit the Virtual Fitness Room.

www.nlm.nih.gov/medlineplus/heartdiseasesprevention.html
This site from the National Institutes of Health is dedicated to presenting information on heart-disease prevention.

www.deliciousdecisions.org
American Heart Association's nutrition Web site includes links to sites on nutrition basics, a cookbook tailored to people with heart disease or those who want to prevent it, and tips for grocery shopping and eating out.

www.americanheart.org
Go to the American Heart Association's home page to find an array of information for consumers on heart disease and stroke.

www.medscape.com
This is an excellent site for health, medical, and nutrition information searches related to heart disease, stroke, and arteriosclerosis.

www.womenheart.org
The National Coalition for women with heart disease provides news, fact sheets, heart hospital ratings, heart healthy recipes, and more at this site.

www.nhlbi.nih.gov
The National Heart, Blood, and Lung Institute offers state-of-the-science information and papers on heart disease and stroke.

Notes

1. Heart disease statistics, NCHS, http://www.cdc.gov/nchs/fastats/heart.htm, accessed 9/06.

2. Caballero E. Dyslipidemia and vascular function: The rationale for early and aggressive intervention. www.medscape.com, 4/18/06.

3. Dietary Reference Intakes. Energy, carbohydrate, fiber, fat, fatty acids, cholesterol, protein, and amino acids. Institute of Medicine, National Academy of Sciences, Washington, DC: National Academies Press, 2002.

4. Detection, evaluation, and treatment of high blood cholesterol in adults. Adult Treatment Panel III. National Heart, Lung, and Blood Institute of the National Institutes of Health, 2001.

5. Kornman KS. Interleukin 1 genetics, inflammatory mechanisms, and nutrigenetic opportunities to modulate diseases of aging. Am J Clin Nutr 2006;83(suppl):475S–83S.

6. Carroll MD et al. Cholesterol levels in US adults. JAMA 2005;294:1773–81.

7. Jellinger PS et al. The American Association of Clinical Endocrinologists medical guidelines for the diagnosis and treatment of dyslipidemia and prevention of atherogenesis. Endocrin Prac 2000; 6:1–213.

8. Dwyer T et al. Differences in HDL cholesterol concentrations in Japanese, American, and Australian children. Circ 1997;76:2830–6.

9. Grundy SM. Cardiovascular and metabolic risk factors: how can we improve outcomes in high-risk patients? www.medscape.com/viewprogram/5700_pnt, accessed 7/06.

10. Davidson MH, New tactics, new targets: the changing landscape of dyslipidemia management in coronary prevention, Medscape CME Activity, www.medscape.com, accessed 5/03.

11. Herron KL et al. The ABCG5 polymorphism contributes to individual responses to dietary cholesterol and carotenoids in eggs. J Nutr 2006;136:1161–5.

12. Ashen MD et al. Low HDL cholesterol levels. N Engl J Med 2005;353:1125260.

13. Robinson SM et al. Combined effects of dietary fat and birth weight on serum cholesterol concentrations: the Hertfordshire Cohort Study. Am J Clin Nutr 2006;84:237–44.

14. Libby P. Inflammation and cardiovascular disease mechanisms. Am J Clin Nutr 2006;83:456S–60S.

15. Broom I. Thinking about abdominal obesity and cardiovascular risk. Br J Vasc Dis 2006;6:58–61.

16. Nettleton JA et al. Dietary patterns are associated with biochemical markers of inflammation and endothelial activation in the Multi-Ethnic Study of Atherosclerosis. Am J Clin Nutr 2006;83:1369–79.

17. Dreher ML et al. The traditional and emerging role of nuts in healthful diets. Nutr Rev 1996;54:241–5.

18. Helmy T et al. Management of dyslipidemia in the elderly. Medscape General Medicine 2005;7(4):8, accessed 10/05.

19. Yajnik C, Interactions of perturbations in intrauterine growth and growth during childhood on the risk of adult-onset disease, Proc Nutr Soc 2000;59:257–65; and Brown JE, Kahn ESB, Maternal nutrition and the outcome of pregnancy, Clin Perinatol 1997;24:433–49.

20. American Heart Association. Guidelines for the secondary prevention of cardiovascular disease. Circulation, posted online May 15, 2006, www.medscape.com/viewarticle/532327.

21. Kereiakes DJ et al., Managing cardiovascular disease in women. Medscape Women's Health, http://womenshealth.medscape.com/CMECircleWomens Health/2000/CME02/pnt-CME02.html, accessed 1/01; and Stampher et al., Primary prevention.

22. Sprecher DL et al. Metabolic coronary risk factors and mortality after bypass surgery. J Am Coll Cardiol 2000;36:1159–65.

23. Knopp RH et al. Saturated fat prevents coronary heart disease? An American paradox. Am J Clin Nutr 2004;80:1102–3.

24. Ernst ND et al. Consistency between US dietary fat intake and serum total cholesterol concentrations: the National Health and Nutrition Examination Surveys. Am J Clin Nutr 1997;66(suppl):965S–72S.

25. Varady KA et al. Combination diet and exercise interventions for the treatment of dyslipidemia: an effective preliminary strategy to lower cholesterol levels? J Nutr 2005;135:1829–35.

26. Fisler JS et al. Dietary fat and genotype: toward individualized prescriptions for lifestyle changes. Am J Clin Nutr 2005;81:1255–6.

27. Patch CS et al. Plant sterol/stanol prescription is an effective treatment strategy for managing hypercholesterolemia. J Am Diet Assoc 2005;105:46–52.

28. Vega C. Diet may lower cholesterol as much as statins, Medscape CME Activity, www.medscape.com, accessed 8/03.

29. Jenkins DJA et al. Assessment of the longer-term effects of a dietary portfolio of cholesterol-lowering foods in hypercholesterolemia. Am J Clin Nutr 2006:83:582–91.

30. Stephensen CP et al. The innate immune system: friend and foe. Am J Clin Nutr 2006;93:187–8.

31. Holick MF. Vitamin D: Important for prevention of osteoporosis, cardiovascular disease, type 1 diabetes, autoimmune disease, and some cancers. www.medscape.com/viewarticle/516238, accessed 9/06.

32. Hope 2 Investigators. Homocysteine lowering with folic and B vitamins in vascular disease. N Engl J Med 2006;354:1567–77.

Nutrition Up Close

Heart Health Quiz

Feedback for Unit 19

Find your score below to learn your estimated risk of developing heart disease:

4–9 very remote
10–15 less than average
16–20 average
21–25 moderately increased
26–30 excessive
31–35 too high—reduce score!

You have no control over certain risk factors for heart disease, such as heredity and age. Other conditions that detract from heart health, such as high blood pressure, diabetes, and high blood cholesterol, should be evaluated by your physician. Some risk factors are within your control, however. These lifestyle changes can help reduce your risk of developing heart disease:

- Keep your weight within a healthy range.
- Exercise regularly.
- Don't smoke.
- Reduce the amount of saturated fat in your diet if it is high.
- Eat at least five servings of vegetables and fruits each day.

Source: Adapted from Heart Health Quiz (Loma Linda University).

Answers to Review Questions

1. True, see page 19-3.
2. False, see page 19-4.
3. True, see page 19-4.
4. True, see page 19-5.
5. True, see page 19-11.
6. False, see page 19-4.
7. True, see page 19-7.
8. False, see page 19-8.
9. False, see page 19-9.
10. True, see page 19-10, Tables 19-4 and 19-5.

UNIT 20

Vitamins and Your Health

NUTRITION SCOREBOARD	TRUE	FALSE
1 The only documented benefit of consuming sufficient amounts of vitamins is protection against deficiency diseases.		
2 Vitamins provide energy.		
3 Vitamin C is found only in citrus fruits.		
4 Nearly all cases of illness due to excessive intake of vitamins result from the overuse of vitamin supplements.		

Key Concepts and Facts

- Vitamins are chemical substances found in food that are required for normal growth and health.

- Adequate intakes of vitamins protect people against deficiency diseases and help prevent a number of chronic diseases.

- Each vitamin has a range of intake in which it functions optimally. Intakes below and above the range impair health.

- Eating five or more servings of fruits and vegetables is a good way to get enough vitamins in your diet each day.

Answers to
NUTRITION SCOREBOARD

	TRUE	FALSE
1 For some vitamins, intake levels above those known to prevent deficiency diseases help protect humans from certain cancers, heart disease, osteoporosis, depression, and other disorders.		✔
2 Nope—only carbohydrates, proteins, and fats provide energy to the body. Vitamins are needed, however, to convert the energy in food into energy the body can use.		✔
3 Citrus fruits are good sources of vitamin C, but so are green peppers, collards, broccoli, strawberries, and a number of other fruits and vegetables.		✔
4 True. Nearly all cases of illness due to vitamin overdoses result from the excessive intake of vitamin supplements.	✔	

Vitamins: They're on Center Stage

These are exciting times for people interested in vitamins. Long relegated to the role of preventing deficiency diseases, vitamins are now being viewed from a very different vantage point. These essential nutrients clearly do more than the important job of protecting us from vitamin deficiency diseases. Vitamins are taking a preeminent position as protectors against a host of ills, ranging from certain birth defects and cataracts to osteoporosis and cancer.[1]

But first, what are vitamins? How much of them do we need? Where do we get them? And what's behind this expanding interest in them? Background information essential to understanding the current interest in the benefits of vitamins is next.

Vitamin Facts

vitamins
Chemical substances that perform specific functions in the body.

Vitamins are chemical substances that perform specific functions in the body. They are essential nutrients because, in general, the body cannot produce them or produce sufficient amounts of them. If we fail to consume enough of any of the vitamins, specific deficiency diseases develop.

Thirteen vitamins have been discovered so far, and they are listed in Table 20.1. It is possible that a few more substances will be added to the list of vitamins in years to come.

Water- and Fat-Soluble Vitamins   Vitamins come in two basic types—those soluble in water (the B-complex vitamins and vitamin C) and those that dissolve

TABLE 20.1

Thirteen vitamins are known to be essential for health. Nine are water soluble—that is, they dissolve completely in water—and four dissolve only in fats.

The Water-Soluble Vitamins[a]	The Fat-Soluble Vitamins
B-complex vitamins	Vitamin A (retinol) (provitamin is beta-carotene)
Thiamin (B_1)	Vitamin D (1,25 dihydroxy-cholecalciferol)
Riboflavin (B_2)	Vitamin E (tocopherol)
Niacin (B_3)	Vitamin K (phylloquinone, menaquinone)
Vitamin B_6 (pyridoxine)	
Folate (folacin, folic acid)	
Vitamin B_{12} (cyanocobalamin)	
Biotin	
Pantothenic acid (pantothenate)	
Vitamin C (ascorbic acid)	

[a]Choline has been assigned Adequate Intake levels in the Dietary Reference Intakes. It is not included here as it has not been shown to be required in the diet during all stages of the life cycle.

Why aren't vitamins named A through F? Vitamins as we know them today don't follow alphabetical order because errors were made when vitamins were discovered. They were named in alphabetical sequence, but some substances turned out not really to be vitamins. The letters assigned to these substances were dropped, so the sequence is interrupted.

Photo Disc

in fat (vitamins D, E, K, and A, or the "deka" vitamins). Their key features are summarized in Table 20.2, which provides an "intensive course" on vitamins. With the exception of vitamin B_{12}, the water-soluble vitamins can be stored in the body only in small amounts. Consequently, deficiency symptoms generally develop within a few weeks to several months after the diet becomes deficient in water-soluble vitamins. Vitamin B_{12} is unique in that the body can build up stores that may last for a year or more after intake of the vitamin stops. Of the water-soluble vitamins, niacin, vitamin B_6, and vitamin C are known to produce ill effects if consumed in excessive amounts.

The fat-soluble vitamins are stored in body fat, the liver, and other parts of the body. Because the body is better able to store these vitamins, deficiencies of fat-soluble vitamins generally take longer to develop than deficiencies of water-soluble vitamins when intake from food is too low.

Bogus Vitamins Some substances that don't belong on the list of vitamins and won't end up there in years to come are listed in Table 20.3. This list includes some of the most common substances claimed to be vitamins by enterprising quacks, misdirected manufacturers of supplements, and some weight-loss and cosmetic product producers. Although these substances may help sales, they aren't essential and therefore cannot be considered vitamins. People do not develop deficiency diseases when they consume too little of the bogus "vitamins."

What Do Vitamins Do?

For starters, vitamins *don't* provide energy or serve as components of body tissues such as muscle and bone. A number of vitamins do play critical roles as *coenzymes* in the conversion of proteins, carbohydrates, and fats into energy. Coenzymes are also involved in reactions that build and maintain body tissues such as bone, muscle, and red blood cells. Thiamin, for example, is needed for reactions that convert glucose into energy. People who are thiamin deficient tire easily and feel weak (among other things). Folate, another B-complex vitamin, is required for reactions that build body proteins. Without enough folate, proteins such as those found in red blood cells form abnormally and function poorly. Vitamin A

coenzymes
Chemical substances, including many vitamins, that activate specific enzymes. Activated enzymes increase the rate at which reactions take place in the body, such as the breakdown of fats or carbohydrates in the small intestine and the conversion of glucose and fatty acids into energy within cells.

TABLE 20.2
An intensive course on vitamins

The Water-Soluble Vitamins

	PRIMARY FUNCTIONS	CONSEQUENCES OF DEFICIENCY
Thiamin (vitamin B$_1$) AI[a] women: 1.1 mg men: 1.2 mg	• Helps body release energy from carbohydrates ingested • Facilitates growth and maintenance of nerve and muscle tissues • Promotes normal appetite	• Fatigue, weakness • Nerve disorders, mental confusion, apathy • Impaired growth • Swelling • Heart irregularity and failure
Riboflavin (vitamin B$_2$) AI women: 1.1 mg men: 1.3 mg	• Helps body capture and use energy released from carbohydrates, proteins, and fats • Aids in cell division • Promotes growth and tissue repair • Promotes normal vision	• Reddened Lips, cracks at both corners of the mouth • Fatigue
Niacin (vitamin B$_3$) RDA women: 14 mg men: 16 mg UL: 35 mg (from supplements and fortfied foods)	• Helps body capture and use energy released from carbohydrates, proteins, and fats • Assists in the manufacture of body fats • Helps maintain normal nervous system functions	• Skin disorders • Nervous and mental disorders • Diarrhea, indigestion • Fatigue
Vitamin B$_6$ (pyridoxine) AI women: 1.3 mg men: 1.3 mg UL: 100 mg	• Needed for reactions that build proteins and protein tissues • Assists in the conversion of tryptophan to niacin • Needed for normal red blood cell formation • Promotes normal functioning of the nervous system	• Irritability, depression • Convulsions, twitching • Muscular weakness • Dermatitis near the eyes • Anemia • Kidney stones

L. V. Bergman & Associates

A look at beriberi, a thiamin-deficiency disease.

Dr. M.A. Ansary/Photo Researchers, Inc.

Pellagra: the niacin-deficiency disease.

[a](Adequate Intakes) and RDAs (Recommended Dietary Allowances) are for 19–30 year olds; UL (Upper Limits) are for 19–70 year olds,

TABLE 20.2

An intensive course on vitamins (continued)

The Water-Soluble Vitamins

CONSEQUENCES OF OVERDOSE	PRIMARY FOOD SOURCES	HIGHLIGHTS AND COMMENTS
• High intakes of thiamin are rapidly excreted by the kidneys. Oral doses of 500 mg/day or less are considered safe.	• Grains and grain products (cereals, rice, pasta, bread) • Ready-to-eat cereals • Pork and ham, liver • Milk, cheese, yogurt • Dried beans and nuts	• Need increases with carbohydrate intake • There is no "e" on the end of thiamin! • Deficiency rare in the U.S.; may occur in people with alcoholism • Enriched grains and cereals prevent thiamin deficiency
• None known. High doses are rapidly excreted by the kidneys.	• Milk, yogurt, cheese • Grains and grain products (cereals, rice, pasta, bread) • Liver, poultry, fish, beef • Eggs	• Destroyed by exposure to light
• Flushing, headache, cramps, rapid heartbeat, nausea, diarrhea, decreased liver function with doses above 0.5 g per day	• Meats (all types) • Grains and grain products (cereals, rice, pasta, bread) • Dried beans and nuts • Milk, cheese, yogurt • Ready-to-eat cereals • Coffee • Potatoes	• Niacin has a precursor– tryptophan. Tryptophan, an amino acid, is converted to niacin by the body. Much of our niacin intake comes from tryptophan. • High doses raise HDL-cholesterol levels.
• Bone pain, loss of feeling in fingers and toes, muscular weakness, numbness, loss of balance (mimicking multiple sclerosis)	• Oatmeal, bread, breakfast cereals • Bananas, avocados, prunes, tomatoes, potatoes • Chicken, liver • Dried beans • Meats (all types), milk • Green and leafy vegetables	• Vitamins go from B_3 to B_6 because B_4 and B_5 were found to be duplicates of vitamins already identified.

(continued)

TABLE 20.2

An intensive course on vitamins (continued)

The Water-Soluble Vitamins

	PRIMARY FUNCTIONS	CONSEQUENCES OF DEFICIENCY	
Folate (folacin, folic acid) RDA: women: 400 mcg men: 400 mcg UL: 1000 mcg (from supplements and fortified foods)	• Needed for reactions that utilize amino acids (the building blocks of protein) for protein tissue formation • Promotes the normal formation of red blood cells	• Megaloblastic anemia • Diarrhea • Red, sore tongue Normal red blood cells.	• Neural tube defects, low birth weight (in pregnancy); increased risk of heart disease and stroke • Elevated blood levels of homocysteine Red blood cells in megaloblastic anemia
Vitamin B₁₂ (cyanocobalamin) AI women: 2.4 mcg men: 2.4 mcg	• Helps maintain nerve tissues • Aids in reactions that build up protein tissues • Needed for normal red blood cell development	• Neurological disorders (nervousness, tingling sensations, brain degeneration) • Pernicious anemia • Fatigue • Deficiency reported in 39% of adults in one study • Elevated blood level of homocysteine	
Biotin AI women: 30 mcg men: 30 mcg	• Needed for the body's manufacture of fats, proteins, and glycogen	• Depression, fatigue, nausea • Hair loss, dry and scaly skin • Muscular pain	
Pantothenic acid (pantothenate) AI women: 5 mg men: 5 mg	• Needed for the release of energy from fat and carbohydrates	• Fatigue, sleep disturbances, impaired coordination • Vomiting, nausea	
Vitamin C (ascorbic acid) RDA women: 75 mg men: 90 mg UL: 2000 mg	• Needed for the manufacture of collagen • Helps the body fight infections, repair wounds • Acts as an antioxidant • Enhances iron absorption Gums that are swollen and bleed easily are signs of scurvy, the vitamin C-deficiency disease.	• Bleeding and bruising easily due to weakened blood vessels, cartilage, and other tissues containing collagen • Slow recovery from infections and poor wound healing • Fatigue, depression • Deficiency reported in 9–24% of adults in one study	

TABLE 20.2

An intensive course on vitamins (continued)

The Water-Soluble Vitamins

CONSEQUENCES OF OVERDOSE	PRIMARY FOOD SOURCES	HIGHLIGHTS AND COMMENTS
• May cover up signs of vitamin B_{12} deficiency (pernicious anemia)	• Fortified, refined grain products (bread, flour, pasta) • Ready-to-eat cereals • Dark green, leafy vegetables (spinach, collards, romaine) • Broccoli, brussels sprouts • Oranges, bananas, grapefruit • Milk, cheese, yogurt • Dried beans	• *Folate* means "foliage." It was first discovered in leafy green vegetables. • This vitamin is easily destroyed by heat. • Synthetic form added to fortified grain products is better absorbed than naturally occurring folates. • Between 6 and 58% of people in various population groups are at risk of folate deficiency due to one or more genetic traits.
• None known. Excess vitamin B_{12} is rapidly excreted by the kidneys or is not absorbed into the bloodstream. • Vitamin B_{12} injections may cause a temporary feeling of heightened energy.	• Animal products: beef, lamb, liver, clams, crab, fish, poultry, eggs • Milk and milk products • Ready-to-eat cereals	• Older people and vegans are at risk for vitamin B_{12} deficiency. • Some people become vitamin B_{12} deficient because they are genetically unable to absorb it. • Vitamin B_{12} is found in animal products and microorganisms only.
• None known. Excesses are rapidly excreted.	• Grain and cereal products • Meats, dried beans, cooked eggs • Vegetables	• Deficiency is extremely rare. May be induced by the overconsumption of raw eggs.
• None known. Excesses are rapidly excreted.	• Many foods, including meats, grains, vegetables, fruits, and milk • Required for the conversion of homocysteine to methionine	• Deficiency is very rare.
• Intakes of 1 g or more per day can cause nausea, cramps, and diarrhea and may increase the risk of kidney stones.	• Fruits: oranges, lemons, limes, strawberries, cantaloupe, honeydew melon, grapefruit, kiwi fruit, mango, papaya • Vegetables: broccoli, green and red peppers, collards, cabbage, tomato, asparagus, potatoes • Ready-to-eat cereals	• Need increases among smokers (to 110–125 mg per day). • Is fragile; easily destroyed by heat and exposure to air • Supplements may decrease duration and symptoms of colds. • Deficiency may develop within 3 weeks of very low intake.

(continued)

TABLE 20.2

An intensive course on vitamins (continued)

The Fat-Soluble Vitamins

	PRIMARY FUNCTIONS	CONSEQUENCES OF DEFICIENCY	
Vitamin A **I. Retinol** RDA women: 700 mcg men: 900 mcg UL: 3000 mcg	• Needed for the formation and maintenance of mucous membranes, skin, bone • Needed for vision in dim light	• Increased susceptibility to infection, increased incidence and severity of infection (including measles) • Impaired vision, blindness • Inability to see in dim light	

Xerophthalmia. *Vitamin A deficiency is the leading cause of blindness in developing countries.*

2. Beta-carotene (a vitamin A precursor or "provitamin") No RDA; suggested intake: 6 mg	• Acts as an antioxidant; prevents damage to cell membranes and the contents of cells by repairing damage caused by free radicals	• Deficiency disease related only to lack of vitamin A

Vitamin E **(alpha-tocopherol)** RDA women: 15 mg men: 15 mg UL: 1000 mg	• Acts as an antioxidant, prevents damage to cell membranes in blood cells, lungs, and other tissues by repairing damage caused by free radicals • Reduces the ability of LDL cholesterol (the "bad" cholesterol) to form plaque in arteries	• Muscle loss, nerve damage • Anemia • Weakness • Many adults may have non-optimal blood levels.	

Vitamin E from foods helps prevent plaque formation in arteries, may improve circulation in people with diabetes, and may decrease symptoms of Alzheimer's disease.

TABLE 20.2
An intensive course on vitamins (continued)

The Fat-Soluble Vitamins

CONSEQUENCES OF OVERDOSE	PRIMARY FOOD SOURCES	HIGHLIGHTS AND COMMENTS
• Vitamin A toxicity (hypervitaminosis A) with acute doses of 500,000 IU, or long-term intake of 50,000 IU per day. Limit retinol use in pregnancy to 5000 IU daily. • Nausea, irritability, blurred vision, weakness, headache • Increased pressure in the skull, hip fracture • Liver damage • Hair loss, dry skin • Birth defects	• Vitamin A is found in animal products only. • Liver, butter, margarine, milk, cheese, eggs • Ready-to-eat cereals	• Symptoms of vitamin A toxicity may mimic those of brain tumors and liver disease. Vitamin A toxicity is sometimes misdiagnosed because of the similarities in symptoms. • 1 IU vitamin A = 0.3 mcg retinol or 0.6 mcg beta-carotene
• High intakes from supplements may increase lung damage in smokers. • With high intakes and supplemental doses (over 12 mg/day for months), skin may turn yellow-orange. • Possibly related to reversible loss of fertility in women	• Deep orange, yellow, and green vegetables and fruits are often good sources. • Carrots, sweet potatoes, pumpkin, spinach, collards, red peppers, broccoli, cantaloupe, apricots, tomatoes, vegetable juice	• The body converts beta-carotene to vitamin A. Other carotenes are also present in food, and some are converted to vitamin A. Beta-carotene and vitamin A perform different roles in the body, however. • May decrease sunburn in certain individuals.
• Intakes of up to 800 IU per day are unrelated to toxic side effects; over 800 IU per day may increase bleeding (blood-clotting time). • Avoid supplement use if aspirin, anticoagulants, or fish oil supplements are taken regularly.	• Oils and fats • Salad dressings, mayonnaise, margarine, shortening, butter • Whole grains, wheat germ • Leafy, green vegetables, tomatoes • Nuts and seeds • Eggs	• Vitamin E is destroyed by exposure to oxygen and heat. • Oils naturally contain vitamin E. It's there to protect the fat from breakdown due to free radicals. • Eight forms of vitamin E exist, and each has different antioxidant strength. • Natural form is better absorbed than synthetic form: 15 IU alpha-tocopherol = 22 IU d-alpha tocopherol (natural form) and 33 IU synthetic vitamin E.

(continued)

Brittle hair and dry, rough, scaly, and cracked skin from vitamin A overdose.

From: American Journal of Clinical Nutrition, Vol 71, No 4, 878–884. April 2000, Robert M. Russell.

TABLE 20.2

An intensive course on vitamins (continued)

The Fat-Soluble Vitamins

	PRIMARY FUNCTIONS	CONSEQUENCES OF DEFICIENCY	

Vitamin D
(1,25 dihydroxy-
cholecalciferol)
AI women: 5 mcg
      (200 IU)
     men: 5 mcg
      (200 IU)
    UL: 50 mcg
      (2000 IU)
	• Needed for the absorption of cal- cium and phosphorus, and for their utilization in bone formation, nerve and muscle activity.	• Weak, deformed bones (children)
		
		• Loss of calcium from bones (adults), osteoporosis
		
		The vitamin D-deficiency disease: rickets	

Vitamin K **(phylloquinone,** **menaquinone)** AI women: 90 mcg men: 120 mcg	• Is an essential component of mechanisms that cause blood to clot when bleeding occurs • Aids in the incorporation of calcium into bones 	• Bleeding, bruises • Decreased calcium in bones • Deficiency is rare. May be induced by the long-term use (months or more) of antibiotics. *The long-term use of antibiotics can cause vitamin K deficiency. People with vitamin K deficiency bruise easily.*	

is needed for reactions that generate new cells to replace worn-out cells lining the mouth, esophagus, intestines, and eyes. Without enough vitamin A, old cells aren't replaced, and the affected tissues are damaged. And vitamin C is required for reactions that build and maintain collagen, a protein found in skin, bones, blood vessels, gums, ligaments, and cartilage. (Approximately 30% of the total amount of protein in the body is collagen.) With vitamin C deficiency, collagen becomes weak, causing tissues that contain collagen to weaken and bleed easily.

The examples just given all relate to the physical effects of vitamins. Vitamins participate in reactions that affect behaviors, too. Alterations in behaviors such as reduced attention span, poor appetite, irritability, depression, or paranoia often precede the physical signs of vitamin deficiency.[2] Vitamins are truly "vital" for health.

TABLE 20.2

An intensive course on vitamins (continued)

The Fat-Soluble Vitamins

CONSEQUENCES OF OVERDOSE	PRIMARY FOOD SOURCES	HIGHLIGHTS AND COMMENTS
• Mental retardation in young children • Abnormal bone growth and formation • Nausea, diarrhea, irritability, weight loss • Deposition of calcium in organs such as the kidneys, liver, and heart	• Vitamin D-fortified milk and margarine • Butter • Fish • Eggs • Mushrooms • Milk products such as cheese, yogurt, and ice cream are generally not fortified with vitamin D.	• Vitamin D is manufactured from cholesterol in cells beneath the surface of the skin upon exposure of the skin to sunlight. • Poor vitamin D status may be common in all age groups. • Breast-fed infants with little sun exposure benefit from vitamin D supplements.
• Toxicity is only a problem when synthetic forms of vitamin K are taken in excessive amounts. That may cause liver disease.	• Leafy, green vegetables • Grain products	• Vitamin K is produced by bacteria in the gut. Part of our vitamin K supply comes from these bacteria. • Newborns are given a vitamin K injection because they have "sterile" guts and consequently no vitamin K–producing bacteria.

TABLE 20.3

Nonvitamins

The "real" vitamins are listed in Table 20.1. But a number of bogus vitamins are also on the market. They turn up in supplements, skin care creams, weight-loss and hair care products, and other items. These are some of the more popular nonvitamins.

Bioflavonoids (vitamin P)
Coenzyme Q_{10}
Gerovital H-3
Hesperidin
Inositol
Laetrile (vitamin B_{17})
Lecithin
Lipoic acid
Nucleic acids
Pangamic acid (vitamin B_{15})
Para-amino benzoic acid (PABA)
Provitamin B_5 complex
Rutin

Illustration 20.1
A baby with spina bifida, a form of neural tube defect associated with poor folate status early in pregnancy.

Protection from Vitamin Deficiencies and More

Current research on vitamins centers around their effects on disease prevention and treatment. It is a very active area of research and, no doubt, new and important results will be announced after this text goes to press. (Nutrition textbooks should really be updated every few weeks. Stay tuned to your instructor for the latest developments.) Here are a few examples of recent developments in research on vitamins and disease prevention and treatment:

Folate, Neural Tube Defects, Dementia, and Cancer   Daily consumption of 400 micrograms of folic acid (the synthetic form of folate added to refined grain products) before and early in pregnancy prevents about two-thirds of cases of

dementia
(Pronounced *di-men-cha*.) A usually progressive condition (such as Alzheimer's disease) marked by the development of memory impairment and an inability to use or comprehend words or to plan and initiate complex behaviors.

neural tube defects in newborns.[3] Neural tube defects are abnormalities of the spinal cord and brain (Illustration 20.1). They are one of the most common types of malformation of newborns in the United States. Adequate intake of folate also reduces the risk of developing ***dementia*** and certain types of cancer such as childhood brain tumors and breast, ovary, and stomach cancer.[4]

In 1998 manufacturers started fortifying refined grain products such as bread, pasta, and rice with folic acid. The addition of folic acid to these foods has produced substantial gains in people's folate status. Blood levels of folate have nearly doubled, and the prevalence of poor folate status has declined by 84%.[5]

Vitamin A—From Measles to "Liver Spots"   Studies undertaken in both developing countries and the United States indicate that adequate vitamin A status decreases the severity of measles and other infectious diseases.[6] (It has been known for decades that adequate vitamin A intake also prevents blindness, an all-too-common consequence of vitamin A deficiency in developing nations.) This vitamin is being used successfully in the treatment of serious cases of acne, skin wrinkles, and "liver" or "aging" spots on the skin due to overexposure to the sun.[7]

Vitamin D, Osteoporosis, and Chronic Inflammation   Vitamin D intakes of 800 to 1000 IU per day are associated with decreased bone loss, and a reduced risk of osteoporosis, and diseases related to ***chronic inflammation***.[8]

chronic inflammation
Inflammation that lasts weeks, months, or years. Inflammation is the first response of the body's immune system to infection or irritation. Inflammation triggers the release of biologically active substances that promote oxidation and other potentially harmful reactions in the body.

Increased rates of breast-feeding in the United States have had the unintended side effect of increased vitamin D deficiency (rickets) in infants.[9] Breast milk contains a very low amount of vitamin D, so infants receiving human milk need to get it from exposure of their skin to sunlight or from supplements. Infants who develop vitamin D deficiency fail to get enough sunlight exposure due to winters in cold climates or from being fully clothed or covered with sunscreen to protect their skin from the sun. Modest exposure to sunshine (about 30 minutes a week with diapers only) is sufficient to supply the required amount of vitamin D (Illustration 20.2). In cold climates, a vitamin D supplement is needed.[10]

Although vitamin D is the best known as the sunshine vitamin that helps build strong bones by facilitating calcium absorption, it does much more than that. Vitamin D plays key roles as a hormone in combating chronic inflammation. Low-grade chronic inflammation is at the core of the development of disorders such as type 1 diabetes, multiple sclerosis, heart disease, and rheumatoid arthritis. Vitamin D reduces inflammation by entering cells and turning genes that produce inflammatory substances "off," and those that produce substances that reduce inflammation "on."[10]

The best way for children and adults to get vitamin D is by exposing the skin to the direct rays of the sun for 5 to 15 minutes a day. Sun exposure builds adequate amounts of vitamin D, and there is no risk of overdose.[10] Foods fortified with vitamin D such and milk, soy milk, and breakfast cereals, as well as supplements, can meet vitamin D needs when sunlight is indirect or weak.

Woodfin Camp, Wally McNamee

Illustration 20.2
Russian children are exposed to a quartz lamp to prevent vitamin D deficiency during the long winter.

Vitamin C and the Common Cold Revisited   Recent research confirms that vitamin C, the popular cold remedy, reduces the symptoms and duration of the common cold modestly but does *not* affect how often colds occur.[11]

Some of the most exciting recent findings on vitamins and disease concern vitamins that function as *antioxidants.*

The Antioxidant Vitamins

Beta-carotene (a *precursor* to vitamin A), vitamin E, and vitamin C function as antioxidants. This means they prevent or repair damage to components of cells caused by exposure to *free radicals.* A free radical is formed when an atom of hydrogen or oxygen loses an electron. Without the electron, there is an imbalance between the atom's positive and negative charges. This makes the atom reactive—it needs to steal an electron from a nearby atom or molecule to reestablish a balance between its positive and negative charges. Atoms and molecules that have lost electrons to free radicals are said to be "oxidized." These oxidized substances are reactive and can damage cell membranes, DNA, and other cell components. Antioxidants such as beta-carotene, vitamin E, and vitamin C donate electrons to stabilize oxidized molecules or repair them in other ways. The effectiveness of antioxidant vitamins appears to be increased when they are consumed in plant foods rather than supplements. Fruits, vegetables, whole grains, and other plant foods contain thousands of naturally occuring antioxidants that work together with the antioxidant vitamins in disease prevention.[12]

Why do free radicals exist in the body? Free radicals play a number of roles in the body, so they are always present. They are produced during energy formation, by breathing, and by the immune system to help destroy bacteria and viruses that enter the body. They can also be formed when the body is exposed to alcohol, radiation emitted by the sun, smoke, ozone, smog, and other environmental pollutants.

Cellular damage due to free radicals appears to play a role in the development of some types of cancer, bronchitis, emphysema, heart disease, cataracts, and premature aging.[12] Food sources of vitamin C and E are associated with slowed progression of Alzheimer's disease, reduced plaque formation in arteries, and improved circulation in people with diabetes.[13, 14] The antioxidant vitamins—along with the mineral selenium, substances found in plants, and other substances produced by the body that function as antioxidants—may help prevent or postpone these disorders by protecting cell components against free radical damage.

Apple slices exposed to air turn brown due to oxidation. Coating the slice with a vitamin C solution or lemon juice prevents the oxidation.

ON THE SIDE

antioxidants
Chemical substances that prevent or repair damage to cells caused by exposure to free radicals. Beta-carotene, vitamin E, and vitamin C function as antioxidants.

precursor
In nutrition, a nutrient that can be converted into another nutrient (also called provitamin). Beta-carotene is a precursor of vitamin A.

free radicals
Chemical substances (usually oxygen or hydrogen) that are missing an electron. The absence of the electron makes the chemical substances reactive and prone to oxidizing nearby atoms or molecules by stealing an electron from them.

Vitamins: Getting Enough without Getting Too Much

Vitamins are widely present in basic foods (Table 20.4). Adequate amounts of them can be obtained from diets that include the variety of foods recommended in the MyPyramid. Most fruits and vegetables are good sources of vitamins, and eating five or more servings a day is one way for Americans to get their vitamins. In the United States, 80% of adults consume two or fewer servings of vegetables and fruits daily, and 10% consume none.[14]

Fortified foods such as ready-to-eat cereals, fruit juices and drinks, and snack bars are not listed in Table 20.4. They can increase vitamin intake substantially, and concern has been expressed that fortified food consumption may increase intake too much. Although an important concern, overdose reactions from the new generation of vitamin-fortified foods have not yet been reported.

Beef and broccoli are good sources of thiamin.

Milk and milk products are good sources of riboflavin.

TABLE 20.4
Food sources of vitamins

Thiamin

FOOD	SERVING SIZE	THIAMIN (MILLIGRAMS)
Meats:		
Pork roast	3 oz	0.8
Beef	3 oz	0.4
Ham	3 oz	0.4
Liver	3 oz	0.2
Nuts and seeds:		
Sunflower seeds	¼ cup	0.7
Peanuts	¼ cup	0.1
Almonds	¼ cup	0.1
Grains:		
Bran flakes	1 cup (1 oz)	0.6
Macaroni	½ cup	0.1
Rice	½ cup	0.1
Bread	1 slice	0.1
Vegetables:		
Peas	½ cup	0.3
Lima beans	½ cup	0.2
Corn	½ cup	0.1
Broccoli	½ cup	0.1
Potato	1 medium	0.1
Fruits:		
Orange juice	1 cup	0.2
Orange	1	0.1
Avocado	½	0.1

Riboflavin

FOOD	SERVING SIZE	RIBOFLAVIN (MILLIGRAMS)
Milk and milk products:		
Milk	1 cup	0.5
2% milk	1 cup	0.5
Yogurt, low-fat	1 cup	0.5
Skim milk	1 cup	0.4
Yogurt	1 cup	0.1
American cheese	1 oz	0.1
Cheddar cheese	1 oz	0.1
Meats:		
Liver	3 oz	3.6
Pork chop	3 oz	0.3
Beef	3 oz	0.2
Tuna	3 oz	0.1
Vegetables:		
Collard greens	½ cup	0.3
Broccoli	½ cup	0.2
Spinach, cooked	½ cup	0.1
Eggs:		
Egg	1	0.2
Grains:		
Macaroni	½ cup	0.1
Bread	1 slice	0.1

TABLE 20.4

Food sources of vitamins (continued)

Niacin

FOOD	SERVING SIZE	NIACIN (MILLIGRAMS)
Meats:		
Liver	3 oz	14.0
Tuna	3 oz	10.3
Turkey	3 oz	9.5
Chicken	3 oz	7.9
Salmon	3 oz	6.9
Veal	3 oz	5.2
Beef (round steak)	3 oz	5.1
Pork	3 oz	4.5
Haddock	3 oz	2.7
Scallops	3 oz	1.1
Nuts and seeds:		
Peanuts	1 oz	4.9
Vegetables:		
Asparagus	½ cup	1.5
Grains:		
Wheat germ	1 oz	1.5
Brown rice	½ cup	1.2
Noodles, enriched	½ cup	1.0
Rice, white, enriched	½ cup	1.0
Bread, enriched	1 slice	0.7
Milk and milk products:		
Cottage cheese	½ cup	2.6
Milk	1 cup	1.9

Salmon is a good source of niacin.

Vitamin B₆

FOOD	SERVING SIZE	VITAMIN B₆ (MILLIGRAMS)
Meats:		
Liver	3 oz	0.8
Salmon	3 oz	0.7
Other fish	3 oz	0.6
Chicken	3 oz	0.4
Ham	3 oz	0.4
Hamburger	3 oz	0.4
Veal	3 oz	0.4
Pork	3 oz	0.3
Beef	3 oz	0.2
Eggs:		
Egg	1	0.3
Legumes:		
Split peas	½ cup	0.6
Dried beans, cooked	½ cup	0.4
Fruits:		
Banana	1	0.6
Avocado	½	0.4
Watermelon	1 cup	0.3
Vegetables:		
Turnip greens	½ cup	0.7
Brussels sprouts	½ cup	0.4
Potato	1	0.2
Sweet potato	½ cup	0.2
Carrots	½ cup	0.2

Bananas are a good source of vitamin B₆.

(continued)

TABLE 20.4

Food sources of vitamins (continued)

Folate

FOOD	SERVING SIZE	FOLATE (MICROGRAMS)
Vegetables:		
Peas	½ cup	0.1
Garbanzo beans	½ cup	141
Navy beans	½ cup	128
Asparagus	½ cup	120
Brussels sprouts	½ cup	116
Black-eyed peas	½ cup	102
Spinach, cooked	½ cup	99
Romaine lettuce	1 cup	86
Lima beans	½ cup	71
Peas	½ cup	70
Collard greens, cooked	½ cup	56
Sweet potato	½ cup	43
Broccoli	½ cup	43
Fruits:		
Cantaloupe	¼ whole	100
Orange juice	1 cup	87
Orange	1	59
Grains:[a]		
Ready-to-eat cereals	1 cup/1 oz	100–400
Oatmeal	½ cup	97
Noodles	½ cup	45
Wheat germ	2 tbs	40
Wild rice	½ cup	37

[a]Fortified, refined grain products such as bread, rice, pasta, and crackers provide approximately 40 to 60 micrograms of folic acid per standard serving.

Vitamin B$_{12}$

FOOD	SERVING SIZE	VITAMIN B$_{12}$ (MICROGRAMS)
Meats:		
Liver	3 oz	6.8
Trout	3 oz	3.6
Beef	3 oz	2.2
Clams	3 oz	2.0
Crab	3 oz	1.8
Lamb	3 oz	1.8
Tuna	3 oz	1.8
Veal	3 oz	1.7
Hamburger, regular	3 oz	1.5
Milk and milk products:		
Skim milk	1 cup	1.0
Milk	1 cup	0.9
Yogurt	1 cup	0.8
Cottage cheese	½ cup	0.7
American cheese	1 oz	0.2
Cheddar cheese	1 oz	0.2
Eggs:		
Egg	1	0.6

Beans are an especially rich source of folate.

Crab is one food source of vitamin B$_{12}$.

TABLE 20.4

Food sources of vitamins (continued)

Vitamin C

FOOD	SERVING SIZE	VITAMIN C (MILLIGRAMS)
Fruits:		
Orange juice, vitamin C-fortified	1 cup	108
Kiwi fruit	1 or ½ cup	108
Grapefruit juice, fresh	1 cup	94
Cranberry juice cocktail	1 cup	90
Orange	1	85
Strawberries, fresh	1 cup	84
Orange juice, fresh	1 cup	82
Cantaloupe	¼ whole	63
Grapefruit	1 medium	51
Raspberries, fresh	1 cup	31
Watermelon	1 cup	15
Vegetables:		
Green peppers	½ cup	95
Cauliflower, raw	½ cup	75
Broccoli	½ cup	70
Brussels sprouts	½ cup	65
Collard greens	½ cup	48
Vegetable (V-8) juice	¾ cup	45
Tomato juice	¾ cup	33
Cauliflower, cooked	½ cup	30
Potato	1 medium	29
Tomato	1 medium	23

Citrus fruits are an excellent source of vitamin C.

Vitamin A (Retinol)

FOOD	SERVING SIZE	VITAMIN A (MICROGRAMS RE)[b]
Meats:		
Liver	3 oz	9124
Salmon	3 oz	53
Tuna	3 oz	14
Eggs:		
Egg	1 medium	84
Milk and milk products:		
Skim milk, fortified	1 cup	149
2% milk	1 cup	139
American cheese	1 oz	82
Whole milk	1 cup	76
Swiss cheese	1 oz	65
Fats:		
Margarine, fortified	1 tsp	46
Butter	1 tsp	38
Vitamin A and beta-carotene		

[b]RE (retinol equivalent) = 3.33 IU.

Eggs are one food source of vitamin A (retinol).

(continued)

Spinach and winter squash are two sources of beta-carotene.

Seafood and asparagus both provide vitamin E.

TABLE 20.4

Food sources of vitamins (continued)

Beta-Carotene

FOOD	SERVING SIZE	VITAMIN A VALUE (MICROGRAMS RE)[b]
Vegetables		
Pumpkin, canned	½ cup	2712
Sweet potato, canned	½ cup	1935
Carrots, raw	½ cup	1913
Spinach, cooked	½ cup	739
Collard greens, cooked	½ cup	175
Broccoli, cooked	½ cup	109
Winter squash	½ cup	53
Green peppers	½ cup	40
Fruits:		
Cantaloupe	¼ whole	430
Apricots, canned	½ cup	210
Nectarine	1 medium	101
Watermelon	1 cup	59
Peaches, canned	½ cup	47
Papaya	½ cup	20

Vitamin E

FOOD	SERVING SIZE	VITAMIN E (IU)[c]
Oils:		
Oil	1 tbs	6.7
Mayonnaise	1 tbs	3.4
Margarine	1 tbs	2.7
Salad dressing	1 tbs	2.2
Nuts and seeds:		
Sunflower seeds	¼ cup	27.1
Almonds	¼ cup	12.7
Peanuts	¼ cup	4.9
Cashews	¼ cup	0.7
Vegetables:		
Sweet potato	½ cup	6.9
Collard greens	½ cup	3.1
Asparagus	½ cup	2.1
Spinach, raw	1 cup	1.5
Grains:		
Wheat germ	2 tbs	4.2
Bread, whole wheat	1 slice	2.5
Bread, white	1 slice	1.2
Seafood:		
Crab	3 oz	4.5
Shrimp	3 oz	3.7
Fish	3 oz	2.4

[b]RE (retinol equivalent) = 3.33 IU.
[c]15 milligrams alpha-tocopherol = 22 IU d-alpha tocopherol (natural form) and 33 IU synthetic vitamin E.

TABLE 20.4

Food sources of vitamins (continued)

Vitamin D

FOOD	SERVING SIZE	VITAMIN D (IU)[d]
Milk:		
Milk, whole, low-fat, or skim	1 cup	100
Fish and seafoods:		
Salmon	3 oz	340
Tuna	3 oz	150
Shrimp	3 oz	127
Organ meats:		
Beef liver	3 oz	42
Chicken liver	3 oz	40
Eggs:		
Egg yolk	1	27

[d]40 IU = 1 microgram.

Shrimp and other shellfish are good sources of vitamin D.

Updated recommendations for vitamin intakes associated with the prevention of deficiency and chronic diseases are represented by standards called Dietary Reference Intakes, or DRIs. DRIs include Recommended Dietary Allowances (RDAs) for vitamins for which convincing scientific data exist for establishing intake standards. Adequate Intakes, or AIs, are assigned to vitamins for which scientific information about levels of intake associated with chronic disease prevention is less convincing. Tolerable Upper Levels of Intake, abbreviated ULs, are also assigned to vitamins and indicate levels of vitamin intake from foods, fortified foods, and supplements that should not be exceeded. The RDAs or AIs and the ULs for the vitamins are given in Table 20.2.

Although people can get all the vitamins they need from supplements, it makes more sense to get them from basic foods. Foods offer fiber, minerals, and other healthful ingredients that don't come in supplements, and they certainly taste better on the way down!

Nutrition Up Close

Antioxidant Vitamins–How Adequate Is Your Diet?

Focal Point: Determine if you eat enough antioxidant-rich foods.

Vitamin C, beta-carotene, and vitamin E, the antioxidant vitamins, help to maintain cellular integrity in the body. Good food sources of these antioxidants reduce the risk of heart disease, certain cancers, and other ailments. Check below to find out how frequently you consume foods containing these important, health-promoting nutrients.

How Often Do You Eat:	Seldom or Never	1–2 Times per Week	3–5 Times per Week	Almost Daily
Vitamin C–rich foods:				
1. Grapefruit, lemons, oranges, or pineapple?	☐	☐	☐	☐
2. Strawberries, kiwi, or honeydew melon?	☐	☐	☐	☐
3. Orange juice, cranberry juice cocktail, or tomato juice?	☐	☐	☐	☐
4. Green, red, or chili peppers?	☐	☐	☐	☐
5. Broccoli, Chinese cabbage, or cauliflower?	☐	☐	☐	☐
6. Asparagus, tomatoes, or potatoes?	☐	☐	☐	☐
Beta-carotene–rich foods:				
7. Carrots, sweet potatoes, pumpkin, or winter squash?	☐	☐	☐	☐
8. Spinach, collard greens, or chard?	☐	☐	☐	☐
9. Cantaloupe, papayas or mangoes?	☐	☐	☐	☐
10. Nectarines, peaches, or apricots?	☐	☐	☐	☐
Vitamin E–rich foods:				
11. Whole-grain breads, whole-grain cereals, or wheat germ?	☐	☐	☐	☐
12. Crab, shrimp, or fish?	☐	☐	☐	☐
13. Peanuts, almonds, or sunflower seeds?	☐	☐	☐	☐
14. Oils, margarine, butter, mayonnaise, or salad dressing?	☐	☐	☐	☐

FEEDBACK (including scoring) can be found at the end of Unit 20.

Key Terms

antioxidants, page 20-13
chronic inflammation, page 20-12
coenzymes, page 20-3

dementia, page 20-12
free radicals, page 20-13

precursor, page 20-13
vitamins, page 20-2

Review Questions

True False

1. Vitamins are essential. Specific deficiency diseases develop if we fail to consume enough of them. ☐ ☐

2. Vitamin D, E, C, and B₆ are fat soluble. ☐ ☐

3. The niacin deficiency disease is called pellagra. ☐ ☐

4. Some people become deficient in vitamin B₁₂ because they are genetically unable to absorb it. ☐ ☐

5. Vitamin A toxicity causes brain tumors. ☐ ☐

True False

6. Three good sources of vitamin D are sunshine, milk, and seafood. ☐ ☐

7. The effectiveness of antioxidant vitamins appears to be increased when they are consumed in plant foods rather than supplements. ☐ ☐

8. Vitamin D acts as a hormone. ☐ ☐

9. Adequate vitamin D status reduces chronic inflammation. ☐ ☐

10. Excessively high intakes of every vitamin have been found to cause toxicity disease. ☐ ☐

Media Menu

www.healthfinder.gov
Search vitamins.

www.ars.usda.gov/ba/bhnrc/ndl
Look up the vitamin and other nutrient content of thousands of foods on this site.

www.iom.edu/fnb
This site provides updated information on the DRIs.

www.merckhomeedition.com
The Merck Manual of Medical Information is free and searchable. It can be used to find out more about health problems related to vitamins.

Notes

1. Hathcock JN. Vitamins and minerals: efficacy and safety. Am J Clin Nutr 1997;66:427–37.

2. Buzina R et al. Workshop on functional significance of mild-to-moderate malnutrition. Am J Clin Nutr 1989;50:172–6.

3. American Academy of Pediatrics. Folic acid and the prevention of neural tube defects. Pediatrics 1999;104: 325–27.

4. Kim Y-I. Will mandatory folic acid fortification prevent or promote cancer? Am J Clin Nutr 2004;80:1123–8.

5. Ganji V et al. Trends in serum folate, RBC folate, and circulating total homocysteine concentrations in the US. J Nutr 2006;136:153–8.

6. Frieden TR et al. Vitamin A levels and severity of measles. Am J Dis Child 1992;146:182–6.

7. Topical retinoids in primary care practice. www.medscape.com, 8/03.

8. Bischoff-Ferrari HA et al. Estimation of optimal serum concentrations of 25-hydroxyvitamin D for multiple health outcomes. Am J Clin Nutr 2006;84:18–28.

9. Welch TR et al. Vitamin D-deficient rickets: the reemergence of a once-conquered disease (editorial). J Pediatr 2000;1137:152–7.

10. Mark BL. The experts speak: multiple sclerosis and vitamin D. J Am Diet Assoc 2006;106:418–24.

11. Douglas RM et al. Vitamin C for preventing and treating the common cold. Cochrane Database Syst Rev. 2004 Oct 18;(4):CD000980.

12. Halvorsen BL et al. Content of redox-active compounds (ie, antioxidants) in foods consumed in the United States. Am J Clin Nutr 2006;84:95–135.

13. Frank B et al. A review of antioxidants and Alzheimer's disease. 2005 Oct–Dec;17(4):269–86.

14. Hercberg S. The history of beta-carotene and cancers: from observational to intervention studies. What lessons can be drawn for future research on polyphenols? Am J Clin Nutr 2005;81 (suppl):218S–22S.

15. Johnston CS, Corte C. People with marginal vitamin C status are at high risk of developing vitamin C deficiency. J Am Diet Assoc 1999;99:854–56.

Answers to Review Questions

1. True, see page 20-2.
2. False, see pages 20-2, 20-3.
3. True, see page 20-4.
4. True, see page 20-7.
5. False, see page 20-9.
6. True, see page 20-11.
7. True, see page 20-13.
8. True, see page 20-12.
9. True, see page 20-12.
10. False, see "Consequences of Overdose" column in Table 20.2.

Nutrition Up Close

Antioxidant Vitamins—How Adequate Is Your Diet?

Feedback for Unit 20

Several responses in the last two columns indicate adequate antioxidant vitamin consumption. If you need to boost your intake, increase the overall amount of fruits, vegetables, nuts, oils, and whole grains in your diet.

Phytochemicals and Genetically Modified Food

NUTRITION SCOREBOARD

	TRUE	FALSE
1 Phytochemicals are found only in plants.		
2 Phytochemicals taken as supplements provide the same health benefits as do phytochemicals consumed in plant foods.		
3 Humans eat lots of DNA every day.		
4 Some chemical substances that occur naturally in food may be harmful to health.		

Key Concepts and Facts

- Plants contain thousands of substances in addition to essential nutrients that affect body processes and health.

- Diets containing lots of vegetables, fruits, whole grains, and other plant foods are strongly associated with the prevention of chronic diseases such as heart disease and cancer.

- Biotechnology is rapidly changing the characteristics and types of foods available to consumers.

- Not all substances that occur naturally in foods are safe to eat.

Answers to NUTRITION SCOREBOARD

		TRUE	FALSE
1	*Phytochemicals* means "plant chemicals."	✔	
2	Beneficial effects of phytochemicals on health appear to result primarily from complex interactions among them in foods. Supplements do not appear to provide these benefits.[1]		✔
3	The average meal contains approximately 250,000 miles (150,000 kilometers) of uncoiled DNA when it's digested.	✔	
4	Some foods contain "naturally occurring toxins" that can be harmful if consumed in excess.	✔	

Photo Disc

Phytochemicals: Nutrition Superstars

Things don't happen by accident in nature. If you observe it, it has a reason for being there.
—NORMAN KRINSKY, TUFTS UNIVERSITY MEDICAL CENTER

As recently as 25 years ago, the science of nutrition focused on the study of the actions and health effects of protein, vitamins, minerals, and the other essential nutrients. Those days are gone. Now nutrition scientists are investigating the effects of thousands of other substances in food. Although we have barely scratched the surface of knowledge about these substances, current research shows that impressive benefits can be gained by diets high in plant foods. People who habitually consume lots of vegetables, fruits, whole grains, and other plant foods are less likely to develop heart disease, cancer, type 2 diabetes, infections, eye disease, premature aging, and a number of other health problems than are people who do not.[2, 3] Research results are sufficiently positive to change the way nutritionists and other scientists are thinking about food and health relationships.

At center stage in this new era in nutrition are the ***phytochemicals,*** or chemical substances found in plants (Illustration 21.1). The number of potentially beneficial substances in plants is mind boggling. There are thousands of individual phytochemicals in plants, and some foods contain hundreds of them. A sampling of the phytochemicals and other substances in two foods is shown in Illustration 21.2. Meats, eggs, dairy products, and other foods of animal origin also contain chemical substances that are not considered essential nutrients but nevertheless affect body processes. Much less is known about these ***zoochemicals*** because their effects on health are not as clear as those of phytochemicals.

Phytochemicals are not considered to be essential nutrients for humans, because we do not develop a deficiency disease if we consume too little of them.

phytochemicals
(phyto = *plant*)
Chemical substances in plants, some of which perform important functions in the body.

zoochemicals
Chemical substances in animal foods, some of which perform important functions in the body.

Illustration 21.1
The foods shown here have star qualities.
They have been featured in recent research reports on the benefits of phytochemicals in food. Foods shown are grape juice, green tea, tomatoes, cabbage, carrots, tofu, crushed garlic; green, red, and yellow peppers; brown rice, whole grain pita bread, and peanut butter. (Food preparation courtesy of The Blue Moon Cafe, Menomonie, WI.)

They are similar to essential nutrients in that the body cannot make them—or enough of them. Consequently, they must be obtained from the diet. The chemical properties of most phytochemicals tend to make them heat and light stable, so they are not easily destroyed by cooking or storage.[4]

Many phytochemicals are excreted within a day or two after ingestion, so intake of vegetables, legumes, nuts, fruits, and other food sources should be maintained.[5] Cooking vegetables, or consuming them with a small amount of fat, increases the body's absorption of some phytochemicals.

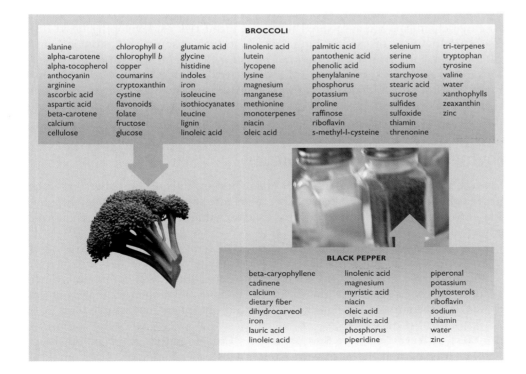

BROCCOLI

alanine	chlorophyll *a*	glutamic acid	linolenic acid	palmitic acid	selenium	tri-terpenes
alpha-carotene	chlorophyll *b*	glycine	lutein	pantothenic acid	serine	tryptophan
alpha-tocopherol	copper	histidine	lycopene	phenolic acid	sodium	tyrosine
anthocyanin	coumarins	indoles	lysine	phenylalanine	starchyose	valine
arginine	cryptoxanthin	iron	magnesium	phosphorus	stearic acid	water
ascorbic acid	cystine	isoleucine	manganese	potassium	sucrose	xanthophylls
aspartic acid	flavonoids	isothiocyanates	methionine	proline	sulfides	zeaxanthin
beta-carotene	folate	leucine	monoterpenes	raffinose	sulfoxide	zinc
calcium	fructose	lignin	niacin	riboflavin	thiamin	
cellulose	glucose	linoleic acid	oleic acid	s-methyl-l-cysteine	threnonine	

BLACK PEPPER

beta-caryophyllene	linolenic acid	piperonal
cadinene	magnesium	potassium
calcium	myristic acid	phytosterols
dietary fiber	niacin	riboflavin
dihydrocarveol	oleic acid	sodium
iron	palmitic acid	thiamin
lauric acid	phosphorus	water
linoleic acid	piperidine	zinc

Illustration 21.2
A sampling of the chemical substances in two foods.

Characteristics of Phytochemicals

Phytochemicals serve a wide variety of functions in plants. They provide color and flavor and protect plants from insects, microbes, and oxidation due to exposure to sunlight and oxygen. Some phytochemicals are components of a plant's energy-making processes, and others act as plant hormones. More than 2000 types of phytochemicals that act as pigments have been identified. There are at least 600 types of carotenoids, for example. These pigments give plants yellow, orange, and red color; they primarily function as antioxidants.

The amount and type of phytochemicals present in plants vary a good deal, depending on the plant. Some plant foods, such as citrus fruits and dried beans, which are considered rich sources of vitamins and minerals, contain an abundance of phytochemicals. Apples, celery, green tea, and garlic, foods considered by nutrient composition tables to be relative "nutrient weaklings," also provide ample amounts of phytochemicals.

Not all of the phytochemicals in plants are beneficial to health, however. Some are "naturally occurring toxins" and can be harmful. This type of phytochemical is discussed later in the unit.

Phytochemicals and Health

It has been known for more than 30 years that diets rich in vegetables and fruits are protective against heart disease and certain types of cancer.[5] For most of this time, the benefits of such diets were attributed to the vitamin, mineral, or dietary fiber content of vegetables and fruits. More recently, scientists discovered that the essential nutrient content of the diet didn't explain all of the differences in disease incidence between people consuming diets low in plants and those eating diets rich in plants. As researchers looked for the reason for the difference, they discovered that other chemical substances in food might be responsible. Many chemically active substances that could contribute to disease prevention were identified and their actions in the body described.

Phytochemicals are associated with a reduced risk of developing heart disease, certain types of cancer (lung, breast, cervical, esophageal, stomach, and colon cancer, for example), *age-related macular degeneration, cataracts,* infectious diseases, osteoporosis, type 2 diabetes, stroke, hypertension, and other disorders.[6,7] Although evidence strongly supports a primary role for phytochemicals in the prevention of these disorders, research studies required to demonstrate *cause-and-effect relationships* in humans have yet to be completed.

Phytochemicals Work in Groups

News about the potential benefits of phytochemicals has sent consumers in search of pills that contain them. The market is replete with such products (Illustration 21.3), even though there is no solid evidence that individual phytochemicals extracted from foods benefit health.[1]

The absorption of phytochemicals appears to depend on the presence of other phytochemicals and nutrients in foods. Most (if not all) phytochemicals act together, producing a desired effect in the body if consumed at the same time. Since the optimal combinations of the different types of phytochemicals are not yet known, it is recommended that foods rather than supplements provide them.[1]

The Case of Beta-Carotene Supplements The lesson that phytochemicals act together was learned the hard way. Numerous studies had shown that smokers

age-related macular degeneration
Eye damage caused by oxidation of the macula, the central portion of the eye that allows you to see details clearly. It is the leading cause of blindness in U.S. adults over the age of 65. Antioxidants provided by the carotenoids in dark green, leafy vegetables such as kale, collard greens, spinach, and Swiss chard may help prevent macular degeneration.[6]

Ralph Eagle/
Photo Researchers, Inc.

cataracts
Complete or partial clouding over the lens of the eye (shown in illustration).

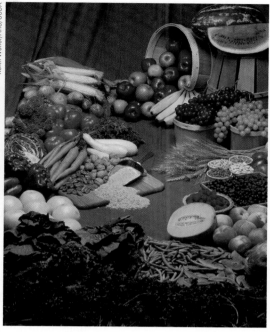

Illustration 21.3
Consumers have their choice of phytochemicals in pills.
The problem is that the supplements have not been shown to benefit health. Phytochemicals taken individually in supplements may actually harm health. They are best consumed in their natural packaging: vegetables, fruits, whole grains, and other plant foods.

consuming diets high in vegetables and fruits and those with high levels of beta-carotene in their blood were much less likely to develop lung cancer than smokers who consumed few vegetables and fruits. Beta-carotene is a powerful antioxidant present in many vegetables and fruits, and researchers theorized that the beta-carotene provided by these plant foods might prevent lung cancer.

Three large and expensive clinical trials of beta-carotene supplementation among male smokers were conducted to test this theory. The results? None of the trials showed a benefit from the beta-carotene supplements, and two studies found a higher rate of lung cancer among the groups using the supplements.[8]

Scientists drew two major conclusions from these studies:

1. High blood levels of beta-carotene found in the group of smokers who did not develop cancer were likely a marker for a high intake of vegetables and fruits and not the cause of the reduced risk.

2. Other phytochemicals in plants, or in a combination of plant foods, are likely responsible for the reduced risk of cancer.[1]

Vegetable Extracts and Essences Dehydrated, powdered extracts of vegetables high in certain phytochemicals and of the parts of vegetables richest in phytochemicals are also widely available. As is the case for supplements of individual phytochemicals, there is no solid evidence that these extracts benefit health. One such product called "Vegetable Essence" claimed to have "200 pounds of vegetables in a bottle." That is, of course, impossible. "Broccoli Concentrate," another vegetable extract, was found to contain primarily sulforaphane, one of the many phytochemicals in vegetables of the ***cruciferous family*** (Illustration 21.4). The problem with this and other vegetable concentrates is that you can fit only so much of a vegetable into a capsule. You would have to consume approximately 100 of the pills to get the amount of sulforaphane present in one serving of broccoli.[9] Even if you did that, you would still be missing out on the other phytochemicals and essential nutrients that are part of those little green trees.

cruciferous vegetables
Sulfur-containing vegetables whose outer leaves form a cross (or crucifix). Vegetables in this family include broccoli, cabbage, cauliflower, brussels sprouts, mustard and collard greens, kale, bok choy, kohlrabi, rutabaga, turnips, broccoflower, and watercress.

Illustration 21.4
Some examples of cruciferous vegetables.

Plants have been on this planet for about a billion years longer than humans. Although the total number of edible plants hasn't changed much over the centuries, most modern humans consume only 150 to 200 different types of plants, or about 0.2% of the total possibilities.[10]

Photo Disc

How Do Phytochemicals Work?

A variety of body processes involved in disease development are affected by phytochemicals. Some of these processes, and the phytochemicals responsible, are listed in Table 21.1. In general, phytochemicals can

1. Act as hormone-inhibiting substances that prevent the initiation of cancer

2. Serve as antioxidants that prevent and repair damage to cells due to oxidation

3. Block or neutralize enzymes that promote the development of cancer and other diseases

4. Modify the absorption, production, or utilization of cholesterol

5. Decrease formation of blood clots

Some plant pigments are powerful antioxidants (see Table 21.2 for the top nine food sources of antioxidants). Zeaxanthin (pronounced *ze-ah-zan-thun*), which gives plants a corn yellow color, anthocyanin (*an-tho-sigh-ah-nin*, the blue in blueberries and grapes), and lycopene (*lie-co-peen*), which helps make tomatoes, strawberries, and guava red, are all strong antioxidants. Dark chocolate contains flavonoids, which act as antioxidants. Some of the phytochemicals that act as antioxidants help reduce plaque formation in arteries and the risk of other inflammation-related disorders such as type 2 diabetes, osteoporosis, and some cancers.[11]

Diets High in Plant Foods

The discovery of a wide array of health-promoting phytochemicals in plants is one more very important reason why vegetables, fruits, whole grains, and other plant foods should be a major part of the diet. They are a minor part now for many Americans. In the United States, 80% of adults consume two or fewer servings of vegetables and fruits daily, and 10% consume none.[9] In reality, people would probably be healthier if they consumed more than the five servings of vegetables and fruits daily.

Naturally Occurring Toxins in Food

All that occurs naturally in foods is not necessarily good. There are many examples of foods containing phytochemicals that can have negative side effects. Spinach, collard greens, rhubarb, and other dark green, leafy vegetables contain oxalic acid. Eating too much of these foods can make your teeth feel as though they are covered with sand and give you a stomachache. Have you ever seen a potato that was partly colored green? (If not, take a look at Illustration 21.5.) The green area contains solanine, a bitter-tasting, insect-repelling phytochemical that is normally found only in the leaves and stalks of potato plants. Small amounts of solanine are harmless, but large quantities (an ounce or so) can interfere with the transmission of nerve impulses.

Phytates are an example of a naturally occurring substance that can have harmful effects. Phytates, which are found in whole grains, tightly bind zinc, iron, and other minerals, making them unavailable for absorption. Although not toxic, phytates do reduce the availability of some minerals in whole grain products. Cassava, a root consumed daily in many parts of tropical Africa, can be very toxic if not prepared properly because it contains cyanide. Soaking cassava roots in water for three nights gets rid of the cyanide, but soaking for shorter periods does not. When the soaking time is cut to one or two nights, as sometimes

PhotoEdit, Tony Freeman

Illustration 21.5
Potatoes grown partly above ground develop a green color in the part exposed to the sun. The green section contains solanine, a naturally occurring, toxic phytochemical.

TABLE 21.1

Examples of phytochemicals, their food sources, and potential mechanisms of action in disease prevention

Phytochemical	Comments	Food Sources	Proposed Action
Indoles, isothiocynates (sulforaphanes)	These sulfur-containing compounds may be particularly protective against breast cancer.	Cruciferous vegetables (broccoli, brussels sprouts, cabbage, cauliflower)	Interfere with the cancer growth-promoting effects of genes on cells; increase the body's ability to neutralize cancer-causing substances.
Allicin	Health benefits are associated with as little as a half clove of garlic per day on average.	Garlic, onions, leeks, shallots, chives, scallions	Interferes with the replication of cancer cells; decreases the production of cholesterol by the liver, reduces blood clotting.
Terpenes (monoterpenes, limonene)	These substances give citrus fruits a slightly bitter taste. Limonene is from the same family of compounds as tamoxifen, a drug used to treat breast cancer.	Oranges, lemons, grapefruit, and their juices	Facilitate the excretion of cancer-causing substances; decrease tumor growth.
Phytoestrogens (plant estrogens; isoflavones, genistein, daidzein, lignans)	May decrease risk of some cancers, heart disease, and osteoporosis. Estrogen effects, if any, are weak. High doses of some individual phytoestrogens may have adverse effects on health.	Soybeans, soy food, chickpeas, other dried beans, peas, peanuts	Interfere with cancer growth–promoting effects of estrogen; block the action of cancer-causing substances; lower blood cholesterol; decrease menopausal symptoms and bone loss, act as antioxidants.
Lignans (phytoestrogens)	Flaxseed is an extremely rich source of lignans. In the gut, lignans are converted to substances that may help prevent breast cancer.	Flaxseed, flaxseed oil, seaweed, soybeans and other dried beans, bran	Interfere with the action of estrogen.
Saponins	Nearly every type of dried beans is an excellent source of saponins.	Dried beans, whole grains, apples, celery, strawberries, grapes, onions, green and black tea, red wine	Neutralize certain potentially cancer-causing enzymes in the gut.
Flavonoids (tannins, phenols)	There are over 4000 flavonoids, some of which are plant pigments. Originally called "vitamin P." Gives red wines and dark teas astringent or bitter taste.	Apples, celery, strawberries, grapes, onions, green and black tea, red wine, soy, purple grape juice, broccoli, dark chocolate	Protect cells from oxidation; decrease plaque formation and blood clotting; increase HDL cholesterol; decrease DNA damage related to cancer development.
Carotenoids (alpha-carotene, beta-carotene, lutein, zeaxanthin, beta-cryptoxanthin, lycopene)	There are more than 600 types of colorful carotenoids in plants. An orange contains at least 20 types. Dark green vegetables are often good sources; the green chlorophyll obscures the colors of carotenoids in these plants. Fat intake increases absorption.	Dark green vegetables, orange, yellow, and red vegetables and fruits	Neutralize oxidation reactions that can damage eyes and promote macular degeneration and cataracts; increase LDL cholesterol and cancer risk.
Plant stanols and sterols	Structurally similar to cholesterol; they block cholesterol absorption.	Edible and nonedible oils	Decrease blood LDL-cholesterol level.

TABLE 21.2

Top 12 food sources of antioxidants in a typical serving size, listed in order[1]

Blackberries	Pecans
Walnuts	Blueberries
Strawberries	Ground cloves
Artichokes	Grape juice
Cranberries	Unsweetened dark
Brewed coffee	chocolate
Raspberries	

Illustration 21.6
Ackee fruit and seeds.

biotechnology
As applied to food products, the process of modifying the composition of foods by biologically altering their genetic makeup. Also called *genetic engineering* of foods. The food products produced are sometimes referred to as "GM" and GMOs (genetically modified organisms).

happens during periods of food shortage, enough of the toxin remains in the root to cause konzo, a disease caused by the cyanide overdose.[12] Konzo is characterized by permanent, spastic paralysis.

Beware of Ackee Fruit   Ackee fruit is another potential hazard to health. If you're from Jamaica or Africa, chances are excellent that you love the taste of the core of ackee—and know the fruit can be deadly. The national fruit of Jamaica, the yellow fleshy part around the seeds tastes like butter and looks like scrambled eggs (Illustration 21.6). The rest of the fruit is not edible. The ackee fruit—and the fruit of unopened, unripe ackee in particular—contains high concentrations of phytochemicals that cause severe vomiting and a drastic drop in blood glucose levels. Ingestion of the fruit has caused hundreds of deaths in Jamaica. Its sale was banned in the United States until 2000.[13]

What's the Scoop on Caffeine?   Caffeine is another phytochemical in plants (primarily, coffee beans) that some people consider to be toxic. To others, it's a gift from the gods every morning. Does caffeine belong in the list of naturally occurring, harmful substances in food?

Caffeine is one of many phytochemicals in coffee beans, cocoa beans, cola nuts, and tea leaves. Coffee, however, is far and away the leading source of caffeine in the diet. Table 21.3 displays the caffeine content of various beverages, chocolate, and nonprescription drugs. Although the effects of caffeine on health are widely debated, it has received a clean bill of health with a couple of exceptions. Too much caffeine causes sleeplessness and "caffeine jitters" in many people.[14] Large amounts of coffee (more than four cups a day) may somewhat delay the time it takes women to become pregnant and may slightly increase the risk of miscarriage.[15] Drinking unfiltered, boiled coffee (as is the custom in Turkey and some Scandinavian countries) raises LDL-cholesterol levels a little. However, oils and other phytochemicals in coffee, and not caffeine, may be responsible for this effect.[16] Suspicions that filtered coffee or caffeine is related to heart disease, cancer, or osteoporosis have not been confirmed.[17, 18] Habitual coffee consumption (one to three cups a day) appears to decrease the risk of cardiovascular disease and type 2 diabetes. The beneficial effects of coffee may be related to its content of phytochemicals that have antioxidant and anti-inflammatory properties and not to caffeine.[7, 17]

Regular coffee drinking can be habit forming. Cessation of coffee intakes that average two-and-a-half cups a day leads to headaches in about 50% of people.[19]

Genetically Modified Foods

Through *biotechnology*, researchers can modify the phytochemical and nutrient makeup of foods by altering the genetic makeup of plants. Resulting plants are called "genetically modified" or "GM" plants (Illustration 21.7). The process of biotechnology usually entails identifying a favorable genetic trait in one plant and transplanting that gene into another plant that lacks the characteristic.

Modifying the genetic makeup of plants has actually been practiced for centuries. Some 8000 years ago, Native Americans increased corn yields by cross-fertilizing corn plants. Over the centuries, thousands of plant hybrids derived from combining the genes of different plant species through cross-pollination have been developed. Advances related to genetic engineering in recent years have refined techniques so that single genes, rather than the full complements of a plant's genetic makeup, can be transferred to another plant.[20]

Genetic engineering is now used to transfer disease-resistant genes from one plant to another plant, conferring upon it an improved ability to resist disease. Flavor genes can be transported from one plant to another, creating new taste sensations like garbanzo beans with a nutty, peppery taste and basil preseasoned with cinnamon. Watermelon and oranges have had their seeds removed through genetic engineering. Colors of vegetables and fruits can be modified by transferring the appropriate genes from one plant to another. Carrots have been modified to be dark red so they would match Texas A&M's school color. Tomatoes have been genetically altered to stay firm during shipment and to produce 10 times the normal amount of lycopene; and rice has been altered to be rich in beta-carotene, the precursor of vitamin A. High beta-carotene rice was produced to make a good source of vitamin A available to people in countries where vitamin A deficiency is widespread. Called "golden rice," it contains two genes extracted from daffodils and one bacterial gene that together lead to the production of beta-carotene within rice seeds.[21]

Genetic Modification of Animals

Food biotechnology applies to foods derived from animals as well as plants. Scientists have engineered Atlantic salmon that grow to market weight in 18 rather than the usual 24 to 30 months. Pigs that produce less smelly stools and gas, cattle with leaner muscles, and hens that lay more eggs are also products of animal genetic engineering. Genetically engineered animals are not, however, approved for entry into the food supply. Approval is being blocked by concerns that genetically engineered fish and other animals could escape, mate with wild animals, and introduce novel and potentially damaging genetic traits in animals.[22]

TABLE 21.3
Caffeine content of foods, beverages, and some drugs

Source	Caffeine (mg)
Coffee, one cup	
Drip	137–153
Decaffeinated (ground or instant)	0.5–4.0
Instant	61–70
Percolated	97–125
Espresso (2 ounces)	100
Tea, one cup	
Black, brewed 5 minutes, U.S. brands	32–144
Black, brewed 5 minutes, imported brands	40–176
Green, brewed 5 minutes	25
Instant	40–80
Soft drinks/other drinks	
Coca-Cola (12 ounces)	47
Cherry Coke (12 ounces)	47
Diet Coca-Cola (12 ounces)	47
Dr. Pepper (12 ounces)	40
Ginger ale (12 ounces)	0
Jolt (12 ounces)	108[a]
Kick (12 ounces)	56
Ripped Force (8 ounces)	120
Red Bull (8 ounces)	80
Mountain Dew (12 ounces)	54
Pepsi-Cola (12 ounces)	38
Diet Pepsi-Cola (12 ounces)	37
7-Up (12 ounces)	0
Surge (8 ounces)	35
Water Joe (8 ounces)	125
Chocolate	
Cocoa, chocolate milk, one cup	10–17
Milk chocolate candy, one ounce	1–15
Chocolate syrup, one ounce (2 tablespoons)	4
Nonprescription drugs, two tablets	
Nodoz	200
Vivarin	200
Dexatrim	200
Excedrin	130
Weight-control pills	150

[a]Estimated from company Web site description of "twice the caffeine."

GM Foods: Are They Safe and Acceptable?

People have a fear of eating DNA. We eat 150,000 kilometers [250,000 miles] of DNA in an average meal.
—DAVID COVE, PhD, LEEDS UNIVERSITY'S GENETICS DEPARTMENT

People generally don't think of DNA as a normal component of food, but since we eat cells when we consume foods, we get hefty helpings of DNA and chromosomes in every meal. People have been consuming foods with novel DNA in hybrid plants for many years without ill effects. Obviously, consuming DNA is

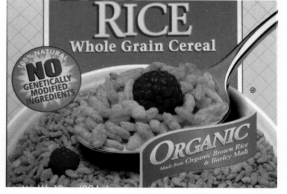

Illustration 21.7
Genetic modification of plants.

Plant, animal, or bacterial chromosome containing desired genes

Desired genes separated from DNA strand

Desired genes

Desired genes incorporated into crop plant DNA and chromosomes

DNA strand separated

DNA strand

Crop plant chromosome

not harmful by itself. Concerns about the safety of GM foods have less to do with consuming them and more to do with their potential impact on people's lives and the environment.

Most Americans are unaware of the broad presence of GM foods in the marketplace. Over 60% of processed foods contain GM ingredients, but only 19% of adults believe they have eaten a GM food.[23,24] Over 80% of all soybeans and a third of corn grown in the United States are from genetically modified seeds. U.S. consumers regularly purchase and eat tomatoes, squash, cantaloupe, and potatoes that have been genetically modified.[23] Although a relatively small percentage of U.S. consumers (27%) are concerned about the safety of GM foods, several baby and snack food producers have excluded them from their products. Individuals preferring GM-free foods in the United States and abroad are opting for organic products (Illustration 21.8) and locally grown produce.[23]

Foods derived from biotechnology are more heavily regulated than any other new foods. They are considered safe to eat, pose little threat of producing allergic reactions, and have real and potential benefits (which are listed in Table 21.4).[23, 24] Important concerns about GM crops and foods exist, and they include the inability of small farmers to afford seed license fees and the fate of locally grown crops (these concerns are also listed in Table 21.4). Serious concerns about GM foods are deeply rooted in some areas of the world. In 1998 the European Union banned GM foods, closing the European market to farmers who produce foods from GM seeds. Reflecting the sentiment of most of the people in countries of the European Union, the ban will stay in effect until products with GM ingredients are labeled as containing genetically modified ingredients and the origins of GM ingredients are stated on packages.[24]

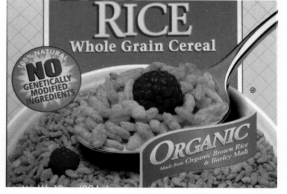

Illustration 21.8
Some organic products in England sport a GM-free label.

Reality Check

Eating Genes

Is it safe to eat genes?

Who gets thumbs up?
Answers on next page

Photo Disc

Hyde:
I have never eaten a gene in my life, and I never will.

Ejay:
I had a huge salad for lunch yesterday and I haven't turned into a head of lettuce yet.

TABLE 21.4

Benefits and concerns related to GM crops and foods[23, 24]

Real and Potential Benefits	**Real and Potential Concerns**
• Reduced herbicide and pesticide use • Increased availability of crops that can grow in dry or salty soils and in hot or cold climates • Increased crop yield • Decreased waste due to spoilage • Improved nutritional content of foods • Improved food flavors • Source of antigens that could help people become immune to food allergens	• GM crop traits may cross-pollinate with local crops and reduce the market for locally grown foods. • GM crop seeds are relatively expensive and perhaps too expensive for small farmers. • Europe does not allow the importation of GM foods. • Nutritionally superior GM foods have yet to reach consumers. • Insects may become resistant to genetic modifications of plants intended to keep them away. • Foods produced may be unacceptable to the consumers they are intended to benefit. • Long-term effects of GM organisms on local plants and crops, insects, and animals are unknown. • A belief by some that humans should not mess with nature by changing the genetic makeup of life-forms.

Leaders in Zimbabwe refused GM corn offered as food aid by the United Nations during a severe famine in 2002. Officials were worried that farmers would plant the corn kernels or feed it to cattle, which would mean neither the crops nor the meat could be sold in Europe. Ground corn was eventually accepted to help avert the famine. The nearby countries of Zambia and Mozambique faced similar famine and refused to accept GM foods.[25]

The Future of GM Foods: The Plot Thickens

> *People want the GM label. Lack of labeling makes people think there is something to hide.*[24]
> —SHARON PALMER, R.D.

Genetically engineered foods remain a controversial subject, and some experts believe that the future of GM crops is at a crossroads. To date, most genetic modifications of crops have benefited producers and not consumers.[21] Sentiment against GM foods is growing in Europe, and people in many countries want GM foods labeled.[24] It is not clear if consumers will accept foods genetically modified to deliver specific nutrients. Populations in need of additional vitamin A may reject yellow rice, preferring short-grained, long-grained, sticky, or dry white rice.[26] GM foods that offer nutritional advantages are not expected to reach the market for several years.[21] The future of GM foods appears to depend on resolution of issues related to cross-pollination and alteration of existing crops, the viability of small farms, consumer benefits, and other concerns. It will take years of experience with GM crops and food to adequately address these issues.[24]

ANSWERS TO *Reality Check*

Eating Genes

Genes in food are primarily protein and don't make it through the digestive tract. Characteristics transferred to foods by genes are not acquired by humans.

Hyde 👎

Ejay 👍

Photo Disc

Nutrition Up Close

Have You Had Your Phytochemicals Today?

FOCAL POINT: Consuming good sources of the "other" beneficial components of food.

Good food sources of beneficial phytochemicals are listed below. Indicate foods you consumed at least twice last week, foods you like, and those you have never tried.

	Foods Eaten at Least Twice Last Week	Foods You Like	Foods You've Never Tried
Broccoli			
Cabbage			
Carrots			
Cauliflower			
Celery			
Collard greens			
Garlic			
Onions			
Tomatoes			
Turnips			
Apple/juice			
Orange/juice			
Grapefruit/juice			
Grapes/juice			
Strawberries			
Brown rice			
Dried beans			
Tofu			
Whole-grain bread, cereal			

FEEDBACK (answers to these questions) can be found at the end of Unit 21.

Key Terms

age-related macular degeneration,
   page 21-4
biotechnology, page 21-8

cataracts, page 21-4
cruciferous vegetables, page 21-5
phytochemicals, page 21-2

zoochemicals, page 21-2

Review Questions

                      True   False

1. Phytochemicals that benefit health are found in high amounts in fish and organ meats. ☐ ☐

2. Some plant foods that contain relatively low amounts of vitamins and minerals are rich sources of beneficial phytochemicals. ☐ ☐

3. Individual supplements of lutein, lycopene, and zeaxanthin more effectively prevent disorders such as macular degeneration and prostate cancer than do foods that naturally contain these phytochemicals. ☐ ☐

4. The top two food sources of antioxidants are dark chocolate and brewed coffee. ☐ ☐

5. The best way to achieve the health benefits of phytochemicals is through the consumption of five to nine servings of vegetables and fruits daily. ☐ ☐

                      True   False

6. If a chemical substance occurs naturally in a plant food, it can be considered harmless to health. ☐ ☐

7. Most of the health effects related to coffee consumption are due to caffeine. ☐ ☐

8. Genetically modified foods have been found to harm human health. ☐ ☐

9. Disease resistance, improved flavor and nutrient content, and reduced pesticide use are benefits of genetically modified plant foods. ☐ ☐

10. Three potential problems related to genetically modified (GM) foods include the loss of small farms, cross-pollination with non-GM plants, and lack of consumer acceptance of these foods. ☐ ☐

Media Menu

www.ers.usda.gov
Receive updates on biotechnology and GM foods by searching these terms.

http://5aday.gov
The National Cancer Institute's 5-a-day program site offers encouragement and tips on selecting colorful vegetables and fruit daily for better health.

www.cqs.com/50harm.htm
This site presents the opponents' view of GM foods.

www.nas.edu
Search "food biotechnology" to get reports on results from the National Academy of Science's subcommittee on genetically modified plants and animals.

Notes

1. Halvorsen BL et al. Content of redox-active compounds (ie, antioxidants) in foods consumed in the United States. Am J Clin Nutr 2006;84:95–135.

2. Thompson HJ, et al. Dietary botanical diversity affects the reduction of oxidative biomarkers in women due to high vegetable and fruit intake. J Nutr 2006;136:2207–12.

3. Kelloff GJ et al., Progress in cancer chemoprevention: development of diet-derived chemopreventive agents, J Nutr 2000;130:467S–71S.

4. Gross M, Flavonoids, platelets, and the risk of cardiovascular disease. Epidemiology Seminar, Minneapolis, 1999 Nov. 10.

5. Craig WT. Phytochemicals: guardians of our health. J Am Diet Assoc 1997;97 (suppl 2):S199–S204.

6. Whitehead AJ, et al. Macular pigment: a review of current knowledge. Arch Ophthalmol 2006;124:1038–45.

7. Hu F et al. Long-term coffee consumption related to reduced risk for type 2 diabetes. Harvard School of Public Health Press release, Jan 5, 2004, www.hsph.harvard.edu/press/releases/press01052004.html. accessed 9/06.

8. Knekt P, Jarvinen R et al. Dietary flavonoids and the risk of lung cancer and other malignant neoplasms. Am J Epidemiol 1997;146:223–30.

9. Johnston CS, Corte C. People with marginal vitamin C status are at high risk of developing vitamin C deficiency. J Am Diet Assoc 1999;99:854–6.

10. Herber D. The stinking rose: organosulfur compounds and cancer. M J Clin Nutr 1997;66:425–6.

11. Hercberg S. The history of beta-carotene and cancers: from observational to intervention studies. What lessons can be drawn for future research on polyphenols? Am J Clin Nutr 2005;81(suppl):218S–22S.

12. Tylleskar T, Banea M et al. Dietary determinants of a non-progressive spastic paparesis (konzo). Int'l J Epidemiol 1995;24:949–56.

13. Holson D. Toxicity, plants: ackee fruit. www.emedicine.com, accessed 9/02.

14. Shirlow MJ, Mathers CD. A study of caffeine consumption and symptoms: indigestion, palpitations, tremor, headache, and insomnia. Int'l J Epidemiol 1985;14:239–48.

15. Grunebaum, A. Coffee and pregnancy: a bad mix? www.webmd.com/content/article/41/3606_696.htm, accessed 9/06.

16. Katan MB et al. Caffeine and arrhythmia. Am J Clin Nutr 2005;81:539–40.

17. Lopez-Garcia E et al. Coffee consumption and the risk for heart disease. Circulation 2006;April 24 (published online before print version).

18. Paynter NP, et al. Coffee and sweetened beverage consumption and the risk of type 2 diabetes mellitus. Am J Epidemiol. 2006 Sep 18 (released ahead of print publication).

19. Silverman K, Evans SM et al. Withdrawal syndrome after the double-blind cessation of caffeine consumption. N Engl J Med 1992;327:1109–14.

20. Babcock BC, Francis CA. Solving global nutrition challenges requires more than new biotechnologies. J Am Diet Assoc 2000;100:1308–11.

21. Falk MC et al. Food biotechnology: benefits and concerns. J Nutr 2002;132:1384–90.

22. Lewis C, A new kind of fish story: the coming of biotech animals. FDA Consumer, Jan.–Feb. 2001.

23. Position of the American Dietetic Association: agricultural and food biotechnology. J Am Diet Assoc 2006;106:285–93.

24. Palmer S. GE foods under the microscope. Today's Dietitian 2005; May: 34–9.

25. Genetic food fight, Wall Street Journal; and Chege W. Africa mulls gene modified food as hunger claws. World Food Program press release, www.medscape.com, accessed 7/02.

26. Merrigan KA et al. Executive summary. Nutr Rev 2003;61:S95–S100.

Answers to Review Questions

1. False, see page 21-2.
2. True, see page 21-4.
3. False, see page 21-4.
4. False, see Table 21.2.
5. True, see page 21-6.
6. False, see page 21-6.
7. False, see page 21-8.
8. False, see page 21-10.
9. True, see page 21-8 and Table 21.4.
10. True, see Table 21-4, pages 21-10, 21-11.

Nutrition Up Close

Have You Had Your Phytochemicals Today?

Feedback for Unit 21

Did you eat four or more of the foods listed at least twice last week? If yes, listen carefully and you'll hear your cells say "thank you."

If you didn't eat four or more twice last week, go for the foods you didn't eat but like. Be adventurous! Try some of the phytochemical-rich foods you have never eaten before.

UNIT 22

Diet and Cancer

NUTRITION SCOREBOARD

	TRUE	FALSE
1 Some types of cancer are contagious.		
2 Cancer is primarily an inherited disease.		
3 People who eat plenty of fruits and vegetables are less likely to get cancer than people who don't.		
4 High levels of body fat contribute to the development of many types of cancer.		
5 Two out of three Americans never get cancer.		

Key Concepts and Facts

- Cancer has many different causes. Diet is a major factor that influences the development of most types of cancer.

- Diets primarily based on plant foods that include lean meats, fish, and low-fat dairy products; regular physical activity; and normal levels of body fat reduce cancer risk.

- Cancer is largely preventable, but there are no absolute guarantees that an individual will not develop cancer.

Answers to
NUTRITION SCOREBOARD

		TRUE	FALSE
1	Cancer doesn't spread from person to person.		✔
2	Cancer is primarily related to environmental factors such as diet, smoking, and overexposure to the sun. Genetic background plays a role by placing some individuals at increased risk for developing cancer.		✔
3	Among the arsenal of cancer-protective measures, the consumption of five or more servings of fruits and vegetables a day stands out.	✔	
4	High levels of body fat *are* related to the development of many types of cancer.	✔	
5	Statistics from the National Cancer Institute demonstrate that most people don't get cancer.[1]	✔	

What Is Cancer?

cancer
A group of diseases in which abnormal cells grow out of control and can spread throughout the body. Cancer is not contagious and has many causes.

Cancer, the second leading cause of death in the United States, is a group of conditions that result from the uncontrolled growth of abnormal cells. Although these cells can begin to grow in any tissue in the body, the lungs, colon, *prostate,* and breasts are the most common sites for cancer development (Illustration 22.1).[1] Some forms of cancer are highly curable.

How Does Cancer Develop?

prostate
A gland located above the testicles in males. The prostate secretes a fluid that surrounds sperm.

Cancer develops by complex processes that are not yet fully understood. Adding to the complexity is the fact that cancer development often does not proceed in a straight line—cancer can progress two steps forward and then take a step or two back.

Illustration 22.2 summarizes the processes involved in the development of cancer, as we currently understand them. Cancer begins when something goes wrong within cells that modifies cell division. Every minute, 10 million cells in the body divide. Usually, the cells divide the right way and on schedule due to a set of regulatory mechanisms that control the replication (duplication) of DNA, the genetic material that becomes part of new cells.

initiation phase
The start of the cancer process; it begins with the alteration of DNA within cells.

promotion phase
The period in cancer development when the number of cells with altered DNA increases.

In the *initiation phase* of cancer development, DNA becomes damaged due to the presence in cells of toxic substances. These toxic substances can be generated by exposure to tobacco smoke, obesity, certain viruses, or iodizing radiation; or by other environmental factors. Damaged DNA can be repaired during the initiation phase, and that stops cancer from developing. If DNA is not repaired, the development of cancer continues to the *promotion phase.*[2] During the *promotion* phase of cancer, cells with altered DNA divide, eventually producing large numbers of abnormal cells. This phase of cancer development commonly takes place over a span of 10 to 30 years. Unless they are hindered by the body or corrected by some other means, abnormal cells continue to divide, leading to the *progression* phase of cancer development.

The progression phase of cancer development is marked by loss of control over the abnormal cells, and their numbers increase rapidly. Eventually, the cells become so numerous that they erode the normal functions of the tissue where they are growing. During this phase, the abnormal cells may migrate to other tissues and cause DNA damage and abnormal cell development in these tissues, too.[3]

Illustration 22.1
Percentage of cancer deaths by selected sites and sex, 2006.
In 2005, 23% of deaths in the United States were due to cancer.[1]

What Causes Cancer?

About 80 to 90% of all cancers are related to environmental factors that modify the structure and function of DNA.[3] Many environmental factors appear to play a role in cancer development, and most are modifiable. The top nine modifiable risk factors for cancer development worldwide are listed in Table 22.1.

Diet is one of the major environmental factors, and it accounts for 40% of cancer risk.[4] Some of the most convincing evidence of the robust relationship between environmental factors and cancer come from studies of cancer rates in people migrating to another country. Rates of breast cancer, for example, are low

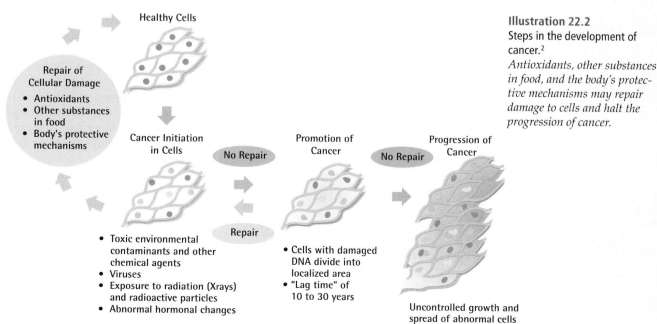

Illustration 22.2
Steps in the development of cancer.[2]
Antioxidants, other substances in food, and the body's protective mechanisms may repair damage to cells and halt the progression of cancer.

The nine leading environmental factors related to cancer development worldwide[9]

1. Obesity
2. Low vegetable and fruit intake
3. Physical inactivity
4. Smoking
5. Excess alcohol intake
6. Unsafe sex
7. Air pollution
8. Indoor use of solid fuels
9. Hepatitis B or C viral infection

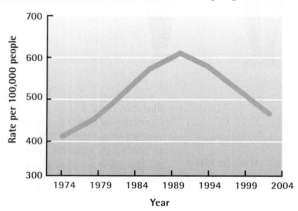

Illustration 22.3

Rates of new cases of cancer in the United States have declined since 1992.[7]

Dietary patterns and lifestyles related to reduced risk of cancer and heart disease[10]

1. Consume a plant-based diet.
   • 5+ servings of vegetables and fruits daily
   • 3+ whole grains/products daily
   • Regular consumption of dried beans, nuts, and seeds
2. Eat foods that are low in saturated fat.
   • Focus on lean meats, low-fat dairy products, fish
3. Exclude charred or nitrate-preserved meats.
4. Exclude smoking.
5. Exclude excess alcohol consumption.
6. Include physical activity.
   • 30 minutes 5+ days a week
7. Lose excess weight.

in rural Asia. When individuals from rural parts of Asia immigrate to the United States, rates of breast cancer become the same or higher than the U.S. rate by the third generation. Rates of prostate cancer similarly increase as people move from countries with low rates to countries with high rates.[4]

Westernization of dietary intake and lifestyles increases the risk of many types of cancer. Rates of breast cancer in Japanese and Alaskan Eskimo women have increased substantially as they have adopted Westernized diets and lifestyles.[5]

Some people have a genetic predisposition toward cancer, which means they have a tendency to develop cancer if regularly exposed to certain substances in the diet or environment. A genetically based susceptibility to cancer can develop in a fetus during pregnancy and in infancy. Exposure to calorie deficits, certain viruses, and specific other substances during these periods of rapid growth and development can modify the function of genes that help protect people from developing cancer.[2] Genetic factors appear to account for 42% of the risk for prostate cancer, 5 to 27% of the risk for breast cancer, and 36% of the risk for pancreatic cancer, for example. Endometrial cancer (cancer of the lining of the uterus), oral, thyroid, and bone cancer do not appear to be caused by genetic factors.[3]

Given the high percentage of cancers related to diet and other environmental factors, cancer is considered a largely preventable disease. Increasing rates of new cases of cancer took a turn for the better after 1992 and correspond to declines in rates of tobacco use (Illustration 22.3)[6] It is anticipated that other improvements in lifestyles and diets will lead to further declines in cancer rates.[1]

Fighting Cancer with a Fork

Leading the list of cancer-promoting diets are those low in vegetables and fruits.[8] High intakes of saturated fat from meats and dairy products, regular intake of charred and nitrate-cured meats, and excessive alcohol consumption are associated with the development of cancer. Low intakes of plant foods such as whole grains, dried beans, nuts, and seeds have been found to increase cancer risk. Other, major risk factors for many types of cancer include smoking, physical inactivity, and excess body fat.[9] Table 22.2 summarizes characteristics of diets and lifestyle that are related to a reduced risk of cancer.

Frequent consumption of certain types of food is sometimes more strongly related to particular cancers than to other types. For example, regular consumption of tomato products is related in particular to decreased risk of prostate cancer, and regular intake of black and green tea appears to contribute to breast and ovarian cancer reduction.[11] Diets and lifestyles that best prevent cancer are represented by a set of characteristics and not by hard rules about specific foods, dietary restrictions, or types of physical activities.[12]

Dietary Risk Factors for Cancer: A Closer Look

Foods contain a variety of vitamins and minerals, as well as fiber and phytochemicals that participate in protecting the body against cancer. These substances in food, particularly plant foods, appear to work together in ways that confer the protection. Attempts to prevent cancer by giving large groups of people individual components of plants that may biologically explain the plant's beneficial

Kumquats and Cancer

"Kumquats Prevent Cancer" reads the headline in the paper. Researchers speculate that "tartaphil," the substance that makes kumquats tangy, may be the reason. Should you eat kumquats?

Who gets thumbs up?
Answers on page 22-7

Sebastian: Kumquats are too sour. I'd take a tartaphil supplement though.

Flora: I already eat about three fruits every day. I'll include kumquats.

effects have not been successful. Particular types of food clearly provide greater levels of protection against cancer than extracts of specific substances in food or other types of food.[13]

Fruits and Vegetables and Cancer People who regularly consume plenty of vegetables and fruits (five or more servings daily) have a lower risk of developing a number of types of cancer than people who eat fewer servings.[8] Because cancer incidence continues to decline as intake of vegetables and fruits increases, some experts are advising people to consume five to nine servings daily.[14] It is not yet clear why fruits and vegetables exert this effect, but a number of possibilities exist.

One possibility is the contribution of the antioxidants vitamin C, beta-carotene, vitamin E, and selenium in vegetables and fruits. Much of the damage to DNA related to cancer initiation is thought to be caused by exposure to oxygen and other oxidizing substances that disrupt molecules within DNA; antioxidants are able to repair damaged DNA.[13] Phytochemicals in vegetables and fruits appear to participate in cancer prevention by protecting cells from damage due to oxidation and by inhibiting the multiplication of abnormal cells.

Consuming three servings per week of vegetables from the cruciferous family (broccoli, cabbage, cauliflower, and brussels sprouts) substantially reduces the risk of lung, bladder, stomach, and prostate cancer (Illustration 22.4).[15] If you enjoy eating these vegetables, consider yourself lucky.

Color-Coding Vegetable and Fruit Choices Many of the phytochemicals participating in mechanisms that likely help to prevent cancer can be identified by their color. This feature of some phytochemicals, and the advantages of eating a variety of vegetables and fruits, has led to the advice to select and consume colorful vegetables and fruits daily.[16] Table 22.3 lists phytochemicals, the color they impart to mature vegetables and fruits, and their sources. The phytochemicals listed all act as antioxidants.

Whole Grains and Cancer Whole grains contain vitamins, minerals, unsaturated fatty acids, fiber, and phytochemicals that play roles in cancer prevention.[25] Effects of whole grains and whole-grain products on cancer risk are related to the combined action of these substances. If you isolate a single substance, you in effect are destroying its ability to function in cancer prevention. Americans are advised to include three or more whole-grain products in their daily diet.[17]

Saturated Fat and Cancer Promotion High intake of saturated fats from meats and dairy products increases the risk of cancer.[18] Consequently, recommenda-

Powerful anti-cancer ingredient is found in broccoli

Illustration 22.4
An example of headline coverage of the cancer-protective effects of vegetables.

TABLE 22.3

Color-Coding Your Vegetables and Fruits

Color	Phytochemical Antioxidant	Vegetable and Fruit Sources
Red	lycopene (*lie-co-peen*)	tomatoes, red raspberries, watermelon, strawberries, red peppers
Yellow-green	lutein zeaxanthin (*lou-te-in, ze-ah-zan-thun*)	leafy greens, avocado, honeydew melon, kiwi fruit
Red-purple	anthocyanins (*antho-sigh-ah-nin*)	grapes, berries, wine, red apples, plums, prunes
Orange	beta-carotene	carrots, mangos, papayas, apricots, pumpkins, yams
Orange-yellow	flavonoids (*fla-von-oids*)	oranges, tangerines, lemons, plums, peaches, cantaloupe
Green	glucosinolates (*glu-co-sin-oh-lates*)	broccoli, brussels sprouts, kale, cabbage
White-green	allyl sulfides (*al-lill sulf-ides*)	onions, leeks, garlic

PhotoEdit, David Young-Wolff

Illustration 22.5
The charred and black, oily coating on grilled or broiled meats is the part you shouldn't eat.

tions for diet and cancer prevention include the use of plant sources of protein because they provide unsaturated fats. There are additional advantages to increased reliance on plant foods in the diet. Plant foods such as dried beans, soy products, nuts, and seeds also provide vitamins, minerals, fiber, and phytochemicals that help ward off cancer progression. Fish are a recommended source of protein, too. Regular consumption of fish is related to lower rates of cancer.[19]

Nitrate-Preserved and Grilled Meats and Cancer Cancer of the stomach and liver appears to be related to the regular consumption of foods such as hot dogs, luncheon meats, bacon, and pickled eggs and vegetables preserved with nitrates.[20] (You can identify foods that contain nitrates by examining food ingredient labels.) Most cases of cancer associated with nitrate use in smoked, salted, and pickled foods now occur in China, parts of the former Soviet Union, and Central and South America, where such foods are commonly consumed.[21]

Certain nitrogen-containing and other substances in beef, chicken, fish, and other meats may become cancer promoting if heated to a high temperature.[22] Such high temperatures can be reached by broiling and grilling food. The potentially cancer-promoting substances formed by meats are found in the charred portions of the meat and the fatty coating that accumulates on meat when fat drips into the heat source and smokes (Illustration 22.5).[23]

The Department of Agriculture concludes that eating broiled or grilled foods several times a week poses no threat to health. People should be careful not to eat the charred portions of meat and to wipe or rinse off any coating on meat that results from fatty flare-ups on the grill.

Diet and Cancer Guidelines

Dietary patterns and lifestyle changes that reduce the risk of cancer (Table 22.1) are similar to, and compatible with, dietary recommendations to reduce the risk of heart disease.[20] Considered together, dietary recommendations for cancer prevention can be transferred to dietary intake by selection of the types of foods shown in the example menu in Table 22.4.

Alcohol Intake and Cancer Consumption of excessive amounts of alcoholic beverages has been linked to cancers of the breast, mouth, throat, and liver.[24] The risk of developing cancers of the mouth and throat increases for people who smoke cigarettes or chew tobacco as well as drink.[3]

Excess Body Fat and Cancer Obesity, and in particular obesity characterized by excess stores of central body fat, increases the risk of cancer at several sites. High levels of central fat appear to alter metabolism of hormones such as estrogen, testosterone, and insulin in ways that promote the growth of abnormal cells. Calorie-reduced diets combined with at least 30 minutes of moderate or vigorous activity five or more days a week are recommended for cancer prevention.[25]

Bogus Cancer Treatments

Unorthodox, purported cancer cures such as macrobiotic diets; hydrogen peroxide ingestion; laetrile tablets; vitamin, mineral, herbal, and phytochemical supplements; and animal gland therapy have not been shown to be effective treatments for cancer. Such remedies have been promoted since the early 1900s. They still exist because, although not proven to work, they offer some cancer patients their last ray of hope. They should not be used as a substitute for conventional cancer treatments.[26]

Eating to Beat the Odds

Two out of every three people in the United States do not develop cancer. You can likely improve your odds of being among the two by not smoking or drinking excessively, by regularly consuming five or more servings of fruits and vegetables each day, by sticking to a low-saturated-fat diet, and by being physically active and achieving or maintaining a normal level of body fat. Although there are no guarantees, people can help themselves prevent cancer by adhering to a good diet and a healthy lifestyle.

TABLE 22.4

Types of foods included in dietary patterns that lower the risk of cancer and heart disease

Breakfast

Oat flakes with banana and skim/low-fat milk
Cracked wheat toast with soft margarine and grape jam
Orange juice
Tea/coffee

Lunch

Skinless chicken sandwich on whole-grain bread with mozzarella cheese
Broccoli, cauliflower, raisin salad
Carrot sticks
Skim or low-fat milk

Dinner

Tomato soup
Broiled tuna with olive oil–lemon sauce
Baked beans
Coleslaw
Ice milk with strawberries
Skim or low-fat milk, coffee, tea, 1 glass wine

Snacks

Nuts, seeds
Yogurt
Raw vegetables
Fruits
Skim or low-fat milk

ANSWERS TO *Reality Check*

Kumquats and Cancer

Phytochemical supplements do not appear to prevent cancer. Diets that help prevent cancer are not based on one or a few foods. They are based on day-to-day intake of a healthy array of foods. P.S. The story about kumquats is bogus.

Sebastian: Flora

Nutrition Up Close

A Cancer Risk Checkup

Focal Point: Reducing cancer risk.

A number of behaviors that help protect people from developing cancer are listed below. Check those that apply to you.

	Yes	No	Don't Know
1. I eat five or more servings of vegetables and fruits daily.	_____	_____	_____
2. I consume whole-grain products.	_____	_____	_____
3. I eat a high-fat diet.	_____	_____	_____
4. I smoke or chew tobacco.	_____	_____	_____
5. I exercise regularly.	_____	_____	_____
6. I have too much body fat.	_____	_____	_____

FEEDBACK (answers to these questions) can be found at the end of Unit 22.

Key Terms

cancer, page 22–2
initiation phase, page 22–2

promotion phase, page 22–2
prostate, page 22–2

Review Questions

	True	False
1. Cancer development is basically related to changes in the structure of DNA.	☐	☐
2. Little can be done to prevent cancer.	☐	☐
3. Most cases of cancer are caused by inherited genetic traits.	☐	☐
4. Diets providing five or more servings per day of vegetables and fruits reduce the risk of cancer.	☐	☐
5. Regular consumption of charred meat and high-saturated-fat foods increases the risk of cancer.	☐	☐

	True	False
6. Tomato sauce and red berries are a good source of lutein, a phytochemical that gives plant foods a red color.	☐	☐
7. Regular consumption of vegetables in the crucifer family reduces the risk of cancer of the lung and stomach.	☐	☐
8. Obesity has *not* been found to be related to the development of cancer.	☐	☐

Media Menu

**www.nlm.nih.gov/medlineplus/
cancergeneral.html**
This address sends you to summarized, scientific information about cancer. Changing the term *cancergeneral* to *breastcancer, coloncancer, alternativemedicine,* and so on takes you to related topics.

www.cancer.gov
The National Cancer Institute's gateway to information about cancer can be found here.

www.cdc.gov/nchs/hus.htm
This site contains the report "Health, United

States." Updated yearly, this report provides statistics on cancer and other disease rates.

www.nci.nih.gov
The Web site for the National Cancer Institute provides evidence-based information about cancer prevention and treatment and research updates.

www.cancer.org
The American Cancer Society's site provides a wealth of information on cancer clinics, treatments, prevention, and statistics.

www.healthfinder.gov
This is the search engine for reliable sources of information on cancer and other health concerns.

www.nal.usda.gov/fnic/
Search "cancer" and find recent results related to diet, nutrition, and cancer development.

Notes

1. Cancer Facts and Statistics, 2006, American Cancer Association, www .cancer.org, accessed 9/06.

2. Jackson AA. Integrating the ideas of life across cellular, individual, and population levels in cancer causation. J Nutr 2006;135:2927S–33S.

3. Lichtenstein P et al. Environmental and heritable factors in the causation of cancer. N Engl J Med 2000;343:78–85.

4. Hoover RN, Cancer-nature, nurture, or both, N Eng J Med 2000;343:135–6.

5. Cowing BC, Saker KE. Polyunsaturated fatty acids and epidermal growth factor receptor. J Nutr 2001;131:1125–8.

6. Cancer facts (www.cancer.org/down loads/STT/CancerFacts&Figures2002 TM.pdf); and Cancer statistics (http:// progressreport.cancer.gov).

7. Death rate from all causes combined fell 32% in 30 years, USA, www .medicalnewstoday.com/medicalnews .php?newsid=30608, accessed 9/06.

8. Pavia M et al. Association between fruit and vegetable consumption and oral cancer: a meta-analysis of observational studies. Am J Clin Nutr 2006;83:1126–34.

9. Danaei G et al. One-third of cancer deaths may be attributable to nine modifiable risk factors. Lancet 2005;366:1784–93.

10. Go VLW et al., Diet, nutrition and cancer prevention, Where do we go from here? J Nutr 2001;131:3121S–26S.

11. Larsson SC et al. Tea consumption and the risk for ovarian cancer. Arch Intern Med 2005;165:2683–6.

12. Campbell TC, Critique of Report on "Food, nutrition and the prevention of cancer: a global perspective," Nutr Today 2001;36:80–4.

13. Hercberg S. The history of beta-carotene and cancers: from observational to intervention studies. What lessons can be drawn for future research on polyphenols? Am J Clin Nutr 2005;81(suppl):218S–22S.

14. Herber D, Bowerman S. Applying science to changing dietary patterns. J Nutr 2001;131:3078S–81S.

15. Kelloff GJ et al. Progress in cancer chemoprevention: development of diet-derived chemopreventive agents. J Nutr 2000;130:467S–71S.

16. Gerber M. The comprehensive approach to diet: a critical review. J Nutr 2001;131:3051S–5S.

17. Adams SM et al. What should we eat? Evidence from observational studies. South Med J 2006;99:744–8.

18. Stoeckli R et al. Nutritional fats and the risk of type 2 diabetes and cancer. Physiol Behav 2004;83:611–5.

19. Tavani A et al. Fish, omega-3 poly-unsaturated fat intake and cancer at select sites. World Rev Nutr Diet 2005;94:166–75.

20. Byers T et al. Guidelines on nutrition and physical activity for cancer prevention: reducing the risk of cancer with healthy food choices and physical activity CA Cancer J 2002;52:92–119.

21. Food, nutrition and the prevention of cander: a global perspective. Washington, DC. World Cancer Research Fund, American Institute for Cancer Research; 1997.

22. Engel LS et al. Risk factors for esophageal and gastric cancer. J Natl Cancer Inst 2003;95:1404–13.

23. de Verdier MG et al. Meat, cooking methods and colorectal cancer. Int'l J Cancer 1991;49:520–5.

24. Mizoue T et al. Alcohol drinking and colorectal cancer risk. Jpn J Clin Oncol 2006;36:682–97.

25. Friendenreich CM et al., Physical activity and cancer prevention: etiologic evidence and biological mechanisms, J Nutr 2002;132:3456S–64S.

26. Straus SE. Complementary and alternative medicine and cancer: what you should know, www. peoplelivingwith cancer.org,accessed 10/03.

A Cancer Risk Checkup

Feedback for Unit 22

Answers to Review Questions

1. True, see page 22-2.
2. False, see page 22-3.
3. False, see page 22-4.
4. True, see page 22-4.
5. True, see pages 22-4, 22-6.
6. False see page 22-6.
7. True, see page 22-5.
8. False, see pages 22-4 (Table 22.1), 22-7.

Best answers:

1. Yes   4. No
2. Yes   5. Yes
3. No   6. No

If you responded "Don't Know" to statements 1, 2, or 3, analyze your diet using the Diet Analysis Plus program. "Regular exercise" in statement 5 means exercising 20 to 30 minutes three times per week. If you didn't know the answer to statement 6, can you pinch an inch of fat above your ribs while you are standing? If yes, you probably have too much body fat.

Good Things to Know about Minerals

NUTRITION SCOREBOARD

	TRUE	FALSE
1 The sole function of minerals is to serve as a component of body structures such as bones, teeth, and hair.		
2 Bones continue to grow and mineralize through the first 30 years of life.		
3 Ounce for ounce, spinach provides more iron than beef does.		
4 Worldwide, the most common nutritional deficiency is iron deficiency.		
5 More than 1 in 4 Americans have hypertension.		

Key Concepts and Facts

- Minerals are single atoms that cannot be created or destroyed by the human body or by any other ordinary means.

- Minerals serve as components of body structures and play key roles in the regulation of body processes.

- Deficiency diseases occur when too little of any of the 15 essential minerals are provided to the body, and overdose reactions occur when too much is provided.

- Inadequate intakes of certain minerals are associated with the development of chronic disorders, including osteoporosis, iron deficiency, and hypertension.

Answers to **NUTRITION SCOREBOARD**	TRUE	FALSE
1 Minerals serve as structural components of the body, but they also play important roles in stimulating muscle and nerve activity and in other functions.		✔
2 Bones continue to grow and mineralize well after we reach adult height. Bone growth and development can continue to about age 30.	✔	
3 Spinach is a nutritious food, providing 0.6 mg iron per ounce. It contains less iron than beef (1 mg iron per ounce). The iron in spinach is poorly absorbed, but the iron in meat is well absorbed by the body. (When several hundred college students were asked this question, 82% voted for spinach.)[1]		✔
4 Iron deficiency is the most common nutritional deficiency in both developed and developing countries. Approximately one-third of the world's population is iron deficient.[2]	✔	
5 An estimated 9% of Americans have hypertension (high blood pressure).[3]		✔

Mineral Facts

What substances are neither animal nor vegetable in origin, cannot be created or destroyed by living organisms (or by any other ordinary means), and provide the raw materials from which all things on Earth are made? The answer is the *mineral* elements, and they are displayed in full in the periodic table presented in Illustration 23.1. Minerals considered "essential," or required in the diet, are highlighted.

The body contains 40 or more minerals. Only 15 are an essential part of our diets; we obtain the others through the air we breathe or from other essential nutrients in the diet such as protein and vitamins.

Minerals are unlike the other essential nutrients in that they consist of single atoms. A single atom of a mineral typically does not have an equal number of protons (particles that carry a positive charge) and electrons (particles that carry a negative charge), and it therefore carries a charge. The charge makes minerals reactive. Many of the functions of minerals in the body are related to this property.

Getting a Charge out of Minerals

The charge carried by minerals allows them to combine with other minerals of the opposite charge and form fairly stable compounds that become part of bones, teeth, cartilage, and other tissues. In body fluids, charged minerals serve as a

minerals
In the context of nutrition, minerals are specific, single atoms that perform particular functions in the body. There are 15 essential minerals—or minerals required in the diet.

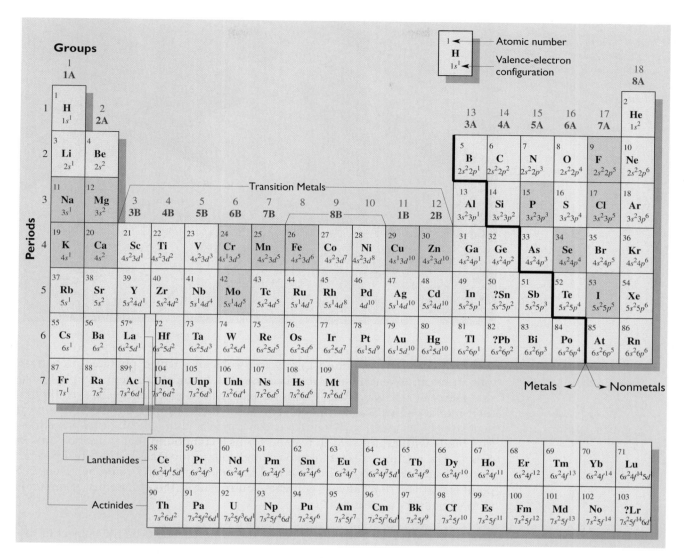

Illustration 23.1
The periodic table lists all known minerals. The highlighted minerals are required in the human diet.

cofactors
Individual minerals required for the activity of certain proteins. For example:

• Iron is needed for hemoglobin's function in oxygen and carbon dioxide transport.
• Zinc is needed to activate or is a structural component of over 200 enzymes.
• Magnesium activates over 300 enzymes involved in the formation of energy and proteins.

source of electrical power that stimulates muscles to contract and nerves to react. The electrical current generated by charged minerals when performing these functions can be recorded by an electrocardiogram (abbreviated EKG or ECG) or an electroencephalogram (EEG). Abnormalities in the pattern of electrical activity in EKGs signal pending or past problems in the heart muscle. An EKG recording is shown in Illustration 23.2. Electroencephalograms similarly record electrical activity in the brain.

The charge minerals carry is related to many other functions. It helps maintain an adequate amount of water in the body and assists in neutralizing body fluids when they become too acidic or basic. Minerals that perform the roles of *cofactors* are components of proteins and enzymes, and they provide the "spark" that initiates enzyme activity.

Charge Problems Because minerals tend to be reactive, they may combine with other substances in food and form highly stable compounds that are not easily absorbed. Absorption of zinc from foods, for example, can vary from zero to 100%, depending on what is attached to it. Zinc in whole-grain products is very poorly

Illustration 23.2

The electrical current measured by an EKG results from the movement of charged minerals across membranes of the muscle cells in the heart.

Index Stock

absorbed because it is bound tightly to a substance called phytate. In contrast, zinc in meats is readily available because it is bound to protein. People whose sole source of zinc was whole grains have developed zinc deficiency, even though their intake of zinc was adequate.[4] The absorption of iron from foods in a meal decreases by as much as 50% if tea is consumed with the meal. In the intestines, iron binds with tannic acid in tea and forms a compound that cannot be broken down.[5] The calcium present in spinach and collard greens is poorly absorbed because it is firmly bound to oxalic acid. Many more examples could be given. The point is that you don't always get what you consume; the availability of minerals in food can vary a great deal.

The Boundaries of This Unit   All of the minerals could be the subject of fascinating stories, but in this unit we will concentrate on just three. In addition, a summary of the main features of all the essential minerals is provided in Table 23.1. This table lists the recommended intake level, functions, consequences of deficiency and overdose, and food sources of each mineral. Table 23.12 at the end of the unit lists food sources for many of the essential minerals.

Minerals highlighted are calcium, iron, and sodium. They have been selected primarily because they play important roles in the development of osteoporosis, iron-deficiency anemia, and hypertension, respectively. These disorders are widespread in the United States and many other countries, and improved diets offer a key to their prevention and treatment.

Selected Minerals: Calcium

What you've heard about calcium is true: It's good for bones and teeth. About 99% of the 3 pounds of calcium in the body is located in bones and teeth. The remaining 1% is found in blood and other body fluids. We don't hear so much about this 1%, but it's very active. Every time a muscle contracts, a nerve sends out a signal, or blood clots to stop a bleeding wound, calcium in body fluids is involved. Calcium's most publicized function, however, is its role in bone formation and the prevention of osteoporosis.

A Short Primer on Bones

Most of the bones we see or study are hard and dead. As a result, people often have the impression that bones in living bodies are that way. Nothing could be further from the truth. The 206 bones in our bodies are slightly flexible living tissues infiltrated by blood vessels, nerves, and cells.

The solid parts of bones consist of networks of strong protein fibers (called the "protein matrix") embedded with mineral crystals (Illustration 23.3). Calcium is the most abundant mineral found in bone, but many other minerals, such as phosphorus, magnesium, and carbon, are also incorporated into the protein matrix. The combination of water, the tough protein matrix, and the mineral crystals makes bone very strong yet slightly flexible and capable of absorbing shocks.

Illustration 23.3
(left) Electron micrograph of healthy bone. (right) Electron micrograph of bone affected by osteoporosis.

Reproduced with permission from Dempster et al, J Bone Min. Res 1, 15–21, 1986.

Teeth Are a Type of Bone   Teeth have the same properties as other bone plus a hard outer covering called enamel, which is not infiltrated by blood vessels or nerves. Enamel serves to protect the teeth from destruction by bacteria and mechanical wear and tear.

Remodeling Your Bones   Bones slowly and continually go through a repair and replacement process known, appropriately enough, as *remodeling*. During remodeling, the old protein matrix is replaced and remineralized. If insufficient calcium is available to complete the remineralization, or if other conditions prevent calcium from being incorporated into the protein matrix, *osteoporosis* results.

Osteoporosis

If you are female, you have a one in four chance of developing osteoporosis in your lifetime. If you are a Caucasian, Hispanic, or Asian female, you have a higher risk of developing osteoporosis than if you are an African American woman or a male. If you are male, your chance of developing osteoporosis is one in eight.[6] Approximately 44 million adults in the United States have osteoporosis, and 1.5 million suffer a broken wrist or hip or crushed spinal vertebrae each year because of the disease.[7]

Osteoporosis is a disabling disease that reduces the quality of life and dramatically increases the need for health care.[7] Because the incidence of osteoporosis increases with age, its importance as a personal and public health problem is intensifying as the U.S. population ages. It currently appears, however, that a

remodeling
The breakdown and buildup of bone tissue.

osteoporosis
(*osteo* = bones; *poro* = porous, *osis* = abnormal condition)
A condition characterized by porous bones; it is due to the loss of minerals from the bones.

Calcium supplements are large because our daily need for calcium is high (1000 milligrams/day). Four pills provide 800 milligrams calcium, the amount in 2²/₃ cups of milk. An aspirin is shown for comparison.

Iron-deficiency anemia is characterized by microcytic anemia and small, pale red blood cells (right photo). Normal red blood cells are shown on the left.

CNRI - Martin M. Rotker

TABLE 23.1

An intensive course on essential minerals

		PRIMARY FUNCTIONS	CONSEQUENCES OF DEFICIENCY
Calcium AI[a] women: 1000 mg men: 1000 mg UL: 2500 mg		• Component of bones and teeth • Needed for muscle and nerve activity, blood clotting	• Poorly mineralized, weak bones (osteoporosis) • Rickets in children • Osteomalacia (rickets in adults) • Stunted growth in children • Convulsions, muscle spasms
Phosphorus RDA women: 700 mg men: 700 mg UL: 4000 mg		• Component of bones and teeth • Component of certain enzymes and other substances involved in energy formation • Needed to maintain the right acid–base balance of body fluids	• Loss of appetite • Nausea, vomiting • Weakness • Confusion • Loss of calcium from bones
Magnesium RDA women: 310 mg men: 400 mg UL: 350 mg (from supplements only)		• Component of bones and teeth • Needed for nerve activity • Activates enzymes involved in energy and protein formation	• Stunted growth in children • Weakness • Muscle spasms • Personality changes
Iron RDA women: 18 mg men: 8 mg UL: 45 mg		• Transports oxygen as a component of hemoglobin in red blood cells • Component of myoglobin (a muscle protein) • Needed for certain reactions involving energy formation	• Iron deficiency • Iron-deficiency anemia • Weakness, fatigue • Pale appearance • Reduced attention span and resistance to infection • Mental retardation, developmental delay in children
Zinc RDA women: 8 mg men: 11 mg UL: 40 mg		• Required for the activation of many enzymes involved in the reproduction of proteins • Component of insulin, many enzymes	• Growth failure • Delayed sexual maturation • Slow wound healing • Loss of taste and appetite • In pregnancy, low-birth-weight infants and preterm delivery

[a]AI (Adequate Intakes) ad RDAs (Recommended Dietary Allowances are for 19–30-year-olds: UL (Upper Limits) are for 19–70-year-olds, 1997–2004.

TABLE 23.1

An intensive course on essential minerals (continued)

CONSEQUENCES OF OVERDOSE	PRIMARY FOOD SOURCES	HIGHLIGHTS AND COMMENTS
• Drowsiness • Calcium deposits in kidneys, liver, and other tissues • Suppression of bone remodeling • Decreased zinc absorption	• Milk and milk products (cheese, yogurt) • Broccoli • Dried beans • Calcium-fortified foods (some juices, breakfast cereals, bread, for example)	• The average intake of calcium among U.S. women is approximately 60% of the DRI. • One in four women and one in eight men in the U.S. develop osteoporosis. • Adequate calcium and vitamin D status must be maintained to prevent bone loss.
• Muscle spasms	• Milk and milk products (cheese, yogurt) • Meats • Seeds, nuts • Phosphates added to foods	• Deficiency is generally related to disease processes.
• Diarrhea • Dehydration • Impaired nerve activity due to disrupted utilization of calcium	• Plant foods (dried beans, tofu, peanuts, potatoes, green vegetables) • Milk • Bread • Ready-to-eat cereals • Coffee	• Magnesium is primarily found in plant foods where it is attached to chlorophyll. • Average intake among U.S. adults is below the RDA.
• Hemochromatosis ("iron poisoning") • Vomiting, abdominal pain • Blue coloration of skin • Liver and heart damage, diabetes • Decreased zinc absorption • Atherosclerosis (plaque buildup) in older adults	• Liver, beef, pork • Dried beans • Iron-fortified cereals • Prunes, apricots, raisins • Spinach • Bread • Pasta	• Cooking foods in iron and stainless steel pans increases the iron content of the foods. • Vitamin C, meat, and alcohol increase iron absorption. • Iron deficiency is the most common nutritional deficiency in the world. • Average iron intake of young children and women in the U.S. is low.
• Over 25 mg/day is associated with nausea, vomiting, weakness, fatigue, susceptibility to infection, copper deficiency, and metallic taste in mouth. • Increased blood lipids	• Meats (all kinds) • Grains • Nuts • Milk and milk products (cheese, yogurt) • Ready-to-eat cereals • Bread	• Like iron, zinc is better absorbed from meats than from plants. • Marginal zinc deficiency may be common, especially in children. • Zinc supplements may decrease duration and severity of the common cold.

(continued)

Phosphates are a common, multipurpose food additive.

Magnesium is to chlorophyll in plants as iron is to hemoglobin in humans.

TABLE 23.1

An intensive course on essential minerals (continued)

		PRIMARY FUNCTIONS	CONSEQUENCES OF DEFICIENCY
Fluoride		• Component of bones and teeth (enamel) • Helps rebuild enamel that is beginning to decay	• Tooth decay and other dental diseases
AI women:	3 mg		
men:	4 mg		
UL:	10 mg		
Iodine		• Component of thyroid hormones that help regulate energy production and growth	• Goiter, thyroid disease • Cretinism (mental retardation, hearing loss, growth failure)
RDA women:	150 mcg		
men:	150 mcg		
UL:	1100 mcg		
Selenium		• Acts as an antioxidant in conjunction with vitamin E (protects cells from damage due to exposure to oxygen) • Needed for thyroid hormone production	• Anemia • Muscle pain and tenderness • Keshan disease (heart failure), Kashin-Beck disease (joint disease)
RDA women:	55 mcg		
men:	55 mcg		
UL:	400 mcg		
Copper		• Component of enzymes involved in the body's utilization of iron and oxygen • Functions in growth, immunity, cholesterol and glucose utilization, brain development	• Anemia • Seizures • Nerve and bone abnormalities in children • Growth retardation
RDA women:	900 mcg		
men:	900 mcg		
UL:	10,000 mcg		
Manganese		• Needed for the formation of body fat and bone	• Weight loss • Rash • Nausea and vomiting
AI women:	2.3 mg		
men:	1.8 mg		
Chromium		• Required for the normal utilization of glucose and fat	• Elevated blood glucose and triglyceride levels • Weight loss
AI women:	35 mcg		
men:	25 mcg		
Molybdenum		• Component of enzymes involved in the transfer of oxygen from one molecule to another	• Rapid heartbeat and breathing • Nausea, vomiting • Coma
RDA women:	45 mcg		
men:	45 mcg		
UL:	2000 mcg		

Custom Medical Stock

Goiter is a highly visible sign of iodine deficiency

TABLE 23.1

An intensive course on essential minerals (continued)

CONSEQUENCES OF OVERDOSE	PRIMARY FOOD SOURCES	HIGHLIGHTS AND COMMENTS
• Fluorosis • Brittle bones • Mottled teeth • Nerve abnormalities	• Fluoridated water and foods and beverages made with it • Tea • Shrimp, crab, fish • Bread	• Toothpastes, mouth rinses, and other dental care products may provide fluoride. • Fluoride overdose has been caused by ingestion of fluoridated toothpaste. • Fluoridated water is not related to cancer.
• Over 1 mg/day may produce pimples, goiter, decreased thyroid function, and thyroid disease.	• Iodized salt • Milk and milk products • Seaweed, seafoods • Bread from commercial bakeries	• Iodine deficiency was a major problem in the U.S. in the 1920s and 1930s. Deficiency remains a major health problem in some developing countries. • Amount of iodine in plants depends on iodine content of soil. • Most of the iodine in our diet comes from the incidental addition of iodine to foods from cleaning compounds used by food manufacturers.
• "Selenosis." Symptoms of selenosis are hair and fingernail loss, weakness, liver damage, irritability, and "garlic" or "metallic" breath.	• Meats and seafoods • Eggs • Whole grains	• Content of foods depends on amount of selenium in soil, water, and animal feeds. • May play a role in the prevention of some types of cancer.
• Wilson's disease (excessive accumulation of copper in the liver and kidneys) • Vomiting, diarrhea • Tremors • Liver disease	• Bread • Potatoes • Grains • Dried beans • Nuts and seeds • Seafood • Ready-to-eat cereals	• Toxicity can result from copper pipes and cooking pans. • Average intake in the U.S. is below the RDA.
• Infertility in men • Disruptions in the nervous system (psychotic symptoms) • Muscle spasms	• Whole grains • Coffee, tea • Dried beans • Nuts	• Toxicity is related to overexposure to manganese dust in miners.
• Kidney and skin damage	• Whole grains • Wheat germ • Liver, meat • Beer, wine • Oysters	• Toxicity usually results from exposure in chrome-making industries or overuse of supplements. • Supplements do not build muscle mass or increase endurance.
• Loss of copper from the body • Joint pain • Growth failure • Anemia • Gout	• Dried beans • Grains • Dark green vegetables • Liver • Milk and milk products	• Deficiency is extraordinarily rare.

(continued)

"Mottled teeth" result from an excessive intake of fluoride in children.

Iodine deficiency during pregnancy produces cretinism in the offspring.

TABLE 23.1

An intensive course on essential minerals (continued)

		PRIMARY FUNCTIONS	CONSEQUENCES OF DEFICIENCY
Sodium		• Needed to maintain the right acid–base balance in body fluids • Helps maintain an appropriate amount of water in blood and body tissues • Needed for muscle and nerve activity	• Weakness • Apathy • Poor appetite • Muscle cramps • Headache • Swelling
AI women:	1.5g		
men:	1.5g		
UL:	2.3g		
Potassium		• Same as for sodium	• Weakness • Irritability, mental confusion • Irregular heartbeat • Paralysis
AI women:	4.7g		
men:	4.7g		
Chloride		• Component of hydrochloric acid secreted by the stomach (used in digestion) • Needed to maintain the right acid–base balance of body fluids • Helps maintain an appropriate water balance in the body	• Muscle cramps • Apathy • Poor appetite • Long-term mental retardation in infants
AI women:	2.3g		
men:	2.3g		
UL:	3.5g		

large percentage of the cases of osteoporosis can be prevented. The key to prevention is to build dense bones during childhood and the early adult years and then keep bones dense as you age.[6]

The Timing of Bone Formation   Bones develop and mineralize throughout the first three decades of life. Even after the growth spurt occurs during adolescence and people think they are as tall as they will ever be, bones continue to increase in width and mineral content for 10 to 15 more years. Peak bone density, or the maximal level of mineral content in bones, is reached somewhere between the ages of 30 and 40 years. After that, bone mineral content no longer increases. The higher the peak bone mass, the less likely it is that osteoporosis will develop. People with higher peak bone mass simply have more calcium to lose before bones become weak and fracture easily. That also is the reason males experience osteoporosis less often than females do: They have more bone mass to lose.[8]

Bone size and density often remain fairly stable from age 30 to the mid-40s, but then bones tend to demineralize with increasing age. By the time women are 70, for example, their bones are 30 to 40% less dense on average than they once were.[9] A woman may lose an inch or more in height with age and develop the "dowager's hump" that is characteristic of osteoporosis in the spine (Illustration 23.4).

What has just been described is the way things are in the United States, but this situation is neither normal nor inevitable. The incidence of osteoporosis is far less in many other countries than in the United States.[10] Although genetic and

Illustration 23.4
This woman's stooped appearance is due to osteoporosis.

PhotoEdit—Robert Brenner

TABLE 23.1

An intensive course on essential minerals (continued)

CONSEQUENCES OF OVERDOSE	PRIMARY FOOD SOURCES	HIGHLIGHTS AND COMMENTS
• High blood pressure in susceptible people • Kidney disease • Heart problems	• Foods processed with salt • Cured foods (corned beef, ham, bacon, pickles, sauerkraut) • Table and sea salt • Bread • Milk, cheese • Salad dressing	• Very few foods naturally contain much sodium. • Processed foods are the leading source of dietary sodium. • High-sodium diets are associated with the development of hypertension in "salt-sensitive" people.
• Irregular heartbeat, heart attack	• Plant foods (potatoes, squash, lima beans, tomatoes, plantains, bananas, oranges, avocados) • Meats • Milk and milk products • Coffee	• Content of vegetables is often reduced in processed foods. • Diuretics (water pills) and other antihypertension drugs may deplete potassium. • Salt substitutes often contain potassium.
• Vomiting	• Same as for sodium. (Most of the chloride in our diets comes from salt.)	• Excessive vomiting and diarrhea may cause chloride deficiency. • Legislation regulating the composition of infant formulas was enacted in response to formula-related chloride deficiency and subsequent mental retardation in infants.

other factors play a role, the causes of osteoporosis are mainly related to diet and other lifestyle behaviors. That means that osteoporosis can largely be prevented.

How Do You Build and Maintain Dense Bones?   Since the late 1980s, a rash of studies have examined the relationships among dietary calcium intake, vitamin D status, and bone density. Results have been quite consistent: Bone mass before the age of 30 is increased by calcium intakes of approximately 1000 milligrams per day along with adequate vitamin D status. In adults over the age of 50 years, bone density tends to be preserved with calcium intakes around 1200 milligrams per day and 400 to 800 IU of vitamin D.[6] In addition, calcium intakes of 1200 milligrams or more per day with adequate vitamin D status decrease the incidence of stress fractures in athletes.[11] Low bone mineral density can be increased somewhat by switching from a low to an adequate intake of calcium and improved vitamin D status. The earlier in life this occurs, the more improvement in bone density is noted. High levels of bone density once achieved do not last forever; bone must be continually renewed by an adequate supply of calcium and vitamin D.[12]

Vitamin D is important for building dense bones because it increases calcium absorption and the deposition of calcium into bone. Our need for vitamin D can be met from vitamin D–fortified milk, vitamin D–fortified breakfast cereals, and seafoods or, preferably, by exposing the skin to the sun. A precursor of vitamin D lies beneath our skin. When skin is exposed to ultraviolet rays from sunlight, this precursor is converted to vitamin D. Exposing the hands and face to direct sunshine for 5 to 15 minutes produces a day's supply of vitamin D. (Windows and sunscreen block ultraviolet rays and prevent the production of vitamin D in the skin.) The "Health Action" on the next page summarizes behaviors that help build dense bones and prevent osteoporosis.

■ **The Current D Controversy** Sun exposure produces much more vitamin D than is consumed in most diets (see the "On the Side" feature), and levels of vitamin D higher than 800 IU per day may be needed for optimal protection against osteoporosis. Currently scientific thinking around vitamin D requirements and the prevention and treatment of osteoporosis are heading in the "we need more vitamin D" direction. Recommendations for daily doses of vitamin D may increase in the future to 1000 IU (25 mcg) per day. Additional safety studies of this dose level are needed before the current recommendation could change.[13]

We Are Consuming Too Little Calcium Unfortunately, about half of U.S. women belong to the "600 Club"; they consume 600 milligrams of calcium per day or less.[8] Only 14% of girls and 36% of boys aged 12 to 19 years consume the recommended amount of calcium during these critical years for building bone density. Low calcium intake during the growing years increases the probability that fractures and osteoporosis will occur at younger ages than is the norm. Men tend to consume more calcium than women but on average still fall short of the recommended intake.[14]

■ **Why Do Women Consume Too Little Calcium?** Women consume too little calcium for several reasons. Although dairy products such as milk, cheese, and yogurt are our richest sources of calcium, some women consider them "fattening" and give them up during the teen years. Skipping breakfast and eating out are common among teens and tend to reduce calcium intake. Many think that milk is a food that only children need, while others stop drinking milk because it "gives them gas." (Such people may be lactose intolerant.)[15]

Increasing intakes of soft drinks, combined with reduced consumption of milk, also appear to be contributing to bone fractures and future osteoporosis in today's youth and women.[16, 17] The wide availability of soft-drink machines in schools is thought to be a factor related to increased consumption by teens. According to some college students, one reason their intake of calcium is low is that they frequently eat at fast-food restaurants that feature soft drinks rather than milk on the menu. But when asked if they would choose specials that included fresh-tasting milk (not the kind that tastes like the milk container) as part of the meal, some students reported they would select the specials (Illustration 23.5).[18]

■ **What Else Influences Osteoporosis?** Table 23.2 summarizes the risk factors identified for osteoporosis. If you were to find all of the various risk factors

TABLE 23.2

Risk factors for osteoporosis[8]

- Female
- Menopause
- Deficient calcium intake
- Caucasian or Asian heritage
- Thinness ("small bones")
- Cigarette smoking
- Excessive alcohol intake
- Ovarectomy (ovaries removed) before age 45
- Physical inactivity
- Deficient vitamin D status
- Genetic factors

HEALTH *Action* Behaviors That Help Prevent Osteoporosis

- Consuming adequate amounts of Vitamin D (400 IU); exposing the hands and face to direct sunlight for 5 to 15 minutes daily from early childhood onward
- Moderating alcohol intake
- Consuming recommended amounts of calcium (800 mg for children, 1000 mg per day for adults)
- Following MyPyramid Food Intake Guide
- Getting regular physical activity
- Not smoking

combined in one person, that individual would be a thin woman with light skin who has consumed too little calcium and vitamin D and is physically inactive, an excessive alcohol drinker, a smoker, and genetically "small-boned." Additionally, she would have had her ovaries surgically removed for medical reasons before the age of 45.[7] Few people meet every aspect of that description, but people who have several of these characteristics are more likely to develop osteoporosis than those who don't.

At one time, researchers thought that a relatively high intake of phosphorus or protein might intensify the effects of low intakes of calcium on the development of osteoporosis. Studies in humans, however, have failed to demonstrate harmful effects of high phosphorus or protein intake on the development of osteoporosis given an average level of calcium intake.[19]

■ **How Is Osteoporosis Treated?** Calcium supplements or adequate calcium intakes (around 1000 to 1200 milligrams per day), vitamin D supplements (800 IU) or sunshine, 5 to 9 fruits and vegetables daily, and weight-bearing exercise like walking and tennis decrease the progression of osteoporosis in many people.[10, 20] Table 23.3 summarizes current recommendations for the treatment of osteoporosis.

Calcium: Where to Find It

Over half of the calcium supplied by the diets of Americans comes from milk and milk products. Milk (including chocolate milk), cheese, and yogurt are all good sources of calcium (see Table 23.12). Some plants such as kale, broccoli, and bok choy provide appreciable amounts of calcium, too. On average, 32% of the calcium content of milk and milk products, calcium-fortified orange juice, and calcium supplements is absorbed, compared to approximately 5 to 60% of the calcium from different plants (Table 23.4).[21] Calcium absorption decreases with age, vitamin D inadequacy, and ingestion of supplemental iron and zinc.[8] The best way to meet the recommended level of intake (1000 milligrams per day for most adults) is to choose rich sources of absorbable calcium.

Many foods rich in calcium aren't loaded with calories or fat. As Table 23.4 shows, low-fat yogurt, skim and soy milk, kale, and broccoli, for example, provide good amounts of calcium at a low cost in calories.

Calcium is increasingly appearing in unexpected places such as in candy, snack bars, waffles, and bread (Illustration 23.6). The array of calcium-fortified foods is increasing dramatically in the United States in response to publicity about our need for more of it. Although development of such products is making it easier to consume more calcium, the availability of calcium in some of these foods has not been established.[22]

Watch for Calcium on the Label Health claims relating to the benefits of calcium in reducing the risk of osteoporosis may appear on the labels of food products that qualify as good sources of calcium. Look for health claims on foods such

MEAL DEALS

ONLY $3.59 — HAMBURGER, FRIES AND MEDIUM SOFT DRINK

ONLY $2.19 — APPLE-CINNAMON MUFFIN AND FRESH MILK

ONLY $3.95 — CHICKEN FILET ON A BUN, COLESLAW AND FRESH MILK

Illustration 23.5
Would you be more likely to drink milk at fast-food restaurants if it was included as part of a bargain special? Some college women think this is a good idea—one they would go for.

TABLE 23.3

Approaches to the treatment of osteoporosis[10, 20]

- Daily intake of 1000–1200 milligrams of calcium
- Regular weight-bearing physical activity
- Adequate vitamin D intake (800 IU daily) or sun exposure
- 5–9 fruits and vegetables daily
- Medication

TABLE 23.4

Caloric content, calcium level, and percentage of available calcium from different foods

Food	Amount	Calories	Calcium (mg)	Calcium Absorbed (%)	Available Calcium (mg)
Yogurt, low fat	1 cup	143	413	32	132
Skim milk	1 cup	85	301	32	96
Soy milk (fortified)	1 cup	79	300	31	93
1% milk	1 cup	163	274	32	88
Tofu	1 cup	188	260	31	81
Cheese	1 ounce	114	204	32	65
Kale, cooked	1 cup	42	94	49	46
Broccoli, cooked	1 cup	44	72	61	44
Bok choy, raw	1 cup	9	73	54	39
Dried beans	1 cup	209	120	24	29
Spinach, cooked	1 cup	42	244	5	12

as milk, yogurt, and calcium-fortified foods such as orange juice, breakfast cereals, and grain products (Illustration 23.7).

Can You Consume Too Much Calcium? Yes, indeed, you can consume too much calcium. Supplemental calcium in doses exceeding 2.5 grams per day may produce drowsiness, lead to constipation, and cause calcium to deposit in tissues such as the liver and kidneys.[12]

Illustration 23.6
The number of calcium-fortified foods is increasing.

Richard Anderson

Illustration 23.7
Food products that are good sources of calcium can be labeled with a health claim.

"Regular exercise and a healthy diet with enough calcium help maintain good bone health and may reduce the risk of osteoporosis later in life."

Nutrition Facts

Serving Size: 1 cup (240ml)
Servings per Container: 16

Amount per Serving

Calories 110   Calories from Fat 20

	% Daily Value*
Total Fat 2.5g	**4%**
Saturated Fat 1.5g	**8%**
Cholesterol 15mg	**4%**
Sodium 135mg	**6%**
Total Carbohydrate 13g	**4%**
Dietary Fiber 0g	**0%**
Sugars 12g	
Protein 8g	

Vitamin A 10%   •   Vitamin C 4%

Calcium 30% Iron 0% Vitamin D 25%

Phosphorus 10%

*Percent Daily Values are based on a 2,000 calorie diet.

MILK
1% LOWFAT MILK

Selected Minerals: Iron

Most of the body's iron supply is found in *hemoglobin*. Small amounts are present in *myoglobin*, and free iron is involved in processes that capture energy released during the breakdown of proteins and fats.

The Role of Iron in Hemoglobin and Myoglobin

What happens to a car when its paint gets scratched? After a while, the exposed metal rusts. The iron in the metal combines with oxygen in the air, and the result is iron oxide, or "rust." Iron readily combines with oxygen, and that property is put to good use in the body. From its location in hemoglobin in red blood cells, iron loosely attaches to oxygen when blood passes near the inner surface of the lungs. The bright red, oxygenated blood is then delivered to cells throughout the body. When the oxygenated blood passes near cells that need oxygen for energy formation or other reasons, oxygen is released from the iron and diffuses into cells (Illustration 23.8). The free iron in hemoglobin then picks up carbon dioxide, a waste product of energy formation. When carbon dioxide attaches to iron, blood turns from bright red to dark bluish red. Blood then circulates back to the lungs, where carbon dioxide is released from the iron and exhaled into the air. The free iron attaches again to oxygen that enters the lungs, and the cycle continues.

Iron in myoglobin traps oxygen delivered by hemoglobin, stores it, and releases it as needed for energy formation for muscle activity. In effect, myoglobin boosts the supply of oxygen available to muscles.

The functions of iron just described operate smoothly when the body's supply of iron is sufficient. Unfortunately, that is often not the case.

Iron Deficiency Is a Big Problem

The most widespread nutritional deficiency in both developing and developed countries is *iron deficiency* (Table 23.5). It is estimated that one out of every three people in the world is iron deficient.[2] For the most part, iron deficiency affects

hemoglobin
The iron-containing protein in red blood cells.

myoglobin
The iron-containing protein in muscle cells.

iron deficiency
A disorder that results from a depletion of iron stores in the body. It is characterized by weakness, fatigue, short attention span, poor appetite, increased susceptibility to infection, and irritability.

Illustration 23.8
Iron's role in carrying oxygen to cells and carbon dioxide away from them by way of the bloodstream.
Fe++ represents charged particles of iron. The charge is neutralized when iron combines with oxygen or carbon dioxide.

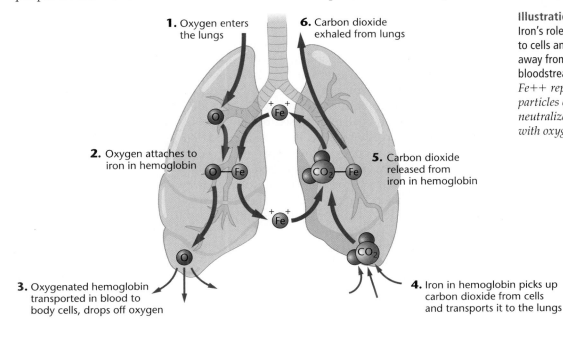

1. Oxygen enters the lungs

6. Carbon dioxide exhaled from lungs

2. Oxygen attaches to iron in hemoglobin

5. Carbon dioxide released from iron in hemoglobin

3. Oxygenated hemoglobin transported in blood to body cells, drops off oxygen

4. Iron in hemoglobin picks up carbon dioxide from cells and transports it to the lungs

TABLE 23.5
Incidence of iron deficiency[2]

	Population with Iron Deficiency (%)
Worldwide (children under 5 years):	
Developing countries	51
Developed countries	12
United States:	
Children, 1–2 years	9
Pregnant women	12
Females, 16–19 years	11
Females, 20–49 years	11
Males, 12–49 years	1 or less

TABLE 23.6
Iron intake of children, women, and men in the United States[24]

	Percentage of Recommended Intake
Children	144%
Women	24%
Men	224%

iron-deficiency anemia
A condition that results when the content of hemoglobin in red blood cells is reduced due to a lack of iron. It is characterized by the signs of iron deficiency plus paleness, exhaustion, and a rapid heart rate.

very young children and women of childbearing age, who have a high need for iron and frequently consume too little of it (Table 23.6). Iron deficiency may develop in people who have lost blood due to injury, surgery, or ulcers. Donating blood more than three times a year can also precipitate iron deficiency.[23]

Consequences of Iron Deficiency Many body processes sputter without sufficient oxygen. People with iron deficiency usually feel weak and tired. They have a shortened attention span and a poor appetite, are susceptible to infection, and become irritable easily. If the deficiency is serious enough, *iron-deficiency anemia* develops, and additional symptoms occur. People with iron-deficiency anemia look pale, are easily exhausted, and have rapid heart rates. Iron-deficiency anemia is a particular problem for infants and young children because it is related to lasting retardation in mental development.[25]

Getting Enough Iron in Your Diet

"Enough" iron according to the RDAs is 8 milligrams for men and 18 milligrams per day for women aged 19 to 50 years. Consuming that much iron can be difficult for women. On average, 1000 calories' worth of food provides about 6 milligrams of iron. Women would have to consume around 2500 calories per day to obtain even 15 milligrams of iron on average. Selection of good sources of iron has to be done on a better-than-average basis if women are to get enough.

Iron is found in small amounts in many foods, but only a few foods such as liver, beef, and prune juice are rich sources. (A list of food sources of iron is given in Table 23.12.) Foods cooked in iron and stainless steel pans, however, can be a significant source of iron because some of the iron in the pan leaches out during cooking. On average, approximately 1 milligram of iron is added to each 3-ounce serving of food cooked in these pans.[26]

Most of the iron in plants and eggs is tightly bound to substances such as phytates or oxalic acid that limit iron absorption, making these foods relatively poor sources of iron even though they contain a fair amount of it. (Differences in the proportions of iron absorbed from various food sources are shown in Illustration 23.9.) A 3-ounce hamburger and a cup of asparagus both contain approximately 3 milligrams of iron, for example. But 20 times more iron can be absorbed from the hamburger than from the asparagus.[27] Absorption of iron from plants can be increased if foods containing vitamin C are included in the same meal.

Iron absorption is also increased by low levels of iron stores in the body (Illustration 23.10). In other words, if you're in need of iron, your body sets off mechanisms that allow more of it to be absorbed from foods or supplements. When iron stores are high, less iron is absorbed.

The body's ability to regulate iron absorption provides considerable protection against iron deficiency and overdose. The protection is not complete, however, as evidenced by widespread iron deficiency and the occurrence of iron toxicity.

Overdosing on Iron Excess iron absorbed into the body cannot be easily excreted. Consequently, it is deposited in various tissues such as the liver, pancreas, and heart. There, the iron reacts with cells, causing damage that can result in liver disease, diabetes, and heart failure. One in 200 people in the United States has an inherited tendency to absorb too much iron (a disorder called hemochroma-

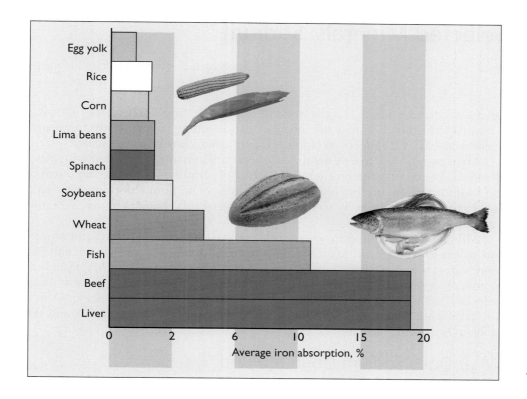

Illustration 23.9
Average percentage of iron absorbed from selected foods by healthy adults.[27]

Egg yolk
Rice
Corn
Lima beans
Spinach
Soybeans
Wheat
Fish
Beef
Liver

Average iron absorption, %

Illustration 23.10
People absorb more iron from foods and supplements when body stores of iron are low than when stores of iron are high.

tosis and pronounced *hem-oh-chrom-ah-toe-sis*). Other people develop iron toxicity from consuming large amounts of iron with alcohol (alcohol increases iron absorption) or from very high iron intakes, usually due to overdoses of iron supplements.

Each year in the United States, more than 10,000 people accidentally overdose on iron supplements.[28] Commonly, victims are young children who mistakenly think iron pills are candy (Table 23.7 and Illustration 23.11). The lethal dose of iron for a 2-year-old child is about 3 grams,[29] the amount of iron present in 25 pills containing 120 milligrams of iron each. Iron supplements that are not being used should be thrown away or stored in a place where toddlers cannot get to them.

TABLE 23.7
Accidental overdoses of chemical substances by children up to age 4 in the United States[30]

Substance	Cases per 100,000 Children
Aspirin and substitutes	12.7
Solvents/petroleum products	11.2
Tranquilizers	9.9
Iron supplements	8.7
Corrosives and caustics	7.4

Illustration 23.11
If you were three years old, could you tell which "pills" are candy and which are the iron supplements? Overdoses of iron supplements are a leading cause of accidental poisoning in young children. (The iron supplements are the lightest green in color.)

Illustration 23.12
The oceans contain enough salt to cover the surface of the Earth to a depth of more than 400 feet.

water balance
The ratio of the amount of water outside cells to the amount inside cells; a proper balance is needed for normal cell functioning.

hypertension
High blood pressure. It is defined as blood pressure exerted inside blood vessel walls that typically exceeds 140/90 millimeters of mercury.

Illustration 23.13
A blood pressure test.
Blood pressure is expressed as two numbers:

$\dfrac{110}{70}$ *Systolic pressure measures the force of the blood when the heart contracts.*
Diastolic pressure measures the force of the blood when the heart is at rest.

Selected Minerals: Sodium

Excess of salty flavor hardens the pulse.
—ABERNATHY, 1000 B.C.

On the one hand, sodium is a gift from the sea (Illustration 23.12); on the other, it is a hazard to health. Sodium's life- and health-sustaining functions are frequently overshadowed by its effects on blood pressure when consumed in excess.

An extremely reactive mineral, sodium occurs in nature combined with other elements. The most common chemical partner of sodium is chloride, and much of the sodium present on this planet is in the form of sodium chloride—table salt. Table salt is 40% sodium by weight; 1 teaspoon of salt contains about 2300 milligrams of sodium.

Although salt is found in abundance in the oceans, humans have not always had access to enough salt to satisfy their taste for it. During times when salt has been scarce, wars have been fought over it. In such periods, a pocketful of salt was as good as a pocketful of cash. (The word *salary* is derived from the Latin word *salarium*, "salt money.") People were said to be "worth their salt" if they put in a full day's work. Today, salt is widely available in most parts of the world, and the problem is limiting human intakes to levels that do not interfere with health.

What Does Sodium Do in the Body?

Sodium appears directly above potassium in the periodic table, and the two work closely together in the body to maintain normal *water balance*. Both sodium and potassium chemically attract water, and under normal circumstances, each draws sufficient water to the outside or inside of cells to maintain an optimal level of water in both places.

Water balance and cell functions are upset when there's an imbalance in the body's supplies of sodium and potassium. You have probably noticed that you become thirsty when you eat a large amount of salted potato chips or popcorn. Salty foods make you thirsty because your body loses water when the high load of dietary sodium is excreted. The thirst signal indicates that you need water to replace what you have lost.

The loss of body water that accompanies ingestion of large amounts of salt explains why seawater neither quenches thirst nor satisfies the body's need for water. When a person drinks seawater, its high concentration of sodium causes the body to excrete more water than it retains. Rather than increasing the body's supply of water, the ingestion of seawater increases the need for water.

In healthy people, the body's adaptive mechanisms provide a buffer against upsets in water balance due to high sodium intakes. It appears that many people are overwhelming the body's ability to cope with high sodium loads, however. High dietary intakes of sodium appear to play an important role in the development of *hypertension* in many people.[31]

A Bit about Blood Pressure   To circulate through the body, blood must exist under pressure in the blood vessels. The amount of pressure exerted on the walls of blood vessels is greatest when pulses of blood are passing through them (that's when "systolic" blood pressure is measured) and least between pulses (that's when "diastolic" blood pressure is taken). Blood pressure measurements note the highest and lowest pressure in blood vessels (Illustration 23.13).

Blood pressure levels less than 120/80 millimeters of mercury (mmHg) are considered normal, whereas levels between 120/80 and 139/89 mmHg are classified as "prehypertension." Values of 140/90 mm Hg and higher qualify as hypertension (Table 23.8). Several blood pressure measurements, taken while a person is relaxed, are needed to obtain an accurate result. (Even going to a clinic or doctor's office to have blood pressure measured can raise it for some people.)

Hypertension is a major public health problem in the United States and other countries. Although not considered a disease by itself, the presence of hypertension substantially increases the risk that a person will develop heart disease or kidney failure or will experience a heart attack or stroke. Hypertension occurs in 30% of U.S. adults and more than 20% of adults worldwide. It is more likely to develop as people become older. About 40% of all cases of hypertension are classified as mild.[3]

What Causes Hypertension?   About 10% of all cases of hypertension can be directly linked to a cause. People who have hypertension with no identifiable cause (over 90% of all cases) are said to have *essential hypertension*.[33]

A number of risk factors for hypertension have been identified, and dietary factors are among the most important (Table 23.9). Foremost among the evidence linking diet to hypertension are population studies that show a relationship between average salt intake and hypertension. As average salt intake rises, so do rates of hypertension. When people with hypertension reduce their salt intake, their blood pressure tends to drop. Further scrutiny of the data on salt and blood pressure, however, reveals that not everyone is equally susceptible to high-salt diets. Some people are genetically susceptible to salt or have subtle forms of kidney disorders that raise their blood pressure in response to high-sodium diets.[34] This condition is known as *salt sensitivity*.

Salt Sensitivity   Approximately 51% of people with hypertension and 26% of people with normal blood pressure are salt sensitive. Reduction in salt intake by people who are salt sensitive, along with weight loss if overweight, substantially improves blood pressure in most cases. Because it is currently difficult to identify who in the population is salt sensitive and who is not; because most Americans consume much more sodium than needed; and because high-sodium foods are often high in fat, official advice for Americans is to limit their sodium intake to 2300 milligrams per day (the equivalent of approximately one level teaspoon of salt from all sources).[35]

Other Risk Factors for Hypertension   Obesity is a major risk factor for hypertension. For obese people with hypertension, the most effective treatment is weight loss. Excessive alcohol intake can prompt the development of hypertension, and moderation of intake (to two or fewer alcoholic drinks per day) can bring blood pressure back down. Low intakes of calcium, vegetables, fruits, and potassium

essential hypertension
Hypertension of no known cause; also called primary or idiopathic hypertension, it accounts for 90 to 95% of all cases of hypertension.

salt sensitivity
A genetically influenced condition in which a person's blood pressure rises when high amounts of salt or sodium are consumed. Such individuals are sometimes identified by blood pressure increases of 10% or more when switched from a low-salt to a high-salt diet.

TABLE 23.8
Hypertension categories[32]

Category	Systolic (mm Hg)[a]	Diastolic (mm Hg)
Normal	<120	80
Prehypertension	120–139	80–89
Hypertension		
Stage 1	140–159	90–99
Stage 2	160+	100+

[a]mm Hg = millimeters of mercury.

TABLE 23.9
Risk factors for hypertension[32]

- Age
- Family history
- High-sodium diet
- Obesity
- Physical inactivity
- Excessive alcohol consumption
- Low vegetable and fruit consumption

HEALTH *Action* **Dietary and Lifestyle Recommendations for Preventing Hypertension**[36]

- Consume a diet rich in vegetables and fruits.
- Consume low-fat dairy products and fish.
- Reduce saturated fat intake.
- Include whole grains, poultry, fish, nuts, and seeds in your diet.

- Reduce sodium intake to about 2300 mg (1 tsp total salt) or less per day.
- Limit alcohol intake to 1 to 2 drinks per day, if any.
- Exercise 30 minutes or more almost every day.

- Lose weight if overweight; maintain a healthy weight.
- Don't smoke.

and high intakes of saturated fat all contribute to the development of hypertension. Physically inactive lifestyles also foster the development of hypertension.[32]

What Is the Best Way to Prevent Hypertension?

The nearby "Health Action" highlights recommendations for decreasing the chances that hypertension will develop. These recommendations contribute to an overall healthy lifestyle whose benefits extend beyond the prevention of hypertension.

How Is Hypertension Treated? The recommended approach to treatment of all cases of hypertension consists of dietary and lifestyle changes and the use of medications if necessary.[32] Weight loss and smoking cessation (if needed), a moderate-sodium diet, regular exercise, moderate alcohol consumption (if any), and the DASH diet are basic components of the approach to treatment (Table 23.10). The DASH diet (Illustration 23.14) is based on vegetables, fruits, low-fat dairy products, whole grains, and poultry and fish. Its composition is similar to that of diets recommended for heart disease and cancer prevention. People who adhere to the DASH diet often bring their blood pressure levels back into the normal range. If blood pressure remains elevated after dietary and lifestyle changes have been implemented, or if blood pressure is quite high when diagnosed, antihypertension drugs are prescribed.[36]

Cutting Back on Salt On average, children and adults in the United States consume twice the recommended amount of sodium.[24] The leading sources of salt (and therefore of sodium) in the U.S. diet are processed foods (Table 23.11). Restricting the use of foods to which salt has been added during processing is the most effective way to lower salt intake and achieve a moderate- to low-salt diet.[39]

High-salt processed foods include salad dressings, pickles, canned soups, corned beef, sausages and other luncheon meats, and snack foods such as potato chips and cheese twists.

Only a small proportion of our total sodium intake enters our diet from fresh foods. Very few foods naturally contain much sodium—at least not until they are processed (Illustration 23.15).

Using spices and lemon juice to flavor foods, consuming fresh vegetables (rather than processed ones) and fresh fruits, and checking out the sodium content of foods and then selecting low-sodium ones can also help reduce sodium intake.

TABLE 23.10

Approaches to the treatment of hypertension[36]

- Weight loss (if needed)
- Moderate salt intake (2300 mg of sodium per day or less)
- Moderate alcohol consumption (if any)
- Regular physical activity
- The DASH diet
- Antihypertension drugs (if hypertension not controlled by above measures)

TABLE 23.11

Major sources of sodium in the U.S. diet[24]

Sodium Source	Contribution to Sodium Intake (%)
Processed foods	77
Fresh foods	12
Salt added at the table	6
Salt added during cooking	5

The DASH diet

In the mid-1990s a revolutionary approach to the control of mild and moderate hypertension was tested, and the results have changed health professionals' thinking about high blood pressure prevention and management. Called the DASH diet (Dietary Approaches to Stop Hypertension), it didn't focus on salt restriction and was related to significant reductions in blood pressure within 2 weeks in most people tested. In some people, reductions in blood pressure were sufficient to erase the need for antihypertension medications, and for others the diet reduced the amount or variety of medications needed. A subsequent study showed that a low-sodium diet boosts the blood pressure lowering effects of the DASH diet, especially in African Americans.[37] This dietary pattern is now recommended in MyPyramid for people in the United States in general.

The DASH diet consists of eating patterns made up of the following food groups:

	DAILY SERVINGS
Vegetables	4–5
Fruits	4–5
Grain products	7–8
Low-fat milk and dairy products	2–3
Lean meats, fish, poultry	2 or less
Nuts, seeds, dried beans	about 1

Although it does not work for all individuals with hypertension, the DASH diet is a mainstay in clinical efforts to control hypertension. Benefits of the diet last as long as the diet does.[38] A very useful Web site is available that describes refinements in the diet based on research studies and that gives practical tips on how to follow the diet.

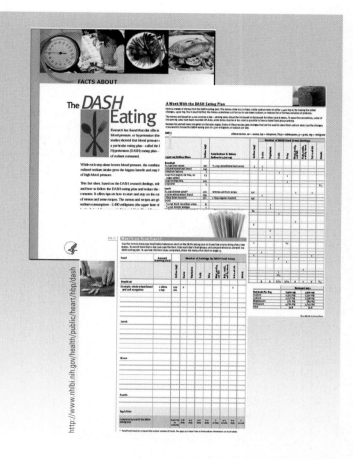

http://www.nhlbi.nih.gov/health/public/heart/hbp/dash

Illustration 23.14
Recommendations for Preventing Hypertension

Label Watch   Not all processed foods with added sodium taste salty. To find out which processed foods are high in sodium, you have to examine the label. Increasingly, low-salt processed foods are entering the market and can be easily identified by the "low-salt" message on the label (Illustration 23.16). Terms used to identify

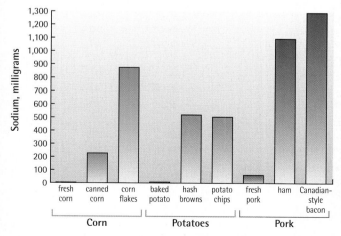

Illustration 23.15
Examples of how processing increases the sodium content of foods. *Sodium values are for a 3-ounce serving of each food shown.*

Term used on label	Definition
LOW SODIUM	Product must contain 140 milligrams or less of sodium per serving.
Reduced Sodium	Product must contain 75% less sodium than the customary amount (comparison data must be given on the label).
Sodium FREE	Product must contain less than 5 milligrams sodium per serving.
LIGHTLY SALTED	Product must contain at least 25% less sodium than the customary amount (comparison data must be given on the label).
UNSALTED No Salt Added	Salt must not have been added to foods customarily salted such as canned green beans, popcorn, and peanuts.

Illustration 23.16
When a label makes a claim about the sodium content of a food product, the label must list the amount of sodium in a serving and adhere to these definitions.

low-salt (or low-sodium) foods are defined by the Food and Drug Administration. To be considered low-sodium, foods must contain 140 milligrams or less of sodium per serving. Food manufacturers must adhere to the definitions when they make claims about the salt or sodium content of a food on the label.

This unit ends having presented the story behind 3 of the 15 essential minerals. Much more information could have been relayed about these three, not to mention the wealth of knowledge that could be shared about minerals such as phosphorus, magnesium, zinc, iodine, selenium, and potassium. Critical information about these and other essential minerals is given in Table 23.1, and Table 23.12 lists food sources of many of the essential minerals. For additional information, look at the comprehensive nutrition books in your college library or take notes on the information presented by your instructor in class.

TABLE 23.12
Food sources of minerals

Magnesium

FOOD	AMOUNT	MAGNESIUM (MG)
Legumes:		
Lentils, cooked	$\frac{1}{2}$ cup	134
Split peas, cooked	$\frac{1}{2}$ cup	134
Tofu	$\frac{1}{2}$ cup	130
Nuts:		
Peanuts	$\frac{1}{4}$ cup	247
Cashews	$\frac{1}{4}$ cup	93
Almonds	$\frac{1}{4}$ cup	80
Grains:		
Bran buds	1 cup	240
Wild rice, cooked	$\frac{1}{2}$ cup	119
Breakfast cereal, fortified	1 cup	85
Wheat germ	2 tbs	45
Vegetables:		
Bean sprouts	$\frac{1}{2}$ cup	98
Black-eyed peas	$\frac{1}{2}$ cup	58
Spinach, cooked	$\frac{1}{2}$ cup	48
Lima beans	$\frac{1}{2}$ cup	32
Milk and milk products:		
Milk	1 cup	30
Cheddar cheese	1 oz	8
American cheese	1 oz	6
Meats:		
Chicken	3 oz	25
Beef	3 oz	20
Pork	3 oz	20

Calcium[a]

FOOD	AMOUNT	CALCIUM (MG)
Milk and milk products:		
Yogurt, low-fat	1 cup	413
Milk shake (low-fat frozen yogurt)	$1\frac{1}{4}$ cup	352
Yogurt with fruit, low-fat	1 cup	315
Skim milk	1 cup	301
1% milk	1 cup	300
2% milk	1 cup	298
3.25% milk (whole)	1 cup	288
Swiss cheese	1 oz	270
Milk shake (whole milk)	$1\frac{1}{4}$ cup	250
Frozen yogurt, low-fat	1 cup	248
Frappuccino	1 cup	220
Cheddar cheese	1 oz	204
Frozen yogurt	1 cup	200
Cream soup	1 cup	186
Pudding	$\frac{1}{2}$ cup	185
Ice cream	1 cup	180
Ice milk	1 cup	180
American cheese	1 oz	175
Custard	$\frac{1}{2}$ cup	150
Cottage cheese	$\frac{1}{2}$ cup	70
Cottage cheese, low-fat	$\frac{1}{2}$ cup	69
Vegetables:		
Spinach, cooked	$\frac{1}{2}$ cup	122
Kale	$\frac{1}{2}$ cup	47
Broccoli	$\frac{1}{2}$ cup	36
Legumes:		
Tofu	$\frac{1}{2}$ cup	260
Dried beans, cooked	$\frac{1}{2}$ cup	60
Foods fortified with calcium:		
Orange juice	1 cup	350
Frozen waffles	2	300
Soy milk	1 cup	200–400
Breakfast cereals	1 cup	150–1000

[a]Actually, the richest source of calcium is alligator meat; $3\frac{1}{2}$ ounces contain about 1231 milligrams of calcium; but just try to find it on your grocer's shelf!

TABLE 23.12

Food sources of minerals (continued)

Selenium

FOOD	AMOUNT	SELENIUM (MG)
Seafood:		
Lobster	3 oz	66
Tuna	3 oz	60
Shrimp	3 oz	54
Oysters	3 oz	48
Fish	3 oz	40
Meats/Eggs:		
Liver	3 oz	56
Egg	1 medium	37
Ham	3 oz	29
Beef	3 oz	22
Bacon	3 oz	21
Chicken	3 oz	18
Lamb	3 oz	14
Veal	3 oz	10

Zinc

FOOD	AMOUNT	ZINC (MG)
Meats:		
Liver	3 oz	4.6
Beef	3 oz	4.0
Crab	1/2 cup	3.5
Lamb	3 oz	3.5
Turkey ham	3 oz	2.5
Pork	3 oz	2.4
Chicken	3 oz	2.0
Legumes:		
Dried beans, cooked	1/2 cup	1.0
Split peas, cooked	1/2 cup	0.9
Grains:		
Breakfast cereal, fortified	1 cup	1.5–4.0
Wheat germ	2 tbs	2.4
Oatmeal, cooked	1 cup	1.2
Bran flakes	1 cup	1.0
Brown rice, cooked	1/2 cup	0.6
White rice	1/2 cup	0.4
Nuts and seeds:		
Pecans	1/4 cup	2.0
Cashews	1/4 cup	1.8
Sunflower seeds	1/4 cup	1.7
Peanut butter	2 tbs	0.9
Milk and milk products:		
Cheddar cheese	1 oz	1.1
Whole milk	1 cup	0.9
American cheese	1 oz	0.8

Sodium

FOOD	AMOUNT	SODIUM (MG)
Miscellaneous:		
Salt	1 tsp	2132
Dill pickle	1 (4 1/2 oz)	1930
Sea salt	1 tsp	1716
Chicken broth	1 cup	1571
Ravioli, canned	1 cup	1065
Spaghetti with sauce, canned	1 cup	955
Baking soda	1 tsp	821
Beef broth	1 cup	782
Gravy	1/4 cup	720
Italian dressing	2 tbs	720
Pretzels	5 (1 oz)	500
Green olives	5	465
Pizza with cheese	1 wedge	455
Soy sauce	1 tsp	444
Cheese twists	1 cup	329
Bacon	3 slices	303
French dressing	2 tbs	220
Potato chips	1 oz (10 pieces)	200
Catsup	1 tbs	155
Meats:		
Corned beef	3 oz	808
Ham	3 oz	800
Fish, canned	3 oz	735
Meat loaf	3 oz	555
Sausage	3 oz	483
Hot dog	1	477
Fish, smoked	3 oz	444
Bologna	1 oz	370
Milk and milk products:		
Cream soup	1 cup	1070
Cottage cheese	1/2 cup	455
American cheese	1 oz	405
Cheese spread	1 oz	274
Parmesan cheese	1 oz	247
Gouda cheese	1 oz	232
Cheddar cheese	1 oz	175
Skim milk	1 cup	125
Whole milk	1 cup	120
Grains:		
Bran flakes	1 cup	363
Cornflakes	1 cup	325
Croissant	1 medium	270
Bagel	1	260
English muffin	1	203
White bread	1 slice	130
Whole-wheat bread	1 slice	130
Saltine crackers	4 squares	125

(continued)

TABLE 23.12

Food sources of minerals (continued)

Iron

FOOD	AMOUNT	IRON (MG)
Meat and meat alternates:		
Liver	3 oz	7.5
Round steak	3 oz	3.0
Hamburger, lean	3 oz	3.0
Baked beans	$\frac{1}{2}$ cup	3.0
Pork	3 oz	2.7
White beans	$\frac{1}{2}$ cup	2.7
Soybeans	$\frac{1}{2}$ cup	2.5
Pork and beans	$\frac{1}{2}$ cup	2.3
Fish	3 oz	1.0
Chicken	3 oz	1.0
Grains:		
Breakfast cereal, iron-fortified	1 cup	8.0 (4-18)
Oatmeal, fortified, cooked	1 cup	8.0
Bagel	1	1.7
English muffin	1	1.6
Rye bread	1 slice	1.0
Whole-wheat bread	1 slice	0.8
White bread	1 slice	0.6
Fruits:		
Prune juice	1 cup	9.0
Apricots, dried	$\frac{1}{2}$ cup	2.5
Prunes	5 medium	2.0
Raisins	1/4 cup	1.3
Plums	3 medium	1.1
Vegetables:		
Spinach, cooked	$\frac{1}{2}$ cup	2.3
Lima beans	$\frac{1}{2}$ cup	2.2
Black-eyed peas	$\frac{1}{2}$ cup	1.7
Peas	$\frac{1}{2}$ cup	1.6
Asparagus	$\frac{1}{2}$ cup	1.5

Phosphorous

FOOD	AMOUNT	PHOSPHORUS (MG)
Milk and milk products:		
Yogurt	1 cup	327
Skim milk	1 cup	250
Whole milk	1 cup	250
Cottage cheese	$\frac{1}{2}$ cup	150
American cheese	1 oz	130
Meats:		
Pork	3 oz	275
Hamburger	3 oz	165
Tuna	3 oz	162
Lobster	3 oz	125
Chicken	3 oz	120
Nuts and seeds:		
Sunflower seeds	$\frac{1}{4}$ cup	319
Peanuts	$\frac{1}{4}$ cup	141
Pine nuts	$\frac{1}{4}$ cup	106
Peanut butter	1 tbs	61

FOOD	AMOUNT	PHOSPHORUS (MG)
Grains:		
Bran flakes	1 cup	180
Shredded wheat	2 large biscuits	81
Whole-wheat bread	1 slice	52
Noodles, cooked	$\frac{1}{2}$ cup	47
Rice, cooked	$\frac{1}{2}$ cup	29
White bread	1 slice	24
Vegetables:		
Potato	1 medium	101
Corn	$\frac{1}{2}$ cup	73
Peas	$\frac{1}{2}$ cup	70
French fries	$\frac{1}{2}$ cup	61
Broccoli	$\frac{1}{2}$ cup	54
Other:		
Milk chocolate	1 oz	66
Cola	12 oz	51
Diet cola	12 oz	45

Potassium

FOOD	AMOUNT	POTASSIUM (MG)
Vegetables:		
Potato	1 medium	780
Winter squash	$\frac{1}{2}$ cup	327
Tomato	1 medium	300
Celery	1 stalk	270
Carrots	1 medium	245
Broccoli	$\frac{1}{2}$ cup	205
Fruits:		
Avocado	$\frac{1}{2}$ medium	680
Orange juice	1 cup	469
Banana	1 medium	440
Raisins	$\frac{1}{4}$ cup	370
Prunes	4 large	300
Watermelon	1 cup	158
Meats:		
Fish	3 oz	500
Hamburger	3 oz	480
Lamb	3 oz	382
Pork	3 oz	335
Chicken	3 oz	208
Grains:		
Bran buds	1 cup	1080
Bran flakes	1 cup	248
Raisin bran	1 cup	242
Wheat flakes	1 cup	96
Milk and milk products:		
Yogurt	1 cup	531
Skim milk	1 cup	400
Whole milk	1 cup	370
Other:		
Salt substitutes	1 tsp	1300–2378

TABLE 23.12

Food sources of minerals (continued)

Fluoride

FOOD	AMOUNT	FLUORIDE (MG)
Grape juice, white	6 oz	350
Instant tea	1 cup	335
Raisins	3.5 oz	234
Wine, white	3.5 oz	202
Wine, red	3.5 oz	105
French fries, McDonald's	1 medium	130
Dannon's Fluoride to Go	8 oz	178
Tap water, U.S.	8 oz	59
Municipal water, U.S.	8 oz	186
Bottled water, store brand	8 oz	37

Iodine

FOOD	AMOUNT	IODINE (MCG)
Iodized salt	1 tsp	400
Haddock	3 oz	125
Cod	3 oz	87
Shrimp	3 oz	30
Bread	1 oz (1 slice)	35–142
Cottage cheese	½ cup	50
Egg	1	22
Cheddar cheese	1 oz	17

Nutrition Up Close

The Salt of Your Diet

Focal Point: Estimate your sodium intake.

Salt added to foods during processing, during home cooking, and at the table is the major source of sodium in the American diet. To determine if you consume too much sodium, answer the following questions.

How Often Do You Usually:	Daily	4–6 Times per Week	1–3 Times per Week	Less than Weekly
1. Eat processed foods such as luncheon meats, bacon, hot dogs, sausage, canned soups, broths, gravy, TV dinners, or smoked fish?	☐	☐	☐	☐
2. Eat pickles, green olives, potato or tortilla chips, pretzels, or salted crackers?	☐	☐	☐	☐
3. Use soy sauce, garlic salt, steak sauce, or catsup in recipes or on foods?	☐	☐	☐	☐
4. Salt foods at the table?	☐	☐	☐	☐
5. Add salt to cooking water or to foods you are preparing?	☐	☐	☐	☐

FEEDBACK (including scoring) can be found at the end of Unit 23.

Source: Adapted from Limit use of sodium. Washington, DC: USDA Home and Garden Publication 232–4; 1986.

Media Menu

www.nal.usda.gov/fnic
It's a quick trip from this site to information on minerals, osteoporosis, hypertension, iron deficiency, and other topics covered in this unit.

www.nhlbi.nih.gov
NIH's Web site on hypertension includes current research and recommendations on prevention and treatment.

www.nof.org
This is the Web address for the National Osteoporosis Foundation.

http://lancaster.unl.edu/food/ osteoporosis.shtml
The University of Nebraska Cooperative Extension Service in Lancaster County offers a "Nutrition and Osteoporosis" online slide show. It is available for viewing from this address.

www.nal.usda.gov/fnic/foodcomp/search
Select "Nutrient lists" from USDA's National Nutrient Database homepage to find lists of food sources of minerals and other nutrients.

Key Terms

cofactors, *page 23-3*
essential hypertension, *page 23-19*
hemoglobin, *page 23-15*
hypertension, *page 23-18*

iron deficiency, *page 23-15*
iron-deficiency anemia, *page 23-16*
minerals, *page 23-2*
myoglobin, *page 23-15*

osteoporosis, *page 23-5*
remodeling, *page 23-5*
salt sensitivity, *page 23-19*
water balance, *page 23-18*

Review Questions

True   False

1. There are 40 essential minerals that perform specific functions in the body. Each must be obtained from the diet. ☐ ☐

2. Essential minerals are a structural component of teeth; they serve as a source of electrical power that stimulate muscles to contract and help maintain an appropriate balance of fluids in the body. ☐ ☐

3. The amount of certain minerals that is absorbed by the body varies extensively based on what else is consumed along with the minerals. ☐ ☐

4. Calcium is the most abundant minerals found in bones, but other minerals such as magnesium, phosphorus, and carbon are also components of bone. ☐ ☐

5. Bones fully develop and mineralize during the first 18 years of life. ☐ ☐

6. Iodine deficiency and overdose are related to goiter and thyroid disease. ☐ ☐

True   False

7. Wilson's disease is related to magnesium deficiency. ☐ ☐

8. An excellent way to develop and maintain a healthy vitamin D status is through brief, daily exposure of the skin to sun. ☐ ☐

9. People with iron deficiency generally feel weak and tired, have a poor appetite, and are susceptible to infection. ☐ ☐

10. Although iron deficiency is a major health problem in America, iron overdose is not. ☐ ☐

11. Blood pressure tends to increase as salt (or sodium) intake increases. This is not true for everyone, however, because some people are insensitive to the effects of salt intake on blood pressure. ☐ ☐

12. The major source of salt (or sodium) in our diet comes from salt added to food at the table. ☐ ☐

Notes

1. Mitchell SJ. Changes after taking a college basic nutrition course. J Am Diet Assoc 1990;90:955–61.

2. Parvanta I. The hidden hunger: micronutrient malnutrition. Centers for Disease Control, www.cdc.gov/epo/mmwr/preview/mmwrhtml/00051880.htm, March 2000.

3. Health, United States, 2005. www.cdc.gov, accessed 9/06.

4. Prasad AS. Discovery and importance of zinc in human nutrition. Federation Proc 1984;43:2829–34.

5. Monsen ER, Balintfy JL. Calculating dietary iron bioavailability: refinement and computerization. J Am Diet Assoc 1982;80:307–11.

6. Raisz LG. Screening for osteoporosis. N Engl J Med 2005;353:164–5. Mark BL. The experts speak: multiple sclerosis and vitamin D. J Am Diet Assoc 2006;106:418–24.

7. Prestwood KM, et al. Prevention and treatment of osteoporosis, Clin Cornerstone 2002;4:31–41.

8. Follin SL, Hansen LB, Current approaches to the prevention and treatment of postmenopausal osteoporosis, Am J Health—Syst Pharm 2003;60:883–901.

9. Maximizing peak bone mass: calcium supplementation increases bone mineral density in children. Nutr Rev 1992;50:335–7.

10. Lanham-New SA. Fruits and vegetables: the unexpected natural answer to the question of osteoporosis? Am J Clin Nutr 2006;83:1254–5.

11. Harris S. Calcium trials. Experimental Biology Annual Meeting, San Diego, Calif. 2000 Apr. 4.

12. Looker AC. Interaction of science, consumer practices, and policy: calcium and bone health as a case study. J Nutr 2003;133:1987S–91S.

13. Calvo MS et al. Overview of the Proceedings from Experimental Biology 2004 Symposium: vitamin D insufficiency: a significant risk factor in chronic diseases and potential disease-specific biomarkers of vitamin D sufficiency. J Nutr 2005;135:301–3.

14. What we eat in America. www.ars .ysda.gov/services/docid/htm?/9098, accessed 3/06.

15. Bone builders. FDA Consumer, 1997;Sept/Oct.

16. Wyshak G. Consumption of carbonated drinks by teenage girls associated with bone fractures, Arch Pediatr Adol Med 2000;154:542–3, 610–13.

17. Tucker K et al, Cola consumption and bone mineral density in women, Medscape Medical News. 2003, www .medscape.com, accessed 10/03.

18. Lewis NM, Hollingsworth M. Food choices of young college women consuming low- or moderate-calcium diets. Nutr Res 1992;12:843–8.

19. Rapuri PB et al. Protein intake: effects on bone mineral density and the rate of bone loss in elderly women. Am J Clin Nutr 2003;77:1517–25.

20. Heaney RP. Role and importance of calcium in preventing and managing osteoporosis, www.medscape.com, March 30, 2006.

21. New SA. Osteoporosis—ask the experts: dairy and bone health. Medscape Ob/Gyn & Women's Health 2003;8(2).

22. Klausner A. EN's guide to calcium in unexpected places. Envir Nutr 1999;May:5.

23. Fairbanks VF. Iron in medicine and nutrition. In: Modern nutrition in health and disease, Shils ME et al, eds. 9th ed. Philadelphia: Lippincott Williams & Wilkins, 1999: pp. 193–221.

24. Ervin RB et al. Dietary intake of selected minerals for the United States population: 1999–2000. Advanced Data, No. 341, April 27, 2004.

25. Fairbanks, Iron in medicine and nutrition; and Hurtado EK et al. Early childhood anemia and mild or moderate mental retardation. Am J Clin Nutr 1999; 69:115–9.

26. Park J, Brittin HC. Increased iron content of food due to stainless steel cookware. J Am Diet Assoc 1997;97:659–61.

27. Monsen ER et al. Estimation of available dietary iron. Am J Clin Nutr 1978;31:134–41.

28. Shannon M. Ingestion of toxic substances by children. N Engl J Med 2000;342:186–89.

29. National Research Council Subcommittee on Iron (Committee on Medical and Biological Effects of Environmental Pollutants). Iron. Washington, DC: Division of Medical Sciences, Assembly of Life Sciences, National Academy of Sciences; 1979.

30. Trinkoff AM, Baker SP. Poisoning hospitalizations and deaths from solids and liquids among children and teenagers. Am J Public Health 1986;76:657–60.

31. Messerli RE et al. Salt: a perpetrator of hypertensive target organ disease? Arch Int Med 1997;157:2449–52.

32. NHLBI issues new high blood pressure clinical practice guidelines. NHLBI Communications Office, 5/14/03, www.nhlbi.nih.gov/index.htm.

33. Messerli et al., Salt; August P, Initial treatment of hypertension, N Engl J Med 2003;348:610–7.

34. Law MR et al., By how much does dietary salt reduction lower blood pressure? I. Analysis of observational data among populations. Br Med J 1991;302:811–15.

35. Dietary Guidelines Executive Summary, http://www.health.gov/ dietaryguidelines/dga2005/document/ html/executivesummary.htm, accessed 7/06.

36. Apple LJ. The science supporting the 2005 Dietary Guidelines for Americans: sodium, potassium, and physical activity. www.healthierus.gov, accessed 9/05.

37. Conlin PR et al., DASH diet can control stage 1 hypertension, Am J Hypertens 2000;13:949–55.

38. Karanja NM et al. Descriptive characteristics of the dietary patterns used in the Dietary Approaches to Stop Hypertension trial. J Am Diet Assoc 1999;99(suppl):S19–S27.

39. Mattes RD, Donnelley D. Relative contribution of dietary sodium sources. J Am Coll Nutr 1991;10:383–93.

Nutrition Up Close

The Salt of Your Diet

Feedback for Unit 23

People in the United States generally consume two to three times more sodium than they need, and you may be doing so too, if you answered "daily" or "4–6 times per week" to several of the questions. If you wish to cut back on sodium, here are some suggestions:

- Choose fresh instead of processed foods more often.
- Read labels to check for sodium content.
- Reduce salt in cooking and at the table.
- When you do choose processed foods, try the lower-sodium versions available.
- To enhance the flavor of foods, try herbs and spices instead of salt.

Answers to Reveiw Questions

1. False, see page 23-2.
2. True, see pages 23-2 and 23-3.
3. True, see pages 23-3, 23-4.
4. True, see page 23-5.
5. False, see page 23-10.
6. True, see page 23-9.
7. False, see page 23-9.
8. True, see page 23-11.
9. True, see page 23-16.
10. False, see page 23-17.
11. True, see page 23-19.
12. False, see page 23-20.

Dietary Supplements and Functional Foods

NUTRITION SCOREBOARD

	TRUE	FALSE
1 Products classified as "dietary supplements" consist of herbs and vitamins and mineral supplements only.		
2 Dietary supplements must be tested for safety and effectiveness before they can be sold.		
3 Herbal remedies have been used for over 100 years in Germany, so they must be safe and effective.		
4 "Probiotics" are "friendly bacteria" that have positive effects on health.		

Key Concepts and Facts

- Dietary supplements include vitamin and mineral pills, herbal remedies, proteins and amino acids, fish oils, and other products.

- Dietary supplements do not have to be shown to be safe or effective prior to being sold.

- Although food is the preferred source of vitamins and minerals, certain people benefit from judiciously selected vitamin or mineral supplements.

Answers to
NUTRITION SCOREBOARD

		TRUE	FALSE
1	Dietary supplements include herbs, vitamin and mineral supplements, protein powders, amino acid and enzyme pills, fish oils and fatty acids, hormone extracts, and other products.		✔
2	Dietary supplements can be sold without proof of their safety or effectiveness.		✔
3	Results of clinical trials, and not historical use, are the gold standard for determining the safety and effectiveness of herbal remedies and other dietary supplements.		✔
4	Yes! There are healthful bacteria in a number of foods.	✔	

dietary supplements
Any products intended to supplement the diet, including vitamin and mineral supplements; proteins, enzymes, and amino acids; fish oils and fatty acids; hormones and hormone precursors; and herbs and other plant extracts. Such products must be labeled "Dietary Supplement."

What do vitamin E supplements, amino acid pills, and herbal remedies all have in common? They are members of the increasingly popular group of products called ***dietary supplements***—and they are all discussed in this unit. Types of dietary supplements available to consumers are presented in Table 24.1. About half of U.S. adults use one or more of these products.[1] Dietary supplements are supposed to "supplement the diet." Many do that and have beneficial health effects in some people, but others neither supplement the diet nor provide health benefits. These

TABLE 24.1
Types of dietary supplements

Type	Example
1. Vitamins and minerals	Vitamins C and E, selenium
2. Herbs (botanicals)	Dong quai, ginseng, saw palmetto
3. Proteins and amino acids	Shark cartilage, chondroitin, creatine
4. Hormones, hormone precursors	DHEA, "Andro"
5. Fats	Fish oils, DHA, lecithin
6. Other plant extracts	Garlic capsules, fiber, cranberry concentrate, bee pollen

Photo Disc

Vitamin, mineral, protein, and amino acid supplements.

Botanical supplements, such as dong quai, often come in gel caps.

Plant extracts can also be taken in liquid form, as tinctures.

Herbal teas are a familiar source of dietary supplements.

disparate products are grouped together because they are regulated by a common set of rules.

You will learn from this unit that taking some of the dietary supplements available on the market can be a gamble. This "buyer beware" situation exists because of the loose rules that govern dietary supplements. You will also learn about "functional foods" and some exciting, new developments related to intestinal fertilizers and friendly bacteria (no kidding).

Regulation of Dietary Supplements

You can call anything a dietary supplement, even something you grow in your back yard.
—Donna Porter, RD, PhD, Congressional Research Service

In 1994 Congress passed the Dietary Supplement Health and Education Act, which started the explosion in the availability of dietary supplements. Under the act, dietary supplements are minimally regulated by the Food and Drug Administration (FDA); they do not have to be tested prior to marketing or shown to be safe or effective.[2] Although often advertised to relieve certain ailments, they are not considered to be drugs. Consequently, dietary supplements are not subjected to vigorous testing to prove safety and effectiveness, as drugs must be. Responsibility for evaluating the safety of dietary supplements lies with manufacturers and not the FDA. Supplements are deemed unsafe when the FDA has proof they are harmful. Since few dietary supplements have been adequately tested, and because results of studies showing negative effects may never see the light of day,[3] it is difficult to prove them to be unsafe. The FDA largely relies on reports of ill effects from manufacturers, health professionals, and consumers to assess supplement safety. Since 1994, the FDA has received thousands of reports of adverse effects of supplements (primarily for herbs), including several hundred deaths.[4]

According to FDA regulations (Table 24.2), dietary supplements must be labeled with a "Supplemental Facts" panel that lists serving size, ingredients, and

---

TABLE 24.2

FDA regulations for dietary supplement labeling

1. Product must be labeled "Dietary Supplement."
2. Product must have a "Supplemental Facts" label that includes serving size, amount of the product per serving, % Daily Value of essential nutrients, a list of other ingredients, and the manufacturer's name and address.
3. Nutrient claims (such as "low in sodium" and "high in fiber") can be made on labels of products that qualify based on nutrition labeling regulations.
4. Structure/function claims about how the product affects normal body structures (such as "helps maintain strong bones") or functions ("enhances normal bowel function") can be made on product labels. If a structure/function claim is made, this FDA disclaimer must appear:

   This statement has not been evaluated by the FDA. This product is not intended to diagnose, treat, cure, or prevent any disease.

Nutrition Facts
Serving size 1 Tablet

Amount Per Serving	% DV
Melatonin 3 mg	*

*Daily Value (DV) not established

Other Ingredients: Dicalcium Phosphate, Cellulose (Plant Origin), Vegetable Stearic Acid, Vegetable Magnesium Stearate, Silica, Croscarmellose.

GUARANTEED FREE OF: wheat, yeast, soy, corn, sugar, starch, milk, eggs. No artificial colors, flavors. No chemical additives. No preservatives. No animal derivatives.

Directions: As a dietary supplement for adults, take one (1) tablet, under the direction of a physician, only at bedtime as Melatonin may produce drowsiness. **DO NOT EXCEED 3 MG IN A 24 HOUR PERIOD.**

Warning: For Adults. Use only at bedtime. This product is not to be taken by pregnant or lactating women. If you are taking medication or have a medical condition such as an autoimmune condition or a depressive disorder, consult your physician before using this product. **NOT FOR USE BY CHILDREN 16 YEARS OF AGE OR YOUNGER.** Do not take this product when driving a motor vehicle, operating machinery or consuming alcoholic beverages.

In case of accidental overdose, seek professional assistance or contact a Poison Control Center immediately.

KEEP OUT OF REACH OF CHILDREN

percent Daily Value (%DV) of essential nutrient ingredients. Products can be labeled with a health claim, such as "high in calcium" or "low fat," if the product qualifies according to the nutrition labeling regulations. Supplements can also be labeled with "structure/function" claims. These claims cannot refer to disease prevention or treatment effects. Claims such as "improves circulation," "prevents wrinkles," "supports the immune system," and "helps maintain mental health" can be used, whereas "prevents heart disease" or "cures depression" cannot be. If a function claim is made on the label or package inserts, the label or insert must include the FDA disclaimer that states the FDA does not support the claim. (This is done to reduce the FDA's liability for problems that may be caused by supplements.) Nonetheless, many people wholeheartedly believe health claims made for supplements.[5]

The Federal Trade Commission (FTC) regulates claims for dietary supplements made in print and broadcast advertisements, including direct marketing, Internet sites, and infomercials. Claims made for dietary supplements in advertisements are supposed to be truthful, but often are not. Although some companies have been prosecuted for making false and misleading claims, neither the FDA nor the FTC has sufficient resources to fully monitor products and enforce laws related to dietary supplements.[6] The FDA has developed an "Adverse Events Reporting System" Web site (www.cfsan.fda.gov) that simplifies recording and tracking of adverse effects of dietary supplements.

Vitamin and Mineral Supplements

Multivitamin and mineral supplements—such as vitamins E or C, calcium, or magnesium—are among the wide variety of vitamin and mineral supplements used by consumers. They represent the most popular type of dietary supplement, with approximately one-third of U.S. adults using them daily.[7] Intake levels of vitamins and minerals below the "Tolerable Upper Limits" of the Dietary Reference Intake values (given on the inside front cover of this book) are safe for the vast majority of people, although lower amounts are related to their optimal functioning in the body.[8] Certain concerns related to vitamin and mineral supplements exist. The *bioavailability* of nutrients contained in some supplements remains uncertain. In general, the wider the assortment of vitamins and minerals in a supplement, the lower the absorption of each.[9] Minerals are particularly prone to forming unabsorbable complexes with each other, reducing the bioavailability of multiple minerals in the supplement.

bioavailability
The amount of a nutrient consumed that is available for absorption and use by the body.

Vitamin and Mineral Supplements: Who Benefits?

Supplements can have positive effects on health, as shown by the examples given in Table 24.3. Vitamin and mineral supplements have come a long way from primarily being used to treat vitamin and mineral deficiency diseases.

Using Vitamin and Mineral Supplements for the Wrong Reason   People often take supplements as a sort of insurance policy against problems caused by poor diets. Although multivitamin and mineral supplements may help fill in some of the nutrient gaps caused by poor food habits, they can't make a bad diet good. Whether a diet is good or bad is determined by more than its vitamin and mineral content. The "goodness" of a diet also depends on its content of essential fatty acid, fiber, water, and other nutrients. Additionally, supplements do not provide phytochemicals found in food such as flavones and antioxidant pigments that have positive influences on health.

One of the most serious consequences of supplements results when they are used as a remedy for health problems that can be treated, but not by vitamins or minerals. Vitamin and mineral supplements have not been found to be an effective treatment for behavioral problems, sexual dysfunction, hair loss, autism, chronic fatigue syndrome, obesity, or stress, for example.

The Rational Use of Vitamin and Mineral Supplements   Like all medications, vitamin and mineral supplements should be taken only if there is a need for them. If they are taken, dosages should not be excessive. Guidelines for the selection and use of vitamin and mineral supplements can be found in the "Health Action" below.

Herbal Remedies

A weed is what we call a plant whose virtues have not yet been discovered.
—RALPH WALDO EMERSON

Herbal remedies have a long history, are used worldwide, and are moving from alternative to mainstream health care in North America. Approximately 20% of U.S. adults use herbal supplements each year.[11]

The herb pharmacopoeia includes over 550 primary herbs known by at least 1800 names. Approximately 30% of all modern drugs are derived from plants,[12] and there's no doubt that many additional plants and plant ingredients benefit health. Plant products known to treat disease are considered drugs, however. Those that have not passed the scientific tests needed to demonstrate safety and effectiveness in disease treatment are often considered herbs. Yet, the truth is that some products sold as herbs in the United States have druglike effects on body functions. An example of an herb that acts powerfully like a drug is given in Illustration 24.1.

TABLE 24.3

Who may benefit from vitamin and mineral supplements? Here are some examples:[7,8,10,13]

- People with vitamin and mineral deficiency diseases
- Newborns (vitamin K)
- People living in areas without a fluoridated water supply (fluoride)
- Vegans (vitamins B_{12} and D)
- Pregnant women (iron and folate)
- People experiencing blood loss (iron)
- Elderly persons on limited diets (multiple vitamins and minerals)
- People on restricted diets (multiple vitamins and minerals)
- Adults with rheumatoid arthritis (EPA, DHA)
- People at risk for osteoporosis due to low calcium intake and poor vitamin D status (calcium, vitamin D)
- People with alcoholism
- People being treated for depression (folic acid)
- Elderly people at risk of vitamin B_{12}, vitamin D, and folate inadequacy

---

HEALTH *Action* Guidelines for Choosing and Using Vitamin and Mineral Supplements

1. Purchase supplements labeled "USP." They are tested for purity and dose.
   - Terms such as "release assured," "laboratory tested," "quality tested," and "scientifically blended" on supplement labels guarantee nothing.
2. Check the expiration date on supplements. Use unexpired supplements.

3. Choose supplements containing 100% of the Daily Value or less.
4. Take supplements with meals.
5. If you have a diagnosed need for a specific vitamin or mineral, take that individual vitamin or mineral and not a multiple supplement.
   - Avoid calcium supplements made from oyster shells, bone, or oral calcium. They may contain lead or aluminum.

6. Store supplements where small children cannot get at them.

Remember! Consult your health care provider about health problems before you start taking supplements to try to treat the problems.

Illustration 24.1
Herb or drug?

"Alternative medicine becomes standard medicine when it is proven true."
—PAUL OKUNIEFF, MD

Statins are a popular type of prescription drug that effectively lower LDL cholesterol, raise HDL cholesterol (the good one), and reduce the risk of heart disease. It turns out that statins are also available in herbal form.

Cholestin is an "herb" made from rice grown with red yeast. If subjected to specific growing conditions, the concentration of naturally occurring statins in the rice-yeast mixture can be increased greatly. In 2000, the FDA ruled that cholestin was a drug, and it was taken off the market. Cholestin has been replaced by similar but somewhat different products, such as Cholestatin, shown here.

Effects of Herbal Remedies

Purported effects and side effects of some herbal and similar remedies are listed in Table 24.4. Herbal remedies, like drugs, have biologically active ingredients that can have positive, negative, and neutral effects on body processes. Basically, an herbal remedy (or a drug) is considered valuable if it has beneficial effects on body processes and if the benefits are not outweighed by the risks. Knowledge of the risks and benefits of many herbal supplements remains incomplete. However, available evidence suggests that some herbal remedies are safe and effective while others appear to be neither.

Which herbal remedies are likely ineffective or unsafe? Human experimentation with various botanicals over the centuries to the present time has helped to identify herbs that lack beneficial effects or have negative side effects. Table 24.5 lists some of these herbs. The extent to which the herbs included in the table pose a risk to health depends on the amount taken and the duration of use, the age and health status of the user, and other factors.

Which Herbal Supplements Are Potentially Beneficial?   Results of many studies on the safety and benefits of specfic herbal remedies have been mixed. The bulk of existing evidence, however, indicates that some may be beneficial:

Illustration 24.2
Rose hips in Tuscany.

- Ginkgo biloba may decrease symptoms in patients with Alzheimer's disease and increase blood flow to the brain.

- Cranberry may relieve urinary tract infection.

- Garlic can reduce elevated blood cholesterol levels and blood pressure.

- Saint John's wort may relieve mild and moderate depression.

- Rose hips have mild laxative and diuretic effects (Illustration 24.2).

- SAMe may relieve mild depression and arthritis pain.

- DHEA supplements may be an effective treatment for midlife-onset depression.

- Senna may relieve constipation.

- Ginger may prevent motion sickness.[14, 15]

TABLE 24.4

Proposed effects and potential side effects of herbal and similar remedies[15]

Herb/Other Remedy	Proposed Effects	Potential Side Effects
Glucosamine-chondroitin sulfate	Slows progression of osteoarthritis and its pain	Gastrointestinal upset, fatigue, headache
Ginseng	Increases energy, normalizes blood glucose, stimulates immune function, relieves impotence in males, cancer prevention	Insomnia, hyperactivity, hypertension, diarrhea, menstrual dysfunction Interacts with blood thinners[a]
SAMe	Relieves mild depression, arthritis pain	May trigger manic excitement, nausea
Garlic	Lowers blood cholesterol, relieves colds and other infections	Heartburn, gas, blood thinner
Cholestin	Maintains desirable blood cholesterol levels	Safety of some ingredients unknown
Echinacea	Prevents and treats colds and sore throat	Allergies to plant components
DHEA	Improves memory, mood, physical well-being	Increases risk of breast cancer
Creatine	Sport supplement (increased performance in short, high-intensity events)	Kidney disease
Saw palmetto	Improves urine flow, reduces urgency of urination in men with prostate enlargement	Nausea, abdominal pain
Ginkgo biloba	Increases mental skills, delays progression of Alzheimer's disease, increases blood flow and sexual performance, decreases depression	Nervousness, headache, diarrhea, nausea. Interacts with blood thinners.
Shark cartilage	Treats lung cancer	Safety unknown
Saint John's wort	Relieves mild-to-moderate depression	Dry mouth, dizziness, sensitivity to light. Interacts with many drugs and chemotherapy.
Kava (kava-kava)	Relaxation, stress relief, sleep aid, mood enhancer	Liver injury
Black cohosh	Improves menopausal and PMS symptoms	Gastric upset, dizziness, headache, low blood pressure. May increase risk of breast cancer. Interacts with anti-hypertension drugs.
Coenzyme Q10 (ubiquinone)	Remedy for heart disease, cancer, Parkinson's disease	Nausea, diarrhea, rash, low blood glucose Interacts with some drugs

[a]Blood thinners include aspirin, warfarin, coumarin.

TABLE 24.5

Examples of dietary supplements that may not be effective or safe[15]

Apricot pits (laetrile)	Ephedra	Pennyroyal oil
Androstenedione (Andro)	Eyebright	Poke weed
Aristolochic acid	Ginkgo seed	Sassafras
Belladonna	Kava	Skullcap
Blue cohosh	Licorice root	Star anise
Bitter orange	Liferoot	Vinca
Borage	Lily of the valley	Wild yam
Broom	Lobelia	Wormwood
Chaparral	Ma huang/ephedra	Yohimbe
Chinese yew	Mandrake	
Comfrey	Mistletoe	
Dong quai	Organ/glandular extracts	

Illustration 24.3
These symbols on the labels of dietary supplements certify quality ingredients and accurate labeling but do not address product safety or effectiveness.

The list of effective herbal remedies may grow, or shrink, with time as additional studies are completed.

Ephedra (Ma Huang)   Ephedra, or Ma huang, is chemically and functionally similar to adrenaline, the "fight-or-flight" hormone. It is traditionally used in China to treat respiratory problems but has gained popularity in the United States as a weight-loss product and athletic performance enhancer. Ephedra delivers limited benefits for these purposes, however, and can produce life-threatening side effects on the heart and nervous system. Side effects and deaths related to ephedra use led to its being banned from the market early in 2004.[15]

A Measure of Quality Assurance for Dietary Supplements

Not all dietary supplements contain the amounts of herbal and other ingredients declared on the label, and some contain contaminants such as bacteria, mold, and lead. Analyses of the composition of 25 ginseng products, for example, found that concentrations of ginseng compounds in the supplements were up to 36 times different than labeled amounts.[16] Similar studies of echinacea products found that 10% of samples contained no echinacea, and half contained the labeled amount. Some male "enlargement" supplements have been found to be contaminated with *E. coli*, mold, lead, and pesticide residues.[17] Oddly enough, dietary supplements are often labeled as "pure," "natural," or "quality assured."

There is no government body that monitors the contents of herbal supplements. Private groups, such as the U.S. Pharmacopeia (USP), the National Formulary (NF), and Consumer Laboratories (CL), offer testing services to ensure that dietary supplements meet standards for disintegration, purity, potency, and labeling (Illustration 24.3). Products that pass these tests can display "USP," "NF," or the CL symbol boldly on product labels. These letters represent quality ingredients and labeling but do not address product safety or effectiveness. New regulations for dietary supplements are being considered by the FDA and other groups.[18]

Due to the lack of studies and potential dangers, the FDA has advised dietary supplement manufacturers not to make claims related to pregnancy for herbs and other products.[19] Considerations for the use of herbal supplements are summarized in the "Health Action" below.

HEALTH *Action* Considerations for the Use of Herbal Remedies

1. Don't use herbal remedies for serious, self-diagnosed conditions such as depression, persistent headaches, and memory loss. (You might benefit more from a different treatment.)
2. Let your doctor know what herbal remedies you take.
3. If you take prescription medications, clear the use of herbal remedies with your doctor.
4. Don't use herbal remedies without medical advice if you are attempting to become pregnant or if you are pregnant or breast-feeding.
5. Don't mix herbal remedies.
6. If you are allergic to certain plants, make sure herbal remedies are not going to be a problem before you use them.
7. If you have a bad reaction to an herbal remedy, stop using it and report the reaction to the FDA from the site www.cfsan.fda.gov.
8. Buy herbs labeled with "USP," "NF," or the CL in a beaker symbol.
9. Investigate brands, herb safety, and effectiveness by checking into one or more of the Web sites and resources listed at the end of this unit.

Reality Check

Herbals on the Web

Can you trust information on herbal products you see on the Web?

Who gets thumbs up?
Answers on next page

Sarah
When I'm sick, the first place I go is to the Web to find an herb that will make me feel better.

Pablo
Herbal products I see advertised on the Web look like they'll work for my problem. But I'm conflicted about buying them, because I'm not sure I can trust the information.

Functional Foods

Leave no aisle unfortified.
—HEADLINE FOR ARTICLE IN A FOOD INDUSTRY MAGAZINE

Also known as "neutraceuticals," *functional foods* include a variety of foods and products that have in reality or theoretically been modified to enhance their contribution to a healthy diet (Table 24.6).[20] All foods are functional in that they provide nutrients. Foods considered "functional," however, are generally specifically formulated to supply one or more dietary ingredients that may improve health, or they are foods containing high amounts of substances that tend to prevent certain diseases.[26] Because there is no statutory definition for what constitutes functional foods, there are no specific regulations that apply to them. Health claims can be made for functional foods given approval by the FDA.[20] Functional foods containing food additives not on the Generally Accepted As Safe (GRAS) list must be tested and approved by the FDA before being sold.

Functional foods that appear to benefit health are listed in Table 24.7. Increasingly, however, the list of functional foods is becoming infiltrated with sports bars, soups, beverages, and cereals spiked with vitamins, minerals, and herbs.

Some of these products carry labels with unsubstantiated health claims and may be of no benefit or are potentially unsafe.[21] For these products, the label "functional food" is a marketing term.

functional foods
Generally taken to mean foods, fortified foods, and enhanced food products that may benefit health beyond the effects of essential nutrients they contain.

TABLE 24.6

How are foods made to be "functional"?

Foods are made to be "functional" by:

1. Taking out potentially harmful components (e.g., cholesterol in egg yolk and lactose in milk)
2. Increasing the amount of nutrients and beneficial nonnutrients (e.g., fiber-fortified liquid meals, calcium-, and vitamin C-fortified orange juice)
3. Using beneficial substances in food production or products (e.g., using "friendly" bacteria in fermented milk and soy products)

Herbals on the Web

Many people use the Web as a source of information about illness remedies. Unfortunately, over half of the sites selling herbal products illegally claim the products prevent or cure specific diseases as if they were really drugs and do not include the required FDA disclaimer.[25]

Take the worry out of decisions about herbal supplements. Check them out using the scientifically reliable Web sites listed at the end of this unit.

Sarah

Pablo

prebiotics
"Intestinal fertilizer." Certain fiberlike forms of nondigestible carbohydrates that support the growth of beneficial bacteria in the gut.

probiotics
"Prolife." Strains of *lactobacillus (lac-toe-bah-sil-us)* and bifido-bacteria *(bif-id-dough bacteria)* that have beneficial effects on the body. Also called "friendly bacteria."

Prebiotics and Probiotics: From "Pharm" to Table

The terms *prebiotics* and *probiotics* were derived from "antibiotics" due to their probable effects on increasing resistance to various diseases. They are in a class of functional foods by themselves. *Prebiotics* are fiberlike, nondigestible carbohydrates that are broken down by bacteria in the colon. The breakdown products foster the growth of beneficial bacteria. For this reason they are considered "intestinal fertilizer." *Probiotics* is the term for live, beneficial—or "friendly"—bacteria that enter food through fermentation and aging processes.[22] Table 24.8 lists food and other sources of pre- and probiotics. Availability of foods and other products containing prebiotics and probiotics is much more common in Japan and European countries than in Canada or the United States.[23] However, availability of such products is increasing as research results shed light on their safety and effectiveness.

The digestive tract, particularly the colon, is home to over 500 species of microorganisms representing 100 trillion bacteria (and billions of viruses and fungi, too). Some species of bacteria such as E. coli may cause disease, whereas others such as lactobacillus and bifidobacteria prevent various diseases.[24] Pre- and probiotics have been credited with important health effects (Table 24.9); the right combination of each fosters the proliferation of healthful bacteria in the colon, nose, and some other internal canals of the body. The concept of the combined benefits of pre- and probiotics has been termed "symbiotics."[21] Because it is difficult to

Taking the coke out of Coke Coca Cola used to be fortified with cocaine. During the early part of the 1900s, the "coke" ingredient was removed.

TABLE 24.7

Example of functional foods with apparent health benefits[20–24]

Functional Food	Benefit
Stanol and sterol fortified margarine, psyllium fiber, soy protein, whole oat products, garlic, nuts	Reduced blood levels of LDL cholesterol
Omega-3 fatty acids	Reduced heart disease risk
Grapes, grape juice	Decreased blood clots in blood vessels
Cranberry juice	Decreased urinary tract infections
Green tea, cooked tomato products, cruciferous vegetables, conjugated linoleic acid	Decreased risk of certain types of cancer
Folic acid–fortified breads and cereals	Decreased risk of neural tube defects, heart disease
Probiotics	Decreased risk of infection, lactose intolerance, food allergies, other disorders

TABLE 24.8

Food and other sources of prebiotics and probiotics[1]

Prebiotics	Probiotics
Chicory	Fermented or aged milk and milk products:
Jerusalem artichokes	• Yogurt with live culture
Wheat	• Buttermilk
Barley	• Kefir
Rye	• Cottage cheese
Onions	• Dairy spreads with added inulin
Garlic	Other fermented products
Leeks	• Soy sauce
Prebiotics tablets and powders and	• Tempeh
nutritional beverages	• Fresh sauerkraut
	• Miso
	Breast milk

TABLE 24.9

Apparent and potential benefits of prebiotics and probiotics[20–24]

Prebiotics	Probiotics
• Prevention and treatment of diarrhea and constipation	• Treatment for chronic diarrhea and traveler's diarrhea
• Prevention of colon cancer	• Treatment of lactose intolerance and some food allergies
• Increased mineral absorption	• Prevention of infections in gastrointestinal tract, ear canals
• Decreased blood triglyceride, glucose, and insulin levels	• Recolonization of colon after antibiotic therapy
	• Increased bacterial production of some B vitamins
	• Decreased blood cholesterol levels
	• Decreased risk of certain types of cancer
	• Treatment of irritable bowel syndrome
	• Decreased dental caries
	• Decreased high blood pressure
	• Prevention of wound infections

A lactobacillus species (blue) taking over harmful E. coli bacteria (red).

Source: Probiotics: Their tiny worlds . . . The Scientist July 22, 2002, pp. 20-22; author: Bob Beale.

recolonize gut bacteria, the benefits of pre- and probiotics last only as long as dietary intake does.[23]

Prebiotics and probiotics are assumed to be safe because they have been part of the human diet for centuries. The primary, negative side effect of their ingestion is flatulence.[23]

Final Thoughts

From dietary supplements to bacteria: The universe of substances considered dietary ingredients is expanding. Knowledge about potential benefits of pre- and probiotics is charging ahead, and advances are catching the attention of consumers and health care professionals. Perhaps you never thought that "intestinal fertilizer" or "friendly bacteria" would ever intentionally pass through your lips. But that may well be the nature of dietary ingredients to come. If you didn't know before, you do now.

Nutrition Up Close

Supplement Use and Misuse

Focal Point: Decide if a dietary supplement is warranted in these situations.

People take dietary supplements for many reasons, but is their use justified? Apply the information from this chapter to determine if you agree with the decisions made in each of the following scenarios.

1. Martha works part-time and takes a full load of classes. Like many college students, she is always on the go, often grabbing something quick to eat at fast-food restaurants or skipping meals altogether. Nevertheless, Martha feels confident her health will not suffer, because she takes a daily vitamin and mineral supplement.

   Is a supplement warranted in this case? _____

   Why or why not? _____

2. Sylvia is a 23-year-old student diagnosed with iron-deficiency anemia. She has learned in her nutrition class that it is preferable to get vitamins and minerals from food instead of supplements. Therefore, instead of taking the iron pills her doctor has prescribed, Sylvia has decided to counteract the anemia by increasing her consumption of iron-rich foods.

   Is a supplement warranted in this case? _____

   Why or why not? _____

3. John is a 21-year-old physical education major involved in collegiate sports. He is very aware that nutrition plays an important role in the way he feels, so he is careful to eat well-balanced meals. In addition, John takes megadoses of vitamins and minerals daily. He is convinced they enhance his physical performance.

   Is a supplement warranted in this case? _____

   Why or why not? _____

4. Roberto, a native Californian, is backpacking through Europe when he is slowed down by constipation. He visits a pharmacy where English is spoken and is given senna by the clerk rather than a fast-acting medicine he expected. Roberto has never taken an herb before and is not sure how his body will react to it, or if it will work.

   Should Roberto try the senna, ask for a nonherbal drug, or take another action? (Assume they cost the same.) _____

   What's the rationale for this decision? _____

5. While shopping at the mall, Yuen notices a kiosk selling "Hypermetabolite," a weight-loss product that guarantees you'll lose 5 pounds a week without dieting. Having gained 10 pounds since she started working full-time, Yuen decides to try it. Her examination of the product's label reveals that an ephedra-derivative and Asian ginseng are major ingredients.

   Should Yuen take Hypermetabolite for weight loss? _____

   Why or why not? _____

FEEDBACK (including answers) can be found at the end of Unit 24.

Key Terms

bioavailability, page 24-4
dietary supplements, page 24-2

functional foods, page 24-9
prebiotics, page 24-10

probiotics, page 24-10

Review Questions

True False

1. Dietary supplement labels must include a "Supplement Facts" panel and may include qualifying nutrient claims and structure/function claims. ☐ ☐

2. Say you read a dietary supplement label that claims the product "helps enhance muscle tone or size." Because the statement appears on the label, it must be true. ☐ ☐

3. The Recommended Dietary Allowance (RDA) for vitamin A for breastfeeding women is 1300 µg per day. A supplement containing 3300 IU vitamin A would supply less than the RDA amount. ☐ ☐

True False

4. Various vitamin and mineral supplements appears to benefit pregnant women, the elderly, and adults at risk of osteoporosis. ☐ ☐

5. Vitamin and mineral supplements available over-the-counter can be safely consumed at any dose levels. ☐ ☐

6. A "USP" label displayed on dietary supplements indicated the product has been tested for safety and effectiveness. ☐ ☐

7. Prebiotics support the growth of beneficial bacteria in the intestines. ☐ ☐

8. Probiotics, or specific strains of bacteria, have been credited with important health effects. ☐ ☐

Media Menu

www.mskcc.org/mskcc/html/11570.cfm
The Memorial Sloan-Kettering Cancer Center provides quick access to reliable information of specific herbs at this site.

www.nccam.nih.gov
Go here for the home page for the National Center for Complementary and Alternative Medicine.

http://dietary-supplements.info.nih.gov
Gain access to state-of-the-science information on vitamin, mineral, herbal, and other

supplements from the National Institutes of Health.

http://ods.od.nih.gov/index.aspx
The National Institutes of Health, Office of Dietary Supplements home page provides descriptions for more than 80 vitamin, mineral, and herbal supplements.

www.cfsan.fda.gov
This site provides news on latest developments, warnings, and health effects of di-

etary supplements. Access the Adverse Events Reporting System from this address.

http://vm.cfsan.fda.gov/~dms/dietsupp.html
This page contains the provisions of the Dietary Supplement Health and Education Act of 1994.

www.mayoclinic.org
Search "herbs" and get the skinny on the helpful and harmful ones.

Notes

1. Walker R et al. Dietary Supplements in the US: pitfalls and safety, www.medscape.com/viewarticle/522892, posted 02/07/2006.

2. ADA/APhA Special Report from the Joint Working Group on Dietary Supplements. A healthcare professional's guide to evaluating dietary supplements. American Dietetic Association, American Pharmaceutical Association, 2000;47;1–40.

3. De Smet P, Herbal remedies, N Engl J Med 2002;347:2046–56.

4. Issues relating to ephedra-containing dietary supplements. http://energycommerce.house.gov/108/Hearings/07232003hearing1021/Zipes1631.htm, accessed 10/06.

5. ADA/APhA Special Report; and Community Nutrition Institute Newsletter, 2000 July 21:2.

6. Herbal remedy sellers on the Web break the rules, www.nlm.nih.gov/medlineplus/news/fullstory14069.html, accessed 10/03.

7. Willett WC, Stampfer MJ. What vitamins should I be taking, doctor? N Engl J Med 2001;345:1819–24.

8. Fletcher RH, Fairfield KM, Vitamins for chronic disease prevention in adults: clinical applications, JAMA 2002;287:3127–9.

9. Lind T et al. A community-based randomized trial of iron and zinc supplementation. Am J Clin Nutr 2003; 77:883–90.

10. Barringer TA et al., Effect of multivitamin and mineral supplement on infection and quality of life: a randomized, double-blind, placebo-controlled trial, Ann Intern Med 2003;93:365–71.

11. Kelly JP et al. Changing trends in herbal supplement use. JAMA, jama.ama-assn.org, 2/22/05.

12. Belew C. Herbs and the childbearing woman. Guidelines for midwives. J Nurse Midwifery 1999;44:231–52.

13. Fairfield KM, Fletcher RH. Vitamins for chronic disease prevention in adults: scientific review. JAMA 2002;9:288:1720–4.

14. Schmidt PJ et al. DHEA may be effective for midlife-onset minor and major depression. Arch Gen Psych 2005;62:154–62.

15. About herbs, botanical, and other products. www.mskcc.org/mskcc/html/11570.cfm, accessed 10/06.

16. Harkey MR et al., Variability in commercial ginseng products: an analysis of 25 preparations, Am J Clin Nutr 2001;73:1101–60

17. Gilroy CM et al. Echinacea and truth in labeling, Arch Intern Med 2003;163:699–704.

18. Rennie D, DeAngelis CD. Stricter regulations of dietary supplements needed. JAMA 2003;289:1568–70.

19. HHS Statement: FDA statement concerning structure/function rule and pregnancy claims. 2000 Jan. 6.

20. Hyman P. Claims for functional foods under the current food regulatory scheme, Nutr Today 2002;37:217–9.

21. Hasler CM. Functional foods: benefits, concerns, and challenges, J Nutr 2002;132:3772–81.

22. Sanders ME. Probiotics: considerations for human health, Nutr Rev 2003;61:91–7.

23. Brannon CB. Prebiotics: feeding friendly bacteria. Today's Dietitian 2003;Sept:12–16.

24. Beale B. Probiotics: their tiny worlds are under scrutiny, The Scientist 2002;July:20–2.

25. Morris et al. Herbal remedy sellers (www.nlm.nih.gov/medlineplus/news/fullstory 14069.html).

26. Halsted CH. Dietary supplements and functional foods: 2 sides of a coin? Am J Clin Nutr 2003;77(suppl);1001S–7S.

Nutrition Up Close

Supplement Use and Misuse

Feedback for Unit 24

1. *Is a supplement warranted in this case?* No

   Why or why not? Martha is deceiving herself! She may be getting the vitamins and minerals she needs by taking a supplement, but this cannot make up for her poor food habits. She needs to improve the overall quality of her diet to ensure optimal health. If Martha follows the MyPyramid recommendations, she should be able to get all the nutrients she needs from what she eats.

2. *Is a supplement warranted in this case?* Yes

   Why or why not? This is one time when a supplement is in order. Sylvia needs to follow her doctor's advice and take the prescribed iron preparation to increase her hemoglobin and replete her iron stores. However, she is correct in consuming more iron-rich foods, too, so that after the anemia has been treated, she will not experience a relapse.

3. *Is a supplement warranted in this case?* No

   Why or why not? There is no scientific evidence that vitamin and mineral megadoses enhance physical performance. In fact, John may be setting himself up for toxicity reactions with prolonged intake of supplements at extremely high dosages.

4. *Should Roberto try the senna, ask for a nonherbal drug, or take another action?* He should take another action—get more information from the pharmacist so he can make a better-informed decision.

5. *Should Yuen take Hypermetabolite for weight loss?* No

   Why or why not? She shouldn't take the product because the ephedra derivative may have serious side effects and might interact with ginseng and because quick weight-loss strategies don't work in the longer run. Plus, there's no guarantee that the product will work or that it isn't mislabeled.

Answers to Review Questions

1. True, see pages 24-3, 24-4.
2. False, see page 24-4.
3. True, pages 24-3, 24-4, and the Glossary of Nutrients Measures table on the inside back cover of this book.
4. True, see page 24-5.
5. False, see page 24-5.
6. False, see page 24-8.
7. True, see page 24-10.
8. True, see pages 24-10, 24-11.

Water Is an Essential Nutrient

NUTRITION SCOREBOARD

	TRUE	FALSE
1 Bottled water is better for health than municipal water from your faucet.		
2 Drinking lots of water helps flush toxins out of your body.		
3 Adequate water intake reduces the risk of some types of cancer and kidney stones.		
4 You can't drink too much water.		

Key Concepts and Facts

- Water is an essential nutrient. It is a required part of the diet. Deficiency symptoms develop when too little is consumed, and toxicity symptoms occur when too much is ingested.

- Functions of water include maintenance of body hydration and temperature, removal of waste products, and participation in energy formation. It is our major source of fluoride.

- Water is a precious resource whose availability and quality are threatened by wasteful use and pollution.

Answers to NUTRITION SCOREBOARD

		TRUE	FALSE
1	Bottled waters have not been shown to be better for health or to be "purer" than tap water obtained from city water supplies.		✔
2	Water does not function as a toxic waste disposal in our bodies.		✔
3	Adequate intake of water from fluids and food (9 to 12 cups a day) reduces the risk of breast, colon, and bladder cancer and of kidney stone formation.[1]	✔	
4	Yes, you can drink too much water, and it results in water intoxication.[1]		✔

Water: Where Would We Be without It?

Photo Disc

Ask any three people you know to name as many essential nutrients as they can. If they mention water, give them a prize. Our need for water is so obvious that it is often taken for granted. Well, that's not going to happen in this text!

Water differs from other essential nutrients in that it is liquid, and our need for it is measured in cups rather than grams or milligrams. Without it, our days are limited to about six. Water is the largest single component of our diet and body. It is a basic requirement of all living things. Now, how could something this important be so easily forgotten?

Water's Roles as an Essential Nutrient

Water qualifies in all respects as an essential nutrient. It is a required part of our diet; it performs specific functions in the body; and deficiency and toxicity signs develop when too little or too much is consumed.

Water is our body's main source of fluoride, an essential mineral needed for the formation and maintenance of enamel and resistance to tooth decay. Fluoride is naturally present in some water supplies and is added to virtually all municipal water supplies in the United States and Canada. Because it is a gas that dissolves in water, fluoride is not removed by water pitcher or faucet filters.

Water is the medium in which many chemical reactions take place within our body. Water plays key roles in energy formation—it is produced as an end product of energy formation from carbohydrates, proteins, and fats. We continue to produce and excrete water even if we quit drinking it for awhile, because energy production is an ongoing process. Water is needed to "carry" nutrients to cells and waste products away from them. Additionally, water acts as the body's cooling system. When our internal temperature gets too high, water transfers heat to the skin and releases it in perspiration. When we're too cool, less water—and heat—is released through the skin. Water's functions are summarized in Table 25.1.

TABLE 25.1

Key functions of water in the body

- Provides a medium for chemical reactions
- Participates in energy formation
- Transports nutrients and waste products
- Helps regulate body temperature

Water has been given credit for other functions in the body, but undeservedly. Drinking more water than is normally needed does not prevent dry, wrinkled skin, lead to weight loss, or flush toxins out of the body.[2] Nor will it cure chronic fatigue, arthritis, migraines, or hypertension.

Water, Water, Everywhere The body of a 160-pound person contains about 12 gallons of water (Illustration 25.1). Adults are approximately 60 to 70% water by weight. Water is distributed in the body in blood, the spaces in between cells, and in all cells. The proportion of water in body tissues varies: blood is 83% water, muscle 75%, and bone 22%. Even fat cells are 10% water.[3]

■ Most Foods Contain Lots of Water, Too Most beverages are more than 85% water, and fruits and vegetables are 75 to 90% (Illustration 25.2). Meats, depending upon their type and how well done they are, contain between 50 and 70% water. Although it is nearly impossible to meet your need for water from solid foods alone, the water content of foods makes an important contribution to our daily intakes. On average, about 31% of water intake comes from plain water, 44% from other beverages, and 25% from foods.[4]

Health Benefits of Water Adequate water consumption may benefit long-term as well as day-to-day health. Consumption of over 10 cups of fluid each day is associated with a decreased risk of bladder, breast, and colon cancer as well as of

Illustration 25.1
The body of a 160-pound person contains approximately 12 gallons of water. That's 96 pounds of water!

Illustration 25.2
The water content of some foods.

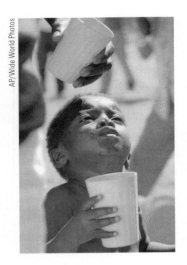

Illustration 25.3
A child waits for water at a refugee camp.

Illustration 25.4
Contaminated water supplies are a threat to the public's health.

kidney stone formation. People feel and perform better when they are adequately hydrated.[3]

The Nature of Our Water Supply

Water covers about three-quarters of the Earth's surface, yet very little of it is drinkable. Nearly 97% of the total supply is salt water, and only 3% is fresh. Of the freshwater supply, only a fourth of the total amount is available for use. The rest is located in polar and glacier ice. Although fresh water is readily available in most locations in the United States, this is not the case in a number of other countries.

Water is sufficiently scarce in parts of Russia that drinking-water dispensers are coin-operated. Fresh water is so highly prized in sections of the arid Middle East that having a decorative fountain in one's home is considered a sign of affluence. Drinking water is sampled, judged, and celebrated in Middle Eastern countries as ceremoniously as fine wines are in France. Although water is not a subject of envy or a cause for celebration in most parts of America, it is being increasingly viewed as a precious resource that must be protected.

Water scarcity has become a primary concern of many more nations in the 21st century (Illustration 25.3). The demand for water is increasing by 100 to 500% per year in countries in Central and South America, Africa, and Europe; but the supply is dwindling. Wasteful use of water, groundwater depletion, pollution, and leaky public water systems are contributing to water shortages and safety concerns (Illustration 25.4).[5] Without more effective protection of the world's water, the quantity of clean water available for agriculture and consumption may be insufficient, limiting food production and increasing the incidence of water-borne illnesses.[6]

The Environmental Protection Agency is responsible for the safety of public water supplies and has set maximum allowable levels of contaminants. Water quality is monitored by local water utilities, and the results are reported to state and federal officials. Problems identified are remedied and attempts are made to prevent future contamination in order to provide safe public water supplies.[5]

Does the Earth Supply "Gourmet" Water?   About 40 years ago, the French made drinking "fine waters" very stylish. In Paris, boulevardiers crowded sidewalk cafes for hours sipping chilled Perrier served with thinly sliced lemon. The popularity of bottled waters skyrocketed in many Western countries. In the United States, "mineral," "spring," and "seltzer" waters are now best sellers (Illustration 25.5). These waters have a strong, positive image. But what *are* these waters? How do they differ from the water we get from the faucet?

True *mineral water* is taken from protected underground reservoirs that are lodged between layers of rock. The water dissolves some of the minerals found in the rocks, and as a result, it contains a higher amount of minerals than most sources of surface water. Actually, most water contains some minerals and could be legitimately considered "mineral water." *Spring water* is taken from freshwater springs that form pools or streams on Earth's surface. True "seltzers" (not the sweetened kind you often find for sale) are *sparkling waters* that are naturally carbonated. Most seltzers, however, become bubbly by the commercial addition of pressurized carbon dioxide.

Bottled waters have their advantages. They are calorie free, are generally sodium free or low in sodium, and quench a thirst better than their major competitor, soft drinks. But they are no "purer" or better for you than tap water.[6] As a matter of fact, some bottled waters contain tap water. The Food and Drug Admin-

Illustration 25.5
Is bottled water better for you than tap water?
Consumers often perceive that bottled waters have fewer impurities and are better for health than tap water. Such beliefs are unfounded.

istration (FDA) estimates that 25% of the bottled water sold in the United States is bottled tap water.

■ **Is Bottled Water Safe?** Every now and then, a story about contaminated bottled water makes the news, raising concerns about the safety of bottled water and the way the industry is monitored. The FDA regulates the bottled water industry. Domestic bottlers must conform to specified standards of water safety (such as allowable levels of chemical contaminants) and labeling requirements. "Mineral" or "spring" water must be just that under FDA regulations. Bottled tap water must be labeled as such.

Bottlers are also subject to unannounced inspections by the FDA. Bottled water is classified as a very low-risk product, however, so domestic bottling plants are inspected, on average, every 5 years.[7] The FDA supplements its monitoring role by requiring that bottlers periodically test their water. Foreign bottling plants are not under the FDA's jurisdiction, but imported waters may be tested when they enter the United States.

Bottling plants that adhere to FDA regulations produce safe water. If you have questions about your favorite brand or want more information, call the Environmental Protection Agency's Safe Drinking Water Hotline (1-800-426-4791) between 8:30 A.M. and 5:00 P.M. Eastern Standard Time.

■ **Water Gimmicks** New waves of bottled water products are finding their way to grocery shelves throughout America. Tapping into consumer interest in vitamins, minerals, herbs, and fitness, producers are marketing "enhanced" bottled water, fortified with everything from ginseng to oxygen (Illustration 25.6). Although mineral waters provide absorbable calcium and magnesium that contribute to nutrient status and potentially to health,[8] the rationale for many of the

Illustration 25.6
"Specialty" bottled water products are of uncertain value to health or fitness.

Richard Anderson

Robin Sterling/Sterling Research

Illustration 25.7
Want to help protect your teeth from cavities? Look for word fluoride on the label of bottled water products.

products (other than their commercial appeal) is elusive. It makes little sense to fortify water with herbs, oxygen, or with nutrients better obtained from foods. Safety of some of the combinations of herbs put into bottled water is unclear, and they do not appear to supplement the diet in meaningful ways.[9] Adding oxygen to beverages and foods is total nonsense. Oxygen our bodies use for energy formation and metabolic processes comes from the air we breathe, and not the beverages we drink or foods we consume.

■ **Fluoridated Bottled Water** Bottled waters have been criticized for their lack of fluoride compared to fluoridated tap water. People of all ages need this mineral for bone health and the prevention of tooth decay. The industry caught on to this shortcoming and responded by offering fluoridated waters (Illustration 25.7). A quick look at the label can tell you if you're getting fluoride along with your water.

Meeting Our Need for Water

In general, individuals require enough water each day to replace water lost in urine, perspiration, stools, and exhaled air. Adequate Intakes (AIs) of water from fluids and food are set at 11 cups per day for women and 15 cups per day for men.[10] No-calorie water is the preferred source for meeting water needs.[11]

Built-in mechanisms that trigger thirst generally protect people from consuming too much or too little water.[4] People who do strenuous work, athletic or otherwise, in hot and humid weather need to consume enough water to replace the amount that is lost in sweat, urine, respiration, and evaporation of water from the skin's surface. How much extra water this takes varies from person to person, but it is often within the ballpark of a 50% increase. You know you are drinking enough water if you haven't lost weight after the physical activity and if your urine is pale yellow and produced in normal volume.[2]

Exposure to both cold weather and high altitudes increases water need. Cold air holds little moisture, so you lose more water when you breathe in cold, dry air than warm, moist air. People exposed to cold weather tend to dress in layers and lose body water in the form of sweat that pools next to the skin. Exposure to cold environmental conditions at high altitudes can increase urine production and water loss. High levels of physical activity in cold, high altitude climates further increases water need (Illustration 25-8).[12]

Prolonged bouts of vomiting, diarrhea, and fever increase water need, and that is why you should drink plenty of fluids when you experience these conditions. High-protein and high-fiber diets and alcohol increase your need for water. Losses in body water that accompany high levels of protein consumption are the reason people on high-protein weight-loss diets are encouraged to "drink a lot of water." Adding fiber to your diet augments water need because fiber increases

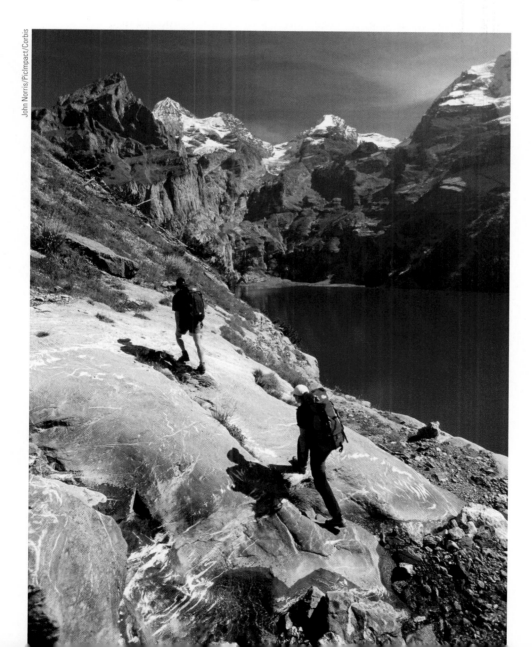

John Norris/PicImpact/Corbis

Illustration 25.8
Cold, high altitude conditions increase water need.

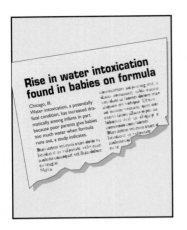

Illustration 25.9
This headline is accurate—overdilution of infant formula can lead to water intoxication in infants.

water loss in stools. The increased loss of water that goes along with alcohol intake explains why people get very thirsty after overindulging in spirits.

Is Caffeine Hydrating? In the past it was widely believed that beverages containing caffeine were not hydrating because caffeine acted as a diuretic. Recent and better research has demonstrated that this conclusion is incorrect. Caffeine does not increase urine output in people accustomed to drinking coffee, tea, and other beverages that contain caffeine.[4] So, count the coffee or tea you consume as contributing to your overall water intake.

Water Deficiency

A deficiency of water can lead to dehydration. Dehydrated people feel very sick. They are generally nauseated, have a fast heart rate and increased body temperature, feel dizzy, and may find it hard to move. The ingestion of fluids produces quick recovery in all but the most serious cases of dehydration. If it is not resolved, however, dehydration can lead to kidney failure and death.[4]

Water Toxicity

People can overdose on water if they drink too much of it. High intake of water can lead to a condition known as hyponatremia—or low blood sodium level and excessive water accumulation in the brain and lungs. The consequences can be devastating and include confusion, severe headache, nausea, vomiting, and even seizure, coma, and death.[13]

Water intoxication is rare, but it has occurred in marathon runners who consumed too much water during an event, infants given too much water or overdiluted formula (Illustration 25.9), and psychotic patients taking medications that produce cravings for water. The drive for water created by antipsychotic medications can be so strong that access to water (even when showering) has to be limited.[14]

Nutrition Up Close

Foods as a Source of Water

Focal Point: Water is primary component of many foods:

Our requirement for water is met by fluids and solid foods. But how much water is in foods? You may be surprised. Circle the food in each set that you think contains the highest percentage of water by weight.

1. avocado; potato, boiled; corn, cooked
2. egg; almonds; ripe (black) olives
3. watermelon; celery; pineapple, fresh
4. 2% milk; Coke; cranberry juice, low calorie
5. cheddar cheese; banana; refried beans
6. hot dog; pork sausage, cooked; ham, extra lean
7. apple; mushrooms, raw; orange
8. onions, cooked; lettuce; okra, cooked
9. peanut butter; butter; mayonnaise
10. cake with frosting; bagel; Italian bread

FEEDBACK (answers to questions) can be found at the end of Unit 25.

Review Questions

True False

1. Water plays key roles in the prevention of acne and dry skin and in the dilution of toxins formed during digestion. ☐ ☐

2. Water is our major source of fluoride. ☐ ☐

3. Municipal water supplies are fluoridated in virtually every county in the United States and province in Canada. ☐ ☐

4. Adult females living in moderate climates require an average of 8 cups of water a day, and men require twice that amount. ☐ ☐

True False

5. Adequate availability of uncontaminated water is a global health concern. ☐ ☐

6. Americans should limit their consumption of bottled waters because they are often contaminated and therefore unsafe to drink. ☐ ☐

7. The Food and Drug Administration requires that all bottled water be fluoridated. ☐ ☐

8. Beverages containing caffeine are not hydrating. ☐ ☐

Media Menu

www.epa.gov
Concerned about the safety of your water supply? Check it out by entering this Environmental Protection Agency Web site and linking with water safety.

www.healthfinder.gov
This is a gateway to reliable information on water. Search "water" and be led to information on the functions of water, water supply concerns, and public health water standards.

www.cdc.gov
The address for the Centers for Disease Control provides access to water safety information and advice for travelers.

www.iom.edu/fnb
You'll be able to access the 2004 DRI Report on water and recommended intake levels from this site.

www.science.gov
Find out more about water and water quality at this Web site by searching "water."

www.pueblo.gsa.gov/cic_text/health/house-wells/six.htm
This EPA (Environmental Protection Agency) site lists and explains the six basic steps you should take to maintain the safety of your drinking water.

www.who.int/water_sanitation_health/en
The World Health Organization's Water, Sanitation, and Health homepage is located at this address. It addresses global health issues related to unsafe water and approaches to improving health by developing safe water supplies.

Notes

1. Kleiner SM. Water: an essential but overlooked nutrient. J Am Diet Assoc 1999;99:200–6.

2. Fiske H. Measuring water's benefits and optimal intake recommendations. Today's Dietitian 2003;Jan:22–4.

3. Askew EW. Water. In: Ziegler EE, Filer LJ Jr., editors. Present knowledge in nutrition. Washington, DC: ILSI Press; 1996: pp. 98–108.

4. Grandjean AC et al. Hydration: issues for the 21st century. Nutr Rev 2003;61:261–71.

5. Water. www.intelihealth.com, accessed 6/03.

6. Haskins J. A third of the world population faces water scarcity today. www.eurekalert.com, 8/21/06.

7. How safe is bottled water? American Institute for Cancer Research Newsletter 1992;36:8.

8. Heaney RP. Absorbability and utility of calcium in mineral waters. Am J Clin Nutr 2006; 84;371–4.

9. Welland D. Drink to good health, especially water: Here's why and how much. Environ Nutr 1999;22(Oct):1,6.

10. Dietary Reference Intakes for Water, Potassium, Sodium Chloride, and Sulfate. Food and Nutrition Board, National Academy of Science. http://books.nap.edu, accessed 5/04.

11. Popkin BM et al. A new proposed guidance system for beverage consumption in the United States. Am J Clin Nutr 2006;83:529–42.

12. Nutritional Needs in Cold and High-Altitude Environments: Applications for Military Personnel in Field Operations (1996), Institute of Medicine, p. 66, darwin.nap.edu/books/0309054842/html/66.html, accessed 10/06.

13. Noakes TD. Too many fluids as bad as too few. BMJ 2003;327:113–4.

14. Goldman MB et al. Mechanisms of altered water metabolism in psychotic patients with polydipsia and hyponatremia. N Engl J Med 1988;318:397–403.

Nutrition Up Close

Foods as a Source of Water

Feedback for Unit 25

The percentage of water is listed after each food.

Answers to Review Questions

1. False, see page 25-5.
2. True, see page 25-2.
3. True, see page 25-2.
4. False, see page 25-6.
5. True, see page 25-4.
6. False, see page 25-5.
7. False, see pages 25-5, 25-6.
8. False, see page 25-8.

1. **avocado (80%);** potato, boiled (77%); corn, cooked (73%)
2. egg (75%); almonds (4%); **ripe (black) olives (80%)**
3. watermelon (91%); **celery (95%);** pineapple, fresh (86%)
4. 2% milk (89%); Coke (89%); **cranberry juice, low calorie (95%)**
5. cheddar cheese (37%); **banana (74%);** refried beans (72%)
6. hot dog (53%); pork sausage, cooked (45%); **ham, extra lean (74%)**
7. apple (84%); **mushrooms, raw (92%);** orange (87%)
8. onions, cooked (88%); **lettuce (96%);** okra, cooked (90%)
9. peanut butter (1%); **butter (16%);** mayonnaise (15%)
10. cake with frosting (22%); bagel (33%); **Italian bread (36%)**

UNIT 26

Nutrient–Gene Interactions in Health and Disease

NUTRITION SCOREBOARD

	TRUE	FALSE
1 Most chronic diseases are genetically caused.		
2 Individual differences in genetic traits are responsible for large differences in nutrient needs between individuals.		
3 Heart disease, cancer, obesity, and hypertension primarily result from interactions among environmental and genetic factors.		

Key Concepts and Facts

- Nutrients interact in important ways with gene functions and thereby affect health status. Nutrients can turn genes on or off, and nutrient intake can compensate for abnormally functioning genes.[1]

- Health problems related to nutrient–gene interactions originate within cells.

- Advances in knowledge of nutrient–gene interactions are dramatically changing nutritional approaches to disease prevention and treatment.

Answers to
NUTRITION SCOREBOARD

		TRUE	FALSE
1	That's incorrect. Only a very small proportion of disease states are caused by genetic traits.[1]		✔
2	Individual genetic traits alter nutrient needs to a small extent in many people, and to a large extent in very few. The Dietary Reference Intakes cover the nutrient needs of nearly all healthy people.		✔
3	How true. Common diseases result from interactions between multiple genetic traits and environmental factors such as dietary intake.[2]	✔	

The science of nutritional genomics is in its infancy, but it has the potential to transform the science of nutrition.
—D. SHATTUCK, JOURNAL OF THE AMERICAN DIETETIC ASSOCIATION

Nutrition and Genomics

In the history of the relatively young science of nutrition, at least two break-throughs have defined the future. The first resulted from experiments in the late 1800s demonstrating that some constituents of foods are essential for life. The second is represented by completion of the draft of the human *genome* and development of the field of nutrigenomics. (Find definitions and explanations of many of the genetic terms used in this unit in Table 26.1.) That breakthrough marked the beginning of a new era in the discovery of contributing factors to health and disease, and the roles of nutritional and other "nurture" factors. It is being made possible by the identification of genetic codes for production of enzymes and other proteins in cells. Enzymes and other proteins produced as a result of genetic codes are central to life and health because they determine which chemical changes will take place within the body. These changes influence growth, digestion, nutrient absorption, disease resistance, blood pressure control, and many other functions of the body.

Diet–Gene Interactions

Research on pinpointing dietary–gene interactions is revolutionizing nutritional approaches to health promotion and to disease prevention and treatment.[3] A number of nutrient-gene interactions identified to date have gotten the revolution

TABLE 26.1

Definitions and explanations of genetic terms

DNA is packed in chromosomes located in cell nuclei

Human chromosomes

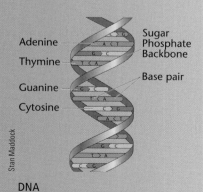

DNA

Stan Maddock

- **Genome** Combined term for "genes" and "chromosomes." It represents all the genes and DNA contained in an organism, which are principally located in chromosomes. The human genome consists of about 25,000 genes.

- **Genes** The basic units of heredity that occupy specific places (loci) on chromosomes. Genes consist of large DNA molecules, each of which contains the code for the manufacture of a specific enzyme or other protein.

- **Genotype** The specific genetic makeup of an individual as coded by DNA.

- **Genomics** The study of the functions and interactions of all genes in the genome. Unlike genetics, it includes the study of genes related to common conditions and their interaction with environmental factors.

- **Nutrigenomics** The study of diet- and nutrient-related functions and interactions of genes and their affects on health and disease.

- **Chromosomes** Structures in the nuclei of cells that contain genes. Humans have 23 pairs of chromosomes (shown below); half of each pair comes from the mother and half from the father.

- **DNA (deoxyribonucleic acid)** Segments of genes that provide instructions for the manufacture of enzymes and other proteins by cells. DNA looks something like an immensely long ladder twisted into a helix, or coil. The sides of the ladder structure of DNA are formed by a backbone of sugar-phosphate molecules, and the "rungs" consist of pairs of bases joined by weak chemical bonds. Bases are the "letters" that spell out the genetic code, and there are over 3 billion of them in human DNA. There are two types of base pairs: adenine-guanine and cytosine-thymine. Each sequence of three base pairs on DNA codes for a specific amino acid. A specific enzyme or other protein is formed when coded amino acids are collected and strung together in the sequence dictated by DNA.

  Some sections of DNA do not transmit genetic information because they don't code for protein production. They appear to signal which genes will turn on and how long they will be activated. Characteristics of these segments of DNA vary among individuals, making it possible to identify individuals based on "DNA fingerprinting."

- **DNA fingerprinting** The process of identifying specific individuals by their DNA. This is possible because no two individuals have the same genetic makeup. Differences among individuals are due to variations in the sections of DNA molecules that do not transmit genetic information.

well under way. The following are examples of the effects gene types (or genotypes) have on the body's response to nutritional factors. In some people:

• Whole oats lower blood cholesterol levels.

• High folate intake decreases the risk of cancer.

• High polyunsaturated fat, low dietary cholesterol, or low-saturated-fat diets lower blood cholesterol levels.

• High-carbohydrate diets increase the risk of type 2 diabetes.

• High alcohol intake during pregnancy produces physical and behavioral abnormalities in the fetus.

• Regular consumption of green tea reduces the risk of prostate cancer.[4,5]

Research into the genetic bases of disease risk is also clarifying which health characteristics have little or nothing to do with genetic traits and everything to do with environmental factors.[1]

Genetic Secrets Unfolded

Our bodies may be new, but our genes have been around for over 40,000 years. Genes replicate themselves exactly over generations, and lasting modifications in them almost never occur. This means that changes in genetic traits do not account for increases or decreases in the incidence of disease. It is why genetic traits cannot be given as the cause of recent increases in rates of obesity and diabetes, nor can they take credit for major declines in heart disease and stroke rates.

Humans are 99.9% alike genetically. It's the 0.1% that is different that makes every individual distinct.[7] Some of the 0.1% of genetic variation represents genes that contribute to disease resistance, disease development, and the way people respond to particular drugs.[6] Many questions about the roles of genetic traits in health and disease remain to be answered. Researchers know the most about disorders related to *single-gene defects.*

Single-Gene Defects

Hundreds of diseases related to one or more defects in a single gene have been identified, and many of these affect nutrient needs. Such defects can alter the absorption or utilization of nutrients such as amino acids, iron, zinc, and the vitamins B_{12}, B_6, or folate.[5,6] PKU, celiac disease, lactose intolerance, and hemochromatosis are four examples of single-gene defects that substantially affect nutrient needs. (These are described in Table 26.2.)

PKU and lactose intolerance are caused by defective genetic codes for enzymes, whereas celiac disease and hemochromatosis result from genetic abnormalities in the formation of other proteins. PKU, celiac disease, and lactose intolerance are treated by diets that limit phenylalanine, gluten, or lactose, respectively. Hemochromatosis is treated by a low iron diet.[1]

Not all single-gene abnormalities produce ill effects. For instance, lack of the gene that codes for an enzyme that helps the body excrete a specific sulfur-containing chemical unique to cruciferous vegetables (such as cabbage, broccoli, and bok choy) may be good. Smokers and nonsmokers who have the gene for the enzyme that helps excrete this beneficial sulfur compound and who consume these vegetables regularly are more likely to develop lung cancer than are people

Are you reminded an hour or two later that you ate asparagus? If you said yes, you are among the 1 out of 10 people born with a gene for detecting the odor of a sulfur-containing compound excreted in urine by 4 in 10 people after consumption of asparagus.

Photo Disc

single-gene defects
Disorders resulting from one abnormal gene. Also called "inborn errors of metabolism." Over 800 single-gene defects have been cataloged, and most are very rare.

TABLE 26.2

Examples of single-gene disorders that affect nutrient need[8,9]

PKU (phenylketonuria)	A very rare disorder caused by the lack of the enzyme phenyl-alanine hydroxylase. Lack of this enzyme causes phenylalanine, an essential amino acid, to build up in the blood. High blood levels of phenylalanine during growth lead to mental retardation, poor growth, and other problems. PKU is treated by low-phenylalanine diets.
Celiac disease	An intestinal malabsorption disorder caused by an inherited intolerance to gluten in wheat, rye, and barley. It causes multiple nutrient deficiencies and is treated with gluten-free diets. Celiac disease is also called "nontropical sprue" and "gluten enteropathy."
Lactose intolerance	A common disorder in adults in many countries resulting from lack of the enzyme lactase. Ingestion of lactose in dairy products causes gas, cramps, and nausea due to the presence of undigested lactose in the gut.
Hemochromatosis	A disorder affecting 1 in 200 people that occurs usually due to a genetic deficiency of a protein that helps regulate iron absorption. Individuals with hemochromatosis absorb more iron than normal and have excessive levels of body iron. High levels of body iron have toxic effects on tissues such as the liver and heart. The disorder can also be produced by excessive levels of iron intake over time and frequent iron injections or blood transfusions.

who eat the vegetables but lack the gene. Without the gene for the enzyme that causes its quick elimination, the beneficial sulfur-containing compound lingers in the body, extending the time it can play a role in cancer prevention.[7] Risk of lung cancer is decreased by not smoking.

Most diseases related to genetic traits are not as straightforward as are single-gene defects. They are more likely to represent an interwoven mesh of genetic and environmental risk factors.

Chronic Disease: Nurture and Nature

> *A very large part of our health story is more genetically influenced than genetically caused.*
> —ELBERT BRANSCOMB, DIRECTOR OF THE JOINT HUMAN GENOME INSTITUTE

Major health problems of our day, including heart disease, cancer, hypertension, obesity, diabetes, and disorders associated with aging, are not due to single-gene defects but result from interactions among multiple genetic traits and environmental factors. Nutrients, a prominent environmental factor, can turn certain genes on or off and, along with other effects on gene function, can compensate for some of the ill effects certain genotypes have on body processes.[3]

Heart disease primarily stems from plaque buildup in arteries near the heart (Illustration 26.1). Plaque buildup "hardens the arteries," narrows artery openings, and increases the risk of artery blockage and subsequent heart attack.

ICI Pharmaceuticals Div.

Illustration 26.1
Healthy (left) versus clogged (right) arteries.
How much plaque builds up in arteries is influenced by multiple environmental and genetic factors.

TABLE 26.3

Estimates of the contribution of environmental and genetic factors to development of some types of cancer[9, 11]

Cancer Site	Environmental Factors (%)	Genetic Factors (%)
Endometrial (uterine wall)	100	0
Ovary	78	22
Lung	74	26
Breast	73–95	5–27
Stomach	72	28
Colon	65	35
Pancreas	64	36
Prostate	58	42

salt sensitivity
A genetically determined condition in which a person's blood pressure rises when high amounts of salt or sodium are consumed. Such individuals are sometimes identified by blood pressure increases of 10% or more when switched from a lower-salt (1 to 3 grams) to a higher-salt (12 to 15 grams) diet.[13]

Illustration 26.2
Obesity is related to complex interactions among nutritional, other environmental, and genetic factors.

Photo Disc

Some people who consume high-saturated-fat diets build up plaque in their arteries, whereas others don't. For others, low fruit and vegetable diets increase the risk that plaque will accumulate. These differences are due to variations in genotypes. Heart disease is also influenced by blood pressure, body weight, and levels of triglycerides and clotting factors in the blood. Each of these is influenced by environmental factors such as high salt or alcohol intake in people with specific genetic traits. Risk for heart disease is the sum of all the pro and con influences of environmental and genetic factors.[12]

Cancer

Most types of cancer are primarily related to environmental factors such as high fat and alcohol intakes, low vegetable and fruit diets, high levels of body fat, smoking, and toxins in the environment.[9] Some cancers have genetic components that interact with environmental exposures. Studies of the roles played by genetic and environmental factors, to the extent they can currently be separated, show differences by cancer site (Table 26.3). As can be seen by the large contributions of environmental factors to cancer development, the hope for cancer prevention primarily lies in our ability to modify harmful environmental exposures.

Hypertension

The primary causes of most cases of hypertension are unknown, but factors such as high body fat, high alcohol intake, physical inactivity, and genetic traits play a role.[14] High salt (sodium, really) intakes are also related, but not in all studies. Differences in results between studies may be related to the proportion of people in study samples who are genetically predisposed to the effects of salt intake on blood pressure. Such people are *salt sensitive;* their blood pressure increases when they consume high amounts of salt. Approximately 51% of people with hypertension, and 26% of people with normal blood pressure, are salt sensitive.[15]

Obesity

All children inherit—along with their parents' genes—their parents, peers, and the community in which they live.
—L. EISENBERG, 1999

Obesity is primarily influenced by excessive-calorie diets and inadequate physical activity. However, over 200 genetic traits have been associated with obesity development, and many of these appear to interact strongly with environmental factors (Illustration 26.2). High-carbohydrate diets may influence gene expression and obesity development in some people, while a genetic predisposition toward inactivity may enhance obesity development in others.[16]

When people in developing countries who have subsisted on low-calorie diets for generations switch to high-calorie intakes, rates of obesity tend to increase dramatically. It appears that genetic traits programmed very early in life, as a result of high food intakes, or body processes that have adapted to survive on low

Reality Check

Changing family history

Elena and Alfredo both come from families with a number of relatives who have died from heart disease.

Who gets thumbs up?
Answers on next page

Elena: Heart disease is in my genes. There's nothing I can do about it.

Alfredo: I already know my LDL cholesterol is high and my HDL cholesterol is low. I'm eating and exercising to beat the odds.

caloric intakes, promote the development of obesity. Diabetes and heart-disease rates in population groups also rise with increasing rates of obesity.[17] Many other nurture and nature interactions on obesity are possible.

Diseases Related to Aging

The incidence of disorders such as diabetes, cancer, heart disease, memory impairment, hypertension, and stroke usually increases with age. Such declines in health are neither inevitable nor largely determined by genetic factors. They are heavily influenced by diet, exercise, and other components of lifestyle (Illustration 26.3).[18]

Genetics of Food Selection

Food preferences are largely learned, but which vegetables we like or don't like may be influenced by genetic traits. There are over 80 genes that help people taste bitter foods, and some people get the set of genes that make them highly sensitive to bitter-tasting foods (which are mainly vegetables). People born with a high sensitivity to bitter tastes tend to dislike cooked cabbage, collard greens, spinach, brussels sprouts, or other vegetables that taste bitter to them. People who tend to like these vegetables generally don't perceive them to be bitter-tasting. A genetic tendency to reject these vegetables is likely to limit intake and therefore may be linked to diseases associated with low vegetable intake.[19]

Illustration 26.3
Diseases related to aging are neither inevitable nor largely determined by genetic factors. They are influenced more by diet, exercise, and other components of lifestyle.

Looking Ahead

In our rush to fit medicine with the genetic mantle, we are losing sight of other possibilities for improving public health.
—N. A. HOLTZMAN, JOHNS HOPKINS MEDICAL INSTITUTE

The promise of advances in knowledge of the genetic bases of disease is longer, healthier lives. No doubt drugs that counter ill effects of genetic traits will continue to be developed, and attempts to fix abnormal genes by gene therapy will

ANSWERS TO *Reality Check*

Changing family history

Alfredo recognizes that a family history of heart disease doesn't mean he is destined to die from it. A small minority of diseases are solely due to genetic factors; many are primarily due to interactions between genetic traits and environmental factors. Disease-promoting traits can be diminished by the right environmental changes.[1]

Elena: 👎   Alfredo: 👍

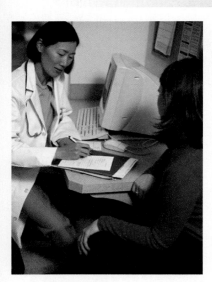

Illustration 26.4
In the future, advice given by registered dietitians may be tailored to the specific genetic traits of individuals.

Dogs have a taste for sweet, but cats don't. Somewhere during genetic evolution the gene in cats that enabled them to recognize sweetness was turned off never to be turned on again.[20]

broaden. Some of the most meaningful breakthroughs in upcoming decades will be in the area of disease prevention and treatment through nutritional changes.[5]

Although used to some extent now, personalized modifications of dietary intake based on genotypes will become standard practice in clinical dietetics and medicine (Illustration 26.4).

Are Gene-Based Designer Diets in Your Future?

In less than 10 years, you'll be able to go to a lab and complete a set of genetic tests to identify your personal disease susceptibilities. When you leave you'll be armed with a list of foods to eat and foods to avoid and recommendations of dietary supplements to help prevent your disease.
—D. Evans, CEO of Wellgen, Inc.

Sound far-fetched? It isn't.[6] In the future, people will have the option of, for example, reducing cancer risk by regularly consuming cruciferous vegetables (cabbage, broccoli, cauliflower, and the like), diabetes risk by switching to a high-protein diet, or heart disease by limiting dietary cholesterol or saturated fat intake (Illustration 26.5). In truth, we have enough knowledge today to implement healthful, disease-preventing dietary changes. The kinds of advances described above will not bring an end to disease or guarantee health, but they will take us a long way down that road.

Illustration 26.5
A menu of the future.

Nutrition Up Close

Nature and Nurture

Focal Point: Gaining insight into your family's and your own "dietary" health history.

Families often share dietary factors that interact with genetic traits to influence chronic disease development. Check out your dietary health history and that of your family by completing the following family food tree activity. If you don't know all of your relatives included in the activity or their dietary behaviors, fill out the parts based on what you do know.

Check each dietary behavior that applies:

Relative	Excess Calorie Consumption		Low Vegetable and Fruit Intake		Low Whole Grains Intake		Low Fiber Intake		Low Fish Intake		High Alcohol Intake	
	yes	no	yes	no	yes	no	yes	no	yes	no	yes	no
A. On your mother's side:												
Grandmother	—	—	—	—	—	—	—	—	—	—	—	—
Grandfather	—	—	—	—	—	—	—	—	—	—	—	—
B. On your father's side:												
Grandmother	—	—	—	—	—	—	—	—	—	—	—	—
Grandfather	—	—	—	—	—	—	—	—	—	—	—	—
C. Your mother	—	—	—	—	—	—	—	—	—	—	—	—
D. Your father	—	—	—	—	—	—	—	—	—	—	—	—
E. Yourself	—	—	—	—	—	—	—	—	—	—	—	—

FEEDBACK (answers to these questions) can be found at the end of Unit 26.

Key Terms

chromosomes, page 26-3
DNA (deoxyribonucleic acid), page 26-3

DNA fingerprinting, page 26-3
genes, page 26-3
genome, page 26-3

genomics, page 26-3
salt sensitivity, page 26-6
single-gene defects, page 26-4

Review Questions

	True	False		True	False
1. Genomics is the study of the functions and interactions of genes.	☐	☐	6. Most chronic diseases result from interactions among multiple genetic traits and environmental factors.	☐	☐
2. Nutrigenomics is the study of the interaction of body weight and genes.	☐	☐	7. How genes function, such as when they are prompted to turn on or off, can be programmed early in life by dietary exposures.	☐	☐
3. Genes provide the codes for the production of enzymes and other proteins in cells.	☐	☐	8. Some of the most meaningful breakthroughs in the prevention and management of chronic diseases in the future will be due to advances in nutrigenomics.	☐	☐
4. Unfavorable changes in genetic traits are the primary cause of the current, global epidemics of obesity and diabetes.	☐	☐			
5. Most chronic diseases result from single-gene defects.	☐	☐			

Media Menu

www.genome.gov
This great site from the National Human Genome Research Institute provides news of the latest developments in genetic research, educational resources, and a written and spoken glossary of genetic terms and useful illustrations.

www.healthfinder.gov
Search any disease of interest to find out current information about it.

www.cdc.gov/genomics
This is CDC's National Office of Public Health Genomics. You'll find information on genomic discoveries and how they can be used to improve health and prevent disease.

Notes

1. Guttmacher AE et al. Welcome to the genomic era. N Engl J Med 2003;349:996–8.

2. Stover PJ. Influence of human genetic variation on nutritional requirements. Am J Clin Nutr 2006;83(suppl):436S–42S.

3. Shattuck D. Nutritional genomics. J Am Diet Assoc 2003;103:16–18.

4. Simopoulos AP. Genetic variation and nutrition, Experimental Biology Annual Meeting, San Diego, CA, 4/12/03.

5. Afman L et al. Nutrigenomics: from molecular nutrition to prevention of disease. J Am Diet Assoc 2006;106:569–78.

6. Debusk RM et al. Nutritional genomics in practice: where do we begin? J Am Diet Assoc 2005;105:589–98.

7. London SJ, Yuan JM, Coetzee GA et al. CYP1A1 1462V genetic polymorphism and lung cancer risk in a cohort of men in Shanghai, China. Cancer Epidemiol Biomarkers Prev 2000;9:987–91.

8. Khoury MJ et al. Population screening in the age of genomic medicine, N Engl J Med 2003;348:50–8.

9. Lichtenstein P, Holm NV, Verkasalo PK et al. Environmental and heritable factors in the causation of cancer—analyses of cohorts of twins from Sweden, Denmark, and Finland. N Engl J Med 2000;343:78–85.

10. Terry P et al. Lifestyle and endometrial cancer risk: a cohort study from the Swedish Twin Registry, Int'l J Cancer 1999;82:38–42.

11. Hunter DJ, Willett WC. Nutrition and breast cancer, Cancer Causes and Control 1996;7:56–68.

12. Ordovas JM. Genetic interactions with diet influence the risk of cardiovascular disease. Am J Clin Nutr 2006;83(suppl):443S–6S.

13. Morimoto A, Uzu T, Fujii T et al. Sodium sensitivity and cardiovascular events in patients with essential hypertension. Lancet 1997;350:1734–7.

14. Wong ZY, Stebbing M, Ellis JA et al. Genetic linkage of beta and gamma subunits of epithelial sodium channel to systolic blood pressure. Lancet 1999;353:1222–5.

15. Kaplan NM. The dietary guideline for sodium: should we shake it up? No. Am J Clin Nutr 2000;71:1020–6.

16. Butte NF et al. Viva la Familia Study: genetic and environmental contributions to childhood obesity and it comorbidities in the Hispanic population. Am J Clin Nutr 2006;84:646–54.

17. Berdanier CD. Nutrient-gene interactions. Nutr Today 2000;35:8–17.

18. Rowe JW. Geriatrics, prevention, and the remodeling of Medicare [editorial]. N Engl J Med 1999;340:720–1.

19. Drewnowski A, Henderson SA, Hann CS et al. Genetic taste markers and preferences for vegetables and fruit of female breast care patients. J Am Diet Assoc 2000;100:191–7.

20. Gorman J. Dogs may laugh, but only cats get the joke. New York Times, 9/5/06.

Answers to Review Questions

1. True, see page 26-3.
2. False, see page 26-3.
3. True, see pages 26-2, 26-3.
4. False, see page 26-4.
5. False, see page 26-5.
6. True, see page 26-5.
7. True, see pages 26-6, 26-7.
8. True, see pages 26-7, 26-8.

Nutrition Up Close

Nature and Nurture

Feedback for Unit 26

The more *yes* responses, the better the odds that some of your family's shared dietary traits may influence disease risk. The most important responses are those you gave yourself.

Nutrition and Physical Fitness for Everyone

NUTRITION SCOREBOARD	TRUE	FALSE
1 Physical fitness means being very muscular.		
2 Overweight people can be physically fit.		
3 Exercise is more effective than diet in preventing heart attacks.		
4 Physical fitness can be achieved only by exercising intensively for at least an hour every day of the week.		

Key Concepts and Facts

- Physical fitness, along with a good diet, confers a number of physical and mental health benefits.

- You don't have to be an athlete or be lean to be physically fit. Fitness depends primarily on muscular strength, endurance, and flexibility.

- People who are physically fit have respiratory and circulatory systems capable of delivering large amounts of oxygen to muscles and muscular systems that can utilize large amounts of oxygen for prolonged periods of time.

- Physical fitness can be achieved by resistance training, aerobic exercises, and stretching.

- The fitness level of most people in the United States is poor.

Answers to
NUTRITION SCOREBOARD

	TRUE	FALSE
1 Muscle strength is one part of physical fitness. The other primary components are endurance and flexibility.		✔
2 You can be fit and fat. Physical fitness is not measured by body fat content.	✔	
3 Regular exercise does help reduce the risk of heart attack—but not as much as a combined program of exercise, weight loss, and smoking cessation (if needed), and consumption of a healthy diet.[1]		✔
4 You don't have to exercise to that extent to become physically fit.[2]		✔

AP Photo/The Denver Post, Cyrus McCrimmon

Physical Fitness: It Offers Something for Everyone

I would not wish to imagine a world in which there were no games to play and no chance to satisfy the natural human impulse to run, to jump, to throw, to swim, to dance. Sport and recreation are, in themselves, eminently worthwhile and desirable. . . . Our aim should be, I suggest, to inspire everyone to become involved in sport and recreation by making the choice irresistible in its scope and variety. . . . I see health as a happy consequence of sporting activity.

—Sir Roger Bannister, 1989

Physical activity has much to offer the athlete and nonathlete alike. It provides recreation good for the body and soul, it doesn't have to cost anything, and it benefits almost everyone. At its best, physical activity is play with happy consequences for health and well-being. As long as there are activities people enjoy doing, there's an "athlete" in everyone.

The "Happy Consequences" of Physical Activity

Ask people who are in good physical condition what they get from exercise, and you're likely to hear a variety of responses. Someone who is 20 years old may say he wants to stay in shape and improve his stamina. A 40-year-old individual may say that exercise helps keep her cholesterol and weight from getting too high.

TABLE 27.1

Benefits of regular physical activity[3-5]

Reduced Risk of Certain Diseases and Disorders	Improved Sense of Well-Being	
• Heart disease • Colon and breast cancer • Hypertension • Stroke • Osteoporosis • Back and other injuries • Obesity, excess abdominal fat • Diabetes (non–insulin dependent) • Bone and joint diseases • Alzheimer's disease	• Increases feeling of well-being • Decreases depression and anxiety • Helps relieve stress • Decreases risk of dementia	

People who are 80 might tell you they exercise so that they won't have to use a walker and will be able to maintain their independence. Ask children why they exercise, however, and they may not understand the question. Children engage in active play; they don't exercise. For them, fitness is truly an unintended consequence of play.

Regular physical activity benefits both physical and psychological health in people of all ages (Table 27.1). By improving a person's physical health, regular exercise may help ward off heart disease, some types of cancer, hypertension and stroke, osteoporosis, back injury, and diabetes. It tends to increase a person's feeling of well-being and helps relieve depression, anxiety, and stress.[3-5]

The Bonus Pack: Exercise plus a Good Diet   Exercise benefits health most when combined with a good diet and other healthy behaviors. The risk of developing heart disease, for example, is meaningfully lower when people combine a diet low in animal fat and high in vegetables, fruits, whole grains, and fiber with regular exercise than when exercise alone is relied upon for protection against heart disease.[6] Regular physical activity helps build bone mass and reduces the risk of osteoporosis. But the risk is lowered to a greater extent if exercise is combined with a diet that supplies adequate amounts of calcium and vitamin D. Some other benefits of exercise, such as a reduced chance of developing colon and breast cancer, may be related to the effect of exercise on body fat content. Regular exercise is one of the few factors yet identified that helps people achieve long-term weight control.[7]

■ **Exercise and Body Weight**   Combined with a moderate decrease in usual caloric intake—on the order of 200 calories per day—exercise helps people lose fat, build muscle mass, and become physically fit.[8] Because the body needs more calories to maintain muscle than fat, exercise that results in an increase in muscle mass leads to an increase in caloric requirements. For some people, this increase in caloric requirement makes it easier to maintain weight within the normal range and to lose weight and keep it off.[9]

Exercise combined with a stable caloric intake can lead to weight loss, but the loss is generally smaller than can be achieved by reducing caloric intake.[10] Even if weight loss is not intended or is achieved at a slow pace, the happy consequences of exercise come to all basically healthy individuals who undertake it on a regular basis.

Physical Activity and Fitness

physical fitness
The health of the body as measured by muscular strength, endurance, and flexibility in the conduct of physical activity.

Many of the benefits of physical activity are related to the **physical fitness** it can produce. Physical fitness is not defined by bulging muscles, thin waistlines, or amount of physical activity. Obese as well as thin people can be physically fit or not. According to the American College of Sports Medicine, physical fitness is a state of health measured by strength, endurance, and flexibility.[11]

The strength component of physical fitness relates to the level of maximum force that muscles can produce. Endurance refers to the length of time muscles can perform physical activities, and flexibility refers to a person's range of motion. Physical fitness results when all three exist.

Muscle Strength   In adults, muscles grow in size and strength when muscle cells increase in size. Muscle cell size—and correspondingly, the strength of muscles—increases in response to weight-bearing or resistance exercise. Activities that require muscles to work harder than usual increase muscular strength.

To increase muscle strength, people should lift or push against a heavy weight; to increase muscle endurance, they should lift lighter weights and push against them repeatedly. Resistance training should include exercises that build both muscle strength and endurance.

Endurance: A Measure of Aerobic Fitness   How long a person can perform an activity depends on inherited traits and conditioning.[12] Both factors contribute to a person's stamina—or the extent of one's ability to deliver oxygen to muscles and the ability of muscles to use the oxygen for work. The amount of oxygen an individual is able to deliver to muscles that can be used by muscles for physical activity corresponds to the level of **aerobic fitness.** The more aerobically fit a person is, the longer and harder she or he is able to exercise.

aerobic fitness
A state of respiratory and circulatory health as measured by the ability to deliver oxygen to muscles and the capacity of muscles to use the oxygen for physical activity.

Aerobic activities include jogging, basketball, swimming, soccer, and other low- and moderate-intensity activities. They give the whole body, or most of it, a continuous workout.

maximal oxygen consumption
The highest amount of oxygen that can be delivered to, and used by, muscles for physical activity. Also called VO$_2$ max and maximal volume of oxygen.

■ **How Is Aerobic Fitness Determined?**   Aerobic fitness is classically assessed by measuring **maximal oxygen consumption** (abbreviated as VO$_2$ max) in a specially equipped laboratory (Illustration 27.1). In the lab, individuals are exercised at increasingly higher intensities, for example, by elevating the grade or speed of

Illustration 27.1
Determining VO$_2$ max.
This man is running on a treadmill. His nose is plugged, so he must breathe through a tube in his mouth. The tube is attached to an apparatus that measures the total amount of air breathed in and out and the difference between the amount of oxygen inhaled and exhaled. Every few minutes the slope or speed of the treadmill is increased, making the exercise more intense. The amount of oxygen he uses at the point when his exercise intensity can go no higher is considered "maximal oxygen consumption," or VO$_2$ max.

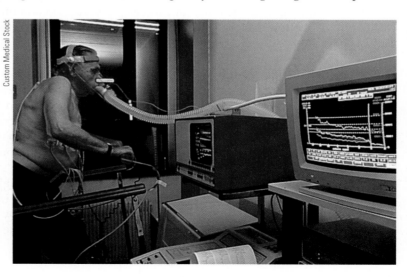

Custom Medical Stock

a treadmill. The individual performing the exercise breathes through a tube that delivers room air. Monitoring equipment measures the amount of oxygen from the air that is breathed and used during the exercise. The maximal amount of oxygen a person delivers to working muscles is the amount used when the intensity of exercise can no longer be increased. The higher the level of oxygen used at the peak level of activity, the higher the level of aerobic fitness and the longer physical activity can be performed.

People can perform physical activity at 100% of VO_2 max for only a few minutes. Consequently, aerobic fitness goals are set below that level. In general, it is recommended that beginners start an aerobic fitness program with a goal of exercising at 40 to 60% of VO_2 max and working up to a higher level.[13] Aerobically fit people may train at 70 to 85% of VO_2 max.

■ **How Do You Know Your VO₂ Max?**   There are several ways to estimate VO_2 max that don't require a supervised laboratory test. One commonly employed method uses the percentage of maximal heart rate. This formula uses heart rate as a proxy measure of oxygen consumption, and it provides a fairly good estimate of VO_2 max because the amount of blood circulated to muscles is roughly equivalent to the number of times the heart sends out a pulse of blood each minute. The more oxygen muscles need to perform an activity, the more oxygen the lungs have to deliver and the faster the heart has to beat (that is, up to the point where the lungs and heart can no longer send an additional supply of oxygen to muscles). The heart rate at which the highest level of oxygen consumption by the body occurs is considered "maximal heart rate." It roughly corresponds to VO_2 max.

The formula for estimating maximal heart rate is quite simple: Subtract your age from 220. To obtain a target heart rate for exercise, multiply the result by the desired percentage of maximal heart rate. (Table 27.2 shows an example.) The resulting figure is your "target heart rate" for aerobic exercise. Table 27.3 lists maximal heart rate, 60% of the maximum, and 75% of the maximum rate for different ages.

■ **How Do You Know Your Heart Rate?**   You can determine if you are exercising at the appropriate level for aerobic fitness by measuring your pulse rate during a break or as soon as you're done exercising. To take your pulse, place a fingertip gently on an artery in your wrist or neck (Illustration 27.2). Count the

TABLE 27.2

Applying the percentage of maximal heart rate (MHR) formula to a 22-year-old individual who will exercise at 60% of MHR

- Target heart rate = (220 − age) × %MHR
- Target heart rate = (220 − 22) × 0.60
  = 198 × 0.60
  = 119

TABLE 27.3

Target heart rates estimated at 60% and 75% of maximum for people at different ages using the percentage of maximal heart rate formula

Age (Years)	Average Maximal Heart Rate (MHR)	60% of MHR	75% of MHR
20	200	120	150
25	195	117	146
30	190	114	142
35	185	111	138
40	180	108	135
45	175	105	131
50	170	102	127
55	165	99	123
60	160	96	120
65	155	93	116
70	150	90	113

Photo Disc

Illustration 27.2
To find out if you're exercising at the right level, gently take your pulse during a break or right after you finish exercising.

number of pulses you feel in a 10-second period and multiply that number by 6. That's your heart rate, measured as heartbeats (or pulses) per minute.

For many years it was assumed that stretching prior to exercise improved flexibility and range of motion and helped prevent injuries. Recent evidence casts doubt on its benefits, however. Stretching muscles to the point of discomfort does not appear to improve flexibility or range of motion, prevent injuries, or enhance performance.[14] Strength exercises that involve major muscle groups, a gentle warm-up of muscle movements prior to exercise and to cool off muscles after exercise, and regular exercise may foster flexibility.[14,15]

Nutrition and Fitness

Your diet can make a strong contribution to your level of physical fitness. That's because good diets contribute to health, and it is easier to become physically fit if you are in good health. Diet also makes a difference to physical fitness because the foods you eat serve as the source of energy for physical activity. For some activities, the body needs glucose as its primary energy source. For others, the fuel of choice is fat.

Muscle Fuel

Muscles use fat, glucose, and amino acids for energy. The proportion of each that is used, as well as the amount, depends on the intensity of activity (Illustration 27.3). When we're inactive, fat supplies between 85 and 90% of the total

Illustration 27.3
Schematic representation showing the proportionate use of sources of energy for physical activities of various intensities. *Most activities are fueled by both fat and glucose.*

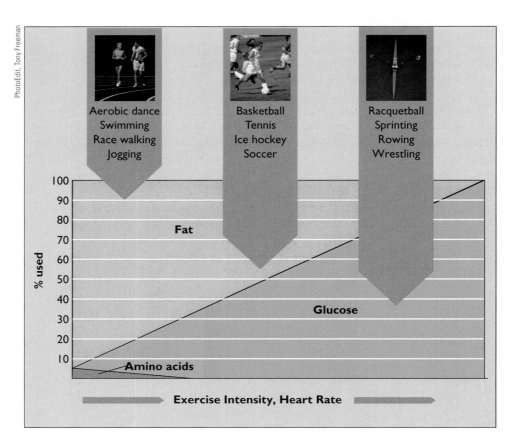

amount of energy needed by muscles. The rest is provided by glucose (about 10%) and amino acids (5% at most).[16] Fat is also the primary source of fuel for activities of low-to-moderate intensity such as walking, running, and swimming. As can be seen from Illustration 27.3, amino acids make a relatively small contribution to meeting energy needs for physical activity.

Oxygen is required to convert fat into energy. That makes activities of low-to-moderate intensity "aerobic"—or "oxygen requiring." Consequently, fat-burning, oxygen-requiring exercises are the type used to increase aerobic fitness. Jogging, long-distance running, swimming, and aerobic dance are all aerobic activities.

High-intensity, short-duration activities like spiking a volleyball, running down a deep fly ball, or sprinting down the block to catch a bus are fueled primarily by glucose. Our supply of glucose for intense activities comes principally from glycogen, the storage form of glucose. Glycogen is stored in muscles and the liver and can be rapidly converted to glucose when needed by working muscles. The conversion of glucose to energy for intense activity doesn't require oxygen (it's *an*aerobic). People can undertake very intensive activity only as long as their stores of glucose last.

Activities such as basketball, hockey, tennis, and soccer that involve walking, running, sprinting, and the occasional high-intensity, quick move use mainly fat or glucose as the energy source, depending on the intensity of the physical effort.

Stiff Muscles in the Out of Shape   People who are out of shape and exercise intensely, or who change the type of exercise performed, may experience stiff muscles 24 to 48 hours after exercise (Illustration 27.4). Although the cause is not clear, the stiffness usually lasts a day or two. The probability of this happening decreases as people get into shape.

Diet and Aerobic Fitness

The best type of diet for physical fitness is the same one that is recommended for people in general. It includes a variety of vegetables and fruits, whole-grain products, lean meats and fish, and low-fat dairy products. Healthy diets promote physical fitness because they facilitate maintenance of normal weight; reduce plaque buildup in arteries; and supply adequate amounts of vitamins, minerals, essential fatty acids, fiber, and other substances that keep the body running like

Illustration 27.4
She's going to feel this tomorrow. People may get stiff muscles if they exercise too hard when they're out of shape.

Reality Check

To drink or not to drink?

Drinking water during exercise upsets your stomach and slows you down. That's what some people say. Are they right?

Who gets thumbs up?
Answers on next page

Tex:
When I played basketball in school, the coach would never let us drink until half-time. Now I know why he did that.

Rose:
I get thirsty when I exercise, so I drink water. It never gave me a stomachache.

To drink or not to drink?

Rose is onto something. Thirst in healthy people signals the need for water. Drinking enough water or other dilute fluids during exercise keeps you hydrated. Not drinking enough slows you down.[19]

Tex

Rose

a well-tuned machine. Diets that promote physical fitness most certainly include well-planned vegetarian diets.[17]

Actually, people who exercise tend to have better diets than people who don't. They are often health conscious, select foods with care, and, because they generally eat more than sedentary people, have a better chance of getting all the nutrients they need.[18]

A Reminder about Water

Physical activity increases the body's need for water, and if the climate is hot and humid, this need increases even more. In general, people should drink in response to thirst and, overall, consume enough water to replace the amount lost in sweat, respiration, and urine during exercise.[19] (The amount of water lost during exercise is equivalent to the amount of weight that is lost during the exercise.) You are consuming the right amount of water if your urine is pale yellow and normal in volume. For exercise that lasts over an hour, alternating water with a sports drink or diluted fruit juice may improve hydration.[20] It's important to keep up fluid intake when exercising in cold weather. Cold air holds less water vapor than hot air, so you lose more water through breathing in cold weather.

People in Washington, DC, and Wyoming tend to be the most physically active, whereas physical activity levels are lowest in Kentucky and Louisiana.[21]

A Personal Fitness Program

Commenting on physical activity for the long run, Sir Roger Bannister has said:

> The best advice to those who are unfit is to take exercise unobtrusively. I am not an enthusiast for fitness and exercise schemes that are boring because they tend to fizzle out. . . . The vast majority of people, I think, can only be attracted for any length of time towards recreational activities if these are rewarding, enjoyable, and satisfying in themselves.

Physical fitness is not something that, once achieved, lasts a lifetime. The beneficial effects of training diminish within 2 weeks after training stops and disappear altogether in 2 to 8 months.[5] That makes it critical to undertake physical activities that "wear well"—those that a person finds rewarding and will reserve a place for in a busy schedule. Too many well-intentioned attempts to get into shape have been abandoned because they were overly ambitious or no fun. As the guidelines for becoming physically fit are presented in the next section, keep firmly in mind that you should seek a realistic and achievable plan—one that feels right and will last because it's good to you and for you.

TABLE 27.4

An overview of the components of physical fitness training

A. Resistance training:
- Goal: To improve muscle strength, flexibility, and endurance
- Components: Two to three resistance training sessions per week
  → Each session includes 10 exercises repeated in sets of 8 to 12.
  → Exercises should involve all major muscle groups.

B. Aerobic training:
- Goal: To increase the ability of the respiratory and circulatory systems to deliver oxygen to working muscles (to build "cardiorespiratory" fitness)
- Components: Three exercise sessions per week (lasting 20 to 30 minutes each) at 60 to 75% of maximal heart rate
  → Aerobic exercise should be a continuous activity (not stop-and-go) and involve all or most major muscle groups.

C. Flexibility:
- Goal: Increase range of motion of muscles and joints.
- Components:
  → Strength training
  → Regular exercise
  → Gentle warm-up exercise
  → Gentle cool down exercise

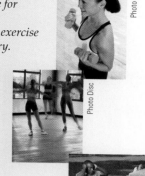

(Top) Moving muscles against resistance for strength.
(Middle) Aerobic exercise for oxygen delivery.
(Bottom) Regular exercise for flexibility.

Photo Disc

Photo Disc

Jason Horowitz/ZeFa/Corbis

Becoming Physically Fit: What It Takes

Physical fitness results from regular physical activity that leads to muscle strength, endurance, and flexibility (Table 27.4). These components of physical fitness can be achieved by resistance training and aerobic exercise.

A Resistance Training Plan   The resistance training program for physical fitness doesn't prepare people for bodybuilding contests. It increases muscle strength, flexibility, and endurance and improves the performance of aerobic exercises.

The American College of Sports Medicine recommends two to three resistance training sessions per week. Two sessions are enough for those who are not interested in "bulking up." Strength training twice weekly will give you 70 to 80% of the benefit of training three times a week.

Each workout session includes about 10 different exercises, repeated 8 to 12 times each. Each set of exercises should leave the muscles exercised and feeling a bit tired. Exercises selected should involve each of the major muscle groups in the legs and buttocks, arms and shoulders, and abdomen and back. Table 27.5 lists examples of exercises for the different muscle groups.

Resistance training can also be used for additional gains in muscular strength. Muscle strength is increased by the use of progressively heavier weights in workouts.

TABLE 27.5

The match between resistance exercises and the body's major muscle groups

Muscle Group Exercised	Activity
Legs and buttocks	Stair climbing, resistance leg pushing, cycling
Arms and shoulders	Free weight lifts, pull-ups, push-ups
Abdomen and back	Abdominal curls, sit-ups, leg lifts (front and side)

The Aerobic Fitness Plan

Aerobic fitness can be achieved by exercising at 60 to 75% of maximal heart rate for at least 30 minutes most of the week.[22] The best kind of aerobic exercise is one that involves all, or most, of the body's major muscle groups in the workout. Beginners should start at 40 to 60% of maximum with low-intensity activities, such as alternating walking with jogging, and increase to a higher level as exercise gets easier. Exercises such as swimming, jogging, running, aerobic dance, and cycling are good choices for subsequent aerobic exercise. Activities such as mowing the lawn, shoveling dirt or snow, and household tasks that elevate heart rate to the target level for 30 minutes are also aerobic exercise. Actually, the most common physical activities of American adults are walking, gardening, and yard work.[5]

Population-Based Physical Activity Recommendations

The American College of Sports Medicine and the Centers for Disease Control and Prevention recommend levels of physical activity for Americans that would benefit health and contribute to a decline in the rising incidence of obesity in the United States.[22] Specifically, it is recommended that moderate-intensity activities be undertaken for at least 30 minutes, five days a week. Moderate-intensity activities include brisk walking, lifting, jogging, tennis playing, lawn mowing (with a push mower), carpet cleaning, child chasing, and fast dancing. Higher, daily levels of moderate-intensity physical activity (60 minutes or more daily) increase weight loss and improve weight control to a greater extent than does less exercise.[23] To complement these aerobic activities, the surgeon general also suggests that people engage in strength-building exercises twice a week.

Some Exercise Is Much Better than None   Sedentary people who take up walking, golfing, biking, gardening, or other similar activities improve their fitness levels. Duration of low- and moderate-intensity exercises counts toward improved blood cholesterol concentrations, decreased body fat, and aerobic fitness.[2]

The Caloric Value of Exercise   Some people prefer to base their aerobic exercise workout on a particular level of calories (such as 200), while others simply like to know how many calories they're burning off. Table 27.6 lists the caloric value of various exercises in case you would like to know.

What's the largest muscle in your body? If your body is like most, its largest muscle is the gluteus maximus (buttocks).

You don't have to be a high-performance athlete to benefit from exercise. Do the things you enjoy—walk, ride a bike, garden, swim, play. Just get out and move 5 times per week for at least 30 minutes or more.

TABLE 27.6

Average energy output per pound of body weight for selected types of exercise

Exercise	Intensity	Calories (pound/hour)	Exercise	Intensity	Calories (pound/hour)
Walking	3 mph (20 min/mi)	1.6	Weight lifting		2.9
	3¹/2 mph (17 min/mi)	1.8	Wrestling		6.2
	4 mph (15 min/mi)	2.7	Handball	Moderate	4.8
	4¹/2 mph (13 min/mi)	2.9		Vigorous	6.2
Jogging	5 mph (12 min/mi)	4.1	Swimming	Resting strokes	1.4
	5¹/2 mph (11 min/mi)	4.5		20 yd/min (mod.)	2.9
	6 mph (10 min/mi)	4.9		40 yd/min (vig.)	4.8
	6¹/2 mph (9 min/mi)	5.2	Rowing (sculling		
	7 mph (8¹/2 min/mi)	5.6	or machine)		4.8
	7¹/2 mph (8 min/mi)	6.0	Downhill skiing		3.8
Running	8 mph (7¹/2 min/mi)	6.3	Cross-country	4 mph (15 min/mi)	4.3
	8¹/2 mph (7 min/mi)	6.7	skiing (level)	6 mph (10 min/mi)	5.7
	9 mph (6²/3 min/mi)	7.1		8 mph (7¹/2 min/mi)	6.7
	9¹/2 mph (6¹/3 min/mi)	7.4		10 mph (6 min/mi)	7.6
	10 mph (6 min/mi)	7.8	Aerobic dancing	Moderate	3.4
	11 mph (5¹/2 min/mi)	8.5		Vigorous	4.3
	12 mph (5 min/mi)	9.5	Racquetball/		
Rebound			squash	Moderate	4.3
trampoline	50–60 steps/min	4.1		Vigorous	4.8
Cycling			Tennis	Moderate	3.4
(stationary)	Mild effort	2.9		Vigorous	4.3
	Moderate effort	3.4	Volleyball	Moderate	3.4
	Vigorous effort	4.3		Vigorous	3.8
Cycling (level)	6 mph (10 min/mi)	1.5	Basketball	Moderate	3.8
	8 mph (7¹/2 min/mi)	1.8		Vigorous	4.8
	10 mph (6 min/mi)	2.0	Football	Moderate	3.8
	12 mph (5 min/mi)	2.8		Vigorous	4.3
	15 mph (4 min/mi)	3.9	Baseball/golf/woodcutting/horseback riding/		
	20 mph (3 min/mi)	5.7	badminton/canoeing		2.4
Skating		2.9	Soccer/hill climbing/fencing/judo/snowshoeing		5.3
Calisthenics	Moderate	2.4	Bowling/archery/pool		1.2
	Vigorous	2.9			
Rope skipping	Moderate	4.8			
Bench stepping	12" high, 24 steps/min	3.2			

Source: Values calculated from Guidelines for graded exercise testing and exercise prescription, 2d ed. Philadelphia: American College of Sports Medicine, Lea & Febiger, 1984.

A Cautionary Note   People who have a chronic disease such as diabetes or hypertension, or are out of shape and over 40, should consult a physician before starting an exercise program. (It may well turn out that exercise will be just what the doctor orders.) Moderate exercise programs (those undertaken at 40 to 60% of VO_2 max) are beneficial and safe for almost everyone.[5]

■ **The Paradox of Death during Exercise**   "Famous athlete dies while jogging" reads the headline. Why would this happen if exercise is good for you? That's the exercise paradox.

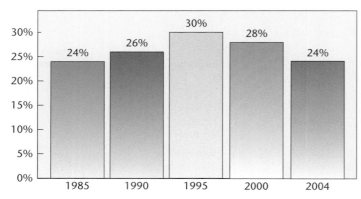

Illustration 27.5
The proportion of adults who engage in no leisure-time physical activity.[25]

Most people who die while shoveling snow, playing basketball, or jogging are those with existing heart disease either due to hardened arteries (atherosclerosis) or an inborn heart problem. It is a very rare event (because exercise really is good for the heart), and the risk can be reduced by regular exercise versus sporadic and intense physical activity.[24]

U.S. Fitness: America Needs to Shape Up

The benefits of regular exercise are sufficient to make it a matter of public health policy in the United States. Americans have much to gain by getting in better shape.

Studies are undertaken periodically to see how well Americans are doing at increasing their physical activity level. Not all of the results are encouraging (Illustration 27.5). Approximately 66% of adults do not achieve recommended levels of physical activity, and one-fourth are inactive. The encouraging news is that recent data indicate that rates of physical activity are beginning to increase. If main-

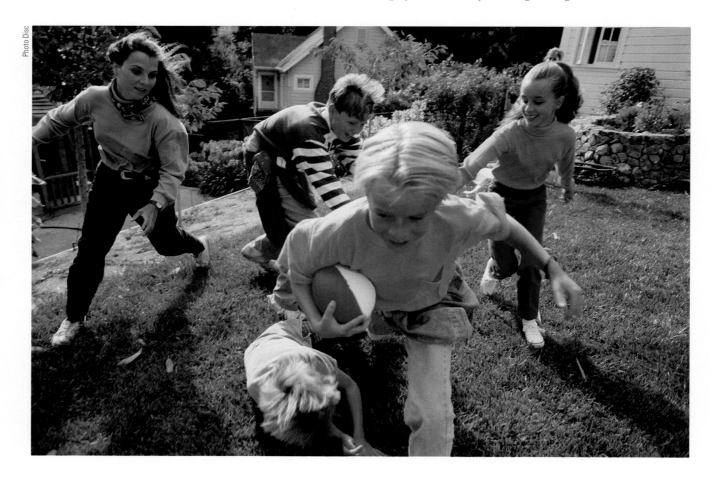

tained, this trend would help lower rates of overweight, obesity, type 2 diabetes, and other disorders.[25]

A Focus on Fitness in Children

Physical fitness advocates are especially concerned about the lack of opportunity for physical activity in schools. They point out that children who are not physically active are more likely to be overweight and that a lifelong love for physical activity and a physically active lifestyle often develop during childhood. Regular exercise benefits obese children by lowering body fat, increasing bone density, and improving aerobic fitness. Fewer than one in ten public schools require students to participate in physical activity daily—and, on average, children undertake vigorous physical activity for only 25 minutes per week.[26]

Illustration 27.6
This child is performing modified pull-ups as part of a fitness assessment.

Source: Greg Merhar, reprinted with permission of the Journal of Physical Education, Recreation and Dance, 59 (9), 1987.

Schools are being encouraged to include physical activity as a daily part of the curriculum for all students in all grades. It is recommended that children and adolescents engage in moderate-intensity physical activities for about an hour each day. Physical fitness goals should be stressed above competitive or sport performance goals.[23] Tests of physical fitness, such as the muscle strength and endurance assessment shown in Illustration 27.6, should be performed periodically on all children and the results used to modify instruction and activities.

Physical activity programs for schoolchildren are most likely to be successful when they offer recreational and sporting activities that are irresistible and fulfill the human desire to run, jump, and play. The happy consequences of physical activity come easiest to those who do it for play.

Nutrition Up Close

Exercise: Your Options

Focal Point: Assess your level of physical activity.

About one-quarter of U.S. adults are physically inactive, and most Americans do not exercise regularly. How much physical activity are you getting?

Physical Activity

Check the category that best describes your usual overall daily activity level. Do not consider activities that are part of an exercise program.

• *Inactive:* Sitting most of the day, with less than 2 hours moving about slowly or standing. ☐

• *Average:* Sitting most of the day, walking or standing 2 to 4 hours each day, but not engaging in strenuous activity. ☐

• *Active:* Physically active 4 or more hours each day. Little sitting or standing, engaging in some physically strenuous activities. ☐

Exercise

Do you exercise five or more days a week for at least 30 minutes?

☐ **Yes**   ☐ **No**

FEEDBACK (including scoring) can be found at the end of Unit 27.

Key Terms

aerobic fitness, page 27-4   *maximal oxygen consumption, page 27-4*   *physical fitness, page 27-4*

Review Questions

	True	False
1. Regular physical activity benefits psychological and physical health.	☐	☐
2. The primary components of physical fitness are body fat composition, strength, and heart rate.	☐	☐
3. Aerobic fitness is related to a person's ability to deliver oxygen to muscles and the capacity of muscles to use the oxygen for physical activity.	☐	☐

	True	False
4. Endurance is related to aerobic fitness.	☐	☐
5. Maximal oxygen consumption is related to endurance.	☐	☐
6. A 44-year-old individual who wants to exercise at 70% of maximal heart rate would aim for a target heart rate of 123 beats per minute.	☐	☐

	True	False

7. Glycogen supplies the fuel to meet most of our energy needs while we are at rest. ☐ ☐

8. Fat is the major fuel for low-to-moderate intensity physical activities. ☐ ☐

9. Resistance training builds endurance. ☐ ☐

10. To build and maintain a state of physical fitness it is recommended that adults engage in vigorous physical activity for 60 minutes seven days each week. ☐ ☐

11. It is recommended that children and adolescents engage in moderate-intensity physical activity for about 60 minutes daily. ☐ ☐

Media Menu

www.healthierus.gov/exercise.html
This site provides guidance for physical activity in children and information about national programs aimed at improving the fitness levels in children.

americanheart.org
Select "Exercise and Fitness" under the Healthy Lifestyles button and find easy ways to introduce physical activity into your life.

www.diabetes.org/weightloss-and-exercise/exercise/overview.jsp
"The journey of a thousand miles begins with a single step. Exercise works the same way." This Web offering from the American Diabetes Association helps you incorporate physical activity into your life.

www.nlm.nih.gov/medlineplus/exerciseforseniors.html
This site provides pictures and diagrams of exercises that improve balance as well as additional information on exercise for seniors.

www.nlm.nih.gov/medlineplus/exercisephysicalfitness.html
This site covers what's new, exercise recommendations, fitness tips, and public health policies related to physical activity.

www.mayoclinic.org
Head to the Mayo Clinic's Health Oasis on Fitness and Exercise to learn about the whys and hows of increasing your physical activity level.

http://presidentschallenge.org
This fun Web site for kids includes a fitness calculator, activity logs, and awards for meeting goals.

www.kidnetic.com
This popular, highly interactive Web site for children focuses on healthy eating and regular physical activity.

Notes

1. Dwyer JD et al. Leisure time physical activity and atherosclerosis. Am J Med 2003;115:19–25.

2. McCaffree J. Physical activity: how much is enough? J Am Diet Assoc 2003; 103:153–4.

3. Kivipelto M et al. Midlife physical activity and rates of dementia and Alzheimer disease. Lancet Neurol, 10/4/2005, www.medscape.com/viewarticle/514204.

4. Bartholomew JB et al. Brief aerobic exercise may improve mood, well-being in major depression. Med Sci Sports Exerc 2005;37:2032–7.

5. Physical activity and health: a report of the surgeon general. Washington, DC: U.S. Department of Health and Human Services; 1996.

6. Risk factors and coronary heart disease. AHA Scientific Position. www.medscape.com, accessed 7/03.

7. Physical activity and good nutrition: essential elements to prevent chronic diseases and obesity. www.cdc.gov/nccdphp/publications/aag/dnpa.htm, accessed 10/06.

8. Broeder C et al., The effects of aerobic fitness on resting metabolic rate, Am J Clin Nutr 1992;55:795–801.

9. Jeffery RW et al. Physical activity and weight loss: does prescribing higher physical activity goals improve outcomes? Am J Clin Nutr 2003;78:684–9.

10. Saris WHM. Physiological aspects of exercise in weight cycling. Am J Clin Nutr 1989;49(suppl):1099–1104.

11. ACSM fitness book, 2nd ed. American College of Sports Medicine, Champaign (IL): Human Kinetics Publishers, 1998.

12. Bouchard C et al. Heredity and trainability of aerobic and anaerobic performances. Sports Med 1988;5:69–73.

13. Position of the American Dietetic Association, Dietitians of Canada, and the American College of Sports Medicine: nutrition and athletic performance. J Am Diet Assoc 2000;100:1543–56.

14. Behm DG et al. Stretching, balance, and movement times. Med Sci Sports Exerc 2004;36:1397–1402.

15. Hart L. Effect of stretching on sport injury risk: a review. Clin J Sport Med. 2005 Mar;15:113.

16. Horton ES. Metabolic fuels, utilization, and exercise. Am J Clin Nutr 1989; 49(suppl):931–2.

17. Barr Sl et al. Nutritional considerations for vegetarian athletes. Nutrition 2004;20:696–703.

18. Nieman DC et al. Nutrient intake of marathon runners. J Am Diet Assoc 1989;89:1273–8.

19. Noakes TD. Too many fluids as bad as too few. BMJ 2003;327:113–4.

20. Peterson C. Ergogenic acids and dietary supplements. Food, Nutrition Conference and Exhibits, St. Louis, 10/23/05.

21. Most US adults not meeting physical activity recommendations. MMWR 2003;53:764–8.

22. Lie D. Moderate intensity exercise effective for weight loss. www.medscape.com, accessed 10/03.

23. Physical activity for everyone: recommendations. www.cdc.gov/nccdphp/dnpa/physical/recommendations/index.htm, accessed 10/9/06.

24. Maron BJ. The paradox of exercise. N Engl J Med 2000;343:1409–11.

25. Trends in leisure-time physical inactivity by age, sex, and race/ethnicity—United States, 1994–2004. MMWR 2005;54:991–4.

26. School study seeks to promote effective interventions, Chronic Disease News and Notes 2002;15:9–10.

Nutrition Up Close

Exercise: Your Options

Feedback for Unit 27

For those of you who answered "Active" to the first question and "Yes" to the second question, congratulations! You qualify as a mover and shaker. Keep up the good work!

If you answered "Inactive" or "Average" to the first question or "No" to the second, now is a good time to think about increasing your level of physical fitness. Besides promoting a feeling of well-being, regular exercise can help beat stress, weight gain, heart disease, and high blood pressure. Maybe it's time you exercised your options by incorporating more physical activity into your daily routine.

Answers to Review Questions

1. True, see Table 27.1, page 27-3.
2. False, see page 27-4.
3. True, see page 27-4.
4. True, see page 27-4.
5. True, see page 27-4.
6. True, see page 27-5.
7. False, see pages 27-6, 27-7.
8. True, see pages 27-6, 27-7.
9. False, see page 27-9.
10. False, see page 27-10.
11. True, see page 27-13.

Nutrition and Physical Performance

NUTRITION SCOREBOARD

	TRUE	FALSE
1 The diet an athlete consumes affects energy substrate availability to muscles during exercise.		
2 Carbohydrate loading is a waste of time.		
3 Supplementation with protein and amino acids increases muscle mass and strength to about the same extent as does resistance training.		
4 Scientists can't explain why, but bee pollen actually improves physical performance.		

Key Concepts and Facts

- Endurance is affected by genetics, training, and nutrition.

- Glycogen stores and endurance can be increased by diet and training.

- Abnormal or absent menstrual cycles in female athletes should not be dismissed but corrected with increased caloric intake.

- Glucose and fat are used as energy sources during exercise but in different proportions. Intense physical activity is primarily fueled by glucose, and fats are the main source of energy for low- to moderate-intensity exercise.

- A few ergogenic aids that claim to improve performance work to some extent, but most do not.

Answers to
NUTRITION SCOREBOARD

		TRUE	FALSE
1	Carbohydrate and fat availability to muscles during exercise varies based on usual diet.	✔	
2	Short-term carbohydrate loading increases endurance in trained athletes.		✔
3	Resistance training increases muscle size and strength. The role of protein and amino acid supplements in building muscle size and strength, if real, is limited.[1,2]		✔
4	In bees, maybe—but not in humans.		✔

Sports Nutrition

> BAM! Nobody heard it, but Lou felt it. She had "hit the wall." She was ahead of her planned pace, but now her legs felt like lead. She would have to finish the last two miles of the marathon at the slow pace her legs would allow.

From her carefully crafted and scrupulously followed training program to her refined shaping of mental attitude, Lou thought she had done everything right. She had left one thing out, however, and that may have cost her the race. Lou failed to pay attention to her diet while training and ran out of *glycogen* too soon (Illustration 28.1).

Three major factors affect physical performance: genetics, training, and nutrition.[1] The first gives some people an innate edge in sprinting or endurance, and nothing can be done about it. The second is acknowledged as a basic truth. Most athletes know a good bit about proper training, and the trick is to follow the right plan. The third is often ignored or, when taken seriously, misunderstood.

Nutrition has important effects on physical performance, but the legitimate role of nutrition is often poorly understood by athletes and coaches alike.[2] Athletes' lack of knowledge may make them vulnerable to phony nutrition claims and bad dietary advice. It leaves some coaches wishing they had taken a college course on nutrition. Only one in ten coaches has had a course in nutrition, but nearly all of them regularly dispense nutritional advice.[3] (To the coaches and future coaches who are taking this nutrition course: Bravo!)

Basic Components of Energy Formation during Exercise

Understanding the role of nutrition and performance, and the potential effects of some *ergogenic aids,* can be fostered by basic knowledge of how energy is formed within muscle cells. Illustration 28.2 summarizes the processes by which energy for muscle movement is formed.

glycogen
The storage form of glucose. Glycogen is stored in muscles and the liver.

ergogenic aids
(ergo = work; genic = producing) Substances that increase the capacity for muscular work.

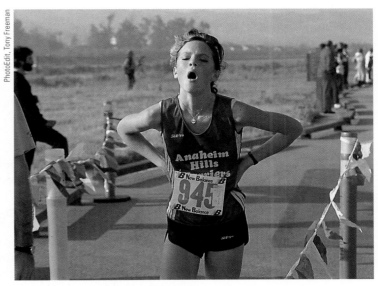

There are two main substrates for energy formation in muscles: glucose from muscle and liver glycogen, and fatty acids released from fat stores. How much of each is used depends on the intensity and duration of the exercise, as well as the body's ability to deliver each along with oxygen to muscle cells. Each substrate is used to form *ATP* from *ADP*. ATP serves as the source of energy for muscle contraction.

Anaerobic Energy Formation   Glucose from the liver and muscle glycogen (which is converted to glucose) form ATP without oxygen. This route of energy formation is "anaerobic," or "without oxygen," and it generates most of the energy used for intense muscular work (70% VO_2 max or higher).[4] Creatine phosphate (abbreviated CrP in Illustration 28.2), an amino acid containing a high-energy phosphate molecule, also converts ADP to ATP to a limited extent. Creatine phosphate stores are limited and decrease rapidly during intensive exercise.

Glucose is converted to "pyruvate" during anaerobic energy formation. In the absence of oxygen, pyruvate is converted to lactate. Lactate can build up in muscles and blood if not reconverted to pyruvate by the addition of oxygen. Pyruvate yields additional energy when it enters "aerobic" energy formation pathways along with fatty acids from fat stores.

Aerobic Energy Formation   The conversion of pyruvate and fatty acids to ATP requires oxygen. Much more ATP is delivered by the breakdown of fatty acids than glucose (fats have 9 calories per gram; glucose has only 4). The rate of energy formation from fatty acids is four times slower than that from glucose, however. It's the reason fatty acids are used to fuel low- and moderate-intensity exercise, or those below 60% VO_2 max.[4] Unlike glucose, energy formation from fatty acids is not limited by availability. Muscle cells can continue to produce energy from fatty acids as long as delivery of oxygen from the lungs and the circulation is sufficient.[5]

Illustration 28.1
Hitting the wall.
This runner ran out of muscle glycogen before the finish line.

ATP, ADP
Adenosine triphosphate *(ah-den-o-scene tri-phos-fate)* and adenosine diphosphate. Molecules containing a form of phosphorous that can trap energy obtained from the macronutrients. ADP becomes ATP when it traps energy and returns to being ADP when it releases energy for muscular and other work.

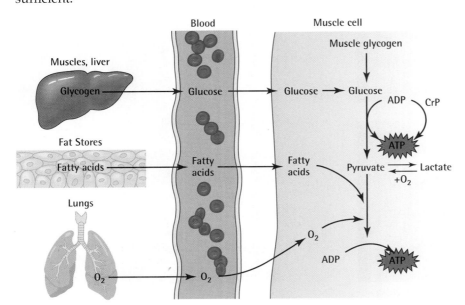

Illustration 28.2
Schematic representation of how ATP is formed for muscular movement.

Nutrition and Physical Performance

When everything else is equal, nutrition can make the difference between winning and losing.
—R. MAUGHAN, "THE ATHLETE'S DIET"[7]

Glycogen stores, foods eaten before and after exercise events, water intake, and ***electrolyte*** balance are all related to energy formation and physical performance.

Glycogen Stores and Performance

The vast majority of energy for muscle activity comes from fat and glucose. For most activities, both types of fuel are used; and the more intense the activity, the higher the amount of fat and glucose used for it. Fats, however, are the principal source of fuel for activities of low to moderate intensity, and glucose is the main source of energy for activities of high intensity (Illustration 28.3).[8]

Glycogen stores in muscles and the liver can deliver about 2000 calories worth of energy, whereas adults have access to over 100,000 calories from fat. Consequently, a person's ability to perform continuous, intense physical activity is limited by the amount of glycogen stored.[5] People who run out of muscle glycogen during an event "hit the wall"—they have to slow down their pace substantially because they can no longer use muscle glycogen as a fuel. The pace they are able to maintain will be dictated by the body's ability to use fat and liver glycogen as fuel for muscular work. If athletes keep pushing themselves after muscle glycogen runs out, they may end up "bonking"—using up the liver's supply of glycogen. That's worse than hitting the wall because hypoglycemia (low blood sugar) develops, and the person becomes dizzy and shaky and may pass out.[7] Obviously, endurance athletes do not want to exhaust their glycogen stores before an event is over.

electrolytes
Minerals such as sodium and potassium that carry a charge when in solution. Many electrolytes help the body maintain an appropriate amount of fluid.

Illustration 28.3
Fat is the main source of energy for low- and moderate-intensity activities, whereas glycogen is the primary fuel for high-intensity events.

Activities primarily fueled by fat.

Activities primarily fueled by glycogen.

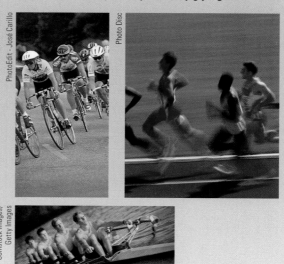

TABLE 28.1
Effects of diet on muscle glycogen stores during training[8]

Carbohydrate Intake Level for Previous 3 Days	Muscle Glycogen (Grams per 100 Grams of Muscle)	Time to Exhaustion at 75% V_{O_2} Max
Low	0.6	60 minutes
Average	1.8	126 minutes
High (65–70% of total calories)	3.5	189 minutes

TABLE 28.2
How many grams of carbohydrate are in a diet that provides 60 or 70% of total calories from carbohydrates? It depends on your caloric intake.

Usual Caloric Intake	Grams of Carbohydrate	
	60% CARB DIET	70% CARB DIET
1600	240 g	280 g
1800	270 g	315 g
2000	300 g	350 g
2200	330 g	385 g
2400	360 g	420 g
2600	390 g	455 g
2800	420 g	490 g
3000	450 g	525 g
3200[a]	480 g	560 g

[a]For each additional 200 calories of usual intake, add 30 grams carbohydrate for a 60% carbohydrate diet and 35 grams for a 70% carbohydrate diet.

Athletes consuming a typical U.S. diet normally have enough glycogen stores to fuel continuous, intense exercise for about an hour or two.[6] Both glycogen stores and endurance can be increased, however, by loading up muscles with glycogen prior to an event (Table 28.1).

Carbohydrate Loading Carbohydrate loading benefits endurance for athletes engaged in sports such as long-distance running and swimming, soccer, canoe racing, and triathlons. It does not improve performance in high-intensity, short-duration events such as sprints or short races, weight lifting, high jump, and baseball.[13] The recommended method for increasing glycogen stores for endurance events is much less stringent than in years past. It basically calls for increasing carbohydrate intake to 60 to 70% of calories for the 24-hour period preceding high-intensity exercise. Loading up on carbohydrates in this 24-hour period can increase glycogen stores to the same extent as previous regimes that lasted up to 6 days.[9] Table 28.2 shows how many grams of carbohydrate correspond to 60 or 70% of total calories for people with different caloric needs. Average diets provide about 50% of calories from carbohydrates. See Illustration 28.4 for an example of a one-day diet that provides 68% of total calories from carbohydrate.

Illustration 28.4
A high-carbohydrate diet. *Carbohydrates provide 68% of the 2420 calories in this day's menus, protein provides 11%, and fats provide 21%. Grams of carbohydrate are shown in parentheses.*

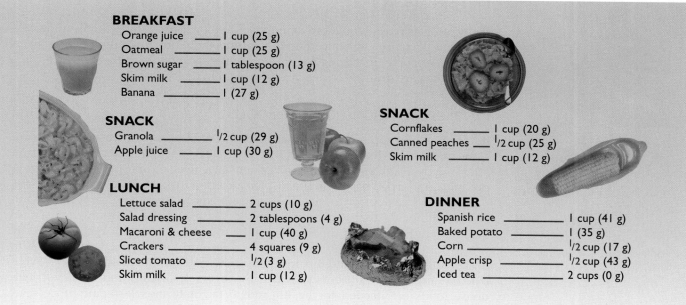

BREAKFAST
Orange juice _____ I cup (25 g)
Oatmeal _____ I cup (25 g)
Brown sugar _____ I tablespoon (13 g)
Skim milk _____ I cup (12 g)
Banana _____ I (27 g)

SNACK
Granola _____ 1/2 cup (29 g)
Apple juice _____ I cup (30 g)

LUNCH
Lettuce salad _____ 2 cups (10 g)
Salad dressing _____ 2 tablespoons (4 g)
Macaroni & cheese _____ I cup (40 g)
Crackers _____ 4 squares (9 g)
Sliced tomato _____ 1/2 (3 g)
Skim milk _____ I cup (12 g)

SNACK
Cornflakes _____ I cup (20 g)
Canned peaches _____ 1/2 cup (25 g)
Skim milk _____ I cup (12 g)

DINNER
Spanish rice _____ I cup (41 g)
Baked potato _____ I (35 g)
Corn _____ 1/2 cup (17 g)
Apple crisp _____ 1/2 cup (43 g)
Iced tea _____ 2 cups (0 g)

TABLE 28.3
High and moderate glycemic index foods and ingredients[10]

High Glycemix Index	Moderate Glycemic Index
Glucose (dextrose, dextrin)	
Maltose (malt)	Popcorn
Coco Pops	Carrots
Corn Chex	Honey
Cornflakes	Sucrose
Corn Pops	Bread
Gatorade	Oatmeal
Doughnuts	Soft drinks
Jelly beans	Orange juice
Potatoes	Muffins
	Bananas

■ Glycemic Index and Carbohydrate Loading

Food sources of carbohydrate that have high or moderate glycemic index (GI) values (Table 28.3) appear to maximize glycogen storage (a longer list of the GI values of food is given in Table 13.2 on page 13-7). These foods increase blood glucose levels more than do low-GI carbohydrates such as dried beans, fructose, apple juice, and oranges.[11] Regular intake of high-GI carbohydrates is generally discouraged because it may increase the risk of insulin resistance and type 2 diabetes. Athletes tend to be very sensitive to the action of insulin, however, and exercise in general lowers the risk of diabetes.[12]

Low glycemic index carbohydrates consumed prior to exercise appear to sustain glycogen stores during endurance exercise. When consumed 3 to 4 hours prior to exercise, low glycemic index carbohydrates increase fat utilization during exercise to a greater extent than do high or moderate glycemic index carbohydrates. Because the use of fat for energy preserves glycogen stores, low glycemic index carbohydrates consumed prior to exercise may improve endurance.[14] Low glycemic index carbohydrates include oatmeal, cracked-wheat bread, orange juice, green peas, dried beans, pasta, yams, and most fruits and vegetables. The Reality Check on the next page addresses this topic.

■ Carbohydrate-Loading Caveats

There is a limit on glycogen storage in muscle. Each gram of glycogen binds with 2 grams of water, limiting the total amount that can "fit" into muscle tissue. Once the capacity of muscles to store glycogen is reached, any additional carbohydrate is largely converted to fat and stored.[15]

Not all athletes tolerate the relatively high-carbohydrate, low-fat, carbohydrate-loading diet equally well. Some athletes dislike the stiff, heavy feeling that may occur when muscles become "loaded" with glycogen. In addition, it should be remembered that carbohydrate loading does not improve performance in athletes participating in physical activities that last less than an hour or two.

Protein Need

Many athletes require no more than their RDA for protein. Individuals undertaking strength training may need up to 15 grams more protein daily, however. Even then, most people (including athletes) consume far more than their RDA for protein, so they are unlikely to need extra protein in their diets.[17]

Preexercise and Recovery Foods

It is generally recommended that athletes engaged in high-intensity events consume meals providing protein and carbohydrate 3 to 4 hours prior to exercise and after exercise.[18] Nutrient-dense foods, such as vegetables and fruits, whole-grain products, and dried beans should also be included in the meals.

Protein intake prior to intense exercise increases the body's supply of available amino acids that can be incorporated into muscles. Carbohydrates raise blood insulin levels, and that facilitates the transfer of amino acids (provided by protein) into muscle cells. Protein intake after hard, physical activity appears to hasten muscle recovery and repair, and carbohydrates augment this effect.[18,19] Antioxidants and other nutrients provided by vegetables, fruits, and other nutrient-dense foods are needed for health and to facilitate muscle repair processes.[19]

Reality Check

Sweet energy?

Can you increase endurance by eating jelly beans, corn pops, doughnuts, or other high-GI foods just before a competition?

Who gets thumbs up?
Answers on next page

Toro:
That's what I do! It's why they call me El Toro!

Lakisha:
That's not what I heard. My coach says it will just make you lose steam sooner.

Illustration 28.5 shows foods that together qualify as a helpful preexercise and recovery meal.

Hydration

Hydration status is a major factor affecting physical performance and health. Adequate hydration during training and competition enhances performance, prevents excessive body temperatures, delays fatigue, and helps prevents injuries. Dehydration has opposite effects on physical performance and health. Hyperhydration, caused by overdrinking water before, during, and after endurance events, can lead to a loss of sodium from blood and body tissues and **hyponatremia.**[20]

Hydration status during exercise is primarily affected by how much a person sweats—or by how much water he or she loses through the skin while exercising.

Muscular activity produces heat that must be eliminated to prevent the body from becoming overheated. To keep the body cool internally, the heat produced by muscles is collected in the blood and then released both through blood that circulates near the surface of the skin and in sweat. Sweating cools the body because heat is released when water evaporates on the skin. People sweat more during physical activity in hot, humid weather because it is harder to add moisture and heat to warm, moist air than to dry, cool air. To stay cool during exercise in hot, humid conditions, the body must release more heat and water than when exercise is undertaken in drier, cooler conditions.

Maintaining Hydration Status during Exercise It is recommended that athletes engaged in events that last an hour or less drink about two cups of water one to two hours before an event and continue to consume water at regular intervals during the event. Athletes undertaking long events (over one hour in duration) should consume beverages that provide sodium (up to 200 mg sodium per

hydration status
The state of the adequacy of fluid in the body tissues. Dehydration indicates the presence of inadequate fluid in body tissues, and hyperhydration means there is too much.

hyponatremia
A deficiency of sodium in the blood (135 mmol/L sodium or less).

Illustration 28.5
An example of a helpful and healthful preexercise and recovery meal

1 PREEXERCISE MEAL
Orange juice
Oatmeal
Honey
Muffin
Coffee, milk

2 PREEXERCISE SNACKS
Bananas   Sherbet
Crackers   Fruit juice
Jell-O   Bagel

3 RECOVERY SNACKS
Low-fat yogurt
Frozen yogurt
Chocolate skim milk
High-carb/protein bar

ANSWERS TO *Reality Check*

Sweet energy burst?

Bravo, Lakisha! Yours is the better answer. Consumption of low-GI carbohydrates before endurance events helps preserve glycogen stores, decreases mental fatigue, and increases endurance somewhat in trained athletes.[21]

Toro

Lakisha

8 ounces) and carbohydrate (about 6% by weight) during the exercise.[22] The sodium is needed to replace that lost in sweat and to prevent hyponatremia, and the carbohydrate helps maintain blood glucose levels and provides muscles with glucose.[23] Hydration status is maintained when athletes do not lose weight during an event and when their urine remains pale yellow and is normal in volume.[24]

■ **Fluids That Don't Hydrate**   Not all fluids helps maintain hydration status. Fluids high in sugar, such as soft drinks, don't quench a thirst and should not be consumed for fluid replacement. Their high-sugar content may draw fluid from the blood into the intestines, thereby increasing the risk of dehydration, nausea, and bloating. Alcohol-containing beverages such as beer, wine, and gin and tonics are not hydrating, either.[20,22]

sweat rate
Fluid loss per hour of exercise. It equals the sum of body weight loss plus fluid intake.

■ **Estimating Fluid Needs: Sweat Rate**   The amount of fluid athletes need during an event can be estimated by calculating an hourly *sweat rate* for a specific activity (Table 28.4). A person preparing for a marathon who loses a pound in an hour of training and who drinks nothing during the hour would have a sweat rate of one pound (16 ounces). If that athlete drank 8 ounces of fluid in the hour and lost a pound—or 16 ounces—his or her sweat rate would be 24 ounces (16 ounces + 8 ounces). The sweat rate amount of fluid should be consumed per hour of the marathon event. Athletes who gain weight during an event have consumed too much fluid.[22]

TABLE 28.4
Calculating sweat rate: an example

1. Determine your body weight one hour before and one hour after exercise.
2. Subtract your postexercise weight from your preexercise weight.
3. Convert the number of pounds lost to ounces. (One pound equals 16 ounces.)
4. Add the number of ounces lost or gained to the number of ounces of fluid you consumed during the hour of exercise. The result is your sweat rate, and that's an approximation of the amount of fluid you need to consume during one hour of that exercise.

Example
1. Terrell weighed 172 pounds an hour before an hour-long bout of exercise and 171 pounds an hour after the exercise:
   172 pounds − 171 pounds = 1 pound lost
   1 pound × 16 ounce per pound = 16 ounces
2. He drank 16 ounces of fluid during the hour of exercise.
   16 ounces + 16 ounces = 32 ounces, or "sweat rate."

Terrell would need to consume about 32 ounces of fluid per hour to remain hydrated.

Dehydration: The Consequences   Loss of more than 2% of body weight (2 to 4 pounds) during an event indicates that the body is becoming dehydrated. Effects of dehydration range from mild to severe, depending on how much body water is lost (Illustration 28.6). Any amount of dehydration impairs physical performance. At the extreme, dehydration can lead to heat exhaustion or heat stroke (Table 28.5). People who overexercise in hot weather when they are out of condition are most likely to suffer heat exhaustion or heat stroke, but these conditions occasionally occur among seasoned athletes, too. Heat exhaustion can be remedied by fluids and electrolytes, but heat stroke requires emergency medical care.

The Health Action feature for this Unit highlights dehydration. The information may help you recognize and avoid it.

TABLE 28.5

A primer on heat exhaustion and heat stroke[25]

Heat exhaustion: A condition caused by low body water and sodium content due to excessive loss of water through sweat in hot weather. Symptoms include intense thirst, weakness, paleness, dizziness, nausea, fainting, and confusion. Fluids with electrolytes and a cool place are the remedy. Also called "heat prostration" and "heat collapse."

Heat stroke: A condition requiring emergency medical care. It is characterized by hot, dry skin, labored and rapid breathing, a rapid pulse, nausea, blurred vision, irrational behavior, and, often, coma. Internal body temperature exceeds 105°F due to a breakdown of the mechanisms for regulating body temperature. Heat stroke is caused by prolonged exposure to environmental heat or strenuous physical activity. The person affected by heat stroke should be kept cool by any means possible, such as removing clothing and soaking the person in ice-cold water. If conscious, the person should be given fluids. Also called "sunstroke."

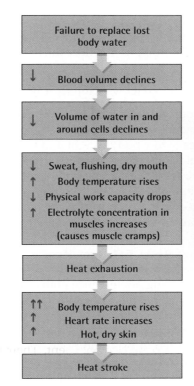

Failure to replace lost body water

↓ Blood volume declines

↓ Volume of water in and around cells declines

↓ Sweat, flushing, dry mouth
↑ Body temperature rises
↓ Physical work capacity drops
↑ Electrolyte concentration in muscles increases (causes muscle cramps)

Heat exhaustion

↑↑ Body temperature rises
↑ Heart rate increases
↑ Hot, dry skin

Heat stroke

Illustration 28.6
Effects of dehydration.[24]
Loss of 9 to 12% of body water can be fatal. Even a low level of dehydration impairs physical performance.

Bob Winslett/ Index Stock

Hyponatremia and Excess Water   Hyponatremia is an important cause of death and life-threatening illness in marathon runners. In a recent Boston Marathon, 13% of runners developed blood sodium levels that qualified as hyponatremia, and 0.6% became very ill due to extremely low levels of sodium.[29] The condition needs to treated promptly. The major signs of hyponatremia are difficulty breathing, bloating, confusion, nausea, vomiting, and swelling around the brain.[30]

Hyponatremia is most likely to occur in athletes who gain weight during an event, consume more than 3 liters (12 cups) of low- or no-sodium fluids during the race, and engage in running for four or more hours. Individuals with a low

HEALTH *Action*   Be Aware of the Signs and Symptoms of Dehydration!

The signs and symptoms of dehydration can include:

- Thirst
- Lightheadedness or fatigue
- Headache
- Inability to concentrate
- Feeing cold or having chills when it is hot out and you are sweating
- Collapse due to heat exhaustion or stroke

Another sign of dehydration is the amount and color of urine. Large amounts and lighter-color urine indicate good hydration versus urine that is dark in color. One should drink until the urine color becomes lighter.

body mass Index (under 20 kg/m²) and women are at higher risk of developing hyponatremia than others.[29] Regular consumption of a beverage containing about 200 mg sodium per 8 ounces during a race decreases the risk of hyponatremia.[22]

Body Fat and Weight: Heavy Issues for Athletes

As a group, athletes tend to be very concerned about their weight and body shape, and they are leaner than the U.S. population in general (Table 28.6).[26] But the advantages of low body weight and body fat disappear when they become too low.[27] Body fat levels of less than 3% in men and less than 12% in women can seriously interfere with health. These percentages represent the obligatory body fat content that people need for the functions of fat unrelated to energy production.[28] Everybody needs some fat to use for hormone production, to maintain normal body temperature in cold weather, and to cushion internal organs.

Exercise, Body Fat, and Health in Women   It is not uncommon for female athletes to experience irregular or absent menstrual periods or the late onset of periods during adolescence (Table 28.7).[31] These aberrations in menstrual period appear to be related to specific effects of high levels of physical activity or caloric intake deficits on hormone production or the body's metabolic function.[32] Female athletes with missing and irregular periods are at risk for developing low bone density, osteoporosis, and bone fractures. Women at particular risk of bone fractures are those with the "female-athlete triad" of disordered eating, amenorrhea (pronounced *a-men-or-re-ah* and meaning "no menstrual periods"), and osteoporosis. Abnormal menstrual cycles should not be dismissed as a normal part of training. Normal menstrual periods should be reinstated through increased caloric intake.[31] Suppressed testosterone levels or other metabolic changes due to caloric deficits in male athletes may be the parallel to menstrual irregularities in female athletes.[33]

Wrestling: The Sport of Weight Cycling   "Making weight" is a common and recurring practice among wrestlers (Illustration 28.7). Most wrestlers will "cut" 1 to 20 pounds over a period of days between 50 and 100 times during a high school or college career.[36] That makes wrestling more than a test of strength and agility. It makes it a contest of rapid weight loss.

For competitive reasons, wrestlers often want to stay in the lowest weight class possible and may go to great lengths to achieve it. They may fast, "sweat

TABLE 28.6
Average body fat content of various athletes

	Body Fat Content (As Percentage of Body Weight)	
	WOMEN	MEN
Bodybuilders	10%	6%
Long-distance runners	17	6–13
Baseball players	—	12–14
Basketball players	21–27	7–10
Football players	—	9–19
Gymnasts	10	5
Soccer players	—	10
Tennis players	20	16
Wrestlers	—	9

TABLE 28.7
Incidence of irregular or absent menstrual cycles in female athletes and sedentary women [31,32]

Joggers (5 to 30 miles per week)	23%
Runners (over 30 miles per week)	34
Long-distance runners (over 70 miles per week)	43
Competitive bodybuilders	86
Noncompetitive bodybuilders	30
Volleyball players	48
Ballet dancers	44
Sedentary women	13

the weight off" in saunas or rubber suits, or vomit after eating to lose weight before the weigh-in. These practices can be dangerous, even life threatening, if taken too far. After the match is over, the wrestlers may binge and regain the weight they lost.

Wrestlers, like other athletes involved in intense exercise, perform better if they have a good supply of glycogen and a normal amount of body water. Fasting before a weigh-in dramatically reduces glycogen stores, and withholding fluids or losing water by sweating puts the wrestler at risk of becoming dehydrated. Trying to stay within a particular weight class too long may also stunt or delay a young wrestler's growth.

The American Medical Association and the Association for Sports Medicine recommend that wrestling weight be determined after six weeks of training and normal eating. In addition, a minimum of 7% body fat should be used as a qualifier for assigning wrestlers to a particular weight class. In addition, weight classes are now based on normal weight for height and age, and weigh-ins are scheduled close to event times.[35]

Wisconsin high school wrestling programs adopted the medical association's recommendations and found that wrestlers, coaches, and parents supported the changes. Participation in wrestling increased because the athletes didn't have to deal with weight cutting.[36]

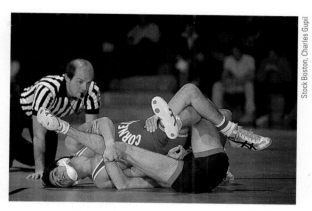

Illustration 28.7
Dropping weight quickly in the days before a match can wreck a wrestler's chances and harm his health.

Iron Status of Athletes

Iron status is an important topic in sports nutrition because iron deficiency (or low iron stores) and iron-deficiency anemia (or low blood hemoglobin level) decrease endurance. Iron is a component of hemoglobin, a protein in blood that carries oxygen to cells throughout the body, and it works with enzymes involved in energy production. When iron stores or hemoglobin levels are low, less oxygen is delivered to cells, and less energy is produced than normal.

Female athletes are at higher risk of iron-deficiency anemia than other females. One study of aerobically fit athletes found that 36% of women and 6% of males were iron deficient.[37] Consequently, it is recommended that female athletes especially pay attention to the amount of iron consumed.[27]

Ergogenic Aids: The Athlete's Dilemma

The quest for ownership of a competitive edge has drawn athletes to ergogenic aids throughout much of history. (See Table 28.8 for a historical review of ergogenic aids use.) Relatively few of the hundreds of available products work, and most are sold as "dietary supplements" so they do not have to be tested for safety. Those found to increase muscle mass, strength, or endurance are usually banned for use by competitive athletes. Several of the aids clearly pose serious risks to health, and others simply represent misguided hopes and misspent money. Table 28.9 summarizes claims made for a variety of ergogenic aids, research results, and what is known about adverse effects.

Protein, Amino Acids, Muscle Size, and Strength Whether protein and amino acid supplements combined with resistance training offer the real life benefits of increased mass size and strength is a red-hot, controversial issue. Some studies indicate protein or amino acid consumption before and after resistance

TABLE 28.8
Faster, higher, stronger, longer: a brief history of performance-enhancing substances used athletes

B.C.	Large quantities of beef consumed by athletes in Greece to obtain "the strength of 10 men"; deer liver and lion heart consumed for stamina
1880s	Morphine used to increase performance in (painful) endurance events
1910s	Strychnine consumed for the same reason as morphine
1930s	Amphetamines used to increase energy levels and endurance; testosterone taken to increase muscle mass
1980s	Blood doping, EPO used to increase endurance

TABLE 28.9

Ergogenic aids: claims and evidence[38]

Ergogenic Aid– Normal Functions	Claims	Evidence
Amphetamines • No normal function.	Reduce fatigue, improve concentration, increase aggressiveness, decrease appetite and weight.	Reduce fatigue, appetite, and weight; increase aggressiveness. **Adverse effects**: Many, including dizziness, tremors, confusion, paranoia, increased heart rate, chest pain, cardiovascular collapse.
Androgenic anabolic steroids • Testosterone and related substances. • Promote protein synthesis, decrease protein tissue breakdown, increase blood volume and red blood cell production.	Gradually increasing doses of several forms of steroids increase muscle mass.	Increase strength, muscle mass; decrease recovery time. **Adverse effects**: Liver, heart, psychiatric disorders, testicle atrophy, reduced HDL and elevated LDL cholesterol.
Bee pollen • Pollen collected by bees from flowers and mixed with bee secretions. Provides food for bee larva.	Detoxification of blood, improves performance.	No such effects found. **Adverse effects**: Allergic reaction in people who are allergic to bee stings and honey and who have asthma.
Boron • A mineral that influences calcium, magnesium, and phosphorus utilization.	Increases testosterone and muscle mass.	No such effects found. May improve bone density, arthritis.
Branched chain amino acids • Isoleucine, leucine, valine. Are essential amino acids. May lower brain serotonin levels.	Improve performance, delay fatigue, build muscle, increase mental concentration.	May delay muscle breakdown, spare glycogen, increase fat utilization and time to exhaustion. **Adverse effects**: High doses (>20 g) may impair performance.
Caffeine • Central nervous system stimulant, diuretic, circulatory and respiratory stimulant.	Decreases fatigue in endurance events.	May reduce fatigue and improve endurance at low to moderate doses (3–6 mg/kg). Increases insulin resistance **Adverse effects**: Nervousness, sleeplessness, irregular heartbeat; fatal dose 3–10 g. Olympic illegal dose equivalent to 6–8 cups coffee 2–3 hours preevent.
Carnitine • A compound required for energy formation from fatty acids.	Enhances fatty acids utilization for energy.	Most studies show no effect on fatty acid utilization during exercise in healthy adults. **Adverse effects**: diarrhea
Cocaine • No normal functions.	Same as for stimulants but much shorter acting.	
Chromium Picolinate • Involved in energy metabolism and insulin utilization.	Increases lean body mass, decreases fat mass.	No effect on body composition. No evidence of ergogenic effect. **Adverse effects**: May increase oxidative damage to muscles.

TABLE 28.9

Ergogenic aids: claims and evidence[38] (*continued*)

Ergogenic Aid– Normal Functions	Claims	Evidence
Creatine • Component of muscle; generates ATP from ADP in muscle.	Increases performance in high-intensity, short-duration exercises.	May slightly increase peak muscle force in short-term, high-intensity exercises, especially in vegetarians with low reserves of creatine. No increase in strength observed, no improvement in aerobic exercise performance. May increase energy available for training. **Adverse effects:** Headache, water retention, muscle cramps.
DHEA, Andro (Dehydroepinandrosterone, Androstendione) • Precursors of testosterone.	Promote testosterone production and increase muscle mass.	Promote testosterone production and increase muscle mass. **Adverse effects:** The same as for anabolic steroids.
Ephedra (ma huang) • No normal function.	Central nervous system stimulant; enhances endurance, strength, and body fat loss.	Increases aerobic capacity when combined with caffeine. **Adverse effects:** Nervousness, anxiety, rapid heart rate, headache, hypertension, death. Not recommended for use; no longer sold.
EPO (Erythropoietin) • A hormone that increases red blood cell production by bone marrow.	Replaces blood doping, or injection of red blood cells for increasing oxygen availability and endurance. Is being replaced by altitude tents and other hypoxic devices.	Increases endurance. **Adverse effects:** Increases blood viscosity, blood pressure, blood clot formation; headache. Banned by the National Olympic Committee.
GBL (Gammabutyrolactone, furanodehydro) • No normal function.	Increases muscle mass and muscle recovery from exercise.	No such effects **Adverse effects:** Vomiting, seizures, death. FDA has called for its voluntary removal from the market.
Ginseng, Siberian • No normal function.	Enhances endurance and strength.	Increases muscular strength and aerobic work capacity.
Glycerol • A clear, sweet fluid attached to fatty acids in fat molecules.	Increases hydration.	Conflicting results. **Adverse effects:** Bloating, nausea.
HGH (Human growth hormone) • Stimulates muscle and bone growth; converted to glucose.	Increases muscle mass and strength; similar effects as anabolic steroids but safer.	Increases muscle mass but probably not strength. **Adverse effects:** Growth of facial bones, heart disease, stroke, hypertension, diabetes. Effects generally irreversible.
HMB (β-hydroxy-β–methylobutyrate) • Derivative of the amino acid leucine.	Improves muscle size and strength.	No performance improvement, may help muscle recovery. **Adverse effects:** Appears safe at doses of 3 grams per day.

(*continued*)

TABLE 28.9

Ergogenic aids: claims and evidence[38] *(continued)*

Ergogenic Aid—Normal Functions	Claims	Evidence
Insulin • Speeds entry of glucose into cells, increases production of glycogen.	Increases muscle mass.	Increases glucose passage into cells and glycogen and protein synthesis. **Adverse effects:** Low blood sugar, seizure, brain damage, death.
Low–GI carbs (low glycemic index carbohydrates)	Increase rate of fat utilization versus high GI carbs; improve endurance.	Low-GI carbs, such as pasta, dried beans, oranges, and peaches, appear to increase fat utilization and endurance compared to high-GI carbs (white bread, corn flakes, candy, and french fries, for example).
Oxygen, oxygen drinks and bars	Increase oxygen availability for muscular work.	No effect on oxygen availability in the body. **Adverse effects:** Burping if consumed.
Protein powders • Can contribute to meeting the body's need for protein.	Increase muscle mass and strength.	No such effects found. **Adverse effects:** Dehydration, liver and kidney problems.
Vitamin and mineral supplements • May contribute to meeting need for vitamins and minerals.	Improve energy and stamina.	Such supplements improve energy and stamina in deficient individuals only. **Adverse effects:** Toxicity reactions.

training increases muscle size but nor strength or performance.[2,39] Other studies report a slight increase in strength and improved performance, as well as gains in muscle size.[1, 40] Protein and amino acids consumed before and after resistance training may improve muscle tissue repair and prevent the decline in strength that comes with postexercise muscle damage.[41] It appears that any effects of protein or amino acid supplementation on muscle size and strength would be minor compared to gains that result from resistance training.

Even *if* protein and amino acid supplements improve muscle size and strength, it makes no sense to buy the supplements. Food that are part of a healthful diet can feed muscles plenty of amino acids. Athletes interested in supplementing their diets with branched-chain amino acids, for example, can eat chicken. You would get more of these amino acids from one chicken breast than seven branch-chain amino acid tablets. A cup of peanuts would supply 11 tablets' worth of these amino acids.[1]

The Path to Improved Performance   Genetics, training, and nutrition: These are the real keys to physical performance. Although other aids will be sought, those that exceed the boundaries of what is considered fair and safe will not be approved for use by athletes. After all, athletic competition is not a test of drugs or performance aids. It's a test of an individual's ability to excel. Anything less wouldn't be sporting.

Nutrition Up Close

Testing Performance Aids

Focal Point: The critical examination of studies on performance aids.

Read the following summary of a study (imaginary) of the effects of a phosphorous supplement on strength and then answer the critical thinking questions.

- *Purpose:* To assess the effect of a phosphorous supplement on strength.
- *Methods:* Twenty volunteers from the crew team were given the phosphorous supplement for a week. Strength, assessed as the maximum number of push-ups a study participant could do in one session, was assessed before and after supplementation. Participants recorded any supplement side effects.
- *Results:* The number of push-ups increased by an average of 5% after supplementation. Diarrhea was the only side effect consistently noted by participants.

Critical Thinking Questions

1. Does the study demonstrate that the supplement is safe?

   Yes _____ No _____ Give reasons for your answer.

   _____

2. Does the study demonstrate that the supplement increases strength?

   Yes _____ No _____ Give reasons for your answer.

   _____

FEEDBACK (answers to these questions) can be found at the end of Unit 28.

Key Terms

ATP, ADP, *page 28-3*
electrolytes, page 28-4
ergogenic aids, page 28-2

glycogen, page 28-2
hydration status, page 28-7

hyponatremia, page 28-7
sweat rate, page 28-8

Review Questions

True False

1. The two main substrates for energy formation in muscles and amino acids and fatty acids. ☐ ☐

2. Glucose is converted to energy by muscles anaerobically. ☐ ☐

3. Much more ATP is produced from the breakdown of fatty acids than from amino acids or glucose. ☐ ☐

4. Hypoglycemia results when athletes use up the liver's supply of glycogen. ☐ ☐

5. Glucose is the major source of energy for low- and moderate-intensity activities. ☐ ☐

6. Carbohydrate loading is rarely used in practice because it takes six days of a special diet and training regime to prepare for it. ☐ ☐

True False

7. Postevent recovery meals should include rich sources of fat because fats facilitate muscle repair. ☐ ☐

8. Hyponatremia is primarily caused by drinking excessive amounts of low- or no-sodium fluids during prolonged exercise. ☐ ☐

9. An athlete who runs for an hour, loses two pounds in that hour, and drinks 12 ounces of fluid during the hour has a sweat rate of 60 ounces. ☐ ☐

10. Signs of dehydration include thirst and lightheadedness or fatigue. ☐ ☐

11. A few ergogenic aids improve performance somewhat, but most do not improve performance. ☐ ☐

12. Athletes attempting to build muscle mass and strength should take protein and amino acid supplements. ☐ ☐

Media Menu

www.mayohealth.com
The Mayo Clinic's Web site offers reliable advice on exercise and sports nutrition.

www.acsm.org
The American College of Sports Medicine Web site offers information on sports nutrition, readily accessible articles on sports nutrition, and the opportunity to join the national Coalition for Promoting Physical Activity.

www.nal.usda.gov/fnic
Search the "sports nutrition" link and head toward information on Healthy People 2010 physical activity and fitness goals for the nation, statistics, and articles on nutrition and physical performance.

www.nutrition.gov
Search "physical performance" and find a wealth of information on study results, evaluation systems, and guidelines.

www.runnersworld.com
This site isn't just for runners! Athletes looking for a competitive edge should check this site out.

www.gssiweb.com
This connects you to the Gatorade Sports Science Institute. The site provides valuable resources for sports nutrition professionals through "Focus on" sections about ergogenic aids, hydration assessment, and other hot sports topics.

Notes

1. Gleason M. Interrelationships between physical activity and branched-chain amino acids. J Nutr 2005;135:1591S–5S.

2. Andersen LL et al. The effect of resistance training combined with timed ingestion of protein on muscle fiber size and muscle strength. Meta Clin Exper 2005;54:165–70.

3. Corley G et al. Nutrition knowledge and dietary practices of college coaches. J Am Diet Assoc 1990;90:705–9.

4. Coleman E. Fat loading for endurance sports. Today's Dietitian. 2003;Mar.:12–16.

5. Brown RC. Nutrition for optimal performance during exercise: carbohydrate and fat. Curr Sports Med 2002;1:222–9.

6. Maughan R. The athlete's diet: Nutritional goals and dietary strategies. Proc Nutr Soc 2002;61:87–96.

7. Coleman E. Eating for endurance. Palo Alto (CA): Bull Publishing Co., 1988.

8. Foster C et al. Effects of preexercise feeding on endurance performance. Med Sci Sports Exer 1979;11:1–5.

9. Fairchild TJ et al. Rapid carbohydrate loading after a short bout of near maximal-intensity exercise. Med Sci Sports Exercise 2002;34:980–6.

10. Foster-Powell K et al. International tables of glycemic index and glycemic load, Am J Clin Nutr 2002;76:5–56.

11. Coleman, Fat loading; and Suzuki M. Glycemic carbohydrates consumed with amino acids or protein right after exercise enhance muscle formation, Nutr Rev 2003;61:S88–S94.

12. Jenkins DJ et al. Glycemic index: overview of implications in health and disease, Am J Clin Nutr 2002;76(suppl):266S–73S.

13. Carbo-loading: Boost your endurance during high-intensity workouts. www.mayoclinic.com/health/carb-loading/HQ00385, 10/8/06.

14. Stevenson EM et al. Influence of high-carbohydrate mixed meals with different glycemic indexes on substrate utilization during subsequent exercise in women. Am J Clin Nutr 2006;84:354–60.

15. Brouns F et al. Utilization of lipids during exercise in human subjects: metabolic and dietary constraints. Brit J Nutr 1998;79:117–28.

16. Atkinson ML. Panama Canal's locomotive "mules" lock ships dead-center in waterway. Newhouse News Service, Sept. 1955.

17. Kleiner SM. Healthy muscle gains. Phys Sports Med 1995;23:21–22.

18. Volek JS et al. Strength nutrition. Curr Sports Med Rep. 2003;2:189–93.

19. Burke LM et al. Carbohydrates and fat for training and recovery. J Sports Sci. 2004;22:15–30.

20. Von Duvillard SP et al. Fluids and hydration in prolonged endurance performance. Nutrition. 2004;20:651–6.

21. Lieberman HR et al. Carbohydrate administration during a day of sustained aerobic activity improves vigilance, as assessed by a novel ambulatory monitoring device, and mood. Am J Clin Nutr 2002;76:120–7.

22. Coleman E. Hydration and hyponatremia in athletes. Today's Dietitian 2005;Oct.:18–25.

23. Winnick JJ et al. Carbohydrate feeding during team sports and preservation of exercise performance, and central nervous system function. Med Sci Sports Exerc 2005;37:306–15.

24. Noakes TD. Too many fluids as bad as too few. BMJ 2003;327:113–4.

25. Bouchama A, Knochel JP. Heatstroke, N Engl J Med 2002;346:1978–88.

26. Wilmore JH. Body composition in sports and exercise: directions for future research. Med Sci Sports Exer 1983;15:21–31.

27. Manson JE, Lee I-M. Exercise for women: how much pain for optimal gain? N Engl J Med 1996;334:1325–6.

28. Elliot DL, Goldberg L. Nutrition and exercise. Med Clin North Am 1985;69:71–81.

29. Almond CSD et al. Hyponatremia in nonelite marathon runners. N Engl J Med 2005;352:1516–8,1550–6.

30. Ayus JC. Hyponatremia in marathon runners. N Engl J Med 2005;353:427.

31. Ireland ML, et al. The female athlete, Philadelphia: W.B. Saunders; 2003.

32. Lloyd T et al. Interrelationships of diet, athletic activity, menstrual status, and bone density in collegiate women, Am J Clin Nutr 1987;46:681–4.

33. Brownell KP et al. Weight regulation practices in athletes: analysis of metabolic and health effects. Med Sci Sports Exer 1987;19:546–56.

34. Steen SN et al. Metabolic effects of repeated weight loss and regain in adolescent wrestlers. JAMA 1988;260:47–50.

35. Cunningham E. Where can I find resources to address weight practices for wrestlers? J Am Diet Assoc 2006;106:761.

36. Oppliger RA et al. The Wisconsin wrestling minimum weight project: a model for weight control among high school wrestlers. Med Sci Sports Exer 1995;27:1220–24.

37. Sinclair LM et al. Prevalence of iron deficiency with and without anemia in recreationally active men and women. J Am Diet Assoc 2005;105:975–8.

38. Ahrendt DM. Ergogenic aids: counseling the athlete. www.aafp.org/afp/20010301/913.html, accessed 10/06.

39. Candow DG et al. Effect of whey and soy protein supplementation combined with resistance training in young adults. Int J Sport Nutr Exerc Metab. 2006;16:233–44.

40. Candow DG et al. Protein supplementation before and after resistance training in older men. Eur J Appl Physiol. 2006;97:548–56.

41. Ratamess NA et al. The effects of amino acids supplementation on muscular performance during resistance training overreaching. J Strength Cond Res. 2003;17:250–8.

Answers to Review Questions

1. False, see page 28-3.
2. True, see page 28-3.
3. True, see page 28-3.
4. True, see page 28-4.
5. False, see page 28-4.
6. False, see page 28-5.
7. False, see page 28-6.
8. True, see page 28-7.
9. False, see page 28-8.
10. True, see page 28-9.
11. True, see Table 28.8 on page 28-11.
12. False, see pages 28-11, 28-14.

Nutrition Up Close

Testing Performance Aids

Feedback for Unit 28

1. Answer: No. Side effects, such as blood pressure changes, diarrhea, abnormalities in kidney function, and weight loss, were not assessed. Safety of long-term use was not addressed.

2. Answer: No. A 5% increase in strength is small and may have been due to usual differences in the number of push-ups a person can do from time to time. Any increase in strength may have been due to increased training during the week the supplement was taken. There was no control group. All participants knew they were taking phosphorous, a supplement that some may believe increases strength. This belief may have changed performance on the push-up test.

UNIT 29

Good Nutrition for Life: Pregnancy, Breast-Feeding, and Infancy

NUTRITION SCOREBOARD

	TRUE	FALSE
1 The fetus is a parasite.		
2 The United States has the lowest rate of infant mortality in the world.		
3 Normal health, growth, and development of infants are defined by the experience of breast-fed infants.		
4 Drinking alcohol-containing beverages is safe during breast-feeding because the alcohol doesn't pass into the milk.		

Key Concepts and Facts

- There are no "maternal instincts" that draw women to select and consume a good diet during pregnancy.

- The fetus is not a parasite.

- Fetal and infant growth are characterized by "critical periods" during which all essential nutrients needed for growth and development must be available or growth and development will not proceed normally.

- An adult's risk of certain chronic diseases may be partially determined by maternal nutrition during pregnancy and the person's own nutrition early in life.

- Breast-feeding is the optimal method of feeding infants.

- Rates of growth and development are higher during infancy than at any other time in life.

Answers to
NUTRITION SCOREBOARD

		TRUE	FALSE
1	The fetus does not harm the mother for its own gain. (A true parasite injures the host while benefiting from it.) If there is a shortage of essential nutrients, the mother generally will get access to them before the fetus does.		✔
2	Over 20 economically developed countries have lower infant mortality rates than the United States.		✔
3	That's the American Academy of Pediatrics's standard.	✔	
4	Alcohol does pass into breast milk. Breast-feeding women should limit their intake of alcohol if they drink at all.		✔

Photo Disc

infant morality rate
Deaths that occur within the first year of life per 1,000 live births.

low-birthweight infants
Infants weighting less than 2500 grams (5.5 pounds) at birth.

preterm infants
Infants born at or before 37 weeks of gestation (pregnancy).

A Healthy Start

The day has finally arrived, the one Crystal and Tyrone have anticipated for a year. Crystal's pregnancy test is positive!

This is a planned pregnancy, and both Crystal and Tyrone have done all they could to prepare. Crystal's diet has been the picture of perfection, she has kept up her regular exercise schedule, and she has quit drinking alcohol entirely. Now they are ready to have the baby of their dreams. They are convinced that not only will this baby be healthy and strong, but it will be above average in every respect.

All parents want to give their children every advantage in life that they can. Although no one can guarantee that a baby will be born healthy and strong—no matter what the parents do—there are steps parents can take to make the best baby possible.

Unhealthy Starts on Life

Unfortunately, many infants born in the United States do not have the advantage of good health at birth. Among countries of the world, the United States ranks fortieth in terms of *infant mortality rate* (Table 29.1). The relatively high rate of infant deaths in the United States is due primarily to the proportion of infants born with *low birthweights* or *preterm* (Illustration 29.1). One in 13 U.S. infants is born too small, and over 1 in 10 is born too early.[1] These infants are at particular

Stock Solutions/Index Stock

Illustration 29.1
In 2004, 7.9% if the infants born in the United States had a low birthweight, and 12.3% were born preterm.[1]

TABLE 29.1

Infant mortality rates (deaths per 1,000 live births) in 2004 for a sampling of countries[2]

Singapore	2.3
Sweden	2.8
Hong Kong	3.0
Japan	3.2
Finland	3.5
Norway	3.7
Czech Republic	3.9
Germany	4.1
France	4.2
Switzerland	4.3
Spain	4.4
Denmark	4.5
Austria	4.6
Australia	4.6
Belgium	4.6
Canada	4.7
Netherlands	5.0
Portugal	5.0
United Kingdom	5.1
Ireland	5.3
Greece	5.4
Italy	5.8
New Zealand	5.8
Cuba	6.2
United States	6.8
Israel	6.9
Slovakia	7.3

risk for requiring intensive and continuing care and of dying within the first year of life. In contrast, infants weighing between 7 pounds, 11 ounces and 8 pounds, 13 ounces (or 3500 to 4000 grams) are least likely to die within the first year. The average weight of U.S. infants is 7 pounds, 7 ounces (3350 grams).[3]

Improving the Health of U.S. Infants

A high proportion of poor infant outcomes in the United States is attributed to a combination of factors including poverty, poor nutrition, limited access to health care, and a maternal lifestyle that includes the use of illicit drugs, cigarettes, and excessive amounts of alcohol.[4] Many infant deaths, preterm infants, and low-birthweight infants can be prevented, however, by optimizing maternal behaviors that influence health. Of the various behaviors that affect maternal health, none offers potentially greater advantage to both pregnant women and their infants than good nutrition.

Nutrition and Pregnancy

Nutrition is important during pregnancy because the *fetus* depends on it. Energy and every nutrient the fetus needs for *growth* and *development* must be supplied by the mother's diet (preferably) or by supplements. If either insufficient or excessive amounts of nutrients are supplied, fetal growth and development may be compromised. The nature and extent of the impairment depend on which organ or tissue is growing most rapidly when the nutritional deficiencies or overdoses occur.[5]

Critical Periods

Fetal growth and development proceed in a series of critical periods. A *critical period* is an interval of time during which cells of a tissue or organ are genetically programmed to multiply. The period is considered critical because if the cells do

fetus
A baby in the womb from the eighth week of pregnancy until birth. (Before then, it is referred to as an *embryo*.)

growth
A process characterized by increases in cell number and size.

development
Processes involved in enhancing functional capabilities. For example, the brain grows, but the ability to reason develops.

critical period
A specific interval of time during which cells of a tissue or organ are genetically programmed to multiply. If the supply of nutrients needed for cell multiplication is not available during the specific time interval, the growth and development of the tissue or organ are permanently impaired.

Illustration 29.2
Baby's first picture.
An ultrasound image of a rapidly growing and developing 16-week-old fetus.

Photo Teke - Barts Medical Library

The ancient Romans believed that if a woman wanted a child with brown eyes, she should eat mice often during pregnancy.

Photo Disc

not multiply as programmed during this set time interval, they cannot make up the deficiency later. The level of nutrients required for cell multiplication to occur normally must be available during this specific time interval. If the nutrients are not available, the developing tissue or organ will contain fewer cells than normal, will form abnormally, or will function less than optimally.

The roof of the mouth (or the hard palate), for example, is formed early in the third month of pregnancy when two developing plates fuse together. This process can only occur early in the third month. If excessive amounts of vitamin A are present in fetal tissues during this period, the two plates may fail to combine, resulting in a cleft palate. (The divided hard palate can be surgically corrected after birth.)

Critical periods of cell multiplication are most intensive in the first few months of pregnancy, when fetal tissues and organs are forming rapidly (Illustration 29.2)—hence the importance of the nutritional status of women at the time of conception and very early in pregnancy. For some organs and tissues, cell multiplication even continues through the first two years after birth. Most of the growth that occurs in the fetus late in pregnancy and throughout the rest of the growing years, however, is due to increases in the size of cells within tissues and organs.

The Fetal Origins Hypothesis

One of the most striking advances in research on pregnancy concerns the potential effects of maternal nutrition on the baby's risk of developing certain chronic diseases later in life. It appears that an important element of increased susceptibility to heart disease, stroke, diabetes, obesity, hypertension, and other disorders may be "programmed," or gene functions modified, by inadequate or excessive supplies of energy or nutrients during pregnancy.[6]

Given optimal conditions, fetal growth and development proceed according to the genetic blueprint established at conception. Organs and tissues are well developed and ready to function optimally after birth. Under less than optimal growing conditions, such as those introduced by maternal weight loss, poor nutrient intakes, or diseases in the mother that alter her ability to supply the fetus with energy or nutrients, fetal growth and development are modified. Fetal tissues undergoing critical phases of development at that time have to make adaptations to cope with the under- or oversupply of nutrients. These adaptations may produce long-term changes in the structure and function of tissues. Low maternal energy intake in the last few months of pregnancy, for example, may hinder the development of cells that produce insulin and limit the body's ability to control blood glucose levels later in life. This could increase the risk of developing diabetes later in life.[6]

A large body of evidence supports the fetal origins hypothesis. Nevertheless, much remains to be discovered about the effects of diet during pregnancy on the risk of later disease.

The Fetus Is Not a Parasite   From early history to current times, every culture has had myths about diet in pregnancy. According to one of the more common pregnancy myths, the fetus is a parasite, capable of drawing whatever nutrients it needs from the mother at the expense of her health. If this were true, there would be no such thing as a small or poorly nourished newborn. Obviously, it is not true.

The fetus is not a parasite because, with few exceptions, it receives an adequate supply of nutrients only if the mother's intake is sufficient to maintain her own health.[5] The situation makes sense in relation to survival of the species. By meeting the mother's needs first, nature protects the reproducer.

Rather striking evidence that the fetus is not a parasite comes from studies that have identified deficiencies of vitamin B_{12}, thiamin, iodine, folate, zinc, and other nutrients in newborns but not in their mothers.[7] Similarly, infants born to women who consumed excessive levels of vitamin or mineral supplements during pregnancy are more likely to display signs of nutrient overdose than are the mothers.[8]

Prepregnancy Weight Status and Prenatal Weight Gain Are Important

Women who enter pregnancy underweight or who fail to gain a certain minimum of weight during pregnancy are much more likely to deliver low-birthweight and preterm infants than are women who enter pregnancy at normal weight or above and gain an appropriate amount of weight.[9] The risk of low birthweight can be reduced substantially by good diets that lead to a desired rate and amount of weight gain during pregnancy. Along with the duration of pregnancy and smoking, prepregnancy weight status and weight gain in pregnancy are the major factors known to influence an infant's birthweight (Table 29.2).

What's the Right Amount of Weight to Gain during Pregnancy?   Weight-gain recommendations for pregnancy vary depending on whether a woman enters pregnancy underweight, normal weight, overweight, or obese. A woman who is obese when she becomes pregnant will need to gain less weight than a woman who enters pregnancy underweight. It is recommended that underweight women gain 28 to 40 pounds, normal-weight women 25 to 35 pounds, overweight women 15 to 25 pounds, and obese women at least 15 pounds during preg-

TABLE 29.2
Major factors that directly influence birthweight[9]

- Duration of pregnancy
- Prenatal weight gain
- Prepregnancy weight status
- Smoking

TABLE 29.3

Prepregnancy weight status and recommended weight gain during pregnancy[1]

Prepregnancy weight status, Body Mass Index[a]	Recommended weight gain
Underweight, <18.5 kg/m²	28–40 pounds
Normal weight, 18.5–24.9 kg/m²	24–35 pounds
Overweight, 25–29.9 kg/m²	15–25 pounds
Obese, 30 kg/m² or higher	15–20 pounds
Twin pregnancy, all weight statuses	35–45 pounds

[a]Body Mass Index categories modified based on 1997 changes from the National Institutes of Health.

Illustration 29.3

An example of a prenatal weight-gain chart for pregnant women.

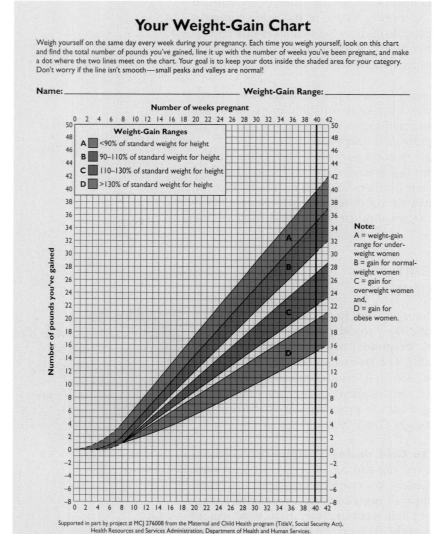

Your Weight-Gain Chart

Weigh yourself on the same day every week during your pregnancy. Each time you weigh yourself, look on this chart and find the total number of pounds you've gained, line it up with the number of weeks you've been pregnant, and make a dot where the two lines meet on the chart. Your goal is to keep your dots inside the shaded area for your category. Don't worry if the line isn't smooth—small peaks and valleys are normal!

Name: _____   Weight-Gain Range: _____

Number of weeks pregnant

Weight-Gain Ranges
A <90% of standard weight for height
B 90–110% of standard weight for height
C 110–130% of standard weight for height
D >130% of standard weight for height

Number of pounds you've gained

Note:
A = weight-gain range for under-weight women
B = gain for normal-weight women
C = gain for overweight women and,
D = gain for obese women.

Supported in part by project # MCJ 276008 from the Maternal and Child Health program (TitleV, Social Security Act), Health Resources and Services Administration, Department of Health and Human Services.

nancy. Women carrying twins should gain 35 to 45 pounds during pregnancy.[9] Table 29.3 can be used to identify prepregnancy weight status by a woman's body mass index (BMI) category and the appropriate recommended weight gain during pregnancy. After the goal is identified, a woman may want to plot her prenatal weight gain on a chart like the one in Illustration 29.3. The weight gained should be the result of a high-quality diet that leads to gradual and consistent gains in weight throughout pregnancy.

Where Does the Weight Gain Go?   If healthy infants tend to weigh about 8 pounds at birth, why do women need to gain more than that? Where does the rest of the weight go? Many pregnant women ask themselves these questions, especially when they don't relish the thought of gaining weight. Illustration 29.4 shows where the weight gain goes.

Fetal growth is accompanied by marked increases in maternal blood volume, fat stores, and breast and uterus size, all of which contribute to weight gain. In addition, water accumulates in the amniotic fluid, which cushions and protects the fetus, and the volume of fluid that exists outside cells increases. The placenta (the tissue that transfers nutrients in the mother's blood supply to the fetus) also accounts for some of the weight that is gained during pregnancy.

Where Does the Weight Gain Go—after Pregnancy?   Delivery is one of the greatest weight-loss plans known to humankind. On average, women lose 15 pounds within the first week after delivery, and weight loss generally continues for up to a year.[10] Women who gain more than the recommended amount of weight, however, and those who gain weight after delivery may have extra pounds to lose to get back to prepregnancy weight.

The Need for Calories and Key Nutrients during Pregnancy

Pregnant women need more calories, protein, and essential nutrients than nonpregnant women need. All of these are important; but calories, folate, vitamin B₆, vitamin A, calcium, vitamin D, iron, iodine, and EPA and DPH are of paramount concern. Although many pregnant women are concerned about protein, low protein intakes are rarely a problem in pregnancy. Most women consume 10 to 30 grams more protein daily than the recommended intake level of 71 grams.[11]

According to dietary intake recommendations, pregnant women need approximately 15% more calories and up to 50% more of various nutrients than do nonpregnant women (Illustration 29.5).[11] Relatively high nutrient requirements mean that pregnant women should increase their intake of nutrient-dense foods more than their consumption of calorie-rich foods.

Calories   On average, women need an additional 340 calories a day in the second *trimester,* and 450 calories in the third trimester of pregnancy.[11] Women entering pregnancy underweight will need more calories than this, and those entering overweight will need fewer. In addition, physically active pregnant women require higher caloric intakes than average. Rather than counting calories, however, it is generally easier and more accurate to monitor the adequacy of caloric intake by tracking weight gain.

Folate   Folate is required for protein tissue construction and therefore is in high demand during pregnancy. It is considered an "at risk" nutrient because some pregnant women do not consume enough to meet their daily need for folate.[12]

Folate deficiency has long been associated with fetal growth failure and malformations, but its link to *neural tube defects* such as spina bifida (Illustration 29.6) is fairly recent. In clinical trials, adequate folate status very early in pregnancy is associated with a 50 to 70% reduction in the occurrence of neural tube defects.[13] This finding, along with the knowledge that many women fail to consume enough folate, led to the fortification of refined grain products with folic acid. (Examples of fortified grain products are shown in Illustration 29.7.) Folic acid is a form of folate used in fortified foods and supplements. It is absorbed from foods about twice as completely as folate is. The vast majority of ready-to-eat breakfast cereals are fortified with at least 100 micrograms (0.1 milligram) of folic acid per serving.

Neural tube defects form before 30 days after conception, so it is important that women are consuming enough folate when they enter pregnancy. "Enough" is 600 micrograms (or 0.6 milligram) daily before and during pregnancy. Women can generally obtain this level of folate by consuming two servings of ready-to-eat breakfast cereal, six servings of grain products, and three servings of vegetables daily.[14] If the need for folate is not met by a diet that includes fortified foods, a 400-microgram folic acid supplement should be taken. Fortification of refined grain products with folic acid is improving folate status in Americans and is decreasing the incidence of neural tube defects.[15]

Vitamin A   Both low and high intakes of vitamin A may cause problems during pregnancy. Too little vitamin A is associated with poor fetal growth. Too much vitamin A in the form of retinol from supplements can cause fetal malformations.[9]

The effects of vitamin A overdoses during pregnancy came to public attention in the early 1990s, when women taking accutane or retinoic acid for acne

Here's where the weight goes for a woman who gains 33 pounds

Breasts 1.1 lb

Fetus 8.3 lb

Placenta 1.6 lb

Amniotic fluid 2.0 lb

Uterus 2.4 lb

Fat stores 8.5 lb

Blood 3.0 lb

Other fluid 6.1 lb

TOTAL = 33 lb

Illustration 29.4
Where does all the weight go?

trimester
One-third of the normal duration of pregnancy. The first trimester is 0 to 13 weeks, the second is 13 to 26 weeks, and the third is 26 to 40 weeks.

neural tube defects
Malformations of the spinal cord and brain. They are among the most common and severe fetal malformations, occurring in approximately one in every 1000 pregnancies. Neural tube defects include spina bifida (spinal cord fluid protrudes through a gap in the spinal cord; shown in Illustration 29.6), anencephaly (absence of the brain or spinal cord), and encephalocele (protrusion of the brain through the skull).

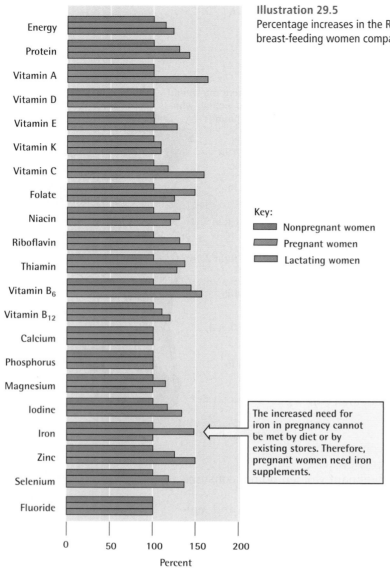

Energy

Protein

Vitamin A

Vitamin D

Vitamin E

Vitamin K

Vitamin C

Folate

Niacin

Riboflavin

Thiamin

Vitamin B$_6$

Vitamin B$_{12}$

Calcium

Phosphorus

Magnesium

Iodine

Iron

Zinc

Selenium

Fluoride

Key:
Nonpregnant women
Pregnant women
Lactating women

The increased need for iron in pregnancy cannot be met by diet or by existing stores. Therefore, pregnant women need iron supplements.

0   50   100   150   200

Percent

Illustration 29.5
Percentage increases in the RDAs or AIs for pregnant and breast-feeding women compared to other women.

Photo Researchers, Inc.

Illustration 29.6
Spina bifida—a neural tube defect.
The interruption in the spinal cord results in paralysis below the injury. This photo shows the back after surgery to close the interruption.

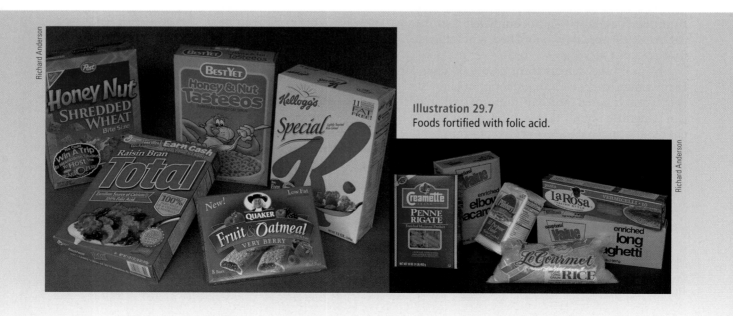

Richard Anderson

Illustration 29.7
Foods fortified with folic acid.

Richard Anderson

were found to deliver far more than the expected number of infants with malformations. Use of these drugs before and very early in pregnancy increases the risk that babies will be born with malformations of facial features and the heart; the intake of more than 10,000 to 15,000 IU of retinol daily during the same period has the same effect. As a precaution, the American College of Obstetrics and Gynecology recommends that women who are or could become pregnant should limit their intake of retinol to less than 5000 IU per day and should not use vitamin A-containing medications. Beta-carotene, a precursor to vitamin A, is not harmful, however.[16]

Calcium   Calcium used to support the mineralization of bones in the fetus is supplied by the mother's diet and, if needed, by the calcium contained in the long bones of the mother's body. Consequently, the fetus has access to as much calcium as it needs. Uptake of calcium by the fetus is especially high during the third trimester, when the fetus's bones are mineralizing.

Pregnant women who regularly consume low-calcium diets lose calcium from their bones during pregnancy but usually regain it after delivery. Because any calcium losses are from the long bones—not from the teeth—the old saying, "for every baby a tooth" has no basis in fact. Several studies have shown that teeth do not demineralize as a result of low-calcium diets during pregnancy.[5]

Vitamin D   Vitamin D during pregnancy supports fetal growth and the addition of calcium to bone and programs genes in ways that can influence the development of chronic diseases such as rheumatoid arthritis and cancer later in life.[17] Lack of it compromises fetal growth and bone development, and this may be happening during many pregnancies. About 42% of African American and 4% of Caucasian women have low blood levels of vitamin D.[18] Vegan women may also be a risk for poor vitamin D status because vitamin D is naturally present only in animal products.

An intake of 5 mcg (200 IU) vitamin D daily is officially recommended for pregnancy. Some credible experts, however, assert that more vitamin D than that is needed. It would not be surprising to see recommended intake levels of vitamin D at least double.[17] Intakes of vitamin D from foods and supplements should not exceed 50 mcg (2000 IU) daily.

Iron   Iron deficiency is the most common nutrient deficiency in pregnant women.[19] It develops when a woman enters pregnancy with low iron stores and fails to consume enough iron during her pregnancy. Iron requirements increase during pregnancy due to increases in hemoglobin production and storage of iron by the fetus.

Because of the large increase in need, it is difficult for pregnant women to get enough iron from foods. Consequently, 30 milligrams of supplemental iron per day are recommended for the second and third trimesters of pregnancy. Women who do not take supplemental iron are more likely to develop iron deficiency and to deliver infants who are small and at risk of developing iron deficiency in their first year than are women who do take supplements.[20]

Iodine   Iodine is needed for normal thyroid function and plays important roles in protein tissue construction and maintenance. A lack of iodine during pregnancy can interfere with fatal development and, in extreme cases, cause mental and growth retardation and malformation in children.[21]

About half of U.S. pregnant women consume less than the recommended 220 mcg of iodine daily, and 7% have moderate iodine deficiency.[21] The most reliable source of iodine is iodized salt. One teaspoon contains 400 mcg iodine. Salt with added iodine is clearly labeled as "iodized." Fish, shellfish, seaweed, and some types of tea also provide iodine. Iodine also ends up in foods manufactured in plants that use iodine-containing solutions to clean equipment. Women who consume iodized salt are not likely to need supplemental iodine. Usual iodine intake should not exceed 1100 mcg daily during pregnancy.

EPA and DHA   These two omega-3 fatty acids are becoming the folic acid of the recent past. New research is quickly informing us of their roles in fetal and infant development and of the benefits of consuming adequate amounts of them. EPA (eicosapentaenoic acid) and DHA (docosahexaenoic acid) are long-chain, highly unsaturated fatty acids that promote maternal health and support the optimal development of vision and the central nervous system of the fetus and infant. Women who have adequate EPA and DHA intake during pregnancy and breast-feeding tend to deliver infants that develop a somewhat higher level of intelligence, better vision, and otherwise more mature central nervous system functioning than do infants born to women with low intake of EPA and DHA. Rates of preterm delivery are significantly lower among women with adequate EPA and DHA status.[22]

It is currently recommended that women consume 300 mg of EPA and DHA daily during pregnancy and breast-feeding.[23] Most pregnant and breast-feeding women in the United States consume less than one-third of this amount. EPA and DHA are found together in fish, fish oils, and seafood (it turns out that fish really is brain food. . .), and DHA is available in omega-3 fatty acid fortified fortified eggs ("Omega Eggs"). Consumption of 12 ounces per week of low-mercury fish during pregnancy and breast-feeding is encouraged. Safe fish that provide high level of EPA and DHA include salmon, herring, anchovies, shrimp, and halibut.

What's a Good Diet for Pregnancy?

Contrary to folklore, women do not instinctively select and consume a healthy diet during pregnancy. A good diet takes planning.

Regardless of a pregnant woman's age and whether or not she is a vegetarian, a healthy diet provides all the nutrients she needs with the possible exception of iron. Foods are recommended over supplements because foods provide fiber, antioxidants, and other beneficial substances that supplements do not. Pregnancy diets should include sufficient fluid and fiber and should consist of regular meals and snacks. A pregnant woman should not consume alcohol and should drink coffee only in moderation. Table 29.4 lists these and other recommendations for the diet during pregnancy. Planning a diet for pregnancy around the basic food groups is the most straightforward approach to meeting nutrient needs (Illustration 29.8). These recommendations also apply to breast-feeding women.

Why Alcohol and Pregnancy Don't Mix   As early as the 1800s, maternal consumption of alcohol during pregnancy was said to cause the birth of "sickly" infants. The ill effects of alcohol on babies were not fully acknowledged, however, until the 1970s, when several research reports described a condition called fetal alcohol syndrome (Illustration 29.9). Women who drank heavily or frequently binged on alcohol during pregnancy were found to be at high risk of delivering infants with specific malformations and retarded physical and mental development. The effect of maternal alcohol consumption on the fetus worsens as intake increases. Heavy drinking in the first half of pregnancy is closely associated with the birth of malformed, small, mentally impaired infants. When excessive drink-

---

TABLE 29.4

Dietary recommendations for pregnancy

- Consume sufficient calories for adequate weight gain.
- Eat a variety of foods from each food group.
- Eat regular meals and snacks.
- Consume sufficient dietary fiber (about 28 grams per day).
- Consume 11 to 12 cups of water each day from fluids and foods.
- Use salt to taste (within reason).
- Do not drink alcoholic beverages.
- Limit coffee to four or fewer cups per day.
- Eat foods you enjoy at pleasant meal times.

PhotoEdit - David Young-Wolff

Illustration 29.9
Children with fetal alcohol syndrome (FAS) have characteristic facial features, mental retardation, heart defects, and other problems. Less severe cases can include small size and slow development.

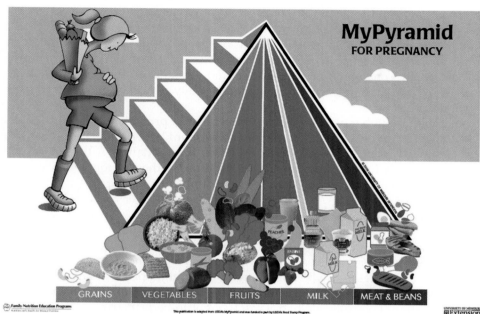

Food group	Ounces/cups recommended per day	Examples of equivalent measures
Grains	8 ounces	• 1 slice bread = 1 ounce • 1 cup cold cereal = 1 ounce • 1 cup cooked rice, pasta, or cereal = 2 ounces
Vegetables	3 cups	• 2 cups tossed salad = 1 cup
Fruits	2 cups	• 1 cup fruit juice = 1 cup
Milk	3 cups	• $^{1}/_{3}$ cup shredded cheese = 1 cup • 2 slices Americans cheese = 1 cup • $1^{1}/_{2}$ ounces hard cheese = 1 cup • $1^{1}/_{2}$ cups ice cream = 1 cup
Meat and beans	6 1/2 ounces	• 1 small egg = 1 ounce • 1 Tbsp peanut butter = 1 ounce • $^{1}/_{4}$ cup dried beans = 1 ounce • $^{1}/_{2}$ ounce nuts = 1 ounce • 1 Tbsp mayonnaise = $2^{1}/_{2}$ teaspoons oil • 1 Tbsp salad dressing = 1 teaspoon oil

[a] Based on MyPyramid food guide for women consuming 2400 calories a day and adjusted for pregnancy nutrient needs. Serving numbers will vary based on actual caloric need.
Source: Judith E. Brown, 2006.

ing occurs only in the second half of the pregnancy, infants are less likely to be malformed but are still likely to be small and to suffer abnormal mental development.[24] These conditions are permanent; they cannot be fully corrected with special treatment, and the child does not outgrow them.

No amount of alcohol has been found to be absolutely safe during pregnancy. When only an occasional drink is consumed, however, the adverse effects of alcohol on fetal development appear to be small and rare. To exclude the possibility of even small impairments in fetal growth and development, it is recommended that women not drink alcohol at all during pregnancy or when they are attempting to become pregnant.[9]

Vitamin and Mineral Supplements   Iron is the only supplement recommended for all pregnant women. About 83% of pregnant women take multiple vitamin and mineral supplements, however. Apparently, supplements are being prescribed as insurance against the possibility of poor diets. Women should be given supplements the same way they are given medications—when they are indicated. Under certain conditions, supplementation with nutrients besides iron is indicated. A multivitamin–mineral preparation is recommended for pregnant women who do not ordinarily consume an adequate diet and for those in high-risk categories, such as women carrying more than one fetus, heavy cigarette smokers, and alcohol and drug abusers.[9]

Because a number of vitamins and minerals can produce overdose reactions when taken in excessive amounts, supplements should not contain more than the recommended intake levels. As little as 15,000 IU of vitamin A (five times the RDA) taken early in pregnancy appears to be linked to the development of central nervous system and bone abnormalities, for example.[25]

Teen Pregnancy

Approximately 4 of every 100 females between the ages of 15 and 19 years in the United States deliver babies each year.[1] Although teens who do not smoke, drink alcohol, or use drugs and who enter pregnancy in good physical and nutritional health tend to deliver healthy infants, many teens do not enter pregnancy in such good condition. Poor lifestyle habits are thought to be primarily responsible for the high rate of low-birthweight and preterm infants born to teens. Few teens (1%) meet all of the food group recommendations. Individualized attention, nutrition intervention, and support all foster healthy outcomes of teen pregnancy.[26]

Breast-Feeding

Food is the first enjoyment of life.
—Lin Yutang

A woman's capacity to nourish a growing infant does not end at birth; it continues in the form of breast-feeding (Illustration 29.10). Breast milk from healthy, well-nourished women is ideally suited for infant nutrition and health.[27]

What's So Special about Breast Milk?

The breastfed infant is the reference model against which all alternative feeding methods must be assessed with regard to growth, health, and development; and other short- and long-term health outcomes.
—HHS Blueprint for Action on Breastfeeding, 2000

Breast milk is like a bonus pack. In addition to serving as a complete source of nutrition for infants for the first 4 to 6 months of life, breast milk contains substances that convey a significant degree of protection against a variety of illnesses (Table 29.5)—including infectious diseases such as polio and the "flu" and ear, respiratory tract, and gastrointestinal tract infections. Evidence indicates that breast milk may confer a degree of protection against the development of cancer of the lymph system (lymphoma), asthma, and diabetes during childhood. In addition,

Illustration 29.10

Photo Disc

TABLE 29.5

A sample of U.S. infants shows fewer significant episodes of illness among those who were breast-fed[27]

Illness	Percentage of Infants Fed:	
	BREAST MILK	FORMULA/OTHER
Ear infections	3.7%	9.1%
Respiratory illness to age 7	17.0	32.0
Diarrhea, vomiting	3.5	6.9
Hospital admissions	1.0	3.0
Average number of episodes of illness per infant	8.2	21.1

breast milk contains essential and nonessential fats and other substances that appear to promote optimal growth and development of the nervous system and eyes. Intelligence, as measured by IQ, tends to be higher in babies receiving breast milk than in those given formula.[28] The disease-preventing and development-promoting components of human milk are lifesaving assets in many developing countries, where a safe water supply and medical care may be unavailable. Although breast-feeding doesn't protect infants from all infectious diseases and food allergies, it's the ounce of prevention that's worth a pound of cure.

Breast-feeding also offers other benefits—ones that may mean a lot to parents. Table 29.6 discusses them.

Is Breast-Feeding Best for All New Mothers and Infants?

Over 96% of women are biologically capable of breast-feeding, and the vast majority of infants thrive on breast milk.[29] But breast-feeding is not best for every woman and infant. Successful breast-feeding involves more than biology; it is heavily influenced by environmental and psychological conditions.

In the United States, about 70% of new mothers breast-feed their infants to some extent (Illustration 29.11).[30] Because rates of breast-feeding were relatively low throughout the 1960s and 1970s, many women giving birth now grew up in an environment where formula feeding was preferred to breast-feeding (Illustration 29.12).

The increase in women returning to work soon after delivery, a lack of health care provider and emotional support for breast-feeding, embarrassment, early hospital discharge, and inadequate knowledge about how to breast-feed all appear to be deterring U.S. women from breastfeeding. If more women are to have the opportunity to breast-feed, these and other barriers must be broken down. Perhaps a good place to start is with the value our

TABLE 29.6

Fifteen reasons to breastfeed

1. The milk container is easy to clean.
2. Breast milk is a renewable resource.
3. There's no packaging to discard.
4. Breast milk comes in an attractive container.
5. The temperature of breast milk is always perfect right out of the container.
6. Breast milk tastes really good.
7. There are no leftovers.
8. You don't have to go to the kitchen in the middle of the night to get breast milk ready.
9. It takes just seconds to get a meal ready.
10. There's no bottle to repeatedly pick up off the floor.
11. The price is right.
12. The meal comes in a perfect serving size.
13. Meals and snacks are easy to bring along on a trip or outing.
14. Feeding units come in an assortment of beautiful colors and sizes.
15. One food makes a complete meal.

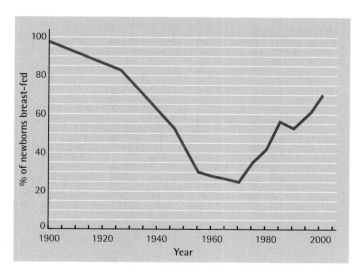

Illustration 29.11
Percentage of breast-fed newborns in the United States.[30,31]

Courtesy of Hans and Libby Schapiro

Illustration 29.12
Said a friend when she saw this photo in 1951: "This is the best advertisement for breast-feeding I've ever seen!"

TABLE 29.7

Actual versus national goals for breast-feeding in the United States[31]

	Newborns	6 Months	12 Months
National goal	75%	50%	25%
Actual (all women)	70	33	25

society places on breast-feeding. A number of European countries that actively promote breast-feeding allow women to stay in the hospital after delivery until breast-feeding is going well. The usual practice in some African countries is to relieve a breast-feeding mother's workload so that she can devote nearly full time to feeding and caring for her young infant. A relative may move in with the family and take over household chores, or the mother and baby may live with her parents for a time.

Public health initiatives are also in place to facilitate breast-feeding. To meet national health objectives for improving rates of breastfeeding in the United States by 2010 (see Table 29.7), it is recommended that:

- Health care workers encourage and facilitate breastfeeding

- "How to" advice and problem solving guidance be available for breast-feeding women from qualified health care staff

- Breast-feeding women returning to work have access to on-site child care facilities and private rooms for breastfeeding

colostrum
The milk produced during the first few days after delivery. It contains more antibodies, protein, and certain minerals than the mature milk that is produced later. It is thicker than mature milk and has a yellowish color.

Breast-feeding is at its best when both the mother and infant benefit from the experience. If mutual benefit is not possible, then formula feeding may be necessary. Infant growth and development are well supported by commercially available infant formulas.[33]

How Breast-Feeding Works

Illustration 29.13
A view of the interior of the breast.

The mother's body prepares for breast-feeding during pregnancy. Fat is deposited in breast tissue, and networks of blood vessels and nerves infiltrate the breasts. Ducts that will channel milk from the milk-producing cells forward to the nipple—the milk collection ducts—also mature (Illustration 29.13). Hormonal changes that occur at delivery signal milk production to begin.

Breast milk produced during the first three days or so after delivery is different from the milk produced later. Called *colostrum*, this early milk contains higher levels of protein, minerals, and antibodies than does "mature" milk.

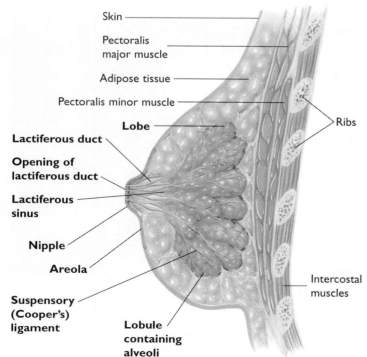

Skin
Pectoralis major muscle
Adipose tissue
Pectoralis minor muscle
Lobe
Lactiferous duct
Opening of lactiferous duct
Lactiferous sinus
Nipple
Areola
Suspensory (Cooper's) ligament
Lobule containing alveoli
Ribs
Intercostal muscles

Breast Milk Production   While an infant is consuming one meal, she or he is "ordering" the next. The pressure produced inside the breast by the infant's sucking and the emptying of the breasts during a feeding cause a hormone to be released from specific cells in the brain. The hormone stimulates the production of milk so that more milk is produced for the next feeding. It generally takes about 2 hours for the milk-producing cells to manufacture enough milk for the next feeding. An important exception to this occurs when an infant enters a growth spurt and consumes more milk than usual. Then milk production takes longer, perhaps a day, to catch up with demand.

Only very rarely is a breast-feeding woman unable to produce enough milk. As long as an infant is allowed to satisfy her or his appetite by breast-feeding as often as desired, milk production will catch up with the baby's need.

Nutrition and Breast-Feeding

A breastfeeding woman needs an adequate and balanced diet to replenish her body's nutrient stores, maintain her health, and produce sufficient milk for her baby. Increases in recommended dietary intakes for breastfeeding are generally higher than those for pregnancy. As during pregnancy, proportionately higher amounts of nutrients than calories are required, indicating the need for a nutrient-dense diet (see Illustration 29.5 on p. 29-8).

Calorie and Nutrient Needs   The RDA for calories is about 15% higher for breast-feeding women than for other women. Actually, a breast-feeding woman needs about 30% more calories than the RDA for women who are not breast-feeding, but she does not have to consume that level of calories from food. Energy supplied from fat stores that normally accumulate during pregnancy contributes to meeting these needs during breastfeeding, so not all of the calories must come from the mother's diet.[34]

A breast-feeding woman has a high requirement for DHA (docohexaenoic acid), an omega-3 fatty acid, because her infant used a good deal of it for central nervous system development. The DHA content of maternal milk directly reflects maternal intake. Adequate intake of DHA by women during breast-feeding is related to gains in intelligence and vision development in infants. Fish and shellfish (the types that are low in mercury should be selected), and "omega eggs" are good sources of DHA.[35]

What's a Good Diet for Breast-Feeding Women?   The calories and nutrients needed by the breast-feeding woman can be obtained from a varied diet that includes foods from each of the basic food groups. The recommended numbers of servings from each food group are shown in Illustration 29.8; they are the same as for pregnant women. Dietary recommendations for breastfeeding women are summarized in Table 29.8.

Increases in hunger and food intake that accompany breast-feeding generally take care of meeting caloric needs. Failure to consume enough calories from food can decrease milk production, however. Low-calorie diets (those providing less than 1500 calories per day) and weight loss that exceeds 1.5 to 2 pounds per week—even in women with a good supply of fat stores—can reduce the amount of milk women produce. Weight loss of about a pound a week starting a month after delivery appears to be safe and helpful in women's attempts to return to prepregnancy weight.[36]

Are Supplements Recommended for Breast-Feeding Women?   Supplements are not recommended for breast-feeding women. Instead, breast-feeding women should try to get all of the nutrients they need from food.[34] Supplements, if needed, should be prescribed on an individual basis.

Dietary Cautions for Breast-Feeding Women   Almost anything a woman consumes may end up in her breast milk. When a breast-feeding woman drinks

Can adopting moms breast-feed their infants? Yes, they can. It takes hormones and motivation, however. Breast-feeding success is higher the younger infants are when they begin to breast-feed with their adoptive moms.[32]

Photo Disc

ON THE SIDE

TABLE 29.8

Dietary recommendations for breast-feeding women[34]

- Diets should supply all of the nutrients breastfeeding women need. (The routine use of vitamin and mineral supplements is not recommended.)
- Fluids should be consumed to thirst (14 cups of fluid each day).
- Weight loss should not exceed 6 to 8 pounds per month after the first month after delivery; caloric intakes should not fall below 1500.
- Alcohol intake should be avoided.

coffee, her infant receives a small dose of caffeine. Breast-fed infants of women who are heavy coffee drinkers (10 or more cups per day) may develop "caffeine jitters." Alcohol also is transferred from a woman's body to breast milk. The development of the brain and nervous system of infants breast-fed by chronic, heavy drinkers appears to be retarded. There is also concern that alcohol ingestion by breast-feeding women may decrease milk production.[37]

Environmental contaminants, such as lead, DDT (dichloro-diphenyl-trichloroethane), chlordane, PCBs (polybrominated biphenyls), and PBBs (polychlorinated biphenyls), for example, are transferred into breast milk. Many environmental contaminants may be stored in a woman's fat or bone tissues. When the fat stores are later broken down for use in breast milk or the calcium in bone is mobilized, the contaminants stored in the fat or bone may enter the breast milk. The ingestion of fish from contaminated waters in Lake Ontario and Lake Michigan was linked to abnormally high levels of PCBs in breast milk. Infants exposed to PCBs can develop rashes, digestive upsets, and nervous system problems. With the exception of breast-feeding women known to be exposed to excess levels of environmental toxins, however, it is concluded that the benefits of breast-feeding outweigh the risks to infants from harmful substances in the environment.[38]

Infants are much smaller than women, and it takes a smaller dose of caffeine, alcohol, drugs, or environmental contaminants to have an effect on them than on an adult. Whereas breast-feeding mothers may show no adverse effects from these substances, their infants may.

Breast-feeding women are advised to limit their consumption of regular coffee to 4 or fewer cups a day and to avoid alcohol or limit it to one drink with a meal per day.[34] Drugs or medications should be taken only on the advice of a health care provider.

Illustration 29.14
Infants develop at a remarkable rate.

Infant Nutrition

He who possesses virtue in abundance may be compared to an infant.
—Lao Tzu, sixteenth century b.c.

At no other time during life outside the womb do growth and development proceed at a faster pace than during infancy. Infants grow out of clothes within weeks, long before they wear them out. Each day infants learn new behaviors, and their minds absorb large chunks of information that will serve them well in the future (Illustration 29.14). The rate at which growth and development proceed in the first year of life is truly amazing. You can get an idea of just how rapidly infants grow and develop by reviewing Illustration 29.15. Just as infants need security, love, and attention to flourish, so too do they need calories and essential nutrients.

Infant Growth

During the first week or two of life, infants generally lose 5 to 10% of their birthweight while adjusting to the new surroundings. After that, infants grow rapidly. Most infants double their birthweight by 4 months and triple it by 1 year. Length usually increases by 50% during the first year. If this rate of growth were to continue, 10-year-old children would be about 10 stories high and weigh over 220 tons! (Look at Illustration 29.15 to get a clear idea of how tall that is.) After infancy, the growth rate declines and remains at a fairly low level until the adolescent growth spurt begins.

Illustration 29.15
This is how tall you would have been as a 10-year-old if you had kept growing at the rate you did during infancy.

Illustration 29.16
Growth and developmental characteristics from birth through one year.

First Month
Generally weighs from 8 to 11 pounds; length is 20 to 23 inches. Head is relatively large and has soft spot on top. Startles and sneezes easily. Jaw may tremble. May hiccup and spit up. Eats every few hours.

One to Three Months
Lifts head briefly when placed on stomach; smiles, coos, and gurgles. Whole body moves when infant is touched or lifted. Eats every three to four hours.

Four to Six Months
Weight nearly doubled. Has grown three to four inches. Follows objects with eyes. Reaches toward objects with both hands; puts fingers and objects into mouth. Turns over, sits unassisted. Awake longer at feeding time. Eats six to seven times per day. Sleeps six to seven hours at night.

Seven to Eleven Months
Gains in weight and height are less rapid, appetite has decreased. Stands up with help, hitches self along the floor. Reaches for, grasps, and examines objects with hands, eyes, and mouth. Has one or two teeth. Takes two naps a day.

Twelve Months
Usually has tripled birthweight and increased length by 50%. Grasps and releases objects with fingers. Holds spoon, but uses it poorly. Begins to walk unassisted.

Growth Charts for Infants   In 2000 the Centers for Disease Control released new growth charts for female and male infants, children, and adolescents. Separate charts were not developed by ethnic group or race, because growth potential is a shared human trait that varies primarily due to environmental factors such as nutritional status and disease. An example of the growth charts that include infants is shown in Illustration 29.17. The charts should be carefully plotted based on accurate, periodic measures of an infant's size. They are best used for screen-

Illustration 29.17
An example of the CDC's growth charts.
The charts can be copied from the Internet using the address www.cdc.gov/growthcharts.

Source: National Center for Chronic Disease Prevention and Health Promotion.

ing growth problems; confirmation of underweight or obesity requires assessment of body fat by skinfold thickness and other measures.[39]

Body Composition Changes with Growth   Humans grow up and out, and their body composition and proportions change as growth progresses. Although generally measured by gains in pounds and inches, growth also reflects changes in bone mass, organ size, body proportions, and composition (the proportionate amounts of water, muscle, and fat). Infants' heads are very large in relation to the rest of their bodies, for example. Brain growth takes precedence over trunk and limb growth early in life, and the body eventually grows to "fit" the head. During the first ten years of life, a child who initially has the shape of a loaf of bread may normally come to resemble a string bean.

Nutrition and Mental Development

Malnutrition has the greatest impact on mental development when it is severe and occurs during the critical period for brain cell multiplication. For humans, this vulnerable period begins during pregnancy and ends after the first year of life. Impairment in mental development is less severe when malnutrition occurs only during pregnancy or only during infancy than if it occurs throughout both pregnancy and infancy.[40]

Children's mental development is also greatly influenced by the social and psychological environment in which they are raised. Because malnutrition is generally accompanied by both social and psychological deprivation, these factors often contribute jointly to poor mental development. For children in the United States, poverty, neglect, illnesses, and psychological problems appear to be the main cause of undernutrition and poor mental development.[41]

Infant Feeding Recommendations

Current dietary recommendations for infants call, preferably, for breast-feeding or, alternatively, for formula feeding for the first 12 months of life (Table 29.9).[42] Infants should be fed "on demand," that is, when they indicate they are hungry, rather than on a rigid schedule set by the clock. In the first few months of life, most infants will get hungry every 3 or 4 hours.

Soft, solid foods should be added to an infant's diet of breast milk or infant formula around 6 months. At one time, it was thought that infants should be offered solid foods within the first few months of life. Although such young infants are unable to swallow much of the food (most of it ends up on their face and bib) or to digest completely what they do swallow, solid foods were thought to help the baby grow and sleep through the night. This belief was incorrect. Infants who receive solids before the age of 6 months are no more likely to grow normally or sleep through the night than are infants who start to receive solid foods around 6 months of age.[43] The age at which an infant begins to sleep for 6 or more hours during the night depends on other factors, including the infant's developmental level and how much he or she has slept during the day. Neither the infant nor the infant's parents are likely to get a full night's sleep for at least 4 months.

Introducing Solid Foods   It is recommended that iron-fortified rice cereal be given to all breast-fed infants and to bottle-fed infants who are not receiving iron-fortified formula (Illustration 29.18). The iron provided by iron-fortified rice cereal or formula helps to restock the infant's iron stores, which have been drawn on since birth. Generally, solid foods prepared for infants should consist of single, basic foods such as strained vegetables, fruits, or meats.

Solid foods for infants can be purchased as commercial baby food or prepared at home. When baby foods are prepared at home, care should be taken to avoid contamination and to achieve the right consistency. Infants should be offered new foods one at a time. The new offerings should be separated by several days, so that any allergic reactions to a food can be identified. Variety is the key to achieving a healthy diet for infants in their second 6 months of life, and an assortment of basic, textured foods should be given.[44,46]

By 9 months of age, infants are ready for mashed foods and foods such as yogurt, applesauce, ripe banana pieces, and grits. Most infants have several teeth by this time, and they are able to bite into and chew soft foods.

Infants graduate to adult-type foods after the age of 12 months. Although most foods still need to be mashed or cut up into small pieces for them, 1-year-olds are able to eat the same types of food as the rest of the family. They can drink from a cup and nearly feed themselves with a spoon. Infants have come a long way in 12 months.

TABLE 29.9
Overview of infant feeding recommendations[44,45]

- Exclusive breast-feeding for the first 6 months of life and continuation of breast-feeding through the first year of life are preferred. Iron-fortified infant formulas may be used as a secondary option to breastfeeding.
- Introduce solid foods around 6 months of age.
- An infant's first food should be iron-fortified, such as iron-fortified baby cereal.
- After 6 months, offer a variety of healthy foods appropriate for baby. Pay attention to the textures of baby's food, going from smooth to mashed to chopped to tiny pieces.
- Keep trying new foods—babies may need to try a new food ten times or more before they like it!
- Use a small baby toothbrush with a tiny dot of fluoride toothpaste when the teeth start to come in.
- Infants should be fed "on demand" and not by a set schedule. Feeding should stop when the baby loses interest in eating.
- Breast-fed infants receiving little sun exposure should be given a vitamin D supplement (200 IU or 5 micrograms per day).
- Infants not receiving fluoridated water should be given fluoride drops (dose depends on age).

Illustration 29.18
Solid foods should be introduced around 6 months.

Illustration 29.19
Baby-bottle tooth decay.

© Dietmar A. J. Kennel

TABLE 29.10

Foods to avoid in the first year of life

Foods that May Cause Allergic Reactions	Foods that May Cause Other Problems	Foods that May Lead to Choking
Cow's milk	Blueberries	Grapes
Egg white	Coffee	Frankfurter pieces
Fish, seafood	Corn	Hard candy
Nuts, peanuts	Fruit drinks	Hard pieces of
Peanut butter	Honey (unpasteurized)	vegetables
Soy protein	Prune juice	Meat chunks
Wheat products	Tea	Nuts and seeds
		Popcorn
		Raw vegetables

TABLE 29.11

Indications for supplementing infants

1. Fluoride supplementation is recommended for infants older than 6 months who live in areas with no or low fluoride in the household water supply (0.25 mg of fluoride/day from 6 months to 3 years of age).[47]

2. Vitamin D supplements (200 IU/day in the United States, 400 IU per day in Canada) should be given to breast-fed infants receiving little exposure to sunshine.[48]

   • Breast-fed infants exposed to about 30 minutes of sunshine each week with only diapers on, or those receiving 2 hours of sun exposure while wearing clothes but no hat, are likely meeting their needs for vitamin D by its manufacture in their skin.[49]

Foods to Avoid   Not all foods can be considered "baby foods." Some foods should be omitted from an infant's diet because they are apt to cause allergic reactions or are too difficult for infants to chew into small pieces and swallow. Table 29.10 provides a list of foods that should not be offered to infants.

Reduced-fat products are not recommended for infants. Infants need a relatively high-fat diet for brain and nervous system development. Unpasteurized honey is not recommended either, because it may cause botulism in infants due to their still maturing gastrointestinal defenses against bacteria. Beverages containing a lot of sugar, such as soft drinks and sweetened fruit juices, should not be given to infants because they promote tooth decay. To help prevent "baby-bottle tooth decay," shown in Illustration 29.19, infants should not be put to sleep with a bottle containing sweet fluids or formula.

Do Infants Need Supplements?   Two situations call for the use of supplements during infancy (see Table 29.11). Breast-fed infants and infants receiving formula from a concentrate that is not diluted with fluoridated water need fluoride supplements after 6 months of age. Since breast milk contains a low amount of vitamin D, and since cases of rickets are increasing among breast-fed infants in the United States, breast-fed infants not exposed regularly to sunshine should receive a vitamin D supplement.[48]

The Development of Healthy Eating Habits Begins in Infancy

Older infants and young children should be offered a wide variety of nutritious foods in a positive eating environment. They alone should make the decision about how much to eat of any food offered. Food preferences are primarily learned and are unique to every individual. The learning process begins early in life and is affected by a variety of influences. "Instincts" that draw infants to select certain, nutritious foods are not one of the influences, however. Infants do not instinctively know what to eat. Rather, they are born with mechanisms that help regulate how much they eat. These basic characteristics of infant food intake were beautifully demonstrated by Clara Davis in experiments with 7- to 9-month-old infants in the 1920s and 1930s.[51] One of her most famous experiments is described in Illustration 29.20.

Subsequent research reinforced earlier conclusions that infants and young children should be offered a wide variety of nutritious foods and that decisions

Illustration 29.20
Infant food preferences.

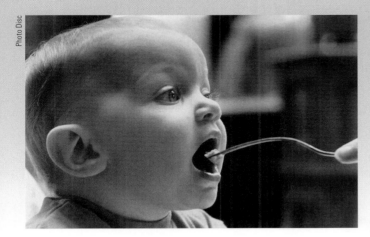

Photo Disc

Clara Davis conducted her most quoted feeding experiment with older infants (7 to 9 months old) living in an orphanage connected to a large hospital in Chicago.

Three times a day a nurse would offer the infants a tray of various foods. The infants would pick up or grab for the foods they wanted, and the nurse would bring the food selected to the infant's hands or mouth.

Altogether 32 different foods were offered, consisting of fresh, unprocessed, unseasoned, and simply prepared basic foods. Foods included milk, beef, kidney, bone marrow, liver, brain, thymus, fish, whole-wheat cereals, raw eggs, sea salt, 15 fresh fruits, and 10 fresh vegetables. Desserts, sugars, syrups, and other sweetened foods were not offered.

Infants quickly formed preferences and narrowed their choices to 14 of the 32 foods. Vegetables were the least liked, whereas milk, bone marrow, eggs, bananas, apples, oranges, and oatmeal were the best liked. They ate enough to grow normally and to remain in good health.

Dr. Davis concluded that appetite is the best guide to how much food an infant or young child needs, and that self-selection will have doubtful value only if the diet is chosen from inferior foods.[50]

about how much to eat should be left up to the infant or child.[52] Beginning at birth, infants who are hungry eat enthusiastically. They stop eating when they are full. Coaxing, cajoling, and pleading by parents or caregivers can override infants' decisions about how much food to eat. That, unfortunately, may upset infants' built-in food intake regulating mechanisms and lead to over- or under-eating, and to "fussy" eaters.

Teaching Infants the Right Lessons about Food   Recommendations for feeding infants are primarily based on energy and nutrient needs, the developmental readiness of infants for solid foods, and the prevention of food allergies. But infant feeding recommendations also include a large educational component. Many of the lessons infants learn about food and eating make an impression that lasts a lifetime. Later food habits and preferences, appetite, and food intake regulation are all influenced by early learning experiences.[53] Table 29.12 provides a lesson plan for teaching your infant the right lessons about food and eating.

Making Feeding Time Pleasurable

More than any other area of development, feeding is an arena in which parents and baby work out the continuing struggle between dependence (being fed) and independence (feeding oneself). Independence must win. Pushing a child to eat is the surest way to create problems. Feeding has got to be pleasurable.
—T. Berry Brazelton, 1993

As Dr. Brazelton, an expert on child development, points out, there is much more to the feeding of infants than meets the eye. Infants learn about communication, relationships, and independence during pleasurable eating experiences. If undertaken in positive and supportive circumstances, mealtimes for infants can provide lessons that will support their physical and psychological health well into the future.

TABLE 29.12

Tips for teaching infants the right lessons about food and eating

1. Infants learn to eat a variety of healthy foods by being offered an assortment of nutritious food choices. There are no inborn mechanisms that direct babies to select a nutritious diet.
2. Infants must be allowed to eat when they are hungry and to stop eating when they are full. Infants, not parents, know when they are hungry and have had enough to eat.
3. Food should be offered in a pleasant environment with positive adult attention.
4. Food should not be used as a reward, punishment, or pacifier.
5. Infants or children should never be coerced into eating anything.
6. Food preferences change throughout infancy. Because an infant rejects a food one time doesn't mean she or he will not accept the food if offered later. Offering a food on a number of occasions often improves acceptance of the food. Babies still may not like strong-flavored vegetables until they are older, however.

Nutrition Up Close

You Be the Judge!

Focal Point: Apply your nutrition knowledge to the diet of a pregnant woman.

Gloria's baby is due in 5 months. Before becoming pregnant, she never gave much thought to what she ate, but now she is trying to eat healthy foods for her baby and herself. Compare her choices with the My-Pyramid recommendations in Illustration 29.8 to find out how well she is doing.

- *Breakfast:* Bran muffin (2 ounces), cold cereal (1 cup) with strawberries ($\frac{1}{2}$ cup), and a glass of milk (1 cup).
- *Lunch:* Vegetable soup (1 cup vegetables), carrot sticks ($\frac{1}{2}$ cup), and a glass of orange juice (1 cup).
- *Afternoon snack:* Yogurt (1 cup).
- *Dinner:* Chicken (3 ounces), turnip greens ($\frac{1}{2}$ cup), a tossed salad (2 cups), and iced tea (2 cups).

Categorize Gloria's choices into the following food groups to determine if she ate the recommended servings from MyPyramid. The grains group has been done for you.

Grains

bran muffin, 2 ounces

cold cereal, 1 cup

_____

Gloria's intake: _____ 3 ounces
Recommended intake: _____ **8 ounces**

Vegetables

_____

_____

_____

Gloria's intake: _____
Recommended intake: _____ **3 cups**

Fruits

_____

_____

_____

Gloria's intake: _____
Recommended intake: _____ **2 cups**

Milk and Milk Products

_____

_____

_____

Gloria's intake: _____
Recommended intake: _____ **3 cups**

Meat and Beans

_____

_____

_____

Gloria's intake: _____
Recommended intake: _____ **6 $\frac{1}{2}$ ounces**

Oils

_____

_____

_____

Gloria's intake: _____
Recommended intake: _____ **7 teaspoons**

FEEDBACK (including answers) can be found at the end of Unit 29.

Key Terms

colostrum, page 29–14
critical period, page 29–3
development, page 29–3
fetus, page 29–3

growth, page 29–3
infant mortality rate, page 29–2
low-birthweight infants, page 29–2

neural tube defects, page 29–7
preterm infants, page 29–2
trimester, page 29–7

Review Questions

True   False

1. Relatively high rates of infant mortality, low-birthweight infants, and preterm delivery represent major public health problems in the United States. ☐ ☐

2. "Critical periods" of growth and development are marked by large increases in the size of cells within developing organs. ☐ ☐

3. Inadequate or excessive levels of nutrient availability during pregnancy can modify the function of fetal genes in ways that increase the risk of chronic disease later in life. ☐ ☐

4. A women entering pregnancy with a BMI (Body Mass Index) of 27 should gain 25 to 35 pounds during pregnancy. ☐ ☐

5. Women need 100% more calories during than before pregnancy because they are "eating for two." ☐ ☐

6. Iodine, EPA and DHA, and vitamin D are considered "risk nutrients" during pregnancy because a sizeable

True   False

portion of women fail to consume enough of them. ☐ ☐

7. Breast-feeding is better for an infant's health than formulas in part because breast milk contains substances that prevent infections and reduce the risk of development of certain chronic disease later in life. ☐ ☐

8. Approximately one in four women is biologically unable to breast-feed successfully. ☐ ☐

9. A woman's milk supply automatically increases as an infant consumes more breast milk. ☐ ☐

10. A daily multivitamin and mineral supplement is recommended for all breast-feeding women. ☐ ☐

11. Infants should be given their first solid food around six months of age and the food should be iron-fortified rice cereal. ☐ ☐

12. Infants are born knowing which foods are good for them. ☐ ☐

Media Menu

www.lalecheleague.org
This is a comprehensive online resource for breastfeeding information and support resources.

www.aap.org
The American Academy of Pediatrics's Web site is loaded with information on breast-feeding and infant and child nutrition.

www.cdc.gov/breastfeeding/
The Centers for Disease Control Web site on breastfeeding can be found here.

www.wicworks.ca.gov/education/ nutrition/infant_Feeditn/InfantFeeding .htm
The "New Infant Feeding Guidelines" for WIC, 2006, are found here. Find these and nutrition educational materials at this site.

www.cia.gov/cia/publications/factbook/ goes/xx.html
The C.I.A. World Factbook home page provides an unbelievable amount of information on each country in the world. Get data

on infant mortality rates, life expectancy, industries, climate, time zones, and much more from this huge resource. (P.S. It really is from the Central Intelligence Agency.)

www.cdc.gov/growthcharts
Need more information about the growth charts, or want copies of them? Get it here.

www.hc-sc.gc.ca/english/
Search pregnancy, breast-feeding, infants, or other topics on Health Canada's Web site.

www.nlm.nih.gov/medlineplus/breastfeeding.html
This site is a good source for updates and advice on breast-feeding, problem solving, alternative therapies, and nutrition.

http://mchlibrary.info
The Maternal and Child Health Bureau's site provides information on resources, hot topics, and announcements of new publications.

www.babycenter.com
A favorite Web site of many consumers, it offers articles, bulletin boards, and features (and advertisements!) related to fertility, pregnancy, and infant feeding.

Notes

1. Final data for 2003: Natality statistics. www.cdc.gov/nchs/fastats/infant_htm, accessed 10/06.

2. The C.I.A. World Factbook. International infant mortality rates.

3. MacDorman MF et al. Annual summary of vital statistics. Pediatrics 2002;110:1037–50.

4. Healthy people 2000: national health promotion and disease prevention objectives. Washington, DC: U.S. Department of Health and Human -Services, DHHS publication No. (PHS) 91-50212; 1991.

5. Rosso P. Nutrition and metabolism in pregnancy. New York: Oxford University Press; 1990.

6. Langley-Evans SC. Developmental programming of health and disease. Proc Nutr Soc. 2006;65:97–105.

7. Winick M. Nutrition and pregnancy, White Plains (NY): March of Dimes Birth Defects Foundation; 1986.

8. Miller DR, Hayes KC. Vitamin excess and toxicity, In: Hathcock JN, editor. Nutritional toxicology, New York: Academic Press; 1982: pp 81–131.

9. National Academy of Science (Institute of Medicine), Nutrition during pregnancy: I. Weight gain, II. Nutrient supplements, Washington, DC: National Academy Press; 1990: p 48.

10. Ohlin A, Rossner S. Maternal body weight development after pregnancy. Int J Obesity 1990;14:159–73.

11. Dietary Reference Intakes: Energy, carbohydrate, fiber, fatty acids, cholesterol, protein, and amino acids. Washington, DC: National Academies Press; 2002.

12. Lawrence JM et al. Racial and ethnic disparities in folate intake. Am J Obstet Gynecol 2006;194:520–6.

13. Daly LE et al. Folate levels and neural tube defects. JAMA 1995;274:1698–1702.

14. Brown JE et al. Predictors of red cell folate level in women attempting pregnancy. JAMA 1997;277:548–52.

15. Tamura T et al. Folate and human reproduction. Am J Clin Nutr 2006; 83:993–1016.

16. Brown JE, Kahn ESB. Maternal nutrition and the outcome of pregnancy, Clin Perinatol 1997;24:433–49

17. Hollis BW et al. Vitamin D deficiency during pregnancy: an ongoing epidemic. Am J Clin Nutr 2006;84:273.

18. Nesby-O'Dell et al. Hypovitaminosis D prevalence and determinants among African American and white women of reproductive age: third National Health and Nutrition Examination Survey, 1998-1994. Am J Clin Nutr. 2007;76:187–92.

19. Recommendations to prevent and control iron deficiency in the United States. MMWR 1998;47:1–28.

20. Preziosi P et al. Effect of iron supplementation on the iron status of pregnant women: consequences for newborns. Am J Clin Nutr 1997;66:1178–82.

21. Utiger RD. Iodine nutrition: more is better. N Engl J Med 2006;354:2819–21.

22. Akabas SR et al. Summary of a workshop on n-3 fatty acids: current status of recommendations and future directions. Am J Clin Nutr 2006;83(suppl): 1536S–8S.

23. Simopoulos AP. Human requirement for N-3 polyunsaturated fatty acids. Poult Sci. 2000;79:961–70.

24. Mattson SN et al. Heavy prenatal alcohol exposure with or without physical features of fetal alcohol syndrome leads to IQ deficits. J Pediatr 1997; 131:718–21.

25. Hathcock JN et al. Evaluation of vitamin A toxicity, Am J Clin Nutr 1990; 52:183–202.

26. Dubois S et al. Ability of the Higgins Nutrition Intervention Program to improve adolescent pregnancy outcome, J Am Diet Assoc 1997;97:871–8.

27. Cummings AS. Morbidity in breast-fed and artificially fed infants, J Pediatr 1979;90:726–9.

28. U.S. Public Health Service. A review of the medical benefits and contraindications to breastfeeding in the United States. Washington, DC: Maternal and Child Health Bureau; 1997: pp 9–10.

29. Simopoulos AP, Grave GD. Factors associated with the choice and duration of infant-feeding practices. Pediatrics 1984;74(suppl):603–14.

30. Health, United States, 2003. www.cdc.gov/nchs/hus.htm, accessed 10/06.

31. Health, United States, 2003. www.cdc.gov/nchs/htm, accessed 10/03.

32. Lakhkar BB. Breastfeeding in adoptive babies. Indian Pediatr 2000;37:1114–16.

33. Committee on Nutrition, American Academy of Pediatrics. Breastfeeding and the use of human milk. Pediatrics 1997;100:1035–9.

34. National Academy of Sciences (Institute of Medicine). Nutrition during lactation. Washington, DC: National Academy Press; 1991.

35. Hoffman DR et al. Maturation of visual acuity is accelerated in breast-fed term infants fed baby food containing DHA-enriched eggs. J Nutr 2004;134:2307–13.

36. Lovelady CA et al. The effect of weight loss in overweight, lactating women on the growth of their infants. N Engl J Med 2000;342:449–53.

37. Menella JA, Beauchamp GK. The transfer of alcohol to human milk, N Engl J Med 1991;325:981–85.

38. Rogan WJ, Pollutants in breast milk, Arch Pediatr Adol Med 1996;150: 981–90.

39. Thompson D et al. New CDC growth charts. WIC National Meeting, Salt Lake City, Utah, 9/8/00.

40. Winick M. Malnutrition and brain development. New York: Oxford University Press; 1976.

41. Lloyd-Still JD et al. Intellectual development after severe malnutrition in infancy. Pediatrics 1974;54:306–11.

42. Rappo PD et al. Pediatrician's responsibility for infant nutrition. Pediatrics 1997;99:749–50.

43. Beal VA. Termination of night feeding in infancy. J Pediatr 1969;75:690–2.

44. New infant feeding guidelines, accessed 10/06. www.wicworks.ca.gov/education/nutrition/Infant_Feeditn/InfantFeeding.htm

45. Briefel R, Reidy K, Karwe V, Devaney B. Feeding Infants and Toddlers Study: Improvements needed in meeting infant feeding recommendations. J Am Diet Assoc. 2004;101(suppl 1):S31–S37.

46. Fiocchi A et al. Consensus document for introducing solid foods into an infant's diet to avoid food allergies. Am Allergy Asthma Immunol 2006;97:10–21.

47. American Academy of Pediatric Dentistry. Guidelines for fluoride therapy. www. aapd.org/media/Policies_Guidelines/G_FluorideTherapy.pdf, accessed 10/06.

48. Ziegler EE. Vitamin D deficiency in breastfed infants in Iowa. Pediatrics. 2006;118:603–10.

49. Specker BL et al. Sunshine exposure and serum 25-hydroxyvitamin D concentrations in exclusively breast-fed infants. J Pediatr 1985;107:372–6.

50. Story M, Brown JE. Do young children instinctively know what to eat? The studies of Clara Davis revisited, N Engl J Med 1987;316:103–6.

51. Davis C. Self-selection of diets by newly weaned infants: an experimental study. Am J Dis Child 1928;36:651–79.

52. Birch LL, Deysher M. Caloric compensation and sensory specific satiety: evidence for self-regulation of food intake by young children. Appetite 1986;7:323–31.

53. Skinner JD et al. Do food-related experiences in the first 2 years of life predict variety in school-aged children? J Nutr Ed Behav 2002;34:310–5.

Nutrition Up Close

You Be the Judge!

Feedback for Unit 29

Gloria's choices:

Grains	3 ounces (bran muffin, cereal)
Vegetables	3 cups (vegetable soup, carrot sticks, turnip greens, tossed salad)
Fruits	2½ cups (strawberries, apple, orange juice)
Milk and milk products	2 cups (milk, yogurt)
Meat and beans	3 ounces (chicken)
Oils	0 teaspoons

All of Gloria's choices are healthy ones, but she is missing some important foods and sufficient calories. While concentrating hard on eating enough fruits and vegetables, she has neglected to consume enough nutrient-dense foods from the remaining food groups. She needs to include in her diet more dairy products, add more meat or protein alternates, and more grains, especially whole-grain products. With a few additions to her existing choices, the quality and quantity of her diet can be greatly improved. Here is a more balanced version of Gloria's menu for a day:

- *Breakfast:* Bran muffin, cereal with strawberries, and a glass of milk.
- *Lunch:* Vegetable soup, carrot sticks, an apple, **burrito,** and a glass of orange juice.
- *Afternoon snack:* Yogurt, **cheese,** and **whole-grain crackers.**
- *Dinner:* Chicken, grits, turnip greens, tossed salad, **with 2 Tbsp ranch dressing, whole grain roll** and iced tea.

Answer to Review Questions

1. True, see pages 29-2, 29-3.
2. False, see pages 29-3, 29-4.
3. True, see page 29-4.
4. False, see page 29-6.
5. False, see page 29-8.
6. True, see pages 29-9, 29-10.
7. True, see page 29-12.
8. False! See page 29-13.
9. True, see page 29-14.
10. False, see page 29-15.
11. True, see page 29-19.
12. False, see page 29-20.

Nutrition for the Growing Years: Childhood through Adolescence

NUTRITION SCOREBOARD

	TRUE	FALSE
1 Parents should not allow their children to eat "junk" foods.		
2 It is up to parents and caretakers to decide how much children should eat.		
3 The incidence of overweight among 6- to 11-year-olds has more than quadrupled since the 1960s.		
4 Childhood and adolescence represent "grace periods" during which the diet consumed does not influence future health.		

Key Concepts and Facts

- There is no evidence to support the notion that children are born knowing what foods they should eat.

- Children are born with regulatory processes that help them decide how much to eat.

- Parents and caretakers should decide *what* foods to offer children. Children should decide *how much* to eat.

- Diet and other behaviors of children and adolescents affect health before and during the adult years.

Answers to
NUTRITION SCOREBOARD

		TRUE	FALSE
1	Withholding sweets and other favorite foods makes kids value and want them more. Such foods should not be given special emphasis.	✔	
2	Children, not parents or caregivers, should decide how much food is consumed.		✔
3	About 4% of 6- to 11-year-olds were overweight in the 1960s. Now over 19% are overweight.[1]	✔	
4	Childhood and adolescent diets can affect future health. Heart disease and diabetes, for example, can have "pediatric" origins that are related to dietary intake patterns.	✔	

Photo Researchers - Juneberg Clark

The Span of Growth and Development

Physical and mental development proceed at a high rate from infancy through adolescence (Illustration 30.1). Young children generally enter this phase of life able to take a step or two and to guide a spoon into their mouths sideways. They will likely leave adolescence with the ability to drive, work for pay, and solve complex problems. These are the formative years that lay the foundation for the rest of life.

The Nutritional Foundation

As the twig is bent, so grows the bough.

Good nutrition takes on particular importance during the growing years, for many reasons. Growth is an energy- and nutrient-requiring process. It will not proceed normally unless the diet supplies enough of both.

Children learn about food and its importance to health and well-being during these early years of life, and they also establish food preferences and physical activity patterns that may endure into the adult years. Food intake regulatory mechanisms are affected by the lessons children learn early in life. Given control over decisions about how much to eat, children generally become responsive to internal cues that signal when they should eat and when they should stop eating.[2] If the lessons go well, they learn to eat for the right reason.

Eating well and learning the right lessons about food and health have implications that transcend the growing years. Early diets may have long-term effects on the risk of developing a number of diseases later in life.

One to Two Years
Gains in height and weight continue at a lower rate; appetite is less. Uses finger and thumb to pick up things. Soft spot grows smaller and then disappears. Baby teeth continue to appear. Usually takes one long nap a day. Drinks from a cup, attempts to feed self with a spoon. Likes to eat with hands. Pulls self up to standing position. Walks alone.

Two to Three Years
Slower and more irregular gains in height and weight. Has all 20 teeth. Runs and climbs, pushes, pulls, lugs, walks upstairs one step at a time. Feeds self using fingers, spoon, and cup; spills a lot. At times has one favorite food. Associates the sensation of hunger with a need for food.

Three to Four Years
Gains 4–6 pounds and grows about 2–3 inches. Feeds self and drinks from a cup quite neatly, carries things without spilling. May give up sleep at nap time, substituting quiet play.

Four to Five Years
Gains in height and weight about same as previous year. Hops and skips, throws ball. Has increasingly good coordination, masters buttons and shoelaces. Can use knife and fork and is a good self-feeder.

Five to Six Years
Growth continues at about the same rate. Legs lengthen. Six-year permanent molars usually appear (new teeth, not replacing baby teeth). Begins to lose front baby teeth. Prefers plain, bland, and unmixed foods.

Illustration 30.1
Growth and development characteristics from 1 through 16 years.

(continued)

Illustration 30.1 (*continued*)
Growth and development characteristics from 1 through 16 years.

Six to Nine Years
Slow gains in height and weight (2–3 inches and 4–6 pounds a year). Some additional permanent teeth appear. Likely to have the childhood communicable diseases. Sleeps 11 to 13 hours.

Nine to Twelve Years
May be long legged and rangy, but health is generally sturdy. Permanent teeth continue to appear. Appetite good. Needs about 10 hours sleep. Growth spurt in girls usually begins. May get very irritable when hungry.

Twelve to Fourteen Years
Wide differences in height and weight in children of either sex of same age. Menstruation usually begins (sometimes earlier). Girls develop breasts, are usually taller and heavier than boys of the same age. Growth spurt of boys begins. Muscle growth rapid. Appetite increases. Likes to spend time with friends.

Fourteen to Sixteen Years
Boys are in period of growth spurt. Voice deepens. Girls have usually achieved maximum growth. Menstruation is established though may still be irregular. Enormous appetite. Pimples a common problem. Sleep reaches its adult pattern. All permanent teeth except wisdom teeth.

Characteristics of Growth in Children

Growth slows substantially after the first year of life. Between the ages of 2 and 10 years, children normally gain somewhere around 5 pounds and 2 to 3 inches in height per year (Illustration 30.2). Gains in weight and height occur in "spurts"

Illustration 30.2
Average yearly growth in weight and height during childhood and adolescence.

rather than continuously and gradually. Prior to a growth spurt, appetite and food intake increase (given an adequate food supply), and the child puts on a few pounds of fat stores (Illustration 30.3). During the growth spurt, these fat stores are used to supply energy needed for growth in height.

CDC's Growth Charts for Children and Adolescents   Growth progress during childhood and adolescence is generally monitored with the use of the Centers for Disease Control (CDC) growth charts. Charts are available for females and males

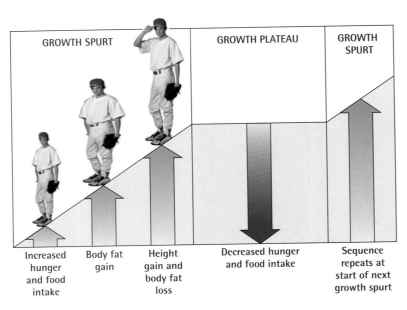

GROWTH SPURT   GROWTH PLATEAU   GROWTH SPURT

Increased hunger and food intake | Body fat gain | Height gain and body fat loss | Decreased hunger and food intake | Sequence repeats at start of next growth spurt

Illustration 30.3
Growth occurs in a series of "spurts." Each spurt follows this sequence.

from 0 to 36 months old and from 2 to 20 years. Growth charts for 2- to 20-year-olds consist of graphs of:

- Weight for age (see Illustration 30.4)
- Height (stature) for age
- Weight for height
- Body mass index (BMI) for age

Each graph provides percentiles that reflect the distribution of these measures in a representative sample of 2- to 20-year-olds in the United States. So, for example, if a child's weight for age is between the 50th and 75th percentiles, it means that this child's weight is somewhat higher than that of most children and similar to that of 25% of children who are represented in the 50th to 75th percentile range. Children and adolescents whose weight status measurements place them in the highest and lowest percentile ranges should be evaluated further for potential underlying nutrition and health problems.

BMI charts for age are a feature of the growth charts. Since BMI increases with age in children and adolescents, ranges of BMI used to classify weight status

Illustration 30.4
An example of the CDC's growth charts: Weight for age for boys and girls.
The charts can be printed off the Internet using the address www.cdc.gov/growthcharts.

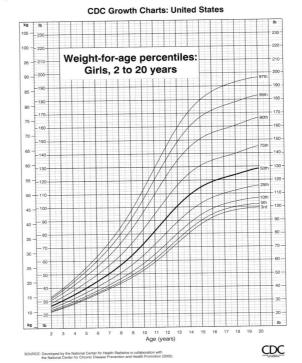

in adults cannot be used. BMIs for age that fall between the 85th and 95th percentiles are considered "at risk" for overweight, and those above the 95th percentile at risk for obesity.[3] Although now recommended for determining weight status in children and adolescents, use of the CDC's new BMI for age charts requires that BMI be calculated. Illustration 30.5 shows you how to calculate BMI.

Food Jags and Normal Appetite Changes

When growth is not occurring, children are often disinterested in food and eat very little at times. They may go on "food jags," insisting on eating only a few favorite foods like peanut butter and jelly sandwiches or breakfast cereal.

The ups and downs of children's food intake can be unnerving for parents. But as long as growth continues normally and children remain in good health, there's little reason to worry about food jags and fluctuations in appetite and food intake. Only children know when they are hungry or full.[4] (Actually, children appear to regulate caloric intake meal by meal better than adults do. Read about it in Illustration 30.6.)

How to calculate BMI

Lucy just turned 7 years old and has had her weight and height carefully measured. She weighs 57 pounds (lb), and her height (stature) is 4 feet (ft), 2 inches (in.). What's her BMI?

1. Convert her height of 4 ft 2 in. to inches:

$$4 \text{ ft} \times 12 \text{ in. per foot} = 48 \text{ in.}$$
$$+2 \text{ in.}$$
$$\overline{50 \text{ in.}}$$

2. Square the inches:

$$50 \text{ in.} \times 50 \text{ in.} = 2500 \text{ in.}^2$$

3. Apply formula:

$$BMI = \frac{kg}{m^2} \quad or \quad \frac{weight \text{ in. lbs} \times 700}{height \text{ in. in.}^2}$$

$$BMI = \frac{57 \text{ lbs} \times 700}{2500 \text{ in.}^2} = \frac{39,900}{2500 \text{ in.}^2} = 15.96 \text{ kg/m}^2$$

In this example, Lucy's BMI for age of 15.96 kg/m² falls between the 50th and 75th percentiles, or well within the normal range.

Illustration 30.5
How to calculate BMI.

Hunger Can Make Kids Irritable!

It was a hot summer day, and 10-year old Kim and her friends had spent it in the pool. The fun went on and on . . . until five o'clock. That's when the argument started and Kim left the pool in a huff.

When Kim arrived home, still in a huff, her mom knew what was going on. Kim was cranky because she was hungry. She was so cranky, in fact, that suggesting she'd feel better if she ate something would only make her more

Illustration 30.6
The pudding study.

Lunch at the day care center the day of the study was a bit different than usual—it started with pudding. Preschoolers were offered a serving of pudding that contained 150 calories or another with 40 calories. Both looked and tasted the same. The amount of pudding consumed was secretly recorded, as was the children's food intake during the rest of lunch. The experiment was repeated on adults.

Children compensated almost perfectly for the calories in the different puddings. Those given the higher-calorie pudding ate less at lunch, and those receiving the low-calorie pudding ate more. The adults, however, didn't fare so well. Their caloric intakes for the rest of lunch bore no relationship to calories consumed in the pudding. The study suggests that children are more sensitive to caloric intake on a short-term basis that adults.[5]

irritated! Kim was ravenously hungry, but she didn't want anybody telling her what her problem was!

Kim's mom had the solution to her daughter's grumpiness on the table. "Dinner's ready," her mom called softly. Kim stomped to the table and grudgingly started to eat. Within minutes, she did feel better, but it wasn't because she was no longer hungry!

Only children know when they are hungry, but sometimes when they stay hungry too long, the signal to eat is overpowered by fun or the effects of not eating. Gently offering food to an irritable child who has skipped a meal or played too long can provide the cure for irritability.

The Adolescent Growth Spurt

It dawned on us that Max was entering his growth spurt when we had to wait for him to finish eating before we could all go out to dinner.
—MAX'S PARENTS

Illustration 30.7
This photo (on the left) was taken when these boys, born 2 months apart, were 12 years old. Ben (on the left) started his growth spurt at age 11, while Max (in the middle) began his at age 13. David (on the right) began to spring up in height after he turned 14. The photo (on the right) shows the same boys at age 19.

The adolescent growth spurt usually occurs in girls between the ages of 9 and 12 years. For boys, this period of growth generally begins around the age of 12 to 14. Actually, though, the age when adolescents start their growth spurt normally varies considerably. Pictured in Illustration 30.7 are three friends, all aged 12. It's not hard to tell which one of them has experienced a growth spurt! The difference was also clear at the dinner table. By the time all three were 19 years old, Max (the one in the middle) had caught up with Ben (on the left), and David had grown taller but less than Max or Ben.

During these years of growth, teenagers gain approximately 50% of their adult weight, 20 to 25% of adult height, and 45% of their total bone mass. In the year of peak growth, girls gain 18 pounds on average, and boys 20 pounds.[6]

Can You Predict or Influence Adult Height?   Ultimate height is difficult to predict. On average, children tend to achieve adult heights that are between the heights of their biological parents.[7] There are many exceptions to this general finding, however, and that means that heredity is not the only influence on height. The dramatic increases in height of Japanese youth since World War II provide clear evidence that nongenetic factors have the strongest influence on height. Since the late 1940s, Japanese youth have grown an average of 2 inches taller each generation. (The increase in height has meant that everything from shoes to beds must be produced in larger sizes.) The increase in size of Japanese people is largely attributed to higher calorie and protein intakes and more nutrient-dense diets. Indeed, people in most economically developed countries continue to grow taller; a maximal genetically determined height has not yet been reached. If children are less well nourished than their parents, though, they tend to be shorter than their parents as adults.[8]

A healthy birthweight and diet during the growing years, and freedom from frequent bouts of illness, support growth in height.[9] There are no supplements, powders, or special diets that can be used to increase growth rate. Growth hormone injections can increase height somewhat, but the side effects of growth hormone are numerous and its use is limited.

Overweight and Type 2 Diabetes: Growing Problems

Rates of overweight in children and adolescents in the United States have risen remarkably since the 1960s (Table 30.1), and overweight-related disorders are also on the rise. Conditions such as type 2 diabetes, bone and joint disorders, abnormal blood lipid levels, and elevated blood pressure that were very rarely observed in children and adolescents in the past are being diagnosed with increasing frequency. Approximately 60% of overweight children have risk factors for cardiovascular disease.[10] The emergence of type 2 diabetes as a problem of the early years is of particular concern because diabetes generally worsens with time and causes long-term health impairments. It is estimated that 4% of children and adolescents have impaired glucose tolerance (a strong risk factor for type 2 diabetes), and that 6 to 17% of overweight and obese children and adolescents have type 2 diabetes.[11]

Causes of Overweight in Young People   Jean Mayer, a noted professor of nutrition, said we shouldn't be surprised that so many children are overweight: "A society where most children sit and watch TV, where nobody walks, and where no domestic chores are required is a society where we may expect to have obese children."[12]

A number of "obesigenic" trends have developed over the past several decades that likely account for rising rates of overweight. Compared to 20 or so years ago, children and adolescents now have fewer opportunities for physical activity and are exposed to a

TABLE 30.1

Changes in the prevalence of overweight in U.S. children and adolescents[1]

	Percent Overweight		
	AGE 2–5 YEARS	AGE 6–11 YEARS	AGE 12–19 YEARS
1960s	—	4%	5%
1992	7	11	11
2000	10	15	16
2004	—	19	17

This fast-food beverage container holds over a quart of soft drink. It's meant for adolescent males and reads: "Do a whole river of icy-cold refreshment. And don't just quench your thirst, stop it in its tracks and make it run for cover. Because showing your thirst who's boss is doing what tastes right."

Author photo

generally plentiful supply of energy-dense foods. Schools offer few physical activity classes and opportunities for exercise, while access to energy-dense, empty-calorie foods in vending machines and cafeteria lines is increasing.[13] Children and adolescents are now exposed to portion sizes of energy-dense fast foods that are up to five times larger than in the past.[14] Many communities lack bike paths or lanes, sidewalks, and safe playgrounds that invite physical activity.[15] Genetically based, inborn processes that regulate appetite and satiety are being overwhelmed by environmental conditions that influence decisions about food intake.

Medicalization of Obesity and Treatment   Stomach stapling and diet drugs are being increasingly used to treat overweight and obesity in children and adolescents, but this trend is not viewed by some as a healthy one. Stomach stapling is forever, and little is known about the psychological and health effects of this type of surgery in children and adolescents. Diet drug use in adults is accompanied by potentially serious side effects and the requirements for long-term use for continued effectiveness.[16] Given that such drastic measures as surgery and diet drugs may not be attractive, how is the problem of the obesity epidemic in children and adolescents to be solved? That's right. The answer is prevention.

Prevention of Overweight   If environmental factors play the predominant role in the overweight epidemic, then the path to prevention is paved with environmental changes. Children and adolescents need more opportunities for physical activity. They should be getting at least 60 minutes of moderate-to-vigorous activity daily.[17] A wider array of healthy food options should be available at schools and in fast-food restaurants, as well as at home. Children who regulate their food intake based on appetite and satiety rather than on environmental cues will have a jump start on overweight prevention.

How Do Food Preferences Develop?

I do not like broccoli, and I haven't liked it since I was a little kid and my mother made me eat it. And I'm President of the United States, and I'm not going to eat any more broccoli.
—GEORGE H. W. BUSH, 1990

Food preferences are a highly individual matter (Illustration 30.8). It's hard to find two people who share the same food likes and dislikes, and many people are very picky about how food is prepared and served. How do food preferences form? Are they "shaped" by early learning experiences, or are they inborn and beyond anyone's control?

With the exception of genetically determined sensitivity to bitter-tasting foods and an inborn preference for sweet-tasting foods, there is no direct evidence that food preferences are inborn.[18] The belief that young children instinctively know what to eat can be hazardous to a child's health. Parents may take an overly relaxed attitude toward poor food habits and inadvertently contribute to the development of health problems later in life. Informed parents should decide what types of food their child should be offered, but the decision about how much to eat should be left to the child.[2]

Food likes and dislikes appear to be almost totally shaped by the environment in which children learn about food. Which foods are offered, the way they are offered, and how frequently particular foods are offered all influence whether a child will like a given food or not.[19]

Richard Anderson (both)

Humans are born with a tendency to be cautious about accepting new things, including foods. They may need to get used to a new food before they trust it. Parents may have to offer a new food on five or more occasions before the child makes a decision.[19] Sometimes children decide they like the food, and sometimes they really don't like it. Often infants and young children do not like strong-flavored vegetables, spicy foods, and mixed foods (Illustration 30.9).

Forcing a young child to eat foods she or he does not like, or totally restricting access to favorite foods, can have lifelong negative effects on food preferences and health. When attempts to get a child to eat a particular food turn the dinner table into a battleground for control, nobody wins. Foods should be offered in an objective, nonthreatening way so that the child has a fair chance to try the food and make a decision about it. Restricting access to, or prohibiting intake of, children's favorite "junk" foods tends to strengthen their interest in the foods and consumption of those foods when they get a chance. Such prohibitions have the opposite effect of that intended because they make kids want the foods even more.[20]

Illustration 30.8
Why select the apple?
Food likes and dislikes appear to be almost totally shaped by the environment in which children learn about food. Which foods are offered, the way they are offered, and how frequently particular foods are offered all influence whether a child will like a given food or not.

Illustration 30.9
MyPyramid for Kids

What's a Good Diet for Children and Adolescents?

> *Children today are tyrants. They contradict their parents, gobble their food, and tyrannize their teachers.*
> —SOCRATES, 470–399 B.C.

Good diets for the growing years can be achieved by following the food group recommendations shown in Table 30.2. The type of foods selected from the food groups is important. Whole-grain breads and cereals should be represented in diets;

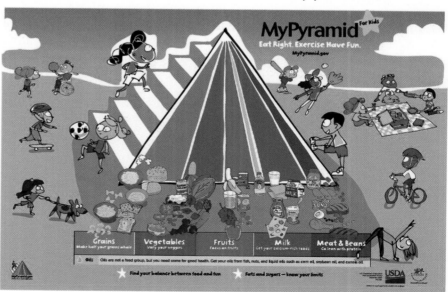

TABLE 30.2

MyPyramid food group recommendations for children and adolescents[a]

	Food Group					
Age in Years	GRAINS (OUNCES)	VEGETABLES (CUPS)	FRUITS (CUPS)	MILK (CUPS)	MEAT AND BEANS (OUNCES)	OIL (TEASPOONS)
6	5	1.5	1.5	2	4	4
12	6–7	2.5–3	2	3	5.5–6	6
16	6–10	2.5–3.5	2–2.5	3	5.5–7	6–8

[a] Higher numbers of ounces and cups apply to males.

and boiled, broiled, and baked foods should be consumed more often than fried foods.

Foods from the "miscellaneous" group that contribute mainly calories may be needed by children and teenagers. Although cakes, candy, cookies, potato chips, and so on are generally considered "sometimes" foods rather than "always" foods, they can serve as sources of needed calories. Snacks also are an important source of calories in the diets of children and teenagers (Illustration 30.10). Good choices for snacks include

Yogurt	Bananas	Carrots
Cheese	Oranges	Cucumbers
Low-fat milk	Apples	Popcorn
Ice milk	Apple juice	Peanuts
Pears	Mangos	Cherry tomatoes
Melons	Raisins	
Peanut butter		

Recommendation for Fat Intake Normal growth and health have been demonstrated in children and adolescents consuming 21 to 35% of total calories from fat. Problems are noted when diets reach 35% or more calories from fat, however. High fat intakes are associated with excessive caloric intake, very low intakes of folate and vitamin C, and high saturated fat intake. Consequently, it is recommended that children older than 3 years and adolescents consume diets that provide 25 to 35% of total calories from fat. Additionally, it is recommended that 3- to 18-year-olds consume as little saturated and *trans* fat as possible, good sources of fiber and omega-3 fatty acids, lean meats and beans, and plenty of vegetables and fruits.[21]

Should Children Drink Milk? Yes! Children certainly should drink milk (Illustration 30.11). If this sounds like a silly question, good. It is, but the notion that children (and adults, for that matter) should not drink milk or consume other animal products erupts in the media from time to time.

Children don't have to consume milk in order to have an adequate diet, because a variety of other foods could

Illustration 30.10
Snacks are an important source of calories and nutrition in the diets of children and teenagers.

Blair Seitz/Photo Researchers

provide the nutrients in milk. However, children who fail to drink milk tend to have low calcium intakes, lower bone density, and more fractures than children who drink milk regularly.[22]

Photo Disc

Status of Children's and Adolescents' Diets

How well are children and teenagers doing at meeting the MyPyramid recommendations? Children over the age of 2 in the United States tend to consume too few vegetables and fruits; too little calcium, zinc, and vitamins E, D, C, and B_6; and too little dietary fiber.[23] Few (9%) children consume five servings of vegetables and fruits a day, and most consume 2.5 servings or less. About 30% of adolescents actually consume their "five a day."[24] Low vegetable and fruit intakes contribute to less-than-optimal intakes of vitamins and minerals by children and teens.

Illustration 30.11
Children should drink milk in order to meet their needs for calcium and vitamin D. *Virtually all school lunch programs include milk as a beverage.*

Approximately half of all children fail to consume the recommended intake of calcium daily. Low calcium intake and poor vitamin D status, along with limited consumption of vegetables and fruits, during childhood and adolescence decrease peak bone mass and increase the risk of osteoporosis.[24] A part of the reason children and teens fail to get enough calcium and vitamin D is that they tend to substitute soft drinks for milk during these years.[25] Children and teens may also have poor vitamin D status due to inadequate exposure to sun.[26]

Diets of children and adolescents generally provide about half of the recommended intake of dietary fiber of 19 to 38 grams per day.[27] This makes constipation a problem for many children and teens.

Early Diet and Later Disease

Children and adolescents generally feel vigorous and healthy even when they are accumulating diet-related risk factors that will influence disease development later in life. Many studies indicate that in populations with a low incidence of heart disease, blood cholesterol values of both children and adults are lower than in countries with high rates of heart disease. Typical cholesterol levels of children in the United States are higher than those of children in countries where the rate of heart disease is lower. In addition, obesity during adolescence is associated with the presence of risk factors for subsequent heart disease: high LDL cholesterol, low HDL cholesterol, and increased blood pressure.[10] High-sodium diets tend to increase blood pressure in children and adolescents who have a hypertensive parent. U.S. children tend to consume far more sodium than is recommended. Adolescents who fail to consume sufficient calcium and to obtain adequate levels of vitamin D are at later risk of osteoporosis.[28] Children and adolescents who become obese are at higher risk not only for obesity later in life but also for disorders related to obesity such as diabetes, heart disease, some cancers, and hypertension.[10] Obesity may also lead to lower self-esteem, a poor body image, and eating disorders.[29]

Nutrition is related to both current and future health and quality of life for people of all ages.

Nutrition Up Close

Overweight prevention close to home

Focal Point: Tips for parents on preventing overweight in their children.

Debra and Dale have a 1-year-old daughter, and their second adopted infant will arrive in a few weeks. Both are acutely aware of the growing problem of obesity and type 2 diabetes in children and don't want their children to become overweight.

Referring back to the information presented on pages 30-10 and 30-11, list four actions Debra and Dale can take to help their children develop healthful food and activity habits:

1. _____

2. _____

3. _____

4. _____

FEEDBACK (including answers) can be found at the end of Unit 30.

Review Questions

True False

1. Children generally become good "self-feeders" between the ages of four to five years. ☐ ☐

2. Gains in weight and height in children and adolescents occur in "spurts" rather than as smooth progressions. ☐ ☐

3. Rates of overweight in children and adolescents have finally stopped increasing in the United States. ☐ ☐

4. Food likes and dislikes appear to be almost entirely shaped by the environment in which children learn about food. ☐ ☐

True False

5. Young children instinctively know what to eat. ☐ ☐

6. According to MyPyramid food intake recommendations, 6-year-old children should consume 1.5 cups of vegetables and fruits daily. ☐ ☐

7. Children have a high need for calories from fat. Therefore, it is not recommended that children limit their fat intake. ☐ ☐

8. A number of chronic diseases such as heart disease, osteoporosis, and type 2 diabetes may begin to develop during childhood and adolescence. ☐ ☐

Media Menu

www.kidnetic.com
This site provides a bunch of fun activities for kids on subjects such as digestion; food, fun, and fitness; healthy and easy recipes; and dancing.

www.healthierus.gov/exercise.html
This site provides guidance for physical activity in children and information about national programs aimed at improving the fitness level of children.

www.nichd.nih.gov/msy
The "Media-Smart Youth" curriculum, developed the National Institutes of Child Health and Human Development, offers ten interactive lesson plans that help kids make healthier choices about food and physical activity.

www.nlm.nih.gov/medlineplus/
childnutrition.html
The National Library of Medicine shares its collection of information on child nutrition through this Web site.

www.nlm.nih.gov/medlineplus/
obesityinchildren.html
Get your news and information on childhood obesity here.

www.nejm.org
Hear an interview with Dr. Marion Nestle on the Institutes of Medicine food marketing to children report and childhood obesity from a link provided at this Web site.

www.mypyramid.gov
Resources offered under the "MyPyramid for Kids" site include lesson plans and worksheets on nutrition and physical activity for teachers; fun, interactive games related to nutrition for children; and healthy eating and exercise tips directed at parents.

www.aap.org
The American Academy of Pediatrics's Web site is loaded with information on child nutrition.

www.cdc.gov/growthcharts
Need more information about the growth charts or printouts of them? Get it here.

www.hc-sc.gc.ca/english/
On Health Canada's Web site, search infant and child nutrition for information and resources.

Notes

1. Centers for Disease Control, fast-stats. Overweight. www.cdc.gov/nchs/fastats/overwt.htm, accessed 10/06. Obesity (Silver Spring). 2006 Mar;14(3):343–4. Links

2. Satter E. The feeding relationship: problems and interventions. J Pediatr 1990;117:S181–S9.

3. Thompson D et al. New CDC growth charts. WIC National Meeting, Salt Lake City, Utah, 9/8/00.

4. Birch LL. Child feeding practices and the etiology of obesity. Obesity (Silver Spring). 2006;14:343–4.

5. Birch LL. Presentation at the National Conference on Nutrition Education Research, Chicago, September 1986.

6. Tanner JM. Fetus into man: physical growth from conception to maturity. Cambridge (MA): Harvard University Press; 1978.

7. Smith GD. Growth and its disorders. Philadelphia: W. B. Saunders; 1979.

8. Bronner F. Adaptation and nutritional needs. (letter) Am J Clin Nutr 1997;65:1570.

9. Gigante DP et al. Early life factors are determinants of female height at age 19 years in a population-based birth cohort. J Nutr 2006;136:473–8.

10. Dietz WH et al. Overweight children and adolescents. N Engl J Med 2005; 352:2100–9.

11. Goran MI. Impaired glucose tolerance in obese children and adolescents (letter), N Engl J Med 2002;347:290.

12. Mayer J. Obesity during childhood. In: Winick M, ed. Childhood obesity. New York: John Wiley & Sons; 1975: pp. 73–80.

13. McGinnis JM et al. Food marketing to children and youth: threat or opportunity? Washington, DC: National Academy Press, Institute of Medicine, 2006.

14. Young LR, Nestle M. Expanding portion sizes in the US marketplace: implications for nutrition counseling. J Am Diet Assoc 2003;103:231–4.

15. American Public Health Association. Surrounding our kids with the opportunity for physical activity. www.medscape.com/viewarticle/527767, 3/23/06.

16. Winslow R, Rundle RL. For obese teens, a radical solution: stomach surgery. Wall Street Journal 2003;Oct.7:A1.

17. Strong WB et al. Evidence-based physical activity for school-aged youth. J Pediatr 2005;146:732–7.

18. Story M, Brown JE. Do young children instinctively know what to eat? The studies of Clara Davis revisited. N Engl J Med 1987;316:103–6.

19. Birch LL. The role of experience in children's food acceptance patterns. J Am Diet Assoc 1987;87(suppl):S36–S40.

20. Fisher JO, Birch LL. Restricting access to palatable foods affects children's behavioral response, food selection, and intake. Am J Clin Nutr 1999;69:1264–72.

21. Gidding SS et al. American Heart Association's Dietary Guidelines for Children. Circulation 2005;112:2061–75.

22. Black RE et al. Children who avoid drinking cow's milk have low dietary calcium intakes and poor bone health. Am J Clin Nutr 2002;76:675–80.

23. Stang J et al., Relationship between vitamin and mineral supplement, dietary intake, and dietary adequacy among adolescents, J Am Diet Assoc 2000; 100:905–10.

24. Vatanparast H et al, Fruit and vegetable intake and bone strength in children. Am J Clin Nutr 2005;82:700–6.

25. Harnack L et al., Soft drink consumption among U.S. children and adolescents: nutritional implications, J Am Diet Assoc 1999;99:436–41.

26. Mughal MZ et al. Vitamin D status in apparently healthy girls. Arch Dis Child 2006;91:549–50, 569–72.

27. Dietary Reference Intakes: Energy, carbohydrate, fiber, fat, fatty acids, cholesterol, protein, and amino acids, Ch. 8, 11. Washington, DC: National Academies Press, 2002.

28. Goulding A et al. Children who avoid drinking cow's milk are at increased risk for prepubertal fractures. J Am Diet Assoc 2004;10:250–3.

29. Parents hold the key to child's healthy weight. www.nlm.nih.gov/medlineplus/news/fullstory_33141.html, 6/3/06.

Nutrition Up Close

Overweight Prevention Close to Home

Feedback for Unit 30

To help their children develop healthful food and activity habits, Debra and Dale can:

1. Provide lots of opportunities for fun physical activities
2. Provide a variety of healthful food choices
3. Limit meals at fast-food restaurants that promote energy-dense, low-nutrient foods to children
4. Work with other parents and school staff to increase physical activity opportunities at school

Other:

- Work with other parents and school staff to increase availability of healthful food choices at school
- Allow their children to eat when they are hungry and to stop eating when they are full
- Never force their children to eat something or totally restrict access to their favorite foods
- Offer foods in an objective, nonthreatening way

Answers to Review Questions

1. True, see page 30-3.
2. True, see pages 30-4, 30-5.
3. False, see page 30-9.
4. True, see page 30-10.
5. False, see page 30-10.
6. True, see page 30-12.
7. False, see page 30-12.
8. True, see page 30-13.

Nutrition and Health Maintenance for Adults of All Ages

NUTRITION SCOREBOARD

	TRUE	FALSE
1 Healthy eating contributes to longevity in part by delaying the age at which chronic disease develops.		
2 Aging is a kind of disease, and people can do little to prevent their health and quality of life from declining as they age.		
3 Calorie restriction and underweight increase longevity.		

Key Concepts and Facts

- Age does not necessarily predict health status. Healthy adults come in all ages.

- Dietary intake, body weight, and physical activity influence changes in health status with age.

- Aging processes begin at the cellular level.

- Medications, diseases, and biological processes associated with aging influence adults' requirements for certain essential nutrients.

Answers to
NUTRITION SCOREBOARD

		TRUE	FALSE
1	Healthy eating pays off.	✔	
2	Aging is not a disease process! Many things can be done to extend life and its quality. (This unit discusses some of those things.)		✔
3	Calorie restriction has not been shown to increase longevity in humans; underweight is associated with shorter life expectancy than normal weight.[1]		✔

Photo Disc

You Never Outgrow Your Need for a Good Diet

One of the most important types of health changes that occurs with age is no change at all.
—Daphne Roe, 1985

Eating right during the adult years is a wise practice. Good nutrition helps adults feel healthy and vigorous as they age and improves their overall sense of well-being. Adults who eat healthfully tend to develop heart disease, cancer, hypertension, and diabetes at older ages and may have more life in their years and years of life than adults who do not.[2]

Aging is a normal process, not a disease. Although the incidence of many diseases increases with age, the causes of the diseases are often unrelated to aging. They may fully or partially result from the cumulative effects of diets high in saturated fat and low in vegetables and fruits, obesity, smoking, physical inactivity, excessive stress, or other habits that insidiously influence health on a day-to-day basis.[3] Aging cannot be prevented, but how healthy we are during aging can be influenced by what we do to our bodies.

Maintaining health as we age is becoming an increasingly important concern. There are more older adults in the United States than ever before, and the numbers are growing.

The Age Wave

The proportion of people in the United States aged 65 years and older is steadily increasing (Illustration 31.1). By the year 2050, approximately 20% of the U.S. population will be 65 years of age or older. Since 1900, **life expectancy** at birth has increased by 63%, from 47.3 to 77.9 years. Although advances in life expectancy are welcomed, there is more progress to be made. The United States ranks twenty-first in life expectancy worldwide, behind countries such as Japan, Sweden, Can-

life expectancy
The average length of life of people of a given age.

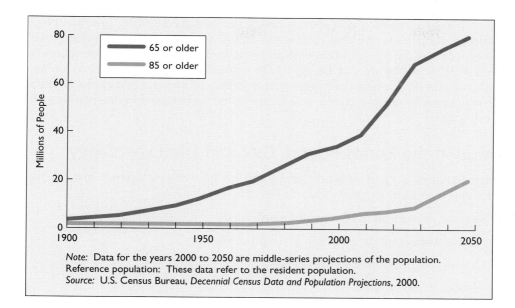

Illustration 31.1
Expected growth in the U.S. population of people aged 65 and 85 years and older.

Note: Data for the years 2000 to 2050 are middle-series projections of the population.
Reference population: These data refer to the resident population.
Source: U.S. Census Bureau, *Decennial Census Data and Population Projections*, 2000.

ada, and Greece.[4] Theoretically, the human life span could reach 130 to 135 years.[5]

All groups in the U.S. population are not benefiting equally from the increase in life expectancy. Life expectancy varies a good deal by sex and race. Table 31.1 shows overall life expectancy for Caucasian and African American men and women in 1900, 1940, and 2003. African American men have the shortest life expectancy, followed by Caucasian men. As in other industrialized countries, life expectancy for females exceeds that for males. In the United States, the gender difference in average life span appears to be largely related to behaviors. U.S. males tend to smoke more, consume more alcoholic beverages, pay less attention to what they eat, and seek medical care less often than do females.[6]

TABLE 31.1

Average length of life for Caucasian and African American males and females in the United States in 1900, 1940, and 2003[4]

	Life Expectancy (Years)		
	1900	1940	2003
Caucasian Americans			
Males	46.6	62.1	75.3
Females	48.7	66.6	80.5
African Americans			
Males	32.5	51.5	69.0
Females	33.5	54.9	76.1

(Data for other ethnic groups were unavailable.)

Why the Gains in Life Expectancy?

"The older an individual gets, the healthier he or she has been."
—T. T. PERLS[9]

Most of the gains in life expectancy in the United States are attributable to decreased infant deaths and deaths from infectious disease, improved nutrition, and medical advances.[7] In 1900, 1 out of 10 newborns died in the first year of life; today the proportion is less than 1 in 100.[11] Since about 1930, the development and mass use of vaccines that prevent common infectious diseases such as pertussis ("whooping cough"), diphtheria, and tetanus have contributed to reductions in infant and child deaths. Advances in medical treatments and surgery have improved the quality of life for many people and have added approximately 5 years to overall life expectancy in the United States since 1900.[8]

Jeanne-Louis Calmet of France had the longest documented life span: 122.5 years. Giant tortoises, by the way, can live up to 177 years.[12]

Photo Disc

A person's genetic background affects her or his longevity. Life insurance company statistics reveal that children of long-lived parents have a 3-year longer life expectancy than do children of short-lived parents. People who inherit genetic traits that favorably affect HDL and LDL cholesterol levels tend to live longer.[9] Since none of us can select our parents, it is perhaps fortunate that environmental factors generally are more important than genetic background in determining life expectancy.

Living in the Bonus Round: Diet and Life Expectancy

Characteristics of diets of adults and older people who experience low disease rates and increased longevity include

- Regular consumption of vegetables and fruits
- Above-average intake of whole grain products
- Lower consumption of saturated fats
- Alcohol in moderation
- Eating breakfast[3, 11]

Taking time to socialize and celebrate with friends and engaging in regular physical activity also appear to contribute bonus years to life (Illustration 31.2).[11]

Calorie Restriction and Longevity

For decades it has been known that laboratory animals fed diets providing 30% or less than normal levels of calories and adequate levels of essential nutrients have increased life expectancies. Why this happens isn't clear, but the reasons could be related to modifications in nutrient utilization, reduced metabolic rate, and decreased aging processes.[13] To date, no human studies have tested the theory that caloric restriction increases longevity. Health effects of starvation have been examined, but those results don't apply. Diets of people who starve are inadequate in many nutrients in addition to calories.[1]

Illustration 31.2
Fun with friends may increase longevity.

Photo Disc

There are some who take the leap and extrapolate results of animal studies to themselves. The number of people practicing calorie restriction is increasing, although their numbers are still small. It's too soon to be adopting a calorie-restricted diet for the purpose of longevity. People who tend to live the longest have normal weights. Life expectancy decreases as body weight decreases below, or increases above, normal.[1]

Nutrition Issues for Adults of All Ages

Researchers and others interested in nutrition and health during middle age and beyond have tended to focus their attention on the cumulative effects of diet on chronic disease. Nutrition exerts its effects on chronic disease development over time, and therefore diseases related to poor diets are most likely to express themselves during the older adult years. The occurrence of diseases related to behavioral traits such as smoking and physical inactivity also increases among adults as they age.

In addition to the health effects of behaviors, people may age biologically. The combined effects of poor diets, other risky lifestyle behaviors, and biological aging increase the rates of serious illness during adulthood. How soon a disease develops largely depends on the intensity and duration of exposure to behavioral risks that contribute to disease development. Behavioral factors also affect the progress of biological aging processes.

Breaking the Chains of Chronic Disease Development

For the most part, the development of chronic disease in middle-aged and older adults can be viewed as a chain that represents the accumulation over time of problems that impair the functions of cells. Each link that is added to the chain, or each additional insult to cellular function, increases the risk that a chronic disease will develop. The presence of a disease indicates that the chain has gotten too long—that the accumulation of problems is sufficient to interfere noticeably with the normal functions of cells and tissues.

It appears that the chain can be shortened by healthful dietary and other behaviors. For example:

- Correcting obesity and stabilizing weight during the adult years may lengthen life expectancy.[14]

- Dietary intakes that correspond to the Dietary Guidelines for Americans are related to decreased death rates in women around the age of 60.[15]

- Regular intake of antioxidant vitamins from plant foods may be protective against declines in cognitive function, Alzheimer's disease, and declines in muscle strength.[16,17]

- Maintaining adequate calcium and vitamin D intake and engaging in regular physical activity during the adult years may prevent, postpone, or lessen the severity of osteoporosis.[21]

- Above-average intakes of fruit and vegetables may delay or prevent the development of a number of types of cancer, heart disease, and stroke, and the formation of cataracts that "cloud over" the lens of the eye.[22,23]

The health status of adults is not necessarily "fixed" by age; it can change for the better or the worse, or not much at all.[24]

TABLE 31.2
Examples of biological changes during aging and nutritional consequences[18–20]

Biological Change	Nutritional Consequences
• Lowered stomach acidity • Decreased lean muscle mass • Reduced production of vitamin D in the skin • Decreased sensation of thirst	• Decreased absorption of vitamin B_{12} and C • Reduced caloric need • Increased dietary requirement for vitamin D, increased risk for inadequate intake of vitamin D and calcium • Dehydration risk

Nutrient Needs of Middle-Aged and Older Adults

Biological processes and lifestyle changes that generally accompany aging affect caloric and nutrient needs (Table 31.2). A person's need for calories generally declines with age as physical activity, muscle mass, and basal metabolic rate decrease. However, people who remain physically active into their older years maintain muscle mass, experience less muscle and bone pain, and gain less body fat than people who are inactive.[25,26]

While caloric need decreases, the requirements for certain nutrients such as protein, vitamin C, vitamin D, vitamin B_{12}, and calcium may increase with aging. The increased need for protein appears to be a result of decreased efficiency of protein utilization. Dietary requirements for vitamin C and calcium may increase due to lower levels of stomach acidity that often occur with advancing age. Decreased stomach acidity reduces the body's ability to absorb vitamins C and B_{12} and calcium. Vitamin D status may become a problem for older adults for two reasons: Their intake of milk and milk products is often low, and exposure of the skin to sunlight produces less than half of the amount of vitamin D as it does in young adults.[27] An estimated 10 to 37% of older adults may be deficient in vitamin D.[28]

The MyPyramid food guide, constructed to reflect the nutrient needs of adults and elderly who exercise 30 to 60 minutes daily is presented in Table 31.3. The food intake plan takes into account the caloric need of most adults and older people. It highlights nutrient-dense foods that are most likely to provide adequate nutrient intakes in this population.

Fluid Needs For reasons that aren't entirely clear, many older people don't get thirsty when their bodies are running low on water. The implication of this change is that older people may become dehydrated and need medical assistance. Approximately 1 million elderly men and women are admitted to hospitals each year due to dehydration.[29] Fluid needs can be met through consumption of water,

TABLE 31.3
MyPyramid food group recommendations for older adults[a]

	Food Group					
Age in Years	GRAINS (OUNCES)	VEGETABLES (CUPS)	FRUITS (CUPS)	MILK (CUPS)	MEAT AND BEANS (OUNCES)	OIL (TEASPOONS)
45	6–9	2.5–3.5	2	3	5.5–6.5	6–8
75	6–7	2.5–3	1.5–2	3	5–6	5–6

[a] Higher numbers of ounces and cups apply to males.

juices, teas, and other beverages. Women of all ages need about 11 cups of water a day from fluids and foods, and men need 15 cups.[19]

Does Taste Change with Age?   Poor diets observed in some middle-aged and older adults have been ascribed to "declining taste" with age. Besides being a dreadful thought, it's not true that taste declines with age to the extent that it makes eating less pleasurable than before. Although taste sensitivity does diminish somewhat with age, chances are your favorite foods will taste as good to you in your older years as they did in your youth (Illustration 31.3). Sight, smell, and hearing senses usually decline with age a good deal more than taste does.[30]

The senses of taste and smell are affected by medications such as antibiotics, antihistamines, some lipid-lowering drugs, and cancer treatments; diseases including Alzheimer's disease, cancer, and allergies; and surgeries that affect parts of the brain and nasal passages. These treatments and conditions are more prevalent in older adults than in middle-aged populations and are primarily responsible for declines in the senses of taste and smell. Changes in these senses increase the likelihood that the person will experience food poisoning or consume an inadequate diet and a low intake of food.[31]

Psychological and Social Aspects of Nutrition in Older Adults

Preparing meals and eating right may not be as simple as it sounds for many older adults. Consuming an adequate and balanced diet may not be easy when you depend on someone else to take you shopping, when mealtimes involve little social life, or when you "don't feel up" to making a meal. Isolation, loneliness, depression, and poor health can be major contributors to poor diets in older adults. The diets of older people are often lacking in nutrients, and because many older adults do not, or cannot, consume enough nutrients to meet their increased need for them, supplementation may be required.

Eating Right during Middle Age and the Older Years

> *I have a theory that chocolate slows down the aging process.*
> *It may not be true, but do I dare take the chance?*
> —AUTHOR UNKNOWN

The best diet for middle-aged adults is one that contains a wide variety of basic foods. A balanced and adequate diet can be obtained by judiciously selecting foods from the basic food groups. It is important to select judiciously because not all foods within the respective food groups are equally desirable. Food choices should emphasize the members of the food groups that are low in saturated fat, trans fat, and sodium, and high in fiber. Diets should highlight vegetables, fruits, low-fat dairy products, and whole grains due to their health benefits.

Altering food habits in favor of healthy diets appears to be a common practice among adults of all ages. A study of 100 women and men over the age of 60 found that 100% had made some change in their food choices in the recent past. The changes had been for health, taste, convenience, and social reasons.[32] Clearly, a person's food choices and intake change throughout life, and the changes can be for the better.

Illustration 31.3
Taste is less affected by age than are some other senses.

PhotoEdit - Jeff Greenberg

Nutrition Up Close

Does He Who Laughs, Last?

Focal Point: Critically thinking about factors that influence longevity.

The following is an adaptation of a letter printed in the "Dear Abby" newspaper column:

> Dear Abby,
>
> Since I've reached my 80s, my mail is full of ads for health products to help me live longer.
>
> I once had many friends, all of whom were health vigilantes. They shook their heads knowingly as I avoided all health food fads and exercise. They made liquid out of good vegetables and spent fortunes buying all the latest supplements. They argued that "organic" was better and "natural" was best. I would tell them that snake venom, poison ivy and manure were "natural." But they wouldn't listen and they didn't laugh.
>
> Now my friends are all dead and I have no one left to argue with.

Critically think about the contents of this letter and identify three alternate explanations for why this individual outlived his friends:

_____

_____

_____

FEEDBACK can be found at the end of Unit 31.

_____

Key Term

life expectancy, page 31-2

Review Questions

True False

1. One of the main reasons the United States ranks first among all countries in life expectancy is the high-quality diet consumed by adults in general. ☐ ☐

2. Three characteristics of dietary intake related to increased longevity are regular consumption of fruits and vegetables, moderate alcohol consumption, and above-average consumption of whole-grain products. ☐ ☐

3. Obesity is associated with decreased life expectancy. ☐ ☐

True False

4. Elderly persons tend to have a lowered sensation of thirst. ☐ ☐

5. The nutrient needs of older adults are the same as those for young adults. ☐ ☐

6. Adults and the elderly have a higher need for water than younger people. ☐ ☐

7. People often lose weight as they age due to normal declines in the sense of taste. ☐ ☐

8. Food habits rarely change among older individuals. ☐ ☐

Media Menu

www.4woman.gov
The National Women's Health Information Center provides information on diabetes, diet and heart health, and other conditions of concern to women. Or call 1-800-994-WOMAN, TDD: 1-888-220-5446.

www.agingwell.state.ny.us
This site provides a "health and wellness village" for mature adults that includes nutrition quizzes, healthy recipes, and diet and exercise tips.

www.nih.seniorhealth.gov
The National Institute on Aging provides health information, exercise videos, and news related to senior health topics.

http://weboflife.nasa.gov/ exerciseandaging/toc.htm
A Guide from the National Institute on Aging presents information on exercise for seniors at this site.

www.nlm.nih.gov/medlineplus/ exerciseforseniors.html
This site provides pictures and diagrams of exercises that improve balance and additional information on exercise for seniors.

www.cdc.gov/men and www.cdc.gov/ women
These CDC sites cover contemporary health information aimed at health concerns of men or women.

Notes

1. Schneider EL. Weighing your longevity part II: should you restrict your calories? www.healthandage.com, accessed 8/03.

2. Dausch JG. Aging issues moving mainstream. J Am Diet Assoc 2003; 103:683–4.

3. Knoops KTB et al. Mediterranean diet and mortality in elderly. JAMA 2004;292:1433–9, 1490–2.

4. Life expectancy. www.cdc.gov/nchs/ fastats/lifexpec.htm, accessed 10/03.

5. US Census Bureau's International Data Base. www.about.com, accessed 10/03.

6. Health United States, 2003. www.cdc. gov/nchs/, accessed 10/03.

7. Bernarducci MP, Owens NJ. Is there a fountain of youth? A review of current life extension strategies. Pharmacotherapy 1996;16:183–200.

8. Guyer B et al. Annual summary of vital statistics: trends in the health of Americans during the 20th century, Pediatrics 2000;106:307–17.

9. Perls TT. The different paths to 100. Am J Clin Nutr 2006;83(suppl):484S–7S.

10. Johnson MA et al. Nutritional patterns of centenarians, Internatl J Aging Hum Dev 1992;34:57–76.

11. Leaf A. Observations of a peripatetic gerontologist. Nutr Today 1973; (Sept./ Oct.):4–12.

12. Mariani SM. Magic potions and the quest for eternal youth. Medscape Molecular Medicine 2003;5:1–7.

13. Heilbronn LK et al. Effect of 6-month calorie restriction on biomarkers of longevity, metabolic adaptation, and oxidative stress in overweight individuals: a randomized controlled trial. JAMA. 2006;295:1539–48.

14. Stevens J et al. The effect of age on the association between body-mass index and mortality. New Engl J Med 1998, 338:1–7.

15. Kant AK et al., Consumption of recommended foods and mortality risk in women, JAMA 2000;283:2109–15.

16. Seshadri S et al. Plasma homocysteine as a risk factor for dementia and Alzheimer's disease, N Engl J Med 2002;346:476–83.

17. Cesari M et al. Antioxidants and physical performance in elderly persons. Am J Clin Nutr 2004;79:289–94.

18. Garry PJ, Vellas BJ. Aging and nutrition. In: Present knowledge in nutrition, Ziegler EE, Filer, LJ eds., ISLI Press: Washington, DC; 1998: pp. 404–19.

19. Bossingham MJ et al. Water balance, hydration status, and fat-free mass hydration in younger and older adults. Am J Clin Nutr 2005;81:1342–50.

20. Francesco VD et al. Unbalanced serum leptin and ghrelin dynamics prolong postprandial satiety and inhibit hunger in healthy elderly: another reason for the "anorexia of aging." Am J Clin Nutr 2006:83:1149–52.

21. Hollick MF. The vitamin D epidemic and its health consequences. J Nutr 2006;135:2739S–48S.

22. Steinmetz KA, Potter JD. Vegetables, fruit, and cancer prevention: a review, J Am Diet Assoc 1996;96:1027–39.

23. Brunce GE, Antioxidant nutrition and cataract in women: a prospective study, Nutr Rev 1993;51: 84–86.

24. Rowe JW, Kahn RL. Human aging: usual and successful, Clin Nutr 1990; 9:26–33.

25. Schmitz KH et al. Strength training and body fat in the elderly. American Heart Associations 46th Annual Conference, Phoenix, AZ, 3/6/06.

26. Bruce B et al. Running and muscle pain in the elderly. Arthritis Res Ther 2005;7: R1263–1270.

27. North American Menopause Society. Osteoporosis guidelines update. Menopause 2006;13:340–67.

28. Visser M et al. Low serum concentrations of 25-hydroxyvitamin D in older persons and the risk of nursing home admission Am J Clin Nutr 2006;84:616–22.

29. Vogelzang JL et al. Overview of fluid maintenance/prevention of dehydration. J Am Diet Assoc 1999;99:605–9.

30. Rolls BJ. Do chemosensory changes influence food intake in the elderly? Physiol Behav 1999;66:193–7.

31. Schiffman SS. Taste and smell losses in normal aging and disease. JAMA 1997;278:1357–62.

32. Bilderbeck N et al., Changing food habits among 100 elderly men and women in the United Kingdom. J Hum Nutr 1981;35:448–55.

Nutrition Up Close

Does He Who Laughs, Last?

Feedback for Unit 31

Alternate explanations:

1. The letter writer's friends may have switched to perceived health foods when they found out they were sick.
2. The letter writer may have consumed a lifelong healthy diet without choosing organic or "natural" foods.
3. The letter writer may come from a family whose members tend to have long lives.

Other possible explanations:

- A sense of humor may help extend life.
- Many factors in addition to diet, exercise, and genetic background influence longevity.
- One individual's experience may not apply to the masses.

Answers to Review Questions

1. False, see page 31-3.
2. True, see page 31-4.
3. True, see page 31-5.
4. True, see page 31-6.
5. False, see 31-6.
6. False, see pages 31-6, 31-7.
7. False, see page 31-7.
8. False, see page 31-7.

The Multiple Dimensions of Food Safety

NUTRITION SCOREBOARD

	TRUE	FALSE
1 The ingestion of pesticides on foods is a major contributor to the high incidence of cancer in humans.		
2 Freezing kills bacteria in food.		
3 Alaska has the highest incidence of botulism in the United States.		
4. You can tell if a food has gone bad by smelling it.		

Key Concepts and Facts

- The leading food safety problem in America is contamination of food by bacteria and viruses.

- Foodborne illnesses primarily result from unsafe methods of producing, storing, and handling food.

- Foodborne illnesses are linked to hundreds of foods but most commonly to raw and undercooked meats and eggs, shellfish, and unpasteurized milk.

- Most cases of foodborne illness are preventable.

Answers to
NUTRITION SCOREBOARD

	TRUE	FALSE
1 Consumers in general are at low risk of developing cancer due to pesticides on food.[1] People at highest risk for pesticide-related illnesses are farmers and others who work with pesticides and do not follow appropriate safety procedures.		✔
2 Freezing causes most bacteria to cease multiplying, but does not kill them. If a food is contaminated with bacteria before you freeze it, bacteria will be present when you thaw it.		✔
3 Here's a question for your nutrition trivia file. Alaska does have the highest incidence of botulism of all 50 states.[2] The reason is explained in this unit. Any guesses?	✔	
4. Smell is not a foolproof indicator of contamination. Bacteria that most commonly cause foodborne illnesses may not change the smell, taste, or appearance of foods.[3]		✔

Photo Disc

foodborne illnesses
An illness related to consumption of foods or beverages containing disease-causing bacteria, viruses, parasites, toxins, or other contaminants.

Threats to the Safety of the Food Supply

Hamburgers contaminated with *E. coli,* chicken tainted with *Salmonella,* fruit coated with organophosphate, mad cow disease—it's enough to give you food fright. Actually, Americans have good reason to be concerned about the safety of some foods (Illustration 32.1). The Centers for Disease Control and Prevention (CDC) estimates that each year in the United States foodborne illnesses cause:

- Sickness in 76 million people

- 325,000 hospitalizations

- Over 5000 deaths[2]

Foodborne illnesses can be caused by bacteria, viruses, marine organisms, fungi, and the toxins they produce, as well as by chemical contaminants in foods or water. They are spread by a wide assortment of foods, including meats, eggs, unpasteurized milk, shellfish, raspberries, sprouts, breakfast cereals, and hummus. Foods most commonly associated with foodborne illness, however, are raw and undercooked meat and eggs, shellfish, and unpasteurized milk.[4]

How Good Foods Go Bad

Bacteria and viruses are the most common causes of *foodborne illnesses* and largely enter the food supply during food processing, storage, or preparation (Illustration 32.2).[2] They are transferred to humans in foods through many different routes, a major one being the contamination of food with animal feces. The lower intestines of many healthy farm animals are colonized by bacteria that may be harmful to humans. These bacteria contaminate food when unsanitary practices are used to prepare and process meats, and when vegetables and fruits are fertilized with animal manure. They can also be transferred to foods by humans who have colonies of harmful bacteria in their gastrointestinal tracts through the use of human sewage on crops and food handling by people carrying the bacteria on their hands (Illustration, 32.2). Humans also transfer certain harmful microorganisms to food when fluids from infected injuries or body secretions contact food.[4] These types of foodborne illnesses are mainly caused by bacteria that are "on" rather than "in" foods.

Although less common, bacteria can be present on the inside of foods. They can enter vegetables and fruits, for example, if the protective skin coatings are broken, allowing an entrance for bacteria. *Salmonella* bacteria can infect the ovaries of hens and cause them to lay normal-looking but infected eggs. Shellfish can concentrate microorganisms present in surrounding water, and although the

Illustration 32.1
Problems with food safety often make the headlines.

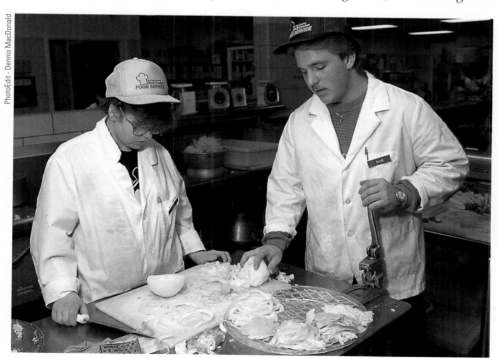

PhotoEdit - Dennis MacDonald

Illustration 32.2
Can you find three potential opportunities for the spread of foodborne illness in this photograph?

bacteria may be harmless to the shellfish, they can provide large doses of harmful bacteria and viruses to humans.[4]

Cross-Contamination of Foods

The source of many cases of foodborne illness is food that has come into contact with contaminated food. This situation, referred to as "cross-contamination," increases the reach of foodborne illnesses.

Microorganisms, including bacteria, viruses, toxins, and other harmful substances, can contaminate safe foods during processing, shipping, preparation, or storage. Opportunities for cross-contamination of foods during processing, for example, are plentiful. The hamburger you eat may contain meat from hundreds of different cows, an omelet in a restaurant may contain the eggs of hundreds of different chickens, and the chicken you baked may have bathed with hundreds of others when they were all washed in the same vat of water at the meat-processing plant. Cross-contamination also occurs on cutting boards across America. The failure to routinely wash cutting boards in between the preparation of different raw foods is a major route to the spread of foodborne illness.[6]

Antibiotics, Hormones, and Other Substances in Foods

Potential sources of food contamination include antibiotics and hormones given to animals, pesticides, and industrial pollutants. Effects of these substances on health vary from potentially life threatening to uncertain.

Antibiotic Resistance   Chickens, cattle, and other farm-raised animals are commonly given antibiotics in feed to prevent infectious disease. In many cases, the antibiotics used are the same ones given to humans to treat infections. Unfortunately, microorganisms are very clever: They can transform themselves to become resistant to antibiotics used to kill them. People can become infected with these new strains of antibiotic-resistant microorganisms when they are consumed in foods. These infections can be difficult to treat because the disease-causing microorganisms are not sensitive to the antibiotics used. Many common forms of foodborne illnesses today are represented by these new forms of bacteria. The FDA is taking action to limit the use of potentially hazardous antibiotics in animals.[7]

Hormones   Hormones are commonly given to farm-raised animals to promote growth or improve milk production. Many consumers are concerned about the effects, however, and would prefer such hormones not be present in food. Hormones given to animals do show up in foods along with other hormones naturally present in animal tissues. Whether animal products free of additional hormones are safer than those without them is unknown but under investigation.[8]

Pesticides and PCBs   Pesticides containing organophosphates, mercury-containing fungicides, and DDT remain causes of foodborne illnesses, as do PCBs. Use of DDT on insects and PCBs in transformers was phased out over 20 years ago due to links with cancer, but these long-lasting chemicals still contaminate some land, lakes, and streams. Fewer than 1 in 10,000 foods on the market contains an excessive level of pesticide, and two out of three contain no trace of agricultural chemicals (Illustration 32.3). Farmers and food

Ed Kashi

Illustration 32.3
Gathering broccoli samples for pesticide testing.
Pesticide residues do not appear to be a major cause of foodborne illnesses.

processing employees are most likely to experience illnesses associated with exposure to agricultural chemicals (Illustration 32.4). Health problems due to agricultural chemicals appear to be rare in other groups of people.[9]

Causes and Consequences of Foodborne Illness

Over 250 types of foodborne illnesses caused by infectious agents (bacteria, viruses, and parasites) and noninfectious agents (toxins and chemical contaminants) have been identified. Their impact on health ranges from a day or two of nausea and diarrhea to death within minutes. Effects of foodborne illnesses are generally most severe in people with weakened immune systems or certain chronic illnesses, pregnant women, young children, and older persons (Table 32.1).[10] Symptoms of foodborne illness most commonly consist of nausea, vomiting, abdominal cramps, and diarrhea. Many cases go unreported, making it difficult to identify the most common causes of foodborne illness. Estimates of the causes are usually based on cases reported by health care professionals and health departments.[2]

The most prevalent reported causes of foodborne illness result from *Salmonella*, *Campylobacter*, and *E. coli* 0157:H7 bacteria and Norwalk-like viruses.[2] Table 32.2 summarizes facts about these top four leading causes of foodborne illness.

Salmonella   Over 37,000 cases of *Salmonella* infection are reported to the CDC yearly, and it is estimated that 38 times that number, or 1,412,498 cases, actually occurred.[2] (A photograph of *Salmonella* bacteria is presented in Illustration 32.5). An outbreak of Salmonella infection that affected approximately 224,000 people

Illustration 32.4
Protective gear—including a chemical-resistant suit, respirator, and goggles—is required for use with some of the most toxic pesticides but is not always worn.

TABLE 32.1

High-risk groups for severe effects of foodborne illness[2]

- People with weakened immune systems due to HIV/AIDS and others with weakened defenses against infections
- People with certain chronic illnesses such as diabetes and cancer
- Pregnant women
- Young children
- Older persons

TABLE 32.2

The top four causes of foodborne illness[11]

Bacteria	Onset	Illness Duration	Symptoms	Foods Most Commonly Affected	Usual Source of Contamination
Salmonella	1–3 days	4–7 days	Diarrhea, abdominal pain, chills, fever, vomiting, dehydration	Uncooked or undercooked eggs, unpasteurized milk, raw meat and poultry, vegetables and fruits	Infected animals, human feces on food, contaminated water
Campylobacter	2–5 days	2–5 days	Diarrhea (may be bloody), abdominal cramps, fever, vomiting	Undercooked poultry, unpasteurized milk	Infected poultry and other animals
E. coli 0157:H7	1–8 days	5–10 days	Watery, bloody diarrhea, abdominal cramps, little or no fever	Raw or undercooked beef, unpasteurized milk, raw vegetables and fruits, contaminated water	Infected cattle
Norwalk-like viruses	1–2 days	1–3 days	Nausea, vomiting, diarrhea	Undercooked seafood	Human feces contamination of oysters and other shellfish beds

Index Stock

Illustration 32.5
A magnified view of *Salmonella* bacteria.

was linked to the transport of an ice cream mixture in a tanker that had previously transported *Salmonella*-infected eggs. The infected ice cream was distributed in 41 states. It is estimated that 5% of the U.S. population experiences a *Salmonella* infection every year.[2]

Campylobacter   This bacterium is a common contaminant on chicken and is the cause of the greatest number of cases of bacteria-related foodborne illness each year. That figure is a startling 2,453,926.[2] The bacteria are found on up to 80% of chickens, and 20% represent a strain of *Campylobacter* resistant to antibiotics. In addition to poultry, *Campylobacter* infection has been traced to unpasteurized milk and contaminated water.[13]

E. Coli   *E. coli* 0157:H7 is a new strain of E. coli that has evolved into a potential killer. As few as 10 of these bacteria can lead to death by kidney failure in vulnerable people. Nearly 80,000 cases of E. coli 0157:H7 occur each year in the United States, and 8% of infected people die from it.[2] E. coli infections are particularly associated with consumption of undercooked ground beef. A major outbreak of E. coli 0157:H7 infection occurred when contaminated ground beef left over from one day's production was added to the next day's batch of beef. Use of the leftover, contaminated meat kept recontaminating subsequent batches.[14] Some 25 million pounds of ground beef had to be recalled as a result.

Norwalk-Like Viruses   Norwalk-like viruses are an extremely common but underreported cause of foodborne illnesses. The viruses go underreported because laboratory tests required for diagnosis are not widely performed. Norwalk-like viruses usually cause an acute bout of vomiting that resolves within two days. They are thought to be primarily spread by infected kitchen workers and fishermen who have dumped sewage waste into waters above oyster beds. An estimated 23,000,000 cases of illnesses related to Norwalk-like viruses occur in the United States each year.[2]

Other Causes of Foodborne Illnesses

Of the hundreds of other causes of foodborne illness, seven have been selected for brief review here. They represent a sampling of foodborne illnesses stemming from seafood consumption, and others that can result from a bacterial toxin, a parasite, and prions (a type of protein).

Photo Disc

Illustration 32.6
Seafoods are an important potential cause of foodborne illness.

Foodborne Illnesses Related to Seafoods   Many types of seafood (Illustration 32.6) may become contaminated due to industrial or human pollution of fresh and ocean waters. Examples presented relate to mercury contamination, ciguatera fish poisoning, and the neurotoxin (or nervous system toxin) responsible for "red tide."

Mercury Contamination   Seafoods have come under fire as a potential source of foodborne illnesses due to mercury contamination of waters and fish by fungicides, fossil fuel exhaust, smelting plants, pulp and paper mills, leather-tanning facilities, and chemical manufacturing plants. High levels of mercury are most likely to be present in large, long-lived fish such as shark, swordfish, and tuna. Because mercury can interfere with fetal brain development, pregnant women should limit their consumption of these types of fish. The Food and Drug Administration (FDA) advises that women who are pregnant or may become pregnant not eat shark, swordfish, king mackerel, or tilefish. Pregnant women should limit consumption of fish caught by family or friends to 6 ounces a week, and they may eat an average of 12 ounces of fish from stores and restaurants weekly. Consumption of fish by adults in general does not appear to pose a health risk.[15]

Ciguatera   Ciguatera poisoning from fish is caused by a neurotoxin (ciguatoxin) present in microorganisms called dinoflagellates that live in reefs (Illustration 32.7). The toxin is transferred through herbivorous reef fish to carnivorous fish and then to humans who eat these fish. Over 200 types of fish may cause ciguatera poisoning, the most common being grouper, red snapper, and barracuda. Primary areas affected by ciguatera poisoning include the Caribbean and South Pacific Islands. Symptoms develop within 1 to 30 hours after ingestion of poisoned fish and cause nausea, vomiting, abdominal cramps, watery diarrhea, and then numbness, shooting pains in the legs, and other symptoms. The toxin causing the poisoning is not destroyed by cooking, freezing, or digestive enzymes, and there is no effective treatment.[16]

Red Tide   "Red tide" may occur between June and October on Pacific and Atlantic coasts. It is due to the accumulation of a microorganism that produces a nerve toxin. Oysters and other shellfish that consume the microorganism become contaminated, as do humans who ingest the shellfish. Resistant to cooking, the toxin produced causes a burning or prickling sensation in the mouth from 5 to 30 minutes after contaminated shellfish are consumed. This symptom is followed by nausea, vomiting, muscle weakness, and a loss of feeling in the hands and feet. Recovery is usually complete, but it's easy to understand that the warning not to eat mussels, clams, oysters, scallops, or other shellfish from red-tide waters is for real.[17]

Botulism   *C. botulinum* bacteria produce a toxin that is one of the deadliest known. It can cause nerve damage and respiratory failure. (An antidote is available but must be given soon after the infection begins.) These bacteria are commonly present in soil and ocean and lake sediment, and they may contaminate crops, honey, animals, and seafood. In humans, botulism usually results from eating underheated, contaminated foods stored in airtight containers. The bacteria thrive without oxygen and produce gases as they grow. The gases expand the food container. (For a real example of a can that exploded due to gas produced by the growth of bacteria, see Illustration 32.8.) Consequently, foods in cans, plastic bags and wraps, and other airtight containers that have bulged-out areas should *not* be eaten.

Alaska has the highest incidence of botulism in the United States. Why does the 49th state rank first? The reason is that some Native Alaskans place uncooked or partially cooked salmon eggs, whale blubber, and other seafoods that harbor *C. botulinum* in plastic bags to ferment. The bags are squeezed to expel the air before sealing, and the no-oxygen environment fosters the growth of botulinum.[18]

Parasites   Various parasitic worms, such as tapeworms, flatworms, and roundworms, may enter food and water through fecal material and soil. They are generally killed by freezing and always killed by high temperature. Once consumed, roundworms may attach to the lining of the intestine and feed on the person's blood. This can lead to anemia. One type of roundworm can bore a hole through the stomach within an hour after the worm's source—raw fish—is eaten. The severe pain that results sends people to their doctors on the double. Hundreds of Japanese people experience this foodborne illness every year.[19]

Illustration 32.7
Magnified view of dinoflagellates that cause ciguatera poisoning.

Illustration 32.8
Pressure caused by bacterial gases in this can of glucose drink made the can explode.

It's a good thing most parasites are destroyed by freezing. One study in Seattle assessed the parasite content of raw fish used in sushi. About 40% of the raw fish samples contained roundworms, but since the fish had been frozen, all the worms were dead.[19]

Mad Cow Disease   Technically called bovine spongiform encephalopathy (BSE), this rare disease in cattle is suspected of causing at least 150 human deaths in Europe. Only one case of the disease in humans—caused by consumption of affected beef—has been diagnosed in the United States, and it originated in England. The disease appears to have started when cows, who are herbivores by nature, were given sheep intestines and parts of the spinal cord in their feed. Some of the sheep harbored a protein, called a prion, that caused a deadly disease when consumed by cows. A prion is not a bacterium or other microorganism; it's a small protein that can transmit disease when consumed by a similar species. Only one other known foodborne illness is spread by the consumption of otherwise healthy body parts. That disease is kuru, and it is transmitted by cannibalism.[20]

Researchers concluded that mad cow disease was transferred to humans who ate the meat of prion-infected cows. The disease represents a new form of Creutzfeldt-Jakob disease (variant Creutzfeldt-Jakob disease, or VCJD) previously identified in humans. It inevitably leads to death in humans, after many years, due to brain damage. Needless to say, it is no longer legal to feed cows animal parts that may transmit the disease. Although the risk of consuming beef from cattle with mad cow disease is extremely small, the possibility that animals may develop this or a similar disease exists. The possibility that mad cow disease may affect cattle in the United States became a reality late in 2003. Stringent monitoring efforts are in place in the United States and many European countries to prevent the occurrence and spread of prions. Rates of BSE are declining due to improved detection and control methods.[21]

Because the disease was first recognized in 1994 and may take 20 or more years to develop, we may not know the full impact of the exposure to contaminated beef for years to come.

Preventing Foodborne Illnesses

There are two major approaches to the prevention of foodborne illnesses. The first relies on food safety regulations that control food processing and handling practices, and the second involves consumer behaviors that lessen the risk of consuming contaminated foods. (Read what can happen when both of these principles are violated in the nearby "Health Action.")

Food Safety Regulations

According to the Federal Food, Drug, and Cosmetic Act, it is illegal to produce or dispense foods that are contaminated with substances that cause illness in humans. Foods are considered "safe" if there is a reasonable certainty that no harm will result from repeated exposure to any substance added to foods. The act governs all substances that are added intentionally or accidentally to foods—except pesticides. According to legislation passed in 1996, pesticides are permitted if their consumption is associated with a "negligible risk" of cancer or other health problems.[23]

HEALTH *Action* **A Double Whammy**

From the true story department of *Nutrition Now* comes the tale of a man who took a night off with his wife to dine at their favorite restaurant. The man, a physician, ordered his favorite item: the luscious chicken burritos. He was served more than he could eat, so he took the leftovers home and promptly put them in the refrigerator. About 8 hours later, he felt like his intestines were exploding; he thought he'd rather die than be so sick. Then the light bulb went off—he had probably consumed *C. perfingens* toxin somehow.

By the third day he began to feel better and noticed he was ravenously hungry. Then he remembered there was a leftover burrito in the refrigerator. . . . Happily, the second round of symptoms wasn't as bad as the first.

The message? It's difficult for consumers to prevent every case of foodborne illness, but we can prevent more cases than we do. Throw out spoiled food.

Irradiation of Foods   Increasing rates of foodborne illness have led some health officials to recommend the expanded use of food irradiation. Food irradiation is a safe process when conducted under specified conditions. It destroys bacteria, parasites, and viruses present in or on foods. Availability of irradiated foods is growing but is still rather limited due to lingering, low consumer acceptance of the process and of the foods.[24]

Although the proposal has merits, food irradiation is not the silver bullet that will prevent all cases of foodborne illness. Prions, toxins, pesticides, mercury, and PCBs are resistant to radiation, so radiation is unlikely to affect the safety of foods with respect to these contaminants. Once irradiated, foods can later become contaminated in such places as a packing plant, grocery store, restaurant, or home. The absence of all microorganisms in irradiated food may enable individual types of bacteria that contact the food after it is sterilized to grow at unusually high rates. Irradiation does not work well on all foods—some fruits and vegetables develop a poor texture and change flavor when irradiated.[24]

The Consumer's Role in Preventing Foodborne Illnesses

Food safety education will never be a substitute for a safe food supply.
—CAROLINE SMITH DEWALL, 1997

Due in large measure to the demands of individuals and consumer groups, many mechanisms are in place to help ensure the safety of the food supply. However, existing safeguards in food production and processing, and government regulations and enforcement efforts, are insufficient to guarantee that all foods purchased by consumers will be free from contamination. This situation, along with the possibility that food can become contaminated in unanticipated ways, means that responsibility for food safety is shared by consumers.

Illustration 32.9
There's a right way to wash your hands.

Food Safety Basics   The first rule of food safety is to wash your hands thoroughly with soap and water before and after handling food (Illustration 32.9).

According to the Centers for Disease Control, this is the single most important means of preventing the spread of foodborne illness caused by bacteria.

■ **There's a Right Way to Wash Your Hands**   Ever watch a TV dramatic series about doctors and hospitals and wondered why they show surgeons scrubbing their hands and upper arms before surgery for so long? TV got it right. It takes about 20 seconds to sanitize your hands. First you need to lather up with soap and very warm water. The soap makes germs lose their grip on your skin. Next you have to make sure to scrub all the crevices between your fingers and under your fingernails. Rinse your hands thoroughly with really warm water and dry them with a paper towel. (If you use a dishcloth, you may reinfect your hands.)[25]

■ **Keep Hot Foods Hot and Cold Foods Cold**   Good foods can go bad if stored improperly. One of the most effective ways to prevent foods from spoiling is to store them at, and heat them to, the right temperatures (see Table 32.3 and Illustrations 32.10 and 32.11). When holding prepared foods before serving, hot foods should be kept hot, and cold foods cold. Freezing foods halts the growth of all of the main types of bacteria that contaminate foods. Once the foods are thawed, however, bacteria growth resumes, and the foods may become contaminated with new bacteria while thawing.

TABLE 32.3

A guide for cooking foods to safe temperatures

Raw Food	Internal Temperature
Ground Meats	
Hamburger	160°F
Beef, veal, lamb, pork	160°F
Chicken, turkey	165°F
Beef, Veal, Lamb	
Roasts and steaks	
Medium rare	145°F
Medium	160°F
Well done	170°F
Pork	
Chops, roasts, ribs	
Medium	160°F
Well done	170°F
Ham, fresh	160°F
Sausage, fresh	160°F
Poultry	
Chicken, whole and pieces	180°F
Duck	180°F
Turkey (unstuffed)	180°F
Whole	180°F
Breast	170°F
Dark meat	180°F
Stuffing (cooked separately)	165°F
Eggs	
Fried, poached	Yolk and white are firm
Casseroles	160°F
Sauces, custards	160°F

Fish should be cooked until the flesh is not clear-looking and flakes easily with a fork; shrimp, lobster, and crab until shells are red and the flesh is opaque and not clear- or raw-looking inside; and clams, oysters, and mussels until the shells open.

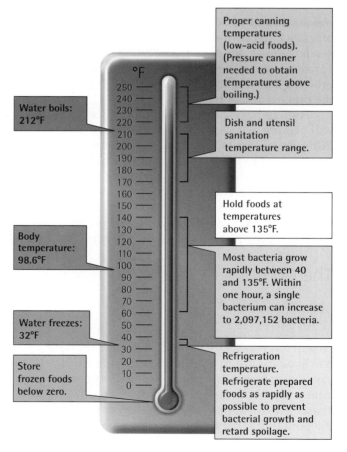

Illustration 32.10

Temperature guide for safe handling of food in the home.

Source: Adapted from Minnesota Extension Service, University of Minnesota; updated in 2004, based on data from USDA.

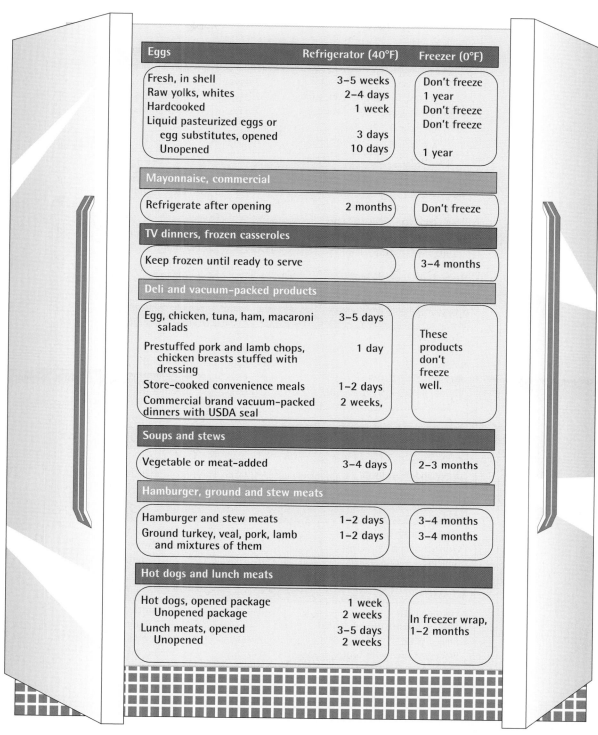

Eggs	Refrigerator (40°F)	Freezer (0°F)
Fresh, in shell	3–5 weeks	Don't freeze
Raw yolks, whites	2–4 days	1 year
Hardcooked	1 week	Don't freeze
Liquid pasteurized eggs or		Don't freeze
egg substitutes, opened	3 days	
Unopened	10 days	1 year

Mayonnaise, commercial		
Refrigerate after opening	2 months	Don't freeze

TV dinners, frozen casseroles		
Keep frozen until ready to serve		3–4 months

Deli and vacuum-packed products		
Egg, chicken, tuna, ham, macaroni salads	3–5 days	These products don't freeze well.
Prestuffed pork and lamb chops, chicken breasts stuffed with dressing	1 day	
Store-cooked convenience meals	1–2 days	
Commercial brand vacuum-packed dinners with USDA seal	2 weeks,	

Soups and stews		
Vegetable or meat-added	3–4 days	2–3 months

Hamburger, ground and stew meats		
Hamburger and stew meats	1–2 days	3–4 months
Ground turkey, veal, pork, lamb and mixtures of them	1–2 days	3–4 months

Hot dogs and lunch meats		
Hot dogs, opened package	1 week	In freezer wrap, 1–2 months
Unopened package	2 weeks	
Lunch meats, opened	3–5 days	
Unopened	2 weeks	

Illustration 32.11
Cold storage.

These safe time limits will help keep refrigerated food from spoiling or becoming dangerous to eat. These time limits will keep frozen food at top quality.

(continued on next page)

Source: U.S. Department of Agriculture, Food Safety and Inspection Service, September 1990 and 1997.

Bacon and Sausage	Refrigerator (40°F)	Freezer (0°F)
Bacon	7 days	1 month
Sausage, raw from pork, beef, turkey	2 days	1–2 months
Smoked breakfast links, patties	7 days	1–2 months
Hard sausage—pepperoni, jerky sticks	2–3 weeks	1–2 months

Ham, corned beef		
Corned beef, in pouch with pickling juices	5–7 days	Drained, wrapped 1 month
Ham, canned— label says keep refrigerated	6–9 months	Don't freeze
opened	3–5 days	1–2 months
Ham, fully cooked—whole	7 days	1–2 months
Ham, fully cooked—half	3–5 days	1–2 months
Ham, fully cooked—slices	3–4 days	1–2 months

Fresh meats		
Steaks, beef	3–5 days	6–12 months
Chops, pork	3–5 days	4–6 months
Chops, lamb	3–5 days	6–9 months
Roasts	3–5 days	4–12 months
Variety meats—tongue, brain, kidneys, liver, heart, chitterlings	2 days	3–4 months

Meat leftovers		
Cooked meat and meat dishes	3–4 days	2–3 months
Gravy and meat broth	1–2 days	2–3 months

Fresh poultry		
Chicken or turkey, whole	1–2 days	1 year
Chicken or turkey pieces	1–2 days	9 months
Giblets	1–2 days	3–4 months

Cooked poultry, leftover		
Fried chicken	3–4 days	4 months
Cooked poultry dishes	3–4 days	4–6 months
Pieces, plain	3–4 days	4 months
Pieces covered with broth, gravy	1–2 days	6 months
Chicken nuggets, patties	1–2 days	1–3 months

Illustration 32.11 (*continued*)
Cold storage: safe time limits

Bacteria grow best at temperatures between 40° and 135°F (see Illustration 32.10). The room temperature of homes is within this range, and that's why foods that spoil should not be left outside the refrigerator for more than an hour. If a food has been improperly stored or kept past the expiration date on the label, it should not be eaten even if it tastes, smells, and looks all right. Bacteria that most commonly cause food borne illnesses usually do not change the taste, smell, or appearance of foods.[3] This general rule should apply: When in doubt, throw it out!

■ **Don't Eat Raw Milk, Eggs, or Meats**   Unpasteurized milk and milk products, raw and partially cooked eggs, raw and undercooked meats and fish, and raw sprouts may be contaminated with microorganisms that cause foodborne illness. Certain raw oysters may be contaminated, too, and should be avoided.[5]

Food Handling and Storage   Particular care needs to be taken when handling and storing foods. Dairy products such as milk and cheese may spoil in about a week or sooner if contaminated with bacteria from the air or hands. If you hold cheese by the wrapper when you take it out of the refrigerator to cut a slice off, it will likely last longer than if you touch it with your hands. Handling foods with clean hands or plastic gloves and using clean utensils on washed surface areas helps to reduce the number of bacteria that come in contact with the food. Raw meats should be separated from other foods, and utensils and surface areas used to prepare meats should be thoroughly cleaned after each use.

The Safety of Canned Foods   Commercially canned foods are heated to the point where all bacteria are killed after the can is sealed. Consequently, the contents of canned foods are sterile and will not spoil if left on the shelf for years. Nevertheless, canned foods may spoil if the can develops a pinhole leak or if errors made during the canning process allow bacteria to enter the food. A cardinal sign that canned foods have spoiled is a buildup of pressure inside the can. The pressure will cause the top of the can to bulge out instead of curving in. Because a buildup of pressure inside a can may be due to bacteria that cause botulism, the food should be thrown away. These and other tips for safe food handling are listed on the USDA's Safe Handling Instructions label (Illustration 32.12) and in the "Health Action" below.

Because so many consumers have questions about food safety, the U.S. Department of Agriculture (USDA) has established a 24-hour hotline for meat and poultry. When in doubt, call the hotline at 1-800-535-4555. If you are planning to visit a foreign country and want to avoid "traveler's diarrhea," contact the Centers for Disease Control for safety tips (www .cdc.gov).

Illustration 32.12
The USDA Safe Handling Instructions label.

Safe Handling Instructions

THIS PRODUCT WAS PREPARED FROM INSPECTED AND PASSED MEAT AND/OR POULTRY. SOME FOOD PRODUCTS MAY CONTAIN BACTERIA THAT CAN CAUSE ILLNESS IF THE PRODUCT IS MISHANDLED OR COOKED IMPROPERLY. FOR YOUR PROTECTION, FOLLOW THESE SAFE HANDLING INSTRUCTIONS.

KEEP REFRIGERATED OR FROZEN.
THAW IN REFRIGERATOR OR MICROWAVE.

KEEP RAW MEAT AND POULTRY SEPARATE FROM OTHER FOODS. WASH WORKING SURFACES (INCLUDING CUTTING BOARDS), UTENSILS, AND HANDS AFTER TOUCHING RAW MEAT OR POULTRY.

COOK THOROUGHLY.   KEEP HOT FOODS HOT. REFRIGERATE LEFTOVERS IMMEDIATELY OR DISCARD.

HEALTH *Action* Tips for Safe Food Handling

- **Tip 1** Wash hands with soap and water before handling foods.
- **Tip 2** Wash hands, surfaces, and utensils with soap and water after contact with raw meat and meat juices
- **Tip 3** Avoid contact between raw and cooked foods. Sanitize cutting boards and utensils in contact with raw meats by washing them in water that contains 1 tablespoon of chlorine (i.e., Clorox) per quart.
- **Tip 4** Cook meats to and store foods at safe temperatures (see Table 32.3 and Illustration 32.10). Eat thoroughly cooked hamburger only.
- **Tip 5** Never leave perishable foods such as potato and egg salad, meats, and dairy products at room temperature for more than an hour. Keep hot foods hot (over 135°F) and cold foods cold (under 40°F).
- **Tip 6** Freeze or refrigerate leftovers promptly in sealed containers.

- **Tip 7** Use flat containers that will allow food to cool quickly if storing large amounts of food.
- **Tip 8** Do not buy or eat food in cans or jars with bulging tops or loose lids.
- **Tip 9** Do not consume foods that are past their expiration date.
- **Tip 10** Follow label instructions for food storage (for example: "Refrigerate after opening").
- **Tip 11** Thaw foods such as turkey and roasts in the refrigerator to reduce bacterial growth.
- **Tip 12** When in doubt, throw it out. If a food has a strange color or odor, throw it away.
- **Tip 13** Wash fresh vegetables and fruits thoroughly with water before eating. Peel vegetables and fruits covered with a waxy film.
- **Tip 14** Avoid "traveler's diarrhea" by getting safety tips from the Centers for Disease Control before you travel (www.cdc.gov).

There Is a Limit to What the Consumer Can Do

In the end, it is up to consumers to demand a safe food supply, up to industry to produce it, up to researchers to develop better ways of doing so, and up to government to see that it happens, to make sure it works, and to identify problems still in need of resolution.
—CDC's Report on Foodborne Illnesses (www.cdc.gov)

Keeping food safe from the farm, lake, or ocean to the table represents a major challenge to industry, government, and consumers. Until contamination of food is prevented, consumers will continue to play an important role in ensuring food's safety.

Nutrition Up Close

Food Safety Detective

Focal Point: Investigating paths to foodborne illness.

Assume an outbreak of foodborne illness occurred among people who ate foods from the onion and lettuce platter shown being prepared in Illustration 32.2. Review the food safety tips in the "Health Action" on page 32-14 and identify three potential causes of the outbreak related to the preparation of the foods as shown in the illustration:

1. _____

2. _____

3. _____

FEEDBACK (answers to these questions) can be found at the end of Unit 32.

Key Terms

foodborne illnesses, page 32-2

Review Questions

	True	False
1. All food borne illnesses are caused by bacteria.	☐	☐
2. Foodborne illnesses often begin with the transfer of harmful bacteria or viruses from feces of an infected animal or person to food.	☐	☐
3. Cross-contamination can result from a failure to wash cutting boards after being used to cut raw meat but not raw vegetables.	☐	☐
4. Antibiotics given to livestock can result in the development of strains of bacteria that resist treatment with antibiotics.	☐	☐

	True	False
5. Symptoms of foodborne illness most commonly consist of headache, blurred vision, and loss of balance.	☐	☐
6. Infection with Norwalk-like viruses is an extremely common cause of foodborne illness. The illness is marked by episodes of vomiting that resolve with two days.	☐	☐
7. The first rule of food safety is to thoroughly wash your hands before and after food preparation.	☐	☐
8. Cheese spoils faster if you touch it with your hands.	☐	☐

Media Menu

www.fightbac.org
The Partnership for Food Safety provides materials for use by the media and consum-

ers. The site gives you access to workbooks, posters about food safety, and "mug shots" of the ten least-wanted foodborne patho-

gens. This site provides a complete cold storage chart.

www.foodsafety.gov
This site is the national gateway to food safety information, resources, outbreak alerts, and consumer advice. Or you can call the Food Safety 24-Hour Hotline at 1-888-SAFEFOOD.

www.cdc.gov
Search foodborne illnesses, food safety, or international travel and find a plethora of information.

www.epa.gov/ost/fish
The place to go for local fish advisories.

www.cdc.gov/mmwr
This URL sends you to the Centers for Disease Control's Morbidity and Mortality Monthly Report. The site includes updates on foodborne illnesses.

www.cdc.gov/foodnet
This address leads you to FoodNet-Foodborne Disease Active Surveillance Network. It belongs to the Centers for Disease Control and provides updates on foodborne illness outbreaks and advances in understanding them.

Notes

1. CDC issues new 5-year report of foodborne illness outbreaks. Food Safety Educator 1997;2:1–6.
2. Mead PS et al. Food-related illness and death in the United States. www.cdc.gov, accessed 10/03.
3. Safe food handling: Myth busters. www.fightbac.org/content/view/151/2, accessed 10/06.
4. Foodborne illnesses. www.cdc.gov, accessed 1/26/01.
5. www.fightbac.org, accessed 10/06.
6. Zhao P et al. Development of a model for evaluation of microbial cross-contamination in the kitchen. J Food Protection 1998;61:960–3.
7. Mathews AW. FDA announces policy designed to curb animal-antibiotic use, Wall Street Journal 2003;Oct. 24:A6.
8. Huffling K. The effects of environmental contaminants in food on women's health. J Midwifery Womens Health 2006;51:19–25.
9. Position of the American Dietetic Association: food and water safety. J Am Diet Assoc 1997;97:184–9.
10. Dietary Guideline for Americans. U.S. Dept. of HHS and USDA, 2005.
11. Food safety: a challenge for everyone in public health. Webcast from the University of North Carolina, 1/26/01.
12. Hennessy TW et al. A national outbreak of salmonella enteritidis infection from ice cream. N Engl J Med 1996;334:1281–6.
13. Wegener HC. The consequences for food safety of the use of fluoroquinolones in food animals, N Engl J Med 1999;340:1581–2.
14. Welland D. *E. coli . . . salmonella . . . Listeria . . .* unwelcome house guests. Envir Nutr 1997;20(Oct.):1, 6.
15. Clarkson TW et al. The toxicology of mercury—current exposures and clinical manifestations, N Engl J Med 2003;349:1731–8.
16. Treatment of ciguatera poisoning with gabapentin. N Engl J Med 2001;344:692–3.
17. The Merck Manual, 17th ed. Whitehouse Station (NJ): Merck Research Laboratories; 1999, p. 262.
18. Lewis R. Botulism from blubber. The Scientist 2003;Mar. 10:11.
19. Tighter reins sought for imported produce. Envir Nutr 1997;20(Nov.):1, 7.
20. Mariani SM. Prions—are they viruses, nucleic acids, or "infectious" proteins? Medscape General Medicine 2003;5:1–8.
21. AMI Foundation Quarterly Report, May 2006. Latest CDC data on foodborne illnesses. www.amif.org/newsletter/AMIFMay2006.pdf, accessed 10/06.
22. Tapeworm, en.wikipedia.org/wiki/Tapeworm, accessed 10/06.
23. Madigan D. After Delaney: how will the new pesticide law work? CNI Weekly Report 1996;Oct. 11:4–5.
24. Shea KM. Technical report: irradiation of food. Pediatrics 2000;106:1505–9.
25. Henneman A. Don't mess with food safety myths. Nutr Today 1999;34:23–8.

Nutrition Up Close

Food Safety Detective

Feedback for Unit 32

Answers to Review Questions

1. False, see page 32-2.
2. True, see pages 32-2, 32-3.
3. False, see page 32-4.
4. True, see page 32-4.
5. False, see page 32-5.
6. True, see page 32-6.
7. True, see page 32-9.
8. True, see page 32-13.

Potential causes of the foodborne illness outbreak:

1. Disease spread from hands to foods.
2. Use of contaminated cutting board, knife, or food platter.
3. Cross-contamination between onions and lettuce or between other foods stored together.

Aspects of Global Nutrition

NUTRITION SCOREBOARD

	TRUE	FALSE
1 Approximately 30% of the world's population does not have access to a safe supply of water.		
2 At most, 25% of deaths of young children in developing countries are related to malnutrition and infection.		
3 The main cause of famine in developing countries is that too little food is available.		

Key Concepts and Facts

- The world produces enough food for all its people.

- Poverty, corrupt governments, the HIV/AIDS epidemic, low rates of breast-feeding, unsafe water supplies, and discrimination against females all contribute to malnutrition.

- Malnutrition early in life has long-term effects on mental and physical development.

- Rates of diseases such as heart disease, diabetes, and hypertension increase in developing countries that adopt Western lifestyles and eating habits.

Answers to NUTRITION SCOREBOARD

		TRUE	FALSE
1	Contaminated water supplies are a major source of infection in many developing countries.	✔	
2	Fifty-three percent of deaths of children in developing countries are associated with malnutrition and infection.[1]		✔
3	Poverty is the primary cause of famine. There are many other important causes as well.		✔

State of the World's Health

The global community consists of countries grouped into the categories of "industrialized nations," "developing nations," and "least-developed nations." (Countries within each category are listed in Table 33.1.) Only 31 countries are considered industrialized, most are developing (113), and, of the developing countries, 49 are least developed.[2] All countries in the global community have interests and issues in common, such as the adequacy of the food supply, availability of health care and education, and safe water. But the countries of the world are also dissimilar in many ways. Differences in financial resources, population growth, and political and other systems translate into large disparities in health status and life expectancy (Table 33.2). People in the least-developed and developing countries are more likely to have substantially shorter life expectancies, to die from infectious diseases, and to experience malnutrition than individuals living in industrialized countries.[4]

The general state of health of populations in various countries is monitored by tracking key environmental, health, and behavioral characteristics. Pregnancy outcome and child growth, rates of breast-feeding, and access to safe drinking water, for example, are key indicators of the health status of a population.[2] As can be seen from Illustration 33.1 and Table 33.3, rates of these key indicators vary substantially worldwide. When percentages of low birthweight in newborns and underweight in young children are high, and rates of breast-feeding and access to a safe water supply are low, one can rightfully assume that malnutrition, infection, and other health problems are common. When these key indicators show improvement, the health status of the population and longevity improve.[5]

If you represented the world's population with a group of 100 people, 70 of the people would be nonwhite, 57 would be Asian, 21 European, 14 North and South American, and 8 African. One person would have a college education, 80 would live in poor housing, and 50 would be malnourished.

Photo Disc

TABLE 33.1

Countries classified by the U.N. International Children's Emergency Fund (Unicef) as industrialized, developing, and least developed[a]

Industrialized Countries	Developing Countries	Least Developed Countries
Andorra; Australia; Austria; Belgium; Canada; Denmark; Finland; France; Germany; Greece; Holy See; Iceland; Ireland; Israel; Italy; Japan; Liechtenstein; Luxembourg; Malta; Monaco; Netherlands; New Zealand; Norway; Portugal; San Marino; Slovenia; Spain; Sweden; Switzerland; United Kingdom; United States.	Afghanistan; Algeria; Angola; Antigua and Barbuda; Argentina; Armenia; Azerbaijan; Bahamas; Bahrain; Bangladesh; Barbados; Belize; Benin; Bhutan; Bolivia; Botswana; Brazil; Brunei Darussalam; Burkina Faso; Burundi; Cambodia; Cameroon; Cape Verde; Central African Rep.; Chad; Chile; China; Colombia; Comoros; Congo; Congo, Dem. Rep.; Cook Islands; Costa Rica; Cote d'Ivoire; Cuba; Cyprus; Djibouti; Dominica; Dominican Rep.; Ecuador; Egypt; El Salvador; Equatorial Guinea; Eritrea; Ethiopia; Fiji; Gabon; Gambia; Georgia; Ghana; Grenada; Guatemala; Guinea; Guinea-Bissau; Guyana; Haiti; Honduras; India; Indonesia; Iran; Iraq; Jamaica; Jordan; Kazakhstan; Kenya; Kiribati; Korea, Dem. People's Rep.; Korea, Rep. of; Kuwait; Kyrgyzstan; Lao People's Dem. Rep.; Lebanon; Lesotho; Liberia; Libya; Madagascar; Malawi; Malaysia; Maldives; Mali; Marshall Islands; Mauritania; Mauritius; Mexico; Micronesia, Fed. States of; Mongolia; Morocco; Mozambique; Myanmar; Namibia; Nauru; Nepal; Nicaragua; Niger; Nigeria; Niue; Oman; Pakistan; Palau; Panama; Papua New Guinea; Paraguay; Peru; Philippines; Qatar; Rwanda; Saint Kitts and Nevis; Saint Lucia; Saint Vincent/Grenadines; Samoa; Sao Tome and Principe; Saudi Arabia; Senegal; Seychelles; Sierra Leone; Singapore; Solomon Islands; Somalia; South Africa; Sri Lanka; Sudan; Suriname; Swaziland; Syria; Tajikistan; Tanzania; Thailand; Togo; Tonga; Trinidad and Tobago; Tunisia; Turkey; Turkmenistan; Tuvalu; Uganda; United Arab Emirates; Uruguay; Uzbekistan; Vanuatu; Venezuela; Viet Nam; Yemen; Zambia; Zimbabwe.	Afghanistan; Angola; Bangladesh; Benin; Bhutan; Burkina Faso; Burundi; Cambodia; Cape Verde; Central African Rep.; Chad; Comoros; Congo, Dem. Rep.; Djibouti; Equatorial Guinea; Eritrea; Ethiopia; Gambia; Guinea; Guinea-Bissau; Haiti; Kiribati; Lao People's Dem. Rep.; Lesotho; Liberia; Madagascar; Malawi; Maldives; Mali; Mauritania; Mozambique; Myanmar; Nepal; Niger; Rwanda; Samoa; Sao Tome and Principe; Sierra Leone; Solomon Islands; Somalia; Sudan; Tanzania; Togo; Tuvalu; Uganda; Vanuatu; Yemen; Zambia.

[a]Data for some eastern European countries, former Soviet republics, and the Baltic states are not available. Least-developed countries represent a subgroup of the developing countries.

TABLE 33.2

Life expectancy in selected countries from lowest to highest[3]

	Life Expectancy (Years)
Zambia	37.2
Mozambique	39.8
Malawi	37.6
Botswana	39.3
Ethiopia	45.2
Afghanistan	45.9
Chad	50.5
Sudan	56.6
Russia	67.1
Philippines	67.5
Iran	70.3
Cuba	76.2
Costa Rica	77.0
Ireland	77.7
United States	77.9
Greece	78.0
Singapore	78.8
France	79.2
Spain	79.3
Italy	79.8
Sweden	80.5
Australia	80.5
Japan	81.2

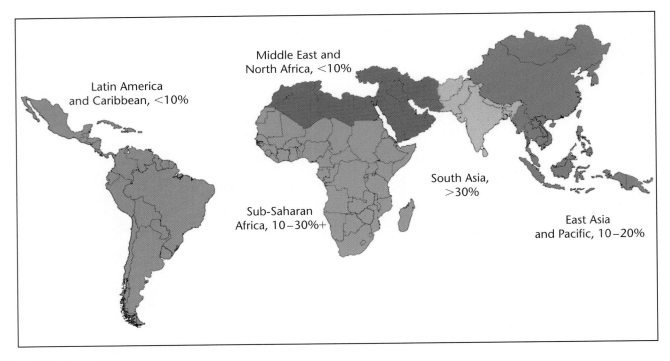

Illustration 33.1
Underweight children are an indicator of malnutrition. This map shows the percentage of children who were underweight in developing regions of the world

Source: WHO global database on child growth and nutrition, 2004, www.who.org.

The health and nutritional status of populations in developing countries is monitored by the World Health Organization (WHO), the Food and Agriculture Organization (FAO), and the United Nations International Children's Emergency Fund (UNICEF). Recent reports have identified leading problem areas related to malnutrition that must be addressed as part of global strategies aimed at improving health and well-being (Table 33.4). Important progress is being made, and rates of malnutrition, poor growth, and infectious disease are decreasing slowly worldwide.[7] Much progress remains to be made, however.

TABLE 33.3

Percentage of low-birthweight infants, exclusive breast-feeding, and access to safe water in different areas of the world.[2]

Region[a]	Low-Birthweight Infants (%)	Exclusive Breast-Feeding 0-3 Months (%)	Access To Safe Water (%)
Sub-Saharan Africa	16	32	49
Middle East and North Africa	11	43	81
South Asia	33	46	80
East Asia and Pacific	11	56	67
Latin America and Caribbean	10	38	77
Developing countries	18	44	70
Least developed countries	23	46	54
Industrialized countries[b]	6	—	100
World	17	44	71

[a]Data for the former Soviet republics and the Baltic states are not available.
[b]Data for exclusive breast-feeding are not available; the figure is approximately 50%.

TABLE 33.4

Priority problem areas related to malnutrition in developing countries[6]

1. Childhood protein-calorie malnutrition
2. Vitamin deficiencies: vitamin A, folate
3. Mineral deficiencies: iodine, iron, zinc
4. Lack of breastfeeding
5. Alcohol abuse
6. Overweight and obesity
7. Insufficient physical activity
8. Low vegetable and fruit intake
9. Malnutrition and increased complications from HIV/AIDS
10. Poor nutritional status of women of child-bearing age

Illustration 33.2
The look of marasmus.
Children with marasmus look like "skin and bones." They don't get enough calories and protein in their diet. This child is suffering from the extreme emaciation of marasmus.

Dr. Robert S. Goodhard

Food and Nutrition: The Global Challenge

He who is healthy has hope; and he who has hope has everything.
—ARAB PROVERB

We have all heard about starvation in Somalia, Ethiopia, Sudan, and Bangladesh. The pictures of starving children with desperation in their big eyes and heads too large for their frail bodies burn an image in our minds. What we don't hear about in the news, however, is the extent of starvation and malnutrition in the world. These are not just problems of isolated areas experiencing civil war or crop failures. In the developing regions, malnutrition is an ongoing problem for 10 to 51% of children under the age of 5, depending on the region. Approximately 50 to 70% of women worldwide are iron deficient, and 2 billion children suffer from iodine deficiency. Vitamin A deficiency exists among 21% of children in developing countries. These children are at risk of frequent and severe infectious disease, decreased growth, and blindness because of the deficiency. It is estimated that one-third of the world's population is mildly to moderately zinc deficient.[6]

The children we usually see in pictures from famine areas are victims of *marasmus,* a disease caused primarily by a lack of calories and protein. These children look starved (Illustration 33.2) and consume far fewer calories and far less protein and other essential nutrients than they need. In victims of marasmus, the body uses its own muscle and other tissues as an energy source.

Sometimes malnutrition is related to an inadequate intake of protein and other nutrients rather than primarily to a deficiency of calories. Children who lack protein and other nutrients develop *kwashiorkor.* These children may appear fat due to massive swelling that occurs with the disease (Illustration 33.3). Most victims of kwashiorkor are somewhat marasmic as well. Children with either type of severe malnutrition are generally deficient in multiple vitamins and minerals and at high risk of dying from infectious diseases due to weakened immune systems.[8]

Survivors of Malnutrition

Malnutrition that occurs before a child's brain has completely grown has lasting effects on development.[5] Brain growth occurs during pregnancy through the age of 2 years. If children suffer severe malnutrition during

marasmus
Malnutrition caused by a lack of calories and protein. Also called protein-calorie malnutrition.

kwashiorkor
A deficiency disease primarily caused by a lack of protein and calories in the diet and abnormal metabolic adaptation to starvation.

Illustration 33.3
The look of kwashiorkor.
This child has the characteristic "moon face" (edema), swollen belly, and patchy dermatitis often seen with kwashiorkor. Children with kwashiorkor may not look starved because of the massive swelling.

Dr. Robert S. Goodhard

these years, they will experience permanent delays in mental development even if refeeding improves their physical growth and health. The severity of the mental delays depends on the timing and duration of the malnutrition. The longer malnutrition exists, the harder it becomes for young children to achieve a brighter future.[9] In addition, the state of hunger and starvation leads people to become self-centered and to lose any sense of well-being. The need for food for survival may prompt unethical behaviors such as stealing and injuring others to obtain food. The devastating psychological effects of starvation may follow children into adulthood and have a lasting effect on behavior throughout life. Persistent malnutrition saps the physical and mental energy of people and, ultimately, compromises the economic and social progress of nations.[4]

Malnutrition and Infection There is a very close relationship between malnutrition and infection: Malnutrition weakens the immune system and increases the likelihood and severity of infection. Simultaneously, repeated infection and the bouts of diarrhea that often accompany it can produce undernutrition. Poor sanitary conditions, contaminated water supplies, and lack of refrigeration contribute to the spread of infectious diseases and thus to the malnutrition the diseases promote. Most deaths of children under the age of 5 years in developing countries are related to the presence of both malnutrition and infection.[10]

Low rates of breast-feeding promote the spread of infection, especially in countries where contaminated water is used to prepare formula. In addition to providing optimal nutrition, breast milk contains a number of substances that protect babies from infection.[6]

Vitamin A deficiency and malnutrition in children play important roles in the spread and severity of infections. Children who are deficient in vitamin A are much more likely to die from a measles infection, for example, than are children whose vitamin A status is sufficient.[9] HIV infection progresses more rapidly into AIDS in malnourished individuals and shortens the time parents can work and support their families. Life expectancy in Botswana, for example, has decreased from 65 years in 1995 to 39 years in 2001. Over 42 million people in the world have HIV infection, and the incidence is increasing.[11]

Why Do Starvation and Malnutrition Happen?

> *It is very much an oversimplification to assert, without qualification, that people starve because they do not have enough to eat.*
> —M. PYKE, 1972

Malnutrition and starvation don't have to happen anywhere. The world's agricultural systems have the capacity to produce enough food to feed everyone on Earth.[12] People become malnourished or starve primarily because of poverty. Human-made disasters, including discrimination against women, the HIV/AIDS epidemic, racism, corrupt governance, and other factors (Table 33.5) heavily contribute to malnutrition. In some instances, starvation and malnutrition are due to natural disasters that lead to crop failures or the inability to distribute food to those in need.[9]

The famine in Somalia that led to the death of over 1.5 million people was due to the collapse of the government, fighting among rival clans, and general lawlessness. Farmers and their families were driven off their land by bandits who stole their crops and possessions. Many had nowhere to go for help. Mass starvation in Ethiopia and Sudan was similarly initiated by violent conflict within the country. Seizing the opposing side's food and preventing them from obtaining

TABLE 33.5
Root causes of malnutrition in developing countries[5]

- Poverty
- Discrimination against females
- HIV/AIDS epidemic
- Racism, ethnocentrism
- Poor and corrupt governance
- Unsafe water

- Low levels of education
- Unequitable distribution of the food supply
- Lack of economic opportunities
- Low agricultural productivity

food aid are among the primary weapons used to fight these civil wars. In Bangladesh, where poor people have no choice but to live on floodplains, cyclones wipe out crops and regularly result in the death of thousands due to floods, starvation, and infectious disease. None of these cases of mass famine was due primarily to the ravages of nature. Starvation and malnutrition are largely initiated by humans, and it is within the power of humans to end them.

Women and female children are at particular risk for malnutrition in some societies because cultural practices call for food to be allocated to men and boys first (Illustration 33.4). If too little food is available, what remains after meals for women and daughters may be too little to support health and growth. In many developing countries, discrimination against women in education and employment, sanctions against the use of birth control, and violence toward women place women at high risk of developing malnutrition and having a low quality of life.[5]

Illustration 33.4
Women and children in some developing countries are particularly vulnerable to malnutrition because they, and their nutritional needs, may be considered less important than men and their needs.
Similarly, boys' needs may take precedence over girls' needs. These children are 2-year-old twins: The child on the left is a girl, and the child on the right is a boy.

Ending Malnutrition

> *Adequate food is the cradle of normal resistance, the playground of normal immunity, the workshop of good health, and the laboratory of long life.*
> —CHARLES MAYO

The long-term solution to malnutrition will depend on the ability of humans to work together to achieve educational and economic development, peace, population growth control, improved sanitation, social equity for women and children, and environmentally sound and productive agricultural policies and practices in developing countries (Illustration 33.5). Bottom-up approaches to problem solving work better than top-down approaches. The people most affected by malnutrition must be active participants in the planning and implementation of improvement programs. Efforts to improve the nutritional status of populations can and have paid off.

Success Stories   Examples of strikingly successful efforts to reduce malnutrition in various countries around the world are available for telling. A repeated theme is the reduction in malnutrition as economic, educational, nutritional, and sanitary conditions improve. Improvements in these same areas were responsible for the dramatic declines in malnutrition and infectious diseases and the

Illustration 33.5
Safe water supplies represent a major advance in quality of life for people in developing countries.

Photo Researchers, Porterfield/Chickering

Approximately 14% of the U.S. population lives in poverty, but there are sizable differences in poverty rates by age and racial group. Twenty percent of children, 30% of Hispanics, 29% of African Americans, and 11% of whites live in poverty.[19]

Photo Disc

increases in life expectancy experienced in the United States, Canada, and many European countries around 100 years ago.[13] In other instances, specific nutrient deficiencies have been greatly reduced or eliminated by food fortification and nutrition education programs. Here are several specific examples:

- Vitamin A supplements and education on vitamin A—rich foods were related to a dramatic decrease in cases of severe and moderate vitamin A deficiency and infection in children in Indonesia.

- Food supplements given to groups of infants in Russia, Brazil, South Africa, and China were associated with higher IQ scores at age 8.

- Iodization of salt has eliminated iodine deficiency in Bolivia and Ecuador. Worldwide, iodization of salt is associated with a drop in the number of iodine-deficient children from 48 million in 1990 to 28 million in 1997.

- Fortification of flour with iron has led to a decrease in iron-deficiency anemia in the Philippines.[14]

In Thailand, a country with a serious iodine-deficiency problem, a highly creative approach to increasing iodine intake was implemented by the king. To celebrate his birthday, salt producers gave the king 10,000 tons of iodized salt. The king had the salt packaged in small plastic bags and, with the help of the Red Cross and the army, delivered a bag to every household in Thailand. A message from the king about the importance of using iodized salt was attached to each bag.[15]

Deaths of young children from malnutrition and its related diseases have dropped substantially in countries experiencing a resurgence of breast-feeding (Illustration 33.6). Breast milk protects infants and young children from a variety of infectious diseases, supports the growth and health of infants, and protects them from the hazards of formulas reconstituted with contaminated water. Worldwide health initiatives have led to the development of the International

Code of Marketing Breast Milk Substitutes and to "Baby Friendly Hospitals." The international marketing code calls for the prohibition of free formula samples as well as the promotion of infant formulas by health care professionals. More than 12,700 hospitals have adopted Baby Friendly Hospital policies that promote and facilitate breast-feeding.[9]

Health status of people in developing countries has also been improved by the widespread use of oral rehydration fluids that protect children with diarrhea from dehydration. Broadly based vaccination programs have protected many from various diseases.[16]

The World Food Summit   The 1996 World Food Summit ended with a pledge that by 2015, countries would reduce by half the number of people in the world who are hungry and undernourished.[9] Progress is being made toward meeting the goal. Access to food and rates of undernutrition are improving, but at a slower pace than envisioned. It is anticipated that it will take until the year 2030 to cut rates of hunger and undernutrition by half.[10]

Illustration 33.6
After a rooming-in policy that enabled mothers to breastfeed their infants on demand was introduced in one hospital in the Philippines, the incidence of early infant deaths due to infection dropped by 95%.

Source: First Call for Children, A UNICEF Quarterly 1992 (January/ March):1.

The "Nutrition Transition": Rising Rates of Obesity and Chronic Diseases
As countries develop economically and food supplies increase, the incidence of undernutrition usually decreases, and chronic health problems such as heart disease, diabetes, and hypertension tend to increase. Elevated rates of chronic disease appear to be closely related to obesity, high saturated fat and low vegetable and fruit intakes, low levels of physical activity, and other characteristics of Western lifestyles and dietary habits.[17]

Countries undergoing the nutrition transition often experience rising rates of obesity.[17] Children born to poorly nourished women may be biologically programmed while in the womb to conserve energy. When exposed to a generous food supply later in life, they may gain fat more readily than height.[18] Excess accumulation of fat later in life may place the children at increased risk for hypertension, diabetes, heart disease, and a number of other "diseases of civilization." Although infections such as malaria and tuberculosis will predominate through the year 2025, the incidence of diseases of "Western civilization" will continue to expand.[17]

The Future

> *Knowing is not enough; we must apply.*
> *Willingness is not enough; we must do.*
> —GOETHE

Benefits derived from improving the nutritional status of the world's population are immense. Adequate nutrition is the bedrock on which present generations secure a future for themselves and the next generation. People who are well nourished are more productive, happier, require less medical care, and are more likely to be self-sufficient than malnourished people. When the malnutrition problem is solved, the world will likely wonder why it didn't happen sooner.[5]

If you would like to be part of the solution to food insecurity, hunger, or malnutrition, consider volunteering at a food or meals program, in community courses for language or self-sufficiency skills, or in other public service programs. The experience will likely enrich your life as well as the lives of those you serve.

Nutrition Up Close

Ethnic Foods Treasure Hunt

FOCAL POINT: Values placed on foods are culturally specific.

The following questions deal with ethnic food preferences. Answer any three of the questions. Resources that will help you find the answers include knowledgeable individuals, the Internet, travel and anthropology books, and encyclopedias.

1. Name three foods commonly served at celebrations by the Igbo in Nigeria (Igbos were formerly referred to as Biafrans):
   a._____
   b._____
   c. _____

2. List two special foods given to Hmong women during pregnancy:
   a. _____
   b. _____

3. List the five foods that make up the traditional Costa Rican breakfast:
   a. _____
   b. _____
   c. _____
   d. _____
   e. _____

4. List three foods commonly sold at soccer matches in Italy:
   a. _____
   b. _____
   c. _____

5. Name two traditional foods of the Zulu in South Africa:
   a. _____
   b. _____

6. Name one of the staple foods of people in Nentsi (it's near Siberia):

7. List two foods considered to be "yin" and two considered to be "yang" in Chinese culture:
   　　　　　Yin　　　　　　　　　　　　Yang
   a. _____　　a. _____
   b. _____　　b. _____

FEEDBACK (answers to these questions) can be found at the end of Unit 33.

Key Terms

kwashiorkor, page 33–5　　　　　*marasmus, page 33–5*

Review Questions

True False

1. Malnutrition and infection are often common health problems in countries with low life expectancy. ☐ ☐

2. Rates of malnutrition are increasing worldwide. ☐ ☐

3. Deficiencies of iodine, iron, and vitamin A remain important public health problems in many developing countries. ☐ ☐

4. The world's agricultural systems cannot produce enough food to feed the world's population. ☐ ☐

True False

5. In many developing countries, discrimination against women places them at high risk of developing malnutrition. ☐ ☐

6. Death rates for young children are decreasing in countries with increasing rates of breast-feeding. ☐ ☐

7. Due to the expanding presence of fast-food restaurants, many developing countries are experiencing increasing rates of chronic diseases. ☐ ☐

Media Menu

www.who.org
This URL leads to causes of illness and death in industrialized and developing countries, information about disease outbreaks, WHO programs, and travelers' health.

www.fao.org
The United Nations' Food and Agriculture Organization address provides updates on worldwide malnutrition and meeting the World Food Summit's goal of reducing hunger by half by 2015.

www.travel.state.gov
Traveling abroad? This site offers travel alerts on health conditions, political instability, crime statistics, and public announcements.

www.census.gov/ipc/www/idbnew.html
The U.S. Census Bureau International Data Base (IDB) is a computerized data bank containing statistical tables of demographic data for 228 countries and areas of the world. The site includes mortality causes, life expectancy, family planning, demographic data, and more.

www.cia.gov/cia/publications/factbook/geos/xx.html
The CIA World Factbook home page provides an unbelievable amount of information on the world and each country in it. Get data on infant mortality rates, life expectancy, industries, climate, time zones, economies, agriculture, and much from this huge resource.

www.who.int/nutrition
The Word Health Organization's main page for international nutrition is at this address.

Notes

1. Nutrition. www.who.int/nutrition, accessed 10/06.

2. Darton-Hill I, Coyne ET. Feast and famine: socioeconomic disparities in global nutrition and health. Public Health Nutr 1998;1:23–31.

3. US Census Bureau International Data Base. www.census.gov/ipc/www/idbnew.html, accessed 10/06.

4. The state of food insecurity in the world in 2000. FAO, United Nations Report, Nov. 2000, www.fao.org, accessed 2/01.

5. Position of American Dietetic Association: addressing world hunger, malnutrition, and food insecurity. J Am Diet Assoc 2003;103:1046–56.

6. Quantifying selected major risks to health, World Health Organization, www.who.int/whr/2002/chapter4/en/print.html, accessed 11/02.

7. Christan P. Micronutrients and reproductive health issues: an international perspective. J Nutr 2003;133:1969S–73S.

8. Scrimshaw NS. Historical concepts of interactions, synergism and antago-nism between nutrition and infection. J Nutr 2003;133:316S–21S.

9. United Nations International Children's Emergency Fund. The state of the world's children, 1998. New York: United Nations; 1998.

10. Underwood BA. Scientific research: essential, but is it enough to combat world food insecurity? J Nutr 2003;133:1434S–7S.

11. Kim JY et al. AIDS in 2006—Moving toward one world, one hope. N Engl J Med 2006;355:645–8.

12. Haddad L, Martorell R. Feeding the world in the coming decades requires improvements in investment, technology and institutions. J Nutr 2002;132:3435S–36S.

13. Gold MR. Presentation at the Tenth Chronic Disease Conference, Washington, DC; 1996.

14. Decimosexta Conferencia Internacional de Nutricion, Montreal. Nutr View 1997; fall:1–36.

15. Food fortification to end micronutrient malnutrition: state of the art. Nutr View 1997; special issue:1–8.

16. Pelletier DL, Frongillo EA. Changes in child survival are strongly associated with changes in malnutrition in developing countries. J Nutr 2003; 133:107–19.

17. Popkin BM. Global nutrition dynamics: the world is shifting rapidly toward a diet linked with noncommunicable diseases. Am J Clin Nutr 2006;84:289–98.

18. De Onis M et al. Prevalence and trends of overweight among preschool children in developing countries. Am J Clin Nutr 2000;72:1032–9.

19. Health, United States, 2000. www.cdc.gov/nchs, accessed 8/00.

Nutrition Up Close

Ethnic Foods Treasure Hunt

Feedback for Unit 33

1. a. yams; b. kola nuts; c. breadfruit with corn and yams; d. cassava.
2. a. chicken; b. rice; c. ripe mango; d. grapes; e. ginger f. milk; g. many more.
3. "Gallo pinto" includes the following:
   a. rice and black beans; b. eggs; c. tortillas; d. coffee; e. fruit; f. sometimes steak.
4. a. Italian ices; b. calzones; c. Italian sodas; d. pizza; e. pasta; f. hot dogs.
5. a. beef; b. cassava; c. milk; d. yams; e. pumpkin; f. millet; g. corn; h. dried beans; i. beer; j. honey; k. yogurt; l. cow's blood.
6. a. reindeer; b. polar bear; c. sour milk; d. caribou; e. tea.
7. Yin: a. raw kelp (seaweed); b. boiled rice; c. raw vegetables; d. fruit; e. dairy products.
   Yang: a. cooked kelp; b. cooked vegetables; c. fried rice; d. fish; e. eggs; f. chicken.

Answers to Review Questions

1. True, see page 33-2.
2. False, see page 33-4.
3. True, see page 33-5.
4. False, see page 33-6.
5. True, see page 33-7.
6. True, see page 33-8.
7. False, see page 33-9.

APPENDIXES

Appendix A
Table of Food Composition

Appendix B
Reliable Sources of Nutrition Information

Appendix C
The U.S. Food Exchange System

Appendix D
Table of Intentional Food Additives

Appendix E
Cells

Appendix F
Canadian Choice System

Table of Food Composition

This edition of the table of food composition includes a wide variety of foods from all food groups. It is updated with each edition to reflect nutrient changes for current foods, remove outdated foods, and add foods that are new to the marketplace.*

The nutrient database for this appendix is compiled from a variety of sources, including the USDA Standard Reference database (Release 16), literature sources, and manufacturers' data. The USDA database provides data for a wider variety of foods and nutrients than other sources. Because laboratory analysis for each nutrient can be quite costly, manufacturers tend to provide data only for those nutrients mandated on food labels. Consequently, data for their foods are often incomplete; any missing information is designated in this table as a blank space. Keep in mind that a blank space means only that the information is unknown and should not be interpreted as a zero.

Whenever using nutrient data, remember that many factors influence the nutrient contents of foods, including the mineral content of the soil, the diet of the animal or the fertilizer of the plant, the season of harvest, the method of processing, the length and method of storage, the method of cooking, the method of analysis, and the moisture content of the sample analyzed. With so many factors involved, users must view nutrient data as a close approximation of the actual amount.

For updates, corrections, and a list of 6000 foods and codes found in the diet analysis software that accompanies this text, visit **www.wadsworth.com/nutrition** and click on *Diet Analysis*.

- *Fats* Total fats, as well as the breakdown of total fats to saturated, monounsaturated, and polyunsaturated fats, are listed in the table. The fatty acids seldom add up to the total due to rounding and to other fatty acid components that are not included in these basic categories, such as *trans*-fatty acids and glycerol. *Trans*-fatty acids can comprise a large share of the total fat in margarine and shortening (hydrogenated oils) and in any foods that include them as ingredients.

- *Vitamin A and Vitamin E* In keeping with the 2001 RDA for vitamin A, which established a new measure of vitamin A activity—retinol activity equivalents (RAE)—this appendix presents data for vitamin A in micrograms (μg) RAE. Similarly, because the 2000 RDA for vitamin E is based only on the alpha-tocopherol form of vitamin E, this appendix reports vitamin E data in milligrams (mg) alpha-tocopherol (listed in the table as mg α).

- *Bioavailability* Keep in mind that the availability of nutrients from foods depends not only on the quantity provided by a food, but also on the amount absorbed and used by the body—the bioavailability. The bioavailability of folate from fortified foods, for example, is greater than from naturally occurring sources. Similarly, the body can make niacin from the amino acid tryptophan, but niacin values in this table (and most databases) report preformed niacin only. Chapter 10 provides conversion factors and additional details.

- *Using the Table* The items in this table have been organized into several categories, which are listed at the head of each right-hand page. Page numbers have been provided, and each group has been color-coded to make it easier to find individual items.

*This food composition table has been prepared for Wadsworth Publishing Company and is copyrighted by ESHA Research in Salem, Oregon—the developer and publisher of the Food Processor and Genesis nutritional software programs. The nutritional data are supported by over 1300 references. Because the list of sources is so extensive, it is not provided here, but is available from the publisher.

In an effort to conserve space, the following abbreviations have been used in the food descriptions and nutrient breakdowns:

- diam = diameter
- ea = each
- enr = enriched
- f/ = from
- frzn = frozen
- g = grams
- liq = liquid
- pce = piece
- pkg = package
- w/ = with
- w/o = without
- t = trace
- 0 = zero (no nutrient value)
- blank space = information not available

INDEX TO APPENDIX A

TABLE A–1

Food Composition   (Computer code number is for West Diet Analysis program)   (For purposes of calculations, use "0" for t, <1, <.1, <.01, etc.)

Computer Code Number	Food Description	Measure	Wt (g)	H₂O (%)	Ener (kcal)	Prot (g)	Carb (g)	Dietary Fiber (g)	Fat (g)	Fat Breakdown (g) Sat	Mono	Poly
	BEVERAGES											
	Alcoholic:											
	Beer:											
22500	Regular (12 fl oz)	1½ c	356	94	117	1	6	<1	<1	0	0	0
22512	Light (12 fl oz)	1½ c	354	95	99	1	5	0	0	0	0	0
22679	Budweiser (12 fl oz)	1½ c	357		145	1	11		0	0	0	0
22616	Miller (12 fl oz)	1½ c	356	96	143	1	13		0	0	0	0
22617	Miller Light (12 fl oz)	1½ c	356		110	1	7		0	0	0	0
	Gin, rum, vodka, whiskey:											
22670	80 proof	1½ fl oz	42	67	97	0	0	0	0	0	0	0
22654	86 proof	1½ fl oz	42	64	105	0	<1	0	0	0	0	0
22661	90 proof	1½ fl oz	42	62	110	0	0	0	0	0	0	0
	Liqueur:											
22519	Coffee liqueur, 53 proof	1½ fl oz	52	31	175	<1	24	0	<1	t	t	t
22520	Coffee & cream liqueur, 34 proof	1½ fl oz	47	46	154	1	10	0	7	4.5	2.1	0.3
22521	Créme de menthe, 72 proof	1½ fl oz	50	28	186	0	21	0	<1	t	t	t
22551	Kahlua	1½ fl oz	30	30	106	<1	13	0	<1	t	t	t
	Wine, 4 fl oz:											
22673	Dessert, sweet	½ c	118	71	189	<1	16	0	0	0	0	0
22501	Red	½ c	118	88	85	<1	2	0	0	0	0	0
22502	Rosé	½ c	118	89	84	<1	2	0	0	0	0	0
22504	White medium	½ c	118	90	80	<1	1	0	0	0	0	0
20077	Nonalcoholic light	1 c	232	98	14	1	3	0	0	0	0	0
22681	Wine cooler	1 c	227	90	113	<1	13	<1	<1	t	t	t
	Carbonated:											
20006	Club soda (12 fl oz)	1½ c	355	100	0	0	0	0	0	0	0	0
20005	Cola beverage (12 fl oz)	1½ c	372	89	156	<1	40	0	0	0	0	0
20030	Diet cola w/aspartame (12 fl oz)	1½ c	355	100	4	<1	<1	0	0	0	0	0
20007	Diet soda pop w/saccharin (12 fl oz)	1½ c	355	100	0	0	<1	0	0	0	0	0
20008	Ginger ale (12 fl oz)	1½ c	366	91	124	0	32	0	0	0	0	0
20031	Grape soda (12 fl oz)	1½ c	372	89	160	0	42	0	0	0	0	0
20032	Lemon-lime (12 fl oz)	1½ c	368	90	147	0	38	0	0	0	0	0
20027	Pepper-type soda (12 fl oz)	1½ c	368	89	151	0	38	0	<1	0.3	0	0
20009	Root beer (12 fl oz)	1½ c	370	89	152	0	39	0	0	0	0	0
20149	Cherry Coke (12 fl oz)	1½ c	375		156	0	42	0	0	0	0	0
20148	Coca Cola Classic (12 fl oz)	1½ c	373		145	0	40	0	0	0	0	0
20150	Diet Coke (12 fl oz)	1½ c	359		1	0	<1	0	0	0	0	0
20207	Diet 7 UP (12 fl oz)	1½ c	360		0	0	0	0	0	0	0	0
20167	Diet Pepsi (12 fl oz)	1½ c	360		0	0	0	0	0	0	0	0
20166	Pepsi cola (12 fl oz)	1½ c	360	88	150	0	41	0	0	0	0	0
20163	Sprite (12 fl oz)	1½ c	373		144	0	39	0	0	0	0	0
	Coffees:											
20012	Brewed	1 c	237	99	9	<1	0	0	2	0	0	0
20592	Cappuccino w/lowfat milk	1½ c	244		110	8	11	0	4	2.5		
20639	Cappuccino w/whole milk	1½ c	244		140	7	11	0	7	4.5		
20668	Latte w/lowfat milk	1½ c	366		170	12	17	0	6	4		
20023	Prepared from instant	1 c	238	99	5	<1	1	0	0			
	Fruit drinks:											
20024	Fruit punch drink, canned	1 c	248	88	117	0	30	<1	0	0	0	0
20142	Gatorade	1 c	241	94	60	0	15	0	0	0	0	0
20026	Grape drink, canned	1 c	250	87	125	<1	32	0	0	0	0	0
20016	Koolade sweetened with sugar	1 c	262	90	97	0	25	0	<1	t	t	t
20017	Koolade sweetened with nutrasweet	1 c	240	95	43	0	11	0	0	0	0	0
20001	Lemonade, frzn concentrate (6-oz can)	¾ c	219	52	396	1	103	<1	<1	t	t	0.1
20000	Lemonade, from concentrate	1 c	248	86	131	<1	34	<1	<1	t	t	t
20003	Limeade, frzn concentrate (6-oz can)	¾ c	218	50	427	<1	106	<1	<1	0	0	0
20002	Limeade, from concentrate	1 c	247	89	104	<1	26	0	<1	0	0	0
20059	Pineapple grapefruit, canned	1 c	250	88	118	<1	29		<1	t	t	t
20025	Pineapple orange, canned	1 c	250	87	125	3	30	<1	0	0	0	0
20559	Powerade	1 c	247		72	0	19	0	0	0	0	0
20737	Snapple, fruit punch	1 c	252	88	110	0	29		0	0	0	0
20761	Snapple, tropical	1 c	252	89	110	0	27		0	0	0	0

PAGE KEY: A–4 = Beverages   A–6 = Dairy   A–10 = Eggs   A–12 = Fat/Oil   A–14 = Fruit   A–20 = Bakery
A–26 = Grain   A–32 = Fish   A–32 = Meats   A–36 = Poultry   A–38 = Sausage   A–40 = Mixed/Fast   A–44 = Nuts/Seeds   A–46 = Sweets
A–48 = Vegetables/Legumes   A–58 = Vegetarian   A–62 = Misc   A–62 = Soups/Sauces   A–64 = Fast   A–78 = Convenience   A–80 = Baby foods

Chol (mg)	Calc (mg)	Iron (mg)	Magn (mg)	Pota (mg)	Sodi (mg)	Zinc (mg)	VT-A (RAE)	Thia (mg)	VT-E (mg α)	Ribo (mg)	Niac (mg)	V-B$_6$ (mg)	Fola (µg)	VT-C (mg)
0	18	0.07	21	89	14	0.04	0	0.02	0	0.09	1.61	0.18	21	0
0	18	0.14	18	64	11	0.11	0	0.03	0	0.11	1.39	0.12	14	0
0	16	<.01	22	111	9	<.01		<.01		0.14	1.4	0.16		0
0					7									
0					6									
0	0	0.02	0	1	<1	0.02	0	<.01	0	<.01	<.01	<.01	0	0
0	0	0.02	0	1	<1	0.02	0	<.01	0	<.01	<.01	<.01	0	0
0	0	0.02	0	1	<1	0.02	0	<.01	0	<.01	<.01	<.01	0	0
0	1	0.03	2	16	4	0.02	0	<.01	0	<.01	0.07	0	0	0
27	8	0.06	1	15	43	0.08	81	<.01	0.21	0.03	0.04	<.01	1	0
0	0	0.04	0	0	2	0.02	0	0	0	0	<.01	0	0	0
0	0	0.02	0	4	2	0.01	0	<.01	0	<.01	0.02	0	0	0
0	9	0.28	11	109	11	0.08	0	0.02	0	0.02	0.25	0	0	0
0	9	0.51	15	132	6	0.11	0	<.01	0	0.03	0.1	0.04	2	0
0	9	0.45	12	117	6	0.07	0	<.01	0	0.02	0.09	0.03	1	0
0	11	0.38	12	94	6	0.08	0	<.01	0	<.01	0.08	0.02	0	0
0	21	0.93	23	204	16	0.19	0	0	0	0.02	0.23	0.05	2	0
0	13	0.62	12	102	19	0.13	<1	0.01	0.01	0.02	0.1	0.03	3	4
0	18	0.04	4	7	75	0.36	0	0	0	0	0	0	0	0
0	11	0.07	4	4	15	0.04	0	0	0	0	0	0	0	0
0	11	0.11	4	21	18	0	0	0.02	0	0.08	0	0	0	0
0	14	0.07	4	14	57	0.11	0	0	0	0	0	0	0	0
0	11	0.66	4	4	26	0.18	0	0	0	0	0	0	0	0
0	11	0.3	4	4	56	0.26	0	0	0	0	0	0	0	0
0	7	0.26	4	4	40	0.18	0	0	0	0	0.06	0	0	0
0	11	0.15	0	4	37	0.15	0	0	0	0	0	0	0	0
0	18	0.18	4	4	48	0.26	0	0	0	0	0	0	0	0
0				0	6	0	0		0		0	0	0	0
0				0	13	0								0
0				18	6	0		0						0
0				52		0		0		0				0
0	0	0		8	35	0					0			0
0	0	0			35	0								0
0				0	34	0								0
0	2	0.02	5	114	2	0.02	0	0	0.05	0.12	0	<.01	5	0
15	250	0			110									2
30	250	0			105									2
25	400	0			170									4
0	10	0.1	7	71	5	0.02	0	0	0	<.01	0.56	0	0	0
0	20	0.22	7	62	94	0.02	5	0.05	0.05	0.06	0.05	0.03	10	73
0	0	0.12	2	27	96	0.05	0	0.01	0	<.01	0	0	0	0
0	8	0.25	8	82	2	0.05	<1	0.02	0	0.03	0.17	0.04	2	40
0	42	0.13	3	3	37	0.08	0	0	0	<.01	<.01	0	0	31
0	17	0.65	5	50	50	0.26	1	0.02	0	0.05	0.05	0	5	78
0	15	1.58	11	147	9	0.18	1	0.06	0.09	0.21	0.16	0.05	11	39
0	10	0.52	5	50	7	0.07	<1	0.02	0.02	0.07	0.05	0.02	2	13
0	28	0.07	11	94	20	0.09	<1	0.02	0	0.03	0.08	0.04	7	25
0	7	0.02	2	22	5	0.02	<1	<.01	0	<.01	0.02	<.01	2	6
0	18	0.78	15	152	35	0.15	<1	0.08	0.02	0.04	0.67	0.1	22	115
0	12	0.68	15	115	8	0.15	2	0.08	0.08	0.05	0.52	0.12	22	56
0	0	0		32	28	0								0
0					10									0
0					10									

TABLE A-1

Food Composition

(Computer code number is for West Diet Analysis program) (For purposes of calculations, use "0" for t, <1, <.1, <.01, etc.)

Computer Code Number	Food Description	Measure	Wt (g)	H₂O (%)	Ener (kcal)	Prot (g)	Carb (g)	Dietary Fiber (g)	Fat (g)	Fat Breakdown (g) Sat	Mono	Poly
	BEVERAGES—Continued											
	Fruit and vegetable juices: see Fruit and Vegetable sections											
	Ultra Slim Fast, ready to drink, can:											
	Chocolate Royale	1 ea	350	83	220	10	40	5	3	1	1.5	0.5
	French Vanilla	1 ea	350	84	220	10	40	5	3	0.5	1.5	0.5
	Strawberries n' cream	1 ea	350	84	220	10	40	5	2	0.5	1.5	0.5
20041	Water, municiple	1 c	237	100	0	0	0	0	0	0	0	0
792	La Croix	1 c	236	100	0	0	0	0	0	0	0	0
20010	Tonic water	1½ c	366	91	124	0	32	0	0	0	0	0
	Tea:											
20014	Brewed, regular	1 c	237	100	2	0	1	0	0	0	0	0
20036	Brewed, herbal	1 c	237	100	2	0	<1	0	0			
20022	From instant, sweetened	1 c	259	91	88	<1	22	0	<1	t	t	t
20020	From instant, unsweetened	1 c	237	100	2	<1	<1	0	0	0	0	0
20443	Green tea bag	1 ea	2		0	0	0	0	0	0	0	0
	DAIRY											
	Butter: see Fats and Oils											
	Cheese:											
	Natural:											
1003	Blue	1 oz	28	42	99	6	1	0	8	5.2	2.2	0.2
1037	Brick	1 oz	28	41	104	7	1	0	8	5.3	2.4	0.2
1004	Brie	1 oz	28	48	94	6	<1	0	8	4.9	2.2	0.2
1006	Camembert	1 oz	28	52	84	6	<1	0	7	4.3	2	0.2
1007	Cheddar:	1 oz	28	37	113	7	<1	0	9	5.9	2.6	0.3
1008	Shredded	1 c	113	37	455	28	1	0	37	23.8	10.6	1.1
1088	Shredded, low fat, sodium	1 oz	28	65	48	7	1	0	2	1.2	0.6	t
1050	Edam	1 oz	28	42	100	7	<1	0	8	4.9	2.3	0.2
1016	Feta	1 oz	28	55	74	4	1	0	6	4.2	1.3	0.2
1054	Gouda	1 oz	28	41	100	7	1	0	8	4.9	2.2	0.2
1073	Gruyere	1 oz	28	33	116	8	<1	0	9	5.3	2.8	0.5
1538	Gorgonzola	1 oz	28	42	97	6	1	0	8	5		
1055	Limburger	1 oz	28	48	92	6	<1	0	8	4.7	2.4	0.1
1017	Monterey Jack	1 oz	28	41	104	7	<1	0	8	5.3	2.5	0.3
1056	Mozzarella whole milk	1 oz	28	50	84	6	1	0	6	3.7	1.8	0.2
1019	Part-skim, low moisture	1 oz	28	46	85	7	1	0	6	3.5	1.6	0.2
1021	Muenster	1 oz	28	42	103	7	<1	0	8	5.4	2.4	0.2
1060	Neufchatel	1 oz	28	62	73	3	1	0	7	4.1	1.9	0.2
1075	Parmesan, grated:	1 oz	28	21	121	11	1	0	8	4.8	2.3	0.3
57	Cup	1 c	100	21	431	38	4	0	29	17.3	8.4	1.2
1023	Provolone	1 oz	28	41	98	7	1	0	7	4.8	2.1	0.2
1064	Ricotta, whole milk	1 c	246	72	428	28	7	0	32	20.4	8.9	0.9
1024	Part-skim milk	1 c	246	74	339	28	13	0	19	12.1	5.7	0.6
1027	Swiss	1 oz	28	37	106	8	2	0	8	5	2	0.3
	Substitute:											
47947	Vegan cheese substitute, slice	1 ea	21		20	2	3		0	0	0	0
47946	Vegetarian cheese substitute topping	2 tsp	5		15	2	<1		<1	0		
	Cottage:											
1099	Low sodium, low fat	1 c	225	84	162	28	6	0	2	1.4	0.6	t
1013	Creamed, large curd	1 c	225	79	232	28	6	0	10	6.4	2.9	0.3
1012	Creamed, small curd	1 c	210	79	216	26	6	0	9	6	2.7	0.3
1049	With fruit	1 c	226	80	219	24	10	<1	9	5.2	2.3	0.3
1014	Low fat 2%	1 c	226	79	203	31	8	0	4	2.8	1.2	0.1
1047	Low fat 1%	1 c	226	82	163	28	6	0	2	1.5	0.7	t
	Cream:											
1015	Regular	1 tbs	15	54	52	1	<1	0	5	3.3	1.5	0.2
47949	Substitute	1 oz	28		49	3	3		3	0		
	Pasteurized processed:											
1456	American	1 oz	28	39	105	6	<1	0	9	5.5	2.5	0.3
1458	Swiss	1 oz	28	42	94	7	1	0	7	4.5	2	0.2
1437	American cheese food, jar	½ c	57	43	188	10	4	0	14	8.5	4.1	0.6

PAGE KEY: A–4 = Beverages   A–6 = Dairy   A–10 = Eggs   A–12 = Fat/Oil   A–14 = Fruit   A–20 = Bakery
A–26 = Grain   A–32 = Fish   A–32 = Meats   A–36 = Poultry   A–38 = Sausage   A–40 = Mixed/Fast   A–44 = Nuts/Seeds   A–46 = Sweets
A–48 = Vegetables/Legumes   A–58 = Vegetarian   A–62 = Misc   A–62 = Soups/Sauces   A–64 = Fast   A–78 = Convenience   A–80 = Baby foods

Chol (mg)	Calc (mg)	Iron (mg)	Magn (mg)	Pota (mg)	Sodi (mg)	Zinc (mg)	VT-A (RAE)	Thia (mg)	VT-E (mg α)	Ribo (mg)	Niac (mg)	V-B$_6$ (mg)	Fola (µg)	VT-C (mg)
5	400	2.7	140	600	220	2.25		0.52	13.64	0.6	7	0.7	120	60
5	400	2.7	140	600	220	2.25		0.52	13.64	0.6	7	0.7	120	60
5	400	2.7	140	600	220	2.25		0.52	13.64	0.6	7	0.7	120	60
0	5	0	2	0	5	0	0	0	0	0	0	0	0	0
0					5									
0	4	0.04	0	0	15	0.37	0	0	0	0	0	0	0	0
0	0	0.05	7	88	7	0.05	0	0	0	0.03	0	0	12	0
0	5	0.19	2	21	2	0.09	0	0.02	0	<.01	0	0	2	0
0	5	0.05	5	49	8	0.03	0	0	0	0.04	0.09	<.01	0	0
0	7	0.05	5	47	7	0.02	0	0	0	<.01	0.09	<.01	0	0
0	0	0			0		0							0
21	148	0.09	6	72	391	0.74	55	<.01	0.07	0.11	0.28	0.05	10	0
26	189	0.12	7	38	157	0.73	82	<.01	0.07	0.1	0.03	0.02	6	0
28	52	0.14	6	43	176	0.67	49	0.02	0.07	0.15	0.11	0.07	18	0
20	109	0.09	6	52	236	0.67	67	<.01	0.06	0.14	0.18	0.06	17	0
29	202	0.19	8	27	174	0.87	74	<.01	0.08	0.1	0.02	0.02	5	0
119	815	0.77	32	111	702	3.51	299	0.03	0.33	0.42	0.09	0.08	20	0
6	197	0.2	8	31	6	0.87		<.01	0.05	<.01	0.03	0.02	5	0
25	205	0.12	8	53	270	1.05	68	0.01	0.07	0.11	0.02	0.02	4	0
25	138	0.18	5	17	312	0.81	35	0.04	0.05	0.24	0.28	0.12	9	0
32	196	0.07	8	34	229	1.09	46	<.01	0.07	0.09	0.02	0.02	6	0
31	283	0.05	10	23	94	1.09	76	0.02	0.08	0.08	0.03	0.02	3	0
30	170	0.18			280									0
25	139	0.04	6	36	224	0.59	95	0.02	0.06	0.14	0.04	0.02	16	0
25	209	0.2	8	23	150	0.84	55	<.01	0.07	0.11	0.03	0.02	5	0
22	141	0.12	6	21	176	0.82	50	<.01	0.05	0.08	0.03	0.01	2	0
15	205	0.07	7	27	148	0.88	38	0.03	0.1	0.09	0.03	0.02	3	0
27	201	0.11	8	38	176	0.79	83	<.01	0.07	0.09	0.03	0.02	3	0
21	21	0.08	2	32	112	0.15	83	<.01	0.26	0.05	0.04	0.01	3	0
25	311	0.25	11	35	428	1.08	34	<.01	0.07	0.14	0.03	0.01	3	0
88	1109	0.9	38	125	1529	3.87	120	0.03	0.26	0.49	0.11	0.05	10	0
19	212	0.15	8	39	245	0.9	66	<.01	0.06	0.09	0.04	0.02	3	0
125	509	0.93	27	258	207	2.85	295	0.03	0.27	0.48	0.26	0.11	30	0
76	669	1.08	37	308	308	3.3	263	0.05	0.17	0.46	0.19	0.05	32	0
26	221	0.06	11	22	54	1.22	62	0.02	0.11	0.08	0.03	0.02	2	0
0	100			10	220									
0	60			50	80									
9	137	0.32	11	194	29	0.86	25	0.04	0.25	0.36	0.29	0.16	27	0
34	135	0.32	11	189	911	0.83	99	0.05	0.09	0.37	0.28	0.15	27	0
32	126	0.29	10	176	850	0.78	92	0.04	0.08	0.34	0.26	0.14	25	0
29	120	0.36	16	203	777	0.75	86	0.07	0.09	0.32	0.34	0.15	25	3
18	156	0.36	14	217	918	0.95	47	0.05	0.05	0.42	0.33	0.17	29	0
9	138	0.32	11	194	918	0.86	25	0.05	0.02	0.37	0.29	0.15	27	0
16	12	0.18	1	18	44	0.08	55	<.01	0.04	0.03	0.02	<.01	2	0
0	99			15	89									
26	155	0.05	8	47	417	0.8	71	<.01	0.08	0.1	0.02	0.02	2	0
24	216	0.17	8	60	384	1.01	55	<.01	0.1	0.08	0.01	0.01	2	0
46	325	0.32	18	166	721	1.82	115	0.04	0.13	0.29	0.1	0.04	4	0

TABLE A-1

Food Composition

(Computer code number is for West Diet Analysis program)   (For purposes of calculations, use "0" for t, <1, <.1, <.01, etc.)

Computer Code Number	Food Description	Measure	Wt (g)	H₂O (%)	Ener (kcal)	Prot (g)	Carb (g)	Dietary Fiber (g)	Fat (g)	Fat Breakdown (g) Sat	Mono	Poly
	DAIRY—Continued											
1002	American cheese spread	1 tbs	15	48	44	2	1	0	3	2	0.9	t
1081	Nonfat cheese (Kraft Singles)	1 oz	28	61	44	6	4	0	0	0	0	0
1094	Velveeta cheese spread, low fat, low sodium, slice	1 pce	34	62	61	8	1	0	2	1.5	0.7	t
	Cream:											
	Sweet:											
69	Half & half (cream & milk)	1 c	242	81	315	7	10	0	28	17.3	8	1
500	Tablespoon	1 tbs	15	81	20	<1	1	0	2	1.1	0.5	t
501	Light, coffee or table	1 tbs	15	74	29	<1	1	0	3	1.8	0.8	0.1
511	Light, whipping cream, liquid	1 tbs	15	64	44	<1	<1	0	5	2.9	1.4	0.1
502	Heavy whipping cream, liquid	1 tbs	15	58	52	<1	<1	0	6	3.5	1.6	0.2
510	Whipped cream, pressurized	1 tbs	4	61	10	<1	<1	0	1	0.6	0.3	t
	Sour, cultured:											
79	Regular	1 c	230	71	492	7	10	0	48	30	13.9	1.8
504	Tablespoon	1 tbs	14	71	30	<1	1	0	3	1.8	0.8	0.1
577	Fat free	1 tbs	15	79	12	1	2	0	0	0	0	0
	Imitation and part-dairy:											
507	Coffee whitener, frozen or liquid	1 tbs	15	77	20	<1	2	0	1	1.4	t	t
506	Coffee whitener, powdered	1 tsp	2	2	11	<1	1	0	1	0.7	t	t
508	Dessert topping, frozen, nondairy	1 tbs	5	50	16	<1	1	0	1	1.1	t	t
509	Dessert topping, mix with whole milk	1 tbs	5	67	9	<1	1	0	1	0.5	t	t
514	Dessert topping, pressurized	1 c	70	60	185	1	11	0	16	13.2	1.3	0.2
505	Sour cream, imitation	1 tbs	14	71	29	<1	1	0	3	2.5	t	t
516	Sour dressing, part dairy	1 tbs	15	75	27	<1	1	0	2	2	0.3	t
	Milk: Fluid											
1	Whole milk	1 c	244	88	146	8	11	0	8	4.6	2	0.5
2	2% lowfat milk	1 c	244	89	122	8	11		5	2.3	2	0.2
147	2% milk solids added	1 c	245	89	125	9	12	0	5	2.9	1.4	0.2
4	1% lowfat milk	1 c	244	90	102	8	12		2	1.5	0.7	t
148	1% milk solids added	1 c	245	90	105	9	12	0	2	1.5	0.7	t
6	Nonfat milk, vitamin A added	1 c	245	91	83	8	12		<1	0.1	t	t
149	Nonfat milk solids added	1 c	245	90	91	9	12	0	1	0.4	0.2	t
7	Buttermilk, skim	1 c	245	90	98	8	12	0	2	1.3	0.6	t
	Canned:											
11	Sweetened condensed	1 c	306	27	982	24	166	0	27	16.8	7.4	1
10	Evaporated, nonfat	1 c	256	79	200	19	29	0	1	0.3	0.2	t
	Dried:											
32	Buttermilk, sweet	1 c	120	3	464	41	59	0	7	4.3	2	0.3
9	Instant, nonfat, vit A added (makes 1 qt)	1 ea	91	4	326	32	47	0	1	0.4	0.2	t
8	Instant nonfat, vit A added, cup	1 c	68	4	243	24	35	0	<1	0.3	0.1	t
23	Goat milk	1 c	244	87	168	9	11	0	10	6.5	2.7	0.4
51	Kefir	1 c	233	88	149	8	11	0	8			
	Milk beverages and powdered mixes:											
	Chocolate:											
20	Whole	1 c	250	82	208	8	26	2	8	5.3	2.5	0.3
18	2% fat	1 c	250	84	180	8	26	1	5	3.1	1.5	0.2
19	1% fat	1 c	250	84	158	8	26	1	2	1.5	0.8	t
	Chocolate-flavored beverages:											
12	Powder containing nonfat dry milk:	1 oz	28	2	111	2	24	1	1	0.7	0.4	t
48	Prepared with water	1 c	275	86	151	2	32	1	2	0.9	0.5	t
14	Powder without nonfat dry milk:	1 oz	28	1	98	1	25	1	1	0.5	0.3	t
39	Prepared with whole milk	1 c	266	81	226	9	32	1	9	4.9	2.2	0.5
	Eggnog:											
17	Commercial	1 c	254	74	343	10	34	0	19	11.3	5.7	0.9
98	2% low-fat	1 c	254	85	191	12	17	0	8	3.7	2.7	0.7
	Instant breakfast:											
24	Envelope, pwd only	1 ea	37	7	131	7	24		1	0.2	0.1	0.1
25	Prepared with whole milk	1 c	281	78	253	15	36	<1	5	3.1		
26	Prepared with 2% milk	1 c	281	77	280	15	36	<1	9	5.3		

PAGE KEY: A–4 = Beverages   A–6 = Dairy   A–10 = Eggs   A–12 = Fat/Oil   A–14 = Fruit   A–20 = Bakery
A–26 = Grain   A–32 = Fish   A–32 = Meats   A–36 = Poultry   A–38 = Sausage   A–40 = Mixed/Fast   A–44 = Nuts/Seeds   A–46 = Sweets
A–48 = Vegetables/Legumes   A–58 = Vegetarian   A–62 = Misc   A–62 = Soups/Sauces   A–64 = Fast   A–78 = Convenience   A–80 = Baby foods

Chol (mg)	Calc (mg)	Iron (mg)	Magn (mg)	Pota (mg)	Sodi (mg)	Zinc (mg)	VT-A (RAE)	Thia (mg)	VT-E (mg α)	Ribo (mg)	Niac (mg)	V-B$_6$ (mg)	Fola (µg)	VT-C (mg)
8	84	0.05	4	36	202	0.39	26	<.01	0.03	0.06	0.02	0.02	1	0
7	221	0		88	398	0.88				0.15				0
12	233	0.15	8	61	2	1.13		0.01	0.17	0.13	0.03	0.03	3	0
90	254	0.17	24	315	99	1.23	235	0.08	0.8	0.36	0.19	0.09	7	2
6	16	0.01	2	20	6	0.08	15	<.01	0.05	0.02	0.01	<.01	<1	0
10	14	<.01	1	18	6	0.04	27	<.01	0.08	0.02	<.01	<.01	<1	0
17	10	<.01	1	15	5	0.04	42	<.01	0.13	0.02	<.01	<.01	1	0
21	10	<.01	1	11	6	0.03	62	<.01	0.16	0.02	<.01	<.01	1	0
3	4	<.01	0	6	5	0.01	8	<.01	0.03	<.01	<.01	<.01	<1	0
101	267	0.14	25	331	122	0.62	407	0.08	1.38	0.34	0.15	0.04	25	2
6	16	<.01	2	20	7	0.04	25	<.01	0.08	0.02	<.01	<.01	2	0
2	28	0			23									0
0	1	<.01	0	29	12	<.01	1	0	0.24	0	0	0	0	0
0	0	0.02	0	16	4	0.01	<1	0	0.01	<.01	0	0	0	0
0	0	<.01	0	1	1	<.01	<1	0	0.05	0	0	0	0	0
0	4	<.01	0	8	3	0.01	1	<.01	0.02	<.01	<.01	<.01	<1	0
0	4	0.01	1	13	43	<.01	3	0	0.6	0	0	0	0	0
0	0	0.05	1	23	14	0.17	0	0	0.1	0	0	0	0	0
1	17	<.01	2	24	7	0.06	<1	<.01	0.2	0.02	0.01	<.01	2	0
24	246	0.07	24	325	105	0.93	68	0.11	0.15	0.45	0.26	0.09	12	0
20	271	0.24	27	342	115	1.17	134	0.1	0.07	0.45	0.22	0.09	12	0
20	314	0.12	34	397	127	0.98	137	0.1	0.1	0.42	0.22	0.11	12	2
12	264	0.85	27	290	122	2.12	142	0.05	0.02	0.45	0.23	0.09	12	0
10	314	0.12	34	397	127	0.98	145	0.1	0.1	0.42	0.22	0.11	12	2
5	223	1.23	22	238	108	2.08	149	0.11	0.02	0.45	0.23	0.09	12	0
5	316	0.12	37	419	130	1	149	0.1	0	0.43	0.22	0.11	12	2
10	284	0.12	27	370	257	1.03	17	0.08	0.12	0.38	0.14	0.08	12	2
104	869	0.58	80	1135	389	2.88	226	0.28	0.49	1.27	0.64	0.16	34	8
10	742	0.74	69	850	294	2.3	302	0.12	0	0.79	0.45	0.14	23	3
83	1421	0.36	132	1910	620	4.82	59	0.47	0.12	1.89	1.05	0.41	56	7
16	1120	0.28	106	1552	500	4.01	645	0.38	<.01	1.59	0.81	0.31	46	5
12	837	0.21	80	1159	373	3	482	0.28	<.01	1.19	0.61	0.23	34	4
27	327	0.12	34	498	122	0.73	139	0.12	0.17	0.34	0.68	0.11	2	3
		0.3	33	373	107									
30	280	0.6	32	418	150	1.02	65	0.09	0.15	0.4	0.31	0.1	12	2
18	285	0.6	32	422	150	1.02	138	0.09	0.1	0.41	0.32	0.1	12	2
8	288	0.6	32	425	152	1.02	145	0.1	0.05	0.42	0.32	0.1	12	2
2	39	0.33	23	199	141	0.41	<1	0.03	0.15	0.16	0.16	0.03	0	1
3	60	0.47	33	270	195	0.58	<1	0.04	0.19	0.21	0.22	0.04	0	1
0	10	0.88	27	165	59	0.43	<1	<.01	0.01	0.04	0.14	<.01	2	0
24	253	0.8	48	458	154	1.28	69	0.11	0.16	0.48	0.38	0.09	13	0
150	330	0.51	48	419	137	1.17	114	0.09	0.51	0.48	0.27	0.13	3	4
194	270	0.71	32	369	155	1.26		0.11	0.58	0.55	0.21	0.15	30	2
4	105	4.74	84	350	142	3.16		0.31	5.31	0.07	5.27	0.42	105	28
24	403	4.87	119	726	264	4.12		0.41	5.48	0.48	5.48	0.53	118	31
38	396	4.87	117	721	262	4.09	0	0.41	5.55	0.47	5.47	0.52	118	31

A

TABLE A-1

Food Composition (Computer code number is for West Diet Analysis program)   (For purposes of calculations, use "0" for t, <1, <.1, <.01, etc.)

Computer Code Number	Food Description	Measure	Wt (g)	H₂O (%)	Ener (kcal)	Prot (g)	Carb (g)	Dietary Fiber (g)	Fat (g)	Fat Breakdown (g) Sat	Mono	Poly
	DAIRY—Continued											
101	Prepared with 1% milk	1 c	281	79	233	15	36	<1	3	1.8		
27	Prepared with nonfat milk	1 c	282	80	216	16	36		1	0.7		
	Malted Milk:											
30	Chocolate Powder	3 tsp	21	1	79	1	18	1	1	0.5	0.2	t
34	Prepared with whole milk	1 c	265	81	225	9	30	1	9	5	2.2	0.6
38	Ovaltine with whole milk	1 c	265	81	223	9	29	1	9	5	2.2	0.5
28	Natural Powder:	3 tsp	21	2	87	2	16	<1	2	0.9	0.4	0.3
29	Prepared with whole milk	1 c	265	81	233	10	27	<1	10	5.4	2.4	0.7
	Milk Shakes:											
2020	Chocolate	1 c	166	72	211	6	34	3	6	3.8	1.8	0.2
2024	Vanilla	1 c	166	75	184	6	30	<1	5	3.1	1.4	0.2
	Milk Desserts:											
2062	Low-fat frozen dessert bars	1 ea	81	72	88	2	19	0	1	0.2	0.1	0.4
	Ice cream, vanilla (about 10% fat):											
2004	Hardened	1 c	132	61	265	5	31	1	15	9	3.9	0.6
2008	Soft serve	1 c	172	60	382	7	38	1	22	12.9	6	0.8
	Ice cream, rich vanilla (16% fat):											
2006	Hardened	1 c	148	57	369	5	33	0	24	15.3	6.6	1
2220	Ben & Jerry's	½ c	108		250	4	22	0	16	11		
	Ice milk, vanilla (about 4% fat):											
2009	Hardened	1 c	132	63	218	7	35	<1	5	3.4	1.1	0.2
2010	Soft serve (about 3.3% fat)	1 c	176	70	222	9	38	0	5	2.9	1.3	0.2
	Pudding, canned (5 oz can = .55 cup):											
2610	Chocolate	1 ea	142	69	197	4	33	1	6	1	2.4	2
2611	Tapioca	1 ea	142	74	169	3	28	<1	5	0.9	2.2	1.9
2612	Vanilla	1 ea	142	71	183	3	31	0	5	0.8	2.2	1.9
	Puddings, dry mix with whole milk:											
2604	Chocolate, regular, cooked	1 c	284	73	338	9	55	2	9	5.1	2.3	0.5
2606	Rice, cooked	1 c	288	72	348	9	60	<1	8	4.8	2.2	0.3
2607	Tapioca, cooked	1 c	282	74	324	8	55	0	8	4.9	2.2	0.3
2609	Vanilla, regular, cooked	1 c	280	75	314	8	52	<1	8	4.6	2	0.6
2011	Sherbet (2% fat)	1 c	198	66	285	2	60	7	4	2.3	1	0.2
	Frozen Yogurt, Low-Fat:											
2064	Cup	1 c	144	65	235	6	35	0	8	4.9	2.3	0.3
2082	Scoop	1 ea	79	74	78	4	16	0	<1	t	t	t
	Milk Substitutes:											
20440	Rice Milk	1 c	245	89	120	<1	25	0	2	0.2	1.3	0.3
20590	Rice/Soy Milk, blend	1 c	241	88	120	7	18	0	3	0.5		
20033	Soy Milk	1 c	245	89	120	9	11	3	5	0.5	0.8	2
7785	Edensoy	1 c	244	89	130	10	13	0	4	0.5	0.8	2.7
7804	Vanilla	1 c	244	87	150	6	23	0	3	0		
20693	Soy Dream	1 c	244		128	7	17	0	4	0.5		
483	Enriched	1 c	244		128	7	17	0	4	0.5		
20404	Veggie original	1 c	227		110	9	13		3	0		
20405	Chocolate	1 c	227		150	9	26		2	0		
	Yogurt:											
	Fat Free:											
2852	Strawberry, container	1 ea	227	86	120	8	22	0	0	0	0	0
2574	Vanilla	1 c	245	76	223	13	43	0	<1	0.3	0.1	t
2012	Plain	1 c	245	85	137	14	19	0	<1	0.3	0.1	t
	Low-Fat:											
2034	Fruit added with low-calorie sweetener	1 c	241	86	122	11	19	1	<1	0.2	t	t
2001	Fruit added	1 c	245	74	250	11	47	0	3	1.7	0.7	t
2000	Plain	1 c	245	85	154	13	17	0	4	2.4	1	0.1
2015	Vanilla or coffee flavor	1 c	245	79	208	12	34	0	3	2	0.8	t
2013	Whole Milk:	1 c	245	88	149	9	11	0	8	5.1	2.2	0.2
	EGGS											
	Raw, Large:											
19501	Whole, without shell	1 ea	50	76	74	6	<1	0	5	1.5	1.9	0.7
19506	White	1 ea	33	88	17	4	<1	0	<1	0	0	0

PAGE KEY: A–4 = Beverages   A–6 = Dairy   A–10 = Eggs   A–12 = Fat/Oil   A–14 = Fruit   A–20 = Bakery
A–26 = Grain   A–32 = Fish   A–32 = Meats   A–36 = Poultry   A–38 = Sausage   A–40 = Mixed/Fast   A–44 = Nuts/Seeds   A–46 = Sweets
A–48 = Vegetables/Legumes   A–58 = Vegetarian   A–62 = Misc   A–62 = Soups/Sauces   A–64 = Fast   A–78 = Convenience   A–80 = Baby foods

Chol (mg)	Calc (mg)	Iron (mg)	Magn (mg)	Pota (mg)	Sodi (mg)	Zinc (mg)	VT-A (RAE)	Thia (mg)	VT-E (mg α)	Ribo (mg)	Niac (mg)	V-B₆ (mg)	Fola (µg)	VT-C (mg)
14	406	4.87	119	731	267	4.12		0.41	5.4	0.48	5.48	0.53	118	31
9	407	4.83	112	755	268	4.14		0.4	5.3	0.42	5.47	0.52	118	31
0	13	0.48	15	130	53	0.17	1	0.04	0.02	0.04	0.42	0.03	11	0
26	260	0.56	40	456	159	1.09	69	0.14	0.16	0.49	0.69	0.12	24	0
26	339	3.76	45	578	231	1.17	904	0.76	0.16	1.32	11.08	1.01	19	32
7	63	0.15	20	159	104	0.21	17	0.11	0.07	0.19	1.1	0.09	10	1
32	310	0.24	45	485	209	1.14	87	0.21	0.32	0.64	1.38	0.17	21	1
22	188	0.51	28	332	161	0.68	43	0.1	0.18	0.41	0.27	0.08	8	1
18	203	0.15	20	289	136	0.6	61	0.07	0.1	0.3	0.31	0.09	8	1
1	81	0.04	9	107	44	0.26	38	0.03	0.07	0.11	0.06	0.03	3	1
58	169	0.12	18	263	106	0.91	156	0.05	0.4	0.32	0.15	0.06	7	1
157	225	0.36	21	304	105	0.89	279	0.08	1.05	0.31	0.16	0.08	15	1
136	173	0.5	16	232	90	0.7	269	0.06	0.75	0.25	0.12	0.07	12	0
75	100	0.36			60									0
33	153	0.11	18	275	98	0.96	182	0.05	0.12	0.23	0.13	0.04	7	0
21	276	0.11	25	389	123	0.93	51	0.09	0.11	0.35	0.21	0.08	11	2
4	128	0.72	30	256	183	0.6	14	0.04	0.41	0.22	0.49	0.04	4	3
1	119	0.33	11	136	226	0.38	<1	0.03	0.43	0.14	0.44	0.03	4	1
10	125	0.18	11	160	192	0.36	9	0.03	0	0.2	0.36	0.02	0	0
26	273	0.99	57	426	278	1.36	68	0.1	0.14	0.55	0.42	0.09	11	0
32	297	1.07	37	369	311	1.09	84	0.21	0.24	0.4	1.27	0.1	12	2
34	290	0.17	34	369	338	0.96	70	0.08	0.23	0.39	0.21	0.09	11	2
25	249	0.11	25	333	437	0.95	70	0.1	0.14	0.45	0.26	0.08	11	0
0	107	0.28	16	190	91	0.95	20	0.06	0.06	0.18	0.15	0.05	14	11
3	206	0.43	20	304	125	0.6	85	0.05	0.16	0.32	0.41	0.12	9	1
1	137	0.07	13	175	53	0.67	1	0.03	<.01	0.16	0.09	0.04	8	1
0	20	0.2	10	69	86	0.24	<1	0.08	1.76	0.01	1.91	0.04	91	1
0	13	1.08	40	270	85	0.9	0	0.09		0.1	0.4	0.08	26	
0	10	1.42	47	345	29	0.56	5	0.39	0.02	0.17	0.36	0.1	5	0
0	80	1.44	60	440	105	0.9	0	0.15		0.07	0.8	0.16	40	0
0	60	0.72	40	290	90	0.6	0	0.12		0.07	1.2	0.12	40	0
0	40	1.78	40	237	138	0.59	0	0.15		0.07	0.79	0.12	59	0
0	295	1.77	39	236	138	0.59		0.15	4.95	0.07	0.79	0.12	59	0
0	400			350	90									
0	401			351	130									
5	350	0		380	160		0							5
4	436	0.21	42	559	168	2.13	4	0.11	<.01	0.52	0.27	0.12	27	2
5	488	0.22	47	625	189	2.38	5	0.12	0	0.57	0.3	0.13	29	2
3	370	0.61	41	550	139	1.83		0.1	0.17	0.45	0.5	0.11	32	26
10	372	0.17	37	478	142	1.81	24	0.09	0.05	0.44	0.23	0.1	22	2
15	448	0.2	42	573	172	2.18	34	0.11	0.07	0.52	0.28	0.12	27	2
12	419	0.17	39	537	162	2.03	29	0.1	0.05	0.49	0.26	0.11	27	2
32	296	0.12	29	380	113	1.45	66	0.07	0.15	0.35	0.18	0.08	17	1
212	26	0.92	6	67	70	0.56	70	0.03	0.48	0.24	0.04	0.07	24	0
0	2	0.03	4	54	55	<.01	0	<.01	0	0.14	0.03	<.01	1	0

A

TABLE A-1

Food Composition   (Computer code number is for West Diet Analysis program)   (For purposes of calculations, use "0" for t, <1, <.1, <.01, etc.)

Computer Code Number	Food Description	Measure	Wt (g)	H₂O (%)	Ener (kcal)	Prot (g)	Carb (g)	Dietary Fiber (g)	Fat (g)	Fat Breakdown (g) Sat	Mono	Poly
	EGGS—Continued											
19508	Yolk	1 ea	17	52	55	3	1	0	5	1.6	2	0.7
	Cooked:											
19509	Fried in margarine	1 ea	46	69	92	6	<1	0	7	2	2.9	1.2
19510	Hard-cooked, shell removed	1 ea	50	75	78	6	1	0	5	1.6	2	0.7
19511	Hard-cooked, chopped	1 c	136	75	211	17	2	0	14	4.4	5.5	1.9
19517	Poached, no added salt	1 ea	50	76	74	6	<1	0	5	1.5	1.9	0.7
19516	Scrambled with milk & margarine	1 ea	61	73	101	7	1	0	7	2.2	2.9	1.3
	Substitute, liquid:											
19525	Egg substitute, liquid:	½ c	126	83	106	15	1	0	4	0.8	1.1	2
19581	Egg Beaters, Fleischmann's	½ c	122		60	12	2	0	0	0	0	0
19552	Egg substitute, liquid, prepared	½ c	105	80	107	12	2	0	6	1.1	2.1	2.1
	FATS AND OILS											
	Butter:											
8000	Tablespoon:	1 tbs	14	16	100	<1	<1	0	11	5.7	4.7	0.4
8025	Unsalted	1 tbs	14	18	100	<1	<1	0	11	7.2	2.9	0.4
8001	Pat (about 1 tsp)	1 ea	5	16	36	<1	<1	0	4	2	1.7	0.1
8142	Whipped	1 tsp	3	16	22	<1	<1	0	2	1.5	0.7	t
	Fats, cooking:											
8004	Beef fat/tallow	1 c	205	0	1849	0	0	0	205	102.1	85.7	8.2
8005	Chicken fat	1 c	205	0	1845	0	0	0	205	61.1	91.6	42.8
8007	Vegetable shortening	1 tbs	13	0	115	0	0	0	13	3.2	5.8	3.4
	Margarine:											
8041	Imitation (about 40% fat), soft	1 tbs	14	58	48	<1	<1	0	5	1.1	2.2	1.9
90234	Regular, hard (about 80% fat)	1 tbs	14	16	101	<1	<1	0	11	2.2	5	3.6
8043	Regular, soft (about 80% fat)	1 tbs	14	16	100	<1	<1	0	11	1.9	4	4.8
8485	Saffola, unsalted	1 tbs	14	20	100	0	0	0	11	2	3	4.5
8486	Saffola, reduced fat	1 tbs	14	37	60	0	0	0	8	1.3	2.7	4.4
	Spread:											
8044	Hard (about 60% fat)	1 tbs	14	37	76	<1	0	0	9	2	3.6	2.5
8045	Soft (about 60% fat)	1 tbs	14	37	76	<1	0	0	9	1.8	4.4	1.9
8602	Touch of Butter (47% fat)	1 tbs	14		60	0	0	0	7	1.5		
	Oils:											
8084	Canola	1 tbs	14	0	124	0	0	0	14	1	8.2	4.1
8009	Corn	1 tbs	14	0	124	0	0	0	14	1.8	3.4	8.2
8008	Olive	1 tbs	14	0	124	0	0	0	14	1.9	10.3	1.4
8361	Olive, extra virgin	1 tbs	14	0	126	0	0	0	14	2	10.8	1.3
8026	Peanut	1 tbs	14	0	124	0	0	0	14	2.4	6.5	4.5
8010	Safflower	1 tbs	14	0	124	0	0	0	14	0.9	2	10.4
8012	Soybean	1 tbs	14	0	124	0	0	0	14	2	3.3	8.1
8028	Soybean/cottonseed	1 tbs	14	0	124	0	0	0	14	2.5	4.1	6.7
8011	Sunflower	1 tbs	14	0	124	0	0	0	14	1.4	2.7	9.2
	Salad Dressings/Sandwich Spreads:											
	Store Brand:											
8013	Blue Cheese	1 tbs	15	32	76	1	1	0	8	1.5	1.8	4.2
	French:											
8015	Regular	1 tbs	16	37	73	<1	2	0	7	0.9	1.3	3.4
8014	Low calorie	1 tbs	16	54	37	<1	5	<1	2	0.2	0.9	0.8
	Italian:											
8020	Regular	1 tbs	15	56	44	<1	2	0	4	0.7	0.9	1.9
8016	Low calorie	1 tbs	15	85	11	<1	1	0	1	t	0.3	0.3
	Ranch:											
8428	Regular	1 tbs	15		80	0	<1	0	8	1.2		
8465	Low calorie	1 tbs	14		22	<1	1	<1	2	0.3		
8022	Russian	1 tbs	15	34	74	<1	2	0	8	1.1	1.8	4.4
8021	Mayo Type	1 tbs	15	40	58	<1	4	0	5	0.7	1.4	2.7
	Mayonnaise:											
8032	Imitation, low calorie	1 tbs	15	63	35	<1	2	0	3	0.5	0.7	1.6
8046	Regular (soybean)	1 tbs	14	15	100	<1	1	0	11	1.7	2.7	6
8148	Regular, low calorie, low sodium	1 tbs	14	63	32	<1	2	0	3	0.5	0.6	1.5
8141	Salad dressing, low calorie, oil free	1 tbs	15	88	4	<1	1	<1	<1	t	0	t

PAGE KEY: A–4 = Beverages   A–6 = Dairy   A–10 = Eggs   A–12 = Fat/Oil   A–14 = Fruit   A–20 = Bakery
A–26 = Grain   A–32 = Fish   A–32 = Meats   A–36 = Poultry   A–38 = Sausage   A–40 = Mixed/Fast   A–44 = Nuts/Seeds   A–46 = Sweets
A–48 = Vegetables/Legumes   A–58 = Vegetarian   A–62 = Misc   A–62 = Soups/Sauces   A–64 = Fast   A–78 = Convenience   A–80 = Baby foods

Chol (mg)	Calc (mg)	Iron (mg)	Magn (mg)	Pota (mg)	Sodi (mg)	Zinc (mg)	VT-A (RAE)	Thia (mg)	VT-E (mg α)	Ribo (mg)	Niac (mg)	V-B$_6$ (mg)	Fola (µg)	VT-C (mg)
210	22	0.46	1	19	8	0.39	65	0.03	0.44	0.09	<.01	0.06	25	0
210	27	0.91	6	68	94	0.55	91	0.03	0.56	0.24	0.04	0.07	23	0
212	25	0.6	5	63	62	0.52	84	0.03	0.52	0.26	0.03	0.06	22	0
577	68	1.62	14	171	169	1.43	230	0.09	1.4	0.7	0.09	0.16	60	0
211	26	0.92	6	66	147	0.55	70	0.03	0.48	0.24	0.04	0.07	24	0
215	43	0.73	7	84	171	0.61	87	0.03	0.52	0.27	0.05	0.07	18	0
1	67	2.65	11	416	223	1.64	23	0.14	0.34	0.38	0.14	<.01	19	0
0	40	2.16		170	250	1.2			1.61	1.7		0.16	64	0
1	83	1.85	11	337	201	1.25		0.09	0.83	0.29	0.12	0.01	11	0
30	3	<.01	0	3	81	0.01	96	<.01	0.32	<.01	<.01	<.01	<1	0
30	3	<.01	0	3	2	0.01	96	<.01	0.32	<.01	<.01	<.01	<1	0
11	1	<.01	0	1	29	<.01	34	<.01	0.12	<.01	<.01	<.01	<1	0
7	1	<.01	0	1	25	<.01	21	<.01	0.07	<.01	<.01	<.01	<1	0
223	0	0	0	0	0	0	0	0	5.54	0	0	0	0	0
174	0	0	0	0	0	0	0	0	5.54	0	0	0	0	0
0	0	0	0	0	0	0	0	0	0.1	0	0	0	0	0
0	3	0	0	4	134	0	115	<.01	0.33	<.01	<.01	<.01	<1	0
0	4	<.01	0	6	132	0	115	<.01	1.26	<.01	<.01	<.01	<1	0
0	4	0	0	5	151	0	102	<.01	0.98	<.01	<.01	<.01	<1	0
0	0			0									0	
0	0			115									0	
0	3	0	0	4	139	0	102	<.01	0.7	<.01	<.01	<.01	<1	0
0	3	0	0	4	139	0	102	<.01	0.7	<.01	<.01	<.01	<1	0
0	0	0		0	110				1.27					0
0	0	0	0	0	0	0	0	0	2.39	0	0	0	0	0
0	0	0	0	0	0	0	0	0	2	0	0	0	0	0
0	0	0.09	0	0	<1	0	0	0	2.01	0	0	0	0	0
									1.74					
0	0	<.01	0	0	0	<.01	0	0	2.2	0	0	0	0	0
0	0	0	0	0	0	0	0	0	4.77	0	0	0	0	0
0	0	<.01	0	0	0	0	0	0	1.29	0	0	0	0	0
0	0	0	0	0	0	0	0	0	1.69	0	0	0	0	0
0	0	0	0	0	0	0	0	0	5.75	0	0	0	0	0
3	12	0.03	0	6	164	0.04	10	<.01	0.9	0.02	0.02	<.01	4	0
0	4	0.13	1	11	134	0.05	4	<.01	0.8	<.01	0.03	0	0	0
0	2	0.14	1	17	129	0.03	4	<.01	0.05	<.01	0.07	<.01	<1	0
0	1	0.09	0	7	248	0.02	<1	<.01	0.75	<.01	0	<.01	0	0
1	1	0.1	1	13	205	0.03	<1	0	0.03	<.01	0	0.01	0	0
5	0	0			105		0							0
3	12	0		16	120		0		0.7					0
3	3	0.09	0	24	130	0.06	2	<.01	0.6	<.01	0.09	<.01	2	1
4	2	0.03	0	1	107	0.03	10	<.01	0.31	<.01	<.01	<.01	1	0
4	0	0	0	2	75	0.02	0	0	0.3	0	0	0	0	0
5	3	0.07	0	5	80	0.02	12	0	0.73	0	<.01	0.08	1	0
3	0	0	0	1	15	0.02	1	0	0.53	<.01	0	0	<1	0
0	1	0.05	2	7	256	<.01	<1	<.01	<.01	<.01	<.01	<.01	<1	0

Table of Food Composition

A

Appendix

TABLE A–1

Food Composition   (Computer code number is for West Diet Analysis program)   (For purposes of calculations, use "0" for t, <1, <.1, <.01, etc.)

Computer Code Number	Food Description	Measure	Wt (g)	H$_2$O (%)	Ener (kcal)	Prot (g)	Carb (g)	Dietary Fiber (g)	Fat (g)	Fat Breakdown (g) Sat	Mono	Poly
	FATS AND OILS—Continued											
8034	Salad dressing, from recipe, cooked	1 tbs	16	69	25	1	2	0	2	0.5	0.6	0.3
53122	Tartar Sauce, low calorie	1 tbs	14	63	31	<1	2	<1	3	0.4	0.6	1.4
	Thousand Island	1 tbs	16									
8024	Regular	1 tbs	16	47	59	<1	2		6	0.8	1.3	2.9
8023	Low Calorie	1 tbs	15	61	31	<1	3	<1	2	0.1	1	0.4
8035	Vinegar and oil	1 tbs	16	47	72	0	<1	0	8	1.5	2.4	3.9
	Kraft, Deliciously Right:											
8563	1000 Island	1 tbs	16		34	0	3	0	2	0.2		
8574	Cucumber ranch	1 tbs	16		31	0	1	0	3	0.5		
	Wishbone:											
8442	Creamy Italian, lite	1 tbs	15	72	26	<1	2		2	0.4	0.9	0.7
8413	Italian, lite	1 tbs	16		6	0	1		<1	0	0.2	0.1
8427	Ranch, lite	1 tbs	15		50	0	2	0	4	0.8		
	FRUITS and FRUIT JUICES											
	Apples:											
	Fresh, w/peel:											
3000	2$^3/_4$" diam (about 3/lb w/cores)	1 ea	138	86	72	<1	19	3	<1	t	t	t
3001	3$^1/_4$" diam (about 2/lb w/cores)	1 ea	212	86	110	1	29	5	<1	t	t	0.1
3004	Slices	1 c	110	87	53	<1	14	1	<1	t	t	t
3005	Dried, sulfured	10 ea	64	32	156	1	42	6	<1	t	t	t
3576	Cup	$^1/_4$ c	40	30	107	<1	26	3	<1	0		
3008	Juice, bottled or canned	1 c	248	88	117	<1	29	<1	<1	t	t	t
3147	Applesauce, sweetened	1 c	255	80	194	<1	51	3	<1	t	t	0.1
3006	Applesauce, unsweetened	1 c	244	88	105	<1	28	3	<1	t	t	t
	Apricots:											
3157	Fresh, w/o pits (about 12 per lb w/pits)	3 ea	105	86	50	1	12		<1	t	0.2	t
	Canned (fruit and liquid):											
3011	Heavy syrup	1 c	240	78	199	1	52	4	<1	t	t	t
3633	Halves	3 ea	120	78	100	1	26	2	<1	t	t	t
3151	Juice pack	1 c	244	87	117	2	30	4	<1	t	t	t
3152	Halves	3 ea	108	87	52	1	13	2	<1	t	t	t
3013	Dried, halves	10 ea	35	31	84	1	22	3	<1	t	t	t
3580	Cup	$^1/_4$ c	40	30	107	1	25	2	<1	0		
3217	Dried, cooked, unsweetened, w/liquid	1 c	250	76	212	3	55		<1	t	t	t
3015	Nectar, canned	1 c	251	85	141	1	36	2	<1	t	t	t
	Avocados, raw, edible part only:											
3210	California	1 ea	173	72	289	3	15	12	27	3.7	17	3.5
3212	Florida	1 ea	304	79	365	7	24		31	6	16.8	5.1
3017	Mashed, fresh, average	1 c	230	73	368	5	20	15	34	4.9	22.5	4.2
	Bananas:											
	Fresh, w/o peel:											
3020	Whole, 8$^3/_4$ long (175g w/peel)	1 ea	118	75	105	1	27	3	<1	0.1	t	t
3021	Slices	1 c	150	75	134	2	34	4	<1	0.2	t	0.1
3023	Dehydrated slices	$^1/_2$ c	50	3	173	2	44	5	1	0.3	t	0.2
	Berries:											
3024	Blackberries, raw	1 c	144	88	62	2	14		1	t	t	0.4
	Blueberries:											
3029	Fresh	1 c	145	84	83	1	21	3	<1	t	t	0.2
3232	Frozen, sweetened	10 oz	284	77	230	1	62	6	<1	t	t	0.2
3231	Frozen, thawed	1 c	230	77	186	1	50	5	<1	t	t	0.1
	Cranberries:											
3042	Juice cocktail, vitamin C added	1 c	253	86	144	0	36	<1	<1	t	t	0.1
3276	Juice, low calorie	1 c	237	95	45	<1	11	0	<1	0	0	0
3223	Cranberry-apple juice, vitamin C added	1 c	245	82	174	<1	44	<1	<1	0	0	0
3040	Sauce, canned, strained	1 c	277	61	418	1	108	3	<1	t	t	0.2
3865	Dried	$^1/_3$ c	40		120	0	33	2	0	0	0	0
	Raspberries:											
3131	Fresh	1 c	123	86	64	1	15	8	1	t	t	0.5

Chol (mg)	Calc (mg)	Iron (mg)	Magn (mg)	Pota (mg)	Sodi (mg)	Zinc (mg)	VT-A (RAE)	Thia (mg)	VT-E (mg α)	Ribo (mg)	Niac (mg)	V-B$_6$ (mg)	Fola (µg)	VT-C (mg)
9	13	0.08	1	19	117	0.06	8	<.01	0.13	0.02	0.04	<.01	3	0
3	1	0.06	0	4	82	0.02	1	<.01	0.84	<.01	<.01	<.01	<1	0
4	3	0.19	1	17	138	0.04	2	0.23	0.18	<.01	0.07	0	0	0
0	2	0.14	1	30	125	0.03	2	<.01	0.15	<.01	0.07	0	0	0
0	0	0	0	1	<1	0	0	0	0.74	0	0	0	0	0
5	0	0		29	165	0								0
0	0	0		10	248	0								0
0	0	0			148	0		0	0.56	0	0			0
0	1	0			255			0	0.24	0	0			0
2	0	0			120	0								0
0	8	0.17	7	148	1	0.06	4	0.02	0.25	0.04	0.13	0.06	4	6
0	13	0.25	11	227	2	0.08	6	0.04	0.38	0.06	0.19	0.09	6	10
0	6	0.08	4	99	0	0.06	2	0.02	0.06	0.03	0.1	0.04	0	4
0	9	0.9	10	288	56	0.13	0	0	0.34	0.1	0.59	0.08	0	2
0	9	0.67		178	251	<1								1
0	17	0.92	7	295	7	0.07	<1	0.05	0.02	0.04	0.25	0.07	0	2
0	10	0.89	8	156	8	0.1	3	0.03	0.54	0.07	0.48	0.07	3	4
0	7	0.29	7	183	5	0.07	2	0.03	0.51	0.06	0.46	0.06	2	3
0	14	0.41	10	272	1	0.21	101	0.03	0.93	0.04	0.63	0.06	9	10
0	22	0.72	17	336	10	0.26	149	0.05	1.44	0.05	0.9	0.13	5	7
0	11	0.36	8	168	5	0.13	74	0.02	0.72	0.03	0.45	0.06	2	4
0	29	0.73	24	403	10	0.27	207	0.04	1.46	0.05	0.84	0.13	5	12
0	13	0.32	11	178	4	0.12	92	0.02	0.65	0.02	0.37	0.06	2	5
0	19	0.93	11	407	4	0.14	63	<.01	1.52	0.03	0.91	0.05	4	0
0	20	1.56		520	1		16							6
0	48	2.35	28	1028	10	0.35	160	0.01	3.82	0.06	2.29	0.13	8	1
0	18	0.95	13	286	8	0.23	166	0.02	0.78	0.04	0.65	0.06	3	2
0	22	1.06	50	877	14	1.18	12	0.13	3.41	0.25	3.31	0.5	107	15
0	30	0.52	73	1067	6	1.22	21	0.06	8.09	0.16	2.04	0.24	106	53
0	28	1.26	67	1116	16	1.47	16	0.15	4.76	0.3	4	0.59	133	23
0	6	0.31	32	422	1	0.18	4	0.04	0.12	0.09	0.78	0.43	24	10
0	8	0.39	40	537	2	0.22	4	0.05	0.15	0.11	1	0.55	30	13
0	11	0.57	54	746	2	0.3	6	0.09	0.2	0.12	1.4	0.22	7	4
0	42	0.89	29	233	1	0.76	16	0.03	1.68	0.04	0.93	0.04	36	30
0	9	0.41	9	112	1	0.23	4	0.05	0.83	0.06	0.61	0.08	9	14
0	17	1.11	6	170	3	0.17	6	0.06	1.48	0.15	0.72	0.17	20	3
0	14	0.9	5	138	2	0.14	5	0.05	1.2	0.12	0.58	0.14	16	2
0	8	0.38	5	46	5	0.18	1	0.02	0	0.02	0.09	0.05	0	90
0	21	0.09	5	59	7	0.05	<1	0	0.12	<.01	<.01	<.01	0	76
0	12	0.29	5	69	17	0.44	<1	0.01	0	0.05	0.15	0.05	0	78
0	11	0.61	8	72	80	0.14	6	0.04	2.3	0.06	0.28	0.04	3	6
0	0	0			0									0
0	31	0.85	27	186	1	0.52	2	0.04	1.07	0.05	0.74	0.07	26	32

A

TABLE A-1

Food Composition

(Computer code number is for West Diet Analysis program)   (For purposes of calculations, use "0" for t, <1, <.1, <.01, etc.)

Computer Code Number	Food Description	Measure	Wt (g)	H₂O (%)	Ener (kcal)	Prot (g)	Carb (g)	Dietary Fiber (g)	Fat (g)	Fat Breakdown (g) Sat	Mono	Poly
	FRUITS and FRUIT JUICES—Continued											
71120	Frozen, sweetened	10 oz	284	73	293	2	74	12	<1	t	t	0.3
3235	Cup, thawed measure	1 c	250	73	258	2	65	11	<1	t	t	0.2
	Strawberries:											
3134	Fresh, whole, capped	1 c	144	91	46	1	11	3	<1	t	t	0.2
3236	Frozen, sliced, sweetened	1 c	255	73	245	1	66	5	<1	t	t	0.2
3663	Breadfruit	1 c	220	71	227	2	60	11	1	0.1	t	0.1
	Cherries:											
3035	Sour, red pitted, canned water pack	1 c	244	90	88	2	22	3	<1	t	t	t
3036	Sweet, red pitted, raw	10 ea	68	82	43	1	11	1	<1	t	t	t
3862	Dried	¼ c	40		120	2	26	3	0	0	0	0
	Dates:											
3044	Whole, without pits	10 ea	83	21	234	2	62	7	<1	t	t	t
3043	Chopped	1 c	178	21	502	4	134	14	1	t	t	t
3162	Figs, dried	10 ea	190	30	473	6	121	19	2	0.3	0.3	0.7
	Fruit cocktail, canned, not drained:											
3045	Heavy syrup pack	1 c	248	80	181	1	47	2	<1	t	t	t
3164	Juice pack	1 c	237	87	109	1	28	2	<1	t	t	t
	Grapefruit:											
	Raw 3¾" diam (half w/rind = 241g)											
3818	Pink/red, half fruit, edible part	1 ea	123	88	52	1	13	2	<1	t	t	t
3047	White, half fruit, edible part	1 ea	118	90	39	1	10	1	<1	t	t	t
3050	Canned sections with light syrup	1 c	254	84	152	1	39	1	<1	t	t	t
	Juice:											
3051	Fresh, white, raw	1 c	247	90	96	1	23	<1	<1	t	t	t
3052	Canned, unsweetened	1 c	247	90	94	1	22	<1	<1	t	t	t
3165	Sweetened	1 c	250	87	115	1	28	<1	<1	t	t	t
3053	Prepared from concentrate	1 c	247	89	101	1	24	<1	<1	t	t	t
	Grapes, European (adherent skin):											
	Fresh:											
3055	Thompson seedless	10 ea	50	81	34	<1	9	<1	<1	t	t	t
3056	Tokay/Emperor, seeded types	10 ea	50	81	34	<1	9	<1	<1	t	t	t
3064	Juice, prepared from frozen, vit C added	1 c	250	87	128	<1	32	<1	<1	t	t	t
3062	Low calorie	1 c	253	84	154	1	38	<1	<1	t	t	t
3636	Jackfruit, fresh, sliced	1 c	165	73	155	2	40	3	<1	0.1	t	0.1
3065	Kiwi fruit, raw, peeled (88g w/peel)	1 ea	76	83	46	1	11	2	<1	t	t	0.2
	Lemons:											
	Fresh:											
3066	Without peel and seeds (about 4/lb)	1 ea	58	89	17	1	5	2	<1	t	t	t
	Juice:											
254	Fresh:	1 c	244	91	61	1	21	1	0	0	0	0
3068	Tablespoon	1 tbs	15	91	4	<1	1	<1	0	0	0	0
3069	Canned or bottled	1 tbs	15	92	3	<1	1	<1	<1	t	t	t
258	Frozen, single strength, Unsweetened:	1 c	244	92	54	1	16	1	1	0.1	t	0.2
3070	Tablespoon	1 tbs	15	92	3	<1	1	<1	<1	t	t	t
	Lime juice:											
3072	Fresh	1 tbs	15	90	4	<1	1	<1	<1	t	t	t
3073	Canned or bottled, unsweetened	1 c	246	93	52	1	16	1	1	t	t	0.2
3758	Pomelos, raw	1 ea	609	89	231	5	59	6	<1			
3221	Mangos, raw, edible part (300g w/skin & seeds)	1 ea	207	82	135	1	35	4	1	0.1	0.2	0.1
	Melons:											
	Raw, without rind and contents:											
3076	Cantaloupe, 5" diam (2⅓ lb whole w/refuse), orange flesh	½ ea	276	90	94	2	23		1	0.1	t	0.2
3081	Honeydew, 6" diam (5/lb whole w/refuse), slice = 1/10 melon	1 pce	160	90	58	1	15	1	<1	t	t	t
3215	Nectarines, raw, w/o pits, 2" diam	1 ea	136	88	60	1	14	2	<1	t	0.1	0.2
	Oranges:											
	Fresh:											

PAGE KEY: A–4 = Beverages   A–6 = Dairy   A–10 = Eggs   A–12 = Fat/Oil   A–14 = Fruit   A–20 = Bakery
A–26 = Grain   A–32 = Fish   A–32 = Meats   A–36 = Poultry   A–38 = Sausage   A–40 = Mixed/Fast   A–44 = Nuts/Seeds   A–46 = Sweets
A–48 = Vegetables/Legumes   A–58 = Vegetarian   A–62 = Misc   A–62 = Soups/Sauces   A–64 = Fast   A–78 = Convenience   A–80 = Baby foods

Chol (mg)	Calc (mg)	Iron (mg)	Magn (mg)	Pota (mg)	Sodi (mg)	Zinc (mg)	VT-A (RAE)	Thia (mg)	VT-E (mg α)	Ribo (mg)	Niac (mg)	V-B$_6$ (mg)	Fola (µg)	VT-C (mg)
0	43	1.85	37	324	3	0.51	9	0.05	2.04	0.13	0.65	0.1	74	47
0	38	1.62	32	285	2	0.45	8	0.05	1.8	0.11	0.57	0.08	65	41
0	23	0.6	19	220	1	0.2	1	0.03	0.42	0.03	0.56	0.07	35	85
0	28	1.5	18	250	8	0.15	3	0.04	0.59	0.13	1.02	0.08	38	106
0	37	1.19	55	1078	4	0.26	0	0.24	0.22	0.07	1.98	0.22	31	64
0	27	3.34	15	239	17	0.17	93	0.04	0.56	0.1	0.43	0.11	20	5
0	9	0.24	7	151	0	0.05	2	0.02	0.05	0.02	0.1	0.03	3	5
0	20	0.36			5		5							0
0	32	0.85	36	544	2	0.24	<1	0.04	0.04	0.05	1.06	0.14	16	0
0	69	1.82	77	1168	4	0.52	1	0.09	0.09	0.12	2.27	0.29	34	1
0	308	3.86	129	1292	19	1.04	1	0.16	0.66	0.16	1.18	0.2	17	2
0	15	0.72	12	218	15	0.2	25	0.04	0.99	0.05	0.93	0.12	7	5
0	19	0.5	17	225	9	0.21	36	0.03	0.95	0.04	0.96	0.12	7	6
0	27	0.1	11	166	0	0.09	71	0.05	0.16	0.04	0.25	0.07	16	38
0	14	0.07	11	175	0	0.08	2	0.04	0.15	0.02	0.32	0.05	12	39
0	36	1.02	25	328	5	0.2	0	0.1	0.23	0.05	0.62	0.05	23	54
0	22	0.49	30	400	2	0.12	5	0.1	0.54	0.05	0.49	0.11	25	94
0	17	0.49	25	378	2	0.22	1	0.1	0.1	0.05	0.57	0.05	25	72
0	20	0.9	25	405	5	0.15	1	0.1	0.1	0.06	0.8	0.05	25	67
0	20	0.35	27	336	2	0.12	1	0.1	0.1	0.05	0.54	0.11	10	83
0	5	0.18	4	96	1	0.04	2	0.03	0.1	0.04	0.09	0.04	1	5
0	5	0.18	4	96	1	0.04	2	0.03	0.1	0.04	0.09	0.04	1	5
0	10	0.25	10	52	5	0.1	1	0.04	0	0.06	0.31	0.1	2	60
0	23	0.61	25	334	8	0.13	1	0.07	0	0.09	0.66	0.16	8	0
0	56	0.99	61	500	5	0.69	25	0.05	0.25	0.18	0.66	0.18	23	11
0	26	0.24	13	237	2	0.11	3	0.02	1.11	0.02	0.26	0.05	19	70
0	15	0.35	5	80	1	0.03	1	0.02	0.09	0.01	0.06	0.05	6	31
0	17	0.07	15	303	2	0.12	2	0.07	0.37	0.02	0.24	0.12	32	112
0	1	<.01	1	19	<1	<.01	<1	<.01	0.02	<.01	0.02	<.01	2	7
0	2	0.02	1	15	3	<.01	<1	<.01	0.02	<.01	0.03	<.01	2	4
0	20	0.29	20	217	2	0.12	2	0.14	0.2	0.03	0.33	0.15	24	77
0	1	0.02	1	13	<1	<.01	<1	<.01	0.01	<.01	0.02	<.01	2	5
0	1	<.01	1	16	<1	<.01	<1	<.01	0.02	<.01	0.02	<.01	1	4
0	30	0.57	17	184	39	0.15	2	0.08	0.3	<.01	0.4	0.07	20	16
0	24	0.67	37	1315	6	0.49	2	0.21	0.55	0.16	1.34	0.22	158	371
0	21	0.27	19	323	4	0.08	79	0.12	2.32	0.12	1.21	0.28	29	57
0	25	0.58	33	737	44	0.5	466	0.11	0.14	0.05	2.03	0.2	58	101
10	10	0.27	16	365	29	0.14	5	0.06	0.03	0.02	0.67	0.14	30	29
0	8	0.38	12	273	0	0.23	23	0.05	1.05	0.04	1.53	0.03	7	7

A

TABLE A–1

Food Composition   (Computer code number is for West Diet Analysis program)   (For purposes of calculations, use "0" for t, <1, <.1, <.01, etc.)

Computer Code Number	Food Description	Measure	Wt (g)	H₂O (%)	Ener (kcal)	Prot (g)	Carb (g)	Dietary Fiber (g)	Fat (g)	Fat Breakdown (g) Sat	Mono	Poly
	FRUITS and FRUIT JUICES—Continued											
3082	Whole w/o peel & seeds, 2⁵/₈" diam (180g w/peel & seeds)	1 ea	131	87	62	1	15	3	<1	t	t	t
3083	Sections, without membranes	1 c	180	87	85	2	21	4	<1	t	t	t
	Juice:											
3090	Fresh, all varieties	1 c	248	88	112	2	26	<1	<1	t	t	t
3093	Canned, unsweetened	1 c	249	89	105	1	25	<1	<1	t	t	t
3480	Calcium fortified	1 c	247		110	1	27	0	0	0	0	0
3092	Chilled	1 c	249	88	110	2	25	<1	1	t	0.1	0.2
3091	Prepared from concentrate	1 c	249	88	112	2	27	<1	<1	t	t	t
20004	Prepared from dry crystals	1 c	248	87	122	0	31	<1	0	0	0	0
3170	Orange and grapefruit juice, canned	1 c	247	89	106	1	25	<1	<1	t	t	t
	Papayas:											
	Fresh:											
3172	¹/₂"slices	1 c	140	89	55	1	14	3	<1	t	t	t
3171	Whole, 3" diam by 5¹/₈"	1 ea	304	89	119	2	30	5	<1	0.1	0.1	t
3095	Nectar, canned	1 c	250	85	142	<1	36	2	<1	0.1	0.1	t
	Peaches:											
	Fresh:											
3096	Whole, 2" diam	1 ea	98	89	38	1	9		<1	t	t	t
3097	Sliced	1 c	170	89	66	2	16		<1	t	0.1	0.1
	Canned, not drained:											
	Heavy syrup pack:											
3098	Cup	1 c	262	79	194	1	52	3	<1	t	t	0.1
3099	Half	1 ea	98	79	73	<1	20	1	<1	t	t	t
	Juice pack:											
3175	Cup	1 c	248	87	109	2	29	3	<1	t	t	t
3176	Half	1 ea	98	87	43	1	11	1	<1	t	t	t
	Dried:											
3100	Uncooked	10 ea	130	32	311	5	80	11	1	0.1	0.4	0.5
3214	Cooked, w/fruit & liquid	1 c	258	78	199	3	51	7	1	t	0.2	0.3
	Frozen, sweetened:											
3234	10-oz package,vitamin C added	1 ea	284	75	267	2	68	5	<1	t	0.1	0.2
57481	Cup, vitamin C added	1 c	250	75	235	2	60	4	<1	t	0.1	0.2
3101	Nectar, canned	1 c	249	86	134	1	35	1	<1	t	t	t
	Pears:											
	Fresh, with skin, cored:											
3103	Bartlett, 2¹/₂" diam (about 2¹/₂/lb)	1 ea	166	84	96	1	26	5	<1	t	t	t
3105	Bosc, 2¹/₅" diam (about 3/lb)	1 ea	139	84	81	1	21	4	<1	t	t	t
3106	D'Anjou, 3" diam (about 2/lb)	1 ea	209	84	121	1	32	6	<1	t	t	t
	Canned, fruit and liquid:											
	Heavy syrup pack:											
3107	Cup	1 c	266	80	197	1	51	4	<1	t	t	t
3108	Half	1 ea	76	80	56	<1	15	1	<1	t	t	t
	Juice pack:											
3179	Cup	1 c	248	86	124	1	32	4	<1	t	t	t
3180	Half	1 ea	76	86	38	<1	10	1	<1	t	t	t
3109	Dried halves	10 ea	175	27	458	3	122	13	1	t	0.2	0.3
3110	Nectar, canned	1 c	250	84	150	<1	39	2	<1	t	t	t
	Pineapple:											
3111	Fresh chunks, diced	1 c	155	86	74	1	20	2	<1	t	t	t
	Canned, not drained:											
	Heavy syrup pack:											
3114	Crushed, chunks, tidbits	¹/₂ c	127	79	99	<1	26	1	<1	t	t	t
3115	Slices	1 ea	49	79	38	<1	10	<1	<1	t	t	t
	Juice pack:											
3183	Crushed, chunks, tidbits	1 c	250	84	150	1	39	2	<1	t	t	t
3184	Slices	1 ea	47	84	28	<1	7	<1	<1	t	t	t
3120	Juice, canned, unsweetened:	1 c	250	86	140	1	34	<1	<1	t	t	t
	Plantains, yellow fleshed, without peel:											
3195	Raw slices (whole = 179g w/o peel)	1 c	148	65	181	2	47	3	1	0.2	t	0.1
3196	Cooked, boiled, sliced	1 c	154	67	179	1	48	4	<1	0.1	t	t

PAGE KEY: A–4 = Beverages A–6 = Dairy A–10 = Eggs A–12 = Fat/Oil A–14 = Fruit A–20 = Bakery
A–26 = Grain A–32 = Fish A–32 = Meats A–36 = Poultry A–38 = Sausage A–40 = Mixed/Fast A–44 = Nuts/Seeds A–46 = Sweets
A–48 = Vegetables/Legumes A–58 = Vegetarian A–62 = Misc A–62 = Soups/Sauces A–64 = Fast A–78 = Convenience A–80 = Baby foods

Chol (mg)	Calc (mg)	Iron (mg)	Magn (mg)	Pota (mg)	Sodi (mg)	Zinc (mg)	VT-A (RAE)	Thia (mg)	VT-E (mg α)	Ribo (mg)	Niac (mg)	V-B$_6$ (mg)	Fola (µg)	VT-C (mg)
0	52	0.13	13	237	0	0.09	14	0.11	0.24	0.05	0.37	0.08	39	70
0	72	0.18	18	326	0	0.13	20	0.16	0.32	0.07	0.51	0.11	54	96
0	27	0.5	27	496	2	0.12	25	0.22	0.1	0.07	0.99	0.1	74	124
0	20	1.1	27	436	5	0.17	22	0.15	0.5	0.07	0.78	0.22	45	86
0	300	0		430	15		0	0	0		0	0	40	78
0	25	0.42	27	473	2	0.1	10	0.28	0.47	0.05	0.7	0.13	45	82
0	22	0.25	25	473	2	0.12	12	0.2	0.5	0.04	0.5	0.11	110	97
0	126	0.02	2	60	10	0.02	191	0	0	0.22	2.54	0.25	0	73
0	20	1.14	25	390	7	0.17	15	0.14	0.35	0.07	0.83	0.06	35	72
0	34	0.14	14	360	4	0.1	77	0.04	1.02	0.04	0.47	0.03	53	87
0	73	0.3	30	781	9	0.21	167	0.08	2.22	0.1	1.03	0.06	116	188
0	25	0.85	8	78	12	0.38	45	0.02	0.6	<.01	0.38	0.02	5	8
0	6	0.24	9	186	0	0.17	16	0.02	0.72	0.03	0.79	0.02	4	6
0	10	0.42	15	323	0	0.29	27	0.04	1.24	0.05	1.37	0.04	7	11
0	8	0.71	13	241	16	0.24	45	0.03	1.28	0.06	1.61	0.05	8	7
0	3	0.26	5	90	6	0.09	17	0.01	0.48	0.02	0.6	0.02	3	3
0	15	0.67	17	317	10	0.27	47	0.02	1.22	0.04	1.44	0.05	7	9
0	6	0.26	7	125	4	0.11	19	<.01	0.48	0.02	0.57	0.02	3	4
0	36	5.28	55	1295	9	0.74	140	<.01	0.25	0.28	5.69	0.09	0	6
0	23	3.38	34	826	5	0.46	26	0.01	0.15	0.05	3.92	0.1	0	10
0	9	1.05	14	369	17	0.14	40	0.04	1.76	0.1	1.85	0.05	9	268
0	8	0.92	12	325	15	0.12	35	0.03	1.55	0.09	1.63	0.04	8	235
0	12	0.47	10	100	17	0.2	32	<.01	0.72	0.03	0.72	0.02	2	13
0	15	0.28	12	198	2	0.17	2	0.02	0.2	0.04	0.26	0.05	12	7
0	13	0.24	10	165	1	0.14	1	0.02	0.17	0.03	0.22	0.04	10	6
0	19	0.36	15	249	2	0.21	2	0.03	0.25	0.05	0.33	0.06	15	9
0	13	0.59	11	173	13	0.21	0	0.03	0.21	0.06	0.64	0.04	3	3
0	4	0.17	3	49	4	0.06	0	<.01	0.06	0.02	0.18	0.01	1	1
0	22	0.72	17	238	10	0.22	1	0.03	0.2	0.03	0.5	0.03	2	4
0	7	0.22	5	73	3	0.07	<1	<.01	0.06	<.01	0.15	0.01	1	1
0	60	3.68	58	933	10	0.68	<1	0.01	0.1	0.25	2.4	0.13	0	12
0	12	0.65	8	32	10	0.18	<1	<.01	0.12	0.03	0.32	0.04	2	3
0	20	0.43	19	178	2	0.16	5	0.12	0.03	0.05	0.76	0.17	23	56
0	18	0.48	20	132	1	0.15	1	0.11	0.01	0.03	0.36	0.09	6	9
0	7	0.19	8	51	<1	0.06	<1	0.04	<.01	0.01	0.14	0.04	2	4
0	35	0.7	35	305	2	0.25	5	0.24	0.02	0.05	0.71	0.18	12	24
0	7	0.13	7	57	<1	0.05	1	0.04	<.01	<.01	0.13	0.03	2	4
0	42	0.65	32	335	2	0.28	1	0.14	0.05	0.06	0.64	0.24	58	27
0	4	0.89	55	739	6	0.21	83	0.08	0.21	0.08	1.02	0.44	33	27
0	3	0.89	49	716	8	0.2	69	0.07	0.2	0.08	1.16	0.37	40	17

TABLE A-1

Food Composition (Computer code number is for West Diet Analysis program) (For purposes of calculations, use "0" for t, <1, <.1, <.01, etc.)

Computer Code Number	Food Description	Measure	Wt (g)	H₂O (%)	Ener (kcal)	Prot (g)	Carb (g)	Dietary Fiber (g)	Fat (g)	Fat Breakdown (g) Sat	Mono	Poly
	FRUITS and FRUIT JUICES—Continued											
	Plums:											
3121	Fresh, medium, 2¹/₈″ diam	1 ea	66	87	30	<1	8	1	<1	t	t	t
	Canned, purple, not drained:											
	Heavy syrup pack:											
3124	Cup	1 c	258	76	230	1	60	2	<1	t	0.2	t
3125	Each	3 ea	138	76	123	<1	32	1	<1	t	t	t
	Juice Pack:											
3185	Cup	1 c	252	84	146	1	38	2	<1	t	t	t
3186	Each	3 ea	138	84	80	1	21	1	<1	t	t	t
3197	Pomegranate, fresh	1 ea	154	81	105	1	26	1	<1	t	t	t
	Prunes:											
	Dried, pitted:											
3126	Uncooked (10 = 97g w/pits, 84g w/o pits)	10 ea	84	31	202	2	54	6	<1	t	t	t
3127	Cooked, unsweetened, fruit & liq, (250 g w/pits)	1 c	248	70	265	2	70	8	<1	t	0.3	t
3128	Juice, bottled or canned	1 c	256	81	182	2	45	3	<1	t	t	t
	Raisins, seedless:											
3130	Cup, not pressed down	1 c	145	15	434	4	115	5	1	t	t	t
3764	One packet, ¹/₂ oz	¹/₂ oz	14	15	42	<1	11	1	<1	t	t	t
3133	Rhubarb, cooked, added sugar	1 c	240	68	278	1	75	5	<1	t	t	t
	Tangerines, without peel and seeds:											
3138	Fresh (2³/₈″ whole) 116g w/refuse	1 ea	84	88	37	1	9	2	<1	t	t	t
3237	Canned, light syrup, fruit and liquid	1 c	252	83	154	1	41	2	<1	t	t	t
3140	Juice, canned, sweetened	1 c	249	87	124	1	30	<1	<1	t	t	t
	Watermelon, raw, without rind and seeds:											
3143	Piece, ¹/₁₆th wedge	1 pce	286	91	86	2	22	1	<1	t	0.1	0.1
3142	Diced	1 c	152	91	46	1	11	1	<1	t	t	t
	BAKED GOODS: BREADS, CAKES, COOKIES, CRACKERS, PIES											
	Bagels:											
42100	Cinnamon raisin, 3¹/₂″ diam.	1 ea	71	32	195	7	39	2	1	0.2	0.1	0.5
42620	Dunkin Donuts	1 ea	125		340	11	72	4	1	0		
42000	Plain, enriched, 3¹/₂″ diam.	1 ea	71	33	195	7	38	2	1	0.2	t	0.5
42619	Dunkin Donuts	1 ea	125		330	12	68	3	1	0		
42596	Oat Bran	1 ea	110	33	280	12	59	4	1	0.2	0.3	0.5
42617	Whole Wheat	1 ea	110	28	291	12	62	10	2	0.3	0.2	0.6
	Biscuits:											
42001	From home recipe	1 ea	60	29	212	4	27	1	10	2.6	4.2	2.5
42002	From mix	1 ea	57	29	191	4	28	1	7	1.6	2.4	2.5
	Bread:											
42672	Cornbread, 2.5 × 2.5 × 1.5″ piece	1 pce	65	49	152	4	23	1	5	1.6	2.4	0.5
42015	Croissants, 4¹/₂ × 4 × 1³/₄″	1 ea	57	23	231	5	26	1	12	6.6	3.1	0.6
42148	Croutons, seasoned	¹/₂ c	20	4	93	2	13	1	4	1	1.9	0.5
42004	Crumbs, dry, grated (see 364, 365 for soft crumbs)	1 c	108	7	427	14	78	5	6	1.3	1.1	2.2
42052	Boston brown, canned, 3¹/₄″ slice	1 pce	45	47	88	2	19	2	1	0.1	t	0.3
	Cracked wheat (¹/₄ cracked-wheat & ³/₄ enriched wheat flour):											
42042	Slice (18 per loaf)	1 pce	25	36	65	2	12	1	1	0.2	0.5	0.2
	French/Vienna, enriched:											
42044	Slice, 4³/₄ × 4¹/₂″	1 pce	25	34	68	2	13	1	1	0.2	0.3	0.2
42043	French, slice, 5 × 2″	1 pce	25	34	68	2	13	1	1	0.2	0.3	0.2
	French toast: see Mixed Dishes, and Fast Foods, #691											
42476	Honey wheatberry	1 pce	38		100	3	18	2	2	0	0.5	1
	Italian, enriched:											
42046	Slice, 4¹/₂ × 3¹/₄ × ³/₄″	1 pce	30	36	81	3	15	1	1	0.3	0.2	0.4
	Mixed grain, enriched:											
42047	Slice (18 per loaf)	1 pce	26	38	65	3	12	2	1	0.2	0.4	0.2
42048	Slice, toasted	1 pce	24	32	65	3	12	2	1	0.2	0.4	0.2

PAGE KEY: A–4 = Beverages   A–6 = Dairy   A–10 = Eggs   A–12 = Fat/Oil   A–14 = Fruit   A–20 = Bakery
A–26 = Grain   A–32 = Fish   A–32 = Meats   A–36 = Poultry   A–38 = Sausage   A–40 = Mixed/Fast   A–44 = Nuts/Seeds   A–46 = Sweets
A–48 = Vegetables/Legumes   A–58 = Vegetarian   A–62 = Misc   A–62 = Soups/Sauces   A–64 = Fast   A–78 = Convenience   A–80 = Baby foods

Chol (mg)	Calc (mg)	Iron (mg)	Magn (mg)	Pota (mg)	Sodi (mg)	Zinc (mg)	VT-A (RAE)	Thia (mg)	VT-E (mg α)	Ribo (mg)	Niac (mg)	V-B$_6$ (mg)	Fola (µg)	VT-C (mg)
0	4	0.11	5	104	0	0.07	11	0.02	0.17	0.02	0.28	0.02	3	6
0	23	2.17	13	235	49	0.18	34	0.04	0.46	0.1	0.75	0.07	8	1
0	12	1.16	7	126	26	0.1	18	0.02	0.25	0.05	0.4	0.04	4	1
0	25	0.86	20	388	3	0.28	126	0.06	0.45	0.15	1.19	0.07	8	7
0	14	0.47	11	213	1	0.15	69	0.03	0.25	0.08	0.65	0.04	4	4
0	5	0.46	5	399	5	0.18	8	0.05	0.92	0.05	0.46	0.16	9	9
0	36	0.78	34	615	2	0.37	33	0.04	0.36	0.16	1.58	0.17	3	1
0	47	1.02	45	796	2	0.47	42	0.06	0.47	0.25	1.79	0.54	0	7
0	31	3.02	36	707	10	0.54	<1	0.04	0.31	0.18	2.01	0.56	0	10
0	72	2.73	46	1086	16	0.32	0	0.15	0.17	0.18	1.11	0.25	7	3
0	7	0.26	4	105	2	0.03	0	0.01	0.02	0.02	0.11	0.02	1	0
0	348	0.5	29	230	2	0.19	10	0.04	0.65	0.06	0.48	0.05	12	8
0	12	0.08	10	132	1	0.2	29	0.09	0.13	0.02	0.13	0.06	17	26
0	18	0.93	20	197	15	0.6	106	0.13	0.25	0.11	1.12	0.11	13	50
0	45	0.5	20	443	2	0.07	32	0.15	0.37	0.05	0.25	0.08	12	55
0	20	0.69	29	320	3	0.29	80	0.09	0.14	0.06	0.51	0.13	9	23
0	11	0.36	15	170	2	0.15	43	0.05	0.08	0.03	0.27	0.07	5	12
0	13	2.7	20	105	229	0.8	15	0.27	0.22	0.2	2.19	0.04	79	0
0	40	3.6			470		0							4
0	53	2.53	21	72	379	0.62	0	0.38	0.21	0.22	3.24	0.04	75	0
0	20	4.5			690		0							4
0	13	3.39	34	126	558	0.99	1	0.36	0.36	0.37	3.26	0.05	108	0
0	32	3.52	116	379	592	2.52		0.34	0.99	0.28	5.75	0.3	66	0
2	141	1.74	11	73	348	0.32	14	0.21	0.78	0.19	1.77	0.02	37	0
2	105	1.17	14	107	544	0.35	15	0.2	0.23	0.2	1.72	0.04	30	0
22	71	0.82	13	105	356	0.38		0.12	0.64	0.16	0.93	0.05	6	0
38	21	1.16	9	67	424	0.43	117	0.22	0.48	0.14	1.25	0.03	50	0
1	19	0.56	8	36	248	0.19	1	0.1	0.08	0.08	0.93	0.02	21	0
0	198	5.22	46	212	791	1.57	0	1.04	0.09	0.44	7.16	0.13	116	0
0	32	0.94	28	143	284	0.22	11	<.01	0.14	0.05	0.5	0.04	5	0
0	11	0.7	13	44	134	0.31	0	0.09	0.15	0.06	0.92	0.08	15	0
0	19	0.63	7	28	152	0.22	0	0.13	0.08	0.08	1.19	0.01	37	0
0	19	0.63	7	28	152	0.22	0	0.13	0.08	0.08	1.19	0.01	37	0
0	20	0.72			200		0	0.12	0.24	0.07	0.8			0
0	23	0.88	8	33	175	0.26	0	0.14	0.09	0.09	1.31	0.01	57	0
0	24	0.9	14	53	127	0.33	0	0.11	0.09	0.09	1.13	0.09	31	0
0	24	0.9	14	53	127	0.33	0	0.08	0.08	0.08	1.02	0.08	28	0

A

TABLE A-1

Food Composition   (Computer code number is for West Diet Analysis program)   (For purposes of calculations, use "0" for t, <1, <.1, <.01, etc.)

Computer Code Number	Food Description	Measure	Wt (g)	H₂O (%)	Ener (kcal)	Prot (g)	Carb (g)	Dietary Fiber (g)	Fat (g)	Fat Breakdown (g) Sat	Mono	Poly
	BAKED GOODS: BREADS, CAKES, COOKIES, CRACKERS, PIES—Continued											
	Oatmeal, enriched:									0.2	0.4	0.5
42049	Slice (18 per loaf)	1 pce	27	37	73	2	13	1	1	0.2	0.4	0.5
42050	Slice, toasted	1 pce	25	31	73	2	13	1	1	t	t	0.3
42007	Pita pocket bread, enr, 6" round	1 ea	60	32	165	5	33	1	1			
	Pumpernickel (²/₃ rye & ¹/₃ enr wheat flr):									0.1	0.2	0.3
42006	Slice, 5 × 4 × ³/₈"	1 pce	26	38	65	2	12	2	1	0.1	0.3	0.4
42054	Slice, toasted	1 pce	29	32	80	3	15	2	1			
	Raisin, enriched:									0.3	0.6	0.2
42051	Slice (18 per loaf)	1 pce	26	34	71	2	14	1	1	0.3	0.6	0.2
42055	Slice, toasted	1 pce	24	28	71	2	14	1	1			
	Rye, light (¹/₃ rye & ²/₃ enr wheat flr):											
42005	1-lb loaf	1 ea	454	37	1176	39	219	26	15	2.8	6	3.6
42005	Slice, 4³/₄ × 3³/₄ × ⁷/₁₆"	1 pce	32	37	83	3	15	2	1	0.2	0.4	0.3
42056	Slice, toasted	1 pce	24	31	68	2	13	2	1	0.2	0.3	0.2
	Wheat (enr wheat & whole-wheat flour):									0.2	0.4	0.2
42012	Slice (18 per loaf)	1 pce	25	37	65	2	12	1	1	0.2	0.4	0.2
42031	Slice, toasted	1 pce	23	32	65	2	12	1	1			
	White, enriched:											
42138	Slice	1 pce	42	35	120	3	21	1	2	0.5	0.5	1.2
	Whole Wheat:									0.3	0.5	0.3
42014	Slice (16 per loaf)	1 pce	28	38	69	3	13	2	1	0.3	0.5	0.3
42029	Slice, toasted	1 pce	25	30	69	3	13	2	1			
	Bread stuffing, prepared from mix:											
42037	Dry type	1 c	200	65	356	6	43	6	17	3.5	7.6	5.2
	Cakes:											
	Prepared from mixes using enrich flour and veg shortening, w/frostings made from margarine:											
46004	Angel Food, ¹/₁₂ of cake	1 pce	28	33	72	2	16	<1	<1	t	t	0.1
46002	Boston cream pie, ¹/₈ of cake	1 pce	92	45	232	2	39	1	8	2.2	4.2	0.9
46005	Coffee Cake, ¹/₆ of cake	1 pce	56	30	178	3	30	1	5	1	2.2	1.8
46013	Devil's food, chocolate frosting, ¹/₁₆ of cake	1 pce	64	23	235	3	35	2	10	3.1	5.6	1.2
	Home recipes w/enrich flour:											
	Fruitcake, dark:											
46205	Piece, ¹/₃₂ of cake, ²/₃" arc	1 pce	43	25	139	1	26	2	4	0.5	1.8	1.4
46015	Sheet, plain, made w/margarine, uncooked white frosting, ¹/₉ of cake	1 pce	64	22	239	2	38	<1	9	1.5	3.9	3.3
	Commercial:											
49004	Cheesecake, ¹/₁₂ of cake	1 pce	80	46	257	4	20	<1	18	7.9	6.9	1.3
46016	Pound cake, ¹/₁₇ of loaf, 2" slice	1 pce	28	25	109	2	14	<1	6	3.2	1.7	0.3
	Yellow, chocolate frosting, 2 layer:											
46012	Slice, ¹/₁₆ of cake	1 pce	64	22	243	2	35	1	11	3	6.1	1.4
	Snack:											
46011	Chocolate w/creme filling, Ding Dong	1 ea	50	20	188	2	30	<1	7	1.4	2.8	2.6
46008	Sponge cake w/creme filling,Twinkie	1 ea	43	20	157	1	27	<1	5	1.1	1.8	1.4
46001	Sponge cake, ¹/₁₂ of 12" cake	1 pce	38	30	110	2	23	<1	1	0.3	0.4	0.2
44032	Chex party mix	1 c	43	4	183	5	28	2	7	2.4	3.9	1.1
	Chips:											
44061	Bagel	5 pce	70	3	298	6	52	4	7	1.3	2.1	3.4
42396	Bagel, onion garlic, toasted	1 oz	28		181	5	30	3	7	1.6	4.9	0
43703	Potato chips	20 pce	28		148	2	15	1	10	3		
4022	Baked	11 pce	28		109	2	23	2	1	0		
	Cookies made with enriched flour:											
	Brownies with nuts:											
47000	Commercial w/frosting, 1¹/₂ × 1³/₄ × ⁷/₈"	1 ea	61	14	247	3	39	1	10	2.6	5.5	1.4
47214	Fat free fudge, Entenmann's	1 pce	40		110	2	27	1	0	0	0	0
	Chocolate chip cookies:											
47001	Commercial, 2¹/₄"diam	4 ea	60	12	275	2	35	2	15	4.4	7.8	2.1
47002	Home recipe, 2¹/₄" diam	4 ea	64	6	312	4	37	2	18	5.2	6.6	5.4

PAGE KEY: A–4 = Beverages A–6 = Dairy A–10 = Eggs A–12 = Fat/Oil A–14 = Fruit A–20 = Bakery
A–26 = Grain A–32 = Fish A–32 = Meats A–36 = Poultry A–38 = Sausage A–40 = Mixed/Fast A–44 = Nuts/Seeds A–46 = Sweets
A–48 = Vegetables/Legumes A–58 = Vegetarian A–62 = Misc A–62 = Soups/Sauces A–64 = Fast A–78 = Convenience A–80 = Baby foods

Chol (mg)	Calc (mg)	Iron (mg)	Magn (mg)	Pota (mg)	Sodi (mg)	Zinc (mg)	VT-A (RAE)	Thia (mg)	VT-E (mg α)	Ribo (mg)	Niac (mg)	V-B$_6$ (mg)	Fola (μg)	VT-C (mg)
0	18	0.73	10	38	162	0.28	1	0.11	0.13	0.06	0.85	0.02	17	0
0	18	0.74	10	38	163	0.28	1	0.09	0.13	0.06	0.77	0.02	13	0
0	52	1.57	16	72	322	0.5	0	0.36	0.18	0.2	2.78	0.02	64	0
0	18	0.75	14	54	174	0.38	0	0.09	0.11	0.08	0.8	0.03	24	0
0	21	0.91	17	66	214	0.47	0	0.08	0.13	0.09	0.89	0.04	25	0
0	17	0.75	7	59	101	0.19	0	0.09	0.07	0.1	0.9	0.02	28	0
0	17	0.76	7	59	102	0.19	0	0.07	0.07	0.09	0.81	0.02	24	0
0	331	12.85	182	754	2996	5.18	2	1.97	1.5	1.52	17.27	0.34	499	2
0	23	0.91	13	53	211	0.36	<1	0.14	0.11	0.11	1.22	0.02	35	0
0	19	0.74	10	44	174	0.3	<1	0.09	0.09	0.08	0.9	0.02	25	0
0	26	0.83	12	50	132	0.26	0	0.1	0.07	0.07	1.03	0.02	23	0
0	26	0.83	12	50	132	0.26	0	0.08	0.07	0.06	0.93	0.02	19	0
1	24	1.25	8	61	151	0.27	9	0.17	0.36	0.16	1.51	0.02	38	
0	20	0.92	24	71	148	0.54	<1	0.1	0.09	0.06	1.07	0.05	14	0
0	20	0.93	24	71	148	0.55	<1	0.08	0.08	0.05	0.97	0.05	10	0
0	64	2.18	24	148	1086	0.56	236	0.27	2.8	0.21	2.95	0.08	78	0
0	39	0.15	3	26	210	0.02	0	0.03	0.03	0.14	0.25	<.01	10	0
34	21	0.35	6	36	132	0.15	22	0.38	0.14	0.25	0.18	0.02	13	0
27	76	0.8	10	63	236	0.25	20	0.09	0.11	0.1	0.85	0.03	27	0
27	28	1.41	22	128	214	0.44	17	0.02	<.01	0.09	0.37	0.03	11	0
2	14	0.89	7	66	116	0.12	3	0.02	0.39	0.04	0.34	0.02	9	0
35	40	0.68	4	34	220	0.16	12	0.06	1.22	0.04	0.32	0.02	17	0
44	41	0.5	9	72	166	0.41	114	0.02	1.26	0.15	0.16	0.04	14	0
62	10	0.39	3	33	111	0.13	42	0.04	0.18	0.06	0.37	0.01	11	0
35	24	1.33	19	114	216	0.4	21	0.08	1.45	0.1	0.8	0.02	14	0
8	36	1.68	20	61	212	0.26	2	0.11	1.09	0.15	1.21	0.01	20	0
7	19	0.55	3	37	157	0.12	2	0.07	0.51	0.06	0.53	0.01	17	0
39	27	1.03	4	38	93	0.19	17	0.09	0.09	0.1	0.73	0.02	18	0
0	15	10.62	27	116	437	0.9	3	0.67	0.11	0.21	7.24	0.67	22	20
0	9	1.39	39	167	419	0.88		0.13	1.71	0.12	1.62	0.15	46	0
0	0	2.37			461		0	0.37	<.01	0.22	3.29			0
0	0	0			178		0							6
0	40	0.36			148		0							1
10	18	1.37	19	91	190	0.44	12	0.16	0.09	0.13	1.05	0.02	29	0
0	0	1.08		90	140		0		<.01					0
0	9	1.45	21	56	196	0.28	<1	0.07	1.74	0.12	0.97	0.1	23	0
20	25	1.57	35	143	231	0.6	92	0.12	1.84	0.11	0.87	0.05	21	0

Table of Food Composition

Appendix   A   Appendix

TABLE A–1

Food Composition   (Computer code number is for West Diet Analysis program)   (For purposes of calculations, use "0" for t, <1, <.1, <.01, etc.)

Computer Code Number	Food Description	Measure	Wt (g)	H₂O (%)	Ener (kcal)	Prot (g)	Carb (g)	Dietary Fiber (g)	Fat (g)	Fat Breakdown (g) Sat	Mono	Poly
	BAKED GOODS: BREADS, CAKES, COOKIES, CRACKERS, PIES—Continued											
47013	From refrigerated dough, 2¼" diam	4 ea	64	13	284	3	39	1	13	4.3	6.7	1.4
47012	Fig bars	4 ea	64	16	223	2	45	3	5	0.7	1.9	1.8
47376	Fruit bar, no fat	1 ea	28		90	2	21	0	0	0	0	0
47324	Fudge, fat free, Snackwell	1 ea	16	14	53	1	12	<1	<1	t	t	t
47154	Nabisco Newtons, fat free, all flavors	1 ea	23		69	1	16		0	0	0	0
47003	Oatmeal raisin, 2⅝" diam	4 ea	60	6	261	4	41	2	10	1.9	4.1	3
47010	Peanut butter, home recipe, 2⅝" diam	4 ea	80	6	380	7	47	2	19	3.6	8.7	5.8
47006	Sandwich-type, all	4 ea	40	2	189	2	28	1	8	1.5	3.4	2.9
47007	Shortbread, commercial, small	4 ea	32	4	161	2	21	1	8	2	4.3	1
47004	Sugar from refrigerated dough, 2" diam	4 ea	48	5	232	2	31	<1	11	2.8	6.2	1.4
47160	Vanilla sandwich, Snackwell's	2 ea	26	4	109	1	21	1	2	0.5	0.8	0.2
47008	Vanilla wafers	10 ea	40	5	176	2	29	1	6	1.5	2.6	1.6
	Crackers:											
43500	Cheese-enriched	10 ea	10	3	50	1	6	<1	3	0.9	1.2	0.2
43501	Cheese with peanut butter-enriched	4 ea	28	3	139	3	16	1	7	1.2	3.6	1.4
	Fat free-enriched:											
43596	Cracked pepper, Snackwell	1 ea	14	2	61	1	10	<1	2	0.3	0.6	0.2
43593	Wheat, Snackwell	7 ea	15	1	60	2	12	1	<1	0.1	0.1	0.1
43590	Whole wheat, herb seasoned	5 ea	14		50	2	11	2	0	0	0	0
43591	Whole wheat, onion	5 ea	14		50	2	11	2	0	0	0	0
43502	Graham-enriched	2 ea	14	4	59	1	11	<1	1	0.2	0.6	0.5
43509	Melba toast, plain-enriched	1 pce	5	5	20	1	4	<1	<1	t	t	t
44064	Rice cakes, unsalted-enriched	2 ea	18	6	70	1	15	1	1	0.1	0.2	0.2
43504	Rye wafer, whole grain	2 ea	22	5	73	2	18	5	<1	t	t	t
43506	Saltine-enriched	4 ea	12	4	52	1	9	<1	1	0.4	0.8	0.2
43586	Saltine, unsalted tops-enriched	2 ea	6		25	1	4	0	<1			
43543	Snack-type, round like Ritz-enriched	3 ea	9	4	45	1	5	<1	2	0.3	1	0.9
43584	Triscuits	7 ea	31	4	150	3	21	3	6	1	2	0.5
43558	Wasa extra crsip crispbread	1 pce	6	6	24	1	4	<1	<1	t	0.2	t
43508	Whole-wheat wafers	2 ea	8	3	35	1	5	1	1	0.3	0.5	0.5
	Pastry:											
	Danish:											
428	Round piece, plain, 4¼" diam, 1" high	1 ea	88	21	349	5	47	<1	17	3.5	10.6	1.6
45512	Ounce, plain	1 oz	28	21	111	2	15	<1	5	1.1	3.4	0.5
45513	Round piece with fruit	1 ea	94	29	335	5	45		16	3.3	10.1	1.6
42094	Pan dulce, sweet roll w/topping	1 ea	79	21	291	5	48	1	9	1.8	4	2.5
49009	Peach crisp, 3 x 3 piece	1 pce	139	73	165	1	30	2	5	0.8		
	Toaster:											
45504	Fortified (PopTarts)	1 ea	52	12	204	2	37	1	5	0.8	2.2	2
	Strudel:											
45647	Cream Cheese	1 ea	54		200	3	24	<1	10	3		
45648	French Toast	1 ea	54		200	3	24	<1	10	3		
	Doughnuts:											
45505	Cake type, plain, 3¼" diam	1 ea	47	21	198	2	23	1	11	1.7	4.4	3.7
45698	Sugared, Dunkin Donuts	1 ea	67		310	4	28	1	20	4		
45506	Yeast-leavened, glazed, 3¾"diam	1 ea	60	25	242	4	27	1	14	3.5	7.7	1.7
45708	Dunkin Donuts	1 ea	46		160	3	23	1	7	2		
	Muffins:											
	English:											
42059	Plain, enriched	1 ea	57	42	134	4	26	2	1	0.1	0.2	0.5
42061	Toasted	1 ea	52	37	133	4	26	2	1	0.1	0.2	0.5
42082	Whole wheat	1 ea	66	46	134	6	27	4	1	0.2	0.3	0.6
44504	Cornmeal	1 ea	50	30	160	4	25	1	5	1.4	2.6	0.6
44614	Banana nut	1 ea	95		340	6	53	2	12	3		
44622	Blueberry	1 ea	95		310	5	51	2	10	2.5		
44616	Chocolate chip	1 ea	95		400	5	63	2	16	6		
	Granola Bars:											
23104	Soft	1 ea	28	6	124	2	19	1	5	2	1.1	1.5
23059	Hard	1 ea	25	4	118	3	16	1	5	0.6	1.1	3
47294	Fat free, all flavors	1 ea	42		140	2	35	3	0	0	0	0

PAGE KEY: A–4 = Beverages   A–6 = Dairy   A–10 = Eggs   A–12 = Fat/Oil   A–14 = Fruit   A–20 = Bakery
A–26 = Grain   A–32 = Fish   A–32 = Meats   A–36 = Poultry   A–38 = Sausage   A–40 = Mixed/Fast   A–44 = Nuts/Seeds   A–46 = Sweets
A–48 = Vegetables/Legumes   A–58 = Vegetarian   A–62 = Misc   A–62 = Soups/Sauces   A–64 = Fast   A–78 = Convenience   A–80 = Baby foods

Chol (mg)	Calc (mg)	Iron (mg)	Magn (mg)	Pota (mg)	Sodi (mg)	Zinc (mg)	VT-A (RAE)	Thia (mg)	VT-E (mg α)	Ribo (mg)	Niac (mg)	V-B$_6$ (mg)	Fola (µg)	VT-C (mg)
15	16	1.44	15	115	134	0.32	12	0.12	1.48	0.12	1.26	0.02	36	0
0	41	1.86	17	132	224	0.25	6	0.1	0.42	0.14	1.2	0.05	22	0
0	0	0.36			95		0		<.01					0
0	3	0.29	5	26	71	0.08		0.02	<.01	0.02	0.26	<.01		0
				77										
20	60	1.59	25	143	323	0.52	86	0.15	1.5	0.1	0.76	0.04	18	0
25	31	1.78	31	185	414	0.66	110	0.18	3.04	0.17	2.81	0.07	44	0
0	10	1.55	18	70	242	0.32	<1	0.03	0.63	0.07	0.83	<.01	20	0
6	11	0.88	5	32	146	0.17	6	0.11	0.11	0.11	1.07	0.03	22	0
15	43	0.88	4	78	225	0.13	6	0.09	0.1	0.06	1.16	0.01	34	0
0	17	0.61	5	28	95	0.16		0.05		0.07	0.69	0.01		0
20	19	0.95	6	39	125	0.14	3	0.11	0.09	0.13	1.24	0.03	24	0
1	15	0.48	4	14	100	0.11	3	0.06	<.01	0.04	0.47	0.06	15	0
0	14	0.76	16	61	199	0.29		0.15	0.66	0.08	1.63	0.04	26	0
0	24	0.51	4	16	117	0.11		0.04	0	0.05	0.75	<.01	11	0
0	28	0.58	7	43	169	0.21		0.04		0.07	0.73	0.02		0
0	0				80									2
0	0	0			80									2
0	3	0.52	4	19	85	0.11	<1	0.03	0.05	0.04	0.58	<.01	6	0
0	5	0.18	3	10	41	0.1	0	0.02	0.02	0.01	0.21	<.01	6	0
0	2	0.27	24	52	5	0.54	<1	0.01	0.02	0.03	1.41	0.03	4	0
0	9	1.31	27	109	175	0.62	<1	0.09	0.18	0.06	0.35	0.06	10	0
0	14	0.65	3	15	156	0.09	0	0.07	0.01	0.06	0.63	<.01	15	0
0		0.36	5	50					0.1					0
0	11	0.32	2	12	76	0.06	0	0.04	0.18	0.03	0.36	<.01	8	0
0		1.44		95	160									
0	5	0.19	3	17	38	0.08		0.03	0.08	0.02	0.2	<.01	1	0
0	4	0.25	8	24	53	0.17	0	0.02	0.07	<.01	0.36	0.01	2	0
27	37	1.8	14	96	326	0.48	5	0.26	0.79	0.19	2.2	0.05	55	3
9	12	0.57	4	31	104	0.15	2	0.08	0.25	0.06	0.7	0.02	17	1
19	22	1.4	14	110	333	0.48	25	0.29	0.85	0.21	1.8	0.06	31	2
26	13	1.84	9	57	75	0.35		0.23	1.22	0.21	2.02	0.03	19	0
0	23	0.95	13	198	69	0.2		0.05	2.46	0.05	1.05	0.03	9	5
0	14	1.81	9	58	218	0.34	150	0.15	0.46	0.19	2.05	0.2	40	0
10	0	1.08			220		0							0
10	0	1.08			220		0							0
17	21	0.92	9	60	257	0.26	18	0.1	0.91	0.11	0.87	0.03	24	0
0	0	1.08			380		0							2
4	26	1.22	13	65	205	0.46	2	0.22	0.21	0.13	1.71	0.03	29	0
0	0	0.36			200		0							1
0	99	1.42	12	75	264	0.4	0	0.25	0.18	0.16	2.21	0.02	54	0
0	98	1.41	11	74	262	0.4	0	0.2	0.17	0.14	1.98	0.02	45	0
0	175	1.62	47	139	420	1.06	<1	0.2	0.27	0.09	2.25	0.11	32	0
31	38	0.97	10	66	398	0.32	20	0.12	0.75	0.14	1.05	0.05	28	0
35	40	1.8			210		0							1
35	40	1.44			190		0							0
35	40	1.8			190		0							0
0	29	0.72	21	91	78	0.42	0	0.08	0.34	0.05	0.14	0.03	7	0
0	15	0.74	24	84	74	0.51	2	0.07	0.33	0.03	0.4	0.02	6	0
0	0	3.6			5									0
														0

A

TABLE A-1

Food Composition   (Computer code number is for West Diet Analysis program)   (For purposes of calculations, use "0" for t, <1, <.1, <.01, etc.)

Computer Code Number	Food Description	Measure	Wt (g)	H₂O (%)	Ener (kcal)	Prot (g)	Carb (g)	Dietary Fiber (g)	Fat (g)	Fat Breakdown (g) Sat	Mono	Poly
	BAKED GOODS: BREADS, CAKES, COOKIES, CRACKERS, PIES—Continued											
	Pancakes, 4″ diam:											
45001	Plain, from home recipe	1 ea	38	53	86	2	11	1	4	0.8	0.9	1.7
45002	Plain, from mix; egg, milk, oil added	1 ea	38	53	74	2	14	<1	1	0.2	0.3	0.3
	Piecrust, enriched flour, vegetable shortening, baked:											
45501	Home recipe, 9″ shell	1 ea	180	10	949	12	86	3	62	15.5	27.3	16.4
45503	From mix, 1 pie shell	1 ea	160	11	802	11	81	3	49	12.3	27.7	6.2
48014	Pie, cherry, commercial fried	1 ea	128	38	404	4	55	3	21	3.1	9.5	6.9
44015	Pretzels, thin twists, 3¼ × 2¼ × ¼″	10 pce	60	3	229	5	48	2	2	0.4	0.8	0.7
	Rolls & buns, enriched, commercial:											
42157	Cloverleaf rolls, 2½″ diam, 2″high	1 ea	28	32	84	2	14	1	2	0.5	1	0.3
42021	Hot dog buns	1 ea	40	35	112	4	20	1	2	0.4	0.4	0.8
42020	Hamburger buns	1 ea	43	35	120	4	21	1	2	0.5	0.5	0.8
42022	Hard roll, white, 3¾″ diam, 2″high	1 ea	57	31	167	6	30	1	2	0.3	0.6	1
42034	Submarine rolls/hoagies, 11¼ × 3 × 2½″	1 ea	135	34	386	11	68	4	7	1.6	3.4	1.2
	Sports/fitness bar:											
	Clif Bars:											
4042	Chocolate brownie	1 ea	68	17	236	10	41	6	4	1		
62709	Chocolate chip	1 ea	68	15	238	10	42	5	4	1		
62711	Chocolate chip peanut	1 ea	68	15	241	12	39	5	5	1.1		
62716	Cookies and cream	1 ea	68	20	225	10	39	5	4	1.4		
62207	Forza energy bar	1 ea	70		231	10	45	4	1			
62205	Tiger sports bar	1 ea	65		260	11	33	2	9	1.9		
	Power bars:											
62275	Apple cinnamon	1 ea	65		230	10	45	3	2	0.5	1.5	0.5
62276	Banana	1 ea	65		230	9	45	3	2	0.5	1	0.5
62823	Chocolate harvest	1 ea	65		240	7	45	4	4	1		
62279	Malt nut	1 ea	65		230	10	45	3	2	0.5	1	1
62280	Mocha	1 ea	65		230	10	45	3	2	1	1	0.5
	Tortillas:											
42023	Corn, enriched, 6″ diam	1 ea	26	44	58	1	12	1	1	t	0.2	0.3
42025	Flour, 10″ diam	1 ea	72	27	234	6	40	2	5	1.3	2.7	0.8
42027	Taco shells	1 ea	14	6	66	1	9	1	3	0.5	1.3	1.2
	Waffles, 7″ diam:											
45003	From home recipe	1 ea	75	42	218	6	25	1	11	2.1	2.6	5.1
45017	Whole grain, prepared from frozen	1 ea	39	43	105	4	13	1	4	1.2	1.8	1.1
	GRAIN PRODUCTS: CEREAL, FLOUR, GRAIN, PASTA and NOODLES, POPCORN											
	Grain:											
38070	Amaranth	1 c	195	10	729	28	129	30	13	3.2	2.8	5.6
	Barley:											
38002	Pearled, dry, uncooked	1 c	200	10	704	20	155	31	2	0.5	0.3	1.1
38003	Pearled, cooked	1 c	157	69	193	4	44	6	1	0.1	t	0.3
38072	Buckwheat, whole grain, dry	1 c	170	10	583	23	122	17	6	1.3	1.8	1.8
38027	Bulgar, dry, uncooked	1 c	140	9	479	17	106	26	2	0.3	0.2	0.8
38028	Bulgar, cooked	1 c	182	78	151	6	34	8	<1	t	t	0.2
38076	Couscous, cooked	1 c	157	73	176	6	36	2	<1	t	t	0.1
38329	Cracked wheat	1 c	120	10	407	16	87	14	2	0.4	0.3	0.9
38052	Millet, cooked	1 c	240	71	286	8	57	3	2	0.4	0.4	1.2
38064	Oat bran, dry	¼ c	24	7	59	4	16	4	2	0.3	0.6	0.7
38079	Quinoa, dry	1 c	170	9	636	22	117	10	10	1	2.6	4
	Rice:											
38010	Brown, cooked	1 c	195	73	216	5	45	4	2	0.4	0.6	0.6
56998	Spanish, cooked	1 c	246		130	3	28	2	1	0		
	White, enriched, all types:											
38012	Regular/long grain, dry	1 c	185	12	675	13	148	2	1	0.3	0.4	0.3
38013	Regular/long grain, cooked	1 c	158	68	205	4	45	1	<1	0.1	0.1	0.1
38019	Instant, prepared without salt	1 c	165	76	162	3	35	1	<1	t	t	t
	Parboiled/converted:											
38015	Raw, dry	1 c	185	10	686	13	151	3	1	0.3	0.3	0.3

Chol (mg)	Calc (mg)	Iron (mg)	Magn (mg)	Pota (mg)	Sodi (mg)	Zinc (mg)	VT-A (RAE)	Thia (mg)	VT-E (mg α)	Ribo (mg)	Niac (mg)	V-B₆ (mg)	Fola (µg)	VT-C (mg)
22	83	0.68	6	50	167	0.21	21	0.08	0.36	0.11	0.6	0.02	14	0
5	48	0.59	8	66	239	0.15	4	0.08	0.32	0.08	0.65	0.03	14	0
0	18	5.2	25	121	976	0.79	0	0.7	0.56	0.5	5.95	0.04	121	0
0	96	3.44	24	99	1166	0.62	0	0.48	8.83	0.3	3.8	0.09	112	0
0	28	1.56	13	83	479	0.29	12	0.18	0.55	0.14	1.82	0.04	23	2
0	22	2.59	21	88	1029	0.51	0	0.28	0.21	0.37	3.15	0.07	103	0
0	33	0.88	6	37	146	0.22	<1	0.14	0.09	0.09	1.13	0.02	27	0
0	55	1.33	8	38	192	0.26	0	0.16	0.03	0.13	1.66	0.03	44	0
0	59	1.43	9	40	206	0.28	0	0.17	0.03	0.14	1.79	0.03	48	0
0	54	1.87	15	62	310	0.54	0	0.27	0.24	0.19	2.42	0.02	54	0
0	188	4.28	27	190	756	0.84	0	0.65	0.62	0.42	5.31	0.06	36	0
0	271	5.75	122	257	149	3.65		0.39	20.18	0.29	3.54	0.42	84	66
0	265	5.22	96	206	76	3.45		0.35	20.26	0.28	3.49	0.39	86	66
0	265	5.37	111	305	274	3.51		0.4	20.34	0.3	6.09	0.41	95	66
0	279	5.23	103	212	179	3.21		0.35	20.26	0.27	3.45	0.43	85	66
0	300	6.3	160	220	65	5.25		1.5	18.35	1.7	20	2	400	60
0	557	5.01	186		139			2.37		1.1	5.57	1.11		11
0	300	6.3	140	110	90	5.25	0	1.5	18.35	1.7	20	2	400	60
0	300	6.3	140	200	90	5.25	0	1.5	18.35	1.7	20	2	400	60
0	150	2.7	60		80	2.25	0	0.75	18.34	0.85	10	1	200	60
0	300	6.3	140	110	90	5.25	0	1.5	18.35	1.7	20	2	400	60
0	300	6.3	140	145	90	5.25	0	1.5	18.35	1.7	20	2	400	60
0	46	0.36	17	40	42	0.24	<1	0.03	0.02	0.02	0.39	0.06	26	0
0	90	2.38	19	94	344	0.51	0	0.38	0.4	0.21	2.57	0.04	75	0
0	22	0.35	15	25	51	0.2	5	0.03	0.42	<.01	0.19	0.05	1	0
52	191	1.73	14	119	383	0.51	49	0.2	1.72	0.26	1.55	0.04	34	0
37	102	0.81	16	90	132	0.44		0.08	0.55	0.13	0.77	0.04	7	0
0	298	14.8	519	714	41	6.2	0	0.16	2.01	0.41	2.51	0.43	96	8
0	58	5	158	560	18	4.26	2	0.38	0.04	0.23	9.21	0.52	46	0
0	17	2.09	35	146	5	1.29	1	0.13	0.02	0.1	3.24	0.18	25	0
0	31	3.74	393	782	2	4.08	0	0.17	1.75	0.72	11.93	0.36	51	0
0	49	3.44	230	574	24	2.7	1	0.32	0.08	0.16	7.16	0.48	38	0
0	18	1.75	58	124	9	1.04	<1	0.1	0.02	0.05	1.82	0.15	33	0
0	13	0.6	13	91	8	0.41	0	0.1	0.2	0.04	1.54	0.08	24	0
0	41	4.67	166	486	6	3.53	0	0.54	0.3	0.26	7.64	0.41	53	0
0	7	1.51	106	149	5	2.18	<1	0.25	0.05	0.2	3.19	0.26	46	0
0	14	1.3	56	136	1	0.75	0	0.28	0.24	0.05	0.22	0.04	12	0
0	102	15.72	357	1258	36	5.61	0	0.34	8.28	0.67	4.98	0.38	83	0
0	20	0.82	84	84	10	1.23	0	0.19	0.06	0.05	2.98	0.28	8	0
0	0	0			1340		0							0
0	52	7.97	46	213	9	2.02	0	1.07	0.2	0.09	7.76	0.3	427	0
0	16	1.9	19	55	2	0.77	0	0.26	0.06	0.02	2.33	0.15	92	0
0	13	1.04	8	7	5	0.4	0	0.12	0.02	0.08	1.45	0.02	116	0
0	111	6.59	57	222	9	1.78	0	1.1	0.06	0.13	6.72	0.65	475	0

A

TABLE A-1

Food Composition   (Computer code number is for West Diet Analysis program)   (For purposes of calculations, use "0" for t, <1, <.1, <.01, etc.)

Computer Code Number	Food Description	Measure	Wt (g)	H₂O (%)	Ener (kcal)	Prot (g)	Carb (g)	Dietary Fiber (g)	Fat (g)	Fat Breakdown (g) Sat	Mono	Poly
	GRAIN PRODUCTS: CEREAL, FLOUR, GRAIN, PASTA and NOODLES, POPCORN—Continued											
38016	Cooked	1 c	175	72	200	4	43	1	<1	0.1	0.1	0.1
38083	Sticky (Glutinous), cooked	1 c	174	77	169	4	37	2	<1	t	0.1	0.1
38021	Wild, cooked	1 c	164	74	166	7	35	3	1	t	t	0.3
38163	Rice and pasta (Rice-a-Roni), cooked	1 c	202	72	246	5	43	5	6	1.1	2.3	1.9
38034	Tapioca-pearl, dry	1 c	152	11	544	<1	135	1	<1	t	t	t
40002	Wheat, rolled, cooked	1 c	240	84	158	5	33	4	1	0.1	0.1	0.5
	Flour & Grain Fractions:											
38053	Buckwheat flour, dark	1 c	120	11	402	15	85	12	4	0.8	1.1	1.1
	Cornmeal:											
38059	Whole-ground, unbolted, dry	1 c	122	10	442	10	94	9	4	0.6	1.2	2
38004	Degermed, enriched, dry	1 c	138	12	505	12	107	10	2	0.3	0.6	1
38041	Degermed, enriched, baked	1 c	138	12	505	12	107	10	2	0.3	0.6	1
38056	Rye flour, medium	1 c	102	10	361	10	79	15	2	0.2	0.2	0.8
7502	Soy flour, low-fat	1 c	88	3	325	45	30	9	6	0.9	1.3	3.3
38024	Wheat bran, crude	1 c	58	10	125	9	37	25	2	0.4	0.4	1.3
38025	Wheat germ, raw	1 c	115	11	414	27	60	15	11	1.9	1.6	6.9
38026	Wheat germ, toasted	1 c	113	6	432	33	56	17	12	2.1	1.7	7.5
38055	Wheat germ, with brown sugar & honey	1 c	113	3	420	30	66	12	9	1.5	1.2	5.5
	Wheat flour:											
38030	All-purpose white flour, enriched	1 c	125	12	455	13	95	3	1	0.2	0.1	0.5
38033	Self-rising, enriched, unsifted	1 c	125	11	442	12	93	3	1	0.2	0.1	0.5
38032	Whole wheat, from hard wheats	1 c	120	10	407	16	87	15	2	0.4	0.3	0.9
	Breakfast Bars:											
47279	Store brand, fat free, all flavors	1 ea	38		110	2	26	3	0	0	0	0
	Snackwell:											
40255	Apple-cinnamon	1 ea	37	16	119	1	29	1	<1	t	t	0.1
40254	Blueberry	1 ea	37	16	121	1	29	1	<1	t	t	0.1
40253	Strawberry	1 ea	37	16	120	1	29	1	<1	t	t	0.1
	Breakfast cereals, hot, cooked: w/o salt added											
	Corn grits (hominy) enriched:											
38007	Regular/quick prep w/o salt, yellow:	1 c	242	85	143	3	31	1	<1	t	0.1	0.2
40089	Instant, prepared from packet, white	1 ea	137	82	93	2	21	1	<1	t	t	t
	Cream of wheat:											
40079	Regular, quick, instant	1 c	239	87	129	4	27	1	<1	t	t	0.3
40087	Mix and eat, plain, packet	1 ea	142	82	102	3	21	<1	<1	t	t	0.2
40006	Farina cereal, cooked w/o salt	1 c	233	88	112	3	24	1	<1	t	t	t
40014	Malt-O-Meal, cooked w/o salt	1 c	240	88	122	4	26	1	<1	0.1	t	t
40239	Maypo	1 c	216	83	153	5	29	4	2	0.4	0.6	0.5
	Oatmeal or rolled oats:											
40000	Cooked w/o salt, nonfortified	1 c	234	85	147	6	25	4	2	0.4	0.7	0.9
492	Plain, from packet, fortified	½ c	118	86	65	3	11	2	1	0.2	0.3	0.4
	Breakfast cereals, ready to eat:											
40095	All-Bran	1 c	62	3	161	8	46	20	2	0.3	0.5	1.3
40258	Alpha Bits	1 c	28	1	110	2	24	1	1	0.1	0.2	0.2
40098	Apple Jacks	1 c	33	3	129	1	30	1	1	0.1	0.2	0.3
40029	Bran Buds	1 c	90	3	225	6	72	39	2	0.4	0.4	1.1
40323	Bran Chex	1 c	49	2	156	5	39	8	1	0.2	0.3	0.7
40084	Honey BucWheat Crisp	1 c	38	5	147	4	31	3	1	0.2	0.2	0.5
40031	C.W. Post, with raisins	1 c	103	4	446	9	74	14	15	11	1.7	1.4
40032	Cap'n Crunch	1 c	37	2	148	2	31	1	2	0.6	0.4	0.3
40033	Cap'n Crunchberries	1 c	35	2	140	2	30	1	2	0.5	0.4	0.3
40034	Cap'n Crunch, peanut butter	1 c	35	2	146	2	28	1	3	0.7	1.4	0.8
40297	Cheerios	1 c	23	3	85	3	17	2	1	0.3	0.5	0.2
40102	Cocoa Krispies	1 c	41	3	156	1	36	1	1	0.8	0.2	t
40257	Cocoa Pebbles	1 c	32	3	127	1	28	1	1	1.2	0.1	t
40036	Corn Bran	1 c	36	2	121	2	31	6	1	0.3	0.3	0.4
40325	Corn Chex	1 c	28	2	104	2	24	1	<1	t	t	t
40195	Corn Flakes, Kellogg's	1 c	28	3	101	2	24	1	<1	t	t	t

PAGE KEY: A–4 = Beverages   A–6 = Dairy   A–10 = Eggs   A–12 = Fat/Oil   A–14 = Fruit   A–20 = Bakery
A–26 = Grain   A–32 = Fish   A–32 = Meats   A–36 = Poultry   A–38 = Sausage   A–40 = Mixed/Fast   A–44 = Nuts/Seeds   A–46 = Sweets
A–48 = Vegetables/Legumes   A–58 = Vegetarian   A–62 = Misc   A–62 = Soups/Sauces   A–64 = Fast   A–78 = Convenience   A–80 = Baby foods

Chol (mg)	Calc (mg)	Iron (mg)	Magn (mg)	Pota (mg)	Sodi (mg)	Zinc (mg)	VT-A (RAE)	Thia (mg)	VT-E (mg α)	Ribo (mg)	Niac (mg)	V-B$_6$ (mg)	Fola (µg)	VT-C (mg)
0	33	1.98	21	65	5	0.54	0	0.44	0.02	0.03	2.45	0.03	133	0
0	3	0.24	9	17	9	0.71	0	0.03	0.07	0.02	0.5	0.05	2	0
0	5	0.98	52	166	5	2.2	<1	0.09	0.39	0.14	2.11	0.22	43	0
2	16	1.9	24	85	1147	0.57	0	0.25	0.27	0.16	3.6	0.2	89	0
0	30	2.4	2	17	2	0.18	0	<.01	0	0	0	0.01	6	0
0	22	1.49	55	170	0	1.18	<1	0.17	0.58	0.13	2.13	0.17	34	0
0	49	4.87	301	692	13	3.74	0	0.5	0.38	0.23	7.38	0.7	65	0
0	7	4.21	155	350	43	2.22	13	0.47	0.51	0.25	4.43	0.37	30	0
0	7	5.7	55	224	4	0.99	15	0.99	0.21	0.56	6.95	0.35	322	0
0	7	5.7	55	224	4	0.99	28	0.74	0.23	0.48	6.6	0.27	52	0
0	24	2.16	76	347	3	2.03	0	0.29	0.81	0.12	1.76	0.27	19	0
0	165	5.27	202	2262	16	1.04	2	0.33	0.17	0.25	1.9	0.46	361	0
0	42	6.13	354	686	1	4.22	<1	0.3	0.86	0.33	7.88	0.76	46	0
0	45	7.2	275	1026	14	14.13	0	2.16	16.1	0.57	7.83	1.5	323	0
0	51	10.27	362	1070	5	18.84	6	1.89	18.07	0.93	6.32	1.11	398	7
0	56	9.1	307	1089	12	15.68	0	1.51	34.07	0.78	5.34	0.56	685	0
0	19	5.8	28	134	2	0.88	0	0.98	0.08	0.62	7.38	0.06	229	0
0	422	5.84	24	155	1588	0.78	0	0.84	0.06	0.52	7.29	0.06	245	0
0	41	4.66	166	486	6	3.52	1	0.54	0.98	0.26	7.64	0.41	53	0
0	20	0.72			25									1
0	17	5	6	68	103	3.88		0.39		0.44	5.2	0.52		0
0	14	4.83	5	44	107	3.85		0.39		0.44	5.2	0.52		0
0	14	4.82	6	47	102	3.83		0.39		0.44	5.2	0.52		2
0	7	1.45	12	51	5	0.17	5	0.2	0.05	0.13	1.75	0.05	80	0
0	8	7.96	10	38	288	0.18	<1	0.16	0.03	0.19	2.21	0.05	47	0
0	50	10.28	12	45	139	0.33	0	0.24	0.02	0	1.43	0.03	108	0
0	20	8.09	7	38	241	0.24	376	0.43	0.01	0.28	4.97	0.57	101	0
0	9	1.16	5	30	5	0.19	0	0.14	0.02	0.1	1.14	0.02	79	0
0	5	9.6	5	31	2	0.17	0	0.48	0.02	0.24	5.76	0.02	5	0
0	117	7.54	48	190	233	1.34	631	0.64	0.15	0.71	8.42	0.83	11	25
0	19	1.59	56	131	2	1.15	0	0.26	0.23	0.05	0.3	0.05	9	0
0	66	5.12	27	63	53	0.54	190	0.17	0.12	0.2	2.4	0.25	51	0
0	205	9.92	236	701	160	3.72	326	0.74	0.76	0.87	9.92	3.72	812	12
0	8	2.66	17	54	178	1.48		0.36	0.02	0.42	4.93	0.5	99	0
0	8	4.59	18	40	157	1.65	51	0.56	0.05	0.43	5.08	0.5	102	15
0	57	13.5	184	900	608	4.5	460	1.08	1.42	1.26	15.3	6.03	1210	18
0	29	13.99	69	216	345	6.48	5	0.64	0.56	0.26	8.62	0.88	173	26
0	54	10.86	43	142	361	0.68	914	0.9	8.99	1.03	12.06	1.88	11	36
0	50	16.38	74	261	161	1.64		1.34	0.72	1.54	18.13	1.85	364	0
0	6	7.07	21	74	277	5.87	3	0.58	0.34	0.66	7.83	0.78	576	0
0	7	6.65	20	72	245	5.54	2	0.55	0.26	0.63	7.38	0.74	545	0
0	4	6.42	24	83	260	5.35	3	0.53	0.42	0.61	7.27	0.71	545	0
0	77	6.21	31	74	209	2.88	115	0.29	0.08	0.33	3.84	0.38	153	5
0	53	6.15	16	66	251	1.97	202	0.49	0.25	0.57	6.56	0.66	135	20
0	4	1.99	12	47	173	1.65		0.41		0.47	5.52	0.55	110	0
0	26	11.1	19	75	309	5.5	3	0.18	0.24	0.62	7.34	0.73	533	0
0	93	8.4	8	23	269	3.5	141	0.35	0.05	0.4	4.68	0.47	187	6
0	2	8.4	3	25	203	0.08	150	0.36	0.04	0.43	5.01	0.5	102	6

A

A Appendix Table of Food Composition

TABLE A-1

Food Composition

(Computer code number is for West Diet Analysis program)   (For purposes of calculations, use "0" for t, <1, <.1, <.01, etc.)

Computer Code Number	Food Description	Measure	Wt (g)	H₂O (%)	Ener (kcal)	Prot (g)	Carb (g)	Dietary Fiber (g)	Fat (g)	Fat Breakdown (g) Sat	Mono	Poly
	GRAIN PRODUCTS: CEREAL, FLOUR, GRAIN, PASTA and NOODLES, POPCORN—Continued											
40263	Corn Flakes, Post Toasties	1 c	28	4	101	2	24	1	<1	0	t	t
40206	Corn Pops	1 c	31	3	118	1	28	<1	<1	t	t	t
40205	Cracklin' Oat Bran	1 c	65	3	266	5	46	8	9	2.7	5.4	1.4
40040	Crispy Wheat `N Raisins	1 c	43	7	143	3	35	4	1	0.2	t	0.3
40041	Fortified Oat Flakes	1 c	48	3	180	8	36	1	1	0.2	0.3	0.4
40202	40% Bran Flakes, Kellogg's	1 c	39	3	124	4	31	7	1	0.2	0.2	0.4
40259	40% Bran Flakes, Post	1 c	47	4	150	4	38	8	1	0.2	0.2	0.7
40218	Froot Loops	1 c	32	3	126	1	28	1	1	0.5	0.1	0.2
40217	Frosted Flakes	1 c	41	3	150	1	37	1	<1	t	t	0.1
40043	Frosted Mini-Wheats	1 c	55	6	187	5	44	6	1	0.2	0.1	0.6
40105	Frosted Rice Krispies	1 c	35	3	133	1	31	<1	<1	t	0.1	0.1
40266	Fruity Pebbles	1 c	32	3	128	1	28	<1	1	0.3	0.6	0.4
40299	Golden Grahams	1 c	39	3	145	2	32	1	1	0.2	0.5	0.5
40048	Granola, homemade	½ c	61	5	299	9	32	5	15	2.8	4.7	6.5
40197	Granola, low fat, commercial	½ c	45	4	165	4	36	2	2	0.7	1.1	0.4
40277	Grape Nuts	½ c	55	4	197	6	45	5	1	0.2	0.2	0.6
40265	Grape Nuts Flakes	1 c	39	3	142	4	32	3	1	0.2	0.3	0.6
40238	Heartland Natural with raisins	1 c	110	5	468	11	76	6	16	4	4.2	6.2
40112	Honey & Nut Corn Flakes	1 c	37	2	150	3	31	1	2	0.3	0.8	0.6
40051	Honey Nut Cheerios	1 c	33	2	123	3	26	2	1	0.3	0.4	0.5
40052	HoneyBran	1 c	35	2	119	3	29	4	1	0.3	t	0.3
40264	HoneyComb	1 c	22	2	87	1	20	1	<1	0.1	0.1	0.2
40054	King Vitaman	1 c	21	2	81	1	18	1	1	0.2	0.2	0.2
40010	Kix	1 c	19	2	72	1	16	1	<1	t	0.1	0.1
40011	Life	1 c	44	4	165	4	34	3	2	0.4	0.7	0.6
40300	Lucky Charms	1 c	32	2	122	2	27	2	1	0.3	0.3	0.3
40124	Mueslix Five Grain	1 c	82	8	289	6	63	6	5	0.7	2	1.8
40008	Nature Valley Granola	1 c	113	4	510	12	74	7	20	2.6	13.3	3.8
40115	Nutri Grain Almond Raisin	1 c	40	6	147	3	31	3	2	t	1	1.2
40281	100% Bran	1 c	66	3	178	8	48	20	3	0.6	0.6	1.9
40063	100% Natural cereal, plain	1 c	104	2	473	11	69	8	20	8.6	4.6	2
40064	100% Natural with apples & cinnamon	1 c	104	2	477	11	70	7	20	15.5	1.8	1.3
40065	100% Natural with raisins & dates	1 c	110	4	496	12	72	7	20	13.6	3.7	1.7
40216	Product 19	1 c	30	3	100	2	25	1	<1	t	0.1	0.2
40066	Quisp	1 c	30	2	122	1	26	1	2	0.5	0.3	0.2
40209	Raisin Bran, Kellogg's	1 c	61	8	195	5	47	7	2	0.3	0.3	0.9
512	Raisin Bran, Post	1 c	59	9	187	5	46	8	1	0.2	0.2	0.7
40117	Raisin Squares	1 c	71	10	239	7	56	7	1	0.2	0.2	0.6
40333	Rice Chex	1 c	33	3	124	2	28	<1	<1	0.1	t	t
40017	Rice Krispies, Kellogg's	1 c	28	2	111	2	25	<1	<1	t	t	t
40018	Rice, puffed	1 c	14	4	54	1	12	<1	<1	t	t	t
40022	Shredded Wheat	1 c	43	5	154	5	35	4	1	0.1	0.1	0.4
40211	Special K	1 c	31	3	117	7	22	1	<1	0.1	0.1	0.2
40261	Super Golden Crisp	1 c	33	2	123	2	30	<1	<1	t	t	0.1
40068	Honey Smacks	1 c	36	2	139	2	32	1	1	0.1	0.2	0.3
40070	Tasteeos	1 c	24	2	94	3	19	3	1	0.2	0.2	0.2
40021	Total, wheat, with added calcium	1 c	40	3	130	3	30	3	1	0.2	0.2	0.4
40306	Trix	1 c	28	2	109	1	25	1	1	0.2	0.6	0.3
40335	Wheat Chex	1 c	46	2	159	5	37	5	1	0.2	0.1	0.4
40023	Wheat cereal, puffed, fortified	1 c	12	4	44	2	9	1	<1	t	t	0.1
40307	Wheaties	1 c	29	3	103	3	23	3	1	0.2	0.3	0.3
7508	Natto	1 c	175	55	371	31	25	9	19	2.8	4.3	10.9
	Pasta:											
	Cellophane Noodles:											
38146	Cooked	1 c	190	79	160	<1	39	<1	<1	t	t	t
7196	Dry	1 c	140	13	491	<1	121	1	<1	t	t	t
38048	Chow Mein, dry	1 c	45	1	237	4	26	2	14	2	3.5	7.8
	Cooked:											
38092	Fresh	2 oz	57	69	75	3	14	1	1	t	t	0.2
38118	Linguini/Rotini	1 c	140	66	197	7	40	2	1	0.1	0.1	0.4

PAGE KEY: A–4 = Beverages   A–6 = Dairy   A–10 = Eggs   A–12 = Fat/Oil   A–14 = Fruit   A–20 = Bakery   A–26 = Grain   A–32 = Fish   A–32 = Meats   A–36 = Poultry   A–38 = Sausage   A–40 = Mixed/Fast   A–44 = Nuts/Seeds   A–46 = Sweets   A–48 = Vegetables/Legumes   A–58 = Vegetarian   A–62 = Misc   A–62 = Soups/Sauces   A–64 = Fast   A–78 = Convenience   A–80 = Baby foods

Chol (mg)	Calc (mg)	Iron (mg)	Magn (mg)	Pota (mg)	Sodi (mg)	Zinc (mg)	VT-A (RAE)	Thia (mg)	VT-E (mg α)	Ribo (mg)	Niac (mg)	V-B$_6$ (mg)	Fola (µg)	VT-C (mg)
0	1	5.4	4	33	266	0.13		0.38	0.67	0.43	5	0.5	100	0
0	5	1.92	2	26	120	1.52	151	0.37	0.03	0.43	4.99	0.5	102	6
0	27	2.4	80	292	186	2.02	298	0.5	0.91	0.57	6.7	0.66	133	21
0	0	5.85	33	178	197	5.85	117	0.58	0.27	0.67	7.83	0.78	157	0
0	68	13.73	58	228	220	2.54		0.62	0.34	0.72	8.45	0.86	169	0
0	21	24.18	55	230	279	20.48	307	2.11	36.14	2.3	26.91	2.73	542	81
0	26	12.69	101	290	344	2.35		0.59		0.67	7.83	0.78	157	0
0	25	4.51	9	35	151	1.5	155	0.38	0.15	0.42	4.99	0.51	100	15
0	2	5.94	3	30	196	0.07	212	0.49	0.02	0.62	6.64	0.66	134	8
0	18	15.95	65	187	6	1.76	0	0.41	0.3	0.46	5.39	0.54	108	0
0	4	2.1	9	33	254	0.32	176	0.46	0.02	0.49	5.95	0.6	234	7
0	2	2.13	6	35	187	1.78		0.44		0.5	5.93	0.59	118	0
0	455	5.85	11	64	349	4.88	195	0.49	0.14	0.55	6.51	0.65	130	8
0	48	2.59	107	328	13	2.51	1	0.45	3.59	0.18	1.29	0.19	51	1
0	19	1.35	34	135	111	2.84	169	0.28	3.78	0.32	3.74	1.48	302	3
0	19	15.36	55	169	336	1.14		0.36		0.4	4.74	0.47	95	0
0	15	10.89	40	133	188	1.61		0.5	0.1	0.57	6.72	0.67	135	0
0	66	4.01	141	415	226	2.83	3	0.32	0.77	0.14	1.54	0.2	44	1
0	4	3.03	3	40	249	0.26	152	0.26	0.09	0.3	3.37	0.33	74	10
0	110	4.95	35	101	296	4.12	165	0.41	0.59	0.47	5.51	0.55	220	7
0	16	5.56	46	150	202	0.9	463	0.46	0.81	0.52	6.16	0.63	23	19
0	4	2.05	8	26	163	1.14		0.28		0.32	3.79	0.38	76	0
0	3	6.09	18	58	176	2.62	210	0.26	1.41	0.3	3.5	0.35	280	8
0	95	5.13	5	22	169	2.38	101	0.24	0.04	0.27	3.17	0.32	127	4
0	154	12.31	42	125	226	5.68	1	0.55	0.24	0.64	7.57	0.76	572	6
0	107	4.8	17	61	217	4	160	0.4	0.1	0.45	5.34	0.53	213	6
0	67	8.94	82	369	107	7.46	747	0.75	8.94	0.84	9.84	0.99	197	1
0	85	3.53	107	375	183	2.27	0	0.35	7.97	0.12	1.25	0.16	17	0
0	122	1	9	143	142	2.72	0	0.28	4	0.32	3.64	0.36	80	0
0	46	8.12	312	652	457	5.74	0	1.58	1.53	1.78	20.92	2.11	47	63
2	124	2.68	115	515	50	2.45	1	0.33	3.34	0.28	2.09	0.19	37	1
0	157	2.89	72	514	52	2		0.33	0.73	0.57	1.87	0.11	17	1
0	160	3.12	124	538	47	2.11		0.31	0.77	0.65	2.09	0.16	45	0
0	5	18.09	16	50	207	15.3	225	1.5	20.13	1.71	20.01	2.07	400	61
0	3	5.51	16	57	222	4.59	12	0.46	0.2	0.52	6.12	0.61	467	3
0	29	4.64	83	372	362	1.55	155	0.39	0.48	0.44	5.18	0.52	104	0
0	27	10.8	88	357	360	2.25		0.38		0.42	5	0.5	100	0
0	28	19.88	56	342	4	1.99	0	0.5	0.37	0.57	6.67	0.67	134	0
0	110	9.9	10	32	311	4.12	165	0.41	0.02	0.47	5.51	0.55	220	7
0	5	0.7	12	27	206	0.46	371	0.52	0	0.59	6.92	0.69	88	15
0	1	0.4	4	16	1	0.15	0	0.06	0.02	0.04	0.49	0	22	0
0	16	1.81	57	155	4	1.42	0	0.11	0.23	0.12	2.26	0.11	22	0
0	9	8.37	19	61	224	0.9	230	0.53	7.07	0.59	7.13	1.98	400	21
0	7	2.08	20	48	51	1.75		0.43	0.12	0.5	5.81	0.59	116	0
0	9	0.47	21	54	67	0.47	204	0.5	0.18	0.58	6.66	0.68	135	8
0	11	6.86	26	71	183	0.69	318	0.31	0.08	0.36	4.22	0.43	85	13
0	1333	24	32	119	256	20	200	2	26.84	2.27	26.68	2.67	533	80
0	93	4.2	3	16	181	3.5	140	0.35	0.56	0.4	4.68	0.47	93	6
0	92	13.34	37	172	410	3.68	138	0.34	0.33	0.39	4.6	0.46	368	6
0	3	0.53	16	44	1	0.37	0	0.08	0	0.05	0.63	0.02	18	0
0	0	7.83	31	107	210	7.25	145	0.72	0.18	0.82	9.66	0.97	193	6
0	380	15.05	201	1276	12	5.3		0.28	0.02	0.33	0	0.23	14	23
0	13	0.86	3	3	8	0.2	0	0.04	0.06	0	0.06	0.02	1	0
0	35	3.04	4	14	14	0.57	0	0.21	0.18	0	0.28	0.07	3	0
0	9	2.13	23	54	198	0.63	<1	0.26	1.57	0.19	2.68	0.05	40	0
19	3	0.65	10	14	3	0.32	3	0.12	0.09	0.09	0.57	0.02	36	0
0	10	1.96	25	43	1	0.74	0	0.29	0.08	0.14	2.34	0.05	108	0

A

TABLE A-1

Food Composition   (Computer code number is for West Diet Analysis program)   (For purposes of calculations, use "0" for t, <1, <.1, <.01, etc.)

Computer Code Number	Food Description	Measure	Wt (g)	H₂O (%)	Ener (kcal)	Prot (g)	Carb (g)	Dietary Fiber (g)	Fat (g)	Fat Breakdown (g) Sat	Mono	Poly
	GRAIN PRODUCTS: CEREAL, FLOUR, GRAIN, PASTA and NOODLES, POPCORN—Continued											
38047	Egg noodles, cooked, enriched	1 c	160	69	213	8	40	2	2	0.5	0.7	0.7
	Macaroni, cooked:											
38102	Enriched	1 c	140	66	197	7	40	2	1	0.1	0.1	0.4
38110	Whole wheat	1 c	140	67	174	7	37	4	1	0.1	0.1	0.3
38117	Vegetable, enriched	1 c	134	68	172	6	36	6	<1	t	t	t
	Spaghetti:											
38121	With salt, enriched	1 c	140	66	197	7	40	2	1	0.1	0.1	0.4
38060	Whole-wheat, cooked	1 c	140	67	174	7	37	6	1	0.1	0.1	0.3
38062	Spinach noodles, dry	3½ oz	100	8	372	13	75	11	2	0.2	0.2	0.6
	Popcorn:											
44012	Air popped, plain	1 c	8	4	31	1	6	1	<1	t	t	0.2
44013	Popped in vegetable oil/salted	1 c	11	3	55	1	6	1	3	0.5	0.9	1.5
44014	Sugar-syrup coated	1 c	35	3	151	1	28	2	4	1.3	1	1.6
	MEATS: FISH AND SHELLFISH											
	Fish:											
17029	Bass, baked or broiled	4 oz	113	69	165	27	0	0	5	1.1	2.1	1.5
17031	Bluefish, baked or broiled	4 oz	113	63	180	29	0	0	6	1.3	2.6	1.5
17126	Catfish, breaded/flour fried	4 oz	113	49	329	21	14	<1	20	4.5	9.2	5.2
	Cod:											
17037	Baked	4 oz	113	76	119	26	0	0	1	0.2	0.1	0.3
17000	Batter fried	4 oz	113	67	197	20	8	<1	9	1.8	3.6	3
17001	Poached, no added fat	4 oz	113	77	116	25	0	0	1	0.1	0.1	0.3
17002	Fish sticks, breaded pollock	2 ea	56	46	152	9	13	1	7	1.8	2.8	1.8
17068	Flounder/sole, baked	4 oz	113	73	132	27	0	0	2	0.4	0.3	0.7
17071	Grouper, baked or broiled	4 oz	113	73	133	28	0	0	1	0.3	0.3	0.5
17007	Haddock, breaded, fried	4 oz	113	60	247	23	10	<1	12	2.6	5.3	3.7
	Halibut:											
17291	Baked	4 oz	113	72	158	30	0	0	3	0.5	1.1	1.1
17259	Smoked	4 oz	113		203	34	0	0	4	0.6	1.2	1.5
17044	Raw	4 oz	113	78	124	24	0	0	3	0.4	0.8	0.8
17012	Herring, pickled	4 oz	113	55	296	16	11	0	20	2.7	13.5	1.9
	Ocean Perch:											
17093	Baked/Broiled	4 oz	113	73	137	27	0	0	2	0.4	0.9	0.6
17015	Breaded/Fried	4 oz	113	58	255	23	10	<1	13	2.7	5.8	3.9
19025	Octopus, raw	4 oz	113	80	93	17	2	0	1	0.3	0.2	0.3
17096	Pollock, baked, broiled, or poached	4 oz	113	74	128	27	0	0	1	0.3	0.2	0.6
17099	Salmon, broiled or baked	4 oz	113	62	244	31	0	0	12	2.2	6	2.7
17060	Sardines, Atlantic, canned, drained, 2=24g	4 oz	113	60	235	28	0	0	13	1.7	4.4	5.8
17022	Snapper, baked or broiled	4 oz	113	70	145	30	0	0	2	0.4	0.4	0.7
19047	Squid, fried in flour	4 oz	113	65	198	20	9	0	8	2.1	3.1	2.4
17080	Surimi	4 oz	113	76	112	17	8	0	1	0.2	0.2	0.5
17065	Swordfish, raw	4 oz	113	76	137	22	0	0	5	1.2	1.7	1
17066	Swordfish, baked or broiled	4 oz	113	69	175	29	0	0	6	1.6	2.2	1.3
17082	Trout, baked or broiled	4 oz	113	70	170	26	0	0	7	1.8	2	2.1
	Tuna:											
	Light, canned, drained solids:											
17025	Oil pack	1 c	145	60	287	42	0	0	12	2.2	4.3	4.2
17027	Water pack	1 c	154	75	179	39	0	0	1	0.4	0.2	0.5
17085	Bluefin, fresh	4 oz	113	68	163	26	0	0	6	1.4	1.8	1.6
	Shellfish:											
	Clams:											
19128	Raw meat only	1 ea	145	82	107	19	4	0	1	0.1	0.1	0.4
19002	Canned, drained	1 c	160	64	237	41	8	0	3	0.3	0.3	0.9
19000	Steamed, meat only	10 ea	95	64	141	24	5	0	2	0.2	0.2	0.5
	Crab, meat only:											
19033	Blue crab, cooked	1 c	118	77	120	24	0	0	2	0.3	0.3	0.8
19037	Imitation, from surimi	4 oz	113	74	115	14	12	0	1	0.3	0.2	0.8
19006	Lobster meat, cooked w/moist heat	1 c	145	76	142	30	2	0	1	0.2	0.2	0.1

PAGE KEY: A–4 = Beverages   A–6 = Dairy   A–10 = Eggs   A–12 = Fat/Oil   A–14 = Fruit   A–20 = Bakery   A–26 = Grain   A–32 = Fish   A–32 = Meats   A–36 = Poultry   A–38 = Sausage   A–40 = Mixed/Fast   A–44 = Nuts/Seeds   A–46 = Sweets   A–48 = Vegetables/Legumes   A–58 = Vegetarian   A–62 = Misc   A–62 = Soups/Sauces   A–64 = Fast   A–78 = Convenience   A–80 = Baby foods

Chol (mg)	Calc (mg)	Iron (mg)	Magn (mg)	Pota (mg)	Sodi (mg)	Zinc (mg)	VT-A (RAE)	Thia (mg)	VT-E (mg α)	Ribo (mg)	Niac (mg)	V-B$_6$ (mg)	Fola (µg)	VT-C (mg)
53	19	2.54	30	45	11	0.99	10	0.3	0.26	0.13	2.38	0.06	102	0
0	10	1.96	25	43	1	0.74	0	0.29	0.08	0.14	2.34	0.05	108	0
0	21	1.48	42	62	4	1.13	<1	0.15	0.42	0.06	0.99	0.11	7	0
0	15	0.66	25	42	8	0.59	7	0.15	0.12	0.08	1.44	0.03	87	0
0	10	1.96	25	43	140	0.74	0	0.29	0.08	0.14	2.34	0.05	108	0
0	21	1.48	42	62	4	1.13	<1	0.15	0.42	0.06	0.99	0.11	7	0
0	58	2.13	174	376	36	2.76	23	0.37	0.64	0.2	4.55	0.32	48	0
0	1	0.21	10	24	<1	0.28	1	0.02	0.02	0.02	0.16	0.02	2	0
0	1	0.31	12	25	97	0.29	1	0.01	0.55	0.01	0.17	0.02	2	0
2	15	0.61	12	38	72	0.2	1	0.02	0.42	0.02	0.77	<.01	2	0
98	116	2.16	43	515	102	0.94	40	0.1	0.84	0.1	1.72	0.16	19	2
86	10	0.7	47	539	87	1.18	156	0.08	0.71	0.11	8.19	0.52	2	0
91	62	1.88	36	391	240	1.17		0.46	2.87	0.21	3.78	0.22	17	1
62	16	0.55	47	276	88	0.66	16	0.1	0.92	0.09	2.84	0.32	9	1
56	33	0.81	28	437	104	0.57		0.08	1.49	0.11	2.58	0.37	10	2
52	10	0.37	31	484	90	0.56	9	0.02	0.33	0.05	2.45	0.45	7	3
63	11	0.41	14	146	326	0.37	17	0.07	0.29	0.1	1.19	0.03	24	0
77	20	0.38	66	389	119	0.71	15	0.09	0.75	0.13	2.46	0.27	10	0
53	24	1.29	42	537	60	0.58	56	0.09	0.71	<.01	0.43	0.4	11	0
88	70	2.02	49	377	194	0.63		0.11	1.93	0.12	4.94	0.31	16	0
46	68	1.21	121	651	78	0.6	61	0.08	1.23	0.1	8.05	0.45	16	0
59	87	1.56	154	833		0.78	86	0.11	1.11	0.14	10.83	0.64	22	0
36	53	0.95	94	508	61	0.47	53	0.07	0.96	0.08	6.61	0.39	14	0
15	87	1.38	9	78	983	0.6	292	0.04	1.93	0.16	3.73	0.19	2	0
61	155	1.33	44	396	108	0.69	16	0.15	1.84	0.15	2.75	0.31	11	1
71	151	1.88	39	335	201	0.75		0.18	2.87	0.2	2.98	0.24	13	1
54	60	5.99	34	396	260	1.9	51	0.03	1.36	0.05	2.37	0.41	18	6
108	7	0.32	82	437	131	0.68	28	0.08	0.89	0.09	1.86	0.08	5	0
98	8	0.62	35	424	75	0.58	71	0.24	1.42	0.19	7.54	0.25	6	0
160	432	3.3	44	449	571	1.48	36	0.09	2.31	0.26	5.93	0.19	14	0
53	45	0.27	42	590	64	0.5	40	0.06	0.71	<.01	0.39	0.52	7	2
294	44	1.14	43	315	346	1.97	12	0.06	2.09	0.52	2.94	0.07	16	5
34	10	0.29	49	127	162	0.37	23	0.02	0.71	0.02	0.25	0.03	2	0
44	5	0.92	31	325	102	1.3	41	0.04	0.56	0.11	10.94	0.37	2	1
56	7	1.18	38	417	130	1.66	46	0.05	0.71	0.13	13.32	0.43	2	1
78	97	0.43	35	506	63	0.58	17	0.17	0.57	0.11	6.52	0.39	21	2
26	19	2.02	45	300	513	1.3	33	0.06	1.26	0.17	17.98	0.16	7	0
46	17	2.36	42	365	521	1.19	26	0.05	0.51	0.11	20.45	0.54	6	0
43	9	1.15	56	285	44	0.68	740	0.27	1.13	0.28	9.78	0.51	2	0
49	67	20.27	13	455	81	1.99	130	0.12	0.45	0.31	2.56	0.09	23	19
107	147	44.74	29	1005	179	4.37	290	0.24	0.99	0.68	5.37	0.18	46	35
64	87	26.56	17	597	106	2.59	162	0.14	1.86	0.4	3.19	0.1	28	21
118	123	1.07	39	382	329	4.98	2	0.12	2.17	0.06	3.89	0.21	60	4
23	15	0.44	49	102	950	0.37	23	0.04	0.11	0.03	0.2	0.03	2	0
104	88	0.57	51	510	551	4.23	38	0.01	1.45	0.1	1.55	0.11	16	0

A

TABLE A–1

Food Composition   (Computer code number is for West Diet Analysis program)   (For purposes of calculations, use "0" for t, <1, <.1, <.01, etc.)

Computer Code Number	Food Description	Measure	Wt (g)	H₂O (%)	Ener (kcal)	Prot (g)	Carb (g)	Dietary Fiber (g)	Fat (g)	Fat Breakdown (g)		
										Sat	Mono	Poly
	MEATS: FISH AND SHELLFISH—Continued											
	Oysters:											
	Raw:											
19026	Eastern	1 c	248	85	169	17	10	0	6	1.9	0.8	2.4
578	Pacific	1 c	248	82	201	23	12	0	6	1.3	0.9	2.2
	Cooked:											
19009	Eastern, breaded, fried, medium	5 ea	73	65	144	6	8	<1	9	2.3	3.4	2.4
19008	Western, simmered	5 ea	125	64	204	24	12	0	6	1.3	1	2.2
	Scallops:											
19030	Breaded, cooked from frozen	6 ea	93	58	200	17	9	<1	10	2.5	4.2	2.7
19046	Imitation, from surimi	4 oz	113	74	112	14	12	0	<1	t	t	0.2
	Shrimp:											
19012	Cooked, boiled, 2 large = 11g	16 ea	88	77	87	18	0	0	1	0.3	0.2	0.4
19016	Canned, drained	¹/₂ c	64	73	77	15	1	0	1	0.2	0.2	0.5
19014	Fried, 2 large = 15g,breaded	12 ea	90	53	218	19	10	<1	11	1.9	3.4	4.6
19032	Raw, large, about 7g each	14 ea	98	76	104	20	1	0	2	0.3	0.2	0.7
19039	Imitation, from surimi	4 oz	113	75	114	14	10	0	2	0.3	0.2	0.8
	MEATS: BEEF, LAMB, AND PORK											
	Beef:											
10008	Corned, canned	4 oz	113	58	282	31	0	0	17	7	6.7	0.7
10009	Dried, cured	1 oz	28	54	43	9	1	0	1	0.3	0.2	t
	Ground beef, broiled, patty:											
10030	Extra lean, about 16% fat	4 oz	113	54	299	32	0	0	18	7	7.8	0.7
10032	Lean, 21% fat	4 oz	113	53	316	32	0	0	20	7.8	8.7	0.7
10002	Rib, whole, rstd, choice, ¹/₄" trim	4 oz	113	46	425	25	0	0	35	14.2	15.2	1.3
	Roast:											
10001	Blade, chuck, lean, brsd, choice, ¹/₄" trim	4 oz	113	55	297	35	0	0	16	6.3	7	0.5
10016	Bottom round, brsd, ¹/₄" trim	4 oz	113	52	311	32	0	0	19	7.2	8.3	0.7
10014	Bottom round, lean, brsd, choice, ¹/₄" trim	4 oz	113	57	249	36	0	0	11	3.6	4.7	0.4
10000	Pot, chuck arm, brsd, choice, ¹/₄" trim	4 oz	113	47	393	30	0	0	29	11.5	12.5	1.1
10017	Round eye, rstd, choice, ¹/₄" trim	4 oz	113	59	272	30	0	0	16	6.2	6.8	0.6
10013	Round eye, lean, rstd, choice, ¹/₄" trim	4 oz	113	65	198	33	0	0	6	2.3	2.7	0.2
	Steak:											
10056	Rib, lean, brld, ¹/₄" trim	4 oz	113	58	250	32	0	0	13	5.1	5.3	0.4
10005	Top sirloin, lean, brld, choice, ¹/₄" trim	4 oz	113	62	228	34	0	0	9	3.5	3.9	0.4
10006	T-bone, brld, choice, ¹/₄" trim	4 oz	113	51	364	26	0	0	28	11	12.7	1
10007	T-bone, lean, brld, choice, ¹/₄" trim	4 oz	113	61	232	30	0	0	11	4.1	5.6	0.3
	Variety meats:											
10010	Liver, fried	4 oz	113	62	198	30	6	0	5	1.7	0.7	0.7
10011	Tongue, cooked	4 oz	113	58	314	22	0	0	25	9.2	11.4	0.7
	Lamb:											
	Chop:											
13508	Shoulder arm, brsd, choice, ¹/₄" trim	1 ea	70	44	242	21	0	0	17	6.9	7.1	1.2
13509	Shoulder arm, lean, brsd, ¹/₄" trim	1 ea	55	49	153	20	0	0	8	2.8	3.4	0.5
13512	Loin, brld, choice, ¹/₄" trim	1 ea	64	52	202	16	0	0	15	6.3	6.2	1.1
13513	Loin chop, lean, brld, choice, ¹/₄" trim	1 ea	46	61	99	14	0	0	4	1.6	2	0.3
13517	Cutlet, avg of lean cuts, cooked	4 oz	113	54	330	28	0	0	23	9.9	9.8	1.7
	Leg:											
13500	Whole, rstd, choice,¹/₄" trim	4 oz	113	57	292	29	0	0	19	7.8	7.9	1.3
13501	Whole, lean, rstd, choice, ¹/₄" trim	4 oz	113	64	216	32	0	0	9	3.1	3.8	0.6
	Rib:											
13618	Rstd, choice, ¹/₄" trim	4 oz	113	48	406	24	0	0	34	14.4	14.1	2.5
13511	Lean, rstd, choice, ¹/₄" trim	4 oz	113	60	262	30	0	0	15	5.4	6.6	1
	Shoulder:											
13502	Whole, rstd, choice, ¹/₄" trim	4 oz	113	56	312	25	0	0	23	9.5	9.2	1.8
13503	Whole, lean, rstd, choice, ¹/₄" trim	4 oz	113	63	231	28	0	0	12	4.6	4.9	1.1

PAGE KEY: A–4 = Beverages   A–6 = Dairy   A–10 = Eggs   A–12 = Fat/Oil   A–14 = Fruit   A–20 = Bakery
A–26 = Grain   A–32 = Fish   A–32 = Meats   A–36 = Poultry   A–38 = Sausage   A–40 = Mixed/Fast   A–44 = Nuts/Seeds   A–46 = Sweets
A–48 = Vegetables/Legumes   A–58 = Vegetarian   A–62 = Misc   A–62 = Soups/Sauces   A–64 = Fast   A–78 = Convenience   A–80 = Baby foods

Chol (mg)	Calc (mg)	Iron (mg)	Magn (mg)	Pota (mg)	Sodi (mg)	Zinc (mg)	VT-A (RAE)	Thia (mg)	VT-E (mg α)	Ribo (mg)	Niac (mg)	V-B$_6$ (mg)	Fola (µg)	VT-C (mg)
131	112	16.52	117	387	523	225.21	74	0.25	2.11	0.24	3.42	0.15	25	9
124	20	12.67	55	417	263	41.22	201	0.17	2.11	0.58	4.98	0.12	25	20
59	45	5.07	42	178	304	63.6	66	0.11	1.66	0.15	1.2	0.05	23	3
125	20	11.5	55	378	265	41.55	182	0.16	1.06	0.55	4.52	0.11	19	16
57	39	0.76	55	310	432	0.99	21	0.04	1.77	0.1	1.4	0.13	34	2
25	9	0.35	49	116	898	0.37	23	0.01	0.12	0.02	0.35	0.03	2	0
172	34	2.72	30	160	197	1.37	60	0.03	1.21	0.03	2.28	0.11	4	2
111	38	1.75	26	134	108	0.81	12	0.02	0.6	0.02	1.76	0.07	1	1
159	60	1.13	36	202	310	1.24	51	0.12	1.35	0.12	2.76	0.09	16	1
149	51	2.36	36	181	145	1.09	53	0.03	1.08	0.03	2.5	0.1	3	2
41	21	0.68	49	101	797	0.37	23	0.03	0.12	0.04	0.19	0.03	2	0
97	14	2.35	16	154	1137	4.03	0	0.02	0.17	0.17	2.75	0.15	10	0
22	1	0.81	6	81	781	1.11	0	0.02	0	0.06	0.93	0.07	2	0
112	10	3.13	28	417	93	7.27	0	0.08	0.2	0.36	6.61	0.36	12	0
114	14	2.77	27	394	101	7.01	0	0.07	0.23	0.27	6.75	0.34	12	0
96	12	2.61	21	334	71	5.92	0	0.08	0.27	0.19	3.8	0.26	8	0
120	15	4.16	26	297	80	11.61	0	0.09	0.16	0.32	3.02	0.33	7	0
108	7	3.53	25	319	56	5.55	0	0.08	0.21	0.27	4.21	0.37	11	0
108	6	3.91	28	348	58	6.19	0	0.08	0.2	0.29	4.61	0.41	12	0
112	11	3.45	21	275	67	7.57	0	0.08	0.26	0.27	3.54	0.32	10	0
81	7	2.07	27	406	67	4.87	0	0.09	0.23	0.18	3.92	0.4	7	0
78	6	2.2	31	446	70	5.36	0	0.1	0.12	0.19	4.24	0.43	8	0
90	15	2.9	31	445	78	7.9	0	0.11	0.16	0.25	5.42	0.45	9	0
101	12	3.8	36	455	75	7.37	0	0.15	0.16	0.33	4.84	0.51	11	0
77	9	3.4	24	311	77	4.73	0	0.1	0.25	0.24	4.37	0.37	8	0
67	7	4.14	29	370	87	5.77	0	0.12	0.16	0.28	5.23	0.44	9	0
431	7	6.97	25	397	87	5.91	8751	0.2	0.52	3.87	19.75	1.16	294	1
149	6	2.95	17	208	73	46.22	0	0.02	0.34	0.33	3.94	0.18	8	1
84	18	1.67	18	214	50	4.26	0	0.05	0.1	0.18	4.66	0.08	13	0
67	14	1.48	16	186	42	4.01	0	0.04	0.1	0.15	3.48	0.07	12	0
64	13	1.16	15	209	49	2.23	0	0.06	0.08	0.16	4.54	0.08	12	0
44	9	0.92	13	173	39	1.9	0	0.05	0.07	0.13	3.15	0.07	11	0
110	12	2.26	25	340	77	4.67	0	0.12	0.15	0.32	7.48	0.16	19	0
105	12	2.24	27	354	75	4.97	0	0.11	0.17	0.31	7.45	0.17	23	0
101	9	2.4	29	382	77	5.58	0	0.12	0.2	0.33	7.16	0.19	26	0
110	25	1.81	23	306	82	3.94	0	0.1	0.11	0.24	7.63	0.12	17	0
99	24	2	26	356	92	5.05	0	0.1	0.17	0.26	6.96	0.17	25	0
104	23	2.23	26	284	75	5.91	0	0.1	0.16	0.27	6.95	0.15	24	0
98	21	2.41	28	299	77	6.83	0	0.1	0.2	0.29	6.51	0.17	28	0

A

TABLE A-1

Food Composition (Computer code number is for West Diet Analysis program)   (For purposes of calculations, use "0" for t, <1, <.1, <.01, etc.)

Computer Code Number	Food Description	Measure	Wt (g)	H₂O (%)	Ener (kcal)	Prot (g)	Carb (g)	Dietary Fiber (g)	Fat (g)	Fat Breakdown (g) Sat	Mono	Poly
	MEATS—Continued											
	Variety meats:											
13624	Brains, pan-fried	4 oz	113	76	164	14	0	0	11	2.9	2.1	1.2
13507	Sweetbreads, cooked	4 oz	113	60	264	26	0	0	17	7.7	6.2	0.8
13527	Tongue, cooked	4 oz	113	58	311	24	0	0	23	8.8	11.3	1.4
	Pork:											
	Cured:											
12000	Bacon, medium slices	3 pce	19	12	103	7	<1	0	8	2.6	3.5	0.9
12009	Breakfast strips, cooked	2 pce	23	27	106	7	<1	0	8	2.9	3.8	1.3
12002	Canadian-style bacon	2 pce	47	62	87	11	1	0	4	1.3	1.9	0.4
	Ham, roasted:											
12211	Reg, 11% fat	4 oz	113	65	201	26	0	0	10	3.5	5	1.6
12212	Extra lean, 5% fat	4 oz	113	68	164	24	2	0	6	2	3	0.6
12209	Extra lean, 4% fat, cnd	4 oz	113	69	154	24	1	0	6	1.8	2.8	0.5
	Chop:											
12029	Whole loin, brsd	1 ea	89	58	213	24	0	0	12	4.5	5.4	1
12033	Whole loin, lean, brsd	1 ea	80	61	163	23	0	0	7	2.7	3.3	0.6
12192	Center loin, w/bone, brld	1 ea	82	58	197	24	0	0	11	3.9	4.8	0.8
12025	Center loin, lean, brld	1 ea	74	61	149	22	0	0	6	2.2	2.7	0.4
12044	Center loin, pan fried	1 ea	78	53	216	23	0	0	13	4.7	5.5	1.5
12040	Blade loin, lean, pan fried	1 ea	63	59	152	16	0	0	10	3.3	3.9	1.2
	Leg:											
12016	Ham, whole, rstd	4 oz	113	55	308	30	0	0	20	7.3	8.9	1.9
12017	Ham, rump, lean, rstd	4 oz	113	61	233	35	0	0	9	3.2	4.3	0.9
	Rib:											
12050	Center loin, w/bone, rstd	4 oz	113	56	288	31	0	0	17	6.7	7.9	1.4
12055	Center loin, lean, w/bone, rstd	4 oz	113	59	252	32	0	0	13	4.9	5.9	1
	Shoulder:											
12003	Picnic, brsd	4 oz	113	48	372	32	0	0	26	9.6	11.7	2.6
12004	Picnic, lean, brsd	4 oz	113	54	280	36	0	0	14	4.7	6.5	1.3
12010	Spareribs, brsd	4 oz	113	40	449	33	0	0	34	12.6	15.2	3.1
14004	Rabbit, roasted (1 cup meat=140g)	4 oz	113	61	223	33	0	0	9	2.7	2.5	1.8
	Veal:											
11519	Short ribs, rstd	4 oz	113	60	258	27	0	0	16	6.1	6.1	1.1
11500	Liver, brsd	4 oz	113	60	217	32	4	0	7	2.2	1.3	1.2
14013	Venison (deer meat), roasted	4 oz	113	65	179	34	0	0	4	1.4	1	0.7
	Chicken:											
15016	Canned, boneless chicken	4 oz	113	69	186	25	0	0	9	2.5	3.6	2
	Fried, batter dipped:											
15013	Breast	1 ea	280	52	728	70	25	1	37	9.9	15.3	8.6
15030	Drumstick	1 ea	72	53	193	16	6	<1	11	3	4.6	2.7
15036	Thigh	1 ea	86	52	238	19	8	<1	14	3.8	5.8	3.4
15034	Wing	1 ea	49	46	159	10	5	<1	11	2.9	4.4	2.5
	Fried, flour coated:											
15003	Breast	1 ea	196	57	435	62	3	<1	17	4.8	6.9	3.8
15057	Breast, without skin	1 ea	172	60	322	58	1	0	8	2.2	3	1.8
15007	Drumstick	1 ea	49	57	120	13	1	<1	7	1.8	2.7	1.6
15009	Thigh	1 ea	62	54	162	17	2	<1	9	2.5	3.6	2.1
15011	Thigh, without skin	1 ea	52	59	113	15	1	0	5	1.4	2	1.3
15029	Wing	1 ea	32	49	103	8	1	<1	7	1.9	2.8	1.6
15902	Patty, breaded, cooked	1 ea	75	49	213	12	11	<1	13	4.1	6.4	1.7
	Roasted:											
15000	All types of meat	1 c	140	64	266	41	0	0	10	2.9	3.7	2.4
15027	Dark meat	1 c	140	63	287	38	0	0	14	3.7	5	3.2
15032	Light meat	1 c	140	65	242	43	0	0	6	1.8	2.2	1.4
15004	Breast, without skin	1 ea	172	65	284	53	0	0	6	1.7	2.1	1.3
15035	Drumstick, without skin	1 ea	44	67	76	12	0	0	2	0.7	0.8	0.6
15156	Leg, without skin	1 ea	95	65	181	26	0	0	8	2.2	2.9	1.9
15010	Thigh	1 ea	62	59	153	16	0	0	10	2.7	3.8	2.1
15012	Thigh, without skin	1 ea	52	63	109	13	0	0	6	1.6	2.2	1.3
15006	Stewed, all types	1 c	140	67	248	38	0	0	9	2.6	3.3	2.2

PAGE KEY: A–4 = Beverages   A–6 = Dairy   A–10 = Eggs   A–12 = Fat/Oil   A–14 = Fruit   A–20 = Bakery
A–26 = Grain   A–32 = Fish   A–32 = Meats   A–36 = Poultry   A–38 = Sausage   A–40 = Mixed/Fast   A–44 = Nuts/Seeds   A–46 = Sweets
A–48 = Vegetables/Legumes   A–58 = Vegetarian   A–62 = Misc   A–62 = Soups/Sauces   A–64 = Fast   A–78 = Convenience   A–80 = Baby foods

Chol (mg)	Calc (mg)	Iron (mg)	Magn (mg)	Pota (mg)	Sodi (mg)	Zinc (mg)	VT-A (RAE)	Thia (mg)	VT-E (mg α)	Ribo (mg)	Niac (mg)	V-B$_6$ (mg)	Fola (µg)	VT-C (mg)
2309	14	1.9	16	232	151	1.54	0	0.12	1.73	0.27	2.79	0.12	6	14
452	14	2.4	21	329	59	3.03	0	0.02	0.78	0.24	2.89	0.06	15	23
214	11	2.97	18	179	76	3.38	0	0.09	0.36	0.47	4.17	0.19	3	8
21	2	0.27	6	107	439	0.66	2	0.08	0.06	0.05	2.11	0.07	<1	0
24	3	0.45	6	107	483	0.85	0	0.17	0.06	0.08	1.75	0.08	1	0
27	5	0.39	10	183	727	0.8	0	0.39	0.16	0.09	3.25	0.21	2	0
67	9	1.51	25	462	1695	2.79	0	0.82	0.35	0.37	6.95	0.35	3	0
60	9	1.67	16	324	1359	3.25	0	0.85	0.28	0.23	4.55	0.45	3	0
34	7	1.04	24	393	1283	2.52	0	1.17	0.29	0.28	5.53	0.51	6	0
71	19	0.95	17	333	43	2.12	2	0.56	0.21	0.23	3.93	0.33	3	1
63	14	0.9	16	310	40	1.98	2	0.53	0.17	0.21	3.67	0.31	3	0
67	27	0.66	20	294	48	1.85	2	0.87	0.27	0.24	4.3	0.35	5	0
61	23	0.63	20	278	44	1.76	1	0.85	0.31	0.23	4.1	0.35	4	0
72	21	0.71	23	332	62	1.8	2	0.89	0.2	0.24	4.37	0.37	5	1
52	14	0.67	16	230	49	2.44	1	0.46	0.14	0.23	2.8	0.26	3	1
106	16	1.14	25	398	68	3.34	3	0.72	0.25	0.35	5.17	0.45	11	0
108	8	1.29	33	442	73	3.4	3	0.91	0.46	0.4	5.56	0.38	3	0
82	32	1.06	24	476	52	2.33	2	0.82	0.41	0.34	6.91	0.37	3	0
80	29	1.11	25	494	53	2.41	2	0.86	0.55	0.36	7.25	0.39	3	0
123	20	1.82	21	417	99	4.72	3	0.61	0.34	0.35	5.89	0.4	5	0
129	9	2.2	25	458	115	5.62	2	0.68	0.33	0.41	6.71	0.46	6	0
137	53	2.09	27	362	105	5.2	3	0.46	0.38	0.43	6.19	0.4	5	0
93	21	2.57	24	433	53	2.57	0	0.1	0.96	0.24	9.53	0.53	12	0
124	12	1.1	25	333	104	4.62	0	0.06	0.4	0.31	7.89	0.28	15	0
577	7	5.77	23	372	88	12.69	23894	0.21	0.77	3.23	14.86	1.04	374	1
127	8	5.05	27	379	61	3.11	0	0.2	0.28	0.68	7.58	0.42	5	0
70	16	1.79	14	156	568	1.59	38	0.02	0.29	0.15	7.15	0.4	5	2
238	56	3.5	67	563	770	2.66	56	0.32	2.97	0.41	29.46	1.2	42	0
62	12	0.97	14	134	194	1.68	19	0.08	0.88	0.15	3.67	0.19	13	0
80	15	1.25	18	165	248	1.75	25	0.1	1.05	0.2	4.91	0.22	16	0
39	10	0.63	8	68	157	0.68	17	0.05	0.52	0.07	2.58	0.15	9	0
174	31	2.33	59	508	149	2.16	29	0.16	1.12	0.26	26.93	1.14	12	0
157	28	1.96	53	475	136	1.86	12	0.14	0.72	0.22	25.43	1.1	7	0
44	6	0.66	11	112	44	1.42	12	0.04	0.41	0.11	2.96	0.17	5	0
60	9	0.92	16	147	55	1.56	18	0.06	0.52	0.15	4.31	0.2	7	0
53	7	0.76	14	135	49	1.45	11	0.05	0.3	0.13	3.7	0.2	5	0
26	5	0.4	6	57	25	0.56	12	0.02	0.18	0.04	2.14	0.13	2	0
45	12	0.94	15	184	399	0.78	11	0.07	1.46	0.1	5.04	0.23	8	0
125	21	1.69	35	340	120	2.94	22	0.1	0.38	0.25	12.84	0.66	8	0
130	21	1.86	32	336	130	3.92	31	0.1	0.38	0.32	9.17	0.5	11	0
119	21	1.48	38	346	108	1.72	13	0.09	0.38	0.16	17.39	0.84	6	0
146	26	1.79	50	440	127	1.72	10	0.12	0.46	0.2	23.58	1.03	7	0
41	5	0.57	11	108	42	1.4	8	0.03	0.12	0.1	2.67	0.17	4	0
89	11	1.24	23	230	86	2.72	18	0.07	0.26	0.22	6	0.35	8	0
58	7	0.83	14	138	52	1.46	30	0.04	0.17	0.13	3.95	0.19	4	0
49	6	0.68	12	124	46	1.34	10	0.04	0.14	0.12	3.39	0.18	4	0
116	20	1.64	29	252	98	2.79	21	0.07	0.38	0.23	8.56	0.36	8	0

A

TABLE A-1

Food Composition   (Computer code number is for West Diet Analysis program)   (For purposes of calculations, use "0" for t, <1, <.1, <.01, etc.)

Computer Code Number	Food Description	Measure	Wt (g)	H₂O (%)	Ener (kcal)	Prot (g)	Carb (g)	Dietary Fiber (g)	Fat (g)	Fat Breakdown (g) Sat	Mono	Poly
	MEATS—Continued											
	Variety meats:											
15025	Gizzards, simmered	1 c	145	68	212	44	0	0	4	1	0.8	0.5
15024	Hearts, simmered	1 c	145	65	268	38	<1	0	11	3.3	2.9	3.3
15215	Liver, simmered: Ounce	3 oz	85	67	142	21	1	0	6	1.8	1.2	1.1
	Duck:											
16295	Whole, w/skin, rstd, about 2.7 cups	½ ea	382	52	1287	73	0	0	108	36.9	49.3	13.9
14000	Whole, w/o skin, rstd, about 1.5 cups	½ ea	221	64	444	52	0	0	25	9.2	8.2	3.2
14003	Goose, whole, w/skin, rstd, about 5.5 cups	½ ea	774	52	2361	195	0	0	170	53.2	79.3	19.5
	Turkey:											
	Roasted, meat only:											
16002	Dark meat	4 oz	113	63	211	32	0	0	8	2.7	1.9	2.4
16158	Light meat	4 oz	113	66	177	34	0	0	4	1.2	0.6	1
16000	All types, chopped or diced	1 c	140	65	238	41	0	0	7	2.3	1.4	2
16003	Ground, cooked	4 oz	113	59	266	31	0	0	15	3.8	5.5	3.6
16010	With gravy, frozen package	3 oz	85	85	57	5	4	0	2	0.7	0.8	0.4
16307	Patty, breaded, fried	2 oz	57	50	161	8	9	<1	10	2.7	4.3	2.7
16308	Roasted, from frozen, seasoned	4 oz	113	68	175	24	3	0	7	2.1	1.4	1.9
16008	Roll, light meat	1 pce	28	72	41	5	<1	0	2	0.6	0.7	0.5
	Lunchmeat:											
	Turkey:											
	Breast:											
13259	Barbecued, Louis Rich	2 oz	56	72	57	11	2		<1	0.2	0.2	0.1
13108	Hickory smoked, Louis Rich	1 pce	80	73	80	16	2	0	1	0		
13110	Honey roasted, Louis Rich	1 pce	80	73	80	16	3	0	1	0.5		
13109	Oven roasted, Louis Rich	1 pce	80		70	16	0	0	1	0		
13114	Fat Free	1 pce	28	76	24	4	1	0	<1	t	t	t
13020	Pastrami	2 pce	57	71	80	10	1	0	4	1	1.2	0.9
13144	Salami	1 pce	28	72	41	4	<1	0	3	0.8	0.9	0.7
13014	Ham	2 pce	57	72	72	10	1	<1	3	0.9	1.1	0.8
	Bologna:											
13002	Beef	1 pce	23	54	72	2	1	0	6	2.6	2.8	0.2
13176	Beef, light, Oscar Mayer	1 pce	28	65	56	3	2	0	4	1.6	2	0.1
13006	Beef & pork	1 pce	28	52	85	4	2	0	7	2.7	2.9	0.3
13218	Healthy Favorites	1 pce	23		22	4	1	0	<1	0		
13032	Pork	1 pce	23	61	57	4	<1	0	5	1.6	2.2	0.5
13174	Regular, light, Oscar Mayer	1 pce	28	65	57	3	2	0	4	1.6	2	0.4
13007	Turkey	1 pce	28	65	59	3	1	<1	4	1.2	1.9	1.1
13149	Turkey, Louis Rich	1 pce	56	67	115	6	1	0	10	2.9	3.6	2.6
	Chicken:											
13222	Chicken breast, Healthy Favorites	4 pce	52		40	9	1	0	0	0	0	0
	Beef:											
13039	Corned beef loaf, jellied	1 pce	28	69	43	6	0	0	2	0.7	0.8	t
	Ham:											
13057	Chopped ham, packaged	2 pce	42	64	96	7	0	0	7	2.4	3.4	0.9
13220	Honey ham, Healthy Favorites	4 pce	52	73	55	9	2	0	1	0.4	0.6	0.1
13168	Oscar Mayer lower sodium ham	1 pce	21	73	23	3	1	0	1	0.3	0.4	t
13048	Mortadella lunchmeat	2 pce	30	52	93	5	1	0	8	2.9	3.4	0.9
13049	Olive loaf lunchmeat	2 pce	57	58	134	7	5	0	9	3.3	4.5	1.1
13051	Pickle & pimento loaf	2 pce	57	57	149	7	3	0	12	4.5	5.5	1.5
	Sausages:											
13035	Beerwurst/beer salami, beef	1 oz	28	57	77	4	1	<1	6	2.4	2.8	0.6
13031	Beerwurst/beer salami, pork	1 oz	28	61	67	4	1	0	5	1.8	2.5	0.7
13001	Berliner sausage	1 oz	28	61	64	4	1	0	5	1.7	2.2	0.4
13066	Braunschweiger sausage	2 pce	57	51	186	8	2	0	16	5.2	7.5	1.6
13036	Bratwurst-link	1 ea	70	51	226	10	2		19	7	9.3	2
13037	Cheesefurter/cheese smokie	2 ea	86	52	281	12	1	0	25	9	11.8	2.6
13070	Chorizo, pork & beef	1 ea	60	32	273	14	1	0	23	8.6	11	2.1
	Frankfurters:											
13008	Beef, large link, 8/package	1 ea	57	52	188	6	2	0	17	6.7	8.2	0.7
13010	Beef and pork, large link, 8/package	1 ea	45	56	137	5	1		12	4.8	6.2	1.2

PAGE KEY: A–4 = Beverages   A–6 = Dairy   A–10 = Eggs   A–12 = Fat/Oil   A–14 = Fruit   A–20 = Bakery
A–26 = Grain   A–32 = Fish   A–32 = Meats   A–36 = Poultry   A–38 = Sausage   A–40 = Mixed/Fast   A–44 = Nuts/Seeds   A–46 = Sweets
A–48 = Vegetables/Legumes   A–58 = Vegetarian   A–62 = Misc   A–62 = Soups/Sauces   A–64 = Fast   A–78 = Convenience   A–80 = Baby foods

Chol (mg)	Calc (mg)	Iron (mg)	Magn (mg)	Pota (mg)	Sodi (mg)	Zinc (mg)	VT-A (RAE)	Thia (mg)	VT-E (mg α)	Ribo (mg)	Niac (mg)	V-B$_6$ (mg)	Fola (µg)	VT-C (mg)
536	25	4.63	4	260	81	6.41	0	0.04	0.29	0.3	4.52	0.1	7	0
351	28	13.09	29	191	70	10.58	12	0.1	2.32	1.07	4.06	0.46	116	3
479	9	9.89	21	224	65	3.38	3384	0.25	0.7	1.69	9.39	0.64	491	24
321	42	10.31	61	779	225	7.11	241	0.66	2.67	1.03	18.43	0.69	23	0
197	27	5.97	44	557	144	5.75	51	0.57	1.55	1.04	11.27	0.55	22	0
704	101	21.9	170	2546	542	20.28	163	0.6	13.47	2.5	32.26	2.86	15	0
96	36	2.63	27	328	89	5.04	0	0.07	0.72	0.28	4.12	0.41	10	0
78	21	1.53	32	345	72	2.31	0	0.07	0.1	0.15	7.73	0.61	7	0
106	35	2.49	36	417	98	4.34	0	0.09	0.46	0.25	7.62	0.64	10	0
115	28	2.18	27	305	121	3.23	0	0.06	0.38	0.19	5.45	0.44	8	0
15	12	0.79	7	52	471	0.6	11	0.02	0.3	0.11	1.53	0.08	3	0
35	8	1.25	9	157	456	0.82	6	0.06	0.72	0.11	1.31	0.11	16	0
60	6	1.84	25	337	768	2.87	0	0.05	0.43	0.18	7.09	0.31	6	0
12	11	0.36	4	70	137	0.44	0	0.02	0.04	0.06	1.96	0.09	1	0
25	14	0.61	16		592	0.58	0							0
35	0	0.72			1060		0							0
35	0	0.72			940		0							0
35	0				910		0							0
9	3	0.31	8	57	334	0.24								0
31	5	0.95	8	148	596	1.23	0	0.03	0.13	0.14	2.01	0.15	3	0
21	11	0.35	6	60	281	0.65								0
41	5	1.33	13	164	635	1.48	4	0.02	0.36	0.08	1.21	0.12	4	0
13	7	0.25	3	40	248	2.09	3	<.01	0.08	0.02	0.58	0.04	2	3
12	4	0.34	4	44	322	0.53	0						4	0
17	24	0.34	5	88	206	0.64	7	0.06	0.1	0.05	0.71	0.08	2	0
8		0.18			255									
14	3	0.18	3	65	272	0.47	0	0.12	0.06	0.04	0.9	0.06	1	0
16	14	0.39	6	46	313	0.45	0	0.04		0.03	0.86	0.05	5	0
21	34	0.84	4	38	351	0.36	3	0.01	0.13	0.03	0.73	0.07	3	4
44	68	0.9	10	103	484	1.14	0	0.03		0.1	2.15	0.1		0
25		0.72			620									
13	3	0.57	3	28	267	1.15	0	0	0.05	0.03	0.49	0.03	2	0
21	3	0.35	7	134	576	0.81	0	0.27	0.1	0.09	1.63	0.15	<1	0
24	6	0.7	18	144	635	1.02	0							0
9	1	0.3	5	197	174	0.42	0							0
17	5	0.42	3	49	374	0.63	0	0.04	0.07	0.05	0.8	0.04	1	0
22	62	0.31	11	169	846	0.79	34	0.17	0.14	0.15	1.05	0.13	1	0
21	54	0.58	10	194	792	0.8	13	0.17	0.14	0.14	1.17	0.11	3	0
17	8	0.48	5	68	205	0.62	<1	0.07	0.05	0.05	0.83	0.06	1	0
17	2	0.21	4	71	347	0.48	0	0.16	0.06	0.05	0.91	0.1	1	0
13	3	0.32	4	79	363	0.69	0	0.11	0.06	0.06	0.87	0.06	1	0
103	5	6.38	6	113	661	1.6	2405	0.14	0.2	0.87	4.77	0.19	25	0
44	34	0.72	11	197	778	1.47	0	0.18	0.19	0.16	2.31	0.09	4	0
58	50	0.93	11	177	931	1.94	40	0.22	0	0.14	2.49	0.11	3	0
53	5	0.95	11	239	741	2.05	0	0.38	0.13	0.18	3.08	0.32	1	0
30	8	0.86	8	89	650	1.4	0	0.02	0.11	0.08	1.35	0.05	3	0
22	5	0.52	4	75	504	0.83	8	0.09	0.11	0.05	1.19	0.06	2	0

A

TABLE A-1

Food Composition   (Computer code number is for West Diet Analysis program)   (For purposes of calculations, use "0" for t, <1, <.1, <.01, etc.)

Computer Code Number	Food Description	Measure	Wt (g)	H₂O (%)	Ener (kcal)	Prot (g)	Carb (g)	Dietary Fiber (g)	Fat (g)	Fat Breakdown (g) Sat	Mono	Poly
	MEATS—Continued											
13012	Turkey frankfurter, 10/package	1 ea	45	63	102	6	1	0	8	2.7	2.5	2.2
13129	Turkey/chicken frank 8/pkg	1 ea	43	67	81	5	2	0	6	1.7	2.4	1.4
13043	Kielbasa sausage	1 pce	26	54	81	3	1	0	7	2.6	3.4	0.8
13044	Knockwurst sausage, link	1 ea	68	55	209	8	2	0	19	6.9	8.7	2
13021	Pepperoni	2 pce	11	31	51	2	<1	<1	4	1.8	2.1	0.3
13022	Polish	1 oz	28	53	91	4	<1	0	8	2.9	3.8	0.9
13024	Salami, pork and beef	2 pce	57	60	142	8	1	0	11	4.6	5.2	1.2
13026	Salami, pork and beef, dry	3 pce	30	35	125	7	1	0	10	3.7	5.1	1
13025	Salami, turkey	2 pce	57	54	126	8	11	<1	5	2	1.8	1.4
13029	Smoked link sausage, beef and pork	1 ea	68	54	218	8	2	0	20	6.6	8.3	2.7
13027	Smoked link sausage, pork	1 ea	68	39	265	15	1	0	22	7.7	10	2.6
13030	Summer sausage	2 pce	46	51	154	7	<1	0	14	5.5	6	0.6
13052	Turkey breakfast sausage	1 pce	28	60	64	6	0		5	1.6	1.8	1.2
13054	Vienna sausage, canned	2 ea	32	60	89	3	1	0	8	3	4	0.5
	Sandwich spreads:											
13034	Ham salad spread	2 tbsp	30	63	65	3	3	0	5	1.5	2.2	0.8
13056	Pork and beef	2 tbsp	30	60	70	2	4	<1	5	1.8	2.3	0.8
	MIXED DISHES											
15907	Almond Chicken	1 c	242	77	280	22	16	3	15	1.9	6.1	5.6
7084	Bean cake	1 ea	32	23	130	2	16	1	7	1	2.9	2.6
56124	Beef fajita	1 ea	223	65	399	23	36	3	18	5.5	7.6	3.5
56119	Beef flauta	1 ea	113	51	354	14	13	2	28	4.8	11.8	9.4
5513	Broccoli, batter fried	1 c	85	74	122	3	9	2	9	1.3	2.1	4.9
15903	Buffalo wings/spicy chicken wings	2 pce	32	53	98	8	<1	<1	7	1.8	2.7	1.8
56649	Cheeseburger deluxe	1 ea	219	52	563	28	38		33	15	12.6	2
56123	Chicken fajita	1 ea	223	65	363	20	44	5	12	2.3	5.5	3.1
56120	Chicken flauta	1 ea	113	55	330	13	12	2	26	4.2	10.7	9.3
50312	Chili con carne	½ c	127	77	128	12	11		4	1.7	1.7	0.3
15915	Chicken teriyaki, breast	1 ea	128	67	178	27	7	<1	4	0.9	1.1	0.9
45557	Chinese Pastry	1 oz	28	46	67	1	13	<1	2	0.2	0.5	0.8
56094	Chop suey with beef & pork	1 c	220	63	421	22	31		24	4.7	8.3	9.2
5461	Coleslaw	1 c	132	74	195	2	17		15	2.1	3.2	8.5
5366	Corn pudding	1 c	250	76	272	11	32	4	13	6.3	4.3	1.7
19539	Deviled egg (½ egg + filling)	1 ea	31	70	63	4	<1	0	5	1.2	1.7	1.5
	Egg Foo Yung Patty:											
56132	Meatless	1 ea	86	77	113	6	3	1	8	2	3.4	2.1
56290	With beef	1 ea	86	76	119	8	3	<1	8	2	2.9	2.2
56287	With chicken	1 ea	86	76	121	8	4	1	8	1.9	2.8	2.3
	Egg Roll:											
56110	Meatless	1 ea	64	70	101	3	10	1	6	1.2	2.9	1.3
57523	With Meat	1 ea	64	66	113	5	9	1	6	1.4	3	1.3
56003	Egg salad	1 c	183	57	584	17	3	0	56	10.5	17.4	23.9
56102	Falafel	1 ea	17	35	57	2	5		3	0.4	1.7	0.7
50182	Hot & Sour Soup (Chinese)	1 c	244	87	162	15	5	1	8	2.7	3.4	1.2
56659	Hamburger deluxe	1 ea	110	49	279	13	27		13	4.1	5.3	2.6
7081	Hummous/hummus	¼ c	62	65	110	3	12	2	5	0.7	3	1.3
16335	Kung Pao Chicken	1 c	162	54	431	29	11	2	31	5.2	13.9	9.7
	Lasagna:											
56108	With meat, homemade	1 pce	245	67	392	23	40		16	8	5.2	0.8
56071	Without meat, homemade	1 pce	218	69	306	16	40		10	5.6	2.5	0.6
56073	Chicken alfredo, w/broccoli	1 ea	340	75	389	24	41		14	6.7	5.5	0.8
57521	Lo mein, meatless	1 c	200	82	135	6	27	4	1	0.1	t	0.3
57522	Lo mein, with meat	1 c	200	72	283	20	21	3	14	2.6	4	6
56080	Moussaka (lamb & eggplant)	1 c	250	82	238	17	13	4	13	4.6		
5514	Mushrooms, batter fried	5 ea	70	63	156	2	11	1	12	1.5	3.6	6
56005	Potato salad with mayonnaise & eggs	½ c	125	76	179	3	14	2	10	1.8	3.1	4.7
56618	Pizza, combination, 1/12 of 12" round	1 pce	79	48	184	13	21		5	1.5	2.5	0.9
56619	Pizza, pepperoni, 1/12 of 12" round	1 pce	71	47	181	10	20		7	2.2	3.1	1.2
56098	Quiche Lorraine 1/8 of 8" quiche	1 pce	176	53	526	15	25	1	41	18.9	14.3	5.2
38067	Ramen noodles-cooked	1 c	227	86	154	3	20	1	7	1.7	1.2	3.3

PAGE KEY: A–4 = Beverages   A–6 = Dairy   A–10 = Eggs   A–12 = Fat/Oil   A–14 = Fruit   A–20 = Bakery   A–26 = Grain   A–32 = Fish   A–32 = Meats   A–36 = Poultry   A–38 = Sausage   A–40 = Mixed/Fast   A–44 = Nuts/Seeds   A–46 = Sweets   A–48 = Vegetables/Legumes   A–58 = Vegetarian   A–62 = Misc   A–62 = Soups/Sauces   A–64 = Fast   A–78 = Convenience   A–80 = Baby foods

Chol (mg)	Calc (mg)	Iron (mg)	Magn (mg)	Pota (mg)	Sodi (mg)	Zinc (mg)	VT-A (RAE)	Thia (mg)	VT-E (mg α)	Ribo (mg)	Niac (mg)	V-B$_6$ (mg)	Fola (µg)	VT-C (mg)
48	48	0.83	6	81	642	1.4	0	0.02	0.28	0.08	1.86	0.1	4	0
40	56	0.94	10	69	488	0.8								0
17	11	0.38	4	70	280	0.53	0	0.06	0.06	0.06	0.75	0.05	1	0
41	7	0.45	7	135	632	1.13	0	0.23	0.39	0.1	1.86	0.12	1	0
13	2	0.16	2	35	197	0.3	0	0.06	0.03	0.03	0.6	0.04	1	0
20	3	0.4	4	66	245	0.54	0	0.14	0.06	0.04	0.96	0.05	1	0
37	7	1.52	9	113	607	1.22	0	0.14	0.13	0.21	2.03	0.12	1	0
24	2	0.45	5	113	558	0.97	0	0.18	0.08	0.09	1.46	0.15	1	0
45	42	0.88	15	226	619	1.77	1	0.24	0.14	0.17	2.27	0.24	6	12
39	8	0.51	9	122	619	0.86	9	0.13	0.09	0.07	2	0.11	1	0
46	20	0.79	13	228	1020	1.92	0	0.48	0.17	0.17	3.08	0.24	3	1
34	6	1.17	6	125	571	1.18	0	0.07	0.1	0.15	1.98	0.12	1	0
23	5	0.51	6	75	188	0.96	0	0.03	0.14	0.08	1.4	0.08	1	0
17	3	0.28	2	32	305	0.51	0	0.03	0.07	0.03	0.52	0.04	1	0
11	2	0.18	3	45	274	0.33	0	0.13	0.52	0.04	0.63	0.04	<1	0
11	4	0.24	2	33	304	0.31	8	0.05	0.52	0.04	0.52	0.04	1	0
40	69	1.97	60	549	526	1.62		0.08	3.8	0.2	9.48	0.44	26	7
0	3	0.67	6	58	1	0.16	0	0.07	1.24	0.05	0.55	0.02	9	0
45	84	3.76	38	479	316	3.52	22	0.39	1.74	0.3	5.4	0.38	23	27
37	51	1.87	28	313	68	3.45		0.06	4.65	0.13	1.88	0.24	10	19
15	66	0.98	20	242	64	0.38		0.08	2.85	0.13	0.73	0.11	36	53
26	5	0.4	6	59	25	0.57		0.01	0.27	0.04	2.06	0.13	1	0
88	206	4.66	44	445	1108	4.6	140	0.39	1.18	0.46	7.38	0.28	81	8
39	101	3.32	48	534	343	1.65		0.43	1.71	0.33	6.12	0.38	42	37
35	50	0.95	27	269	71	1.13		0.05	4.36	0.09	3.1	0.22	8	18
67	34	2.6	23	347	505	1.79	42	0.06	0.81	0.57	1.24	0.17	23	1
82	27	1.71	35	309	1683	1.96		0.08	0.35	0.19	8.75	0.47	12	3
0	6	0.18	7	25	3	0.16	<1	0.02	0.26	<.01	0.27	0.04	1	0
43	39	4.19	54	519	950	3.48		0.36	1.8	0.37	5.73	0.39	44	20
7	45	0.96	12	236	356	0.26	48	0.05	5.28	0.04	0.11	0.15	51	11
250	100	1.4	38	402	138	1.25	135	1.03	0.52	0.32	2.47	0.3	62	7
122	15	0.35	3	37	50	0.3	50	0.02	0.61	0.15	0.02	0.05	13	0
185	31	1.04	12	117	317	0.7		0.04	1.22	0.26	0.43	0.09	30	5
166	25	1.01	11	139	131	1.01		0.05	1.06	0.23	0.65	0.13	22	3
167	27	0.82	11	136	132	0.76		0.05	1.1	0.23	0.89	0.12	22	3
30	14	0.81	9	97	274	0.25		0.08	0.85	0.11	0.8	0.05	13	3
37	15	0.83	10	124	274	0.46		0.16	0.8	0.13	1.28	0.09	10	2
581	74	1.81	13	181	464	1.45	263	0.09	7.66	0.66	0.09	0.46	61	0
0	9	0.58	14	99	50	0.26	<1	0.02	0.19	0.03	0.18	0.02	16	0
34	29	1.9	29	384	1011	1.51		0.27	0.15	0.25	5	0.2	13	1
26	63	2.63	22	227	504	2.06	4	0.23	0.82	0.2	3.68	0.12	52	2
0	30	0.97	18	107	150	0.68	<1	0.06	0.46	0.03	0.25	0.25	37	5
64	49	1.96	63	428	907	1.5		0.15	3.9	0.15	13.23	0.59	43	8
58	270	3.07	50	460	391	3.33		0.24	1.16	0.34	4.2	0.25	20	14
32	265	2.35	44	373	365	1.82		0.23	1.09	0.28	2.51	0.17	17	14
55	265	3.82	64	759	840	3.7		0.29	3.45	0.39	5.07	0.32	28	13
0	46	2.03	33	386	564	0.92	65	0.23	0.35	0.24	2.82	0.19	48	12
42	29	2.07	42	332	142	1.83		0.41	2.06	0.28	5.02	0.36	53	11
97	75	1.74	40	565	460	2.57		0.16	0.98	0.31	4.14	0.23	46	6
2	15	1.22	7	154	112	0.42		0.11	2.34	0.26	2.25	0.04	8	1
85	24	0.81	19	318	661	0.39	40	0.1	2.33	0.08	1.11	0.18	9	12
21	101	1.53	18	179	382	1.11	58	0.21		0.17	1.96	0.09	32	2
14	65	0.94	9	153	267	0.52	53	0.13		0.23	3.05	0.06	37	2
221	231	1.88	24	239	221	1.5		0.26	2.02	0.49	2.01	0.1	19	1
0	13	0.39	10	49	802	0.18	2	0.02	2.34	0.01	0.25	0.01	3	0

TABLE A-1

Food Composition (Computer code number is for West Diet Analysis program) (For purposes of calculations, use "0" for t, <1, <.1, <.01, etc.)

Computer Code Number	Food Description	Measure	Wt (g)	H₂O (%)	Ener (kcal)	Prot (g)	Carb (g)	Dietary Fiber (g)	Fat (g)	Fat Breakdown (g) Sat	Mono	Poly
	MIXED DISHES—Continued											
56302	Ravioli, meat	½ c	125	69	197	11	18	1	9	3	3.7	1
38145	Fried rice (meatless)	1 c	166	68	271	5	34	1	12	1.8	3.2	6.7
	Spaghetti (enriched) in tomato sauce:											
56097	With cheese, home recipe	1 c	250	77	260	9	37		9	2		
56100	With meatballs, home recipe	1 c	248	71	362	18	28	3	18	4.8		
56076	Spinach souffle	1 c	136	74	219	11	3		18	7.1	6.8	3.1
2995	Spring roll, vegetable	1 ea	63	49	158	4	20	1	7	0.9		
	Sushi:											
56313	Fish and vegetable	1 c	166	65	232	9	47	2	1	0.2	0.2	0.2
56314	Vegetable seaweed	1 c	166	71	194	4	43	1	<1	0.1	0.1	0.1
12900	Sweet & sour pork	1 c	226	77	231	15	25	2	8	2.1	3.2	2.3
15921	Sweet & sour chicken breast	1 ea	131	79	118	8	15	1	3	0.5	0.9	1.4
56916	Tabouli	1 c	160	77	199	3	16	4	15	2	10.8	1.4
	Thai dishes											
1984	Beef peanut satay, svg	1 ea	129	60	286	26	5	1	18	5.9		
1988	Drunkard noodles, svg	1 ea	366	80	344	5	51	4	14	1.8		
1994	Lemongrass vegetables, svg	1 ea	187	75	238	8	19	4	16	2.6		
1998	Peanut chicken, svg	1 ea	309	81	272	19	26	3	11	2.3		
2994	Spicy noodles, svg	1 ea	310	58	626	34	65	3	26	4.4		
2994	Sweet noodles, svg	1 ea	211	58	426	23	44	2	18	3		
56118	Three bean salad	1 c	150	81	140	4	15	5	8	1.1	1.7	4.4
56006	Waldorf salad	1 c	137	58	411	4	12	3	41	4.3		
56111	Wonton, meat filled	1 ea	19	45	55	3	5	<1	3	0.8	1.2	0.3
	FAST FOODS and SANDWICHES											
	(see end of this appendix for additional Fast Foods)											
	Burritos:											
66026	Beef & bean	1 ea	116	52	255	11	33	3	9	4.2	3.5	0.6
66025	Bean	1 ea	109	53	225	7	36	4	7	3.5	2.4	0.6
16255	Chicken con queso	1 ea	299		350	14	60	6	6	2.5		
	Cheeseburgers:											
66013	With bun, 4-oz patty	1 ea	166	51	417	21	35		21	8.7	7.8	2.7
66015	With bun, regular	1 ea	154	55	359	18	28		20	9.2	7.2	1.5
56668	Corndog	1 ea	175	47	460	17	56		19	5.2	9.1	3.5
15181	Chicken	1 ea	113		271	11	23	2	15			
66021	Enchilada	1 ea	163	63	319	10	29		19	10.6	6.3	0.8
66032	English muffin with egg, cheese, bacon	1 ea	146	57	308	18	28	2	13	5	5	1.7
	Fish sandwich:											
66010	Large, no cheese	1 ea	158	47	431	17	41	<1	23	5.2	7.7	8.2
66011	Regular, with cheese	1 ea	183	45	523	21	48	<1	29	8.1	8.9	9.4
	Hamburgers:											
56658	With bun, regular	1 ea	107	45	275	12	35	2	10	3.6	3.4	1
66006	With bun, 4-oz patty	1 ea	215	51	576	32	39		32	12	14.1	2.8
66004	Hotdog/frankfurter with bun	1 ea	98	54	242	10	18		15	5.1	6.9	1.7
	Lunchables:											
56938	Bologna & American cheese	1 ea	128		450	18	19	0	34	15		
56939	Ham & cheese	1 ea	128		320	22	19	0	17	8		
56930	Honey ham & Amer. w/choc pudding	1 ea	176		390	18	34	<1	20	9		
56931	Honey turkey & cheddar w/Jello	1 ea	163		320	17	27	<1	16	9		
56940	Pepperoni & American cheese	1 ea	128		480	20	19	0	36	17		
56936	Salami & American cheese	1 ea	128		430	18	18	0	32	15		
56937	Turkey & cheddar cheese	1 ea	128		360	20	20	1	22	11		
	SANDWICHES:											
	Avocado, chesse, tomato & lettuce:											
56021	On white bread, firm	1 ea	210	62	429	14	35	5	27	7.7		
56022	On part whole wheat	1 ea	201	63	402	14	30	6	27	7.8		
56023	On whole wheat	1 ea	214	63	424	15	33	8	28	8.2		
	Bacon, lettuce & tomato sandwich:											
56009	On white bread, soft	1 ea	124	52	318	10	29	2	18	4.1		
56011	On part whole wheat	1 ea	124	53	314	11	26	3	19	4.6		

PAGE KEY: A–4 = Beverages   A–6 = Dairy   A–10 = Eggs   A–12 = Fat/Oil   A–14 = Fruit   A–20 = Bakery
A–26 = Grain   A–32 = Fish   A–32 = Meats   A–36 = Poultry   A–38 = Sausage   A–40 = Mixed/Fast   A–44 = Nuts/Seeds   A–46 = Sweets
A–48 = Vegetables/Legumes   A–58 = Vegetarian   A–62 = Misc   A–62 = Soups/Sauces   A–64 = Fast   A–78 = Convenience   A–80 = Baby foods

Chol (mg)	Calc (mg)	Iron (mg)	Magn (mg)	Pota (mg)	Sodi (mg)	Zinc (mg)	VT-A (RAE)	Thia (mg)	VT-E (mg α)	Ribo (mg)	Niac (mg)	V-B$_6$ (mg)	Fola (µg)	VT-C (mg)
85	35	2.15	20	264	90	1.7		0.16	1.32	0.22	2.99	0.14	14	4
43	28	1.94	23	128	261	0.92		0.21	2.51	0.11	2.24	0.15	22	4
8	80	2.25	26	408		1.3		0.25	2.75	0.18	2.25	0.2	8	12
65	92	3.33	44	479	1133	3.4		0.25	2.46	0.3	4.38	0.29	68	16
184	230	1.35	38	201	763	1.29	267	0.09	1.22	0.3	0.48	0.12	80	
3	58	1.78	12	74	262	0.38		0.18	1.29	0.14	1.88	0.03	32	1
11	25	2.33	27	218	93	0.84		0.28	0.62	0.07	2.96	0.16	15	4
0	22	1.65	21	106	5	0.75	33	0.21	0.13	0.04	1.99	0.15	11	3
39	28	1.44	34	386	839	1.47		0.55	1.09	0.21	3.63	0.41	10	20
23	15	0.84	21	185	506	0.67		0.06	0.67	0.08	3.09	0.18	6	12
0	29	1.25	36	246	799	0.48	34	0.08	2.16	0.05	1.14	0.11	31	29
71	15	2.77	39	408	1173	5.81		0.1	1.23	0.24	4.75	0.41	16	3
0	37	1.77	37	524	1227	0.77		0.16	3.83	0.13	1.89	0.39	42	159
0	146	2.88	49	453	741	1.09		0.18	2.41	0.12	1.45	0.29	48	92
36	42	2.65	48	344	900	1.21		0.22	1.73	0.12	8.24	0.48	66	67
212	73	4.77	95	623	1660	2.31		0.26	1.97	0.32	9.47	0.55	90	20
144	49	3.25	65	424	1130	1.57		0.17	1.34	0.22	6.45	0.38	61	14
0	35	1.48	27	247	520	0.58		0.08	1.74	0.1	0.44	0.04	56	4
21	44	0.97	37	258	235	0.7		0.09	8.62	0.05	0.54	0.36	34	5
20	4	0.4	4	51	10	0.32		0.09	0.18	0.06	0.63	0.04	3	0
24	53	2.46	42	329	670	1.93	16	0.27	0.7	0.42	2.71	0.19	58	1
2	57	2.27	44	328	495	0.76	9	0.32	0.87	0.31	2.04	0.15	44	1
35	40	1.8			590									6
60	171	3.42	30	335	1051	3.49	71	0.35		0.28	8.05	0.18	61	2
52	182	2.65	26	229	976	2.62	82	0.32	1.34	0.23	6.38	0.15	65	2
79	102	6.18	18	262	973	1.31	60	0.28	0.7	0.7	4.16	0.09	103	0
53	60	2.71			738		0							0
44	324	1.32	51	240	784	2.51	99	0.08	1.47	0.42	1.91	0.39	65	1
250	161	2.6	25	212	777	1.66	188	0.53	0.6	0.48	3.55	0.16	73	2
55	84	2.61	33	340	615	1	33	0.33	0.87	0.22	3.4	0.11	85	3
68	185	3.5	37	353	939	1.17	130	0.46	1.83	0.42	4.23	0.11	92	3
30	127	2.74	24	254	539	2.27	4	0.29	0.43	0.24	3.95	0.12	52	2
103	92	5.55	45	527	742	5.8	2	0.34	1.61	0.41	6.73	0.37	84	1
44	24	2.31	13	143	670	1.98	0	0.24	0.27	0.27	3.65	0.05	48	0
85	300	2.7			1620									0
60	300	1.8			1770									
55	250	2.7			1540									
50	20	6			1360									
95	250	2.7			1840									
80	250	2.7			1740									
70	300	1.8			1650									
30	283	6.1	48	548	507	1.45		0.37	3.15	0.43	3.76	0.3	102	12
29	272	5.96	56	576	454	1.59		0.3	3.2	0.37	3.47	0.31	85	12
30	270	6.36	83	636	499	2.21		0.3	3.51	0.36	3.67	0.38	77	13
21	68	2.21	22	240	632	0.98		0.41	2.33	0.26	3.7	0.16	66	6
22	61	2.22	32	288	625	1.21		0.37	2.56	0.21	3.68	0.18	53	6

A

TABLE A–1

Food Composition   (Computer code number is for West Diet Analysis program)   (For purposes of calculations, use "0" for t, <1, <.1, <.01, etc.)

Computer Code Number	Food Description	Measure	Wt (g)	H₂O (%)	Ener (kcal)	Prot (g)	Carb (g)	Dietary Fiber (g)	Fat (g)	Fat Breakdown (g)		
										Sat	Mono	Poly
	MIXED DISHES—Continued											
56010	On whole wheat	1 ea	137	52	339	12	29	5	20	4.9		
	Cheese, grilled:											
56013	On white bread, soft	1 ea	119	37	399	17	30	1	23	11.9		
56015	On part whole wheat	1 ea	119	37	402	18	26	2	25	13.1		
56014	On whole wheat	1 ea	132	38	432	20	30	4	27	13.8		
56000	Chicken fillet	1 ea	182	47	515	24	39		29	8.5	10.4	8.4
	Chicken salad:											
56017	On white bread, soft	1 ea	110	41	366	10	31	2	22	2.7		
56019	On part whole wheat	1 ea	110	41	369	11	27	3	24	3.1		
56018	On whole wheat	1 ea	123	41	399	13	32	5	26	3.4		
56020	Corned beef & swiss on rye	1 ea	156	47	427	28	22	6	26	9.5		
	Egg salad:											
56025	On white bread, soft	1 ea	117	43	379	9	31	1	24	3.8		
56027	On part whole wheat	1 ea	116	44	378	10	27	2	26	4.3		
56026	On whole wheat	1 ea	130	44	410	11	31	5	28	4.6		
	Ham:											
56032	On rye bread	1 ea	150	52	345	22	29	4	15	3.2		
56029	On white bread, soft	1 ea	157	52	365	24	30	2	16	3.3		
56031	On part whole wheat	1 ea	156	54	355	25	26	2	17	3.6		
56030	On whole wheat	1 ea	169	53	378	27	29	4	18	3.9		
	Ham & cheese:											
56035	On white bread, soft	1 ea	157	48	423	24	31	2	23	8.1		
56037	On part whole wheat	1 ea	156	49	417	25	26	2	24	8.7		
56036	On whole wheat	1 ea	170	48	446	27	30	4	25	9.2		
56033	Ham & swiss on rye	1 ea	150	48	386	22	30	4	19	6.5		
	Ham salad:											
56064	On white bread, soft	1 ea	131	47	361	10	37	1	19	4.2		
56066	On part whole wheat	1 ea	131	48	358	11	33	2	21	4.7		
56065	On whole wheat	1 ea	144	48	383	12	37	4	22	5		
56038	Patty melt: Ground beef & cheese on rye	1 ea	182	46	561	37	22	6	37	12.7		
	Peanut butter & jelly:											
56040	On white bread, soft	1 ea	101	27	348	11	47	3	14	2.7		
56042	On part whole wheat	1 ea	101	27	339	12	45	6	15	3		
56041	On whole wheat	1 ea	114	27	398	13	51	5	17	3.6		
	Roast beef:											
66003	On a bun	1 ea	139	49	346	22	33		14	3.6	6.8	1.7
56044	On white bread, soft	1 ea	157	46	405	29	35	1	16	2.9		
56046	On part whole wheat	1 ea	156	47	398	30	30	2	17	3.2		
56045	On whole wheat	1 ea	169	47	423	32	34	4	18	3.4		
	Tuna salad:											
56048	On white bread, soft	1 ea	122	46	326	13	35	1	14	1.9		
56050	On part whole wheat	1 ea	122	47	322	14	32	2	16	2.2		
56049	On whole wheat	1 ea	135	47	347	16	36	4	17	2.4		
	Turkey:											
56052	On white bread, soft	1 ea	156	54	346	24	30	1	14	1.9		
56054	On part whole wheat	1 ea	155	55	336	25	25	2	15	2.1		
56053	On whole wheat	1 ea	169	54	360	27	29	4	16	2.3		
	Turkey ham:											
56103	On rye bread	1 ea	150	60	280	21	20	6	14	2.5		
56104	On white bread, soft	1 ea	156	55	331	21	30	2	14	2.5		
56106	On part whole wheat	1 ea	156	56	346	21	25	2	18	3.6		
56105	On whole wheat	1 ea	169	56	344	24	29	4	15	3		
57531	Taco	1 ea	171	58	369	21	27		21	11.4	6.6	1
	Tostada:											
66017	With refried beans	1 ea	144	66	223	10	27		10	5.4	3.1	0.7
66018	With beans & beef	1 ea	225	70	333	16	30		17	11.5	3.5	0.6
56062	With beans & chicken	1 ea	156	70	242	19	16	3	11	4.5		
	NUTS, SEEDS and PRODUCTS											
	Almonds:											
4571	Dry roasted, salted	1 c	138	3	824	30	27	16	73	5.6	46.4	17.5
4503	Slivered, packed, unsalted	1 c	108	5	624	23	21	13	55	4.2	34.7	13.2

PAGE KEY: A–4 = Beverages   A–6 = Dairy   A–10 = Eggs   A–12 = Fat/Oil   A–14 = Fruit   A–20 = Bakery
A–26 = Grain   A–32 = Fish   A–32 = Meats   A–36 = Poultry   A–38 = Sausage   A–40 = Mixed/Fast   A–44 = Nuts/Seeds   A–46 = Sweets
A–48 = Vegetables/Legumes   A–58 = Vegetarian   A–62 = Misc   A–62 = Soups/Sauces   A–64 = Fast   A–78 = Convenience   A–80 = Baby foods

Chol (mg)	Calc (mg)	Iron (mg)	Magn (mg)	Pota (mg)	Sodi (mg)	Zinc (mg)	VT-A (RAE)	Thia (mg)	VT-E (mg α)	Ribo (mg)	Niac (mg)	V-B$_6$ (mg)	Fola (µg)	VT-C (mg)
23	51	2.54	61	346	690	1.87		0.37	2.91	0.2	3.97	0.25	45	7
53	407	2	26	162	1155	2.03		0.29	1.01	0.4	2.37	0.08	60	0
57	431	2	38	208	1197	2.37		0.24	1.15	0.37	2.25	0.1	46	0
60	440	2.33	68	264	1293	3.13		0.24	1.44	0.36	2.46	0.16	37	0
60	60	4.68	35	353	957	1.87	31	0.33		0.24	6.81	0.2	100	9
30	76	2.21	20	146	483	0.79		0.3	5.48	0.24	4.09	0.27	63	1
33	70	2.25	32	194	468	1.04		0.25	6.08	0.2	4.15	0.31	49	1
35	59	2.6	63	252	528	1.76		0.25	6.65	0.18	4.47	0.38	39	1
83	267	3.03	28	232	1470	3.59		0.2	2.59	0.33	2.76	0.18	32	0
154	87	2.36	18	122	500	0.76		0.31	4.24	0.38	2.45	0.21	74	0
166	80	2.38	29	165	479	0.99		0.25	4.65	0.34	2.3	0.24	60	0
177	71	2.75	60	222	543	1.71		0.26	5.17	0.33	2.54	0.32	52	0
50	54	2.89	40	390	1245	2.76		0.84	2.36	0.42	6.42	0.37	54	0
54	76	3.1	32	384	1237	2.61		0.9	2.52	0.44	6.83	0.39	60	0
56	68	3.12	43	438	1252	2.92		0.88	2.72	0.4	6.91	0.42	44	0
59	57	3.46	72	500	1336	3.66		0.9	3.06	0.39	7.29	0.49	34	0
64	246	2.82	33	329	1366	2.71		0.7	2.58	0.46	5.36	0.31	61	0
67	248	2.82	44	379	1388	3.03		0.67	2.78	0.42	5.35	0.34	46	0
71	247	3.17	74	441	1487	3.8		0.69	3.14	0.41	5.7	0.41	36	0
56	241	2.68	43	370	1397	3		0.65	2.47	0.45	5.06	0.29	55	0
29	72	2.25	21	167	935	1.06		0.55	3.34	0.28	3.69	0.18	59	0
30	65	2.26	32	213	950	1.31		0.52	3.67	0.23	3.65	0.21	44	0
32	54	2.59	62	269	1029	2.02		0.53	4.08	0.21	3.92	0.28	34	0
113	222	4.17	39	391	714	7.04		0.25	3.52	0.46	6.14	0.37	37	0
1	76	2.29	52	240	429	1.05		0.29	2.55	0.23	5.44	0.15	79	2
0	54	2.41	88	320	418	1.82		0.22	2.9	0.15	5.43	0.22	53	2
0	80	2.66	75	335	465	1.5		0.28	3.24	0.21	6.41	0.2	76	2
51	54	4.23	31	316	792	3.39	11	0.38	0.19	0.31	5.87	0.26	57	2
43	76	4.13	30	436	1606	3.73		0.35	3.39	0.36	6.78	0.4	67	0
45	67	4.21	41	493	1643	4.11		0.29	3.63	0.32	6.86	0.44	51	0
47	57	4.6	70	557	1743	4.9		0.29	4.01	0.3	7.24	0.51	42	0
13	76	2.41	25	168	589	0.68		0.3	2.77	0.24	5.89	0.13	63	1
13	69	2.44	36	215	578	0.9		0.25	3.07	0.19	6.07	0.16	48	1
14	59	2.79	67	273	642	1.6		0.25	3.46	0.17	6.47	0.23	38	1
43	72	2.19	31	307	1589	1.33		0.31	3.28	0.29	9.29	0.42	60	0
45	63	2.15	42	356	1625	1.56		0.25	3.52	0.24	9.52	0.45	45	0
47	53	2.46	71	417	1736	2.25		0.25	3.92	0.22	10.07	0.53	35	0
55	51	4.04	27	343	1179	2.94		0.22	2.86	0.33	4.32	0.31	29	0
53	77	4.22	29	351	1252	2.85		0.32	2.83	0.41	5.29	0.29	62	0
57	68	4.22	39	395	1264	3.12		0.26	1.23	0.36	5.16	0.34	46	0
58	59	4.72	69	466	1361	3.95		0.26	3.41	0.36	5.63	0.39	37	0
56	221	2.41	70	474	802	3.93	108	0.15	1.88	0.44	3.21	0.24	68	2
30	210	1.89	59	403	543	1.9	45	0.1	1.15	0.33	1.32	0.16	43	1
74	189	2.45	68	490	871	3.17	101	0.09	1.8	0.5	2.86	0.25	86	4
55	146	1.57	41	263	386	1.93		0.09	0.66	0.15	4.26	0.31	25	5
0	367	6.22	395	1029	468	4.89	<1	0.1	35.88	1.19	5.31	0.17	46	0
0	268	4.64	297	786	1	3.63	<1	0.26	27.94	0.88	4.24	0.14	31	0

A

TABLE A–1

Food Composition   (Computer code number is for West Diet Analysis program)   (For purposes of calculations, use "0" for t, <1, <.1, <.01, etc.)

Computer Code Number	Food Description	Measure	Wt (g)	H₂O (%)	Ener (kcal)	Prot (g)	Carb (g)	Dietary Fiber (g)	Fat (g)	Sat	Mono	Poly
	NUTS, SEEDS and PRODUCTS—Continued											
4502	Whole, dried, unsalted	1 oz	28	5	162	6	6	3	14	1.1	9	3.4
4534	Almond butter:	1 tbs	16	1	101	2	3	1	9	0.9	6.1	2
4572	Salted	1 tbs	16	1	101	2	3	1	9	0.9	6.1	2
4750	Brazil nuts, dry (about 7)	1 c	140	3	918	20	17	10	93	21.2	34.4	28.8
	Cashew nuts:											
4519	Dry roasted, salted	1 oz	28	2	161	4	9	1	13	2.6	7.6	2.2
4621	Dry roasted, unsalted	1 oz	28	2	161	4	9	1	13	2.6	7.6	2.2
4596	Oil roasted, salted	1 oz	28	2	163	5	8	1	13	2.4	7.3	2.4
4622	Oil roasted, unsalted:	1 c	130	3	754	22	39	4	62	11	33.7	11.1
4622	Ounce	1 oz	28	3	162	5	8	1	13	2.4	7.3	2.4
4537	Cashew butter, unsalted	1 tbs	16	3	94	3	4	<1	8	1.6	4.7	1.3
4662	Salted	1 tbs	16	3	94	3	4	<1	8	1.6	4.7	1.3
4538	Chestnuts, European, roasted, (1 cup = approx 17 kernels)	1 c	143	40	350	5	76	7	3	0.6	1.1	1.2
	Coconut, raw:											
4508	Piece 2 × 2 × ¹/₂"	1 pce	45	47	159	1	7	4	15	13.4	0.6	0.2
4507	Shredded/grated, unpacked	¹/₂ c	40	47	142	1	6	4	13	11.9	0.6	0.1
	Coconut, dried, shredded/grated:											
4510	Unsweetened	1 c	78	3	515	5	18	13	50	44.6	2.1	0.6
4511	Sweetened	1 c	93	13	466	3	44	4	33	29.3	1.4	0.4
4559	Coconut milk, canned	1 c	226	73	445	5	6	3	48	42.7	2	0.5
4514	Filberts/hazelnuts, chopped	1 oz	28	5	176	4	5	3	17	1.2	12.8	2.2
	Mixed Nuts:											
4592	Dry roasted, salted	1 c	137	2	814	24	35	12	70	9.5	43	14.8
4593	Oil roasted, salted	1 c	142	2	876	24	30	13	80	12.4	45	18.9
4533	Oil roasted, unsalted	1 c	142	2	876	24	30	14	80	12.4	45	18.9
4762	Peanuts, oil roasted, salted	1 oz	28	1	168	8	4	3	15	2.4	7.3	4.3
4578	Pecan halves, dried, unsalted	1 oz	28	4	193	3	4	3	20	1.7	11.4	6.1
4583	Dry roasted, salted	¹/₄ c	28	1	199	3	4	3	21	1.8	12.3	5.8
4554	Pine nuts/pinons, dried	1 oz	28	6	176	3	5	3	17	2.6	6.4	7.2
4520	Pistachios, dried, shelled	1 oz	28	4	156	6	8	3	12	1.5	6.5	3.8
4540	Dry roasted, salted, shelled	1 c	128	2	727	27	34	13	59	7.1	31	17.8
4522	Pumpkin kernels, dried, unsalted	1 oz	28	7	151	7	5	1	13	2.4	4	5.9
4625	Roasted, salted	1 c	227	7	1185	75	30	9	96	18.1	29.7	43.6
4524	Sesame seeds, hulled, dried	¹/₄ c	38	5	225	8	6	5	21	2.9	7.9	9.1
8878	Soy nuts, BBQ	5 pce	28		119	12	9	4	4	1		
8877	Salted	5 pce	28		119	12	9	5	4	1		
	Sunflower seed kernels:											
4545	Dry	¹/₄ c	36	5	205	8	7	4	18	1.9	3.4	11.8
4546	Oil roasted	¹/₄ c	34	3	209	7	5	2	20	2	3.7	12.9
4532	Tahini (sesame butter)	1 tbs	15	3	91	3	3	1	8	1.2	3.2	3.7
44059	Trail Mix w/chocolate chips	1 c	146	7	707	21	66	8	47	8.9	19.8	16.5
4525	Black walnuts, chopped	1 oz	28	5	173	7	3	2	17	0.9	4.2	9.8
4556	English walnuts, chopped	1 oz	28	4	183	4	4	2	18	1.7	2.5	13.2
	SWEETENERS and SWEETS (see also Dairy (milk desserts) and Baked Goods)											
23000	Apple butter	2 tbs	36	56	62	<1	15	1	0	0	0	0
1124	Butterscotch topping	2 tbs	41	32	103	1	27	<1	<1	t	t	0
23069	Caramel topping	2 tbs	41	32	103	1	27	<1	<1	t	t	0
	Cake frosting, creamy vanilla:											
46009	Canned	2 tbs	39	15	164	0	26	<1	6	1.2	1.9	3.1
46018	From mix	2 tbs	39	12	161	<1	29	<1	5	0.7	1.5	1.1
	Candy:											
23405	Almond Joy candy bar	1 oz	28	8	134	1	17	1	8	4.9	1.5	0.3
23184	Butterscotch morsels	¹/₄ c	43	8	243	0	27	0	12	10.6		
23015	Caramel, plain or chocolate	1 pce	10	8	38	<1	8	<1	1	0.7	t	t
90712	Chewing gum	1 pce	3	3	7	0	2		<1	t	t	t
25125	Sugarless	1 pce	3		6	0	2		0	0	0	0
	Chocolate:											
23016	Milk chocolate	1 oz	28	2	150	2	17	1	8	4	3.7	0.2

PAGE KEY: A–4 = Beverages A–6 = Dairy A–10 = Eggs A–12 = Fat/Oil A–14 = Fruit A–20 = Bakery
A–26 = Grain A–32 = Fish A–32 = Meats A–36 = Poultry A–38 = Sausage A–40 = Mixed/Fast A–44 = Nuts/Seeds A–46 = Sweets
A–48 = Vegetables/Legumes A–58 = Vegetarian A–62 = Misc A–62 = Soups/Sauces A–64 = Fast A–78 = Convenience A–80 = Baby foods

Chol (mg)	Calc (mg)	Iron (mg)	Magn (mg)	Pota (mg)	Sodi (mg)	Zinc (mg)	VT-A (RAE)	Thia (mg)	VT-E (mg α)	Ribo (mg)	Niac (mg)	V-B6 (mg)	Fola (μg)	VT-C (mg)
0	69	1.2	77	204	<1	0.94	<1	0.07	7.24	0.23	1.1	0.04	8	0
0	43	0.59	48	121	2	0.49	0	0.02	3.25	0.1	0.46	0.01	10	0
0	43	0.59	48	121	72	0.49	<1	0.02	4.16	0.1	0.46	0.01	10	0
0	224	3.4	526	923	4	5.68	0	0.86	8.02	0.05	0.41	0.14	31	1
0	13	1.68	73	158	179	1.57	0	0.06	0.26	0.06	0.39	0.07	19	0
0	13	1.68	73	158	4	1.57	0	0.06	0.26	0.06	0.39	0.07	19	0
0	12	1.69	76	177	86	1.5	0	0.1	0.26	0.06	0.49	0.09	7	0
0	56	7.86	355	822	17	6.96	0	0.47	1.2	0.28	2.26	0.42	32	0
0	12	1.69	76	177	4	1.5	0	0.1	0.26	0.06	0.49	0.09	7	0
0	7	0.8	41	87	2	0.83	0	0.05	0.25	0.03	0.26	0.04	11	0
0	7	0.8	41	87	98	0.83	0	0.05	0.15	0.03	0.26	0.04	11	0
0	41	1.3	47	847	3	0.82	1	0.35	0.72	0.25	1.92	0.71	100	37
0	6	1.09	14	160	9	0.5	0	0.03	0.11	<.01	0.24	0.02	12	1
0	6	0.97	13	142	8	0.44	0	0.03	0.1	<.01	0.22	0.02	10	1
0	20	2.59	70	424	29	1.57	0	0.05	0.34	0.08	0.47	0.23	7	1
0	14	1.79	46	313	244	1.69	0	0.03	0.36	0.02	0.44	0.25	7	1
0	41	7.46	104	497	29	1.27	0	0.05	1.47	0	1.44	0.06	32	2
0	32	1.32	46	190	0	0.69	<1	0.18	4.21	0.03	0.5	0.16	32	2
0	96	5.07	308	818	917	5.21	<1	0.27	14.99	0.27	6.44	0.41	68	1
0	153	4.56	334	825	926	7.21	<1	0.71	10.22	0.32	7.19	0.34	118	1
0	153	4.56	334	825	16	7.21	1	0.71	8.52	0.32	7.19	0.34	118	1
0	17	0.43	49	203	90	0.92	0	0.02	1.94	0.02	3.87	0.13	34	0
0	20	0.71	34	115	0	1.27	1	0.18	0.39	0.04	0.33	0.06	6	0
0	20	0.78	37	119	107	1.42	2	0.13	0.36	0.03	0.33	0.05	4	0
0	2	0.86	66	176	20	1.2	<1	0.35	0.98	0.06	1.22	0.03	16	1
0	30	1.16	34	287	<1	0.62	8	0.24	0.64	0.04	0.36	0.48	14	1
0	141	5.38	154	1334	518	2.94	17	1.08	2.47	0.2	1.82	1.63	64	3
0	12	4.19	150	226	5	2.09	5	0.06	0	0.09	0.49	0.06	16	1
0	98	33.91	1212	1830	1305	16.89	43	0.48	0	0.72	3.95	0.2	129	4
0	50	2.96	132	155	15	3.9	1	0.27	0.1	0.03	1.78	0.06	36	0
0	59	1.07			415		0							0
0	59	1.07			148		0							0
0	42	2.44	127	248	1	1.82	1	0.82	12.42	0.09	1.62	0.28	82	1
0	19	2.28	43	164	1	1.77	<1	0.11	13.25	0.1	1.4	0.27	80	0
0	21	0.95	53	69	<1	1.57	<1	0.24	0.34	0.02	0.85	0.02	15	0
6	159	4.95	235	946	177	4.58	3	0.6	15.62	0.33	6.43	0.38	95	2
0	17	0.87	56	146	1	0.94	1	0.02	0.5	0.04	0.13	0.16	9	0
0	27	0.81	44	123	1	0.87	<1	0.1	0.2	0.04	0.32	0.15	27	0
0	5	0.11	2	33	5	0.02	<1	<.01	<.01	<.01	0.02	0.01	<1	0
0	22	0.08	3	34	143	0.08	11	<.01	0	0.04	0.02	<.01	1	0
0	22	0.08	3	34	143	0.08	11	<.01	0	0.04	0.02	<.01	1	0
0	1	0.06	0	13	72	0.03	0	<.01	0.82	0.12	0.09	0	3	0
0	2	<.01	1	4	44	<.01	31	<.01	0.76	<.01	<.01	<.01	0	0
1	18	0.36	18	71	40	0.22		<.01	<.01	0.04	0.13	0.02		0
0	0	0		80	45		0	0.03		0.03	0.03			0
1	14	0.01	2	21	24	0.04	<1	<.01	0.28	0.03	0.02	<.01	<1	0
0	0	0	0	0	<1	0	0	0	0	0	0	0	0	0
				0	0									
6	53	0.66	18	104	22	0.56	14	0.03	0.57	0.08	0.11	0.01	3	0

TABLE A-1

Food Composition   (Computer code number is for West Diet Analysis program)   (For purposes of calculations, use "0" for t, <1, <.1, <.01, etc.)

SWEETENERS and SWEETS (see also Dairy (milk desserts) and Baked Goods)—Continued

Computer Code Number	Food Description	Measure	Wt (g)	H₂O (%)	Ener (kcal)	Prot (g)	Carb (g)	Dietary Fiber (g)	Fat (g)	Sat	Mono	Poly
	SWEETENERS and SWEETS (see also Dairy (milk desserts) and Baked Goods)—Continued											
23018	Milk chocolate with almonds	1 oz	28	2	147	3	15	2	10	4.8	3.8	0.6
23058	Milk chocolate with rice cereal	1 oz	28	2	139	2	18	1	7	4.4	2.4	0.2
23012	Semisweet chocolate chips	1 c	168	1	805	7	106	10	50	29.8	16.7	1.6
23057	Sweet Dark chocolate (candy bar)	1 ea	41	1	218	2	24	3	13	7.9	2.1	0.2
23024	Fondant candy, uncoated (mints, candy corn, other)	1 pce	16	7	60	0	15	0	<1	0	0	0
23409	Gumdrops	1 c	182	1	721	0	180	<1	0	0	0	0
23031	Hard candy-all flavors	1 pce	6	1	24	0	6	0	<1	0	0	0
23033	Jellybeans	10 pce	11	6	41	0	10	<1	<1	0	0	0
23025	Fudge, chocolate	1 pce	17	10	70	<1	13	<1	2	1	0.5	t
23046	M&M's plain chocolate candy	10 pce	7	2	34	<1	5	<1	1	0.9	0.2	t
23048	M&M's peanut chocolate candy	10 pce	20	2	103	2	12	1	5	2.1	2.2	0.8
23037	MARS almond bar	1 ea	50	4	234	4	31	1	12	3.6	5.3	2
23038	MILKY WAY candy bar	1 ea	60	6	254	3	43	1	10	4.7	3.6	0.4
23021	Milk chocolate-coated peanuts	1 c	149	2	773	20	74	7	50	21.8	19.3	6.5
23081	Peanut brittle, recipe	1 c	147	1	711	11	104	4	28	6.1	11.9	6.7
23036	Skor English toffee candy bar	1 ea	39	2	209	1	24	1	13	7.3	3.6	0.5
1131	Snickers candy bar (2.2oz)	1 ea	62	5	297	5	37	2	15	5.6	6.5	3
23086	Fruit Roll-Up (small)	1 ea	14	10	52	<1	12	<1	<1	t	0.2	t
23174	Fruit juice bar (2.5 fl oz)	1 ea	77	78	63	1	16	1	<1	t	0	t
23052	Gelatin dessert/Jello, prepared	½ c	135	84	84	2	19	0	0	0	0	0
23093	SugarFree	½ c	117	95	23	1	5	0	0	0	0	0
25001	Honey	1 tbs	21	17	64	<1	17	<1	0	0	0	0
23003	Jellies:	1 tbs	19	30	51	<1	13	<1	<1	t	t	t
23004	Packet	1 ea	14	30	37	<1	10	<1	<1	t	t	t
23005	Marmalade	1 tbs	20	33	49	<1	13	<1	0	0	0	0
23007	Marshmallows	1 ea	7	16	22	<1	6	<1	<1	t	t	t
23071	Marshmallow creme topping	2 tbs	38	20	122	<1	30	<1	<1	t	t	t
25003	Molasses	2 tbs	41	22	119	0	31	0	<1	t	t	t
23050	Popsicle/ice pops	1 ea	128	80	92	0	24	0	0	0	0	0
23171	Rice crispie bar	1 ea	28	13	107	1	20	<1	3	0.6	1.3	0.8
	Sugars:											
25005	Brown sugar	1 c	220	2	829	0	214	0	0	0	0	0
25006	White sugar, granulated	1 tbs	12	0	46	0	12	0	0	0	0	0
25007	Packet	1 ea	6	0	23	0	6	0	0	0	0	0
	Sweeteners:											
25038	Equal, packet	1 ea	1	12	4	<1	1	0	<1	t	t	t
25208	Sweet 'N Low, packet	1 ea	1		4	0	1	0	0	0	0	0
	Syrups											
23014	Chocolate, hot fudge type	2 tbs	43	22	150	2	27	1	4	1.7	1.7	0.1
23056	Thin type	2 tbs	38	29	93	1	25	1	<1	0.3	0.2	t
23042	Pancake table syrup (corn and maple)	2 tbs	40	38	94	0	25	<1	0	0	0	0
	VEGETABLES AND LEGUMES											
	Amaranth leaves:											
5375	Raw, chopped	1 c	28	92	6	1	1	<1	<1	t	t	t
5376	Raw, each	1 ea	14	92	3	<1	1	<1	<1	t	t	t
5377	Cooked	1 c	132	91	28	3	5	2	<1	t	t	0.1
6033	Arugula, raw, chopped	½ c	10	92	2	<1	<1	<1	<1	t	t	t
5000	Artichokes, cooked globe (300 g with refuse)	1 ea	120	84	60	4	13	6	<1	t	t	t
	Artichoke hearts:											
5192	Cooked from frozen	1 c	168	86	76	5	15	8	1	0.2	t	0.4
5191	Marinated	1 c	130		116	5	14	5	7	0		
7962	In water	½ c	100		38	2	6	0	0	0	0	0
	Asparagus, green:											
5003	Cooked from fresh, cuts and tips	½ c	90	93	20	2	4	2	<1	t	t	0.1
5004	Spears, ½" diam at base	4 ea	60	93	13	1	2	1	<1	t	t	t
5005	Cooked from frozen, cuts and tips	½ c	90	94	16	3	2	1	<1	t	t	0.2
5006	Spears, ½" diam at base	4 ea	60	94	11	2	1	1	<1	t	t	0.1
5007	Canned, spears, ½" diam at base	4 pce	72	94	14	2	2		<1	0.1	t	0.2

PAGE KEY: A–4 = Beverages   A–6 = Dairy   A–10 = Eggs   A–12 = Fat/Oil   A–14 = Fruit   A–20 = Bakery
A–26 = Grain   A–32 = Fish   A–32 = Meats   A–36 = Poultry   A–38 = Sausage   A–40 = Mixed/Fast   A–44 = Nuts/Seeds   A–46 = Sweets
A–48 = Vegetables/Legumes   A–58 = Vegetarian   A–62 = Misc   A–62 = Soups/Sauces   A–64 = Fast   A–78 = Convenience   A–80 = Baby foods

Chol (mg)	Calc (mg)	Iron (mg)	Magn (mg)	Pota (mg)	Sodi (mg)	Zinc (mg)	VT-A (RAE)	Thia (mg)	VT-E (mg α)	Ribo (mg)	Niac (mg)	V-B$_6$ (mg)	Fola (µg)	VT-C (mg)
5	63	0.46	25	124	21	0.38	12	0.02	1.26	0.12	0.21	0.01	4	0
5	48	0.21	14	96	41	0.31	17	0.02	0.56	0.08	0.13	0.02	4	0
0	54	5.26	193	613	18	2.72	0	0.09	0.39	0.15	0.72	0.06	5	0
2	12	0.87	13	206	2	<.01		0	0.08	<.01	0	0	0	0
0	0	<.01	0	1	3	0	0	<.01	0	<.01	<.01	<.01	0	0
0	5	0.73	2	9	80	0	0	0.01	0	0.02	0.02	<.01	0	0
0	0	0.02	0	0	2	<.01	0	<.01	0	<.01	<.01	<.01	0	0
0	0	0.01	0	4	6	<.01	0	<.01	0	<.01	<.01	<.01	0	0
2	8	0.3	6	22	8	0.19	7	<.01	0.03	0.01	0.03	<.01	1	0
1	7	0.08	2	14	4	0.08	2	<.01	0.08	0.01	0.01	<.01	<1	0
2	20	0.23	15	69	10	0.48	5	0.02	0.51	0.03	0.82	0.02	8	0
8	84	0.55	36	162	85	0.56	8	0.02	3.88	0.16	0.47	0.03	4	0
8	78	0.46	20	145	144	0.43	11	0.02	0.75	0.13	0.21	0.03	4	1
13	155	1.95	143	748	61	3.56	51	0.17	5.16	0.26	6.33	0.31	12	0
18	40	1.79	62	247	654	1.28	57	0.2	3.76	0.06	3.89	0.12	68	0
21	51	0.22	4	60	124	0.07		<.01	0.02	0.04	0.05	0.01	1	0
8	58	0.47	35	185	165	0.97	26	0.08	0.68	0.12	0.98	0.04	19	0
0	4	0.14	3	41	44	0.03	1	0.01	0.08	<.01	0.01	0.04	1	17
0	4	0.15	3	41	3	0.04	1	<.01	0.1	0.01	0.12	0.02	5	7
0	4	0.03	1	1	101	0.01	0	0	0	<.01	<.01	0	1	0
0	4	0.01	1	1	56	0	0	0	0	0	0	0	0	0
0	1	0.09	0	11	1	0.05	0	0	0	<.01	0.03	<.01	<1	0
0	1	0.04	1	10	6	<.01	<1	<.01	0	<.01	<.01	<.01	<1	0
0	1	0.03	1	8	4	<.01	<1	<.01	0	<.01	<.01	<.01	<1	0
0	8	0.03	0	7	11	<.01	1	<.01	0.01	<.01	0.01	<.01	2	1
0	0	0.02	0	0	6	<.01	0	<.01	0	<.01	<.01	<.01	<1	0
0	1	0.08	1	2	30	0.02	<1	<.01	0	<.01	0.03	<.01	<1	0
0	84	1.94	99	600	15	0.12	0	0.02	0	<.01	0.38	0.27	0	0
0	0	0	1	5	15	0.03	0	0	0	0	0	0	0	0
0	2	0.51	4	12	123	0.15		0.1	0.41	0.11	1.31	0.13	28	4
0	187	4.2	64	761	86	0.4	0	0.02	0	0.02	0.18	0.06	2	0
0	0	<.01	0	0	0	0	0	0	0	<.01	0	0	0	0
0	0	<.01	0	0	0	0	0	0	0	<.01	0	0	0	0
0	0	<.01	0	0	<1	0	0	0	0	0	0	0	0	0
0	0	0			0		0							0
1	43	0.68	28	194	149	0.36	2	0.03	1.07	0.12	0.16	0.03	2	0
0	5	5.15	25	183	58	0.28		<.01	0.01	0.31	12.76	<.01	2	0
0	1	0.01	1	6	33	0.03		<.01	0	<.01	<.01	<.01	0	0
0	60	0.65	15	171	6	0.25	0	<.01	0.22	0.04	0.18	0.05	24	12
0	30	0.32	8	86	3	0.13	0	<.01	0.11	0.02	0.09	0.03	12	6
0	276	2.98	73	846	28	1.16	183	0.03	0.66	0.18	0.74	0.23	75	54
0	16	0.15	5	37	3	0.05	12	<.01	0.04	<.01	0.03	<.01	10	2
0	54	1.55	72	425	114	0.59	11	0.08	0.23	0.08	1.2	0.13	61	12
0	35	0.94	52	444	89	0.6	13	0.1	0.27	0.27	1.54	0.15	200	8
0	0	0			488		0							46
0	0	1.35		0	250		6							4
0	21	0.82	13	202	13	0.54	45	0.15	0.34	0.13	0.98	0.07	134	7
0	14	0.55	8	134	8	0.36	30	0.1	0.23	0.08	0.65	0.05	89	5
0	16	0.5	9	155	3	0.37	36	0.06	1.08	0.09	0.93	0.02	122	22
0	11	0.34	6	103	2	0.25	24	0.04	0.72	0.06	0.62	0.01	81	15
0	12	1.32	7	124	207	0.29	30	0.04	0.22	0.07	0.69	0.08	69	13

A

TABLE A-1

Food Composition   (Computer code number is for West Diet Analysis program)   (For purposes of calculations, use "0" for t, <1, <.1, <.01, etc.)

Computer Code Number	Food Description	Measure	Wt (g)	H$_2$O (%)	Ener (kcal)	Prot (g)	Carb (g)	Dietary Fiber (g)	Fat (g)	Fat Breakdown (g) Sat	Mono	Poly
	VEGETABLES AND LEGUMES—Continued											
	Bamboo shoots:											
5401	Canned, drained slices	1 c	131	94	25	2	4		1	0.1	t	0.2
6736	Raw slices	1 c	151	91	41	4	8	3	<1	0.1	t	0.2
5249	Cooked slices	1 c	120	96	14	2	2	1	<1	t	t	0.1
	Beans (see also alphabetical listing this section):											
7034	Adzuki beans, cooked	¹/₂ c	115	66	147	9	28	8	<1	t		
7012	Black beans, cooked	¹/₂ c	86	66	114	8	20	7	<1	0.1		0.2
	Canned beans (white/navy):											
7004	With pork and tomato sauce	¹/₂ c	127	73	124	7	25	6	1	0.5	0.6	0.2
7023	With sweet sauce	¹/₂ c	130	71	144	7	27	7	2	0.7	0.8	0.2
56101	With frankfurters	¹/₂ c	130	69	185	9	20	9	9	3.1	3.7	1.1
	Lima beans:											
5247	Fordhooks, cooked from frozen	¹/₂ c	85	73	88	5	16	5	<1	t	t	0.1
5019	Baby, cooked from frozen	¹/₂ c	90	72	94	6	18	5	<1	t	t	0.1
7010	Cooked from dry, drained	¹/₂ c	94	70	108	7	20	7	<1	t	t	0.2
7352	Red Mexican, cooked f/dry	¹/₂ c	112	70	127	8	24	9	<1	t	t	0.2
	Snap bean/green string beans:											
5011	Cooked from fresh	¹/₂ c	63	89	22	1	5	2	<1	t	t	t
5013	Cooked from frozen	¹/₂ c	68	91	19	1	4	2	<1	t	t	t
5015	Canned, drained	¹/₂ c	68	93	14	1	3	1	<1	t	t	t
5194	Snap bean, yellow, cooked f/fresh	¹/₂ c	63	89	22	1	5	2	<1	t	t	t
	Bean sprouts (mung):											
5020	Raw	¹/₂ c	52	90	16	2	3	1	<1	t	t	t
5246	Cooked, stir fried	¹/₂ c	62	84	31	3	7	1	<1	t	t	t
5021	Cooked, boiled, drained	¹/₂ c	62	93	13	1	3	<1	<1	t	t	t
5197	Canned, drained	¹/₂ c	63	96	8	1	1	1	<1	t	t	t
	Beets:											
5022	Cooked from fresh, sliced or diced	¹/₂ c	85	87	37	1	8	2	<1	t	t	t
5023	Whole beets, 2" diam	2 ea	100	87	44	2	10	2	<1	t	t	t
5024	Canned, sliced or diced	¹/₂ c	79	91	24	1	6	1	<1	t	t	t
5310	Pickled slices	¹/₂ c	114	82	74	1	19	3	<1	t	t	t
5025	Beet greens, cooked, drained	¹/₂ c	72	89	19	2	4	2	<1	t	t	t
	Broccoli:											
5027	Broccoli, raw, spear	1 ea	31	89	11	1	2	1	<1	t	t	t
5029	Cooked from fresh, spears	1 ea	180	89	63	4	13	6	1	0.1	t	0.3
5028	Chopped	¹/₂ c	78	89	27	2	6	3	<1	t	t	0.1
5234	Cooked from frozen, spear, small piece	¹/₂ c	92	91	26	3	5	3	<1	t	t	t
5030	Chopped	¹/₂ c	92	91	26	3	5	3	<1	t	t	t
5679	Broccoflower-steamed	¹/₂ c	78	90	25	2	5	2	<1	t	t	t
5033	Brussels sprouts, cooked from fresh	¹/₂ c	78	89	28	2	6	2	<1	t	t	0.2
5035	Cooked from frozen	¹/₂ c	78	87	33	3	6	3	<1	t	t	0.2
5036	Cabbage, common, raw, chopped	1 c	70	92	17	1	4		<1	t	t	t
5038	Cooked, drained	1 c	150	94	33	2	7		1	t	t	0.3
	Cabbage, Chinese:											
5041	Bok Choy, raw, shredded	1 c	70	95	9	1	2	1	<1	t	t	t
5237	Cooked, drained	1 c	170	96	20	3	3		<1	t	t	0.1
5535	Kim chee style	1 c	150	92	31	2	6	2	<1	t	t	0.1
5040	Pe Tsai, raw, chopped	1 c	76	94	12	1	2		1	<1	t	t
5235	Cooked	1 c	119	95	17	2	3	2	<1	t	t	t
6766	Cabbage, red, raw, chopped	1 c	89	90	28	1	7	2	<1	t	t	t
5238	Cooked, drained	1 c	150	91	44	2	10	4	<1	t	t	t
5043	Cabbage, savoy, raw, chopped	1 c	70	91	19	1	4	2	<1	t	t	t
5044	Cooked	1 c	145	92	35	3	8	4	<1	t	t	t
5511	Capers	1 ea	5		0	<1	<1		<1			
	Carrots:											
5045	Raw, whole, 7¹/₂ × 1¹/₈"	1 ea	72	88	30	1	7	2	<1	t		t
5046	Grated	¹/₂ c	55	88	23	1	5	2	<1	t	t	t
5047	Cooked from raw	¹/₂ c	78	90	27	1	6	2	<1	t	t	t
5358	From frozen	¹/₂ c	73	90	27	<1	6	2	<1	t	t	0.2
5199	Canned	¹/₂ c	73	93	18	<1	4	1	<1	t	t	t

PAGE KEY: A–4 = Beverages   A–6 = Dairy   A–10 = Eggs   A–12 = Fat/Oil   A–14 = Fruit   A–20 = Bakery
A–26 = Grain   A–32 = Fish   A–32 = Meats   A–36 = Poultry   A–38 = Sausage   A–40 = Mixed/Fast   A–44 = Nuts/Seeds   A–46 = Sweets
A–48 = Vegetables/Legumes   A–58 = Vegetarian   A–62 = Misc   A–62 = Soups/Sauces   A–64 = Fast   A–78 = Convenience   A–80 = Baby foods

Chol (mg)	Calc (mg)	Iron (mg)	Magn (mg)	Pota (mg)	Sodi (mg)	Zinc (mg)	VT-A (RAE)	Thia (mg)	VT-E (mg α)	Ribo (mg)	Niac (mg)	V-B$_6$ (mg)	Fola (μg)	VT-C (mg)
0	10	0.42	5	105	9	0.85	1	0.03	0.83	0.03	0.18	0.18	4	1
0	20	0.76	5	805	6	1.66	2	0.23	1.51	0.11	0.91	0.36	11	6
0	14	0.29	4	640	5	0.56	0	0.02	0.8	0.06	0.36	0.12	2	0
0	32	2.3	60	612	9	2.04	<1	0.13	0.12	0.07	0.82	0.11	139	0
0	23	1.81	60	305	1	0.96	<1	0.21	0.07	0.05	0.43	0.06	128	0
9	71	4.17	44	381	559	7.44	5	0.07	0.69	0.06	0.63	0.09	29	4
9	79	2.16	44	346	437	1.95	1	0.06	0.7	0.08	0.46	0.11	48	4
8	62	2.25	36	306	559	2.43	5	0.08	0.6	0.07	1.17	0.06	39	3
0	26	1.55	36	258	59	0.63	8	0.06	0.25	0.05	0.91	0.1	18	11
0	25	1.76	50	370	26	0.5	7	0.06	0.58	0.05	0.69	0.1	14	5
0	16	2.25	40	478	2	0.89	0	0.15	0.17	0.05	0.4	0.15	78	0
0	42	1.87	48	371	6	0.87	<1	0.13	0.08	0.07	0.38	0.12	94	2
0	28	0.41	11	92	1	0.16	22	0.05	0.28	0.06	0.39	0.04	21	6
0	33	0.6	16	86	6	0.33	19	0.02	0.24	0.06	0.26	0.04	16	3
0	18	0.61	9	74	178	0.2	15	0.01	0.19	0.04	0.14	0.03	22	3
0	29	0.81	16	188	2	0.23	3	0.05	0.28	0.06	0.39	0.04	21	6
0	7	0.47	11	77	3	0.21	1	0.04	0.05	0.06	0.39	0.05	32	7
0	8	1.18	20	136	6	0.56	1	0.09	<.01	0.11	0.74	0.08	43	10
0	7	0.4	9	63	6	0.29	1	0.03	0.04	0.06	0.51	0.03	18	7
0	9	0.27	6	17	88	0.18	<1	0.02	0.03	0.04	0.14	0.02	6	0
0	14	0.67	20	259	65	0.3	2	0.02	0.03	0.03	0.28	0.06	68	3
0	16	0.79	23	305	77	0.35	2	0.03	0.04	0.04	0.33	0.07	80	4
0	12	1.44	13	117	153	0.17	1	<.01	0.02	0.03	0.12	0.05	24	3
0	13	0.47	17	169	301	0.3	1	0.01	0.15	0.05	0.29	0.06	31	3
0	82	1.37	49	654	174	0.36	276	0.08	1.3	0.21	0.36	0.1	10	18
0	15	0.23	7	98	10	0.13	10	0.02	0.24	0.04	0.2	0.05	20	28
0	72	1.21	38	527	74	0.81	176	0.11	2.61	0.22	1	0.36	194	117
0	31	0.52	16	229	32	0.35	76	0.05	1.13	0.1	0.43	0.16	84	51
0	47	0.56	18	166	22	0.28	52	0.05	1.21	0.07	0.42	0.12	28	37
0	30	0.56	12	131	10	0.26	52	0.05	1.21	0.07	0.42	0.12	52	37
0	25	0.55	16	251	18	0.39	3	0.06	0.23	0.07	0.59	0.14	38	49
0	28	0.94	16	247	16	0.26	30	0.08	0.34	0.06	0.47	0.14	47	48
0	20	0.37	14	226	12	0.19	36	0.08	0.4	0.09	0.42	0.23	79	36
0	33	0.41	10	172	13	0.13	6	0.04	0.1	0.03	0.21	0.07	30	23
0	46	0.26	12	146	12	0.14	10	0.09	0.18	0.08	0.42	0.17	30	30
0	74	0.56	13	176	46	0.13	156	0.03	0.06	0.05	0.35	0.14	46	32
0	158	1.77	19	631	58	0.29	360	0.05	0.15	0.11	0.73	0.28	70	44
0	145	1.28	27	375	995	0.36	213	0.07	0.24	0.1	0.75	0.34	88	80
0	59	0.24	10	181	7	0.17	12	0.03	0.09	0.04	0.3	0.18	60	21
0	38	0.36	12	268	11	0.21	57	0.05	0.14	0.05	0.6	0.21	63	19
0	40	0.71	14	216	24	0.2	50	0.06	0.1	0.06	0.37	0.19	16	51
0	63	0.99	26	393	12	0.38	3	0.11	0.18	0.09	0.57	0.34	36	16
0	24	0.28	20	161	20	0.19	35	0.05	0.12	0.02	0.21	0.13	56	22
0	44	0.55	35	267	35	0.33	64	0.07	0.15	0.03	0.03	0.22	67	25
0	2	0.05			105		1							0
0	24	0.22	9	230	50	0.17	433	0.05	0.48	0.04	0.71	0.1	14	4
0	18	0.16	7	176	38	0.13	331	0.04	0.36	0.03	0.54	0.08	10	3
0	23	0.27	8	183	45	0.16	671	0.05	0.8	0.03	0.5	0.12	11	3
0	26	0.39	8	140	43	0.26	607	0.02	0.74	0.03	0.3	0.06	8	2
0	18	0.47	6	131	177	0.19	407	0.01	0.54	0.02	0.4	0.08	7	2

A

TABLE A-1

Food Composition   (Computer code number is for West Diet Analysis program)   (For purposes of calculations, use "0" for t, <1, <.1, <.01, etc.)

Computer Code Number	Food Description	Measure	Wt (g)	H₂O (%)	Ener (kcal)	Prot (g)	Carb (g)	Dietary Fiber (g)	Fat (g)	Fat Breakdown (g) Sat	Mono	Poly
	VEGETABLES AND LEGUMES—Continued											
5226	Juice, canned	1 c	236	89	94	2	22	2	<1	t	t	0.2
5625	Cassava, cooked	1 c	137	59	221	2	53	2	<1	0.1	0.1	t
5049	Cauliflower, flowerets, raw:	½ c	50	92	12	1	3	1	<1	t	t	t
5051	Cooked from fresh, drained	½ c	62	93	14	1	3	2	<1	t	t	0.1
5053	From frozen, drained	½ c	90	94	17	1	3	2	<1	t	t	t
5055	Celery, raw, large outer stalk, 8 × 1½" (root end)	1 ea	40	95	6	<1	1		<1	t	t	t
5054	Diced	1 c	120	95	17	1	4		<1	t	t	t
5200	Celeriac/celery root, cooked	1 c	155	92	42	1	9	2	<1	t	t	0.2
5057	Chard, swiss, raw, chopped	1 c	36	93	7	1	1	1	<1	t	t	t
5059	Cooked	1 c	175	93	35	3	7	4	<1	t	t	t
5413	Chayote fruit, raw	1 ea	203	95	35	2	8	3	<1	t	t	0.1
5414	Cooked	1 c	160	93	38	1	8	4	1	0.1	t	0.3
	Chickpeas (see Garbanzo Beans #854)											
5061	Collards, cooked from raw	½ c	95	92	25	2	5	3	<1	t	t	0.2
5062	From frozen	½ c	85	88	31	3	6	2	<1	t	t	0.2
	Corn, yellow:											
5364	Cooked from frozen, on cob, 3½" long	1 ea	63	73	59	2	14	2	<1	t	0.1	0.2
5065	Kernels, cooked from frozen	½ c	82	77	66	2	16	2	1	t	0.2	0.3
5068	Canned, cream style	½ c	128	79	92	2	23	2	1	t	0.2	0.3
5067	Whole kernel, vacuum pack	½ c	105	77	83	3	20	2	1	t	0.2	0.2
	Cowpeas (see Black-eyed peas #814-816)											
5610	Cucumber, kim chee style	1 c	150	91	32	2	7	2	<1	t	t	t
5241	Dandelion greens, raw	1 c	55	86	25	1	5	2	<1	t	t	0.2
5242	Chopped, cooked, drained	1 c	105	90	35	2	7	3	1	0.2	t	0.3
5072	Eggplant, cooked	1 c	99	90	35	1	9	2	<1	t	t	t
5202	Endive, fresh, chopped	1 c	50	94	8	1	2		2	<1	t	t
856	Escarole/curly endive-chopped	1 c	50	94	8	1	2	2	<1	t	t	t
7001	Garbanzo beans (Chickpeas), cooked	1 c	164	60	269	15	45	12	4	0.4	1	1.9
7961	Grape leaf, raw:	1 ea	3	73	3	<1	1	<1	<1	t	t	t
7914	Cup	1 c	14	73	13	1	2	2	<1	t	t	0.1
7021	Great northern beans, cooked	1 c	177	69	209	15	37	12	1	0.2	t	0.3
5077	Jerusalem artichoke, raw slices	1 c	150	78	114	3	26	2	<1	0	t	t
5224	Jicama	1 c	120	90	46	1	11		<1	t	t	t
5075	Kale, cooked from raw	1 c	130	91	36	2	7	3	1	t	t	0.3
5076	From frozen	1 c	130	90	39	4	7	3	1	t	t	0.3
7292	Kidney beans, canned	1 c	256	77	218	13	40	16	1	0.1	t	0.5
5078	Kohlrabi, raw slices	1 c	135	91	36	2	8	5	<1	t	t	t
5079	Cooked	1 c	165	90	48	3	11	2	<1	t	t	t
5205	Leeks, raw, chopped	1 c	89	83	54	1	13	2	<1	t	t	0.1
5203	Cooked, chopped	1 c	104	91	32	1	8	1	<1	t	t	0.1
7006	Lentils, cooked from dry	1 c	198	70	230	18	40	16	1	0.1	0.1	0.3
5390	Lentils, sprouted, stir fried	1 c	124	69	125	11	26	5	1	Sat	0.1	Poly
5389	Raw	1 c	77	67	82	7	17	3	<1	t	t	0.2
	Lettuce:											
5082	Butterhead/Boston, head, 5" diameter	¼ ea	41	96	5	1	1	<1	<1	t	t	t
5081	Leaves, inner or outer	4 ea	30	96	4	<1	1	<1	<1	t	t	t
5083	Iceberg/crispshead, chopped	1 c	55	96	6	<1	1		<1	t	t	t
5085	Head, 6" diameter	1 ea	539	96	54	4	11		1	t	t	0.3
866	Wedge, ¼ head	1 ea	135	96	14	1	3		<1	t	t	t
5086	Looseleaf, chopped	¼ c	28	95	4	<1	1	<1	<1	t	t	t
5088	Romaine, chopped	¼ c	28	95	5	<1	1	1	<1	t	t	t
5089	Romaine, inner leaf	3 pce	30	95	5	<1	1	1	<1	t	t	t
5528	Luffa, cooked (Chinese okra)	1 c	178	90	57	3	13	4	<1	t	t	t
6778	Manioc, raw	1 c	206	60	330	3	78	4	1	0.2	0.2	t
	Mushrooms:											
5090	Raw, sliced	½ c	35	92	8	1	1	<1	<1	t	t	t
5092	Cooked from fresh, pieces	½ c	78	91	22	2	4	2	<1	t	t	0.1
1962	Stir fried, shitake slices	½ c	73	83	40	1	10	2	<1	t	t	t
5094	Canned, drained	½ c	78	91	20	1	4	2	<1	t	t	t
5612	Mushroom caps, pickled	8 ea	47	92	11	1	2	<1	<1	t	t	t

PAGE KEY: A–4 = Beverages A–6 = Dairy A–10 = Eggs A–12 = Fat/Oil A–14 = Fruit A–20 = Bakery
A–26 = Grain A–32 = Fish A–32 = Meats A–36 = Poultry A–38 = Sausage A–40 = Mixed/Fast A–44 = Nuts/Seeds A–46 = Sweets
A–48 = Vegetables/Legumes A–58 = Vegetarian A–62 = Misc A–62 = Soups/Sauces A–64 = Fast A–78 = Convenience A–80 = Baby foods

Chol (mg)	Calc (mg)	Iron (mg)	Magn (mg)	Pota (mg)	Sodi (mg)	Zinc (mg)	VT-A (RAE)	Thia (mg)	VT-E (mg α)	Ribo (mg)	Niac (mg)	V-B$_6$ (mg)	Fola (µg)	VT-C (mg)
0	57	1.09	33	689	68	0.42	2256	0.22	2.74	0.13	0.91	0.51	9	20
0	21	0.35	28	338	18	0.45	1	0.1	0.26	0.06	1.06	0.11	24	19
0	11	0.22	8	152	15	0.14	<1	0.03	0.04	0.03	0.26	0.11	28	23
0	10	0.2	6	88	9	0.11	1	0.03	0.04	0.03	0.25	0.11	27	27
0	15	0.37	8	125	16	0.12	<1	0.03	0.05	0.05	0.28	0.08	37	28
0	16	0.08	4	104	32	0.05	9	<.01	0.11	0.02	0.13	0.03	14	1
0	48	0.24	13	312	96	0.16	26	0.03	0.32	0.07	0.38	0.09	43	4
0	40	0.67	19	268	95	0.31	0	0.04	0.31	0.06	0.66	0.16	5	6
0	18	0.65	29	136	77	0.13	110	0.01	0.68	0.03	0.14	0.04	5	11
0	102	3.96	150	961	313	0.58	536	0.06	3.31	0.15	0.63	0.15	16	32
0	35	0.69	24	254	4	1.5	0	0.05	0.24	0.06	0.95	0.15	189	16
0	21	0.35	19	277	2	0.5	3	0.04	0.19	0.06	0.67	0.19	29	13
0	133	1.1	19	110	15	0.22	386	0.04	0.84	0.1	0.55	0.12	88	17
0	178	0.95	26	213	42	0.23	489	0.04	1.06	0.1	0.54	0.1	65	22
0	2	0.38	18	158	3	0.4	8	0.11	0.05	0.04	0.96	0.14	20	3
0	2	0.39	23	191	1	0.52	8	0.02	0.06	0.05	1.08	0.08	29	3
0	4	0.49	22	172	365	0.68	5	0.03	0.09	0.07	1.23	0.08	58	6
0	5	0.44	24	195	286	0.48	4	0.04	0.04	0.08	1.23	0.06	51	9
0	14	7.23	12	176	1532	0.76	25	0.04	0.24	0.04	0.69	0.16	34	5
0	103	1.7	20	218	42	0.23	136	0.1	2.63	0.14	0.44	0.14	15	19
0	147	1.89	25	244	46	0.29	521	0.14	3.57	0.18	0.54	0.17	14	19
0	6	0.25	11	122	1	0.12	2	0.08	0.41	0.02	0.59	0.09	14	1
0	26	0.42	8	157	11	0.4	54	0.04	0.22	0.04	0.2	<.01	71	3
0	26	0.42	8	157	11	0.4	54	0.04	0.22	0.04	0.2	<.01	71	3
0	80	4.74	79	477	11	2.51	2	0.19	0.57	0.1	0.86	0.23	282	2
0	11	0.08	3	8	<1	0.02	41	<.01	0.06	0.01	0.07	0.01	2	0
0	51	0.37	13	38	1	0.09	193	<.01	0.28	0.05	0.33	0.06	12	2
0	120	3.77	88	692	4	1.56	<1	0.28	0.53	0.1	1.21	0.21	181	2
0	21	5.1	26	644	6	0.18	2	0.3	0.28	0.09	1.95	0.12	20	6
0	14	0.72	14	180	5	0.19	1	0.02	0.55	0.03	0.24	0.05	14	24
0	94	1.17	23	296	30	0.31	885	0.07	1.1	0.09	0.65	0.18	17	53
0	179	1.22	23	417	20	0.23	956	0.06	1.2	0.15	0.87	0.11	18	33
0	61	3.23	72	658	873	1.41	0	0.27	1.54	0.23	1.17	0.06	131	3
0	32	0.54	26	472	27	0.04	3	0.07	0.65	0.03	0.54	0.2	22	84
0	41	0.66	31	561	35	0.51	3	0.07	0.86	0.03	0.64	0.25	20	89
0	53	1.87	25	160	18	0.11	74	0.05	0.82	0.03	0.36	0.21	57	11
0	31	1.14	15	90	10	0.06	2	0.03	0.63	0.02	0.21	0.12	25	4
0	38	6.59	71	731	4	2.51	1	0.33	0.22	0.14	2.1	0.35	358	3
0	17	3.84	43	352	12	1.98	2	0.27	0.11	0.11	1.49	0.2	83	16
0	19	2.47	28	248	8	1.16	2	0.18	0.07	0.1	0.87	0.15	77	13
0	14	0.51	5	98	2	0.08	68	0.02	0.07	0.03	0.15	0.03	30	2
0	10	0.37	4	71	2	0.06	50	0.02	0.05	0.02	0.11	0.02	22	1
0	11	0.19	4	84	5	0.09	9	0.02	0.02	0.01	0.07	0.03	31	2
0	108	1.89	43	819	49	0.86	86	0.2	0.16	0.11	0.67	0.25	302	21
0	27	0.47	11	205	12	0.22	22	0.05	0.04	0.03	0.17	0.06	76	5
0	10	0.24	4	54	8	0.05	104	0.02	0.08	0.02	0.1	0.03	11	5
0	9	0.27	4	69	2	0.06	81	0.02	0.04	0.02	0.09	0.02	38	7
0	10	0.29	4	74	2	0.07	87	0.02	0.04	0.02	0.09	0.02	41	7
0	112	0.8	101	573	9	0.98	52	0.23	1.23	0.1	1.55	0.33	81	29
0	33	0.56	43	558	29	0.7	2	0.18	0.39	0.1	1.76	0.18	56	42
0	1	0.18	3	110	1	0.18	0	0.03	<.01	0.15	1.35	0.04	6	1
0	5	1.36	9	278	2	0.68	0	0.06	<.01	0.23	3.48	0.07	14	3
0	2	0.32	10	85	3	0.97	0	0.03	0.01	0.12	1.1	0.12	15	0
0	9	0.62	12	101	332	0.56	0	0.07	<.01	0.02	1.24	0.05	9	0
0	2	0.51	5	140	2	0.28	0	0.03	0.05	0.16	1.42	0.03	6	1

A

TABLE A–1

Food Composition   (Computer code number is for West Diet Analysis program)   (For purposes of calculations, use "0" for t, <1, <.1, <.01, etc.)

Computer Code Number	Food Description	Measure	Wt (g)	H₂O (%)	Ener (kcal)	Prot (g)	Carb (g)	Dietary Fiber (g)	Fat (g)	Sat	Mono	Poly
	VEGETABLES AND LEGUMES—Continued											
5096	Mustard greens, cooked from raw	½ c	70	94	10	2	1	1	<1	t	t	t
5097	From frozen	½ c	75	94	14	2	2		<1	t	t	t
7022	Navy beans, cooked from dry	1 c	182	63	258	16	48	12	1	0.3	t	0.4
	Okra, cooked:											
5098	From fresh pods	8 ea	85	93	19	2	4		<1	t	t	t
5100	From frozen slices	1 c	184	91	52	4	11	5	1	0.1	t	0.1
5644	Batter fried from fresh	1 c	92	67	175	2	14	2	13	1.7	3.1	7.1
	Onions, yellow/white:											
7499	Raw, chopped	½ c	80	89	34	1	8	1	<1	t	t	t
7808	Raw, sliced	½ c	58	89	24	1	6	1	<1	t	t	t
7812	Cooked, drained, chopped	½ c	105	88	46	1	11	1	<1	t	t	t
5113	Dehydrated flakes	¼ c	14	4	49	1	12	1	<1	t	t	t
5530	Onions, pearl, cooked	½ c	93	88	41	1	9	1	<1	t	t	t
5114	Spring/green onions, bulb and top, chopped	½ c	50	90	16	1	4	1	<1	t	t	t
5190	Onion rings, breaded, heated from frozen	2 ea	20	28	81	1	8	<1	5	1.7	2.2	1
5522	Palm hearts, cooked slices	1 c	146	70	150	4	39	2	<1	t	t	0.1
26012	Parsley, raw, chopped	½ c	30	88	11	1	2	1	<1	t	t	t
5212	Parsnips, sliced, cooked	½ c	78	80	55	1	13	3	<1	t	t	t
	Peas:											
7018	Black-eyed, cooked from dry, drained	½ c	86	70	100	7	18	6	<1	0.1	t	0.2
5213	From fresh, drained	½ c	82	75	80	3	17	4	<1	t	t	0.1
5115	From frozen, drained	½ c	85	66	112	7	20	5	1	0.1	t	0.2
5122	Edible pod peas, cooked	½ c	80	89	34	3	6	2	<1	t	t	t
5119	Green, canned, drained:	½ c	85	82	59	4	11	3	<1	t	t	0.1
5267	Unsalted	½ c	124	86	66	4	12	4	<1	t	t	0.2
5118	Green, cooked from frozen	½ c	80	80	62	4	11	4	<1	t	t	0.1
6836	Snow peas, raw	½ c	49	89	21	1	4	1	<1	t	t	t
5121	each	10 ea	34	89	14	1	3	1	<1	t	t	t
7020	Split, green, cooked from dry	½ c	98	69	116	8	21	8	<1	t	t	0.2
5123	Peas & carrots, cooked from frozen	½ c	80	86	38	2	8	2	<1	t	t	0.2
5281	Canned w/liquid	½ c	128	88	49	3	11	3	<1	t	t	0.2
	Peppers, hot:											
5063	Hot green chili, canned	½ c	68	92	14	1	3	1	<1	t	t	t
5400	Raw	1 ea	45	88	18	1	4	1	<1	t	t	t
5288	Hot red chili, raw, diced	1 tbs	9	88	4	<1	1	<1	<1	t	t	t
5293	Jalapeno, chopped, canned	½ c	68	89	18	1	3	2	1	t	t	0.3
	Peppers, sweet:											
6846	Green, whole, raw	1 ea	119	94	24	1	6	2	<1	t	t	t
5126	Cooked, chopped	½ c	68	92	19	1	5	1	<1	t	t	t
5128	Red, raw, chopped	½ c	75	92	20	1	5		<1	t	t	0.1
5294	Raw, each	1 ea	74	92	19	1	4		<1	t	t	0.1
5278	Cooked, chopped	½ c	68	92	19	1	5	1	<1	t	t	t
5441	Yellow, raw, whole	1 ea	186	92	50	2	12	2	<1	t	t	0.2
5442	Strips	10 pce	52	92	14	1	3	<1	<1	t	t	t
7013	Pinto beans, cooked from dry	½ c	85	64	119	8	21	7	1	t	t	0.2
5229	Poi - two finger	½ c	120	72	134	<1	33	<1	<1	t	t	t
	Potatoes:											
	Baked in oven, 4¾" × 2⅓" diam											
5129	Flesh only	1 ea	156	75	145	3	34	2	<1	t	t	t
5339	Skin only	1 ea	58	47	115	2	27	5	<1	t	t	t
	Baked in microwave, 4¾" × 2⅓" dm:											
5340	With skin	1 ea	202	72	212	5	49	5	<1	t	t	t
5345	Flesh only	1 ea	156	74	156	3	36	2	<1	t	t	t
5350	Skin only	1 ea	58	64	77	3	17	3	<1	t	t	t
	Boiled, about 2½" diam:											
5133	Peeled after boiling	1 ea	136	77	118	3	27	2	<1	t	t	t
5135	Peeled before boiling	1 ea	135	77	116	2	27	2	<1	t	t	t
5139	French fried, strips, oven heated	10 ea	50	35	166	2	20	2	9	3	5.7	0.7
5140	Hashed browns from frozen	1 c	156	56	340	5	44	3	18	7	8	2.1

PAGE KEY: A–4 = Beverages   A–6 = Dairy   A–10 = Eggs   A–12 = Fat/Oil   A–14 = Fruit   A–20 = Bakery
A–26 = Grain   A–32 = Fish   A–32 = Meats   A–36 = Poultry   A–38 = Sausage   A–40 = Mixed/Fast   A–44 = Nuts/Seeds   A–46 = Sweets
A–48 = Vegetables/Legumes   A–58 = Vegetarian   A–62 = Misc   A–62 = Soups/Sauces   A–64 = Fast   A–78 = Convenience   A–80 = Baby foods

Chol (mg)	Calc (mg)	Iron (mg)	Magn (mg)	Pota (mg)	Sodi (mg)	Zinc (mg)	VT-A (RAE)	Thia (mg)	VT-E (mg α)	Ribo (mg)	Niac (mg)	V-B$_6$ (mg)	Fola (µg)	VT-C (mg)
0	52	0.49	10	141	11	0.08	221	0.03	0.85	0.04	0.3	0.07	51	18
0	76	0.84	10	104	19	0.15	266	0.03	1.01	0.04	0.19	0.08	52	10
0	127	4.51	107	670	2	1.93	<1	0.37	0.73	0.11	0.97	0.3	255	2
0	65	0.24	31	115	5	0.37	12	0.11	0.23	0.05	0.74	0.16	39	14
0	177	1.23	94	431	6	1.14	31	0.18	0.59	0.23	1.44	0.09	269	22
2	61	1.26	36	190	122	0.5		0.18	3.04	0.14	1.44	0.12	38	10
0	18	0.15	8	115	2	0.13	<1	0.04	0.02	0.02	0.07	0.12	15	5
0	13	0.11	6	84	2	0.09	<1	0.03	0.01	0.01	0.05	0.09	11	4
0	23	0.25	12	174	3	0.22	<1	0.04	0.02	0.02	0.17	0.14	16	5
0	36	0.22	13	227	3	0.26	<1	0.07	0.03	0.01	0.14	0.22	23	10
0	20	0.22	10	154	3	0.2	0	0.04	0.12	0.02	0.15	0.12	14	5
0	36	0.74	10	138	8	0.2	25	0.03	0.28	0.04	0.26	0.03	32	9
0	6	0.34	4	26	75	0.08	2	0.06	0.14	0.03	0.72	0.02	13	0
0	26	2.47	15	2637	20	5.45	5	0.07	0.73	0.25	1.25	1.06	30	10
0	41	1.86	15	166	17	0.32	126	0.03	0.22	0.03	0.39	0.03	46	40
0	29	0.45	23	286	8	0.2	0	0.06	0.78	0.04	0.56	0.07	45	10
0	21	2.16	46	239	3	1.11	1	0.17	0.24	0.05	0.43	0.09	179	0
0	105	0.92	43	343	3	0.84	33	0.08	0.18	0.12	1.15	0.05	104	2
0	20	1.8	42	319	4	1.21	3	0.22	0.26	0.05	0.62	0.08	120	2
0	34	1.58	21	192	3	0.3	43	0.1	0.31	0.06	0.43	0.12	23	38
0	17	0.81	14	147	214	0.6	23	0.1	0.03	0.07	0.62	0.05	37	8
0	22	1.26	21	124	11	0.87	89	0.14	0.02	0.09	1.04	0.08	36	12
0	19	1.22	18	88	58	0.54	84	0.23	0.02	0.08	1.18	0.09	47	8
0	21	1.02	12	98	2	0.13	26	0.07	0.19	0.04	0.29	0.08	21	29
0	15	0.71	8	68	1	0.09	18	0.05	0.13	0.03	0.2	0.05	14	20
0	14	1.26	35	355	2	0.98	<1	0.19	0.03	0.05	0.87	0.05	64	0
0	18	0.75	13	126	54	0.36	374	0.18	0.42	0.05	0.92	0.07	21	6
0	29	0.96	18	128	333	0.74	370	0.09	0.24	0.07	0.74	0.11	23	8
0	5	0.34	10	127	798	0.12	24	0.01	0.47	0.03	0.54	0.1	7	46
0	8	0.54	11	153	3	0.14	27	0.04	0.31	0.04	0.43	0.13	10	109
0	1	0.09	2	29	1	0.02	4	<.01	0.06	<.01	0.11	0.05	2	13
0	16	1.28	10	131	1136	0.23	58	0.03	0.47	0.03	0.27	0.13	10	7
0	12	0.4	12	208	4	0.15	21	0.07	0.44	0.03	0.57	0.27	13	96
0	6	0.31	7	113	1	0.08	10	0.04	0.36	0.02	0.32	0.16	11	51
0	5	0.32	9	158	2	0.19	118	0.04	1.18	0.06	0.73	0.22	14	142
0	5	0.32	9	156	1	0.18	116	0.04	1.17	0.06	0.72	0.22	13	141
0	6	0.31	7	113	1	0.08	187	0.04	1.12	0.02	0.32	0.16	11	116
0	20	0.86	22	394	4	0.32	19	0.05	1.28	0.05	1.66	0.31	48	341
0	6	0.24	6	110	1	0.09	5	0.01	0.36	0.01	0.46	0.09	14	95
0	36	1.77	35	246	9	0.86	0	0.08	0.8	0.06	0.17	0.08	146	1
0	19	1.06	29	220	14	0.26	4	0.16	2.76	0.05	1.32	0.33	25	5
0	8	0.55	39	610	8	0.45	0	0.16	0.06	0.03	2.18	0.47	14	20
0	20	4.08	25	332	12	0.28	1	0.07	0.02	0.06	1.78	0.36	13	8
0	22	2.5	55	903	16	0.73	0	0.24	0.1	0.06	3.46	0.69	24	31
0	8	0.64	39	641	11	0.51	0	0.2	0.06	0.04	2.54	0.5	19	24
0	27	3.45	21	377	9	0.3	0	0.04	0.02	0.04	1.29	0.29	10	9
0	7	0.42	30	515	5	0.41	<1	0.14	0.01	0.03	1.96	0.41	14	18
0	11	0.42	27	443	7	0.36	<1	0.13	0.01	0.03	1.77	0.36	12	10
0	6	0.83	12	270	306	0.2	0	0.04	0.25	0.02	1.33	0.11	11	3
0	23	2.36	27	680	53	0.5	0	0.17	0.3	0.03	3.78	0.2	11	10

TABLE A-1

Food Composition   (Computer code number is for West Diet Analysis program)   (For purposes of calculations, use "0" for t, <1, <.1, <.01, etc.)

Computer Code Number	Food Description	Measure	Wt (g)	H₂O (%)	Ener (kcal)	Prot (g)	Carb (g)	Dietary Fiber (g)	Fat (g)	Fat Breakdown (g) Sat	Mono	Poly
	VEGETABLES AND LEGUMES—Continued											
	Mashed:											
5137	Home recipe with whole milk	¹/₂ c	105	79	87	2	18	2	1	0.3	0.1	t
5272	Home recipe with milk and marg	¹/₂ c	105	75	119	2	18	2	4	1	1.8	1.3
5138	Prepared from flakes with milk and marg	¹/₂ c	110	76	124	2	17	3	6	1.6	2.5	1.7
	Potato products, prepared:											
5276	Au gratin from dry mix	¹/₂ c	123	79	114	3	16	1	5	3.2	1.4	0.2
5275	From home recipe, using butter	¹/₂ c	122	74	161	6	14	2	9	4.3	3.2	1.3
5271	Scalloped from dry mix	¹/₂ c	122	79	113	3	16	1	5	3.2	1.5	0.2
5270	From home recipe, using butter	¹/₂ c	123	81	106	4	13	2	5	1.7	1.7	0.9
	Potato Salad (see Mixed Dishes #715)											
5265	Potato Puffs, cooked from frozen	¹/₂ c	64	53	142	2	20	2	7	3.3	2.8	0.5
5396	Pumpkin, cooked from fresh, mashed	¹/₂ c	123	94	25	1	6	1	<1	t	t	t
5142	Canned	¹/₂ c	123	90	42	1	10	4	<1	0.2	t	t
5451	Radicchio, raw, shredded	¹/₂ c	20	93	5	<1	1	<1	<1	t	t	t
5452	Leaf	10 ea	80	93	18	1	4	1	<1	t	t	t
5143	Red radishes	10 ea	45	95	7	<1	2		<1	t	t	t
1793	Daikon radishes (Chinese) raw	¹/₂ c	44	95	8	<1	2	1	<1	t	t	t
7024	Refried beans, canned	¹/₂ c	126	76	118	7	20	7	2	0.6	0.7	0.2
7225	Rutabaga, cooked cubes	¹/₂ c	85	89	33	1	7	2	<1	t	t	t
5145	Sauerkraut, canned with liquid	¹/₂ c	118	93	22	1	5	3	<1	t	t	t
6857	Seaweed, kelp, raw	¹/₂ c	40	82	17	1	4	1	<1	t	t	t
5260	Seaweed, spirulina, dried	¹/₂ c	8	5	23	5	2	<1	1	0.2	t	0.2
5427	Shallots, raw, chopped	1 tbs	10	80	7	<1	2	<1	<1	t	t	t
5666	Snow Peas, stir fried	¹/₂ c	83	89	35	2	6	2	<1	t	t	t
7015	Soybeans, cooked from dry	¹/₂ c	86	63	149	14	9	5	8	1.1	1.7	4.4
7063	dry roasted	¹/₂ c	86	1	388	34	28	7	19	2.7	4.1	10.5
	Soybean products:											
7503	Miso	¹/₂ c	138	41	284	16	39	7	8	1.2	1.9	4.7
	Soy milk (see Dairy)											
	Tofu (soybean curd):											
	Silken:											
7540	Extra firm	¹/₂ c	126	88	69	9	3	<1	2	0.4	0.4	1.3
7542	Firm	¹/₂ c	126	87	78	9	3	<1	3	0.5	0.7	1.9
7500	Regular	¹/₂ c	124	87	76	8	2	<1	5	0.7	1	2.6
7721	Block	1 pce	116	85	88	9	2	<1	6	0.8	1.2	3.1
7720	Firm	1 pce	81	70	117	13	3	2	7	1	1.6	4
7519	Dried, frozen	1 pce	17	6	82	8	2	1	5	0.7	1.1	2.9
90039	Fried	1 pce	13	51	35	2	1	1	3	0.4	0.6	1.5
7541	Soft	¹/₂ c	124	89	68	6	4	<1	3	0.4	0.6	1.9
7518	Prepared with nigari	¹/₂ c	126	84	97	10	4	1	6	0.8	1.2	3.2
	Spinach:											
5146	Raw, chopped	¹/₂ c	15	91	3	<1	1	<1	<1	t	t	t
5147	Cooked, from fresh, drained	¹/₂ c	90	91	21	3	3	2	<1	t	t	t
5148	From frozen (leaf)	¹/₂ c	95	89	30	4	5	4	<1	t	0	0.2
5149	Canned, unsalted	¹/₂ c	107	92	25	3	4	3	1	t	t	0.2
	Spinach souffle (see Mixed Dishes)											
	Squash, summer varieties, cooked w/skin:											
5152	Varieties averaged	¹/₂ c	90	94	18	1	4	1	<1	t	t	0.1
5322	Crookneck	¹/₂ c	90	94	18	1	4	1	<1	t	t	0.1
5327	Zucchini	¹/₂ c	90	95	14	1	4	1	<1	t	t	t
	Squash, winter varieties, cooked:											
5303	Average of all varieties, baked, cubes	1 c	205	89	76	2	18	6	1	0.2	t	0.4
5314	Acorn, baked, mashed	¹/₂ c	123	83	69	1	18	5	<1	t	t	t
5316	Boiled, mashed	¹/₂ c	122	90	41	1	11	3	<1	t	t	t
	Butternut squash:											
5317	Baked, mashed	¹/₂ c	103	88	41	1	11	3	<1	t	t	t
5274	Cooked from frozen	¹/₂ c	120	88	47	1	12	3	<1	t	t	t
5453	Hubbard, baked, mashed	¹/₂ c	120	85	60	3	13	3	1	0.2	t	0.3
5454	Boiled, mashed	¹/₂ c	118	91	35	2	8	3	<1	t	t	0.2
5455	Spaghetti, baked or boiled	¹/₂ c	77	92	21	1	5	1	<1	t	t	t

Chol (mg)	Calc (mg)	Iron (mg)	Magn (mg)	Pota (mg)	Sodi (mg)	Zinc (mg)	VT-A (RAE)	Thia (mg)	VT-E (mg α)	Ribo (mg)	Niac (mg)	V-B$_6$ (mg)	Fola (µg)	VT-C (mg)
2	23	0.28	19	311	317	0.3	4	0.09	0.02	0.05	1.18	0.24	8	7
1	21	0.27	20	342	350	0.32	43	0.1	0.44	0.05	1.23	0.26	9	11
4	54	0.24	20	256	365	0.2	51	0.12	0.77	0.06	0.74	<.01	8	11
18	102	0.39	18	269	540	0.3	64	0.02	1.48	0.1	1.15	0.05	9	4
18	145	0.78	24	483	528	0.84	78	0.08	0.64	0.14	1.21	0.21	13	12
13	44	0.46	17	248	416	0.3	43	0.02	0.18	0.07	1.26	0.05	12	4
7	70	0.7	23	465	412	0.49	41	0.08	0.4	0.11	1.3	0.22	14	13
0	19	1	12	243	477	0.19	<1	0.13	0.03	0.05	1.38	0.15	11	4
0	18	0.7	11	283	1	0.28	308	0.04	0.98	0.1	0.51	0.05	11	6
0	32	1.71	28	253	6	0.21	957	0.03	1.3	0.07	0.45	0.07	15	5
0	4	0.11	3	60	4	0.12	<1	<.01	0.45	<.01	0.05	0.01	12	2
0	15	0.46	10	242	18	0.5	1	0.01	1.81	0.02	0.2	0.05	48	6
0	11	0.15	4	105	18	0.13	<1	<.01	0	0.02	0.11	0.03	11	7
0	12	0.18	7	100	9	0.07	0	<.01	0	<.01	0.09	0.02	12	10
10	44	2.09	42	336	377	1.47	0	0.03	0	0.02	0.4	0.18	14	8
0	41	0.45	20	277	17	0.3	<1	0.07	0.27	0.03	0.61	0.09	13	16
0	35	1.73	15	201	780	0.22	1	0.02	0.12	0.03	0.17	0.15	28	17
0	67	1.14	48	36	93	0.49	2	0.02	0.35	0.06	0.19	<.01	72	1
0	10	2.28	16	109	84	0.16	2	0.19	0.4	0.29	1.03	0.03	8	1
0	4	0.12	2	33	1	0.04	6	<.01	<.01	<.01	0.02	0.03	3	1
0	36	1.73	20	166	3	0.22	5	0.11	0.32	0.06	0.47	0.13	28	42
0	88	4.42	74	443	1	0.99	<1	0.13	0.3	0.25	0.34	0.2	46	1
0	120	3.4	196	1173	2	4.1	0	0.37	3.96	0.65	0.91	0.19	176	4
0	91	3.78	58	226	5033	4.58	6	0.13	0.01	0.34	1.19	0.3	46	0
0	39	1.5	34	194	79	0.76	0	0.1	0.18	0.04	0.3	0.01		0
0	40	1.3	34	244	45	0.77	0	0.13	0.24	0.05	0.31	0.01		0
0	138	1.38	33	149	10	0.79	<1	0.06	0.01	0.05	0.66	0.06	55	0
0	406	6.22	35	140	8	0.93	5	0.09	0.01	0.06	0.23	0.05	17	0
0	553	2.15	47	192	11	1.27	6	0.13	0.02	0.08	0.31	0.07	23	0
0	62	1.65	10	3	1	0.83	4	0.08	0.01	0.05	0.2	0.05	16	0
0	125	0.63	12	19	2	0.26	0	0.02	<.01	<.01	0.01	0.01	4	0
0	38	1.02	36	223	6	0.64	0	0.12	0.25	0.05	0.37	0.01		0
0	204	1.83	58	222	10	1.27	0	0.12	0.03	0.13	0.01	0.08	42	
0	15	0.41	12	84	12	0.08	70	0.01	0.3	0.03	0.11	0.03	29	4
0	122	3.21	78	419	63	0.68	472	0.09	1.87	0.21	0.44	0.22	131	9
0	145	1.86	78	287	92	0.47	573	0.07	3.36	0.17	0.42	0.13	115	2
0	136	2.46	81	370	29	0.49	524	0.02	2.08	0.15	0.42	0.11	105	15
0	24	0.32	22	173	1	0.35	10	0.04	0.13	0.04	0.46	0.06	18	5
0	20	0.33	14	153	0	0.2	7	0.04	0.11	0.04	0.46	0.08	18	5
0	12	0.32	20	228	3	0.16	50	0.04	0.11	0.04	0.39	0.07	15	4
0	45	0.9	27	896	2	0.45	535	0.03	0.25	0.14	1.01	0.33	41	20
0	54	1.14	53	538	5	0.21	26	0.21	0.15	0.02	1.08	0.24	23	13
0	32	0.68	32	321	4	0.13	50	0.12	0.15	<.01	0.65	0.14	13	8
0	42	0.62	30	293	4	0.13	575	0.07	1.33	0.02	1	0.13	20	16
0	23	0.7	11	160	2	0.14	200	0.06	0.16	0.05	0.56	0.08	19	4
0	20	0.56	26	430	10	0.18	362	0.09	0.14	0.06	0.67	0.21	19	11
0	12	0.33	15	253	6	0.12	236	0.05	0.14	0.03	0.39	0.12	12	8
0	16	0.26	8	90	14	0.15	5	0.03	0.09	0.02	0.62	0.08	6	3

A

Appendix A — Table of Food Composition

TABLE A–1

Food Composition (Computer code number is for West Diet Analysis program) (For purposes of calculations, use "0" for t, <1, <.1, <.01, etc.)

Computer Code Number	Food Description	Measure	Wt (g)	H₂O (%)	Ener (kcal)	Prot (g)	Carb (g)	Dietary Fiber (g)	Fat (g)	Fat Breakdown (g) Sat	Mono	Poly
	VEGETABLES AND LEGUMES—Continued											
5154	Succotash, cooked from frozen	½ c	85	74	79	4	17	3	1	0.1	0.1	0.4
	Sweet potatoes:											
5155	Baked in skin, peeled, 5 × 2″ diam	1 ea	114	76	103	2	24	4	<1	t	t	t
5159	Boiled without skin, 5 × 2″ diam	1 ea	151	80	115	2	27	4	<1	t	0	t
5166	Candied pieces, 2½ × 2″	1 pce	105	67	144	1	29	3	3	1.4	0.7	0.2
5162	Canned, solid pack	½ c	128	74	129	3	30	2	<1	t	t	0.1
5164	Vacuum pack, mashed	½ c	127	76	116	2	27	2	<1	t	t	0.1
5543	Taro shoots, cooked slices	1 c	140	95	20	1	4	1	<1	t	t	t
5544	Taro, tahitian, cooked slices	1 c	137	86	60	6	9	1	1	0.2	t	0.4
5445	Tomatillos, raw, each	1 ea	34	92	11	<1	2	1	<1	t	t	0.1
5444	Chopped	1 c	132	92	42	1	8	3	1	0.2	0.2	0.6
	Tomatoes:											
5169	Raw, whole, 2⅗″ diam	1 ea	123	94	22	1	5	1	<1	t	t	0.1
6492	Whole, small	1 ea	62	94	11	1	2	1	<1	t	t	t
5170	Chopped	1 c	180	94	32	2	7	2	<1	t	t	0.2
5178	Cooked from raw	1 c	240	94	43	2	10	2	<1	t	t	0.1
5179	Canned, solids and liquid:	1 c	240	94	41	2	9	2	<1	t	t	0.1
6992	Unsalted	1 c	240	94	46	2	10	2	<1	t	t	0.1
5446	Sundried:	1 c	54	15	139	8	30	7	2	0.2	0.3	0.6
5447	Pieces	10 pce	20	15	52	3	11	2	1	t	t	0.2
5448	Oil pack, drained	10 pce	30	54	64	2	7	2	4	0.6	2.6	0.6
5188	Tomato juice, canned:	1 c	243	94	41	2	10	1	<1	t	t	t
5397	Unsalted	1 c	243	94	41	2	10	1	<1	t	t	t
	Tomato products, canned:											
5181	Paste-no added salt	1 c	262	74	215	11	50	12	1	0.3	0.2	0.4
5225	Puree-no added salt	1 c	250	88	95	4	22	5	1	t	t	0.2
5180	Sauce-with salt	1 c	245	89	78	3	18	4	1	t	t	0.2
5183	Turnips, cubes, cooked from fresh	1 c	156	94	34	1	8	3	<1	t	t	t
5185	Turnip greens, cooked from fresh, leaves and stems	1 c	144	93	29	2	6	5	<1	t	t	0.1
5186	From frozen, chopped	1 c	164	90	48	5	8	6	1	0.2	t	0.3
20080	Vegetable juice cocktail, canned	1 c	242	94	46	2	11	2	<1	t	t	t
5305	Vegetables, mixed, canned, drained	½ c	81	87	40	2	8	2	<1	t	t	t
5187	Frozen, cooked, drained	½ c	91	83	59	3	12	4	<1	t	t	t
5386	Water chestnuts, Chinese, raw	½ c	62	73	60	1	15	2	<1	t	t	t
5387	Canned, slices	½ c	70	86	35	1	9	2	<1	t	t	t
5388	Whole	4 ea	28	86	14	<1	3	1	<1	t	t	t
5222	Watercress, fresh, chopped	½ c	17	95	2	<1	<1	<1	<1	t	t	t
	VEGETARIAN FOODS:											
7509	Bacon strips, meatless	3 ea	15	49	46	2	1	<1	4	0.7	1.1	2.3
7038	Baked beans, canned	½ c	127	73	118	6	26	6	1	0.1	t	0.2
7526	Bakon Crumbles	¼ c	7	8	31	2	2	1	2	0		
7723	Chicken, fillet	3 oz	85		90	15	8	4	2	0		
7557	Chili w/meat substitute	½ c	107	65	141	19	15	5	2	0.3	0.6	0.9
	Frankfurters:											
7550	Meatless	1 ea	51	58	102	10	4	2	5	0.8	1.2	2.6
8127	Deli	1 ea	76	72	80	16	4	1	0	0	0	0
8835	Tofu wiener	1 ea	38		45	9	2	0	<1	0		
8839	Jumbo	1 ea	76		100	16	7	2	2	0		
8840	Hot and spicy	1 ea	52		70	13	3	2	1	0		
	Garden Burger patties:											
7504	Regular	1 ea	71		110	6	16	3	3	1.5		
8811	Mushroom	1 ea	71		120	6	18	4	2	1		
8813	Roasted vegetable	1 ea	71		120	6	18	4	2	1		
7652	Veggie medley	1 ea	71		90	5	18	3	0	0	0	0
7505	Garden Sausage, patty	1 ea	71		83	8	3	3	6	0		
57433	Nuteena	1 ea	55	58	162	6	6	2	13	5.2	5.8	1.7
7556	Pot pie, meatless	1 ea	227	60	510	14	41	5	32	8.6	12.4	9.6
8831	Soyburger mix	⅓ c	55	73	60	10	4	3	0	0	0	0

PAGE KEY: A–4 = Beverages   A–6 = Dairy   A–10 = Eggs   A–12 = Fat/Oil   A–14 = Fruit   A–20 = Bakery   A–26 = Grain   A–32 = Fish   A–32 = Meats   A–36 = Poultry   A–38 = Sausage   A–40 = Mixed/Fast   A–44 = Nuts/Seeds   A–46 = Sweets   A–48 = Vegetables/Legumes   A–58 = Vegetarian   A–62 = Misc   A–62 = Soups/Sauces   A–64 = Fast   A–78 = Convenience   A–80 = Baby foods

Chol (mg)	Calc (mg)	Iron (mg)	Magn (mg)	Pota (mg)	Sodi (mg)	Zinc (mg)	VT-A (RAE)	Thia (mg)	VT-E (mg α)	Ribo (mg)	Niac (mg)	V-B$_6$ (mg)	Fola (µg)	VT-C (mg)
0	13	0.76	20	225	38	0.38	8	0.06	0.15	0.06	1.11	0.08	28	5
0	43	0.79	31	542	41	0.36	1096	1.65	0.81	0.12	1.7	0.33	7	22
0	41	1.09	27	347	41	0.3	1190	0.08	1.42	0.07	0.81	0.25	9	19
8	27	1.19	12	198	74	0.16	219	0.02	3.99	0.04	0.41	0.04	12	7
0	38	1.7	31	269	96	0.27	968	0.03	0.35	0.12	1.22	0.3	14	7
0	28	1.13	28	396	67	0.23	507	0.05	1.27	0.07	0.94	0.24	22	34
0	20	0.57	11	482	3	0.76	4	0.05	1.4	0.07	1.13	0.16	4	26
0	204	2.14	70	854	74	0.14	121	0.06	3.7	0.27	0.66	0.16	10	52
0	2	0.21	7	91	<1	0.07	2	0.01	0.13	0.01	0.63	0.02	2	4
0	9	0.82	26	354	1	0.29	8	0.06	0.5	0.05	2.44	0.07	9	15
0	12	0.33	14	292	6	0.21	52	0.05	0.66	0.02	0.73	0.1	18	16
0	6	0.17	7	147	3	0.11	26	0.02	0.33	0.01	0.37	0.05	9	8
0	18	0.49	20	427	9	0.31	76	0.07	0.97	0.03	1.07	0.14	27	23
0	26	1.63	22	523	26	0.34	58	0.09	1.34	0.05	1.28	0.19	31	55
0	74	2.33	26	451	307	0.34	14	0.11	1.7	0.11	1.76	0.22	19	22
0	72	1.32	29	545	24	0.38	17	0.11	1.92	0.07	1.76	0.22	19	34
0	59	4.91	105	1851	1131	1.07	24	0.29	<.01	0.26	4.89	0.18	37	21
0	22	1.82	39	685	419	0.4	9	0.11	<.01	0.1	1.81	0.07	14	8
0	14	0.8	24	470	80	0.23	19	0.06	0.16	0.11	1.09	0.1	7	31
0	24	1.04	27	556	654	0.36	56	0.11	0.78	0.08	1.64	0.27	49	44
0	24	1.04	27	556	24	0.36	56	0.11	0.78	0.08	1.64	0.27	49	44
5	94	7.81	110	2657	257	1.65	199	0.16	11.27	0.4	8.06	0.57	31	57
5	45	4.45	58	1098	70	0.9	65	0.06	4.92	0.2	3.66	0.32	28	26
0	32	2.5	39	811	1284	0.49	42	0.06	5.1	0.16	2.39	0.24	22	17
0	51	0.28	14	276	25	0.19	0	0.04	0.03	0.04	0.47	0.1	14	18
0	197	1.15	32	292	42	0.2	549	0.06	2.71	0.1	0.59	0.26	170	39
0	249	3.18	43	367	25	0.67	882	0.09	4.36	0.12	0.77	0.11	64	36
0	27	1.02	27	467	653	0.48	189	0.1	12.1	0.07	1.76	0.34	51	67
0	22	0.85	13	236	121	0.33	471	0.04	0.28	0.04	0.47	0.06	19	4
0	23	0.75	20	154	32	0.45	195	0.06	0.4	0.11	0.77	0.07	17	3
0	7	0.04	14	362	9	0.31	0	0.09	0.74	0.12	0.62	0.2	10	2
0	3	0.61	4	83	6	0.27	0	<.01	0.35	0.02	0.25	0.11	4	1
0	1	0.24	1	33	2	0.11	0	<.01	0.14	<.01	0.1	0.04	2	0
0	20	0.03	4	56	7	0.02	40	0.02	0.17	0.02	0.03	0.02	2	7
0	3	0.36	3	26	220	0.06	1	0.66	1.03	0.07	1.13	0.07	6	0
0	64	0.37	41	376	504	1.78	6	0.19	0.67	0.08	0.54	0.17	30	4
0	16	0.28			163		0							0
0	80	1.8			170		0							1
0	54	4.38	36	366	355	1.27	37	0.12	1.28	0.07	1.22	0.15	82	6
0	17	0.92	9	76	219	0.61	0	0.56	0.98	0.61	8.16	0.5	40	0
0	40	0.72			590									1
0	20	2.16	8	90	240	0.6	0	0.15						0
0	20	4.5	16	170	480	3.75	0	0.38						0
0	20	3.6	16	135	400	3	38	0.3						0
20	60	0			560		0							0
20	100	0.36			370									0
15	100	0.36			330		0							0
0	40	0			280									0
0	33	1.19			198		0							0
0	9	0.27		166	119	0.46	0	0.1		0.35	1.04	0.45		0
20	68	2.96	32	378	486	1.08		0.82	4.23	0.44	5.17	0.41	58	10
0	40	2.7		240	270	3	0	0.22		0.14	3	0.2		0

A

Table of Food Composition

Appendix A

TABLE A-1

Food Composition (Computer code number is for West Diet Analysis program)   (For purposes of calculations, use "0" for t, <1, <.1, <.01, etc.)

Computer Code Number	Food Description	Measure	Wt (g)	H₂O (%)	Ener (kcal)	Prot (g)	Carb (g)	Dietary Fiber (g)	Fat (g)	Fat Breakdown (g) Sat	Mono	Poly
	VEGETARIAN FOODS:—Continued											
7751	Griller patty	1 ea	113		150	21	11	6	4	0		
7562	Patty, with cheese	1 ea	135	51	308	20	30	4	12	3.6	3.6	3.6
7517	Soy protein isolate	1 oz	28	5	95	23	2	2	1	0.1	0.2	0.5
7564	Tempeh	1 c	166	60	320	31	16		18	3.7	5	6.4
7670	Vegan burger, patty	1 ea	78	71	83	13	7	4	<1	t	0.3	0.1
8842	Veggie slices, soy	1 pce	15		19	3	1	<1	<1	0		
8830	Veggie ground soy	⅓ c	55	73	60	10	4	3	0	0	0	0
7902	Vegetarian burger mix	½ c	47		170	8	30	5	3	0		
7511	Vegetarian sausage link	1 ea	25	50	64	5	2	1	5	0.7	1.1	2.3
7512	Patty	1 ea	38	50	98	7	4	1	7	1.1	1.7	3.5
	Vegetarian foods, Green Giant harvest burger											
7673	Italian, patty	1 ea	90		140	17	8	5	4	1.5	0.5	0.5
7674	Original, patty	1 ea	90	65	138	18	7	6	4	1	2.1	0.3
6658	Mix, frozen	⅔ c	54		90	15	7	3	0	0	0	0
7675	Southwestern, patty	1 ea	90		140	16	9	5	4	1.5	0	0.5
	Vegetarian Foods, Loma Linda											
7727	Chik nuggets, frozen	5 pce	85	47	245	12	13	5	16	2.5	4	8.8
7767	Fakin Bakin, bacon bits	1 tsp	3		12	1	1	0	1	0		
7860	Smokey tempeh strip	1 pce	19		27	3	2	<1	1	0.2		
7744	Franks, big, canned	1 ea	51	58	118	12	2		7	0.8	1.5	3.7
7997	Gimme Lean! Meatless ground beef	2 oz	57		70	9	8	1	0	0	0	0
7998	Meatless sausage	2 oz	57		70	9	8	1	0	0	0	0
8159	Italian sausage, link	1 ea	40	68	60	5	5	0	2	1		
7747	Linketts, canned	1 ea	35	60	72	7	3	2	4	0.7	1.2	2.5
57434	Redi-burger, patty	1 ea	85	59	172	16	7	5	10	1.5	2.4	5.8
7755	Swiss stake w/gravy, canned	1 pce	92	71	120	9	8	4	6	0.8	1.5	3.3
57435	Vege-Burger, patty	1 ea	55	71	66	11	3	2	2	0.4	0.6	0.5
7999	Light	3 oz	85	64	130	16	12	2	1	0		
	Vegetarian foods, Morningstar Farms:											
7766	Better-n-eggs	¼ c	57	88	26	5	<1	0	<1	t	0.1	0.1
57436	Breakfast links	2 pce	45	68	64	9	2	1	2	0.4	0.5	1.1
7752	Breakfast strips	2 pce	16	43	56	2	2	1	4	0.7	0.9	2.6
477	Buffalo wings	5 ea	85	50	204	13	18	3	9	1.2	2.2	4.8
7725	Burger crumbles, svg	1 ea	55	60	116	11	3	3	6	1.6	2.3	2.5
7726	Burger, spicy black bean	1 ea	78	60	115	12	15	5	1	0.2	0.2	0.4
62541	Chik nuggets	4 pce	86	53	183	14	17	4	7	0.6	1.1	2.7
7665	Chik pattie	1 ea	71	56	148	10	14	3	6	0.9	1.6	3.6
7724	Frank, deli	1 ea	57	52	141	13	5	3	8	1.1	2.5	4.2
7722	Garden vege pattie	1 ea	67	60	119	11	10	4	4	0.5	1.1	2.2
7746	Grillers	1 ea	64	56	139	15	5	2	6	1.1	1.5	3.1
62545	Griller Prime	1 ea	71	56	160	17	4	2	8	1	3.6	3.8
7789	Ground burger	½ c	55	71	62	10	4	1	<1	t	t	0.3
62359	Sausage breakfast patty	1 ea	38	54	79	10	4	2	3	0.5	0.7	1.3
	Vegetarian foods, Worthington:											
7634	Beef style, meatless, frzn	3 pce	55	58	113	9	4	3	7	1.2	2.7	2.6
7732	Burger, meatless, patty	¼ c	55	71	60	9	2	1	2	0.3	0.5	1.1
7610	Choplets, slices, canned	2 pce	92	72	93	17	3	2	2	0.9	0.3	0.3
7608	Corned beef style, meatless, frzn	4 pce	57	55	138	10	5	2	9	1.9	4.1	3.1
7607	Country stew, canned	1 c	240	81	208	13	20	5	9	1.6	2.3	4.8
7612	Numete, slices, canned	1 pce	55	58	132	6	5	3	10	2.4	4.4	2.7
7613	Prime stakes, slices, canned	1 pce	92	71	136	9	4	4	9	1.4	2.9	4.9
7617	Protose, slices, canned	1 pce	55	53	131	13	5	3	7	1	3	2.4
7606	Roast, dinner, meatless, frzn	1 ea	85	63	180	12	5	3	13	2.2	5	5.2
7619	Saucette links, canned	1 pce	38	62	86	6	2	1	6	1.1	1.6	3.8
7620	Savory slices, canned	1 pce	28	66	48	3	2	1	3	1.2	1.3	0.6
7735	Stakelets, frzn	1 pce	71	58	145	12	6	2	8	1.4	2.7	3.9
7625	Turkee slices, canned	1 pce	33	64	68	5	1	1	5	0.8	1.9	2.1
7782	Vegan meatless patty	1 ea	71		90	10	12	2	0	0	0	0
7765	Vita Burger chunks	¼ c	21	6	75	10	6	4	1	0.3	0.2	0.6

Chol (mg)	Calc (mg)	Iron (mg)	Magn (mg)	Pota (mg)	Sodi (mg)	Zinc (mg)	VT-A (RAE)	Thia (mg)	VT-E (mg α)	Ribo (mg)	Niac (mg)	V-B$_6$ (mg)	Fola (µg)	VT-C (mg)
0					475									
9	158	2.94	27	242	922	1.91		0.8	1.58	0.59	8.32	0.84	65	1
0	50	4.06	11	23	281	1.13	0	0.05	0	0.03	0.4	0.03	49	0
0	184	4.48	134	684	15	1.89	0	0.13	0.03	0.59	4.38	0.36	40	0
0	80	2.66	15	398	351	0.69		0.23	<.01	0.51	3.77	0.18	225	0
0	10	0.87	4	31	104	0.73	2	0.07						0
0	40	2.7		240	270	3	0	0.22		0.14	3	0.2		0
0	60	1.8			320									2
0	16	0.93	9	58	222	0.36	0	0.59	0.52	0.1	2.8	0.21	6	0
0	24	1.41	14	88	337	0.55	0	0.89	0.8	0.15	4.25	0.31	10	0
0	80	2.7			370	6.75	0	0.3		0.14	4	0.3		0
0	102	3.85	70	432	411	8.07		0.32	1.56	0.2	6.3	0.39	22	0
0	100	1.8			370	4.5	0	0.22		0.1	3	0.16		0
0	80	2.7			370	6.75	0	0.3		0.14	4	0.3		0
2	40	1.4		153	709	0.43	0	0.67		0.3	2.89	0.45		0
0	0	0			30		0							0
0	30	0.3			77									0
0	10	0.99		61	224	1.2		0.28		0.68	5.78	0.67		
0	40	1.81			241		0							2
0	40	1.81			292		0							2
0	20	1.08			160		0							2
1	4	0.39		29	160	0.46	0	0.13		0.22	0.64	0.29		0
1	12	1.06		121	455	1.1	0	0.14		0.3	1.9	0.51		0
2	24	0.31		225	433	0.41	0	1.25		0.65	5.41	1		0
0	8	0.5		30	114	0.58	0	0.2		0.25	0.78	0.31		0
0	110	3.96			410		0							0
1	24	0.83		60	98	0.6		0.05		0.36	0	0.11		0
1	9	1.77		46	355	0.35	0	5.43		0.15	2.5	0.38		0
0	3	0.33		16	228	0.06	0	0.67		0.05	0.75	0.08		0
2	39	2.75		299	628	0.65		0.31		0.26	2.82	0.37		0
0	40	3.2	1	89	238	0.82		4.96	0.35	0.18	1.49	0.27		0
1	56	1.84	44	269	499	0.93		8.06	0.36	0.14	0	0.21		0
1	37	1.95		297	632	0.83	0	1.28		0.29	2.64	0.24		0
1	24	2.31		174	514	0.58	0	1.39		0.23	2.26	0.18		0
1	22	0.77	5	63	545	0.48		0.18	1.59	0.03	0	0.01		0
1	48	1.21	29	180	382	0.58	134	6.47	0.55	0.1	0	0	59	0
2	22	2.5		122	269	0.67	0	11.81		0.2	4.07	0.48		0
1	25	1.43		159	365	0.87	0	0.23		0.15				0
0	21	0.62		135	262	0.72	0	0.35		0.15	6.39	0.36		0
1	18	1.92	1	102	259	0.37		5.38	0.3	0.13	1.84	0.19		0
0	4	2.63		44	624	0.22	0	0.89		0.34	6.46	0.56		0
0	4	1.73		25	269	0.38	0	0.13		0.1	1.96	0.24		0
0	6	0.37		40	500	0.65	0	0.05		0.06	0	0.06		0
1	6	1.17		58	524	0.26	0	10.61		0.07	1.36	0.3		0
2	51	5.09		270	826	1.03	108	1.85		0.29	4.22	0.86		0
0	10	1.12		155	272	0.56	0	0.08		0.06	0.54	0.2		0
2	12	0.38		82	445	0.38	0	0.12		0.13	1.98	0.38		0
0	1	1.84		50	283	0.7	0	0.18		0.13	1.34	0.24		0
2	36	2.87		38	566	0.64	0	2.13		0.26	6.02	0.6		0
1	9	1.15		25	205	0.26	0	0.59		0.08	0.1	0.13		0
0	0	0.47		14	179	0.08	0	0.08		0.06	0.48	0.1		0
2	49	0.99		95	484	0.5	0	1.51		0.12	3.1	0.26		0
1	3	0.48		16	203	0.11	0	1.13		0.05	0.39	0.09		0
0	40	4.5			230		0							0
0	27	1.83		499	353	4.79	0	1.23		0.17	3.06	0.12		0

A

TABLE A–1

Food Composition   (Computer code number is for West Diet Analysis program)   (For purposes of calculations, use "0" for t, <1, <.1, <.01, etc.)

Computer Code Number	Food Description	Measure	Wt (g)	H₂O (%)	Ener (kcal)	Prot (g)	Carb (g)	Dietary Fiber (g)	Fat (g)	Fat Breakdown (g) Sat	Mono	Poly
	MISCELLANEOUS											
28006	Baking powder, low sodium	1 tsp	5	6	5	<1	2	<1	<1	t	t	t
28003	Baking soda	1 tsp	5	0	0	0	0	0	0	0	0	0
26001	Basil, dried	1 tbsp	5	6	13	1	3	2	<1	t	t	0.1
38063	Carob flour	1 c	103	4	229	5	92	41	1	t	0.2	0.2
27000	Catsup	1 tbsp	15	70	14	<1	4	<1	<1	t	t	t
26027	Cayenne/red pepper	1 tbsp	5	8	16	1	3	1	1	0.2	0.1	0.4
26040	Celery seed	1 tsp	2	6	8	<1	1	<1	1	t	0.3	t
26002	Chili powder	1 tbsp	8	8	25	1	4	3	1	0.2	0.3	0.6
23010	Chocolate, baking, unsweetened	1 oz	28	1	140	4	8	5	15	9.1	4.5	0.4
	(For other chocolate items, see Sweeteners & Sweets)											
26038	Cilantro/Coriander, fresh	1 tbsp	1	92	0	<1	<1	<1	<1	t	t	t
26003	Cinnamon	1 tsp	2	10	5	<1	2	1	<1	t	t	t
30197	Cornstarch	1 tbsp	8	8	30	<1	7	<1	<1	t	t	t
26004	Curry powder	1 tsp	2	10	6	<1	1	1	<1	t	0.1	t
26021	Dill weed, dried	1 tbsp	3	7	8	1	2	<1	<1	t	t	t
26005	Garlic cloves	1 ea	3	59	4	<1	1	<1	<1	t	t	t
26007	Garlic powder	1 tsp	3	6	10	1	2	<1	<1	t	t	t
23009	Gelatin, dry, unsweetened: Envelope	1 ea	7	13	23	6	0	0	<1	t	t	t
26043	Ginger root, slices, raw	2 pce	5	79	4	<1	1	<1	<1	t	t	t
27004	Horseradish, prepared	1 tbsp	15	85	7	<1	2	<1	<1	t	t	t
7081	Hummous/hummus	1 c	246	65	435	12	49	10	21	2.8	12.1	5.1
27072	Mustard, country dijon	1 tsp	5		5	1	1	0	0	0	0	0
6313	Mustard, gai choy chinese	1 tbsp	16	97	1	<1	<1	<1	0	0	0	0
	Miso (see Vegetables and Legumes, Soybean products)											
27009	Olives, ripe, pitted	5 ea	22	80	25	<1	1	1	2	0.3	1.7	0.2
26008	Onion powder	1 tsp	2	5	7	<1	2	<1	<1	t	t	t
26009	Oregano, ground	1 tsp	2	7	6	<1	1	1	<1	t	t	0.1
26010	Paprika	1 tsp	2	10	6	<1	1	1	<1	t	t	0.2
26011	Parsley, freeze dried	¼ c	1	2	3	<1	<1	<1	<1	t	t	t
26016	Pepper, black	1 tsp	2	11	5	<1	1	1	<1	t	t	t
27012	Pickles, dill, medium, 3¾ × 1¼" diam	1 ea	65	92	12	<1	3		<1	t	t	t
27016	Sweet, medium	1 ea	35	65	41	<1	11	<1	<1	t	t	t
	Popcorn (see Grain Products)											
44006	Potato chips:	10 pce	20	2	107	1	11	1	7	2.2	2	2.4
44076	Unsalted	1 oz	28	2	150	2	15	1	10	3.1	2.8	3.4
26031	Sage, ground	1 tsp	1	8	3	<1	1	<1	<1	t	t	t
27020	Salsa, from recipe	1 tbsp	15	95	3	<1	1	<1	<1	t	t	t
26014	Salt 1 tsp	6	0	0	0	0	0	0	0	0	0	
	Salt Substitutes:											
26090	Morton, salt substitute	1 tsp	6	0	0	0	<1		0	0	0	0
26048	Morton, light salt	1 tsp	6	0	0	0	<1		<1			
26098	Seasoned salt, no MSG	1 tsp	5	5	0	0	0	0	0	0	0	0
27007	Vinegar, cider	½ c	120	94	17	0	7	0	0	0	0	0
27148	Balsamic	1 tbsp	15	64	21	0	5	0	0	0	0	0
27154	Malt	1 tbsp	15	90	5	0	1	0	0	0	0	0
27158	Tarragon	1 tbsp	15	95	3	0	<1	0	0	0	0	0
27156	White wine	1 tbsp	15	89	5	0	2	0	0	0	0	0
28001	Yeast, baker's, dry, active, package	1 ea	7	8	21	3	3	1	<1	t	0.2	t
	SOUPS, SAUCES, AND GRAVIES											
	SOUPS, canned:											
	Unprepared, condensed:											
50650	Cream of celery	1 c	251	85	181	3	18	2	11	2.8	2.6	5
50654	Cream of chicken	1 c	251	82	233	7	19	1	15	4.2	6.6	3
50666	Cream of mushroom	1 c	251	81	259	4	19	1	19	5.1	3.6	8.9
50668	Onion	1 c	246	86	113	8	16	2	3	0.5	1.5	1.3
	Prepared w/equal volume of whole milk:											
50008	Clam chowder, New England	1 c	248	85	164	9	17	1	7	3	2.3	1.1
50015	Cream of celery	1 c	248	86	164	6	15	1	10	3.9	2.5	2.7

PAGE KEY: A–4 = Beverages   A–6 = Dairy   A–10 = Eggs   A–12 = Fat/Oil   A–14 = Fruit   A–20 = Bakery
A–26 = Grain   A–32 = Fish   A–32 = Meats   A–36 = Poultry   A–38 = Sausage   A–40 = Mixed/Fast   A–44 = Nuts/Seeds   A–46 = Sweets
A–48 = Vegetables/Legumes   A–58 = Vegetarian   A–62 = Misc   A–62 = Soups/Sauces   A–64 = Fast   A–78 = Convenience   A–80 = Baby foods

Chol (mg)	Calc (mg)	Iron (mg)	Magn (mg)	Pota (mg)	Sodi (mg)	Zinc (mg)	VT-A (RAE)	Thia (mg)	VT-E (mg α)	Ribo (mg)	Niac (mg)	V-B$_6$ (mg)	Fola (µg)	VT-C (mg)
0	217	0.41	1	505	4	0.04	0	0	0	0	0	0	0	0
0	0	0	0	0	1368	0	0	0	0	0	0	0	0	0
0	106	2.1	21	172	2	0.29	23	<.01	0.37	0.02	0.35	0.12	14	3
0	358	3.03	56	852	36	0.95	1	0.05	0.65	0.47	1.95	0.38	30	0
0	3	0.08	3	57	167	0.04	7	<.01	0.22	0.07	0.23	0.02	2	2
0	7	0.39	8	101	2	0.12	104	0.02	1.49	0.05	0.44	0.12	5	4
0	35	0.9	9	28	3	0.14	<1	<.01	0.02	<.01	0.06	0.02	<1	0
0	22	1.14	14	153	81	0.22	119	0.03	2.32	0.06	0.63	0.29	8	5
0	28	4.87	92	232	7	2.7	0	0.04	0.11	0.03	0.38	<.01	8	0
0	1	0.02	0	5	<1	<.01	3	<.01	0.02	<.01	0.01	<.01	1	0
0	25	0.76	1	10	1	0.04	<1	<.01	0.02	<.01	0.03	<.01	1	1
0	0	0.04	0	0	1	<.01	0	0	0	0	0	0	0	0
0	10	0.59	5	31	1	0.08	1	<.01	0.44	<.01	0.07	0.02	3	0
0	54	1.46	14	99	6	0.1	9	0.01		<.01	0.08	0.05		2
0	5	0.05	1	12	1	0.03	0	<.01	<.01	<.01	0.02	0.04	<1	1
0	2	0.08	2	33	1	0.08	0	0.01	0.02	<.01	0.02	0.09	<1	1
0	4	0.08	2	1	14	<.01	0	<.01	0	0.02	<.01	<.01	2	0
0	1	0.03	2	21	1	0.02	0	<.01	0.01	<.01	0.04	<.01	1	0
0	8	0.06	4	37	47	0.12	<1	<.01	<.01	<.01	0.06	0.01	9	4
0	121	3.86	71	426	595	2.68	1	0.22	1.84	0.13	0.98	0.98	145	19
0				10	120									
5	0.07			1			14						4	
0	19	0.73	1	2	192	0.05	4	<.01	0.36	0	<.01	<.01	0	0
0	7	0.05	2	19	1	0.05	0	<.01	<.01	<.01	0.01	0.02	3	0
0	32	0.88	5	33	<1	0.09	7	<.01	0.38	<.01	0.12	0.02	5	1
0	4	0.47	4	47	1	0.08	53	0.01	0.6	0.03	0.31	0.08	2	1
0	2	0.54	4	63	4	0.06	32	0.01	0.06	0.02	0.1	0.01	2	1
0	9	0.58	4	25	1	0.03	<1	<.01	0.01	<.01	0.02	<.01	<1	0
0	6	0.34	7	75	833	0.09	6	<.01	0.06	0.02	0.04	<.01	1	1
0	1	0.21	1	11	329	0.03	3	<.01	0.03	0.01	0.06	<.01	<1	0
0	5	0.33	13	255	119	0.22	0	0.03	1.82	0.04	0.77	0.13	9	6
0	7	0.46	19	357	2	0.31	0	0.05	2.55	0.06	1.07	0.18	13	9
0	17	0.28	4	11	<1	0.05	3	<.01	0.07	<.01	0.06	0.03	3	0
0	1	0.05	1	23	1	0.02	3	<.01	0.04	<.01	0.06	0.01	2	2
0	1	0.02	0	0	2325	<.01	0	0	0	0	0	0	0	0
33		0	3018	1										
2		4	1518	1182										
0					1583									
0	7	0.72	26	120	1	0	0	0	0	0	0	0	0	0
2	0.15		10	3			0.15			0.15	0.15		0	
2	0.15		14	4			0.15			0.15	0.15		2	
0	0.15		2	1			0.15			0.15	0.15		0	
1	0.15		12	1			0.15			0.15	0.15		0	
0	4	1.16	7	140	4	0.45	0	0.17	0	0.38	2.78	0.11	164	0
28	80	1.25	13	246	1900	0.3	55	0.06	3.49	0.1	0.67	0.03	5	1
20	68	1.2	5	176	1973	1.25	108	0.06	1.36	0.12	1.64	0.03	3	0
3	65	1.05	10	168	1737	1.18	20	0.06	2.03	0.17	1.62	0.03	8	2
0	54	1.35	5	138	2116	1.23	7	0.07	0.54	0.05	1.21	0.1	30	2
22	186	1.49	22	300	992	0.79	57	0.07	0.45	0.24	1.03	0.13	10	3
32	186	0.69	22	310	1009	0.2	114	0.07	0.97	0.25	0.44	0.06	7	1

A

TABLE A-1

Food Composition   (Computer code number is for West Diet Analysis program)   (For purposes of calculations, use "0" for t, <1, <.1, <.01, etc.)

Computer Code Number	Food Description	Measure	Wt (g)	H₂O (%)	Ener (kcal)	Prot (g)	Carb (g)	Dietary Fiber (g)	Fat (g)	Sat	Mono	Poly
	SOUPS, SAUCES, AND GRAVIES—Continued											
50006	Cream of chicken	1 c	248	85	191	7	15	<1	11	4.6	4.5	1.6
50011	Cream of mushroom	1 c	248	85	203	6	15	<1	14	5.1	3	4.6
50026	Cream of potato	1 c	248	87	149	6	17	<1	6	3.8	1.7	0.6
50024	Oyster stew	1 c	245	89	135	6	10	0	8	5	2.1	0.3
50012	Tomato	1 c	248	85	161	6	22	3	6	2.9	1.6	1.1
	Prepared with equal volume of water:											
50000	Bean with bacon	1 c	253	84	172	8	23	9	6	1.5	2.2	1.8
50003	Beef noodle	1 c	244	92	83	5	9	1	3	1.1	1.2	0.5
50005	Chicken noodle	1 c	241	92	75	4	9	1	2	0.7	1.1	0.6
50020	Chicken rice	1 c	241	94	60	4	7	1	2	0.5	0.9	0.4
50007	Chili beef	1 c	250	85	170	7	21	10	7	3.4	2.8	0.3
50021	Clam chowder, Manhattan	1 c	244	92	78	2	12	1	2	0.4	0.4	1.3
50018	Cream of chicken	1 c	244	91	117	3	9	<1	7	2.1	3.3	1.5
50049	Cream of mushroom	1 c	244	90	129	2	9	<1	9	2.4	1.7	4.2
50009	Minestrone	1 c	241	91	82	4	11	1	3	0.6	0.7	1.1
50022	Onion	1 c	241	93	58	4	8	1	2	0.3	0.7	0.7
50025	Split pea & ham	1 c	253	82	190	10	28	2	4	1.8	1.8	0.6
50028	Tomato	1 c	244	90	85	2	17	<1	2	0.4	0.4	1
50014	Vegetable beef	1 c	244	92	78	6	10	<1	2	0.9	0.8	0.1
50013	Vegetarian vegetable	1 c	241	92	72	2	12	<1	2	0.3	0.8	0.7
50052	Ready to serve, Chunky chicken soup	1 c	251	84	178	13	17	2	7	2	3	1.4
	SOUPS, dehydrated:											
	Prepared with water:											
50032	Beef broth/bouillon	1 c	244	97	20	1	2	0	1	0.3	0.3	t
50034	Chicken broth	1 c	244	97	22	1	1	0	1	0.3	0.4	0.4
50037	Chicken noodle	1 c	252	94	58	2	9	<1	1	0.3	0.5	0.4
50036	Cream of chicken	1 c	261	91	107	2	13	<1	5	3.4	1.2	0.4
50040	Onion	1 c	246	96	27	1	5	1	1	0.1	0.3	t
50041	Split pea	1 c	255	87	125	7	21	3	1	0.4	0.7	0.3
50043	Tomato vegetable	1 c	253	94	56	2	10	1	1	0.4	0.3	t
	Unprepared, dry products:											
50051	Beef bouillon, packet	1 ea	6	3	14	1	1	0	1	0.3	0.2	t
50054	Onion soup, packet	1 ea	39	4	115	5	21	4	2	0.5	1.4	0.3
	SAUCES											
53097	From home recipe, lowfat cheese sauce	¼ c	61	74	81	6	4	<1	5	1.8	1.8	0.8
	Ready to serve:											
53396	Alfredo sauce, reduced fat	¼ c	69		144	5	9	0	9	6.2		
53000	Barbeque sauce	1 tbsp	16	81	12	<1	2	<1	<1	t	0.1	0.1
53355	Creole sauce	¼ c	62	89	25	1	4	1	1	t	0.2	0.3
53392	Pesto sauce	2 tbsp	16		83	2	1	<1	8	1.5		
53002	Soy sauce	1 tbsp	16	71	8	1	1	<1	<1	t	t	t
53349	Szechuan sauce	1 tbsp	16	71	21	<1	3	<1	1	0.1	0.3	0.4
53004	Teriyaki sauce	1 tbsp	18	68	15	1	3	<1	0	0	0	0
53011	Spaghetti with meat sauce	1 c	250	85	178	7	19	4	8	1.8	3.3	1.8
	GRAVIES											
53023	Beef	1 c	233	87	123	9	11	1	5	2.7	2.2	0.2
53022	Chicken	1 c	238	85	188	5	13	1	14	3.4	6.1	3.6
53026	Mushroom	1 c	238	89	119	3	13	1	6	1	2.8	2.4
	FAST FOOD RESTAURANTS											
	ARBY'S											
	Roast beef sandwiches:											
56336	Regular	1 ea	155		326	21	35	2	14	6.9		
56337	Junior	1 ea	89		200	11	23	1	8	3.4		
56338	Super	1 ea	254		467	23	50	3	22	8.3		
69056	Beef 'n cheddar	1 ea	194		451	23	42	2	23	8.8		
56341	Chicken breast sandwich	1 ea	204		539	24	46	2	29	4.9		
56342	Ham'n cheese sandwich	1 ea	169		338	23	35	1	13	4.5		
69048	Italian sub sandwich	1 ea	297		742	28	47	3	50	14.3		
56343	Turkey sandwich, deluxe	1 ea	218		292	26	37	3	6	0.6	2.5	2.6
69044	Turkey sub sandwich	1 ea	277		570	24	46	2	33	8.1		

PAGE KEY: A–4 = Beverages A–6 = Dairy A–10 = Eggs A–12 = Fat/Oil A–14 = Fruit A–20 = Bakery
A–26 = Grain A–32 = Fish A–32 = Meats A–36 = Poultry A–38 = Sausage A–40 = Mixed/Fast A–44 = Nuts/Seeds A–46 = Sweets
A–48 = Vegetables/Legumes A–58 = Vegetarian A–62 = Misc A–62 = Soups/Sauces A–64 = Fast A–78 = Convenience A–80 = Baby foods

Chol (mg)	Calc (mg)	Iron (mg)	Magn (mg)	Pota (mg)	Sodi (mg)	Zinc (mg)	VT-A (RAE)	Thia (mg)	VT-E (mg α)	Ribo (mg)	Niac (mg)	V-B$_6$ (mg)	Fola (µg)	VT-C (mg)
27	181	0.67	17	273	1047	0.67	179	0.07	0.25	0.26	0.92	0.07	7	1
20	179	0.6	20	270	918	0.64	35	0.08	1.24	0.28	0.91	0.06	10	2
22	166	0.55	17	322	1061	0.67	52	0.08	0.1	0.24	0.64	0.09	10	1
32	167	1.05	20	235	1041	10.34	56	0.07	0.49	0.23	0.34	0.06	10	4
17	159	1.81	22	449	744	0.3	64	0.13	1.24	0.25	1.52	0.16	17	68
3	81	2.05	46	402	951	1.04	46	0.09	0.76	0.03	0.57	0.04	33	2
5	15	1.1	5	100	952	1.54	7	0.07	0.68	0.06	1.07	0.04	20	0
7	17	0.77	5	55	1106	0.39	36	0.05	0.1	0.06	1.39	0.03	22	0
7	17	0.75	0	101	815	0.27	22	0.02	0.1	0.02	1.13	0.02	0	0
12	42	2.12	30	525	1035	1.4	75	0.06	1.52	0.08	1.07	0.16	18	4
2	27	1.63	12	188	578	0.98	56	0.03	0.34	0.04	0.82	0.1	10	4
10	34	0.61	2	88	986	0.63	163	0.03	0.2	0.06	0.82	0.02	2	0
2	46	0.51	5	100	881	0.59	15	0.05	0.95	0.09	0.72	0.01	5	1
2	34	0.92	7	313	911	0.75	118	0.05	0.07	0.04	0.94	0.1	36	1
0	27	0.67	2	67	1053	0.6	<1	0.03	0.19	0.02	0.6	0.05	14	1
8	23	2.28	48	400	1007	1.32	23	0.15	0.15	0.08	1.47	0.07	3	2
0	12	1.76	7	264	695	0.24	29	0.09	2.32	0.05	1.42	0.11	15	66
5	17	1.12	5	173	791	1.54	95	0.04	0.37	0.05	1.03	0.08	10	2
0	22	1.08	7	210	822	0.46	116	0.05	0.41	0.05	0.92	0.06	10	1
30	25	1.73	8	176	889	1	68	0.09	0.33	0.17	4.42	0.05	5	1
0	10	0.02	7	37	1362	0.07	0	<.01	0.1	0.02	0.36	0	0	0
0	15	0.07	5	24	1484	0	2	<.01	0.07	0.03	0.2	0	2	0
10	5	0.5	8	33	577	0.2	3	0.2	0.13	0.08	1.09	0.03	18	0
3	76	0.26	5	214	1185	1.57	21	0.1	0.6	0.2	2.61	0.05	5	1
0	12	0.15	5	64	849	0.05	0	0.03	0	0.06	0.48	0	2	0
3	20	0.94	43	224	1148	0.56	3	0.21	0.03	0.14	1.26	0.05	41	0
0	8	0.63	20	104	1146	0.18	10	0.06	0.35	0.05	0.79	0.05	10	6
1	4	0.06	3	27	1019	0	0	<.01	0.13	0.01	0.27	0.01	2	0
2	55	0.58	25	260	3493	0.23	<1	0.11	0.03	0.24	1.99	0.04	6	1
9	167	0.24	10	101	307	0.74		0.03	0.51	0.14	0.16	0.03	4	0
31	103	0	8	93	618			0		0.11	0		0	0
0	3	0.14	3	28	130	0.03	<1	<.01	<.01	<.01	0.14	0.01	1	1
0	35	0.31	9	187	339	0.1		0.03	0.61	0.02	0.53	0.07	9	0
4	65	0.09		26	137			0		0.02	0		2	0
0	3	0.32	5	29	914	0.06	0	<.01	0	0.02	0.54	0.03	3	0
0	2	0.12	2	13	218	0.02		<.01	0.07	<.01	0.1	<.01	1	0
0	4	0.31	11	40	690	0.02	0	<.01	0	0.01	0.23	0.02	4	0
15	54	2.1	43	742	982	1.27		0.13	2.98	0.12	3.46	0.31	25	19
7	14	1.63	5	189	1305	2.33	2	0.07	0.05	0.08	1.54	0.02	5	0
5	48	1.12	5	259	1373	1.9	2	0.04	0.31	0.1	1.05	0.02	5	0
0	17	1.57	5	252	1357	1.67	0	0.08	0.19	0.15	1.6	0.05	29	0
44	59	3.55	16	422	879	3.75	0							0
28	41	1.86	8	201	483	1.5		0.18		0.26	6.6	0.1	7	
47	62	3.73	25	533	1099	3.73		0.39		0.58	12.4	0.3	21	1
49	98	3.53		321	1146	2.94		0.42		0.63	9.8			1
88	78	1.77	30	330	1138	0.15		0.22		0.54	8.99	0.38	18	4
89	149	2.68	31	380	1441	0.89		0.82		0.37	7.75	0.31	26	1
114	238	2.57			565	2323		0.91		0.49	8.19			2
45	90	2.02	33	394	1157	1.69		0.09		0.46	17.22	0.58	22	1
91	181	0.33		500	1964			13.2		0.54	18.8			2

A

TABLE A-1

Food Composition (Computer code number is for West Diet Analysis program)   (For purposes of calculations, use "0" for t, <1, <.1, <.01, etc.)

Computer Code Number	Food Description	Measure	Wt (g)	H₂O (%)	Ener (kcal)	Prot (g)	Carb (g)	Dietary Fiber (g)	Fat (g)	Fat Breakdown (g)		
										Sat	Mono	Poly
	FAST FOOD RESTAURANTS—Continued											
	Milkshakes:											
2124	Chocolate	1 ea	340	71	411	9	72	0	14	6.9		
2125	Jamocha	1 ea	326	72	386	8	67	0	12	5.7		
2126	Vanilla	1 ea	312	72	369	8	65	0	12	5.5		
6431	Salad, roast chicken	1 ea	400		152	19	14	6	2	0		
	Source: Arby's											
	BOSTON MARKET											
	Chicken											
15247	Half chicken with skin, svg	1 ea	277		590	70	4	0	33	10		
15246	Dark meat, svg	1 ea	95		190	22	1	0	10	3		
15249	Dark meat with skin, svg	1 ea	125		320	30	2	0	21	6		
15244	White meat, svg	1 ea	140		170	33	2	0	4	1		
15245	White meat, with skin, svg	1 ea	152	64	280	40	2	0	12	3.5		
57530	Chicken pot pie	1 ea	425		780	32	61	4	46	13		
15248	Turkey breast, svg	1 ea	142		170	36	1	0	1	0.5		
	Sandwiches											
69084	Ham and turkey club	1 ea	266		430	29	64	4	6	2		
69081	Ham	1 ea	266		440	25	66	4	8	2.5		
69076	Chicken	1 ea	281	64	430	34	62	4	4	1		
69075	Chicken with cheese and sauce	1 ea	352		750	41	72	5	33	12		
69079	Turkey	1 ea	266		400	45	61	4	4	1		
69078	Turkey with cheese and sauce	1 ea	337		710	45	68	4	28	10		
69083	Turkey and ham with cheese and sauce	1 ea	379		890	47	79	4	43	20		
69074	Meatloaf	1 ea	351		690	40	86	6	21	7		
69082	Meatloaf with cheese	1 ea	383		860	46	95	6	33	16		
	Side dishes											
7393	BBQ baked beans	³/₄ c	201		270	8	48	12	5	2		
7387	New Potatoes	³/₄ c	131		130	3	25	2	2	0		
7390	Mashed potatoes	²/₃ c	161		190	3	24	1	9	6		
28257	Red beans and rice	1 c	227		260	8	45	4	5	0		
7388	Whole kernel corn	³/₄ c	146		180	5	30	2	4	0.5		
7386	Steamed vegetables	²/₃ c	105		35	2	7	3	<1	0		
52111	Caesar side salad	4 oz	113		199	7	7	1	17	4.5		
52103	Chunky chicken salad	³/₄ c	158		370	28	3	1	27	4.5		
52108	Mediterranean pasta salad	³/₄ c	129		170	4	16	2	10	2.5		
	Source: Boston Chicken, Inc.											
	BURGER KING											
56346	Croissant sandwich, egg, sausage & cheese	1 ea	176	45	575	22	30	1	41	15		
	Sandwiches:											
56354	Whopper	1 ea	270		660	28	51	4	38	11.7		
56355	Whopper with cheese	1 ea	294		757	33	53	4	46	16.5		
57002	BK broiler chicken sandwich	1 ea	248		529	29	50	3	24	4.8		
56352	Cheeseburger	1 ea	138		375	20	31	2	18	9.1		
56360	Chicken sandwich	1 ea	229		675	26	54	3	40	8.2		
56356	Double beef	1 ea	351		915	48	53	4	57	19.9		
56357	Double beef & cheese	1 ea	375	52	1012	53	55	4	64	24.8		
56353	Double cheeseburger with bacon	1 ea	218		649	40	34	2	39	19.1		
56351	Hamburger	1 ea	126		328	18	31	2	14	6.1		
56362	Ocean catch fish fillet	1 ea	255		688	23	65	4	37	13.6		
15158	Chicken tenders	1 ea	88		251	16	15	1	14	3.4		
6141	French fries (salted), svg	1 ea	116		357	4	46	4	18	5		
42429	French toast sticks, svg	1 ea	141		491	8	58	3	25	5.7		

PAGE KEY: A–4 = Beverages   A–6 = Dairy   A–10 = Eggs   A–12 = Fat/Oil   A–14 = Fruit   A–20 = Bakery
A–26 = Grain   A–32 = Fish   A–32 = Meats   A–36 = Poultry   A–38 = Sausage   A–40 = Mixed/Fast   A–44 = Nuts/Seeds   A–46 = Sweets
A–48 = Vegetables/Legumes   A–58 = Vegetarian   A–62 = Misc   A–62 = Soups/Sauces   A–64 = Fast   A–78 = Convenience   A–80 = Baby foods

Chol (mg)	Calc (mg)	Iron (mg)	Magn (mg)	Pota (mg)	Sodi (mg)	Zinc (mg)	VT-A (RAE)	Thia (mg)	VT-E (mg α)	Ribo (mg)	Niac (mg)	V-B$_6$ (mg)	Fola (µg)	VT-C (mg)
39	428	0.62	48	410	317	1.5		0.12		0.68	0.8	0.14	14	2
37	411	0.59	36	525	320	1.48		0.12		0.68	0.8	0.14	14	2
35	393	0.85	36	686	283	1.49		0.12		0.68	4	0.14	37	2
38	76	1.71		877	667			0.31		0.54	5.6			0
290	0	2.7			1010		0							0
115	0	1.08			440		0							0
155	0	1.8			500		0							0
85	0	0.72			480		0							0
175	0	1.08			510		0							0
135	40	3.6			1480									4
100	20	1.8			850		0							0
55	80	2.7			1330									9
45	80	1.8			1450									9
65	100	1.8			910									9
135	400	2.7			1860									9
60	100	2.7			1070									9
110	500	2.7			1390									9
150	800	4.5			2310									9
120	100	4.5			1610									15
165	400	4.5			2270									15
0	100	3.6			540									6
0	0	0.72			150		0							12
25	60	0.36			570									6
5	60	2.7			1050									12
0	0	0.36			170		10							5
0	20	0.36			35		200							21
15	199	1.08			448									12
120	20	0.72			800									4
10	60	1.08			490									9
219	173	3.11			1173									0
78	97	5.24			913									9
102	243	5.24			1349									9
101	58	3.46			1067									6
56	152	2.74			761									0
72	82	2.76			1360									0
149	149	7.16			1014									9
169	298	7.14			1448									9
128	266	4.79			1244									0
46	82	2.77			543									0
48	78	3.49			1163									0
34	0	0.41			606		0							0
0	20	0.71			684		0							15
0	76	2.27			554		0							0

A

TABLE A-1

Food Composition   (Computer code number is for West Diet Analysis program)   (For purposes of calculations, use "0" for t, <1, <.1, <.01, etc.)

Computer Code Number	Food Description	Measure	Wt (g)	H₂O (%)	Ener (kcal)	Prot (g)	Carb (g)	Dietary Fiber (g)	Fat (g)	Fat Breakdown (g) Sat	Mono	Poly
	FAST FOOD RESTAURANTS—Continued											
56363	Onion rings, svg	1 ea	124		436	5	55	4	22	5.5		
2127	Milk shakes, chocolate	1 ea	284		315	9	57	3	6	3.6		
2129	Vanilla	1 ea	284		315	9	57	1	6	3.6		
48134	Fried apple pie	1 ea	113		340	2	52	1	14	3		
	Source: Burger King Corporation											
	CARL'S JR											
	Sandwiches:											
91413	Carl's catch	1 ea	201	51	510	18	50	1	27	7		
91407	Charbroiled BBQ Chicken	1 ea	199		280	25	37	2	3	1		
91411	Bacon swiss chicken	1 ea	291		720	32	66	3	36	10		
91410	Ranch crispy chicken	1 ea	266		730	29	77	4	34	7.1		
91406	Hamburger	1 ea	134		330	18	34	1	13	5		
91402	Famous star hamburger	1 ea	254		580	25	49	2	32	9		
91403	Super star hamburger	1 ea	345		790	42	49	2	46	14		
91404	Western bacon cheeseburger	1 ea	225		650	32	63	2	30	12		
91405	Western bacon double cheeseburger	1 ea	308		900	51	64	2	49	21		
91424	Charbroiled chicken salad to go	1 ea	350		200	25	12	3	7	3.5		
	Side dishes:											
91437	Bacon, 2 slices, svg	1 ea	9	15	50	3	0	0	4	1.5		
91418	Criss-cut french fries, svg	1 ea	139		410	5	43	4	24	5		
91414	French fries, svg	1 ea	92	36	290	5	37	3	14	3		
91415	Onion rings, svg	1 ea	127		430	7	53	3	21	5		
	Source: Carl Karcher Enterprises, Inc.											
	CHICK-FIL-A											
	Sandwiches:											
69153	Chargrilled chicken	1 ea	150		290	26	30	1	7	2.1		
69154	Chicken salad club	1 ea	232		413	34	36	2	15	5.7		
	Salads:											
52139	Carrot and raisin	1 ea	76		109	1	18	2	4	0.8		
69181	Chargrilled chicken caesar	1 ea	241		240	31	6	2	10	6		
52134	Chicken garden, charbroiled	1 ea	397	86	257	33	11	4	9	4.3		
52135	Chick-n-strips	1 ea	451		487	43	27	4	23	7.2		
52138	Cole slaw	1 ea	79		158	1	11	2	13	1.9		
69184	Chargrilled chicken cool wrap	1 ea	240		390	31	53	3	7	3		
69183	Chicken caesar cool wrap	1 ea	227		460	38	51	3	11	6		
69188	Chicken breast fillet, breaded	1 ea	105		230	23	10	0	11	2.5		
15263	Chicken nuggets, svg	1 ea	110	54	253	25	12	1	12	2.4		
15262	Chicken-n- strips, svg	1 ea	119	54	275	28	13	0	12	2.8		
50885	Hearty breast of chicken soup, svg	1 ea	215	88	100	9	13	1	2	0		
7973	Waffle potato fries, svg	1 ea	85	34	280	3	37	5	14	5		
46489	Cheesecake, svg	1 ea	88		322	6	28	2	20	11.4		
49134	Fudge nut brownie, svg	1 ea	74	13	330	4	45	2	15	3.5		
20601	Icedream, svg	1 ea	127		151	4	26	0	4	1.9		
48214	Lemon pie, svg	1 ea	99	39	280	6	45	3	9	3.1		
20602	Lemonade, svg	1 ea	255		170	0	41	0	<1	0		
	Source: Chick-Fil-A											
	DAIRY QUEEN											
	Ice cream cones:											
2144	Small vanilla	1 ea	142		230	6	38	0	7	4.5		
2143	Regular vanilla	1 ea	213		355	9	57	0	10	6.5		
2142	Large vanilla	1 ea	253		410	10	65	0	12	8		
2136	Chocolate dipped	1 ea	220		490	8	59	1	24	13		

Chol (mg)	Calc (mg)	Iron (mg)	Magn (mg)	Pota (mg)	Sodi (mg)	Zinc (mg)	VT-A (RAE)	Thia (mg)	VT-E (mg α)	Ribo (mg)	Niac (mg)	V-B$_6$ (mg)	Fola (µg)	VT-C (mg)
0	136	0			627		0							0
25	250	1.29			193									0
18	286	0			243									4
0	0	1.44			470									0
80	150	1.8			1030									2
60	80	2.7			830									5
75	250	3.6			1610									6
59	177	4.24			1436									6
45	60	3.6			480		0							2
70	100	4.5			910									6
130	100	5.4			910									9
80	200	4.5			1430									1
155	300	6.3			1770									1
75	200	1.8			440									27
10	0	0			140		0							0
0	20	1.8			950		0							12
0	0	1.08			170		0							21
0	20	0.36			700		0							0
62	83	1.86			1034		0							0
92	172	3.1			1573									4
0	17	0.3			75									3
85	350	0.36			1170									0
100	214	0.51			1042									9
122	215	1.55			974									9
15	30	0.27			135									20
70	200	3.6			1120									5
85	400	3.6			1540									0
60	40	1.08			990									0
68	39	1.05			1061		0							0
77	44	1.19			628		0							0
20	40	0			940		0							0
15	20	0			105		0							21
85	57	0			255									0
20	20	1.8			210									0
14	94	0.34			75									0
96	131	0			193									4
0	0	0.36			10		0							15
20	200	1.08			115									1
32	269	1.94			172									3
40	350	1.8			200									2
30	250	1.8			190									2

A

TABLE A-1

Food Composition   (Computer code number is for West Diet Analysis program)   (For purposes of calculations, use "0" for t, <1, <.1, <.01, etc.)

Computer Code Number	Food Description	Measure	Wt (g)	H₂O (%)	Ener (kcal)	Prot (g)	Carb (g)	Dietary Fiber (g)	Fat (g)	Fat Breakdown (g) Sat	Mono	Poly
	FAST FOOD RESTAURANTS—Continued											
2154	Chocolate sundae	1 ea	234		400	8	71	0	10	6		
2131	Banana split	1 ea	369		510	8	96	3	12	8		
2151	Peanut buster parfait	1 ea	305		730	16	99	2	31	17		
2133	Buster bar	1 ea	149		450	10	41	2	28	12		
2135	Dilly bar	1 ea	85	55	210	3	21	0	13	7		
	Milkshakes:											
2152	Large	1 ea	461	72	600	13	101	<1	16	10	2	2
2145	Malted	1 ea	418	68	610	13	106	<1				
2351	Misty slush, small	1 ea	454		220	0	56	0	0	0	0	0
2359	Starkiss	1 ea	85		80	0	21	0	0	0	0	0
	Sandwiches:											
56372	Cheeseburger, double	1 ea	219		540	35	30	2	31	16		
56371	Single	1 ea	152	55	340	20	29	2	17	8		
56379	Chicken	1 ea	191		466	18	45	4	24	4.2		
69029	Chicken fillet, grilled	1 ea	184		310	24	30	3	10	2.5		
56381	Fish fillet sandwich	1 ea	170	58	370	16	39	2	16	3.5		
56382	With cheese	1 ea	184	56	420	19	40	2	21	6		
56368	Hamburger, single	1 ea	138	56	290	17	29	2	12	5		
56369	Hamburger, double	1 ea	212		440	30	29	2	22	10		
	Hotdogs:											
56374	Regular	1 ea	99	55	240	9	19	1	14	5		
56375	With cheese	1 ea	113		290	12	20	1	18	8	8	2
56376	With chili	1 ea	128		297	13	20	2	19	8.1		
6143	French fries, small	1 ea	112	41	347	4	42	3	18	3.5		
56383	Onion rings	1 ea	113		320	5	39	3	16	4		
	Source: International Dairy Queen											
	HARDEES'S											
	Sandwiches:											
69061	Frisco burger hamburger	1 ea	242		793	36	41	2	54	15.3		
56422	Chicken sandwich	1 ea	187		444	23	39	2	17	4.6		
6146	French fries, svg	1 ea	96		289	3	38	0	14	1.7		
	Source: Hardees Food Systems											
	IN-N-OUT BURGERS											
81110	Hamburger with spread	1 ea	243		390	16	39	3	19	5		
81111	Hamburger w/mustard and ketchup	1 ea	243		310	16	41	3	10	4		
81113	Cheeseburger with spread	1 ea	268		480	22	39	3	27	10		
81114	Cheeseburger w/mustard and ketchup	1 ea	268		400	22	41	3	18	9		
81115	Cheeseburger, protein style, without bun	1 ea	300		330	18	11	2	25	9		
81116	Double cheeseburger with spread	1 ea	328		670	37	40	3	41	18		
81117	Double cheeseburger with mustard and ketchup	1 ea	328		590	37	42	3	32	17		
81118	Double cheeseburger, protein style, without bun	1 ea	361		520	33	11	2	39	17		
	Source: In-N-Out Burgers											
	JACK IN THE BOX											
	Breakfast items:											
56430	Breakfast jack sandwich	1 ea	126		280	17	28	1	12	5		
56431	Sausage crescent	1 ea	181		660	20	37	0	48	15		
56432	Supreme crescent	1 ea	164		530	21	37	0	34	10		
	Sandwiches:											
69032	Bacon cheeseburger	1 ea	274		760	39	39	2	50	17		

Chol (mg)	Calc (mg)	Iron (mg)	Magn (mg)	Pota (mg)	Sodi (mg)	Zinc (mg)	VT-A (RAE)	Thia (mg)	VT-E (mg α)	Ribo (mg)	Niac (mg)	V-B₆ (mg)	Fola (µg)	VT-C (mg)
30	250	1.44			210									0
30	250	1.8			180									15
35	300	1.8			400									1
15	150	1.08			280									0
10	100	0.36			75									0
50	450	1.44		660	260			0.15		0.68	0.8			0
45				570										
0	0	0			20	0								0
0	0	0			10	0								0
115	250	4.5			1130									4
55	150	3.6			850									4
28	75	4.19			1016									2
50	200	2.7			1040	0								0
45	40	1.8		280	630	0		0.3		0.22	3			0
60	100	1.8		290	850			0.3		0.26	5			0
45	60	2.7			630									4
90	60	4.5			680									6
25	60	1.8			730									4
40	150	1.8		180	950			0.22		0.17	2			4
41	135	1.62			983									3
0	20	1.07			872	0								4
0	20	1.44			180	0								0
111					1192									
61					1081									
0					331									
40	40	3.6			640									15
35	40	3.6			720									15
60	200	3.6			1000									15
55	200	3.6			1080									15
60	200	1.08			720									18
120	350	5.4			1430									15
115	350	5.4			1510									15
120	350	1.08			1160									18
190	150	3.6		120	750									10
240	100	1.8		160	860									0
225	100	1.8		165	1060									4
135	250	4.5		530	1570			0.27		0.54	9.96	0.44		9

A-72   APPENDIX A

TABLE A-1

Food Composition (Computer code number is for West Diet Analysis program)   (For purposes of calculations, use "0" for t, <1, <.1, <.01, etc.)

Computer Code Number	Food Description	Measure	Wt (g)	H₂O (%)	Ener (kcal)	Prot (g)	Carb (g)	Dietary Fiber (g)	Fat (g)	Sat	Mono	Poly
	FAST FOOD RESTAURANTS—Continued											
1655	Chicken sandwich	1 ea	164		400	15	38	3	21	3		
56443	Chicken supreme	1 ea	305		830	33	66	3	49	7		
56366	Double cheeseburger	1 ea	158		440	24	31	2	24	11		
69033	Grilled sourdough burger	1 ea	233		690	34	37	2	45	15		
56436	Jumbo jack burger	1 ea	271		550	27	43	2	30	10		
56437	Jumbo jack burger with cheese	1 ea	296		640	31	44	2	38	15		
56377	Tacos, regular	1 ea	90	66	170	7	12	2	10	3.5		
56378	Super	1 ea	138	63	270	12	19	4	17	6		
57014	Chicken teriyaki bowl	1 ea	502		670	26	128	3	4	1		
6150	French fries	1 ea	113	40	350	4	46	3	16	4		
6149	Hash browns	1 ea	57		170	1	14	1	12	2		
2164	Milkshake, strawberry	1 ea	382		640	10	85	0	28	15		
48135	Apple turnover	1 ea	107	40	340	4	41	2	18	4		
	Source: Jack in the Box Restaurant, Inc											
	KENTUCKY FRIED CHICKEN											
	Original Recipe chicken:											
15163	Side breast	1 ea	153		400	29	16	1	24	6	14.4	3.6
15165	Drumstick	1 ea	61	56	140	13	4	0	9	2	5.3	1.7
15166	Thigh	1 ea	91	54	250	16	6	1	18	4.5	10.2	3.4
15167	Wing	1 ea	47	48	140	9	5	0	10	2.5	5.8	1.7
	Hot & spicy chicken:											
15185	Center breast	1 ea	180	48	505	38	23	1	29	8		
15183	Thigh	1 ea	107		355	19	13	1	26	7		
15187	Wing	1 ea	55	37	210	10	9	1	15	4		
	Extra Crispy Recipe chicken:											
15169	Center breast	1 ea	168	48	470	39	17	1	28	8	16.7	3.3
15170	Drumstick	1 ea	67	48	195	15	7	1	12	3	7.4	1.6
15171	Thigh	1 ea	118		380	21	14	1	27	7	15.8	4.2
15172	Wing	1 ea	55	35	220	10	10	1	15	4	8.9	2.1
7139	Baked beans	½ c	167		203	6	35	6	3	1.1	1.4	0.7
42331	Buttermilk biscuit	1 ea	56		180	4	20	1	10	2.5	5.4	2.1
56451	Coleslaw, svg	1 ea	142		232	2	26	3	14	2	3.5	7
6152	Corn-on-the-cob	1 ea	162		150	5	35	2	2	0	0.6	0.9
15177	Chicken, hot wings, svg	1 ea	135		471	27	18	2	33	8		
56681	Macaroni & cheese, svg	1 ea	153		180	7	21	2	8	3		
56453	Mashed potatoes & gravy, svg	1 ea	136	82	120	1	17	2	6	1	3.6	1.4
56454	Potato salad	½ c	160		230	4	23	3	14	1.8	4.5	7.8
6188	Potato wedges, svg	1 ea	135		278	5	28	5	13	4		
	Source: Kentucky Fried Chicken Corp											
	McDONALD'S											
	Sandwiches:											
69010	Big Mac	1 ea	216		590	24	47	3	34	11		
69013	Filet-o-fish	1 ea	156	43	470	15	45	1	26	5		
69011	Quarter-pounder	1 ea	171		428	23	37	2	21	8		
69012	Quarter-pounder with cheese	1 ea	199		527	28	38	2	30	12.9		
15174	Chicken McNuggets	4 pce	71		187	10	13	1	11	2.5		
	Sauces (packet):											
27070	Hot mustard	1 ea	28		60	1	7	1	4	0		
53176	Barbecue	1 ea	28		45	0	10	0	0	0	0	0
53177	Sweet & sour	1 ea	28		50	0	11	0	0			
	Low-fat (frozen yogurt) milk shakes:											
2167	Chocolate	1 ea	295		360	11	60	1	9	6		
2168	Strawberry	1 ea	294	72	360	11	60	0	9	6		
2169	Vanilla	1 ea	293		360	11	59	0	9	6		
	Low-fat (frozen yogurt) sundaes:											

PAGE KEY: A–4 = Beverages A–6 = Dairy A–10 = Eggs A–12 = Fat/Oil A–14 = Fruit A–20 = Bakery
A–26 = Grain A–32 = Fish A–32 = Meats A–36 = Poultry A–38 = Sausage A–40 = Mixed/Fast A–44 = Nuts/Seeds A–46 = Sweets
A–48 = Vegetables/Legumes A–58 = Vegetarian A–62 = Misc A–62 = Soups/Sauces A–64 = Fast A–78 = Convenience A–80 = Baby foods

Chol (mg)	Calc (mg)	Iron (mg)	Magn (mg)	Pota (mg)	Sodi (mg)	Zinc (mg)	VT-A (RAE)	Thia (mg)	VT-E (mg α)	Ribo (mg)	Niac (mg)	V-B₆ (mg)	Fola (µg)	VT-C (mg)
40	100	2.7		200	770									5
65	200	3.6		250	2140									9
80	250	4.5		290	1100									1
105	200	4.5		480	1180									9
75	150	4.5		490	880									9
105	250	4.5		530	1340									9
15	100	1.08	40	235	390	1.38								0
30	200	1.44	49	365	630	1.8								2
15	100	4.5		620	1730									24
0	10	0.72		590	710		0							6
0	10	0.18		100	250		0							0
85	350	0		620	300									0
0	10	1.8		85	510									10
135	40	1.08			1116									1
75	20	0.72			422									1
95	20	0.72			747									1
55	20	0.36			414									1
162	60	1.08			1170									1
126	20	0.72			630									1
55	20	0.72			350									1
160	20	1.08			874									1
77	20	0.72			375									1
118	20	1.08			625									1
55	20	0.36			415									1
5	86	1.93			814									1
0	20	1.08			560									1
8	30	0.36			285									34
0	20	0.36			20		5							4
150	40	1.44			1230									1
10	150	0.36			860									1
1	20	0.36			440									1
15	20	2.7			540									1
5	20	1.79			744		5							1
85	300	4.5	46	455	1090	4.8		0.49	1.01	0.44	6.07	0.25	49	4
50	200	1.8	34	286	890	0.76			1.64					0
70	199	4.47	34	405	835	4.66		0.39	0.36	0.32	6.78	0.24	27	2
95	348	4.48			1303				0.81		6.78		33	2
35	9	0.7	16	199	355	0.65	0		0.93		4.87			0
5	7	0.72		27	240			<.01		<.01	0.14			0
0	3	0		45	250		0	<.01		<.01	0.15			4
0	2	0.14		7	140			0		<.01	0.07			0
40	350	0.72		543	250									1
40	350	0.72		542	180									6
40	350	0.36		533	250									1

A

TABLE A-1

Food Composition   (Computer code number is for West Diet Analysis program)   (For purposes of calculations, use "0" for t, <1, <.1, <.01, etc.)

Computer Code Number	Food Description	Measure	Wt (g)	H₂O (%)	Ener (kcal)	Prot (g)	Carb (g)	Dietary Fiber (g)	Fat (g)	Fat Breakdown (g) Sat	Mono	Poly
	FAST FOOD RESTAURANTS—Continued											
2170	Hot caramel	1 ea	182		360	7	61	0	10	6		
2171	Hot fudge	1 ea	179	59	340	8	52	1	12	9		
2172	Strawberry	1 ea	178		290	7	50		7	5		
2166	Vanilla	1 ea	90		150	4	23	0	4	3		
47147	Cookies, McDonaldland	1 ea	42	13	169	2	28	1	6	1.5		
47146	Chocolaty chip	1 ea	35		175	2	23	1	9	5		
42333	Muffin, apple bran, fat-free	1 ea	114		300	6	61	3	3	0.5		
48136	Pie, apple	1 ea	77		260	3	34	1	13	3.5		
	Breakfast items:											
42064	English muffin with spread	1 ea	63	33	189	5	30	2	6	2.4	1.5	1.3
69005	Egg McMuffin	1 ea	137		292	17	27	1	12	4.5		
45069	Hotcakes with marg & syrup	1 ea	228		600	9	104	0	17	3		
19579	Scrambled eggs	1 ea	102		160	13	1	0	11	3.5		
12230	Pork sausage	1 ea	43		170	6	0	0	16	5		
69006	Sausage McMuffin	1 ea	112		360	13	26	1	23	8		
69007	Sausage McMuffin with egg	1 ea	163	52	443	19	27	1	28	10.1		
42332	Biscuit with biscuit spread	1 ea	84	33	292	5	37	1	13	3		
69003	Biscuit with sausage	1 ea	127	36	465	11	34	1	32	9.1		
69004	Biscuit with sausage & egg	1 ea	178	49	538	18	34	1	36	11		
69002	Biscuit with bacon, egg, cheese	1 ea	157	44	496	21	32	1	32	10.3		
56479	Garden salad	1 ea	149		100	7	4	2	6	3		
	Source: McDonald's Corporation											
	PIZZA HUT											
	Pan pizza:											
56481	Cheese	2 pce	216		569	24	55	4	27	11.8		
56482	Pepperoni	2 pce	208		549	22	55	4	27	9.8		
56483	Supreme	2 pce	273		657	27	60	6	35	12.3		
56484	Super supreme	2 pce	286		680	28	60	6	36	12		
	Thin 'n crispy pizza:											
56485	Cheese	2 pce	174		409	20	45	4	18	10.2		
56486	Pepperoni	2 pce	168		394	19	44	4	19	8.3		
56487	Supreme	2 pce	232		496	24	46	4	26	11.9		
56488	Super supreme	2 pce	247		532	25	44	4	28	11.4		
	Hand tossed pizza:											
56489	Cheese	2 pce	216		489	24	57	4	20	10.2		
56490	Pepperoni	2 pce	208		502	23	50	4	23	10.8		
56492	Super supreme	2 pce	286		599	29	64	7	27	11.1		
56493	Pepperoni personal pan pizza	1 ea	255		615	26	69	5	28	10.9		
	Source: Pizza Hut.											
	SUBWAY											
	Deli style sandwich:											
69102	Ham	1 ea	171		253	13	42	4	5	1.8		
69103	Roast beef	1 ea	180		262	15	42	4	5	2.4		
69107	Tuna with light mayo	1 ea	178		350	14	38	3	17	4.8		
69101	Turkey	1 ea	180		262	15	43	4	4	1.8		
	Sandwiches, 6 inch:											
69117	Club on white bread	1 ea	246		309	23	44	4	6	1.9		
69113	Cold cut trio on white bread	1 ea	246		421	20	45	4	20	6.7		
69115	Ham on white bread	1 ea	232		290	18	46	4	5	1.5		
69139	Italian B.M.T. on white bread	1 ea	246		476	23	46	4	24	8.9		
69129	Meatball on white bread	1 ea	260		480	22	48	5	24	9.1		
69127	Melt with turkey, ham, bacon, cheese, on white bread	1 ea	251		397	23	46	4	16	5.8		
69121	Roast Beef on white bread	1 ea	232		303	20	47	4	5	2.1		

Chol (mg)	Calc (mg)	Iron (mg)	Magn (mg)	Pota (mg)	Sodi (mg)	Zinc (mg)	VT-A (RAE)	Thia (mg)	VT-E (mg α)	Ribo (mg)	Niac (mg)	V-B6 (mg)	Fola (µg)	VT-C (mg)
35	250				180									1
30	250	0.72			170								1	
30	200	0.36			95									1
20	100	0.36			75									1
0	15	1.33	8	46	184	0.29	0		0.74		1.5			0
25	12	0.9	15	89	106	0.25		0.09	0.58	0.1	0.92			0
0	100	1.44	20	117	380	0.5	0	0.22	0	0.22	2.01	0.04	8	1
0	20	1.08	7	63	200	0.21		0.18	1.38	0.11	1.42	0.03	8	24
13	103	1.59	13	69	386	0.42	32	0.25	0.13	0.32	2.61	0.04	57	1
237	201	2.72	24	199	796	1.56		0.49	0.85	0.44	3.32	0.15	33	1
20	100	4.5	28	292	770	0.55			1.23				<1	0
425	40	1.08	10	126	170	1.06	150	0.07	0.92	0.51	0.06	0.12	44	0
35	7	0.36	7	102	290	0.78	0	0.18	0.26	0.06	1.7	0.09		0
45	200	1.8	22	191	740	1.51			0.66				16	0
257	252	2.72	26	251	895	2.06			1.11				30	0
0	49	2.19	10	116	779	0.33			0.89		2.46		5	0
40	45	2.7	16	221	1055	1.08		0.51	1.13	0.31	4.19	0.13	5	0
269	88	2.97	21	283	1110	1.68			1.59		4.14		28	0
258	155	2.79	21	253	1456	1.69			1.54		3.43		31	0
75	150	1.08			120	75								15
20	393	3.53			1159									5
29	196	3.53			1197									5
41	308	3.69			1375									12
50	300	3.6			1560									12
20	409	2.95			1208									5
31	207	2.99			1265									5
40	297	3.57			1408									18
48	285	3.42			1596									17
20	408	2.93			1325									5
36	359	3.23			1417									43
44	333	3.99			1618									13
30	298	4.46			1419									6
12	72	4.34			927									14
18	72	6.44			787									14
26	159	3.81			879									13
18	72	4.29			870									14
34	58	5.21			1254									20
53	144	5.17			1608									23
25	60	3.6			1270									21
55	149	3.57			1885									24
50	136	4.89			1232									24
44	145	3.49			1677									23
21	63	6.58			951									22

A

TABLE A–1

Food Composition (Computer code number is for West Diet Analysis program) (For purposes of calculations, use "0" for t, <1, <.1, <.01, etc.)

Computer Code Number	Food Description	Measure	Wt (g)	H₂O (%)	Ener (kcal)	Prot (g)	Carb (g)	Dietary Fiber (g)	Fat (g)	Fat Breakdown (g) Sat	Mono	Poly
	FAST FOOD RESTAURANTS—Continued											
69125	Roasted chicken breast on white bread	1 ea	246		334	24	49	5	5	2.1		
69147	Seafood and crab with light mayo on white bread	1 ea	246		396	15	50	5	15	4.3		
69119	Steak and cheese on white bread	1 ea	257		392	24	48	5	14	5		
69143	Tuna with light mayo on white bread	1 ea	246		434	19	44	4	21	5.8		
69111	Turkey on white bread	1 ea	232		293	19	48	4	5	1.6		
69137	Turkey and ham on white bread	1 ea	232		290	20	46	4	5	1.5		
69109	Veggie delite on white bread	1 ea	175		242	9	46	4	3	1.1		
	Salads:											
52124	B.M.T., classic italian	1 ea	331		302	17	12	3	20	8.6		
52115	Club	1 ea	331		154	17	12	3	4	1.5		
52120	Cold cut trio	1 ea	330		240	15	11	3	16	6.3		
52123	Ham	1 ea	316		120	12	12	3	3	1.1		
52129	Meatball	1 ea	345		319	17	17	4	20	9		
52131	Melt	1 ea	336		211	18	12	3	11	4.8		
52126	Roast beef	1 ea	316		131	13	11	3	3	1.6		
52119	Roasted chicken breast	1 ea	331		153	17	13	3	3	1.1		
52116	Seafood and crab with light mayo	1 ea	331		211	9	18	4	12	3.7		
52130	Steak and cheese	1 ea	342		195	18	13	4	9	3.8		
52118	Tuna with light mayo	1 ea	331		253	14	11	3	17	4.2		
52113	Veggie delite	1 ea	260		56	2	10	3	1	0		
	Cookies:											
47658	With M&M's	1 ea	48	5	235	2	32	1	11	4.3		
47659	Chocolate chunk	1 ea	48		235	2	32	1	11	4.3		
47656	Oatmeal raisin	1 ea	48		213	3	32	2	9	2.7		
47657	Peanut butter	1 ea	48	5	235	4	28	1	13	4.3		
47660	Sugar	1 ea	48		245	2	30	0	13	4.3		
47661	White chip macademia	1 ea	48		235	2	30	1	12	4.3		
	Source: Subway International											
	TACO BELL											
	Burritos:											
56519	Bean with red sauce	1 ea	198		370	14	55	8	10	3.5		
56688	Chicken burrito	1 ea	171		283	14	34	3	10	4.1		
56522	Supreme with red sauce	1 ea	255		452	19	52	7	19	8.2		
56691	7 layer burrito	1 ea	283		530	18	67	10	22	8		
	Tacos:											
56525	Soft taco	1 ea	90		191	9	19	2	9	4.1		
56526	Soft taco supreme	1 ea	142		276	12	23	3	15	7.4		
56689	Soft taco, chicken	1 ea	121		232	17	23	1	7	3.1		
56693	Soft taco, steak	1 ea	128		282	12	21	1	17	4.5		
56528	Tostada with red sauce	1 ea	177		260	11	30	7	10	4.2		
56531	Mexican pizza	1 ea	220		560	21	47	7	32	11.2		
56537	Taco salad with salsa	1 ea	539		799	31	74	13	42	15.2		
56533	Nachos, regular	1 ea	99	40	320	5	33	2	19	4.5		
56534	Nachos, bellgrande	1 ea	312		790	20	81	12	44	13.2		
56536	Pintos & cheese with red sauce	1 ea	120	69	169	9	19	6	7	3.3		
53186	Taco sauce, packet	1 ea	11		4	0	<1	0	0	0	0	0
45585	Cinnamon twists	1 ea	28		128	1	22	0	4	0.8		
	Source: Taco Bell Corporation											
	WENDY'S											
56571	Cheeseburger, bacon	1 ea	166		382	20	34	2	19	7		
69059	Chicken sandwich, grilled	1 ea	189		302	24	36	2	7	1.5		
	Baked potatoes:											
6167	Plain	1 ea	284		310	7	72	7	0	0	0	0

PAGE KEY: A–4 = Beverages   A–6 = Dairy   A–10 = Eggs   A–12 = Fat/Oil   A–14 = Fruit   A–20 = Bakery
A–26 = Grain   A–32 = Fish   A–32 = Meats   A–36 = Poultry   A–38 = Sausage   A–40 = Mixed/Fast   A–44 = Nuts/Seeds   A–46 = Sweets
A–48 = Vegetables/Legumes   A–58 = Vegetarian   A–62 = Misc   A–62 = Soups/Sauces   A–64 = Fast   A–78 = Convenience   A–80 = Baby foods

Chol (mg)	Calc (mg)	Iron (mg)	Magn (mg)	Pota (mg)	Sodi (mg)	Zinc (mg)	VT-A (RAE)	Thia (mg)	VT-E (mg α)	Ribo (mg)	Niac (mg)	V-B$_6$ (mg)	Fola (μg)	VT-C (mg)
47	63	5.63			1042									22
24	145	3.47			1235									23
35	151	8.13			1215									24
39	14	3.47			1148									23
21	63	3.76			1055									22
25	60	3.6			1220									21
0	63	3.8			538									22
59	108	1.16			1714									32
36	41	18.5			1141									31
57	157	1.88			1431									31
27	44	1.18			1170									33
55	100	1.79			1047									36
48	106	1.14			1490									32
22	44	1.97			787									33
49	44	1.18			874									33
26	105	1.14			1023									32
38	109	3.91			966									33
42	105	1.14			928									32
0	45	1.21			346									33
16	0	1.15			112		0							0
11	0	1.15			112		0							0
16	0	1.15			192		0							0
11	0	1.15			213									0
16	0	1.15			144		0							0
16	0	1.15			171									0
10	200	2.7			1200									5
31	138	1.86			876									6
41	206	2.78			1368									9
25	300	3.6			1360									5
23	91	1.64			564									2
42	159	1.91			668									5
37	122	1.32			672									1
30	101	1.45			655									4
16	156	1.5			739									5
46	356	3.67			1049									6
66	405	6.37			1689									21
5	80	0.72			530		0							0
35	203	2.74			1317									6
14	141	1.01			656									3
0	0	0			102									0
0	0	0.29			120		0							0
55	151	3.62		322	895									9
55	80	2.71		432	744									9
0	20	3.6		1190	25		0							36

TABLE A-1

Food Composition (Computer code number is for West Diet Analysis program)   (For purposes of calculations, use "0" for t, <1, <.1, <.01, etc.)

Computer Code Number	Food Description	Measure	Wt (g)	H₂O (%)	Ener (kcal)	Prot (g)	Carb (g)	Dietary Fiber (g)	Fat (g)	Fat Breakdown (g) Sat	Mono	Poly
	FAST FOOD RESTAURANTS—Continued											
56579	With bacon & cheese	1 ea	380		580	18	79	7	22	6		
56580	With broccoli & cheese	1 ea	411		480	9	81	9	14	3		
56582	With sour cream & chives	1 ea	439		521	10	103	10	8	5.6		
50311	Chili	1 ea	227		200	17	21	5	6	2.5		
2176	Frosty dairy dessert	1 ea	227		330	8	56	0	8	5		
	Source: Wendy's International											
	CONVENIENCE FOODS & MEALS											
	HAAGEN DAZS											
70642	Ice cream bar, vanilla almond	1 ea	87		304	5	21	1	22	11.5		
70645	Lemon sorbet	½ c	113		120	0	31	<1	0	0	0	0
70646	Raspberry	½ c	105		120	0	30	2	0	0	0	0
	Source: Pillsbury											
	HEALTHY CHOICE											
	Entrees:											
18825	Fish, lemon pepper	1 ea	303		320	14	50	5	7	2		
81039	Lasagna	1 ea	383		420	26	59	6	9	3		
11119	Meatloaf, traditional	1 ea	340	78	316	15	52	6	5	2.5	1.9	0.6
	Low-Fat ice cream:											
2184	Brownie	½ c	71		120	3	22	1	2	1	0.3	0.7
2185	Chocolate chip	½ c	71		120	3	21	1	2	1	1	0
2105	Cookie & cream	½ c	71		120	3	21	1	2	1	1	0
2123	Rocky road	½ c	71		140	3	28	1	2	1	1	0
	Source: ConAgra Frozen Foods, Omaha, NE											
	HEALTH VALLEY											
	Soups, fat-free:											
50355	Beef broth, no salt added	1 c	240		18	5	0	0	0	0	0	0
50366	Beef broth, w/salt	1 c	240		30	5	2	0	0	0	0	0
50363	Black bean & vegetable	1 c	240		110	11	24	12	0	0	0	0
50364	Chicken broth	1 c	240		30	6	0	0	0	0	0	0
50365	14 garden vegetable	1 c	240		80	6	17	4	0	0	0	0
50362	Lentil & carrot	1 c	240		90	10	25	14	0	0	0	0
50361	Split pea & carrot	1 c	240		110	8	17	4	0	0	0	0
50360	Tomato vegetable	1 c	240		80	6	17	5	0	0	0	0
	Source: Health Valley											
	LA CHOY											
83016	Egg rolls, mini, chicken, svg	1 ea	106		108	3	13	1	5	1.3		
83013	Egg rolls, mini, shrimp, svg	1 ea	106		98	3	14	1	3	0.8		
	Source: Beatrice/Hunt-Wesson											
	LEAN CUISINE											
	Dinners:											
56901	Baked cheese ravioli	1 ea	241		260	12	38	4	7	3.5	1.5	0.5
15964	Chicken chow mein	1 ea	255		240	14	37	3	3	1	1.5	0.5
56740	Lasagna	1 ea	291		293	19	37	4	8	4.4	2	1
56702	Macaroni & cheese	1 ea	255		261	13	38	2	6	3.6	1.3	0.4
56732	Spaghetti w/meatballs	1 ea	269	75	299	18	40	5	8	2.1	2.7	1.3
56734	French bread sausage pizza	1 ea	170		210	8	24	1	9	3.5		
	Source: Stouffer's Foods Corp, Solon OH											

Chol (mg)	Calc (mg)	Iron (mg)	Magn (mg)	Pota (mg)	Sodi (mg)	Zinc (mg)	VT-A (RAE)	Thia (mg)	VT-E (mg α)	Ribo (mg)	Niac (mg)	V-B₆ (mg)	Fola (µg)	VT-C (mg)
40	200	3.6		1410	950									
5	200	4.5		1400	510									42
21	84	5.07		1731	56									72
35	80	1.8		470	870									51
35	300	1.08		590	200									2
														0
74	123	0.59		180	66					0.15				0
0	0	0		30	5	0								4
0	0	0		56	0	0								2
30	20	1.08			480									30
35	150	3.6		500	580			0.3		0.26	2			6
37	48	2.24			459									55
5	100	0		268	55									0
5	100	0		240	50									0
5	100			254	90			0.03		0.15				0
5	100	0		168	60			0.03		0.15				0
0				196	74						0.98			
0	0	0			160	0								5
0	40	3.6			280									9
0	20	1.8			170	0								1
0	40	1.8			250									15
0	60	5.4			220									2
0	40	5.4			230									9
0	40	5.4			240									9
8	10	0.56			335									
5	10	0.56			377									0
														0
35	150	1.44	42	450	590	1.5		0.06		0.26	1.2	0.2	48	5
35	40	0.72	30	300	590	1.1		0.15		0.17	5			0
29	244	1.41	44	596	576	2.9		0.15		0.25	3	0.32		6
18	180	0.65		423	567		0	0.12		0.25	1.2			0
5	94	2.37		539	465									6
10	100	2.7	39	165	630	2.2		0.45		0.51	0.05	0.07		1

TABLE A-1

Food Composition (Computer code number is for West Diet Analysis program)   (For purposes of calculations, use "0" for t, <1, <.1, <.01, etc.)

Computer Code Number	Food Description	Measure	Wt (g)	H$_2$O (%)	Ener (kcal)	Prot (g)	Carb (g)	Dietary Fiber (g)	Fat (g)	Fat Breakdown (g) Sat	Mono	Poly
	CONVENIENCE FOODS & MEALS—Continued											
	TASTE ADVENTURE SOUPS											
50325	Black bean	1 c	242		807	51	148	36	4	0.9	0.3	1.5
50324	Curry lentil	1 c	241		795	66	135	71	3	0.4	0.5	1.2
50326	Lentil chili	1 c	242		411	24	75	15	2			
50323	Split pea	1 c	244		807	58	143	60	3	0.4	0.6	1.2
	Source: Taste Adventure Soups											
	BABY FOODS											
60465	Apple juice	½ c	125	88	59	0	15	<1	<1	t	t	t
60475	Applesauce, strained	1 tbsp	16	87	8	<1	2	<1	<1	t	t	t
60502	Carrots, strained	1 tbsp	14	92	4	<1	1	<1	<1	t	t	t
60563	Cereal, mixed, milk added	1 tbsp	15	75	17	1	2	<1	1	0.3		
60622	Cereal, rice, milk added	1 tbsp	15	75	17	1	3	<1	1	0.3		
60515	Chicken and noodles, strained	1 tbsp	16	86	11	<1	1	<1	<1	t	0.1	t
60603	Peas, strained	1 tbsp	15	88	6	1	1	<1	<1	t	t	t
60634	Teething biscuits	1 ea	11	6	43	1	8	<1	<1	0.2	0.2	t

PAGE KEY: A–4 = Beverages   A–6 = Dairy   A–10 = Eggs   A–12 = Fat/Oil   A–14 = Fruit   A–20 = Bakery
A–26 = Grain   A–32 = Fish   A–32 = Meats   A–36 = Poultry   A–38 = Sausage   A–40 = Mixed/Fast   A–44 = Nuts/Seeds   A–46 = Sweets
A–48 = Vegetables/Legumes   A–58 = Vegetarian   A–62 = Misc   A–62 = Soups/Sauces   A–64 = Fast   A–78 = Convenience   A–80 = Baby foods

Chol (mg)	Calc (mg)	Iron (mg)	Magn (mg)	Pota (mg)	Sodi (mg)	Zinc (mg)	VT-A (RAE)	Thia (mg)	VT-E (mg α)	Ribo (mg)	Niac (mg)	V-B$_6$ (mg)	Fola (µg)	VT-C (mg)
0	296	12.18	405	3521	1978	8.67	39	2.12	0.12	0.48	4.8	0.84	1043	1
0	140	22.04	256	2170	2182	8.51	48	1.12	0.72	0.6	6.33	1.25	1005	16
			1476	1016										
0	140	10.65	272	2324	1729	7.14		1.71	2.75	0.51	6.84	0.42	646	5
0	5	0.71	4	114	4	0.04	1	<.01	0.75	0.02	0.1	0.04	0	72
0	1	0.04	0	11	<1	<.01	<1	<.01	0.03	<.01	<.01	<.01	<1	6
0	3	0.05	1	27	5	0.02	80	<.01	0.15	<.01	0.06	0.01	2	1
2	33	1.56	4	30	7	0.11	4	0.06		0.09	0.87	<.01	2	1
2	36	1.83	7	28	7	0.1	3	0.07		0.08	0.78	0.02	1	0
3	4	0.1	2	22	4	0.09	18	<.01	0.03	<.01	0.12	0.01	2	0
0	3	0.14	2	17	1	0.05	3	0.01	0.01	<.01	0.15	0.01	4	1
0	29	0.39	4	36	40	0.1	3	0.03	0.03	0.06	0.48	0.01	5	1

A

Reliable Sources of Nutrition Information

Many sources of nutrition information are available to consumers, but the quality of the information they provide varies widely. All of the sources listed here provide scientifically based information.

Expert Advice

Registered dietitians (hospitals and the yellow pages) Public health nutritionists (public health departments) College nutrition instructors/professors (colleges and universities) Extension Service home economists (state and county U.S. Department of Agriculture Extension Service offices) Consumer affairs staff of the Food and Drug Administration (national, regional, and state FDA offices)

You can find hundreds of toll-free telephone numbers for health information through the following Website: www.healthfinder.gov. After you are connected, search the term toll-free numbers.

U.S. Government

- **Federal Trade Commission (FTC)**
  Public Reference Branch
  (202) 326-2222
  www.ftc.gov

- **Food and Drug Administration (FDA)**
  Office of Consumer Affairs, HFE 1
  Room 16-85
  5600 Fishers Lane
  Rockville, MD 20857
  (301) 443-1544
  www.fda.gov

- **FDA Consumer Information Line**
  (301) 827-4420

- **FDA Office of Food Labeling, HFS 150**
  Washington, DC 20204
  (202) 205-4561; fax (202) 205-4564
  www.cfsan.fda.gov

- **FDA Office of Plant and Dairy Foods and Beverages**
  HFS 300
  200 C Street SW
  Washington, DC 20204
  (202) 205-4064; fax (202) 205-4422

- **FDA Office of Special Nutritionals,**
  HFS 450
  200 C Street SW
  Washington, DC 20204
  (202) 205-4168; fax (202) 205-5295

- **Food and Nutrition Information Center**
  National Agricultural Library,
  Room 304
  10301 Baltimore Avenue
  Beltsville, MD 20705-2351
  (301) 504-5719; fax (301) 504-6409
  www.nal.usda.gov/fnic

- **Food Research Action Center (FRAC)**
  1875 Connecticut Avenue NW,
  Suite 540
  Washington, DC 20009
  (202) 986-2200; fax (202) 986-2525

- **Superintendent of Documents**
  U.S. Government Printing Office
  Washington, DC 20402
  (202) 512-1071
  www.access.gpo.gov/su_docs

- **U.S. Department of Agriculture (USDA)**
  14th Street SW and Independence
  Avenue
  Washington, DC 20250
  (202) 720-2791
  www.usda.gov/fcs

- **USDA Center for Nutrition Policy and Promotion**
  1120 20th Street NW, Suite 200
  North Lobby
  Washington, DC 20036
  (202) 208-2417
  www.usda.gov/fcs/cnpp.htm

- **USDA Food Safety and Inspection Service**
  Food Safety Education Office,
  Room 1180-S
  Washington, DC 20250
  (202) 690-0351
  www.usda.gov/fsis

- **U.S. Department of Education (DOE)**
  Accreditation Agency Evaluation
  Branch
  7th and D Street SW
  ROB 3, Room 3915
  Washington, DC 20202-5244
  (202) 708-7417

- **U.S. Department of Health and Human Services**
  200 Independence Avenue SW
  Washington, DC 20201
  (202) 619-0257
  www.os.dhhs.gov

- **U.S. Environmental Protection Agency (EPA)**
  401 Main Street SW
  Washington, DC 20460
  (202) 260-2090
  www.epa.gov

- **U.S. Public Health Service**
  Assistant Secretary of Health
  Humphrey Building, Room 725-H
  200 Independence Avenue SW
  Washington, DC 20201
  (202) 690-7694

Health Canada

Headquarters

- **Health Canada**
  A.L. 0900C2
  Ottawa, Canada
  K1A 0K9
  Telephone: (613) 957-2991
  TTY: 1-800-267-1245
  http://www.hc-sc.gc.ca/

Regional Headquarters

- **British Columbia/Yukon**
  Suite 405, Winch Building
  757 West Hastings Street
  Vancouver, BC
  V6C 1A1
  Tel: (604) 666-2083
  Fax: (604) 666-2258

- **Alberta/NWT**
  Suite 710, Canada Place
  9700 Jasper Avenue
  Edmonton, AB
  T5J 4C3
  Tel: (780) 495-2651
  Fax: (780) 495-3285
- **Manitoba/Saskatchewan**
  391 York Avenue, Suite 425
  Winnipeg, MB
  R3C 0P4
  Tel: (204) 983-2508
  Fax: (204) 983-3972
- **Ontario/Nunavut**
  25 St. Clair Avenue East, 4th Floor
  Toronto, ON
  M4T 1M2
  Tel: (416) 973-4389
  Toll free: 1-866-999-7612
  Fax: (416) 973-1423
- **Quebec**
  Room 218, Complexe Guy-Favreau
  East Tower
  200 René Lévesque Blvd. West
  Montreal, QC
  H2Z 1X4
  Tel: (514) 283-2306
  Fax: (514) 283-6739
- **Atlantic**
  Suite 1525, 15th Floor,
  Maritime Centre
  1505 Barrington Street
  Halifax, NS B3J 3Y6
  Tel: (902) 426-2700
  Fax: (902) 426-9689

International Agencies

- **Food and Agriculture Organization of the United Nations (FAO)**
  Liaison Office for North America
  2175 K Street, Suite 300
  Washington, DC 20437
  (202) 653-2400
  www.fao.org
- **International Food Information Council Foundation**
  1100 Connecticut Avenue NW,
  Suite 430
  Washington, DC 20036
  (202) 296-6540
  ificinfo.health.org
- **UNICEF**
  3 United Nations Plaza
  New York, NY 10017
  (212) 326-7000
  www.unicef.com
- **World Health Organization (WHO)**
  Regional Office
  525 23rd Street NW
  Washington, DC 20037
  (202) 974-3000
  www.who.org

Professional Nutrition Organizations

- **American Dietetic Association (ADA)**
  216 West Jackson Boulevard,
  Suite 800
  Chicago, IL 60606-6995
  (800) 877-1600; (312) 899-0040
  www.eatright.org
- **ADA, The Nutrition Hotline**
  (800) 366-1655
- **American Society for Clinical Nutrition**
  9650 Rockville Pike
  Bethesda, MD 20814-3998
  (301) 530-7110; fax (301) 571-1863
  www.faseb.org/ascn
- **Dietitians of Canada**
  480 University Avenue, Suite 604
  Toronto, Ontario M5G 1V2, Canada
  (416) 596-0857; fax (416) 596-0603
  www.dietitians.ca
- **Human Nutrition Institute (INACG)**
  1126 Sixteenth Street NW
  Washington, DC 20036
  (202) 659-0789
  www.ilsi.org
- **National Academy of Sciences/ National Research Council (NAS/NRC)**
  2101 Constitution Avenue, NW
  Washington, DC 20418
  (202) 334-2000
  www.nas.edu
- **National Institute of Nutrition**
  265 Carling Avenue, Suite 302
  Ottawa, Ontario K1S 2E1
  (613) 235-3355; fax (613) 235-7032
  www.nin.ca
- **Society for Nutrition Education**
  7101 Wisconsin Avenue, Suite 901
  Bethesda, MD 20814-4805
  (301) 656-4938

Aging

- **Administration on Aging**
  330 Independence Avenue SW
  Washington, DC 20201
  (202) 619-0724
  www.aoa.dhhs.gov
- **American Association of Retired Persons (AARP)**
  601 E Street NW
  Washington, DC 20049
  (202) 434-2277
  www.aarp.org
- **National Aging Information Center**
  330 Independence Avenue SW
  Washington, DC 20201
  (202) 619-7501
  www.aoa.dhhs.gov/naic

- **National Institute on Aging**
  Public Information Office
  31 Center Drive, MSC 2292
  Bethesda, MD 20892
  (301) 496-1752
  www.nih.gov/nia

Alcohol and Drug Abuse

- **Al-Anon Family Group Headquarters, Inc.**
  1600 Corporate Landing Parkway
  Virginia Beach, VA 23454-5617
  (800) 356-9996
  www.al-anon.alateen.org
- **Alateen**
  1600 Corporate Landing Parkway
  Virginia Beach, VA 23454-5617
  (800) 356-9996
  www.al-anon.alateen.org
- **Alcoholics Anonymous (AA)**
  General Service Office
  475 Riverside Drive
  New York, NY 10115
  (212) 870-3400
  www.aa.org
- **Narcotics Anonymous (NA)**
  P.O. Box 9999
  Van Nuys, CA 91409
  (818) 773-9999; fax (818) 700-0700
  www.wsoinc.com
- **National Clearinghouse for Alcohol and Drug Information (NCADI)**
  P.O. Box 2345
  Rockville, MD 20847-2345
  (800) 729-6686
  www.health.org
- **National Council on Alcoholism and Drug Dependence (NCADD)**
  12 West 21st Street
  New York, NY 10010
  (800) NCA-CALL or (800) 622-2255
  (212) 206-6770; fax (212) 645-1690
  www.ncadd.org
- **U.S. Center for Substance Abuse Prevention**
  1010 Wayne Avenue, Suite 850
  Silver Spring, MD 20910
  (301) 459-1591 ext. 244;
  fax (301) 495-2919
  www.covesoft.com/csap.html

Consumer Organizations

- **Center for Science in the Public Interest (CSPI)**
  1875 Connecticut Avenue NW,
  Suite 300
  Washington, DC 20009-5728
  (202) 332-9110; fax (202) 265-4954
  www.cspinet.org

- **Choice in Dying, Inc.**
  1035 30th Street NW
  Washington, DC 20007
  (202) 338-9790; fax (202) 338-0242
  www.choices.org
- **Consumer Information Center**
  Pueblo, CO 81009
  (888) 8 PUEBLO or (888) 878-3256
  www.pueblo.gsa.gov
- **Consumers Union of US Inc.**
  101 Truman Avenue
  Yonkers, NY 10703-1057
  (914) 378-2000
  www.consunion.org
- **National Council Against Health Fraud, Inc. (NCAHF)**
  P.O. Box 1276
  Loma Linda, CA 92354
  (909) 824-4690
  www.ncahf.org

Fitness

- **American College of Sports Medicine**
  P.O. Box 1440
  Indianapolis, IN 46206-1440
  (317) 637-9200
  _www.acsm.org/sportsmed
- **American Council on Exercise (ACE)**
  5820 Oberlin Drive, Suite 102
  San Diego, CA 92121
  (800) 529-8227
  www.acefitness.org
- **President's Council on Physical Fitness and Sports**
  Humphrey Building, Room 738
  200 Independence Avenue SW
  Washington, DC 20201
  (202) 690-9000; fax (202) 690-5211
  _www.indiana.edu/~preschal
- **Shape Up America!**
  6707 Democracy Boulevard, Suite 306
  Bethesda, MD 20817
  (301) 493-5368
  www.shapeup.org
- **Sport Medicine and Science Council of Canada**
  1600 James Naismith Drive, Suite 314
  Gloucester, Ontario K1B 5N4, Canada
  (613) 748-5671; fax (613) 748-5729
  www.smscc.ca

Food Safety

- **Alliance for Food & Fiber Food Safety Hotline**
  (800) 266-0200

- **FDA Center for Food Safety and Applied Nutrition**
  200 C Street SW
  Washington, DC 20204
  (800) FDA-4010 or (800) 332-4010
  vm.cfsan.fda.gov
- **National Lead Information Center**
  (800) LEAD-FYI or (800) 532-3394
  (800) 424-LEAD or (800) 424-5323
- **National Pesticide Telecommunications Network (NPTN)**
  Oregon State University
  333 Weniger Hall
  Corvallis, OR 97331-6502
  (541) 737-6091
  _www.ace.orst.edu/info/nptn
- **USDA Meat and Poultry Hotline**
  (800) 535-4555
- **U.S. EPA Safe Drinking Water Hotline**
  (800) 426-4791

Health and Disease

- **Alzheimer's Disease Education and Referral Center**
  P. O. Box 8250
  Silver Spring, MD 20907-8250
  (800) 438-4380
  www.alzheimers.org
- **Alzheimer's Disease Information and Referral Service**
  919 North Michigan Avenue, Suite 1000
  Chicago, IL 60611
  (800) 272-3900
  www.alz.org
- **American Academy of Allergy, Asthma, and Immunology**
  611 East Wells Street
  Milwaukee, WI 53202
  (414) 272-6071; fax (414) 276-3349
  www.aaaai.org
- **American Cancer Society National Home Office**
  1599 Clifton Road NE
  Atlanta, GA 30329-4251
  (800) ACS-2345 or (800) 227-2345
  www.cancer.org
- **American Council on Science and Health**
  1995 Broadway, 2nd Floor
  New York, NY 10023-5860
  (212) 362-7044; fax (212) 362-4919
  www.acsh.org
- **American Dental Association**
  211 East Chicago Avenue
  Chicago, IL 60611
  (312) 440-2800
  www.ada.org

- **American Diabetes Association**
  1660 Duke Street
  Alexandria, VA 22314
  (800) 232-3472 or (703) 549-1500
  www.diabetes.org
- **American Heart Association**
  Box BHG, National Center
  7320 Greenville Avenue
  Dallas, TX 75231
  (800) 275-0448 or (214) 373-6300
  www.amhrt.org
- **American Institute for Cancer Research**
  1759 R Street NW
  Washington, DC 20009
  (800) 843-8114 or (202) 328-7744; fax (202) 328-7226
  www.aicr.org
- **American Medical Association**
  515 North State Street
  Chicago, IL 60610
  (312) 464-5000
  www.ama-assn.org
- **American Public Health Association (APHA)**
  1015 Fifteenth Street NW, Suite 300
  Washington, DC 20005
  (202) 789-5600
  www.apha.org
- **American Red Cross**
  National Headquarters
  8111 Gatehouse Road
  Falls Church, VA 22042
  (703) 206-7180
  www.redcross.org
- **Canadian Diabetes Association**
  15 Toronto Street, Suite 800
  Toronto, ON M5C 2E3
  (800) BANTING or (800) 226-8464
  (416) 363-3373
  www.diabetes.ca
- **Canadian Public Health Association**
  400-1565 Carling Avenue
  Ottawa, Ontario K1Z 8R1
  (613) 725-3769; fax (613) 725-9826
  www.cpha.ca
- **Centers for Disease Control and Prevention (CDC)**
  1600 Clifton Road NE
  Atlanta, GA 30333
  (404) 639-3311
  www.cdc.gov
- **The Food Allergy Network**
  10400 Eaton Place, Suite 107
  Fairfax, VA 22030-2208
  (800) 929-4040 or (703) 691-3179
  www.foodallergy.org
- **Internet Health Resources**
  www.ihr.com

- **National AIDS Hotline (CDC)**
  (800) 342-AIDS (English)
  (800) 344-SIDA (Spanish)
  (800) 2437-TTY (Deaf)
  (900) 820-2437
- **National Cancer Institute**
  Office of Cancer Communications
  Building 31, Room 10824
  Bethesda, MD 20892
  (800) 4-CANCER or (800) 422-6237
  www.nci.nih.gov
- **National Diabetes Information Clearinghouse**
  1 Information Way
  Bethesda, MD 20892-3560
  (301) 654-3327
  www.niddk.nih.gov
- **National Digestive Disease Information Clearinghouse (NDDIC)**
  2 Information Way
  Bethesda, MD 20892-3570
  (301) 654-3810
  www.niddk.nih.gov
- **National Health Information Center (NHIC)**
  Office of Disease Prevention and
  Health Promotion
  (800) 336-4797
  nhic-nt.health.org
- **National Heart, Lung, and Blood Institute**
  Information Center
  P.O. Box 30105
  Bethesda, MD 20824-0105
  (301) 251-1222
  _www.nhlbi.nih.gov/nhlbi/nhlbi
  .htm
- **National Institute of Allergy and Infectious Diseases**
  Office of Communications
  Building 31, Room 7A50
  31 Center Drive, MSC2520
  Bethesda, MD 20892-2520
  (301) 496-5717
  www.niaid.nih.gov
- **National Institute of Dental Research (NIDR)**
  National Institute of Health
  Bethesda, MD 20892-2190
  (301) 496-4261
  www.nidr.nih.gov
- **National Institutes of Health (NIH)**
  9000 Rockville Pike
  Bethesda, MD 20892
  (301) 496-2433
  www.nih.gov
- **National Osteoporosis Foundation**
  1150 17th Street NW, Suite 500
  Washington, DC 20036
  (202) 223-2226
  www.nof.org

- **Office of Disease Prevention and Health Promotion**
  odphp.osophs.dhhs.gov
- **Office on Smoking and Health (OSH)**
  _www.americanheart.org/heart.org/
  Heart_and_stroke_A_Z_Guide/osh
  .html

Infancy and Childhood

- **American Academy of Pediatrics**
  141 Northwest Point Boulevard
  Elk Grove Village, IL 60007-1098
  (847) 228-5005
  www.aap.org
- **Association of Birth Defect Children, Inc.**
  930 Woodcock Road, Suite 225
  Orlando, FL 32803
  (407) 245-7035
  www.birthdefects.org
- **Canadian Paediatric Society**
  100-2204 Walkley Road
  Ottawa, ON K1G 4G8
  (613) 526-9397; fax (613) 526-3332
  www.cps.ca
- **National Center for Education in Maternal & Child Health**
  2000 15th Street North, Suite 701
  Arlington, VA 22201-2617
  (703) 524-7802
  www.ncemch.org

Pregnancy and Lactation

- **American College of Obstetricians and Gynecologists Resource Center**
  409 12th Street SW
  Washington, DC 20024-2188
  (202) 638-5577
  www.acog.org
- **La Leche International, Inc.**
  1400 N. Meacham Road
  Schaumburg, IL 60173
  (847) 519-7730
  www.lalecheleague.org
- **March of Dimes Birth Defects Foundation**
  1275 Mamaroneck Avenue
  White Plains, NY 10605
  (914) 428-7100
  www.sunkist.com

World Hunger

- **Bread for the World**
  1100 Wayne Avenue, Suite 1000
  Silver Spring, MD 20910
  (301) 608-2400
  www.bread.org

- **Center on Hunger, Poverty and Nutrition Policy**
  Tufts University School of Nutrition
  11 Curtis Avenue
  Medford, MA 02155
  (617) 627-3956
- **Freedom from Hunger**
  P.O. Box 2000
  1644 DaVinci Court
  Davis, CA 95617
  (530) 758-6200
  www.freefromhunger.org
- **Oxfam America**
  26 West Street
  Boston, MA 02111
  (617) 482-1211
  www.oxfamamerica.org
- **SEEDS Magazine**
  P.O. Box 6170
  Waco, TX 76706
  (254) 755-7745
  _www.helwys.com/seedhome.htm
- **Worldwatch Institute**
  1776 Massachusetts Avenue NW,
  Suite 800
  Washington, DC 20036
  (202) 452-1999
  www.worldwatch.org

Scientific Literature

Nutrition Journals

American Journal of Clinical Nutrition

British Journal of Nutrition

Human Nutrition, Applied Nutrition

Journal of the American College of Nutrition

Journal of the American Dietetic Association

Journal of the Canadian Dietetic Association

Journal of Food Composition and Analysis

Journal of Nutrition

Journal of Nutrition Education

Nutrition Abstracts and Reviews

Nutrition and Metabolism

Nutrition Reports International

Nutrition Research

Nutrition Reviews

Nutrition Today

B

Other Journals

American Journal of Epidemiology
American Journal of Nursing
American Journal of Public Health
Annals of Internal Medicine
Annals of Surgery
Canadian Journal of Public Health
Caries Research
Food Technology

Gastroenterology
International Journal of Obesity
Journal of the American Dental Association
Journal of the American Medical Association
Journal of Clinical Investigation
Journal of Food Science
Journal of Home Economics
Journal of Pediatrics

Lancet
New England Journal of Medicine
Nutrition Today
Pediatrics
Science
The Scientist

The U.S. Food Exchange System

The U.S. exchange system divides the foods suitable for use in planning a healthy diet into six lists—the starch/bread, meat/meat alternate, vegetable, fruit, milk, and fat lists.a These lists are shown in Tables C.1 through C.6. Following these lists are three other sets of foods: free foods, combination foods, and foods for occasional use (Tables C.7, C.8, and C.9).

[a]The Exchange Lists are the basis of a meal planning system designed by a committee of the American Diabetes Association and The American Dietetic Association. While designed primarily for people with diabetes and others who must follow special diets, the Exchange Lists are based on principles of good nutrition that apply to everyone. © 1989 American Diabetes Association, Inc., The American Dietetic Association.

TABLE C.1

The U.S.. Exchange System: Starch/Bread List

15 g carbohydrate, 3 g protein, trace fat, 80 cal

Amount	Food	Amount	Food
Cereals/Grains/Pasta		**Bread**	
$1/2$ c	Bran cereals, concentrated 🌾	$1/2$ (1 oz)	Bagels
$1/2$ c	Bran cereals, flaked 🌾	2 ($2/3$ oz)	Bread sticks, crisp, 4" × $3^1/_2$"
$1/2$ c	Bulgur, cooked	1 c	Croutons, low-fat
$1/2$ c	Cooked cereals	$1/2$	English muffins
$2^1/2$ tbs	Cornmeal, dry	$1/2$ (1 oz)	Frankfurter or hamburger buns
3 tbs	Grapenuts	$1/2$ loaf	Pita, 6" across
$1/2$ c	Grits, cooked	1 (1 oz)	Plain rolls, small
$3/4$ c	Other ready-to-eat unsweetened cereals	1 slice (1 oz)	Raisin, unfrosted
$1/2$ c	Pasta, cooked	1 slice (1 oz)	Rye, pumpernickel 🌾
$1^1/2$ c	Puffed cereals	1	Tortillas, 6" across
$1/3$ c	Rice, white or brown, cooked	1 slice (1 oz)	White (including French, Italian)
$1/2$ c	Shredded wheat	1 slice (1 oz)	Whole-wheat
3 tbs	Wheat germ 🌾	**Crackers/Snacks**	
Dried Beans/Peas/Lentils		8	Animal crackers
$1/4$ c	Baked beans 🌾	3	Graham crackers, $2^1/2$" square
$1/3$ c	Beans and peas, cooked, such as kidney, white, split, black-eyed 🌾	$3/4$ oz	Matzoth
		5 slices	Melba toast
$1/3$ c	Lentils, cooked 🌾	24	Oyster crackers
Starchy Vegetables		3 c	Popcorn, popped, no fat added
$1/2$ c	Corn 🌾	$3/4$ oz	Pretzels
1 cob	Corn on the cob, 6" long 🌾	4	Rye crisp, 2" × $3^1/2$"
$1/2$ c	Lima beans 🌾	6	Saltine-type crackers
$1/2$ c	Peas, green, canned or frozen 🌾	2 to 4 ($3/4$ oz)	Whole-wheat crackers, no fat added (crisp breads)
$1/2$ c	Plantains 🌾		
1 small (3 oz)	Potatoes, baked	**Starch Foods Prepared with Fat**	
$1/2$ c	Potatoes, mashed	(Count as 1 starch/bread serving, plus 1 fat serving.)	
$3/4$ c	Squash, winter (acorn, butternut)	1	Biscuits, $2^1/2$" across
$1/3$ c	Yams, sweet potatoes, plain	$1/2$ c	Chow mein noodles
		1 (2 oz)	Cornbread, 2" cube
		6	Crackers, round butter type
		10 ($1^1/2$ oz)	French fries, 2" to $3^1/2$" long
		1	Muffins, plain, small
		2	Pancakes, 4" across
		$1/4$ c	Stuffing, bread, prepared
		2	Taco shells, 6" across
		1	Waffles, $4^1/2$" square
		4 to 6 (1 oz)	Whole-wheat crackers, fat added

🌾 3 grams or more dietary fiber per serving. Average fiber contents of whole-grain products is 2 grams per serving. For starch foods not on this list, the general rule is that $1/2$ cup cereal, grain, or pasta is 1 serving: 1 ounce of a bread product is 1 serving.

TABLE C.2

U.S. Exchange System: Meat/Meat Alternate Lists

Lean meat = 7 g protein, 3 g fat, 55 cal; medium-fat meat = 7 g protein, 5 g fat, 75 cal; high-fat meat = 7 g protein, 8 g fat, 100 cal.

Category	Amount	Food
Lean Meat and Alternates		
Beef	1 oz	USDA Good or Choice grades of lean beef, such as round, sirloin, and flank steak; tenderloin; chipped beef ✏
Pork	1 oz	Leak pork, such as fresh ham; canned, cured, or boiled ham ✏, Canadian bacon ✏, tenderloin
Veal	1 oz	All cuts are lean except veal cutlets (ground or cubed); examples of lean veal: chops and roasts
Poultry	1 oz	Chicken, turkey, Cornish hen (without skin)
Fish	1 oz	All fresh and frozen fish
	2 oz	Crab, lobster, scallops, shrimp, clams (fresh or canned in water ✏)
	6 medium	Oysters
	¼ c	Tuna ✏, canned in water
	1 oz	Herring uncreamed or smoked
	2 medium	Sardines, canned
Wild game	1 oz	Venison, rabbit, squirrel
	1 oz	Pheasant, duck, goose (without skin)
Cheese	¼ c	Any cottage cheese
	2 tbs	Grated parmesan
	1 oz	Diet cheeses ✏ (with less than 55 cal/oz)
Other	1 oz	95% fat-free lunch meats
	3 whites	Egg whites
	¼ c	Egg substitutes with less than 55 cal per ¼ c
Medium-Fat Meat and Alternates		
Beef	1 oz	Most beef products fall into this category; examples: all ground beef, roasts (rib, chuck, rump), steak (cubed, porterhouse, T-bone), meatloaf
Pork	1 oz	Most pork products fall into this category; examples: chops, loin roast, Boston butt, cutlets
Lamb	1 oz	Most lamb products fall into this category; examples: chops, leg, roast

Category	Amount	Food
Medium-Fat Meat and Alternates (*continued*)		
Veal	1 oz	Cutlet, ground or cubed, unbreaded
Poultry	1 oz	Chicken (with skin), domestic duck or goose (well-drained of fat), ground turkey
Fish	¼ c	Tuna ✏, canned in oil and drained
	¼ c	Salmon ✏, canned
Cheese		Skim or part-skim milk cheeses, such as:
	¼ c	Ricotta
	1 oz	Mozzarella
	1 oz	Diet cheeses ✏ (with 56 to 80 cal/oz)
Other	1 oz	86% fat-free lunch meat ✏
	1	Eggs (high in cholesterol, limit to 3 per week)
	¼ c	Egg substitutes with 56 to 80 cal per ¼ c
	4 oz	Tofu, 2½" × 2¾" × 1"
	1 oz	Liver, hearts, kidneys, sweetbreads (high in cholesterol)
High-Fat Meat and Alternates[a]		
Beef	1 oz	Most USDA Prime cuts of beef, such as ribs, corned beef ✏
Pork	1 oz	Spareribs, ground pork, pork sausages ✏ (patties or links)
Lamb	1 oz	Patties, ground lamb
Fish	1 oz	Any fried fish product
Cheese	1 oz	All regular cheeses ✏, such as American, blue, Cheddar, Monterey, Swiss
Other	1 oz	Lunch meats ✏, such as bologna, salami, pimento loaf
	1 oz	Sausage ✏, such as Polish, Italian
	1 oz	Knockwurst, smoked
	1 oz	Bratwurst ✏
	1 (10/lb)	Frankfurters ✏ (turkey or chicken)
	1 tbs	Peanut butter (contains unsaturated fat)
Count as 1 high-fat meat plus 1 fat exchange:		
1 frank	(10/lb)	Frankfurters ✏ (beef, pork, or combination)

✏ 400 milligrams or more sodium per exchange. Meats contribute no fiber to the diet.
[a]These items are high in saturated fat, cholesterol, and calories and should be used no more than three times per week.

TABLE C.3

U.S. Exchange System: Vegetable List

5 g carbohydrate, 2 g protein, 25 cal
All portion sizes, except as otherwise noted, are $^1/_2$ c of any cooked vegetable or vegetable juice, 1 c of any raw vegetable.

Artichokes, $^1/_2$ medium	Cabbage, cooked	Leeks	Spinach, cooked
Asparagus	Carrots	Mushrooms, cooked	Summer squash (crookneck)
Bean sprouts	Cauliflower	Okra	Tomatoes, 1 large
Beans (green, wax, Italian)	Eggplant	Onions	Tomato/vegetable juice ✎
Beets	Green peppers	Pea pods	Turnips
Broccoli	Greens (collard, mustard,	Rutabagas	Water chestnuts
Brussels sprouts	turnip)	Sauerkraut ✎	Zucchini, cooked
	Kohlrabi		

Starchy vegetables such as corn, peas, and potatoes are found on the Starch/Bread List.
For free vegetables, see the Free Food List (Table D.7).

✎ 400 milligrams or more sodium per serving. Most vegetable servings contain 2 to 3 grams dietary fiber.

TABLE C.4

U.S. Exchange System: Fruit List

15 g carbohydrate, 60 cal
All portion sizes, unless otherwise noted, are $^1/_2$ c fresh fruit or fruit juice, $^1/_4$ c dried fruit.

Amount	Food	Amount	Food
Fresh, Frozen, and Unsweetened Canned Fruit		$^1/_2$ c (2 halves)	Pears, canned
		2	Persimmons, medium, native
1	Apples, raw, 2" across	$^3/_4$ c	Pineapple, raw
$^1/_2$ c	Applesauce, unsweetened	$^1/_3$ c	Pineapple, canned
4	Apricots, medium, raw	2	Plums, raw, 2" across
$^1/_2$ c (4 halves)	Apricots, canned	$^1/_2$	Pomegranates ✄
$^1/_2$	Bananas, 9" long	1 c	Raspberries, raw ✄
$^3/_4$ c	Blackberries, raw ✄	$1^1/_4$ c	Strawberries, raw, whole
$^3/_4$ c	Blueberries, raw ✄	2	Tangerines, $2^1/_2$" across
$^1/_3$	Cantaloupe, 5" across	$1^1/_4$ c	Watermelon, cubes
1 c	Cantaloupe, cubes		
12	Cherries, large, raw	**Dried Fruit**	
$^1/_2$ c	Cherries, canned	4 rings	Apples ✄
2	Figs, raw, 2" across	7 halves	Apricots ✄
$^1/_2$ c	Fruit cocktail, canned	$2^1/_2$ medium	Dates
$^1/_2$	Grapefruit, medium	$1^1/_2$	Figs ✄
$^3/_4$ c	Grapefruit, segments	3 medium	Prunes ✄
15	Grapes, small	2 tbs	Raisins
$^1/_8$	Honeydew melon, medium		
1 c	Honeydew melon, cubes	**Fruit Juice**	
1	Kiwis, large		
$^3/_4$ c	Mandarin oranges	$^1/_2$ c	Apple juice/cider
$^1/_2$	Mangoes, small	$^1/_3$ c	Cranberry juice cocktail
1	Nectarines, $1^1/_2$" across ✄	$^1/_3$ c	Grape juice
1	Oranges, $2^1/_2$" across	$^1/_2$ c	Grapefruit juice
1 c	Papayas	$^1/_2$ c	Orange juice
1 ($^3/_4$ c)	Peaches, $2^3/_4$" across	$^1/_2$ c	Pineapple juice
$^1/_2$ c (2 halves)	Peaches, canned	$^1/_3$ c	Prune juice
$^1/_2$ large or 1 small	Pears		

✄ 3 grams or more dietary fiber per serving. Average fiber contents of fresh, frozen, and dry fruits: 2 grams per serving.

TABLE C.5
U.S. Exchange System: Milk List

Nonfat and very low-fat milk = 12 g carbohydrate, 8 g protein, trace fat, 90 cal; low-fat milk = 12 g carbohydrate, 8 g protein, = g fat, 120 cal; whole milk = 12 g carbohydrate, 8 g protein, 8 g fat, 150 cal.

Amount	Food	Amount	Food
Nonfat and Very Low-Fat Milk		**Low-Fat Milk**	
1 c	Nonfat milk	1 c fluid	2% milk
1 c	1/4% milk	8 oz	Plain low-fat yogurt, with added nonfat milk solids
1 c	1% milk		
1/3 c	Dry nonfat milk	**Whole Milk**	
1/4 c	Evaporated nonfat milk	1 c	Whole milk
1 c	Low-fat buttermilk	1/4 c	Evaporated whole milk
8 oz	Plain nonfat yogurt	8 oz	Whole plain yogurt

TABLE C.6
U.S. Exchange System: Fat List

5 g fat, 45 cal

Amount	Food	Amount	Food
Unsaturated Fats		**Saturated Fats**	
1/8 medium	Avocados	1 slice	Bacon[a]
1 tsp	Margarine	1 tsp	Butter
1 tbs	Margarine, diet[a]	1/2 oz	Chitterlings
1 tsp	Mayonnaise	2 tbs	Coconut, shredded
1 tbs	Mayonnaise, reduced calorie[a]	2 tbs	Coffee whitener, liquid
	Nuts and seeds:	4 tsp	Coffee whitener, powder
6 whole	Almonds, dry roasted	1 tbs	Cream (heavy, whipping)
1 tbs	Cashews, dry roasted	2 tbs	Cream (light, coffee, table)
20 small or 10 large	Peanuts	2 tbs	Cream (sour)
2 whole	Pecans	1 tbs	Cream cheese
2 tsp	Pumpkin seeds	1/4 oz	Salt pork[a]
1 tbs	Other nuts		
1 tbs	Seeds, pine nuts, sunflower seeds (without shells)		
2 whole	Walnuts		
1 tsp	Oil (corn, cottonseed, safflower, soybean, sunflower, olive, peanut)		
10 small or 5 large	Olives[a]		
1 tbs	Salad dressing, all varieties[a]		
2 tsp	Salad dressing, mayonnaise type		
1 tbs	Salad dressing, mayonnaise type, reduced calorie		
2 tbs	Salad dressing, reduced calorie		

Two tablespoons of low-calorie salad dressing is a free food.

[a]If more than one or two servings are eaten, these foods provide 400 milligrams or more sodium.

400 milligrams or more sodium per serving.

TABLE C.7
U.S. Exchange System: Free Foods

A free food is any food or drink that contains less than 20 cal/serving. People with diabetes are advised to eat as much as they want of those items that have no serving size specified. They may eat two or three servings per day of those items that have a specific serving size. It is suggested that they spread the servings out through the day.

Amount	Food	Amount	Food
Drinks	Bouillon, low-sodium	**Condiments**	
	Bouillon 🖋 or broth without fat	1 tbs	Catsup
	Carbonated drinks, sugar-free		Horseradish
	Carbonated water		Mustard
	Club soda		Pickles 🖋, dill, unsweetened
1 tbs	Cocoa powder, unsweetened	2 tbs	Salad dressing, low-calorie
	Coffee/tea	1 tbs	Taco sauce
	Drink mixes, sugar-free		Vinegar
	Tonic water, sugar-free	**Seasonings**	Basil, fresh
Nonstick Pan Spray			Celery seeds
Fruit			Chili powder
1/2 c	Cranberries, unsweetened		Chives
1/2 c	Rhubarb, unsweetened		Cinnamon
Vegetables (raw, 1 c)	Cabbage		Curry
	Celery		Dill
	Chinese cabbage 🌿		Flavoring extracts (almond, butter, lemon,
	Cucumbers		peppermint, vanilla,
	Green onions		walnut, etc.)
	Hot peppers		Garlic
	Mushrooms		Garlic powder
	Radishes		Herbs
	Zucchini 🌿		Hot pepper sauce
Salad Greens	Endive		Lemon
	Escarole		Lemon juice
	Lettuce		Lemon pepper
	Romaine		Lime
	Spinach		Lime juice
Sweet Substitutes	Candy, hard, sugar-free		Mint
	Gelatin, sugar-free		Onion powder
	Gum, sugar-free		Oregano
2 tsp	Jam/jelly, sugar-free		Paprika
1 to 2 tbs	Pancake syrup, sugar-free		Pepper
	Sugar substitutes (saccharin, aspartame,		Pimento
	acesulfame-K)		Soy sauce 🖋
2 tbs	Whipped topping		Soy sauce, low-sodium ("lite")
			Spices
		1/4 c	Wine, used in cooking
			Worcestershire sauce

🌿 3 grams or more dietary fiber per serving.

🖋 400 milligrams or more sodium per serving.

Appendix C   The U.S. Food Exchange System

TABLE C.8

U.S. Exchange Combination Foods

Much of the food we eat is mixed together in various combinations. These combination foods do not fit into only one exchange list. It can be quite hard to tell what is in a certain casserole dish or baked food item. This is a list of average values for some typical combination foods. This list will help you fit these foods into your meal plan. Ask your dietitian for information about any other foods you'd like to eat. The *American Diabetes Association/American Dietetic Association Family Cookbooks* and the *American Diabetes Associates Holiday Cookbook* have many recipes and further information about many foods, including combination foods. Check your library or local bookstore.

Food	Amount	Exchanges
Casseroles, homemade	1 c (8 oz)	2 starch, 2 medium-fat meat, 1 fat
Cheese pizza 🖉, thin crust	3/4 of 15 oz or 1/4 of 10"	2 starch, 1 medium-fat meat, 1 fat
Chili with beans 🌾 🖉, commercial	1 c (8 oz)	2 starch, 2 medium-fat meat, 2 fat
Chow mein 🌾 🖉, without noodles or rice	2 c (16 oz)	1 starch, 2 vegetable, 2 lean meat
Macaroni and cheese 🖉	1 c (8 oz)	2 starch, 1 medium-fat meat, 2 fat
Soups:		
Bean 🌾 🖉	1 c (8 oz)	1 starch, 1 vegetable, 1 lean meat
Chunky, all varieties 🖉	10 3/4-oz can	1 starch, 1 vegetable, 1 medium-fat meat
Cream 🖉, made with water	1 c (8 oz)	1 starch, 1 fat
Vegetable 🖉 or broth 🖉	1 c (8 oz)	1 starch

Food	Amount	Exchanges
Spaghetti and meatballs 🖉, canned	1 c (8 oz)	2 starch, 1 medium-fat meat, 1 fat
Sugar-free pudding, made with nonfat milk	1/2 c	1 starch

If beans are used as a meat substitute:

Food	Amount	Exchanges
Dried beans 🌾, peas 🌾, lentils 🌾	1 c (cooked)	2 starch, 1 lean meat

🖉 400 milligrams or more sodium per serving.

🌾 3 grams or more dietary fiber per serving.

TABLE C.9

U.S. Exchange Foods for Occasional Use

The following list includes average exchange values for some foods high in sugar and fat. People are advised to use them only occasionally and in moderate amounts.

Food	Amount	Exchanges	Food	Amount	Exchanges
Angel food cake	1/12 cake	2 starch	Granola bars	1 small	1 starch, 1 fat
Cake, no icing	1/12 cake or a 3" square	2 starch, 2 fat	Ice cream, any flavor	1/2 c	1 starch, 2 fat
			Ice milk, any flavor	1/2 c	1 starch, 1 fat
Cookies	2 small, 1 3/4" across	1 starch, 1 fat	Sherbet, any flavor	1/4 c	1 starch
Frozen fruit yogurt	1/3 c	1 starch	Snack chips 🖍, all varieties	1 oz	1 starch, 2 fat
Gingersnaps	3	1 starch	Vanilla wafers	6 small	1 starch, 1 fat
Granola	1/4 c	1 starch, 1 fat			

🖍 If more than one serving is eaten, these foods have 400 milligrams or more sodium.

C

Table of Intentional Food Additives

TABLE D.1

A guide to Intentional Food Additives

Abbreviation	Type of Additive	Uses
AA	Anticaking agents	Keeps dry powders and crystals from clumping together (e.g., salt, powdered sugar).
C	Colors	Synthetic (laboratory-made) vegetable and fruit concentrates and other substances used to color foods (e.g., soft drinks, frosting).
E	Emulsifiers	Used to make oil and water mix (e.g., salad dressings, sauces).
EX	Extenders	"Fillers" such as fruit pulp and texturized protein (e.g., fruit drinks, hamburger).
FA	Flavoring agents	Used to add particular flavors to food (e.g., pudding, rye bread).
N	Nutrients	Used to add vitamins or minerals to foods (e.g., breakfast cereals, skim milk).
P	Preservatives	Used to keep food from spoiling (e.g., breads, breakfast cereals).
S	Sweeteners	Used to sweeten foods. Some are "artificial," such as aspartame and saccharin, and some are extracted from plants, such as sugar cane and sugar beets (e.g., soft drinks, catsup).
T	Texturizers	Used to improve the texture of food by stabilizing moisture content, dryness, volume, tenderness, or hardness (e.g., cakes, breads).
TH	Thickeners	Used to improve the consistency of foods (e.g., low-fat salad dressings, low-calorie jams).

Common Food Additives and Their Primary Function

Alginates (T)
Alpha tocopherol (vitamin E) (N)
Alpha tocopheryl acetate (vitamin E) (N)
Ascorbic acid (vitamin C) (P)
Aspartame (NutraSweet™) (S)
Baking powder (T)
Beet juice (C)
Beet sugar (S)
Beta-carotene (N)
BHA (P)
BHT (P)
Calcium carbonate (calcium) (N)
Calcium pantothenate (pantothenic acid, a B vitamin) (N)
Calcium propionate (P)
Calcium silicate (AA)
Cane sugar (S)

Carotene (C)
Carrageenan (E, TH)
Cellulose gum (TH)
Chromium chloride (N)
Citric acid (FA, P)
Corn syrup (S)
Cupric oxide (copper) (N)
Cyanocobalamin (vitamin B_{12}) (N)
Dextrin (TH)
Dextrose (S)
Dibasic calcium phosphate (calcium phosphorus) (N)
Diglycerides (E)
EDTA (P)
Extracts (FA)
FD&C blue no. 1 (C)
FD&C red no. 3 (use is being phased out) (C)

(continued)

TABLE D.1
A guide to Intentional Food Additives, continued

Common Food Additives and Their Primary Function

FD&C yellow no. 5 (C)
Ferrous fumarate (iron) (N)
Ferrous sulfate (iron) (N)
Fructose (S)
Fruit pulp (EX)
Gelatin (TH)
Glycerol (T)
Glyceryl abietate (E, T)
Guar gum (T, TH)
High-fructose corn syrup (S)
Honey (S)
Hydrolyzed protein (EX)
Lecithin (E)
Magnesium oxide (magnesium) N
Maltodextrin (S, TH)
Manganese sulfate (manganese) (N)
Modified food starch (TH)
Monoglycerides (E)
Monosodium glutamate (MSG) (FA)
Natural flavorings (FA)
Niacinamide (niacin, vitamin B_3) (N)
Nitrates (P)
Nitrites (P)
Paprika (C)
Pectin (TH)
Phosphoric acid (P)
Phytonadione (vitamin K) (N)
Polysorbates (P)
Potassium benzoate (P)
Potassium bicarbonate (T)
Potassium chloride (N)
Potassium iodite (iodine) (N)
Potassium metabisulfite (P)
Potassium sorbate (P)
Propyl gallate (P)
Propyleneglycol (T)
Pyridoxine hydrochloride (vitamin B_6) (N)
Reduced iron (iron) (N)
Retinol (vitamin A) (N)
Riboflavin (vitamin B_2) (N)
Saccharin (S)

Salt (FA, P)
Silicondioxide (AA)
Sodium aluminum phosphate (T)
Sodium ascorbite (vitamin C) (N)
Sodium benzoate (P)
Sodium bicarbonate (baking soda) (T)
Sodium bisulfite (P)
Sodium chloride (P, FA)
Sodium citrate (P)
Sodium erythorbate (P)
Sodium hexametaphosphate (P, T)
Sodium metabisulfite (P)
Sodium molydate (molybdenum) (N)
Sodium nitrate (P)
Sodium nitrite (P)
Sodium propionate (P)
Sodium saccharin (S)
Sodium selenate (selenium) (N)
Sodium stearolyn-2-lactylate (P, E)
Sodium sulfite (P)
Sorbitan monostearate (E)
Sorbitol (S, T)
Sorghum (S, P)
Starches (TH)
Sucrose (S)
Sugar (P)
Sulfur dioxide (P)
Sweeteners (FA)
Texturized protein (EX)
Thiamin hydrochloride (thiamin, B_1) (N)
Thiamin mononitrate (thiamin, B_1) (N)
Tocopherols (P, N)
Turmeric (C)
Vitamin A palmitate (vitamin A) (N)
Vitamin C (ascorbic acid) (N, P)
Vitamin E (N, P)
Xanthan gum (T)
Xylitol (S)
Yeast (T)
Zinc oxide (zinc) (N)

Cells

This appendix presents an overview of the basic structure and functions of cells in the human body. Cell structure and function are central to discussions of nutrition for it is within cells that nutrients are utilized to sustain life and health. The life-sustaining processes that take place within each of the more than one hundred trillion cells in the human body are maintained by the nutrients we consume in our diet. Chemical reactions within the cells produce energy from carbohydrates, proteins, and fats; cause proteins to be broken down or built up; and in thousands of other ways keep us in a state of health.

A cell is a basic unit of life. Any substance that does not consist of one or more cells cannot be alive. Cells are the "building blocks" of tissues (such as muscles and bones), organs (such as the kidneys and liver), and systems (the respiratory and digestive systems, for example). Normal cell health and functioning are maintained when a state of nutritional and environmental utopia exists within and around the cells. A disruption in the availability of nutrients or the presence of harmful substances in the cell's environment can initiate disorders that eventually affect our health or growth. Health problems in general begin with disruptions in the normal activity of cells.

The types and amounts of food and supplements people consume affect the cells' environment and their ability to function normally. Excessive or inadequate supplies of nutrients and other chemical substances disrupt cell functions and result in health problems. Humans remain in a state of health as long as their cells do.

■ **Cells**
The basic unit of life, of which all living things are composed. Every cell is surrounded by a membrane and contains cytoplasm, within which are organelles and a nucleus; the cell nucleus contains chromosomes.

Cell Structures and Functions

A generalized diagram of a human cell is shown in Illustration E.1. Although all cells have some structures and functions in common, the specific functions performed, and the structures that support those functions, can vary a good deal from cell to cell. Cells lining the esophagus, stomach, and intestines, for example, are specialized to produce and secrete mucus that helps food pass through the digestive tract. Red blood cells are specially formed to transport oxygen and carbon dioxide.

Every cell is surrounded by a cell membrane that helps move nutrients into and out of the cell. Inside the cell membrane lies the cytoplasm, a fluid material that contains many organelles, tubes, and particles. Among the organelles in the cytoplasm are ribosomes, mitochondria, and lysosomes. Each of these "little organs" is encased in a cell membrane and performs specific functions. Ribosomes assemble amino acids into proteins following the instructions of DNA and its messenger RNA. The mitochondria are made of intricately folded membranes that bear thousands of highly organized sets of enzymes on their surfaces. These enzymes are actively involved in the production of energy and are found in particularly dense quantities in muscle cells. The lysosomes are like packets of enzymes. The enzymes are used to break down old cell particles that are being recycled and to destroy substances that are harmful to the body.

Cytoplasm also contains a highly organized system of membranes called the endoplasmic reticulum. When these membranes are dotted with ribosomes, they are called "rough endoplasmic reticulum." When ribosomes are absent, the endoplasmic reticulum is referred to as "smooth." Some membranes within the cytoplasm form tubes that collect certain types of cellular material and transport it out of the cell. These membranous tubes are called "Golgi apparatus." The rough and smooth endoplasmic reticulum are continuous with the Golgi apparatus, so secretions produced throughout the cell can be collected and transported to its exterior.

Within each cell is a nucleus covered by a two-layer membrane. The nucleus contains chromosomes and the genetic material DNA. DNA encodes all of the instructions a cell needs to conduct protein synthesis and to replicate life.

All cells within the body are part of a complex communication system that uses hormones, electrical impulses, and other chemical messengers to link each cell to the others. No cell is an island that operates independently from the others.

Cell membrane
Cytoplasm
Smooth endoplasmic reticulum
Golgi apparatus
Lysosome
Nucleus
Chromosomes
Rough endoplasmic reticulum
Mitochondrion
Ribosomes

Illustration E.1
Generalized structure of a human cell.

E

cell membrane: the membrane that surrounds the cell and encloses its contents; made primarily of lipid and protein.

chromosomes: a set of structures within the nucleus of every cell that contain the cell's genetic material, DNA, associated with other materials (primarily proteins).

cytoplasm (SIGH-toe-plazm): the cell contents, except for the nucleus.

> *cyto* = cell
>
> *plasm* = a form

Golgi (GOAL-gee) **apparatus:** a set of membranes within the cell where secretory materials are packaged for export.

lysosomes: cellular organelles; membrane-enclosed sacs of degradative enzymes.

> *lysis* = dissolution

mitochondria (my-toe-KON-dree-uh; *singular* **mitochondrion**): the cellular organelles responsible for producing ATP aerobically; made of membranes (lipid and protein) with enzymes mounted on them.

> *mitos* = thread (referring to their slender shape)

> *chondros* = cartilage (referring to their external appearance)

nucleus: a major membrane-enclosed body within every cell, which contains the cell's genetic material, DNA, embedded in chromosomes.

> *nucleus* = a kernel

organelles: membrane-bound subcellular structures such as ribosomes, mitochondria, and lysosomes.

> *organelle* = little organ

ribosomes: protein-making organelles in cells; composed of RNA and protein.

> *ribo* = containing the sugar ribose
>
> *some* = body

rough endoplasmic reticulum (en-doh-PLAZ-mic reh-TIC-you-lum): intracellular membranes dotted with ribosomes, where protein synthesis takes place.

> *endo* = inside
>
> *plasm* = the cytoplasm

smooth endoplasmic reticulum: smooth intracellular membranes bearing no ribosomes.

WHO: Nutrition Recommendations
Canada: Choice System and Guidelines

This appendix first presents nutrition recommendations from the World Health Organization (WHO) and then provides details for Canadians on Canada's *Food Guide to Healthy Eating* and on the exchange system (called the choice system).

Nutrition Recommendations from WHO

The World Health Organization (WHO) has assessed the relationships between diet and the development of chronic diseases. Its recommendations include:

- Total energy: sufficient to support normal growth, physical activity, and healthy body weight (body mass index = 20 to 22).

- Total fat: 15 to 30 percent of total energy.

- Saturated fat: less than 10 percent of total energy.

- Total carbohydrate: 55 to 75 percent of total energy.

- Added sugars: less than 10 percent of total energy.

- Protein: 10 to 15 percent of total energy.

- Salt: less than 5 grams/day, preferably iodized.

- Fruit and vegetables: at least 400 grams (almost 1 pound) daily.

- Physical activity: one hour per day of moderate intensity on most days of the week.

Canada's *Food Guide to Healthy Eating*

Figure F-1 presents the 1992 Canada's *Food Guide to Healthy Eating,* which interprets Canada's *Guidelines for Healthy Eating* (see Table 2-1 on p. 42) for consumers and recommends a range of servings to consume daily from each of the four food groups. The following publications, which are available from Health Canada, through its website, explain how to use the *Guide: Using the Food Guide; Food Guide Facts: Background for Educators and Communicators; Canada's Food Guide to Healthy Eating—Focus on Preschoolers: Background for Educators and Communicators;* and *Canada's Food Guide to Healthy Eating—Focus on Children Six to Twelve Years: Background for Educators and Communicators.* Figure F-2 presents Canada's Physical Activity Guide.

Canada's *Guidelines for Healthy Eating* and Canada's *Food Guide to Healthy Eating* are being reviewed for consistency with the new Dietary Reference Intakes. Check the website for the Health Canada Office of Nutrition Policy and Promotion, www.hc-sc.gc.ca/hpfb-dgpsa/onpp-bppn, for the status of the review.

Illustration F-1
Canadian Food Guide to Healthy Eating

 Health and Welfare Canada   Santé et Bien-être social Canada

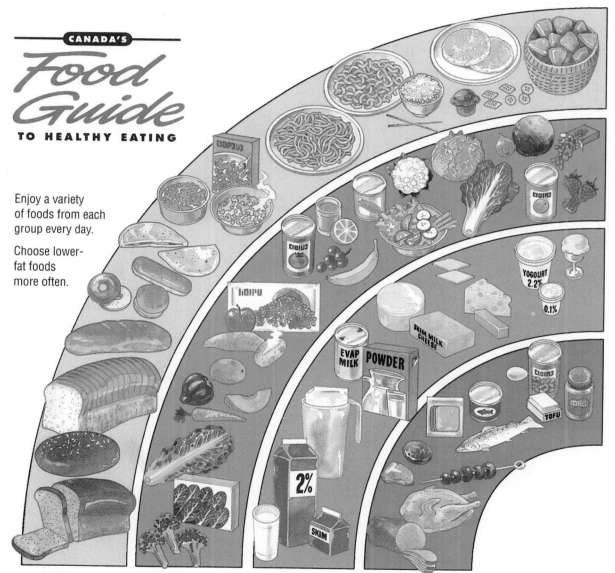

Enjoy a variety of foods from each group every day.

Choose lower-fat foods more often.

Grain Products
Choose whole grain and enriched products more often.

Vegetables & Fruit
Choose dark green and orange vegetables and orange fruit more often.

Milk Products
Choose lower-fat milk products more often.

Meat & Alternatives
Choose leaner meats, poultry and fish, as well as dried peas, beans and lentils more often.

Canadian Choice System

F

Appendix

Illustration F-1
Canadian Food Guide to Healthy Eating (*continued*)

Different People Need Different Amounts of Food

The amount of food you need every day from the 4 food groups and other foods depends on your age, body size, activity level, whether you are male or female and if you are pregnant or breast-feeding. That's why the Food Guide gives a lower and higher number of servings for each food group. For example, young children can choose the lower number of servings, while male teenagers can go to the higher number. Most other people can choose servings somewhere in between.

Grain Products 5-12 SERVINGS PER DAY

1 Serving — 1 Slice; Cold Cereal 30 g; Hot Cereal 175 mL 3/4 cup
2 Servings — 1 Bagel, Pita or Bun; Pasta or Rice 250 mL 1 cup

Vegetables & Fruit 5-10 SERVINGS PER DAY

1 Serving — 1 Medium Size Vegetable or Fruit; Fresh, Frozen or Canned Vegetables or Fruit 125 mL 1/2 cup; Salad 250 mL 1 cup; Juice 125 mL 1/2 cup

Milk Products SERVINGS PER DAY
Children 4–9 years: 2–3
Youth 10–16 years: 3–4
Adults: 2–4
Pregnant & Breast-feeding Women: 3–4

1 Serving — MILK 250 mL 1 cup; Cheese 3"x1"x1" 50 g; 2 Slices 50 g; Yogourt 175 g 3/4 cup

Other Foods

Taste and enjoyment can also come from other foods and beverages that are not part of the 4 food groups. Some of these foods are higher in fat or Calories, so use these foods in moderation.

Meat & Alternatives 2-3 SERVINGS PER DAY

1 Serving — Meat, Poultry or Fish 50–100 g; Fish 1/3–2/3 Can 50–100 g; 1-2 Eggs; Beans 125–250 mL; Tofu 100 g 1/3 cup; Peanut Butter 30 mL 2 tbsp

Enjoy eating well, being active and feeling good about yourself. That's VITALIT*e*

© Minister of Supply and Services Canada 1992 Cat. No. H39-252/1992E No changes permitted. Reprint permission not required.
ISBN 0-662-19648-1

Illustration F-1
Canadian Food Guide to Healthy Eating (*continued*)

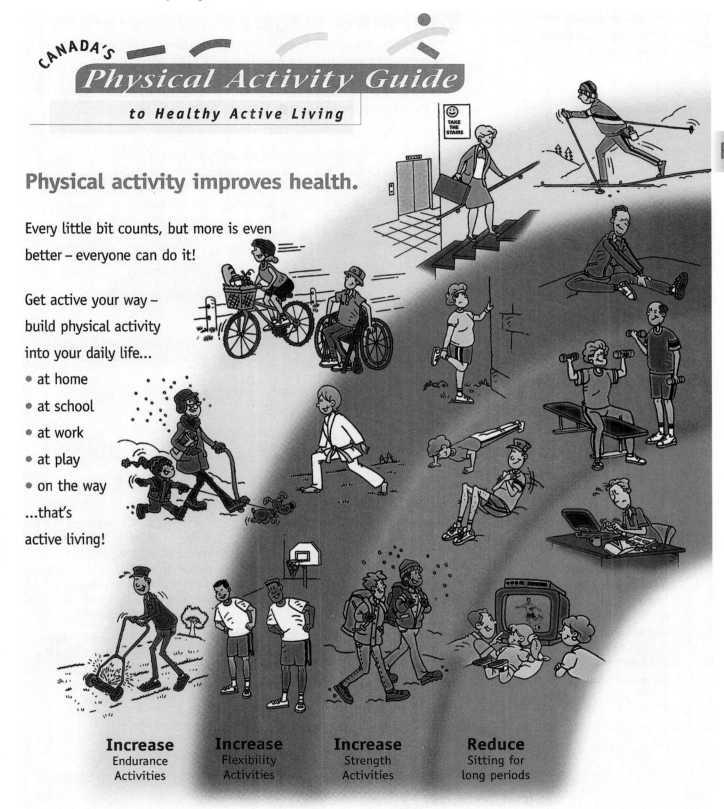

CANADA'S *Physical Activity Guide*
to Healthy Active Living

Physical activity improves health.

Every little bit counts, but more is even better – everyone can do it!

Get active your way – build physical activity into your daily life...
- at home
- at school
- at work
- at play
- on the way
...that's active living!

Increase Endurance Activities

Increase Flexibility Activities

Increase Strength Activities

Reduce Sitting for long periods

Health Canada   Santé Canada

CSEP SCPE Canadian Society for Exercise Physiology

F

Illustration F-1
Canadian Food Guide to Healthy Eating (*continued*)

Canadian Choice System

F

Appendix

Choose a variety of activities from these three groups:

Endurance

4-7 days a week
Continuous activities for your heart, lungs and circulatory system.

Flexibility

4-7 days a week
Gentle reaching, bending and stretching activities to keep your muscles relaxed and joints mobile.

Strength

2-4 days a week
Activities against resistance to strengthen muscles and bones and improve posture.

Starting slowly is very safe for most people. Not sure? Consult your health professional.

For a copy of the *Guide Handbook* and more information:
1-888-334-9769, or
www.paguide.com

Eating well is also important. Follow *Canada's Food Guide to Healthy Eating* to make wise food choices.

Get Active Your Way, Every Day–For Life!

Scientists say accumulate 60 minutes of physical activity every day to stay healthy or improve your health. As you progress to moderate activities you can cut down to 30 minutes, 4 days a week. Add-up your activities in periods of at least 10 minutes each. Start slowly... and build up.

Time needed depends on effort

Very Light Effort	Light Effort *60 minutes*	Moderate Effort *30-60 minutes*	Vigorous Effort *20-30 minutes*	Maximum Effort
• Strolling • Dusting	• Light walking • Volleyball • Easy gardening • Stretching	• Brisk walking • Biking • Raking leaves • Swimming • Dancing • Water aerobics	• Aerobics • Jogging • Hockey • Basketball • Fast swimming • Fast dancing	• Sprinting • Racing

Range needed to stay healthy

You Can Do It – Getting started is easier than you think

Physical activity doesn't have to be very hard. Build physical activities into your daily routine.

- Walk whenever you can– get off the bus early, use the stairs instead of the elevator.
- Reduce inactivity for long periods, like watching TV.
- Get up from the couch and stretch and bend for a few minutes every hour.
- Play actively with your kids.
- Choose to walk, wheel or cycle for short trips.

- Start with a 10 minute walk– gradually increase the time.
- Find out about walking and cycling paths nearby and use them.
- Observe a physical activity class to see if you want to try it
- Try one class to start, you don't have to make a long-term commitment.
- Do the activities you are doing now, more often.

Benefits of regular activity:

- better health
- improved fitness
- better posture and balance
- better self-esteem
- weight control
- stronger muscles and bones
- feeling more energetic
- relaxation and reduced stress
- continued independent living in later life

Health risks of inactivity:

- premature death
- heart disease
- obesity
- high blood pressure
- adult-onset diabetes
- osteoporosis
- stroke
- depression
- colon cancer

ACTIVE LIVING

CANADA'S *Physical Activity Guide* to Healthy Active Living

Canada's Choice System for Meal Planning

The *Good Health Eating Guide* is the Canadian choice system of meal planning.[1] It contains several features similar to those of the U.S. exchange system including the following:

- Foods are divided into lists according to carbohydrate, protein, and fat content.

- Foods are interchangeable within a group.

- Most foods are eaten in measured amounts.

- An energy value is given for each food group.

Tables F-1 through F-8 present the Canadian choice system.

---

[1]The tables for the Canadian choice system are adapted from the *Good Health Eating Guide Resource,* copyright 1994, with permission of the Canadian Diabetes Association.

TABLE F-1
Canadian Choice System: Starch Foods

1 starch choice = 15 g carbohydrate (starch), 2 g protein, 290 kJ (68 kcal)

Food	Measure	Mass (Weight)
Breads		
Bagels	$1/2$	30 g
Bread crumbs	50 mL ($1/4$ c)	30 g
Bread cubes	250 mL (1 c)	30 g
Bread sticks	2	20 g
Brewis, cooked	50 mL (¼ c)	45 g
Chapati	1	20 g
Cookies, plain	2	20 g
English muffins, crumpets	$1/2$	30 g
Flour	40 mL ($2^1/2$ tbs)	20 g
Hamburger buns	$1/2$	30 g
Hot dog buns	$1/2$	30 g
Kaiser rolls	$1/2$	30 g
Matzo, 15 cm	1	20 g
Melba toast, rectangular	4	15 g
Melba toast, rounds	7	15 g
Pita, 20 cm (8") diameter	$1/4$	30 g
Pita, 15 cm (6") diameter	$1/2$	30 g
Plain rolls	1 small	30 g
Pretzels	7	20 g
Raisin bread	1 slice	30 g
Rice cakes	2	30 g
Roti	1	20 g
Rusks	2	20 g
Rye, coarse or pumpernickel	$1/2$ slice	30 g
Soda crackers	6	20 g
Tortillas, corn (taco shell)	1	30 g
Tortilla, flour	1	30 g
White (French and Italian)	1 slice	25 g
Whole-wheat, cracked-wheat, rye, white enriched	1 slice	30 g
Cereals		
Bran flakes, 100% bran	125 mL ($1/2$ c)	30 g
Cooked cereals, cooked	125 mL ($1/2$ c)	125 g
Dry	30 mL (2 tbs)	20 g

(continued on the next page)

TABLE F-1
Canadian Choice System: Starch Foods—continued

1 starch choice = 15 g carbohydrate (starch), 2 g protein, 290 kJ (68 kcal)

Food	Measure	Mass (Weight)
Cornmeal, cooked	125 mL ($^1/_2$ c)	125 g
Dry	30 mL (2 tbs)	20 g
Ready-to-eat unsweetened cereals	125 mL ($^1/_2$ c)	20 g
Shredded wheat biscuits, rectangular or round	1	20 g
Shredded wheat, bite size	125 mL ($^1/_2$ c)	20 g
Wheat germ	75 mL ($^1/_3$ c)	30 g
Cornflakes	175 mL ($^2/_3$ c)	20 g
Rice Krispies	175 mL ($^2/_3$ c)	20 g
Cheerios	200 mL ($^3/_4$ c)	20 g
Muffets	1	20 g
Puffed rice	300 mL (1$^1/_4$ c)	15 g
Puffed wheat	425 mL (1$^2/_3$ c)	20 g

Grains

Food	Measure	Mass (Weight)
Barley, cooked	125 mL ($^1/_2$ c)	120 g
Dry	30 mL (2 tbs)	20 g
Bulgur, kasha, cooked, moist	125 mL ($^1/_2$ c)	70 g
Cooked, crumbly	75 mL ($^1/_3$ c)	40 g
Dry	30 mL (2 tbs)	20 g
Rice, cooked, brown & white (short & long grain)	125 mL ($^1/_2$ c)	70 g
Rice, cooked, wild	75 mL ($^1/_3$ c)	70 g
Tapioca, pearl and granulated, quick cooking, dry	30 mL (2 tbs)	15 g
Couscous, cooked moist	125 mL ($^1/_2$ c)	70 g
Dry	30 mL (tbs)	20 g
Quinoa, cooked moist	125 mL ($^1/_2$ c)	70 g
Dry	30 mL (2 tbs)	20 g

Pastas

Food	Measure	Mass (Weight)
Macaroni, cooked	125 mL ($^1/_2$ c)	70 g
Noodles, cooked	125 mL ($^1/_2$ c)	80 g
Spaghetti, cooked	125 mL ($^1/_2$ c)	70 g

Starchy Vegetables

Food	Measure	Mass (Weight)
Beans and peas, dried, cooked	125 mL ($^1/_2$ c)	80 g
Breadfruit	1 slice	75 g
Corn, canned, whole kernel	125 mL ($^1/_2$ c)	85 g
Corn on the cob	$^1/_2$ medium cob	140 g
Cornstarch	30 mL (2 tbs)	15 g
Plantains	$^1/_3$ small	50 g
Popcorn, air-popped, unbuttered	750 mL (3 c)	20 g
Potatoes, whole (with or without skin)	$^1/_2$ medium	95 g
Yams, sweet potatoes (with or without skin)	$^1/_2$	75 g

Note: Food items found in this category provide more than 1 starch choice:

Food	Choices per Serving	Measure	Mass (Weight)
Bran flakes	1 starch + $^1/_2$ sugar	150 mL ($^2/_3$ c)	24 g
Croissant, small	1 starch + 1$^1/_2$ fats	1 small	35 g
Large	1 starch + 1$^1/_2$ fats	$^1/_2$ large	30 g
Corn, canned creamed	1 starch + $^1/_2$ fruits and vegetables	12 mL ($^1/_2$ c)	113 g
Potato chips	1 starch + 2 fats	15 chips	30 g
Tortilla chips (nachos)	1 starch + 1$^1/_2$ fats	13 chips	20 g
Corn chips	1 starch + 2 fats	30 chips	30 g

(*continued on the next page*)

TABLE F-1

Canadian Choice System: Starch Foods—continued

1 starch choice = 15 g carbohydrate (starch), 2 g protein, 290 kJ (68 kcal)

Food	Choices per Serving	Measure	Mass (Weight)
Cheese twists	1 starch + 1^1/$_2$ fats	30 chips	30 g
Cheese puffs	1 starch + 2 fats	27 chips	30 g
Tea biscuit	1 starch + 2 fats	1	30 g
Pancakes, homemade using 50 mL (^{1}/$_4$ c) batter (6″ diameter)	1^1/$_2$ starches + 1 fat	1 medium	50 g
Potatoes, french fried (homemade or frozen)	1 starch + 1 fat	10 regular size	35 g
Soup, canned* (prepared with equal volume of water)	1 starch	250 mL (1 c)	260 g
Waffles, packaged	1 starch + 1 fat	1	35 g

*Soup can vary according to brand and type. Check the label for Food Choice Values and Symbols or the core nutrient listing.

TABLE F-2

Canadian Choice System: Fruits and Vegetables

1 fruits and vegetables choice = 10 g carbohydrate, 1 g protein, 190 kJ (44 kcal)

Food	Measure	Mass (Weight)
Fruits (fresh, frozen, without sugar, canned in water)		
Apples, raw (with or without skin)	^{1}/$_2$ medium	75 g
Sauce unsweetened	125 mL (^{1}/$_2$ c)	120 g
Sweetened	see *Combined Food Choices (Table I-8)*	
Apple butter	20 mL (4 tsp)	20 g
Apricots, raw	2 medium	115 g
Canned, in water	4 halves, plus 30 mL (2 tbs) liquid	110 g
Bake-apples (cloudberries), raw	125 mL (^{1}/$_2$ c)	120 g
Bananas, with peel	^{1}/$_2$ small	75 g
Peeled	^{1}/$_2$ small	50 g
Berries (blackberries, blueberries, boysenberries, huckleberries, loganberries, raspberries)		
Raw	125 mL (^{1}/$_2$ c)	70 g
Canned, in water	125 mL (^{1}/$_2$ c), plus 30 mL (2 tbs) liquid	100 g
Cantaloupe, raw, with rind	^{1}/$_4$ wedge	240 g
Cubed or diced	250 mL (1 c)	160 g
Cherries, raw, with pits	10	75 g
Raw, without pits	10	70 g
Canned, in water, with pits	75 mL (^{1}/$_3$ c), plus 30 mL (2 tbs) liquid	90 g
Canned, in water, without pits	75 mL (^{1}/$_3$ c), plus 30 mL (2 tbs) liquid	85 g
Crabapples, raw	1 small	55 g
Cranberries, raw	250 mL (1 c)	100 g
Figs, raw	1 medium	50 g
Canned, in water	3 medium, plus 30 mL (2 tbs) liquid	100 g
Foxberries, raw	250 mL (1 c)	100 g
Fruit cocktail, canned, in water	125 mL (^{1}/$_2$ c), plus 30 mL (2 tbs) liquid	120 g
Fruit, mixed, cut-up	125 mL (^{1}/$_2$ c)	120 g
Gooseberries, raw	250 mL (1 c)	150 g
Canned, in water	250 mL (1 c), plus 30 mL (2 tbs) liquid	230 g
Grapefruit, raw, with rind	^{1}/$_2$ small	185 g
Raw, sectioned	125 mL (^{1}/$_2$ c)	100 g
Canned, in water	125 mL (^{1}/$_2$ c), plus 30 mL (2 tbs) liquid	120 g

(continued on the next page)

Canadian Choice System

F

Appendix

TABLE F-2

Canadian Choice System: Fruits and Vegetables—continued

1 fruits and vegetables choice = 10 g carbohydrate, 1 g protein, 190 kJ (44 kcal)

Food	Measure	Mass (Weight)
Grapes, raw, slip skin	125 mL (1/2 c)	75 g
Raw, seedless	125 mL (1/2 c)	75 g
Canned, in water	75 mL (1/3 c), plus 30 mL (2 tbs) liquid	115 g
Guavas, raw	1/2	50 g
Honeydew melon, raw, with rind	1/2	225 g
Cubed or diced	250 mL (1 c)	170 g
Kiwis, raw, with skin	2	155 g
Kumquats, raw	3	60 g
Loquats, raw	8	130 g
Lychee fruit, raw	8	120 g
Mandarin oranges, raw, with rind	1	135 g
Raw, sectioned	125 mL (1/2 c)	100 g
Canned, in water	125 mL (1/2 c), plus 30 mL (2 tbs) liquid	100 g
Mangoes, raw, without skin and seed	1/3	65 g
Diced	75 mL (1/3 c)	65 g
Nectarines	1/2 medium	75 g
Oranges, raw, with rind	1 small	130 g
Raw, sectioned	125 mL (1/2 c)	95 g
Papayas, raw, with skin and seeds	1/4 medium	150 g
Raw, without skin and seeds	1/4 medium	100 g
Cubed or diced	125 mL (1/2 c)	100 g
Peaches, raw, with seed and skin	1 large	100 g
Raw, sliced or diced	125 mL (1/2 c)	100 g
Canned in water, halves or slices	125 mL (1/2 c), plus 30 mL (2 tbs) liquid	120 g
Pears, raw, with skin and core	1/2	90 g
Raw, without skin and core	1/2	85 g
Canned, in water, halves	1 half plus 30 mL (2 tbs) liquid	90 g
Persimmons, raw, native	1	30 g
Raw, Japanese	1/4	50 g
Pineapple, raw	1 slice	75 g
Raw, diced	125 mL (1/2 c)	75 g
Canned, in juice, diced	75 mL (1/3 c), plus 15 mL (1 tbs) liquid	55 g
Canned, in juice, sliced	1 slice, plus 15 mL (1 tbs) liquid	55 g
Canned, in water, diced	125 mL (1/2 c), plus 30 mL (2 tbs) liquid	100 g
Canned, in water, sliced	2 slices, plus 15 mL (1 tbs) liquid	100 g
Plums, raw	2 small	60 g
Damson	6	65 g
Japanese	1	70 g
Canned, in apple juice	2, plus 30 mL (2 tbs) liquid	70 g
Canned, in water	3, plus 30 mL (2 tbs) liquid	100 g
Pomegranates, raw	1/2	140 g
Strawberries, raw	250 mL (1 c)	150 g
Frozen/canned, in water	250 mL (1 c), plus 30 mL (2 tbs) liquid	240 g
Rhubarb	250 mL (1 c)	150 g
Tangelos, raw	1	205 g
Tangerines, raw	1 medium	115 g
Raw, sectioned	125 mL (1/2 c)	100 g

(*continued* on the next page)

TABLE F-2
Canadian Choice System: Fruits and Vegetables—continued

1 fruits and vegetables choice = 10 g carbohydrate, 1 g protein, 190 kJ (44 kcal)

Food	Measure	Mass (Weight)
Watermelon, raw, with rind	1 wedge	310 g
Cubed or diced	250 mL (1 c)	160 g
Dried Fruit		
Apples	5 pieces	15 g
Apricots	4 halves	15 g
Banana flakes	30 mL (2 tbs)	15 g
Currants	30 mL (2 tbs)	15 g
Dates, without pits	2	15 g
Peaches	$1/_2$	15 g
Pears	$1/_2$	15 g
Prunes, raw, with pits	2	15 g
Raw, without pits	2	10 g
Stewed, no liquid	2	20 g
Stewed, with liquid	2, plus 15 mL (1 tbs) liquid	35 g
Raisins	30 mL (2 tbs)	15 g
Juices (no sugar added or unsweetened)		
Apricot, grape, guava, mango, prune	50 mL ($1/_4$ c)	55 g
Apple, carrot, papaya, pear, pineapple, pomegranate	75 mL ($1/_3$ c)	80 g
Cranberry	see *Sugars (Table I-4)*	
Clamato	see *Sugars (Table I-4)*	
Grapefruit, loganberry, orange, raspberry, tangelo, tangerine	125 mL ($1/_2$ c)	130 g
Tomato, tomato-based mixed vegetables	250 mL (1 c)	255 g
Vegetables (fresh, frozen, or canned)		
Artichokes, French, globe	2 small	50 g
Beets, diced or sliced	125 mL ($1/_2$ c)	85 g
Carrots, diced, cooked or uncooked	125 mL ($1/_2$ c)	75 g
Chestnuts, fresh	5	20 g
Parsnips, mashed	125 mL ($1/_2$ c)	80 g
Peas, fresh or frozen	125 mL ($1/_2$ c)	80 g
Canned	75 mL ($1/_3$ c)	55 g
Pumpkin, mashed	125 mL ($1/_2$ c)	45 g
Rutabagas, mashed	125 mL ($1/_2$ c)	85 g
Sauerkraut	250 mL (1 c)	235 g
Snow peas	250 mL (1 c)	135 g
Squash, yellow or winter, mashed	125 mL ($1/_2$ c)	115 g
Succotash	75 mL ($1/_3$ c)	55 g
Tomatoes, canned	250 mL (1 c)	240 g
Tomato paste	50 mL ($1/_4$ c)	55 g
Tomato sauce*	75 mL ($1/_3$ c)	100 g
Turnips, mashed	125 mL ($1/_2$ c)	115 g
Vegetables, mixed	125 mL ($1/_2$ c)	90 g
Water chestnuts	8 medium	50 g

*Tomato sauce varies according to brand name. Check the label or discuss with your dietitian.

TABLE F-3
Canadian Choice System: Milk

Type of Milk	Carbohydrate (g)	Protein (g)	Fat (g)	Energy
Nonfat (0%)	6	4	0	170 kJ (40 kcal)
1%	6	4	1	206 kJ (49 kcal)
2%	6	4	2	244 kJ (58 kcal)
Whole (4%)	6	4	4	319 kJ (76 kcal)

Food	Measure	Mass (Weight)
Buttermilk (higher in salt)	125 mL ($^1/_2$ c)	125 g
Evaporated milk	50 mL ($^1/_4$ c)	50 g
Milk	125 mL ($^1/_2$ c)	125 g
Powdered milk, regular	30 mL (2 tbs)	15 g
Instant	50 mL ($^1/_4$ c)	15 g
Plain yogurt	125 mL ($^1/_2$ c)	125 g

Food	Choices per Serving	Measure	Mass (Weight)
Note: Food items found in this category provide more than 1 milk choice:			
Milk shake	1 milk + 3 sugars + $^1/_2$ protein	250 mL (1 c)	300 g
Chocolate milk, 2%	2 milks 2% + 1 sugar	250 mL (1 c)	300 g
Frozen yogurt	1 milk + 1 sugar	125 mL ($^1/_2$ c)	125 g

TABLE F-4

Canadian Choice System: Sugars

1 sugar choice = 10 g carbohydrate (sugar), 167 kJ (40 kcal)

Food	Measure	Mass (Weight)
Beverages		
Condensed milk	15 mL (1 tbs)	
Flavoured fruit crystals*	75 mL (1/3 c)	
Iced tea mixes*	75 mL (1/3 c)	
Regular soft drinks	125 mL (1/2 c)	
Sweet drink mixes*	75 mL (1/3 c)	
Tonic water	125 mL (1/2 c)	
*These beverages have been made with water.		
Miscellaneous		
Bubble gum (large square)	1 piece	5 g
Cranberry cocktail	75 mL (1/3 c)	80 g
Cranberry cocktail, light	350 mL (1 1/3 c)	260 g
Cranberry sauce	30 mL (2 tbs)	
Hard candy mints	2	5 g
Honey, molasses, corn & cane syrup	10 mL (2 tsp)	15 g
Jelly bean	4	10 g
Licorice	1 short stick	10 g
Marshmallows	2 large	15 g
Popsicle	1 stick (1/2 popsicle)	
Powdered gelatin mix (Jello®) (reconstituted)	50 mL (1/4 c)	
Regular jam, jelly, marmalade	15 mL (1 tbs)	
Sugar, white, brown, icing, maple	10 mL (2 tsp)	10 g
Sweet pickles	2 small	100 g
Sweet relish	30 mL (2 tbs)	

Food	Choices per Serving	Measures	Mass (Weight)
The following food items provide more than 1 sugar choice:			
Brownie	1 sugar + 1 fat	1	20 g
Clamato juice	1 1/2 sugars	175 mL (2/3 c)	
Fruit salad, light syrup	1 sugar + 1 fruits & vegetables	125 mL (1/2 c)	130 g
Aero® bar	2 1/2 sugars + 2 1/2 fats	1 bar	43 g
Smarties®	4 1/2 sugars + 2 fats	1 box	60 g
Sherbet	3 sugars + 1/2 fat	125 mL (1/2 c)	95 g

TABLE F-5

Canadian Choice System: Protein Foods

1 protein choice = 7 g protein, 3 g fat, 230 kJ (55 kcal)

Food	Measure	Mass (Weight)
Cheese		
Low-fat cheese, about 7% milk fat	1 slice	30 g
Cottage cheese, 2% milkfat or less	50 mL (¹/₄ c)	55 g
Ricotta, about 7% milkfat	50 mL (¹/₄ c)	60 g
Fish		
Anchovies	see *Extras (Table I-7)*	
Canned, drained (e.g., mackerel, salmon, tuna packed in water)	(¹/₃ of 6.5 oz can)	30 g
Cod tongues, cheeks	75 mL (¹/₃ c)	50 g
Fillet or steak (e.g., Boston blue, cod, flounder, haddock, halibut, mackerel, orange roughy, perch, pickerel, pike, salmon, shad, snapper, sole, swordfish, trout, tuna, whitefish)	1 piece	30 g
Herring	¹/₃ fish	30 g
Sardines, smelts	2 medium or 3 small	30 g
Squid, octopus	50 mL (¹/₄ c)	40 g
Shellfish		
Clams, mussels, oysters, scallops, snails	3 medium	30 g
Crab, lobster, flaked	50 mL (¹/₄ c)	30 g
Shrimp, fresh	5 large	30 g
Frozen	10 medium	30 g
Canned	18 small	30 g
Dry pack	50 mL (¹/₄ c)	30 g
Meat and Poultry (e.g., beef, chicken, goat, ham, lamb, pork, turkey, veal, wild game)		
Back, peameal bacon	3 slices, thin	30 g
Chop	¹/₂ chop, with bone	40 g
Minced or ground, lean or extra-lean	30 mL (2 tbs)	30 g
Sliced, lean	1 slice	30 g
Steak, lean	1 piece	30 g
Organ Meats		
Hearts, liver	1 slice	30 g
Kidneys, sweetbreads, chopped	50 mL (¹/₄ c)	30 g
Tongue	1 slice	30 g
Tripe	5 pieces	60 g
Soyabean		
Bean curd or tofu	¹/₂ block	70 g
Eggs		
In shell, raw or cooked	1 medium	50 g
Without shell, cooked or poached in water	1 medium	45 g
Scrambled	50 mL (¹/₄ c)	55 g

Note: The following choices provide more than 1 protein exchange:

Food	Choices per Serving	Measures	Mass (Weight)
Cheese			
Cheeses	1 protein + 1 fat	1 piece	25 g
Cheese, coarsely grated (e.g., cheddar)	1 protein + 1 fat	50 mL (¹/₄ c)	25 g

(*continued* on the next page)

TABLE F-5

Canadian Choice System: Protein Foods—continued

1 protein choice = 7 g protein, 3 g fat, 230 kJ (55 kcal
Note: The following choices provide more than 1 protein exchange:

Food	Choices per Serving	Measures	Mass (Weight)
Cheese, dry, finely grated (e.g., parmesan)	1 protein + 1 fat	45 mL	15 g
Cheese, ricotta, high fat	1 protein + 1 fat	50 mL (1/4 c)	55 g
Fish			
Eel	1 protein + 1 fat	1 slice	50 g
Meat			
Bologna	1 protein + 1 fat	1 slice	20 g
Canned lunch meats	1 protein + 1 fat	1 slice	20 g
Corned beef, canned	1 protein + 1 fat	1 slice	25 g
Corned beef, fresh	1 protein + 1 fat	1 slice	25 g
Ground beef, medium-fat	1 protein + 1 fat	30 mL (2 tbs)	25 g
Meat spreads, canned	1 protein + 1 fat	45 mL	35 g
Mutton chop	1 protein + 1 fat	1/2 chop, with bone	35 g
Paté	see *Fats and Oils (Table I-6)*		
Sausages, garlic, Polish or knockwurst	1 protein + 1 fat	1 slice	50 g
Sausages, pork, links	1 protein + 1 fat	1 link	25 g
Spareribs or shortribs, with bone	1 protein + 1 fat	1 large	65 g
Stewing beef	1 protein + 1 fat	1 cube	25 g
Summer sausage or salami	1 protein + 1 fat	1 slice	40 g
Wieners, hot dog	1 protein + 1 fat	1/2 medium	25 g
Miscellaneous			
Blood pudding	1 protein + 1 fat	1 slice	25 g
Peanut butter	1 protein + 1 fat	15 mL (1 tbs)	15 g

F

Canadian Choice System

Appendix

F

TABLE F-6
Canadian Choice System: Fats and Oils

1 fat choice = 5 g fat, 190 kJ (45 kcal)

Food	Measure	Mass (Weight)	Food	Measure	Mass (Weight)
Avocado*	1/8	30 g	Nuts (continued):		
Bacon, side, crisp*	1 slice	5 g	Sesame seeds	15 mL (1 tbs)	10 g
Butter*	5 mL (1 tsp)	5 g	Sunflower seeds		
Cheese spread	15 mL (1 tbs)	15 g	Shelled	15 mL (1 tbs)	10 g
Coconut, fresh*	45 mL (3 tbs)	15 g	In shell	45 mL (3 tbs)	15 g
Coconut, dried*	15 mL (1 tbs)	10 g	Walnuts	4 halves	10 g
Cream, Half and half			Oil, cooking and salad	5 mL (1 tsp)	5 g
(cereal), 10%*	30 mL (2 tbs)	30 g	Olives, green	10	45 g
Light (coffee), 20%*	15 mL (1 tbs)	15 g	Ripe black	7	57 g
Whipping, 32 to 37%*	15 mL (1 tbs)	15 g	Pâté, liverwurst,	15 mL (1 tbs)	15 g
Cream cheese*	15 mL (1 tbs)	15 g	meat spreads		
Gravy*	30 mL (2 tbs)	30 g	Salad dressing: blue,	10 mL (2 tsp)	10 g
Lard*	5 mL (1 tsp)	5 g	French, Italian,		
Margarine	5 mL (1 tsp)	5 g	mayonnaise,		
Nuts, shelled:			Thousand Island	5 mL (1 tsp)	5 g
Almonds	8	5 g	Salad dressing,	30 mL (2 tbs)	30 g
Brazil nuts	2	10 g	low-kcalorie		
Cashews	5	10 g	Salt pork, raw	5 mL (1 tsp)	5 g
Filberts, hazelnuts	5	10 g	or cooked*		
Macadamia	3	5 g	Sesame oil	5 mL (1 tsp)	5 g
Peanuts	10	10g	Sour cream		
Pecans	5 halves	5 g	12% milkfat	30 mL (2 tbs)	30 g
Pignolias, pine nuts	25 mL (5 tsp)	10 g	7% milkfat	60 mL (4 tbs)	60 g
Pistachios, shelled	20	10 g	Shortening*	5 mL (1 tsp)	
Pistachios, in shell	20	20 g			
Pumpkin and squash seeds	20 mL (4 tsp)	10 g			

*These items contain higher amounts of saturated fat.

TABLE F-7
Canadian Choice System: Extras

Extras have no more than 2.5 g carbohydrate, 60 kJ (14 kcal).

Vegetables 125 mL (1/2 c)

Artichokes
Asparagus
Bamboo shoots
Bean sprouts, mung or soya
Beans, string, green, or yellow
Bitter melon (balsam pear)
Bok choy
Broccoli
Brussels sprouts
Cabbage
Cauliflower
Celery
Chard
Cucumbers
Eggplant
Endive
Fiddleheads
Greens: beet, collard, dandelion, mustard, turnip, etc.
Kale
Kohlrabi
Leeks
Lettuce
Mushrooms
Okra
Onions, green or mature
Parsley
Peppers, green, yellow or red
Radishes
Rapini
Rhubarb
Sauerkraut
Shallots
Spinach
Sprouts: alfalfa, radish, etc.
Tomato wedges
Watercress
Zucchini

Free Foods (may be used without measuring)

Artificial sweetener, such as cyclamate or aspartame
Baking powder, baking soda
Bouillon from cube, powder, or liquid
Bouillon or clear broth
Chowchow, unsweetened
Coffee, clear
Consommé
Dulse
Flavorings and extracts
Garlic
Gelatin, unsweetened
Ginger root
Herbal teas, unsweetened
Horseradish, uncreamed
Lemon juice or lemon wedges
Lime juice or lime wedges
Marjoram, cinnamon, etc.
Mineral water
Mustard
Parsley
Pimentos
Salt, pepper, thyme
Soda water, club soda
Soya sauce
Sugar-free Crystal Drink
Sugar-free Jelly Powder
Sugar-free soft drinks
Tea, clear
Vinegar
Water
Worcestershire sauce

Condiments

Food	Measure
Anchovies	2 fillets
Barbecue sauce	15 mL (1 tbs)
Bran, natural	30 mL (2 tbs)
Brewer's yeast	5 mL (1 tsp)
Carob powder	5 mL (1 tsp)
Catsup	5 mL (1 tsp)
Chili sauce	5 mL (1 tsp)
Cocoa powder	5 mL (1 tsp)
Cranberry sauce, unsweetened	15 mL (1 tbs)
Dietetic fruit spreads	5 mL (1 tsp)
Maraschino cherries	1
Nondairy coffee whitener	5 mL (1 tsp)
Nuts, chopped pieces	5 mL (1 tsp)
Pickles	
unsweetened dill	2
sour mixed	11
Sugar substitutes, granular	5 mL (1 tsp)
Whipped toppings	15 mL (1 tbs)

F

TABLE F-8
Canadian Choice System: Combined Food Choices

Food	Choices per Serving	Measure	Mass (Weight)
Angel food cake	$1/2$ starch + $2^1/2$ sugars	$1/12$ cake	50 g
Apple crisp	$1/2$ starch + $1^1/2$ fruits & vegetables + 1 sugar + 1–2 fats	125 mL ($1/2$ c)	
Applesauce, sweetened	1 fruits & vegetables + 1 sugar	125 mL ($1/2$ c)	
Beans and pork in tomato sauce	1 starch + $1/2$ fruits & vegetables + $1/2$ sugar + 1 protein	125 mL ($1/2$ c)	135 g
Beef burrito	2 starches + 3 proteins + 3 fats		110 g
Brownie	1 sugar + 1 fat	1	20 g
Cabbage rolls*	1 starch + 2 proteins	3	310 g
Caesar salad	2–4 fats	20 mL dressing (4 tsp)	
Cheesecake	$1/2$ starch + 2 sugars + $1/2$ protein + 5 fats	1 piece	80 g
Chicken fingers	1 starch + 2 proteins + 2 fats	6 small	100 g
Chicken and snow pea Oriental	2 starches + $1/2$ fruits & vegetables + 3 proteins + 1 fat	500 mL (2 c)	
Chili	$1^1/2$ starches + $1/2$ fruits & vegetables + $3^1/2$ protein	300 mL ($1^1/4$ c)	325 g
Chips			
Potato chips	1 starch + 2 fats	15 chips	30 g
Corn chips	1 starch + 2 fats	30 chips	30 g
Tortilla chips	1 starch + $1^1/2$ fats	13 chips	
Cheese twist	1 starch + $1^1/2$ fats	30 chips	30 g
Chocolate bar			
Aero®	$2^1/2$ sugars + $2^1/2$ fats	bar	43 g
Smarties®	$4^1/2$ sugars + 2 fats	package	60 g
Chocolate cake (without icing)	1 starch + 2 sugars + 3 fats	$1/10$ of a 8″ pan	
Chocolate devil's food cake (without icing)	2 starches + 2 sugars + 3 fats	$1/12$ of a 9″ pan	
Chocolate milk	2 milks 2% + 1 sugar	250 mL (1 c)	300 g
Clubhouse (triple-decker) sandwich	3 starches + 3 proteins + 4 fats		
Cookies			
Chocolate chip	$1/2$ starch + $1/2$ sugar + $1^1/2$ fats	2	22 g
Oatmeal	1 starch + 1 sugar + 1 fat	2	40 g
Donut (chocolate glazed)	1 starch + $1^1/2$ sugars + 2 fats	1	65 g
Egg roll	1 starch + $1/2$ protein + 1 fat		75 g
Four bean salad	1 starch + $1/2$ protein + 1 fat	125 mL ($1/2$ c)	
French toast	1 starch + $1/2$ protein + 2 fats	1 slice	65 g
Fruit in heavy syrup	1 fruits & vegetables + $1^1/2$ sugars	125 mL ($1/2$ c)	
Granola bar	$1/2$ starch + 1 sugar + 1–2 fats		30 g
Granola cereal	1 starch + 1 sugar + 2 fats	125 mL ($1/2$ c)	45 g
Hamburger	2 starches + 3 proteins + 2 fats	junior size	

*If eaten with sauce, add $1/2$ fruits & vegetables exchange.

(*continued on the next page*)

TABLE F-8

Canadian Choice System: Combined Food Choices—continued

Food	Choices per Serving	Measure	Mass (Weight)
Ice cream and cone, plain flavour			
Ice cream	$^1/_2$ milk + 2–3 sugars + 1–2 fats		100 g
Cone	$^1/_2$ sugar		4 g
Lasagna			
Regular cheese	1 starch + 1 fruits & vegetables + 3 proteins + 2 fats	30 × 40 piece	
Low-fat cheese	1 starch + 1 fruits & vegetables + 3 proteins	30 × 40 piece	
Legumes			
Dried beans (kidney, navy, pinto, fava, chick peas)	2 starches + 1 protein	250 mL (1 c)	180 g
Dried peas	2 starches + 1 protein	250 mL (1 c)	210 g
Lentils	2 starches + 1 protein	250 mL (1 c)	210 g
Macaroni and cheese	2 starches + 2 proteins + 2 fats	250 mL (1 c)	210 g
Minestrone soup	$1^1/_2$ starches + $^1/_2$ fruits & vegetables + $^1/_2$ fat	250 mL (1 c)	
Muffin	1 starch $1^1/_2$ sugar + 1 fat	1 small	45 g
Nuts (dry or roasted without any oil added):			
Almonds, dried sliced	$^1/_2$ protein + 2 fats	50 mL ($^1/_4$ c)	22 g
Brazil nuts, dried unblanched	$^1/_2$ protein + $2^1/_2$ fats	5 large	23 g
Cashew nuts, dry roasted	$^1/_2$ starch + $^1/_2$ protein + 2 fats	50 mL ($^1/_4$ c)	28 g
Filbert hazelnut, dry	$^1/_2$ protein + $3^1/_2$ fats	50 mL ($^1/_4$ c)	30 g
Macadamia nuts, dried	$^1/_2$ protein + 4 fats	50 mL ($^1/_4$ c)	28 g
Peanuts, raw	1 protein + 2 fats	50 mL ($^1/_4$ c)	30 g
Pecans, dry roasted	$^1/_2$ fruits & vegetables + 3 fats	50 mL ($^1/_4$ c)	22 g
Pine nuts, pignolia dried	1 protein + 3 fats	50 mL ($^1/_4$ c)	34 g
Pistachio nuts, dried	$^1/_2$ fruits & vegetables + $^1/_2$ protein + $2^1/_2$ fats	50 mL ($^1/_4$ c)	27 g
Pumpkin seeds, roasted	2 proteins + 2½ fats	50 mL ($^1/_4$ c)	47 g
Sesame seeds, whole dried	$^1/_2$ fruits & vegetables + $^1/_2$ protein + $2^1/_2$ fats	50 mL ($^1/_4$ c)	30 g
Sunflower kernel, dried	$^1/_2$ protein + $1^1/_2$ fats	50 mL ($^1/_4$ c)	17 g
Walnuts, dried chopped	$^1/_2$ protein + 3 fats	50 mL ($^1/_4$ c)	26 g
Perogies	2 starches + 1 protein + 1 fat	3	
Pie, fruit	1 starch + 1 fruits & vegetables + 2 sugars + 3 fats	1 piece	120 g
Pizza, cheese	1 starch + 1 protein + 1 fat	1 slice ($^1/_8$ of a 12″)	50 g
Pork stir fry	$^1/_2$ to 1 fruits & vegetables + 3 proteins	200 mL ($^3/_4$ c)	
Potato salad	1 starch + 1 fat	125 mL ($^1/_2$ c)	130 g
Potatoes, scalloped	2 starches + 1 milk + 1–2 fats	200 mL ($^3/_4$ c)	210 g
Pudding, bread or rice	1 starch + 1 sugar + 1 fat	125 mL ($^1/_2$ c)	
Pudding, vanilla	1 milk + 2 sugars	125 mL ($^1/_2$ c)	
Raisin bran cereal	1 starch + $^1/_2$ fruits & vegetables + $^1/_2$ sugar	175 mL ($^2/_3$ c)	40 g
Rice krispie squares	$^1/_2$ starch + $1^1/_2$ sugars + $^1/_2$ fat	1 square	30 g
Shepherd's pie	2 starches + 1 fruits & vegetables + 3 proteins	325 mL ($1^1/_3$ c)	
Sherbet, orange	3 sugars + $^1/_2$ fat	125 mL ($^1/_2$ c)	
Spaghetti and meat sauce	2 starches + 1 fruits & vegetables + 2 proteins + 3 fats	250 mL (1 c)	
Stew	2 starches + 2 fruits & vegetables + 3 proteins + $^1/_2$ fat	200 mL ($^3/_4$ c)	
Sundae	4 sugars + 3 fats	125 mL ($^1/_2$ c)	
Tuna casserole	1 starch + 2 proteins + $^1/_2$ fat	125 mL ($^1/_2$ c)	
Yogurt, fruit bottom	1 fruits & vegetables + 1 milk + 1 sugar	125 mL ($^1/_2$ c)	125 g
Yogurt, frozen	1 milk + 1 sugar	125 mL ($^1/_2$ c)	125 g

F

GLOSSARY

Note to the reader: If you have a customized book, this glossary may include some entries that will not be found in your book.

absorption
The process by which nutrients and other substances are transferred from the digestive system into body fluids for transport throughout the body.

adequate diet
A diet consisting of foods that together supply sufficient protein, vitamins, and minerals and enough calories to meet a person's need for energy.

Adequate Intakes (AIs)
Provisional RDAs developed when there is insufficient evidence to support a specific level of intake.

aerobic fitness
A state of respiratory and circulatory health as measured by the ability to deliver oxygen to muscles, and the capacity of muscles to use the oxygen for physical activity.

age-related macular degeneration
Eye damage caused by oxidation of the macula, the central portion of the eye that allows you to see details clearly. It is the leading cause of blindness in U.S. adults over the age of 65. Antioxidants provided by the carotenoids in dark green, leafy vegetables such as kale, collard greens, spinach, and Swiss chard may help prevent macular degeneration.

alcohol sugars
Simple sugars containing an alcohol group in their molecular structure. The most common are xylitol, mannitol, and sorbitol.

alcoholism
An illness characterized by a dependence on alcohol and by a level of alcohol intake that interferes with health, family and social relations, and job performance.

amylophagia (am-e-low-phag-ah)
Laundry starch or cornstarch eating.

anaphylactic shock (an-ah-fa-lac-tic)
Reduced oxygen supply to the heart and other tissues due to the body's reaction to an allergen in food. Symptoms of anaphylactic shock include paleness, a flush discoloration of the lips and fingertips, a "faceless" expression, weak and rapid pulse, and difficulty breathing.

anorexia nervosa
An eating disorder characterized by extreme weight loss, poor body image, and irrational fears of weight gain and obesity.

antibodies
Blood proteins that help the body fight particular diseases. They help the body develop an immunity, or resistance, to many diseases.

antioxidants
Chemical substances that prevent or repair damage to cells caused by exposure to oxidizing agents such as environmental pollutants, smoke, ozone, and oxygen. Oxidation reactions are also a normal part of cellular processes. Beta-carotene, vitamin E, and vitamin C function as antioxidants.

appetite
The desire to eat; a pleasant sensation that is aroused by thoughts of the taste and enjoyment of food.

association
The finding that one condition is correlated with, or related to another condition, such as a disease or disorder. For example, diets low in vegetables are associated with breast cancer. Associations do *not* prove that one condition (such as a diet low in vegetables) *causes* an event (such as breast cancer). They indicate that a statistically significant relationship between a condition and an event exists.

atherosclerosis
"Hardening of the arteries" due to a build-up of plaque.

ATP, ADP
Adenosine triphosphate (ah-den-o-scene tri-phos-fate) and adenosine diphosphate. Molecules containing a form of phosphorous that can trap energy obtained from the macronutrients. ADP becomes ATP when it traps energy, and returns to being ADP when it releases energy for muscular and other work.

balanced diet
A diet that provides neither too much nor too little of nutrients and other components of food such as fat and fiber.

bariatrics
The field of medicine concerned with weight loss.

basal metabolism
Energy used to support body processes such as growth, health, tissue repair and maintenance, and other functions. Assessed while at rest, basal metabolism includes energy the body expends for breathing, the pumping of the heart, the maintenance of body temperature, and other life-sustaining, ongoing functions. Also called resting metabolism.

bile
A yellowish-brown or green fluid produced by the liver, stored in the gallbladder, and secreted into the small intestine. It acts like a detergent, breaking down globs of fat entering the small intestine to droplets, making the fats more accessible to the action of lipase.

binge eating
The consumption of a large amount of food in a small amount of time.

binge-eating disorder
An eating disorder characterized by periodic binge eating, which normally is not followed by vomiting or the use of laxatives. People must experience eating binges twice a week on average over a period of 6 months to qualify for the diagnosis.

bioavailability
The amount of a nutrient consumed that is available for absorption and use by the body.

biotechnology
As applied to food products, the process of modifying the composition of foods by biologically altering their genetic makeup. Also called *genetic engineering* of foods. The food products produced are sometimes referred to as "GM" and GMOs (genetically modified organisms).

body mass index (BMI)
An indicator of the appropriateness of a person's weight for their height. It is calculated by dividing weight in kilograms by height in meters. It can also be calculated using the method shown in Table 9.2.

bulimia nervosa
An eating disorder characterized by recurrent episodes of rapid, uncontrolled eating of large amounts of food in a short period of time. Episodes of binge eating are often followed by purging.

calorie (calor = heat)
A unit of measure used to express the amount of energy produced by foods in the form of heat. The calorie used in nutrition is the large "Calorie," or the "kilocalorie" (kcal). It equals the amount of energy needed to raise the temperature of 1 kilogram of water (about 4 cups) from 15 to 16°C (59 to 61°F). The term *kilocalorie,* or "calorie" as used in this text, is gradually being replaced by the "kilojoule" (kJ) in the United States; 1 kcal = 4.2 kJ.

cancer
A group of diseases in which abnormal cells grow out of control and can spread throughout the body. Cancer is not contagious and has many causes.

carbohydrates
Chemical substances in foods that consist of a simple sugar molecule or multiples of them in various forms.

cardiovascular disease
Disorders related to plaque build-up in arteries of the heart, brain, and other organs and tissues.

cataracts
Complete or partial clouding over the lens of the eye.

cause and effect
A finding that demonstrates that a condition causes a particular event. For example, vitamin C deficiency causes the deficiency disease scurvy.

celiac disease
A disease characterized by inflammation of the small intestine lining resulting from a genetically based intolerance to gluten. The inflammation produces diarrhea, fatty stools, weight loss, and vitamin and mineral deficiencies. (Also called *celiac sprue* and *gluten-sensitive enteropathy.*)

cholesterol
A fat-soluble, colorless liquid found in many animal products but not in plants. High cholesterol intake is related to the development of heart disease.

chronic diseases
Slow-developing, long-lasting diseases that are not contagious (e.g., heart disease, cancer, diabetes). They can be treated but not always cured.

chronic inflammation
Inflammation that lasts weeks, months, or years. Inflammation is the first response of the body's immune system to infection or irritation. Inflammation triggers the release of biologically active substances that promote oxidation and other potentially harmful reactions in the body.

circulatory system
The heart, arteries, capillaries, and veins responsible for circulating blood throughout the body.

cirrhosis of the liver
Degeneration of the liver, usually caused by excessive alcohol intake over a number of years.

clinical trial
A study design in which one group of randomly assigned subjects (or subjects selected by the "luck of the draw") receives an active treatment and another group receives an inactive treatment, or "sugar pill," called the placebo.

coenzymes
Chemical substances, including many vitamins, that activate specific enzymes. Activated enzymes increase the rate at which reactions take place in the body, such as the breakdown of fats or carbohydrates in the small intestine and the conversion of glucose and fatty acids into energy within cells.

cofactors
Individual minerals required for the activity of certain proteins. For example: • iron is needed for hemoglobin's function in oxygen and carbon dioxide transport, • zinc is needed to activate or is a structural component of over 200 enzymes, and • magnesium activates over 300 enzymes involved in the formation of energy and proteins.

colostrum
The milk produced during the first few days after delivery. It contains more antibodies, protein, and certain minerals than the mature milk that is produced later. It is thicker than mature milk and has a yellowish color.

complementary protein sources
Plant sources of protein that together provide sufficient quantities of the nine essential amino acids.

complete proteins
Proteins that contain all of the essential amino acids in amounts needed to support growth and tissue maintenance.

complex carbohydrates
The form of carbohydrate found in starchy vegetables, grains, and dried beans and in many types of dietary fiber. The most common form of starch is made of long chains of interconnected glucose units.

control group
Subjects in a study who do not receive the active treatment or who do *not* have the condition under investigation. Control periods, or times when subjects are not receiving the treatment, are sometimes used instead of a control group.

critical period
A specific interval of time during which cells of a tissue or organ are genetically programmed to multiply. If the supply of nutrients needed for cell multiplication is not available during the specific time interval, the growth and development of the tissue or organ are permanently impaired.

cruciferous vegetables
Sulfur-containing vegetables whose outer leaves form a cross (or crucifix). Vegetables in this family include broccoli, cabbage, cauliflower, brussels sprouts, mustard and collard greens, kale, bok choy, kohlrabi, rutabaga, turnips, broccoflower, and watercress.

Daily Values (DVs)
Scientifically agreed-upon daily dietary intake standards for fat, saturated fat, cholesterol, carbohydrate, dietary fiber, and protein intake compatible with health. DVs are intended for use on nutrition labels only and are listed in the Nutrition Facts panel. "% Daily Value" on Nutrition Facts panels is calculated as the percentage of each DV supplied by a serving of the labeled food.

development
Processes involved in enhancing functional capabilities. For example, the brain grows, but the ability to reason develops.

diabetes
A disease characterized by abnormal utilization of carbohydrates by the body and elevated blood glucose levels. There are three main types of diabetes: type 1, type 2, and gestational diabetes. The word *diabetes* in this unit refers to type 2, which is by far the most common form of diabetes.

diarrhea
The presence of three or more liquid stools in a 24-hour period.

dietary fiber
Naturally occurring, intact forms of nondigestible carbohydrates in plants and "woody" plant cell walls. Oat and wheat bran, and raffinose in dried beans, are examples of this type of fiber.

dietary supplements
Any products intended to supplement the diet, including vitamin and mineral supplements; proteins, enzymes, and amino acids; fish oils and fatty acids; hormones and hormone precursors; and herbs and other plant extracts. Such products must be labeled "Dietary Supplement."

dietary thermogenesis
Thermogenesis means "the production of heat." Dietary thermogenesis is the energy expended during food ingestion, the digestion of food, and the absorption and utilization of nutrients. Some of the energy escapes as heat. It accounts for approximately 10% of the body's total energy need. Also called *diet-induced thermogenesis* and *thermic effect of foods or feeding.*

digestion
The mechanical and chemical processes whereby ingested food is converted into substances that can be absorbed by the intestinal tract and utilized by the body.

disaccharides (di = two, saccharide = sugar)
Simple sugars consisting of two molecules of monosaccharides linked together. Sucrose, maltose, and lactose are disaccharides.

DNA (deoxyribonucleic acid)
Genetic material contained in cells that initiates and directs the production of proteins in the body.

double blind
A study in which neither the subjects participating in the research nor the scientists performing the research know which subjects are receiving the treatment and which are getting the placebo. Both subjects and investigators are "blind" to the treatment administered.

double-blind, placebo-controlled food challenge
A test used to determine the presence of a food allergy or other adverse reaction to a food. In this test, neither the patient nor the care provider knows whether a suspected offending food or a placebo is being tested.

duodenal (do-odd-en-all) and stomach ulcers
Open sores in the lining of the duodenum (the uppermost part of the small intestine) or the stomach.

edema
Swelling due to an accumulation of fluid in body tissues.

electrolytes
Minerals such as sodium and potassium that carry a charge when in solution. Many electrolytes help the body maintain an appropriate amount of fluid.

empty-calorie foods
Foods that provide an excess of calories in relation to nutrients. Soft drinks, candy, sugar, alcohol, and fats are considered empty-calorie foods.

endothelium
(Pronounced *n-dough-theil-e-um*) The layer of cells lining the inside of blood vessels.

enrichment
The replacement of thiamin, riboflavin, niacin, and iron lost when grains are refined.

environmental trigger
An environmental factor, such as inactivity, a high-fat diet, or a high sodium intake, that causes a genetic tendency toward a disorder to be expressed.

enzymes
Protein substances that speed up chemical reactions. Enzymes are found throughout the body but are present in particularly large amounts in the digestive system.

epidemiological studies
Research that seeks to identify conditions related to particular events within a population. This type of research does *not* identify cause-and-effect relationships. For example, much of the information known about diet and cancer is based on epidemiological studies that have found that diets low in vegetables and fruits are associated with the development of heart disease.

ergogenic aids (ergo = work; genic = producing)
Substances that increase the capacity for muscular work.

essential amino acids
Amino acids that cannot be synthesized in adequate amounts by humans and therefore must be obtained from the diet. They are sometimes referred to as "indispensable amino acids."

essential fatty acids
Components of fats (linoleic acid–pronounced lynn-oh-lay-ick and alpha-linolenic acid–lynn-oh-len-ick) required in the diet.

essential hypertension
Hypertension of no known cause; also called primary or idiopathic hypertension, it accounts for 90–95% of all cases of hypertension.

essential nutrients
Substances required for normal growth and health that the body cannot generally produce, or produce in sufficient amounts; they must be obtained in the diet.

experimental group
Subjects in a study who receive the treatment being tested or have the condition that is being investigated.

fermentation
The process by which carbohydrates are converted to ethanol by the action of the enzymes in yeast.

fetus
A baby in the womb from the eighth week of pregnancy until birth. (Before then, it is referred to as an embryo.)

flatulence (flat-u-lens)
Presence of excess gas in the stomach and intestines.

food additives
Any substances added to food that become part of the food or affect the characteristics of the food. The term applies to substances added both intentionally and unintentionally to food.

food allergen
A substance in food (almost always a protein) that is identified as harmful by the body and elicits an allergic reaction from the immune system.

food allergy
Adverse reaction to a normally harmless substance in food that involves the body's immune system. (Also called food hypersensitivity.)

food insecurity
Limited or uncertain availability of safe, nutritious foods.

food intolerance
Adverse reaction to a normally harmless substance in food that does not involve the body's immune system.

food security
Access at all times to a sufficient supply of safe, nutritious foods.

foodborne illnesses
An illness related to consumption of foods or beverages containing disease-causing bacteria, viruses, parasites, toxins, or other contaminants.

fortification
The addition of one or more vitamins and/or minerals to a food product.

free radicals
Chemical substances (usually oxygen or hydrogen) that are missing an electron. The absence of the electron makes the chemical substances reactive and prone to oxidizing nearby atoms or molecules by stealing an electron from them.

fruitarian
A form of vegetarian diet in which fruits are the major ingredient. Such diets provide inadequate amounts of a variety of nutrients.

functional fiber
Specific types of nondigestible carbohydrates and connective tissues that have beneficial effects on health. Two examples of functional fibers are psyllium and pectin.

functional foods
Generally taken to mean foods, fortified foods, and enhanced food products that may benefit health beyond the effects of essential nutrients they contain.

genotype
The specific genetic makeup of an individual as coded by DNA.

geophagia (ge-oh-phag-ah)
Clay or dirt eating.

gestational diabetes
Diabetes first discovered during pregnancy.

glycemic index (GI)
A measure of the extent to which blood glucose level is raised by a 50-gram portion of a carbohydrate-containing food compared to 50 grams of glucose or white bread.

glycemic load (GL)
A measure of the extent to which blood glucose level is raised by a given amount of a carbohydrate-containing food. GL is calculated by multiplying a food's GI by its carbohydrate content.

glycogen
The body's storage form of glucose. Glycogen is stored in the liver and muscles.

growth
A process characterized by increases in cell number and size.

heart disease
One of a number of disorders that result when circulation of blood to parts of the heart is inadequate. Also called coronary heart disease. ("Coronary" refers to the blood vessels at the top of the heart. They look somewhat like a crown.)

heartburn
A condition that results when acidic stomach contents are released into the esophagus, usually causing a burning sensation.

hemoglobin
The iron-containing protein in red blood cells.

hemorrhoids (hem-or-oids)
Swelling of veins in the anus or rectum.

histamine (hiss-tah-mean)
A substance released in allergic reactions. It causes dilation of blood vessels, itching, hives, and a drop in blood pressure and stimulates the release of stomach acids and other fluids. Antihistamines neutralize the effects of histamine and are used in the treatment of some cases of allergies.

homocysteine
A compound produced when the amino acid methionine is converted to another amino acid, cysteine. High blood levels of homocysteine increase the risk of hardening of the arteries, heart attack, and stroke.

hormone
A substance, usually a protein or steroid (a cholesterol-derived chemical), produced by one tissue and conveyed by the bloodstream to another. Hormones affect the body's metabolic processes such as glucose utilization and fat deposition.

hunger
Unpleasant physical and psychological sensations (weakness, stomach pains, irritability) that lead people to acquire and ingest food.

hydrogenation
The addition of hydrogen to unsaturated fatty acids.

hypertension
High blood pressure. It is defined as blood pressure exerted inside blood vessel walls that typically exceeds 140/90 millimeters of mercury.

hypoglycemia
A disorder resulting from abnormally low blood glucose levels. Symptoms of hypoglycemia include irritability, nervousness, weakness, sweating, and hunger. These symptoms are relieved by consuming glucose or foods that provide carbohydrate.

hypothesis
A statement made prior to initiating a study of the relationship sought to be proved (found to be true) by the research.

immunoproteins
Blood proteins such as phagocytes (pronounced fag-o-sites) that play a role in the functioning of the immune system. White blood cells are a type of phagocyte. They engulf and absorb foreign bodies and harmful organisms in blood and tissues.

immune system
Body tissues that provide protection against bacteria, viruses, and other substances identified by cells as harmful.

incomplete proteins
Proteins that are deficient in one or more essential amino acids.

infant mortality rate
Deaths that occur within the first year of life per 1,000 live births.

initiation
The start of the cancer process; it begins with the alteration of DNA within cells.

insulin resistance
A condition in which cell membrane have reduced sensitivity to insulin so that more insulin than normal is required to transport a given amount of glucose into cells. It is characterized by elevated levels of serum insulin, glucose, and triglycerides, and increased blood pressure.

iron deficiency
A disorder that results from a depletion of iron stores in the body. It is characterized by weakness, fatigue, short attention span, poor appetite, increased susceptibility to infection, and irritability.

iron-deficiency anemia
A condition that results when the content of hemoglobin in red blood cells is reduced due to a lack of iron. It is characterized by the signs of iron deficiency plus paleness, exhaustion, and a rapid heart rate.

irritable bowel syndrome (IBS)
A disorder of bowel function characterized by chronic or episodic gas, abdominal pain, diarrhea or constipation, or both.

kwashiorkor (kwa-she-or-kor)
A deficiency disease related to a lack of calories and protein and abnormal metabolic adaptations to starvation.

lactose intolerance
The term for gastrointestinal symptoms (flatulence, bloating, abdominal pain, diarrhea, and "rumbling in the bowel") resulting from the consumption of more lactose than can be digested with available lactase.

lactose intolerance
The term for gastrointestinal symptoms (flatulence, bloating, abdominal pain, diarrhea, and "rumbling in the bowel") resulting from the consumption of more lactose than can be digested with available lactase.

lactose maldigestion
A disorder characterized by reduced digestion of lactose due to the low availability of the enzyme lactase.

life expectancy
The average length of life of people of a given age.

lipids
Compounds that are insoluble in water and soluble in fat. Triglycerides, saturated and unsaturated fats, and essential fatty acids are examples of lipids, or "fats."

low-birthweight infants
Infants weighing less than 2,500 grams (5.5 pounds) at birth.

lymphatic system
A network of vessels that absorb some of the products of digestion and transport them to the heart, where they are mixed with the substances contained in blood.

macronutrients
The group name for the energy-yielding nutrients of carbohydrate, protein, and fat. They are called macronutrients because we need relatively large amounts of them in our daily diet.

malnutrition
Poor nutrition resulting from an excess or lack of calories or nutrients.

marasmus
A condition of severe body wasting due to deficiencies of both protein and calories. Also called protein-energy malnutrition and protein-calorie malnutrition.

maximal oxygen consumption
The highest amount of oxygen that can be delivered to, and utilized by, muscles for physical activity. Also called VO$_2$ max and maximal volume of oxygen.

meta-analysis
An analysis of data from multiple studies. Results are based on larger samples than the individual studies and are therefore more reliable. Differences in methods and subjects among the studies may bias the results of meta-analyses.

metabolic syndrome
A constellation of metabolic abnormalities that increase the risk of heart disease, hypertension, type 2 diabetes, and other disorders. Metabolic syndrome is characterized by insulin resistance, abdominal obesity, high blood pressure and triglycerides levels, low levels of HDL cholesterol, and impaired glucose tolerance. It is also called *Syndrome X* and *insulin resistance syndrome*.

metabolism
The chemical changes that take place in the body. The conversion of glucose to energy or to body fat is an example of a metabolic process.

minerals
In the context of nutrition, minerals are specific, single atoms that perform particular functions in the body. There are 15 essential minerals—or minerals required in the diet.

monosaccharides
(mono = one, saccharide = sugar): Simple sugars consisting of one sugar molecule. Glucose, fructose, and galactose are monosaccharides.

myoglobin
The iron-containing protein in muscle cells.

neural tube defects
Malformations of the spinal cord and brain. They are among the most common and severe fetal malformations, occurring in approximately 1 in every 1000 pregnancies. Neural tube defects include spina bifida (spinal cord fluid protrudes through a gap in the spinal cord; shown in Illustration 29.6), anencephaly (absence of the brain or spinal cord), and encephalocele (protrusion of the brain through the skull).

nonessential amino acids
Amino acids that can be readily produced by humans from components of the diet. Also referred to as "dispensable amino acids."

nonessential nutrients
Nutrients required for normal growth and health that the body can manufacture in sufficient quantities from other components of the diet. We do not require a dietary source of nonessential nutrients.

nutrigenomics
The study of diet- and nutrient-related functions and interactions of genes and their affects on health and disease.

nutrient-dense foods
Foods that contain relatively high amounts of nutrients compared to their calorie value. Broccoli, collards, bread, cantaloupe, and lean meats are examples of nutrient-dense foods.

nutrients
Chemical substances found in food that are used by the body for growth and health. The six categories of nutrients are carbohydrates, proteins, fats, vitamins, minerals, and water.

nutrition
The study of foods, their nutrients and other chemical constituents, and the effects of food constituents on health.

obesity
A condition characterized by excess body fat.

osteoporosis (osteo = bones; poro = porous, osis = abnormal condition)
A condition in which bones become fragile and susceptible to fracture due to a loss of calcium and other minerals.

overweight
A high weight-for-height.

pagophagia (pa-go-phag-ah)
Ice eating.

peer review
Evaluation of the scientific merit of research or scientific reports by experts in the area under review. Studies published in scientific journals have gone through peer review prior to being accepted for publication.

% Daily Value (%DV)
Scientifically agreed-upon standards of daily intake of nutrients from the diet developed for use on nutrition labels. The "% Daily Values" listed in nutrition labels represent the percentages of the standards obtained from one serving of the food product.

phenylketonuria (feen-ol-key-tone-u-re-ah), PKU
A rare genetic disorder related to the lack of the enzyme phenylalanine hydroxylase. Lack of this enzyme causes the essential amino acid phenylalanine to build up in blood.

physical fitness
The health of the body as measured by muscular strength, endurance, and flexibility in the conduct of physical activity.

phytochemicals (phyto = plant)
Chemical substances in plants, some of which perform important functions in the human body. Phytochemicals give plants color and flavor, participate in processes that enable plants to grow, and protect plants against insects and diseases.

pica (pike-eh)
The regular consumption of nonfood substances such as clay or laundry starch.

placebo
A "sugar pill," an imitation treatment given to subjects in research.

placebo effect
Changes in health or perceived health that result from expectations that a "treatment" will produce an effect on health.

plant stanols or sterols
Substances in corn, wheat, oats, rye, olives, wood, and some other plants that are similar in structure to cholesterol but that are not absorbed by the body. They decrease cholesterol absorption.

plaque (dental)
A soft, sticky, white material on teeth; formed by bacteria.

plaque (arterial)
Deposits of cholesterol, other fats, calcium, and cell materials in the lining of the inner wall of arteries.

plumbism
Lead (primarily from old paint flakes) eating.

polysaccharides (poly = many, saccharide = sugar)
Carbohydrates containing many molecules of sugar linked together. Starch, glycogen, and dietary fiber are the three major types of polysaccharides.

prebiotics
"Intestinal fertilizer." Certain fiber-like forms of nondigestible carbohydrates that support the growth of beneficial bacteria in the gut.

precursor
In nutrition, a nutrient that can be converted into another nutrient. (Also called provitamin.) Beta-carotene is a precursor of vitamin A.

prediabetes
A condition in which blood glucose levels are higher than normal but not high enough for the diagnosis of diabetes. It is characterized by impaired glucose tolerance, or fasting blood glucose levels between 110 and 126 mg/dl.

preterm
Infants born at or before 37 weeks of gestation (pregnancy).

probiotics
Non-harmful bacteria and some yeasts that help colonize the intestinal tract with beneficial microorganisms and that sometimes replace colonies of harmful microorganisms. Most common probiotic strains are Lactobacilli and Bifidobacteria.

progression
The uncontrolled growth of abnormal cells.

promotion
The period in cancer development when the number of cells with altered DNA increases.

prostate
A gland located above the testicles in males. The prostate secretes a fluid that surrounds sperm.

protein
Chemical substance in foods made up of chains of amino acids.

purging
The use of self-induced vomiting, laxatives, or diuretics (water pills) to prevent weight gain.

Recommended Dietary Allowances (RDAs)
Intake levels of essential nutrients that meet the nutritional needs of practically all healthy people while decreasing the risk of certain chronic diseases.

remodeling
The breakdown and buildup of bone tissue.

restrained eating
The purposeful restriction of food intake below desired amounts in order to control body weight.

salt sensitivity
A genetically determined condition in which a person's blood pressure rises when high amounts of salt or sodium are consumed. Such individuals are sometimes identified by blood pressure increases of 10% or more when switched from a low-salt (1–3 grams) to a high-salt (12–15 grams) diet.

satiety
A feeling of fullness or of having had enough to eat.

saturated fats
The type of fat that tends to raise blood cholesterol levels and the risk for heart disease. They are solid at room temperature and are found primarily in animal products such as meat, butter, and cheese.

serotonin (pronounced sare-uh-tone-in)
A neurotransmitter, or chemical messenger, for nerve cell activities that excite or inhibit various behaviors and body functions. It plays a role in mood, appetite regulation, food intake, respiration, pain transmission, blood vessel constriction, and other body processes.

simple sugars
Carbohydrates that consist of a glucose, fructose, or galactose molecule; or a combination of glucose and either fructose or galactose. High-fructose corn syrup and alcohol sugars are also considered simple sugars. Simple sugars are often referred to as "sugars."

single-gene defects
Disorders resulting from one abnormal gene. Also called "inborn errors of metabolism." Over 800 single-gene defects have been cataloged, and most are very rare.

starch
Complex carbohydrates made up of complex chains of glucose molecules. Starch is the primary storage form of carbohydrate in plants. The vast majority of carbohydrate in our diet consists of starch, monosaccharides, and disaccharides.

statistically significant
Research findings that likely represent a true or actual result and not one due to chance.

stroke
The event that occurs when a blood vessel in the brain suddenly ruptures or becomes blocked, cutting off blood supply to a portion of the brain. Stroke is often associated with "hardening of the arteries" in the brain. (Also called a *cerebral vascular accident*.)

subcutaneous fat
(Pronounced *sub-q-tain-e-ous*) Fat located under the skin.

tooth decay
The disintegration of teeth due to acids produced by bacteria in the mouth that feed on sugar. Also called dental caries or cavities.

total fiber
The sum of functional and dietary fiber.

trans fats
A type of unsaturated fat present in hydrogenated oils, margarine, shortening, pastries, and some cooking oils that increase the risk of heart disease. Fats containing fatty acids in the trans form are generally referred to as trans fats.

trimester
One-third of the normal duration of pregnancy. The first trimester is 0 to 13 weeks, the second is 13 to 26 weeks, and the third is 26 to 40 weeks.

tryptophan (pronounced trip-tuh-fan)
An essential amino acid that is used to form the chemical messenger serotonin (among other functions). Tryptophan is generally present in lower amounts in food protein

than most other essential amino acids. It can be produced in the body from niacin, a B vitamin.

type 1 diabetes
A disease characterized by high blood glucose levels resulting from destruction of the insulin-producing cells of the pancreas. This type of diabetes was called juvenile-onset diabetes and insulin-dependent diabetes in the past, and its official medical name is type 1 diabetes mellitus.

type 2 diabetes
A disease characterized by high blood glucose levels due to the body's inability to use insulin normally, or to produce enough insulin. This type of diabetes was called adult-onset diabetes and non-insulin-dependent diabetes in the past, and its official medical name is type 2 diabetes mellitus.

underweight
Usually defined as a low weight-for-height. May also represent a deficit of body fat.

unsaturated fats
The type of fat that tends to lower blood cholesterol level and the risk of heart disease. They are liquid at room temperature and found in foods such as nuts, seeds, fish, shellfish, and vegetable oils.

visceral fat
(Pronounced *vis-sir-el*) Fat located under the skin and muscle of the abdomen.

vitamins
Chemical substances that perform specific functions in the body.

water balance
The ratio of the amount of water outside cells to the amount inside cells; this balance is needed for normal cell functioning.

zoochemicals
Chemical substances in animal foods, some of which perform important functions in the body.

CREDITS

CR-1

This page constitutes an extension of the copyright page. We have made every effort to trace the ownership of all copyrighted material and to secure permission from copyright holders. In the event of any question arising as to the use of any material, we will be pleased to make the necessary corrections in future printings. Thanks are due to the following authors, publishers, and agents for permission to use the material indicated.

Chapter Opener 1. 1-01: bottom, Photo Disc **1-01:** top, Photo Disc **Chapter Opener 2. 2-01:** bottom, Photo Disc **2-01:** top, Photo Disc **Chapter Opener 3. 3-01:** bottom, Photo Disc **3-01:** top, Photo Disc **Chapter Opener 4. 4-01:** bottom, Photo Disc **4-01:** top, Photo Disc **Chapter Opener 5. 5-01:** bottom, Photo Disc **5-01:** top, Photo Disc **Chapter Opener 6. 6-01:** By William Carlos Williams, from Collected Poems, 1909-1939, Volume 1, copyright 1938 by New Directions Publishing Corp. Reprinted by permission of New Directions Publishing Corp. **6-01:** bottom, Photo Disc **6-01:** top, Photo Disc **Chapter Opener 7. 7-01:** bottom, Photo Disc **7-01:** top, Photo Disc **Chapter Opener 8. 8-01:** bottom, Photo Disc **8-01:** top, Photo Disc **Chapter Opener 9. 9-01:** bottom, Photo Disc **9-01:** top, Photo Disc **Chapter Opener 10. 10-01:** bottom, Photo Disc **10-01:** top, Photo Disc **Chapter Opener 11. 11-01:** bottom, Photo Disc **11-01:** top, Photo Disc **Chapter Opener 12. 12-01:** bottom, Photo Disc **12-01:** top, Photo Disc **Chapter Opener 13. 13-01:** bottom, Photo Disc **13-01:** top, Photo Disc **Chapter Opener 14. 14-01:** top, Photo Disc **14-01:** bottom, Photo Disc **Chapter Opener 15. 15-01:** bottom, Photo Disc **15-01:** top, Photo Disc **Chapter Opener 16. 16-01:** bottom, Photo Disc **16-01:** top, Photo Disc **Chapter Opener 17. 17-01:** bottom, Photo Disc **17-01:** top, Photo Disc **Chapter Opener 18. 18-01:** bottom, Photo Disc **18-01:** top, Photo Disc **Chapter Opener 19. 19-01:** bottom, Photo Disc **19-01:** top, Photo Disc **Chapter Opener 20. 20-01:** bottom, Photo Disc **20-01:** top, Photo Disc **Chapter Opener 21. 21-01:** bottom, Photo Disc **21-01:** top, Photo Disc **Chapter Opener 22. 22-01:** bottom, Photo Disc **22-01:** top, Photo Disc **Chapter Opener 23. 23-01:** bottom, Photo Disc **23-01:** top, Photo Disc **Chapter Opener 24. 24-01:** bottom, Photo Disc **24-01:** top, Photo Disc **Chapter Opener 25. 25-01:** bottom, Photo Disc **25-01:** top, Photo Disc **Chapter Opener 26. 26-01:** bottom, Photo Disc **26-01:** top, Photo Disc **Chapter Opener 27. 27-01:** bottom, Photo Disc **27-01:** top, Photo Disc **Chapter Opener 28. 28-01:** bottom, Photo Disc **28-01:** top, Photo Disc **Chapter Opener 29. 29-01:** bottom, Photo Disc **29-01:** top, Photo Disc **Chapter Opener 30. 30-01:** bottom, Photo Disc **30-01:** top, Photo Disc **Chapter Opener 31. 31-01:** bottom, Photo Disc **31-01:** top, Photo Disc **Chapter Opener 32. 32-01:** bottom, Photo Disc **Chapter Opener 33. 33-01:** bottom, Photo Disc **33-01:** top, Photo Disc

Chapter 1. 1-02: Gary Conner/Photo Edit **1-03:** top right, Phil Schermeister/Corbis **1-03:** bottom right, Photo Disc **1-03:** top left, Royalty free/CORBIS **1-03:** bottom left, Royalty free/CORBIS **1-04:** top right, David Young-Wolff/Photo Edit **1-04:** top left, Photo Disc **1-04:** bottom, AP/Wide World Photos **1-05:** top right, Photo Disc **1-05:** top left, Royalty Free/CORBIS **1-06:** Richard Anderson **1-07:** bottom left, Photo Disc **1-07:** bottom right, Photo Disc **1-08:** bottom, Brand X Pictures **1-08:** top, W.P. Armstrong **1-09:** top, Photo Disc **1-09:** top center, Photo Disc **1-09:** top right, Photo Disc **1-09:** center left, Photo Disc **1-09:** bottom left, Photo Disc **1-10:** bottom, Photo Disc **1-10:** top left, Photo Disc **1-11:** Ann Borman, Wadsworth/Thomson **1-12:** bottom left, Food and Nutrition Board **1-12:** left, Photo Disc **1-12:** center right, Photo Disc **1-12:** right, Photo Disc **1-12:** center left, Photo Disc **1-13:** Ann Borman, Wadsworth/Thomson **1-14:** Index Stock **1-15:** top right, Photo Disc **1-15:** top right, Ann Borman, Wadsworth/Thomson **1-16:** bottom, Chad Johnston/Masterfile **1-16:** center left, ICI Pharmaceuticals **1-16:** top left, Photo Disc **1-16:** center right, Photo Disc **1-16:** top right, Photo Disc **1-16:** top center, Photo Disc **1-16:** top center, Stan Maddock **1-18:** Erika McConnell/Getty Images **1-19:** Stan Maddock **1-20:** Courtesy, USDA,

Chapter 2. 2-03: bottom right, Biophoto/Photo Researchers, Inc. **2-03:** top right, Photo Disc **2-03:** center right, Photo Disc **2-03:** center left, Photo Disc **2-03:** bottom left, Photo Researchers, Inc. **2-03:** top right, Stan Maddock **2-04:** Ann Borman, Wadsworth/Thomson **2-04:** top, Stan Maddock **2-05:** top left, Photo Disc **2-05:** bottom left, Photo Disc **2-05:** bottom right, Photo Disc **2-05:** top right, Photo Disc **2-06:** right, Anthro-Photo File **2-06:** left, Anthro-Photo File **2-07:** left, Irven DeVore, Anthro-Photo File **2-07:** right, Richard Anderson **2-08:** Ann Borman, Wadsworth/Thomson **2-08:** bottom right, Photo Disc **2-09:** bottom, Ann Borman, Wadsworth/Thomson **2-09:** top right, Stan Maddock **2-10:** top, Photo Disc **2-10:** bottom, USDA **2-11:** center, Courtesy, USDA

Chapter 3. 3-02: Photo Disc **3-03:** Photo Disc **3-04:** Stan Maddock **3-05:** top, Stan Maddock **3-05:** bottom, Stan Maddock **3-06:** bottom, Richard Anderson **3-07:** bottom, Richard Anderson **3-07:** top, Stan Maddock **3-09:** These materials developed and distributed by the Nutrition Screening Initiative, a project of: American Academy of Family Physicians, The American Dietetic Association, and the National Council of the Aging. Reprinted by permission. **3-09 & 03-10:** Photo Disc **3-11:** Mug Shots/Corbis **3-12:** Photo Disc **3-15:** top, Stan Maddock **3-15:** bottom right, Stan Maddock **3-16:** right, Herbert Spichtinger/Zefa/Corbis **3-16:** left, Photo Disc **3-17:** Stan Maddock

Chapter 4. 4-02: Richard Anderson **4-03:** Stan Maddock **4-05:** top, Photo Disc **4-05:** bottom, Stan Maddock **4-08:** right, Richard Anderson **4-08:** left, Richard Anderson **4-09:** left, Richard Anderson **4-09:** right, Richard Anderson **4-10:** Copyright, Judith E. Brown, 2006 **4-11:** top, Courtesy of Health Canada **4-11:** bottom, Richard Anderson **4-12:** bottom, Richard Anderson **4-12:** top, Richard Anderson **4-13:** center, Erik Freeland/Corbis **4-13:** top, Richard Anderson **4-13 & 04-15:** bottom, Photo Disc

Chapter 5. 5-02: Jorge Uzon/AFP/Getty Images **5-03:** top, Photo Disc **5-03:** bottom right, Photo Disc **5-04:** top, J. E. Steiner, Taste and Development: The Genesis of Sweet Preference, ed J.M. Weiffenbach, HHS Publication no. NIH 77-1068 (Bethesda, Md.: US Department of Health and Human Services, 1977). **5-04:** bottom, Michael Newman/Photo Edit **5-05:** Photo Disc **5-06:** bottom, Photo Disc **5-06:** top, Photo Disc **5-07:** bottom, Photo Disc **5-07:** top, Stan Maddock **5-09 & 05-10:** Photo Disc **5-10:** center, Enrique Marcarian, Reuters NewMedia, Corbis **5-11:** PhotoEdit, Felicia Martinez **5-12:** Ed Bock/Corbis

Chapter 6. 6-02: bottom left, Photo Disc **6-02:** top right, Photo Disc **6-02:** bottom right, Photo Disc **6-02:** top left, Susie Eising/Getty Images **6-03:** top left, Photo Disc **6-03:** bottom left, Photo Disc **6-03:** top right, Photo Disc **6-03:** bottom right, Photo Disc **6-04:** Photo Disc **6-05:** Photo Disc **6-07:** USDA **6-08:** Courtesy, USDA **6-10:** Courtesy, USDA **6-10:** Courtesy, USDA **6-11:** Courtesy, USDA **6-12:** Photo Disc **6-13:** top, Courtesy, Media Partners, Inc., www.mediapartners.com **6-13:** bottom left, Oldways Preservation and Exchange Trust **6-13:** bottom right, Oldways Preservation and Exchange Trust **6-14:** Photo Disc **6-15:** top, Photo Disc **6-15 & 06-17:** bottom, Photo Disc **6-16:** top right, Envision/Corbis **6-16:** bottom left, Photo Disc **6-16:** top left, Royalty-Free/Corbis **6-17:** top right, Photo Disc **6-17:** top left, Royalty-free **6-18:** bottom, Richard Anderson **6-18:** top, www.slowfood.com

Chapter 7. 7-02: Photo Disc **7-03:** Index Stock **7-06:** top, Michael Newman, PhotoEdit **7-06:** bottom left, Photo Researchers, Inc., Meckes/Ottawa **7-06:** bottom right, Prof. P. Motta/Photo Researchers, Inc. **7-07:** McMahon **7-08:** Stan Maddock **7-09:** Richard Anderson **7-10:** Photo Disc **7-11:** Stan Maddock **7-12:** top, Photo Disc **7-12:** bottom, Richard Anderson

Chapter 8. 8-02: Stan Maddock **8-03:** bottom right, Alfred Pasieka/Photo Researchers, Inc. **8-03:** top left, Copyright 2007, Custom Medical Stock Photo, All Rights Reserved **8-03:** bottom center, Photo Disc **8-03:** bottom left, Royalty-Free/Corbis **8-04:** top left, Felicia Martinez/Photo Edit **8-04:** top right, Michael Newman/Photo Edit **8-07:** top, Felicia Martinez, PhotoEdit **8-07:** bottom left, Richard Anderson **8-07:** bottom right, Richard Anderson **8-07:** bottom center, Richard Anderson **8-08:** bottom left, Richard Anderson **8-08:** bottom right, Richard Anderson **8-08:** bottom center, Richard Anderson **8-09:** Photo Disc **8-10:** Stan Maddock **8-11:** center, Photo Disc

Chapter 9. 9-02: Photo Disc **9-03:** top, Ann Borman, Wadsworth/Thomson **9-04:** Public Domain **9-05:** left, Centers for Disease Control, chart is in the public domain **9-05:** bottom right, Stan Maddock **9-06:** Stan Maddock **9-07:** bottom, Photo Disc **9-07:** top, www.medscape.com, 2006 **9-08:** bottom, Photo Disc **9-08:** top, Photo Disc **9-09:** Photo Disc **9-10:** bottom, Dr. Alon Eliakim **9-10:** top center, Rudolph J. Liedtke/www.rjlsystems.com **9-10:** top, Tom Pantages **9-10:** bottom center, Yoav Levy/Photo Take **9-11:** top, Andy Gillum, ARS, USDA, public domain **9-11:** bottom, Courtesy, Life Measurement, Inc., www.bodpod.com **9-14:** Richard Hutchings/Photo Researchers, Inc. **9-15:** bottom, Chad Johnston/Masterfile **9-15:** top left, Michael Newman/Photo Edit **9-15:** top right, Michael Newman/Photo Edit **9-16:** top, Photo Disc

Chapter 10. 10-02: Photo Disc **10-03:** Richard Anderson **10-04 & 10-05:** top left, Photo Disc **10-04 & 10-05:** top right, Photo Disc **10-06:** top left, Stan Maddock **10-06:** top center, Stan Maddock **10-06:** bottom, Stan Maddock **10-06:** top right, Stan Maddock **10-07:** top, Photo Disc **10-07:** bottom, Richard Anderson **10-07:** center, Susan Van Etten/Photo Edit **10-09:** Stan Maddock **10-10:** top, Courtesy, American Society of Plastic Surgeons **10-10:** top, Custom Medical Stock, I. Richard Toranto, M.D. **10-12:** Photo Disc **10-14:** Stan Maddock

Chapter 11. 11-03: bottom, Felicia Martinez/PhotoEdit **11-03:** top left, Richard Anderson **11-04:** Photo Disc **11-04:** Reprinted with permission from the Diagnostic and Statistical Manual of Mental Disorders, Fourth Edition, Text Revision, (Copyright 2000). American Psychiatric Association. **11-05:** Richard Anderson **11-07:** Reprinted with permission from the Diagnostic and Statistical Manual of Mental Disorders, Fourth Edition, Text Revision, (Copyright 2000). American Psychiatric Association. **11-09:** bottom right, Stan Maddock **11-09 & 11-10:** top left, Photo Disc **11-09 & 11-10:** top right, Photo Disc **11-11:** Richard Anderson **11-12:** bottom, Photo Disc **11-13:** PhotoEdit Page **11-14:** Garner, D.M., Olmsted, M.P., Bohr, Y., and Garfinkel, P.E. (1982). The Eating Attitudes Test: Psychometric features and clinical correlates. Psychological Medicine, 12, 871-878.

Chapter 12. 12-03: right, Felicia Martinez/PhotoEdit **12-03:** left, Richard Anderson **12-04:** top left, Don W. Fawcett, Visuals Unlimited **12-04:** top right, Stan Maddock **12-05:** bottom, Stan Maddock **12-07:** top, Photo Disc **12-07:** bottom right, Photo Disc **12-07:** center, Richard Anderson **12-08:** Ann Borman, Wadsworth/Thomson **12-09:** Richard Anderson **12-10 & 12-13:** top left, Photo Disc **12-10 & 12-13:** top right, Photo Disc **12-11:** top left, Photo Disc **12-11:** top right, Photo Disc **12-11:** center right, PhotoEdit **12-11:** center left, Richard Anderson **12-11:** bottom right, Richard Anderson **12-11:** bottom left, Richard Anderson **12-12:** Photo Disc **12-13:** bottom, Stan Maddock **12-14:** center, Photo Disc **12-15:** bottom, Photo Disc **12-15:** top, Reprinted by permission of Toothfriendly International. **12-16:** Custom Medical Stock, All Rights Reserved

Chapter 13. 13-02: Ann Borman, Wadsworth/Thomson **13-04:** top left, Photo Disc **13-05:** bottom, Photo Disc **13-06 & 13-07:** Richard Anderson **13-08:** bottom left, PhotoDisc **13-09:** Royalty-Free/Corbis **13-10:** top, Elisa Cicinelli/Index Stock **13-10:** bottom, Jan Bengtsson/Corbis **13-11:** Ann Borman, Wadsworth/Thomson

Chapter 14. 14-03: bottom, National Institute on Alcohol Abuse and Alcoholism, 2001, 2006 **14-03:** top, Photo Disc **14-04:** top left, Photo Disc **14-04:** top right, Jonathan Nourok/Photo Edit **14-05:** University of Oklahoma Police Department, Blood Alcohol Calculator. www.ou.edu/oupd/bac.htm. Reprinted by permission of the Board of Regents of the University of Oklahoma. **14-05 & 14-06:** top left, Photo Disc **14-05 & 14-06:** top right, Photo Disc **14-06:** bottom left,

Louis B. Wallach, Inc/Getty Images **14-06**: center left, Photo Disc **14-07**: Photo Disc

Chapter 15. 15-02: Photo Disc **15-03**: top right, Felicia Martinez/Photo Edit **15-03**: bottom, Felicia Martinez/Photo Edit **15-03**: bottom left, Felicia Martinez/Photo Edit **15-03**: center right, Felicia Martinez/Photo Edit **15-03**: center, Felicia Martinez/Photo Edit **15-03**: center left, Felicia Martinez/Photo Edit **15-03**: bottom, Index Stock **15-03**: bottom, Richard Anderson **15-03**: bottom right, Richard Anderson **15-03**: top left, Tony Freeman/Photo Edit **15-03**: top, Tony Freeman/Photo Edit **15-03**: top, Tony Freeman/Photo Edit **15-04**: Stan Maddock **15-05**: center left, Photo Disc **15-05**: bottom right, Photo Disc **15-05**: bottom center, PhotoEdit, David Young-Wolff **15-05**: top right, Richard Anderson **15-06**: top, Richard Anderson **15-06**: bottom, Richard Anderson **15-06**: center, Taxi/Getty Images **15-07**: Photo Disc **15-08 & 15-09**: top left, Photo Disc **15-09**: top left, Dr. Robert S. Goodhard, M.D. **15-09**: top right, Dr. Robert S. Goodhard, M.D.

Chapter 16. 16-02: Photo Disc **16-03**: Richard Anderson **16-04**: Photo Disc **16-05 & 16-06**: top left, Photo Disc **16-05 & 16-06**: top right, Photo Disc **16-09**: Photo Disc **16-10**: bottom, David Young-Wolff/Photo Edit **16-10**: bottom, Michael Newman/Photo Edit **16-10**: bottom, Photo Disc **16-10**: bottom right, Photo Disc **16-10**: bottom top left, Photo Edit **16-10**: bottom, PhotoEdit, Amy C. Etra **16-10**: bottom, PhotoEdit, Michael Newman **16-10**: bottom, Richard Anderson

Chapter 17. 17-03: top, Felicia Martinez/Photo Edit **17-03**: bottom right, Richard Anderson **17-03**: bottom right, Richard Anderson **17-03**: bottom right, Stan Maddock **17-03**: bottom right, Stan Maddock **17-04**: Felicia Martinez/Photo Edit **17-05**: top, Photo Disc **17-07**: bottom right, Biophoto/Photo Researchers, Inc. **17-07 & 17-08**: top left, Photo Disc **17-07 & 17-08**: top right, Photo Disc **17-09**: Photo Disc **17-10**: Photo Disc

Chapter 18. 18-03: Biophoto/Photo Researchers, Inc. **18-04**: top, AP/Wide World Photos **18-04**: center left, Stan Maddock **18-05**: Stan Maddock **18-06**: bottom, Photo Disc **18-07**: Eric Risberg/AP Photo **18-10**: Richard Anderson **18-11**: top right, Photo Disc **18-14**: Photo Disc **18-15**: left, Photo Disc **18-15**: right, Photo Disc

Chapter 19. 19-02: Photo Disc **19-03**: top left, Ann Borman, Wadsworth/Thomson **19-03**: bottom left, ICI Pharmaceutical Division, Cheshire, England **19-03**: bottom right, ICI Pharmaceutical Division, Cheshire, England **19-03**: top right, Stan Maddock **19-04**: top, Stan Maddock **19-04**: bottom, Stan Maddock **19-05**: center right, Photo Disc **19-05**: bottom, Stan Maddock **19-05 & 19-06**: top left, Photo Disc **19-05 & 19-06**: top right, Photo Disc **19-07**: top, From Dyslipidemia and Vascular Function: The Rationale for Early and Aggressive Intervention, by E. Caballero. www.medscape.com. Reprinted by permission. **19-07**: bottom, From Dyslipidemia and Vascular Function: The Rationale for Early and Aggressive Intervention, by E. Caballero. www.medscape.com. Reprinted by permission. **19-08**: top left, Courtesy of California Walnuts/Walnut Marketing Board **19-08**: bottom, Photo Disc **19-09**: Keith Weller/ARS/USDA **19-11**: Photo Disc

Chapter 20. 20-02: Photo Disc **20-04**: bottom, Dr. M.A. Ansary/Photo Researchers, Inc. **20-04**: top, L.V. Bergman & Associates **20-06**: top right, Photo Researchers, Inc. **20-06**: bottom center, Photo Researchers, Inc. **20-06**: top left, Photo Take **20-08**: ICI Pharmaceuticals **20-08**: bottom, ICI Pharmaceuticals Division, Cheshire, England **20-09**: American Journal of Clinical Nutrition, Vol. 71, No 4, 878-884. April 2000, Robert M. Russell **20-10**: top, Photo Researchers, Inc. **20-10**: bottom, Photo Researchers, Inc. **20-11**: Photo Researchers, Inc. **20-12**: Wally McNamee/Woodfin Camp **20-13**: PhotoDisc **20-14**: top, Holly Hitzeman/Index Stock **20-14**: bottom, Photo Disc **20-15**: top, Photo Disc **20-15**: bottom, Photo Disc **20-16**: top, Photo Disc **20-16**: bottom, Photo Disc **20-17**: top, Photo Disc **20-17**: bottom, Photo Disc **20-18**: top, Photo Disc **20-18**: bottom, Photo Disc **20-19**: Photo Disc

Chapter 21. 21-02: Photo Disc **21-03**: bottom left, Photo Disc **21-03**: bottom right, Photo Disc **21-03**: top, PhotoDisc **21-04**: Ralph Eagle/Photo Researchers, Inc. **21-05**: top right, Keith Weller/ARS/USDA **21-05**: top left, Richard Anderson **21-05**: bottom right, Richard Anderson **21-06**: Photo Disc **21-06**: bottom, Tony Freeman/Photo Edit **21-08**: top, Conde

Nast Archive/Corbis **21-08**: bottom, Photo Disc **21-10**: center left, Paul Sakuma/AP Photo **21-10**: top, Stan Maddock **21-10 & 21-11**: bottom left, Photo Disc **21-10 & 21-11**: bottom right, Photo Disc

Chapter 22. 22-03: top left, Photo Disc **22-03**: top right, Photo Disc **22-03**: bottom, Stan Maddock **22-04**: center left, Ann Borman, Wadsworth/Thomson **22-05**: bottom, Stan Maddock **22-05 & 22-07**: top left, Photo Disc **22-05 & 22-07**: top right, Photo Disc **22-06**: bottom, David Young-Wolff/Photo Edit **22-06**: top, Photo Disc

Chapter 23. 23-03: Stan Maddock **23-04**: Index Stock **23-05**: left, Dempster et al, J Bone Min. Res. 1, 15-21, 1986 **23-05**: right, Dempster et al, J Bone Min. Res. 1, 15-21, 1986 **23-06**: bottom left, Martin M. Rotker/CNRI **23-06**: bottom right, Martin M. Rotker/CNRI **23-06**: top, Richard Anderson **23-07**: bottom, Photo Disc **23-07**: top, Tony Freeman/Photo Edit **23-08**: Custom Medical Stock **23-09**: top, Ray Stanyard **23-09**: bottom, Ray Stanyard **23-10**: Robert Brenner/Photo Edit **23-13**: top right, Stan Maddock **23-14**: bottom left, Richard Anderson **23-14**: bottom right, Stan Maddock **23-15**: Stan Maddock **23-17**: top left, Stan Maddock **23-17**: center right, Stan Maddock **23-17**: bottom, Tony Freeman/PhotoEdit **23-18**: bottom, Photo Disc **23-18**: top, Photo Researchers, Inc. **23-21**: bottom left, Stan Maddock **23-21**: bottom right, Stan Maddock **23-21**: top right, http://www.nhlbi.nih.gov/health/public/heart/hbp/dash

Chapter 24. 24-02: Photo Disc **24-03**: bottom right, Stan Maddock **24-06**: bottom left, Judy Brown **24-06**: top right, Richard Anderson **24-08**: center left, Government Agency **24-08**: top left, Richard Anderson **24-09 & 24-10**: top left, Photo Disc **24-09 & 24-10**: top right, Photo Disc **24-10**: bottom left, Photo Disc **24-11**: top right, Probiotics: Their tiny worlds...The Scientist July 22, 2002, pp. 20-22; author: Bob Beale

Chapter 25. 25-02: Photo Disc **25-03**: bottom left, Photo Disc **25-03**: top right, Stan Maddock **25-04**: top, AP/Wide World Photos **25-04**: bottom, Photo Disc **25-05**: right, Richard Anderson **25-05**: left, Tony Freeman/Photo Edit **25-06**: top, Richard Anderson **25-06**: bottom, Robin Sterling/Sterling Research **25-07**: John Norris/PicImpact/Corbis **25-08**: Stan Maddock

Chapter 26. 26-02: Photo Disc **26-03**: Stan Maddock **26-04**: Photo Disc **26-05**: left, ICI Pharmaceuticals Division **26-06**: right, ICI Pharmaceuticals Division **26-06**: bottom right, Photo Disc **26-07 & 26-08**: top left, Photo Disc **26-07 & 26-08**: top right, Photo Disc **26-08**: center left, Photo Disc **26-08**: bottom right, Photo Disc **26-08**: bottom left, Royalty-Free/Corbis

Chapter 27. 27-02: Cyrus McCrimmon, The Denver Post, AP Photo **27-03**: Photo Disc **27-04**: Custom Medical Stock **27-05**: bottom right, Photo Disc **27-06**: bottom center, Photo Disc **27-07**: top right, Photo Disc **27-07 & 27-08**: bottom left, Photo Disc **27-07 & 27-08**: bottom right, Photo Disc **27-09**: bottom right, Jason Horowitz/Zefa/Corbis **27-09**: top right, Photo Disc **27-10**: top left, Photo Disc **27-10**: bottom, PhotoEdit **27-12**: top, Ann Borman, Wadsworth/Thomson **27-12**: bottom, Photo Disc **27-13**: Greg Merhar, reprinted with permission of the Journal of Physical Education, Recreation and Dance, 59 (9), 1987

Chapter 28. 28-03: bottom, Stan Maddock **28-03**: top, Tony Freeman/Photo Edit **28-04**: bottom right, Comstock Images/Getty Images **28-04**: bottom center, Jose Carillo/Photo Edit **28-04**: bottom left, Photo Disc **28-04**: bottom center, Photo Disc **28-04**: bottom right, Photo Disc **28-04**: center left, Thinkstock **28-05**: bottom, Photo Disc **28-06**: bottom left, AP Photo **28-07 & 28-08**: top left, Photo Disc **28-07 & 28-08**: top right, Photo Disc **28-09**: top right, Bob Winslett/Index Stock **28-11**: top right, Charles Gupil/Stock, Boston

Chapter 29. 29-02: Photo Disc **29-03**: top left, Stock Solution/Index Stock **29-04**: top right, Barts Medical Library/Photo Take **29-04**: left, Photo Disc **29-06**: bottom left, U.S. Department of Health and Human Services **29-07**: Stan Maddock **29-08**: top right, Biophoto Associates/Science Source/Photo Researchers, Inc. **29-08**: bottom left, Richard Anderson **29-08**: bottom right, Richard Anderson **29-08**: top left, Stan Maddock **29-10**: bottom left, David Young-Wolff/Photo Edit **29-11**: Reprinted by permission of the University of Missouri Extension. muextension.missouri.edu/xplor/hesguide/foodnut/n00944.htm. **29-12**: Photo Disc **29-13**: top right, Hans and Libby Schapiro **29-13**: bottom left, Stan Maddock **29-14**: bottom, Stan Maddock **29-15**: top right, Photo Disc

29-16: top, Photo Disc **29-16**: bottom, Robin Sterling/Sterling Research **29-17**: top center, Mary Kate Denny/Photo Edit **29-17**: center, Myrleen Ferguson/Photo Edit **29-17**: bottom, Photo Disc **29-17**: bottom center, Photo Disc, Shirley Zieberg **29-17**: top, Suzanne Szasz/Photo Researchers, Inc. **29-18**: National Center for Chronic Disease Prevention and Health Promotion **29-19**: bottom right, Norbert Schaefer/Corbis **29-20**: top left, Dietmar A. J. Kennel **29-21**: top left, Photo Disc **29-22**: bottom, Stan Maddock

Chapter 30. 30-02: Juneberg Clark/Photo Researchers, Inc. **30-03**: bottom, Mary Kate Denny/Photo Edit **30-03**: top, Photo Disc **30-03**: top center, PhotoEdit **30-03**: bottom center, Robert Brenner/Photo Edit **30-03**: center, Tony Freeman/Photo Edit **30-04**: top, Carl Purcell/Photo Researchers, Inc. **30-04**: top center, David Young-Wolff/Photo Edit **30-04**: bottom, David Young-Wolff/Photo Edit **30-04**: bottom center, David Young-Wolff/Photo Edit **30-05**: top, Photo Disc **30-05**: bottom, Photo Disc **30-06**: bottom, National Center for Chronic Disease Prevention **30-06**: top, Photo Disc **30-07**: top right, Photo Disc **30-07**: bottom left, Richard Anderson **30-08**: bottom left, Judy Brown, author of Nutrition Now, Fourth Edition **30-08**: bottom right, Judy Brown, author of Nutrition Now, Fourth Edition **30-10**: Judith E. Brown **30-11**: top left, Richard Anderson **30-11**: bottom right, USDA **30-12**: bottom left, Blair Seitz/Photo Researchers, Inc. **30-12**: top right, USDA **30-13**: Photo Disc

Chapter 31. 31-02: Photo Disc **31-03**: top left, Stan Maddock **31-04**: bottom left, Photo Disc **31-04**: bottom center, Photo Disc **31-04**: bottom right, Photo Disc **31-04**: top left, Photo Disc **31-06**: bottom, USDA **31-07**: Jeff Greenberg/Photo Edit

Chapter 32. 32-02: Photo Disc **32-03**: bottom, Dennis MacDonald/Photo Edit **32-03**: top right, Stan Maddock **32-04**: Ed Kashi Photography **32-05**: top right, Martha Cooper Arnold **32-06**: top, Index Stock **32-06**: bottom, Photo Disc **32-07**: top right, Biophoto Associates/Photo Researchers, Inc. **32-07**: bottom right, Richard Anderson **32-08**: Centers for Disease Control and Prevention, US Department of Health and Human Services **32-09**: top, Photo Disc **32-09**: bottom, Photo Disc **32-10**: center right, Stan Maddock **32-11**: Stan Maddock **32-12**: Stan Maddock **32-13**: USDA

Chapter 33. 33-02: Photo Disc **33-03**: bottom left, Photo Disc **33-03**: bottom right, Photo Disc **33-04**: top, Stan Maddock **33-05**: top right, Dr. Robert S. Goodhard, MD **33-05**: bottom, Dr. Robert S. Goodhard, MD **33-07**: center right, WHO/Unicef **33-08**: bottom left, Photo Disc **33-08**: top, Porterfield/Chickering/Photo Researchers, Inc. **33-09**: S. Yabas/UNICEF

Chapter FM. FM: Copyright 2000 by the National Academy Press. All rights reserved. **Cover:** Image Shop/Corbis **Table of Contents ix:** bottom right, Morton Salt Company **Table of Contents ix:** top right, Photo Disc **Table of Contents ix:** bottom left, PhotoDisc **Table of Contents v:** top right, Photo Disc **Table of Contents v:** bottom left, PhotoDisc **Table of Contents v:** bottom center, PhotoDisc **Table of Contents v:** bottom right, PhotoDisc **Table of Contents vi:** top left, Photo Disc **Table of Contents vi:** bottom left, PhotoDisc **Table of Contents vi:** bottom right, PhotoDisc **Table of Contents vi:** bottom right, Bettmann/Corbis **Table of Contents vii:** top right, Photo Disc **Table of Contents vii:** bottom left, Stefano Bianchetti/Corbis **Table of Contents viii:** bottom center, Bettman/Corbis **Table of Contents viii:** top left, Photo Disc **Table of Contents viii:** bottom left, PhotoDisc **Table of Contents viii:** bottom right, PhotoDisc **Table of Contents x:** top left, Photo Disc **Table of Contents x:** bottom left, Photo Disc **Table of Contents x:** bottom right, Photo Disc **Table of Contents xi:** bottom left, FDR Library **Table of Contents xi:** top right, Photo Disc **Table of Contents xi:** bottom center, Photo Disc **Table of Contents xi:** bottom right, Photo Disc **Table of Contents xii:** bottom, Photo Disc **Table of Contents xiii:** top left, Photo Disc **Table of Contents xiii:** bottom, Photo Disc **Table of Contents xiv:** top, Photo Disc **Table of Contents xiv:** bottom, Richard Anderson **Table of Contents xv:** Photo Disc **Table of Contents xvi:** Photo Disc **Table of Contents xvii:** Photo Disc **Table of Contents xviii:** Photo Disc **Table of Contents xx:** Photo Disc

INDEX

Note to the reader: If you have a customized book, this index may include some entries that will not be found in your book.

Daily Values for Food Labels

The Daily Values are standard values developed by the Food and Drug Administration (FDA) for use on food labels. The values are based on 2000 kcalories a day for adults and children over 4 years old.

Proteins, Vitamins, and Minerals

Nutrient	Amount
Protein[a]	50 g
Thiamin	1.5 mg
Riboflavin	1.7 mg
Niacin	20 mg NE
Biotin	300 μg
Pantothenic acid	10 mg
Vitamin B$_6$	2 mg
Folate	400 μg
Vitamin B$_{12}$	6 μg
Vitamin C	60 mg
Vitamin A	5000 IU[b]
Vitamin D	400 IU[b]
Vitamin E	30 IU[b]
Vitamin K	80 μg
Calcium	1000 mg
Iron	18 mg
Zinc	15 mg
Iodine	150 μg
Copper	2 mg
Chromium	120 μg
Selenium	70 μg
Molybdenum	75 μg
Manganese	2 mg
Chloride	3400 mg
Magnesium	400 mg
Phosphorus	1000 mg

[a]The Daily Values for protein vary for different groups of people: pregnant women, 60 g; nursing mothers, 65 g; infants under 1 year, 14 g; children 1 to 4 years, 16 g.
[b]Equivalent values for nutrients expressed as IU are: vitamin A, 1500 RAE (assumes a mixture of 40% retinol and 60% beta-carotene); vitamin D, 10 μg; vitamin E, 20 mg.

Nutrients and Food Components

Food Component	Amount	Calculation Factors
Fat	65 g	30% of kcalories
Saturated fat	20 g	10% of kcalories
Cholesterol	300 mg	Same regardless of kcalories
Carbohydrate (total)	300 g	60% of kcalories
Fiber	25 g	11.5 g per 1000 kcalories
Protein	50 g	10% of kcalories
Sodium	2400 mg	Same regardless of kcalories
Potassium	3500 mg	Same regardless of kcalories

GLOSSARY OF NUTRIENT MEASURES

kcal: kcalories; a unit by which energy is measured.

g: grams; a unit of weight equivalent to about 0.03 ounces.

mg: milligrams; one-thousandth of a gram.

μg: micrograms; one-millionth of a gram.

IU: international units; an old measure of vitamin activity determined by biological methods (as opposed to new measures that are determined by direct chemical analyses). Many fortified foods and supplements use IU on their labels.
- For vitamin A, 1 IU = 0.3 μg retinol, 3.6 μg β-carotene, or 7.2 μg other vitamin A carotenoids.
- For vitamin D, 1 IU = 0.025 μg cholecalciferol.
- For vitamin E, 1 IU = 0.67 natural α-tocopherol (other conversion factors are used for different forms of vitamin E).

mg NE: milligrams niacin equivalents; a measure of niacin activity.
- 1 NE = 1 mg niacin.
  = 60 mg tryptophan (an amino acid).

μg DFE: micrograms dietary folate equivalents; a measure of folate activity
- 1 μg DFE = 1 μg food folate.
  = 0.6 μg fortified food or supplement folate.
  = 0.5 μg supplement folate taken on an empty stomach.

μg RAE: micrograms retinol activity equivalents; a measure of vitamin A activity.
- 1 μg RAE = 1 μg retinol.
  = 12 μg β-carotene.
  = 24 μg other vitamin A carotenoids.

mmol: millimoles; one-thousanth of a mole, the molecular weight of a substance. To convert mmol to mg, multiply by the atomic weight of the substance.
- For sodium, mmol × 23 = mg Na.
- For chloride, mmol × 35.5 = mg Cl.
- For sodium chloride, mmol × 58.5 = mg NaCl.